AF570750

Adhesion Molecules In Health And Disease

Adhesion Molecules In Health And Disease

edited by

Leendert C. Paul
University of Toronto at St. Michael's Hospital
Toronto, Ontario, Canada

Thomas B. Issekutz
University of Toronto
Toronto, Ontario, Canada

Marcel Dekker, Inc. New York • Basel • Hong Kong

Library of Congress Cataloging-in-Publication Data

Adhesion molecules in health and disease / edited by Leendert C. Paul and Thomas B. Issekutz
p. cm.
Includes bibliographical references and index.
ISBN 0-8247-9824-4 (hardcover : alk. paper)
1. Cell adhesion molecules—Physiological effect. 2. Cell adhesion molecules—Pathophysiology. I. Paul, Leendert C. II. Issekutz, Thomas B.
[DNLM: 1. Cell Adhesion Molecules—physiology. 2. Ingegrins—physiology. 3. Inflammation—immunology. QU 55 A2349 1997]
QP552.C42A345 1997
171'.2—dc21
DNLM/DLC
for Library of Congress 97-12034
CIP

The publisher offers discounts on this book when ordered in bulk quantities. For more information, write to Special Sales/Professional Marketing at the address below.

This book is printed on acid-free paper.

MARCEL DEKKER, INC.
270 Madison Avenue, New York, New York 10016
http://www.dekker.com

Current printing (last digit):
10 9 8 7 6 5 4 3 2 1

PRINTED IN THE UNITED STATES OF AMERICA

Preface

Cell adhesion is essential for normal development and regulation of cellular interactions necessary for an organism to maintain its structure and function. This is perhaps best illustrated by the way the body regulates immune and inflammatory defense processes and tissue repair in response to injury, infections, or ischemia. Simple or complex glycoproteins involved in these reactions act by mediating cell-cell interaction or cellular adhesion to the extracellular matrix. Several families of cell adhesion molecules have been identified as therapeutic targets for promoting tissue repair, inhibition of tumor progression, alleviation of inflammatory processes, and transplant acceptance. This monograph introduces the reader to the various adhesion molecules and their involvement in biological processes such as morphogenesis, inflammation, immunological reactions, blood coagulation, tumor metastasis, bone tissue remodeling, and transplant rejection.

We begin with a review of cell adhesion molecules involved in the trafficking of heterogeneous populations of leukocytes throughout tissues in response to localized inflammatory stimuli. Specific interactions among adhesion molecules regulate and stabilize leukocyte adhesion to other cells and components of the extracellular matrix. This results in an orderly sequence of leukocyte invasion into inflammatory lesions. Chapter 2 reviews potential strategies to inhibit cellular adhesion molecules, including the use of antibodies or antibody fragments, soluble receptors or counterreceptors, receptor peptide homologues, antisense oligonucleotides, and inhibitors of cell activation. Chapters 3 and 4 discuss the measurement of soluble adhesion molecules in biological fluids in health and disease.

Cell adhesion, or adhesion receptor engagement, results not only in specific cellular interactions, but also in the induction of a variety of cellular signals. These signals follow an organized cascade of events from the initial receptor engagement all the way to gene induction and cellular activation. Chapter 5 reviews the unique and common pathways of signal transduction following engagement of adhesion molecules.

The transmembrane topography and direct cytoskeletal binding of cell adhesion molecules enable them to transmit changes in mechanical force equilibrium to the cell and its surroundings, resulting in alterations in cell function. The role of adhesion molecules as mechanoreceptors is reviewed in Chapter 6. Chapter 7 explains the role of adhesion molecules in morphogenesis, motility, and invasiveness. The principles and mechanisms involved are discussed using epithelial cells as a model system. Chapter 8 summarizes the role of adhesion molecules in the development of T and B lymphocytes and the three-dimensional structure in which immune and inflammatory cells migrate in health and disease. The issue of reversibility of tissue damage is considered in the context of the three-dimensional structure of tissue adhesion molecules.

Chapter 9 details the regulation of the endothelial selectins and their ligands, as well as their cell type-specific expression, and the specificity of the various receptor interactions. The role of chemoattractants in neutrophil recruitment and tissue inflammation along with strategies aimed at blocking them is reviewed in Chapter 10.

Chapter 11 reviews the role of various adhesion receptors in lymphocyte extravasation in vivo. Although T and B lymphocytes appear to utilize the same adhesion molecules as other leukocytes for interaction with endothelial cells, the extensive differentiation of lymphocytes in the secondary lymphoid tissues, the role of many adhesion receptors as costimulatory molecules in lymphocyte activation, and the heterogeneity of lymphocytes make analysis of the pathways of lymphocyte migration under various conditions in vivo a particular challenge.

P-selectin and its ligand PDGL-1 stand at the crossroads of inflammation and thrombosis. They hold promise as potential targets for therapy to retard or interrupt the cycle of vascular perturbation, inflammation, and thrombosis. The structure and function of P-selectin are examined in detail in Chapter 12.

Adhesion is a central event in microbial pathogenesis. Chapter 13 describes host-pathogen interactions in which host adhesion molecules are utilized as receptors for a number of pathogens. Chapter 14 reviews the mechanisms of leukocyte trafficking to sites infected with gram-negative or gram-positive microorganisms, the rationale to modulate leukocyte trafficking in these conditions, and the steps amenable to down-regulate the inflammatory response.

The pathogenesis of atherosclerosis is classically understood as a complex set of interactions among blood cells, platelets, cells of the vessel wall, and various determinants of thrombosis. These interactions are now being understood at the molecular level, as the factors that promote atherosclerosis are directly linked to expression of cell adhesion molecules involved in atherogenesis (Chapter 15). Chapter 16 reviews cell-cell and cell-matrix interactions in bone, the role of adhesion molecules in the bone remodeling processes, and possible applications of anti-adhesion strategies to prevent or treat selected bone diseases.

Chapter 17 discusses the molecular aspects of the structure and function of the major classes of adhesion receptors as they pertain to processes relevant to tumor progression and metastasis, and the mechanisms whereby neoplastic cells uncouple these adhesion-mediated processes from normal regulatory influences. In vitro studies of cell adhesion and signaling are often relevant to an understanding of tumor behavor in vivo and, conversely, studies of human tumors can lead to important new insights into adhesion molecules as potential therapeutic agents.

Many manifestations of ischemia-reperfusion injury result from neutrophil-mediated tissue damage. Chapter 18 reviews the cell adhesion molecules involved in this multistep process and the therapeutic efficacy of monoclonal antibodies against various cell adhesion molecules in experimental models. Chapter 19 summarizes the current literature of cell adhesion molecules in the nervous system and their role in disease processes like multiple sclerosis and neuro-oncological disorders. Chapters 20 and 21 review the cytokine and adhesion molecule requirements for the induction of inflammatory lung and intestinal lesions, respectively, in various experimental models. Diverse cytokine and cell adhesion pathways have been identified; the pattern of adhesion molecules involved depends on the stimulus used to initiate the inflammatory response. Chapter 22 reviews the pathogenesis of rheumatoid arthritis and the cell adhesion pathways involved in the influx of cells into the joint. Chapter 23 reviews the molecular basis of leukocyte trafficking through the skin, the role of various adhesion molecules in this process and the potential efficacy of anti-adhesion treatments in some skin diseases. Chapter 24 highlights some recent studies on the role of adhesion molecules in inflammatory kidney diseases and in signal transduction in glomerular epithelial cells. Chapter 25 describes the role of adhesion molecules in the interactions of leukocytes and platelets, with artificial membranes during hemodialysis and cardio-pulmonary bypass, two commonly used clinical modalities involving extracorporeal circulation of the blood. Chapter 26 reviews the cell-cell and cell-extracellular matrix interactions in organ transplant rejection, the adhesion molecules involved, and the potential of interference in these interactions to prevent or treat graft rejection. The final chapter provides an overview of peritoneal host defense mechanisms in the setting of peritoneal dialysis, with particular emphasis on the involvement of adhesion molecules, cytokines, and chemokines in the host's response to peritoneal infection.

It is clear that stimulus and organ specific differences exist in the use of cell adhesion molecules. Blocking strategies that target these adhesion pathways or interfere with intra-cellular signaling by these receptors may be one clinical benefit.

We hope that this book will be useful to biomedical researchers and students, clinicians, physiologists and immunologists, and that it may serve as a reference for medical and science educators.

Leendert C. Paul
Thomas B. Issekutz

Contents

Contributors

Anthony R. Berendt Adhesion and Infection Group, Nuffield Department of Medicine, University of Oxford, Oxford, England

Carmen Birchmeier Department of Medical Genetics, Max-Delbrueck Center for Molecular Medicine, Berlin, Germany

Elisabeth Bloemena Department of Pathology, Free University Hospital, Amsterdam, The Netherlands

Volker Brinkman Department of Cell Biology, Max-Delbrueck Center for Molecular Medicine, Berlin, Germany

M. Büchler Unité d'Immunologie Clinique et de Transplantation et Groupe Interactions Hôte-greffon, Hôpital Bretonneau, Tours, France

Alfred K. Cheung, M.D. Departments of Medicine and Medical Service, Veterans Affairs Medical Center and University of Utah, Salt Lake City, Utah

Carol J. Cornejo, M.D. Department of Surgery, University of Washington, Seattle, Washington

Colleen Sweeney Crovello New England Medical Center, Tufts University School of Medicine, Boston, Massachusetts

Andrey V. Cybulsky Department of Medicine, Royal Victoria Hospital, McGill University, Montreal, Quebec, Canada

Pranab K. Das Departments of Cardiovascular Pathology and Dermatology, Academic Medical Centre, University of Amsterdam, Amsterdam, The Netherlands

Mark G. Davies, M.D., Ph.D., FRSCI Vascular Biology and Atherosclerosis Research Laboratory, Department of Surgery, Duke University Medical Center, Durham, North Carolina

Onno J. de Boer Department of Cardiovascular Pathology, Academic Medical Centre, University of Amsterdam, Amsterdam, The Netherlands

Shoukat Dedhar, Ph.D. Professor, Department of Medical Biophysics, University of Toronto, and Division of Cancer Biology Research, Sunnybrook Health Science Center, North York, Ontario, Canada

Gregory J. Fulton, M.D., FRCSI Vascular Biology and Atherosclerosis Research Laboratory, Department of Surgery, Duke University Medical Center, Durham, North Carolina

Barbara C. Furie New England Medical Center, Tufts University School of Medicine, Boston, Massachusetts

Bruce Furie Professor of Medicine, New England Medical Center, Tufts University School of Medicine, Boston, Massachusetts

Simon A. L. Gibbs, FRCS Department of Surgery, Brigham and Women's Hospital and Harvard Medical School, Boston, Massachusetts

Michael A. Gimbrone, Jr. Department of Pathology, Brigham and Women's Hospital and Harvard Medical School, Boston, Massachusetts

D. Neil Granger Department of Physiology, Louisiana State University Medical Center, Shreveport, Louisiana

Per-Otto Hagen, Ph.D. Department of Surgery, Duke University Medical Center, Durham, North Carolina

Gregory E. Hannigan, Ph.D. Assistant Professor, Department of Pathology, University of Toronto and Hospital for Sick Children, Toronto, Ontario, Canada

John M. Harlan, M.D. Department of Medicine, Division of Hematology, University of Washington, Seattle, Washington

Herbert B. Hechtman, M.D. Department of Surgery, Brigham and Women's Hospital and Harvard Medical School, Boston, Massachusetts

Clifford J. Holmes, Ph.D. Renal Division, Baxter Healthcare, McGraw Park, Illinois

Shervanthi Homer-Vanniaskinkam Vascular Surgical Research Unit, The General Infirmary at Leeds, Leeds, England

David E. Hughes, M.D., Ph.D. Lecturer and Honorary Senior Registrar, Department of Pathology, University of Sheffield Medical School, Sheffield, England

Thomas B. Issekutz, M.D., FRCPC Department of Medicine and Immunology, University of Toronto, Toronto, Ontario, Canada

Brent Johnson Department of Medical Physiology, Univerisity of Calgary, Calgary, Alberta, Canada

Paul Kubes Department of Medical Physiology, University of Calgary, Calgary, Alberta, Canada

Jerzy Kupiec-Weglinski Department of Surgery, Surgical Research Laboratory, Brigham and Women's Hospital and Harvard Medical School, Boston, Massachusetts

Y. Lebranchu Unité d'Immunologie Clinique et de Transplantation et Groupe Interactions Hôte-greffon, Hôpital Bretonneau, Tours, France

John Kenneth Leypoldt, Ph.D. Departments of Medicine, Bioengineering, and Research Service, Veterans Affairs Medical Center and University of Utah, Salt Lake City, Utah

Joseph Loscalzo, M.D., Ph.D. Department of Pathology, Brigham and Women's Hospital and Harvard Medical School, Boston, Massachusetts

Francis W. Luscinskas Department of Pathology, Brigham and Women's Hospital and Harvard Medical School, Boston, Massachusetts

Paul J. Marchetti Division of Neurology, University of Toronto at St. Michael's Hospital, Toronto, Ontario, Canada

Christopher J. McCormick Adhesion and Infection Group, Nuffield Department of Medicine, University of Oxford, Oxford, England

Carol T. Mei New England Medical Center, Tufts University School of Medicine, Boston, Massachusetts

Chris J. L. M. Meijer Department of Pathology, Free University Hospital, Amsterdam, The Netherlands

Syed Fazal Mohammad, Ph.D. Department of Pathology and Artificial Heart Laboratory, University of Utah, Salt Lake City, Utah

Michael S. Mulligan, Ph.D. Department of Pathology, University of Michigan Medical School, Ann Arbor, Michigan

Paul W. O'Connor Division of Neurology, University of Toronto at St. Michael's Hospital, Toronto, Ontario, Canada

Quirino Orlandi, M.D. Whitaker Cardiovascular Institute, Boston University School of Medicine, Boston, Massachusetts

Aiyappa Palecanda Department of Medicine and Immunology, University of Toronto, Toronto, Ontario, Canada

Leendert C. Paul, M.D., Ph.D., FRCPC Keenan Professor of Medicine, Director, Renal Division, Department of Medicine, University of Toronto at St. Michael's Hospital, Toronto, Ontario, Canada

Dieter Riethmacher Department of Medical Genetics, Max-Delbrueck Center for Molecular Medicine, Berlin, Germany

Daniel R. Saloman, M.D. Departments of Molecular and Experimental Medicine, and Immunology, The Scripps Research Institute, La Jolla, California

Donald M. Salter, BSc, MB, ChB, M.D., MRCPath Senior Lecturer and Honorary Consultant, Department of Pathology, University of Sheffield Medical School, Sheffield, England

Oliver Spertini, M.D. Division of Hematology, University of Lausanne, Lausanne, Switzerland

Timothy A. Springer Center for Blood Research, Harvard Medical School, Boston, Massachusetts

Peter Tan Vascular Surgical Research Unit, The General Infirmary at Leeds, Leeds, England

Nicholas Topley Institute of Nephrology, University of Wales College of Medicine, Cardiff Royal Infirmary, Cardiff, Wales, United Kingdom

Elaine Tuomanen, M.D. Lab of Molecular Infectious Diseases, Rockefeller University, New York, New York

J. F. Valentin Unité d'Immunologie Clinique et de Transplantation et Groupe Interactions Hôte-greffon, Hôpital Bretonneau, Tours, France

Anna C. H. M. van Dinther-Janssen Department of Pathology, Free University Hospital, Amsterdam, The Netherlands

Dietmar Vestweber Institute for Cell Biology, ZMBE, University of Münster, Münster, Germany

Peter A. Ward, M.D. Professor and Chairman, Department of Pathology, University of Michigan Medical School, Ann Arbor, Michigan

Martin R. Weiser, M.D. Department of Surgery, Brigham and Women's Hospital and Harvard Medical School, Boston, Massachusetts

Robert K. Winn, M.D. Department of Physiology and Biophysics, University of Washington, Seattle, Washington

Andrew D. Yurochko, Ph.D. Lineberger Comprehensive Cancer Center, University of North Carolina at Chapel Hill, Chapel Hill, North Carolina

Adhesion Molecules In Health And Disease

1

Traffic Signals on Endothelium for Lymphocyte Recirculation and Leukocyte Emigration

Timothy A. Springer
Center for Blood Research, Harvard Medical School, Boston, Massachusetts

I. INTRODUCTION

The circulatory and migratory properties of white blood cells have evolved to allow efficient surveillance of tissues for infectious pathogens and rapid accumulation at sites of injury and infection. Lymphocytes continually patrol the body for foreign antigen by recirculating from blood, through tissue, into lymph, and back to blood. Lymphocytes acquire a predilection, based on the environment in which they first encounter foreign antigen, to home to or recirculate through that same environment (1,2). Granulocytes and monocytes cannot recirculate, but emigrate from the bloodstream in response to molecular changes on the surface of blood vessels that signal injury or infection. Lymphocytes can similarly accumulate in response to inflammatory stimuli. The nature of the inflammatory stimulus determines whether lymphocytes, monocytes, neutrophils, or eosinophils predominate, and thus exerts specificity in the molecular signals or "area codes" that are displayed on endothelium and control traffic of particular leukocyte classes.

Recent findings show that the "traffic signals" for lymphocyte recirculation and for neutrophil and monocyte localization in inflammation are strikingly similar at the molecular level. These "traffic signal" or "area code" molecules are displayed together on endothelium but act on leukocytes in a sequence that was first defined for neutrophils and appears to hold true with slight modification for lymphocyte homing as well (Fig. 1). The selectin, or green light, allows cells to tether and roll; the chemoattractant, or yellow light, tells cells to activate integrin adhesiveness and put on the brakes; and the Ig family member, or red light, binds integrins and causes cells to come to a full stop. These three steps, with multiple molecular choices at each step, provide great combinatorial diversity in signals. Accordingly, the selective responses of different leukocyte classes to inflammatory agents, as well as the preferential recirculation patterns of distinct lymphocyte subpoplations, can be explained by their distinct receptivity to combinations of molecular signals. Following an overview of leukocytes and endothelium, and of the molecules important in their interactions, I will review the traffic signals that enable selective emigratory behavior of monocytes and neutrophils, and then elaborate how a paradigm of three or four sequential signals can be extended to lymphocyte recirculation. This review updates and extends about twofold a previous one (3). For recent reviews see (4–20).

II. FUNCTION OF LEUKOCYTE CLASSES CORRELATES WITH CIRCULATORY BEHAVIOR

Neutrophilic granulocytes are among the most abundant leukocytes in the bloodstream, and the first to appear at sites of bacterial infection or injury. Neutrophils are produced at the prodigious rate of 10^9 cells/kg body weight/day in the bone marrow, and have a half-life in the circulation of 7 hours. Their lifespan after extravasation is hours or less (21). Their primary function is to phagocytose and eliminate foreign microorganisms and damaged tissue.

Monocytes are far less numerous in the blood than neutrophils, where their half-life is about 24 hours (22). Like neutrophils, they are phagocytic and accumulate in response to traumatic injury or bacterial infection. However, monocytes differ from neutrophils, in that they accumulate at sites where T lymphocytes have recognized antigen, as in delayed-type hypersensitivity reactions and graft rejection. Monocytes are important effector cells in antigen-specific T cell immunity, are activated by T cell products such as γ-interferon, and can organize around parasites into protective structures called granulomas. After extravasation, monocytes may also differentiate into longer-lived tissue macrophages or mononuclear phagocytes such as the Kupffer cells of the liver, which have a half-life of weeks to months.

In contrast to the neutrophil and monocyte, a lymphocyte may emigrate and recirculate many thousands of times during its life history. Recirculation of lymphocytes correlates with their role as antigen receptor-bearing surveillance cells. Lymphocytes function as the reservoir of "immunological

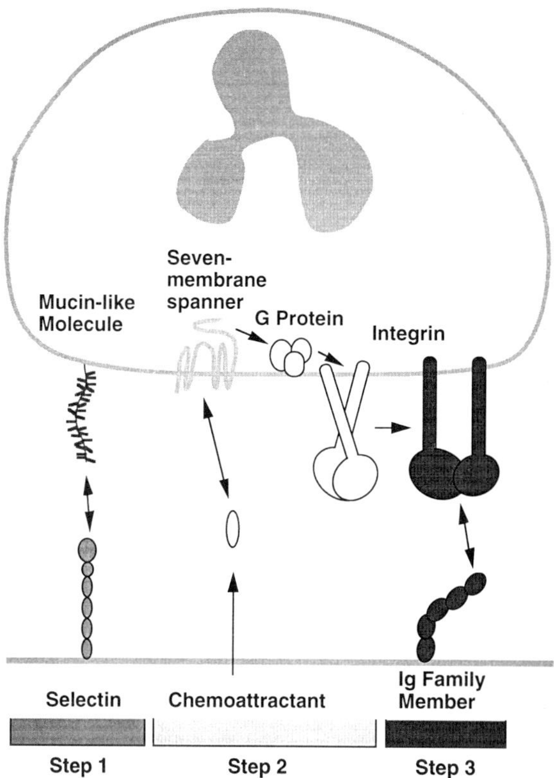

Figure 1 Three sequential steps provide the traffic signals that regulate leukocyte localization in the vasculature. Selectin molecules that bind carbohydrate ligands, often displayed on mucinlike molecules, are responsible for the initial tethering of a flowing leukocyte to the vessel wall and labile, rolling adhesions (green light). Tethering brings leukocytes into proximity with chemoattractants that are displayed on or released from the endothelial lining of the vessel wall. Chemoattractants bind to receptors that span the membrane seven times on the surface of leukocytes. These couple to G proteins, which transduce signals that activate integrin adhesiveness (yellow light). The integrins can then bind to immunoglobulin superfamily (IgSF) members on the endothelium, increasing adhesiveness and resulting in arrest of the rolling leukocyte (red light). Following directional cures from chemoattractants and using integrins for traction, leukocytes then cross the endothelial lining of the blood vessel and enter the tissue.

memory," and recirculate through tissues to provide systemic memory. Few of the body's lymphocytes are present at any one time in the bloodstream, where their half-life is 2 hours. Distinct subsets of lymphocytes extravasate through the microvasculature in tissues such as skin and gut, and through specialized high endothelial venules (HEV) in lymphoid organs (1,6,17). After migrating through tissue, lymphocytes find their way into the lymphatics. They percolate through draining lymph nodes in the lymphatic system and finally enter the thoracic duct, through which they return to the bloodstream. This journey is completed roughly every 1 to 2 days.

III. ENDOTHELIUM

By displaying specific signals, the endothelium is the most active player in controlling leukocyte traffic. Vascular endothelium is diversified at a number of levels. Large vessels differ from small vessels and capillaries, venular endothelium differs from arterial endothelium, and endothelial phenotype varies between tissues. The preferential migration of leukocytes from postcapillary venules may be related to factors such as shear stress, which is lower there and hence more favorable for leukocyte attachment than in capillaries or arterioles, or to events that occur when leukocytes pass through capillaries. However, when flow is controlled so that shear stress is equivalent in arterioles and venules (23), or when the direction of blood flow is reversed (24), attachment and emigration is far greater from venules, suggesting molecular differences in their endothelial surfaces. In agreement with this, P-selectin is much more abundant on postcapillary venules than on large vessels, arterioles, or capillaries (25), and induction of E-selectin and vascular cell adhesion molecule-1 (VCAM-1) expression in inflammation is most prominent on postcapillary venules (9). The mucin-like cluster of differentiation 34 (CD34) molecule is well expressed on capillaries and is absent from most large vessels (26), and CD36 is expressed on microvascular but not large vessel endothelium (27). The extracellular matrix may exert an influence on endothelial differentiation, as exemplified by modulation of adhesiveness (28). The high endothelium in lymphoid tissue, which expresses addressins for lymphocyte recirculation, is one of the most dramatic examples of endothelial specialization (6).

Inflammatory cytokines dramatically and selectively modulate the transcription and expression of adhesion molecules and chemoattractants in endothelial cells (29). Tumor necrosis factor (TNF) and interleukin-1 (IL-1) increase adhesiveness of endothelium for both neutrophils and lymphocytes and induce ICAM-1, E-selectin, and VCAM-1. IL-4, synergistically with other cytokines, increases adhesion of lymphocytes and induces VCAM-1 (30,31). It is likely that the precise mixture of chemoattractants and cytokines produced at inflammatory sites in vivo determines which types of leuckocytes emigrate. Thus, injection into skin of IL-1α induces emigration of neutrophils and monocytes, as do lipopolysaccharide (LPS) and TNF-α, but with more prolonged emigration of the monocytes. INF-γ induces emigration of monocytes but not neutrophils (22). INF-γ and TNF-α, but not IL-1α or LPS recruit lymphocytes, and IL-4 is ineffective by itself but synergizes with TNF (32–34).

Acting more quickly than cytokines, vasoactive substances such as histamine and thrombin modulate endothelial function in seconds or minutes. They induce secretion of the storage granules of endothelial cells and platelets. Furthermore, they dilate arterioles, increase plasma leakage and thereby raise the hematocrit within microvessels, and thus alter the rheology of blood so as to increase the collision of leukocytes with the vessel wall (35). Furthermore, arteriolar dilation and the ensuing increased blood flow in inflammatory sites are responsible for two of the cardinal signs of inflammation, rubor (redness) and calor (heat), and greatly enhance the discharge and, thus, accelerate the accumulation of leukocytes.

IV. AREA CODE MOLECULES

A. Selectins

Multiple protein families, each with a distinct function, provide the traffic signals for leukocytes. The selectin family of adhesion molecules (Fig. 2) has a N-terminal domain homologous to Ca^{2+}-dependent lectins (9,18, 19,36,37). The name selectin capitalizes on the derivation of "lectin" and "select" from the same Latin root, meaning to separate by picking out. Selectins are limited in expression to cells of the vasculature (Fig.2) L-selectin is expressed on all circulating leukocytes, except for a subpopulation of lymphocytes (38–40). P-selectin in stored preformed in the Weibel-Palade bodies of endothelial cells and the α granules of platelets. In response to mediators of acute inflammation such as thrombin or histamine, P-selectin is rapidly mobilized to the plasma membrane to bind neutrophils and monocytes (25,41,42). E-selectin is induced on vascular endothelial cells by cytokines such as IL-1, LPS, or TNF and requires de novo mRNA and protein synthesis (43).

B. Carbohydrates and Mucin-Like Molecules

All selectins appear to recognize a sialylated carbohydrate determinant on their counterreceptors (7,8,19). E-selectin and P-selectin recognize carbohydrate structures that are distinct, but are both closely related to the tetrasaccharide sialyl Lewis^{x} and its isomer sialyl Lewis^{a} (Fig. 2). The actual

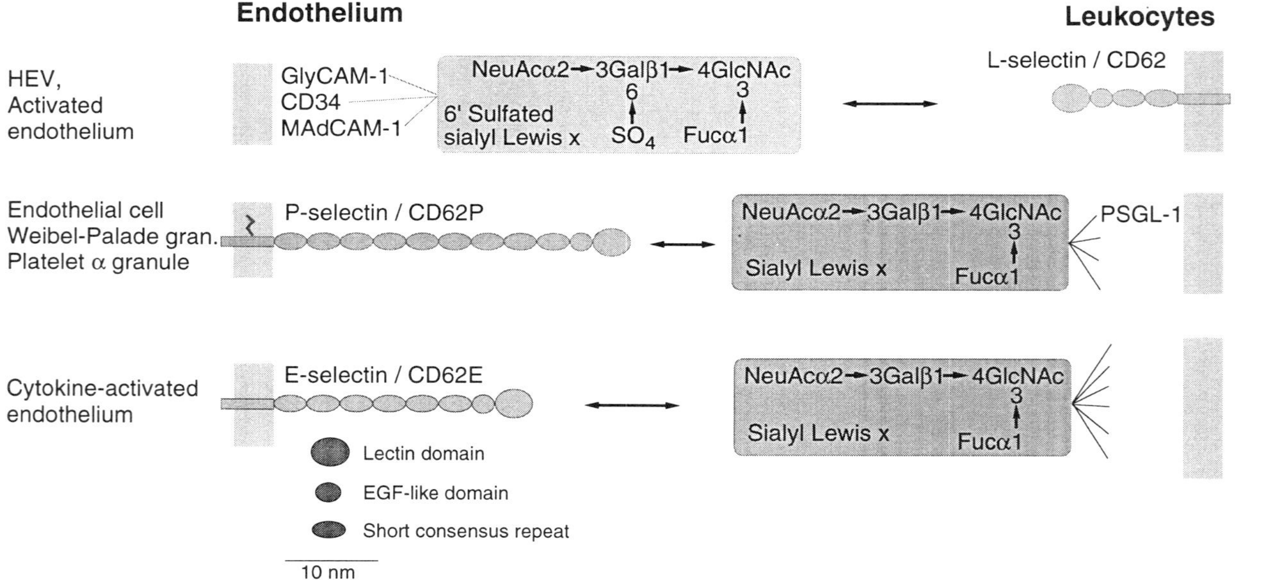

Figure 2 Selectins and their ligands. The selectins are shown to scale, based on electron micrographs of P-selectin (47), the X-ray structure of E-selectin lectin and EGF domains (49), and estimates of the sizes of the short consensus repeats (SCR) (36). P-selectin is shown palmitylated on a transmembrane cysteine (259). The carbohydrates are not to scale. Sialyl Lewis^a and Lewis^x contain Galβ1-3(Fucα1-4)GlcNAc and Galβ1-4(Fucα1-3)GlcNAc linkages, respectively.

ligand structures for E-and P-selectin are more complex, as shown by display of the ligand for E-selectin but not P-selectin on fucosyl transferase-transfected cells that express sialyl Lewis^x (44). The affinity of E-selectin for soluble sialyl Lewis^x is quite low, with K_d = 0.2-0.8 mM (45), which suggests that a higher-affinity ligand may yet be identified. P-selectin is specific for carbohydrate displayed on the P-selectin glycoprotein ligand (PSGL-1), suggesting either that PSGL-1 expresses a specific carbohydrate structure or that PSGL-1 protein forms part of the ligand binding site (46). The affinity of P-selectin for PSGL-1 is high with a K_d = 70 nM (47). Structure-function studies suggest that the Ca^{2+} binding site and a cluster of basic residues on E-selectin coordinate with the fucosyl and sialic acid carboxylate moieties, respectively, of sialyl Lewis^x (48,49).

The carbohydrate ligands for L- and P-selectin are O-linked to specific mucinlike molecules. Mucins are serine-and threonine-rich proteins that are heavily O-glycosylated and have an extended structure. L-selectin recognizes at least two mucins in HEV (Fig. 3)—glycosylation-dependent cell adhesion molecule-1 (GlyCAM-1), which is secreted (37), and CD34, which is on the cell surface (50). The carbohydrate ligand for L-selectin is related to sialyl Lewis^a and Lewis^x (51,52), contains sialic acid and sulfate, and is O-linked to mucinlike structures of HEV (19). Structural studies on the carbohydrates of GlyCAM-1 show that 6′ sulfated sialyl Lewis^x (Fig. 2) is a major oligosaccharide capping group, and is a candidate for the ligand structure (53).

The mucinlike P-selectin glycoprotein ligand (PSGL-1) is a disulfide-linked dimer of 120 kD_a subunits (46) that is sensitive to O-glycoprotease, which selectively cleaves mucinlike domains (54,55). PSGL-1 (Fig. 3) was isolated by screening for cDNA that expressed ligand activity (56). COS cells must be transfected both with the PSGL-1 cDNA and α-¾-fucosyl transferase cDNA to express P-selectin ligand activity. By contrast, COS cells cotransfected with cDNA for α-¾-fucosyl transferase and another mucinlike molecule that is expressed by neutrophils, CD43, lack P-selectin ligand activity.

C. Function of Selectins and Their Ligands

Selectins mediate functions unique to the vasculature, tethering of flowing leukocytes to the vessel wall, and formation of labile adhesions with the wall that permit leukocytes subsequently to roll in the direction of flow. One study demonstrated this with purified P-selectin incorporated into supported planar lipid bilayers on one wall of a flow chamber (57). At wall shear stresses within the range of those found in postcapillary venules, neutrophils formed labile attachments to the P-selectin in the bilayer and

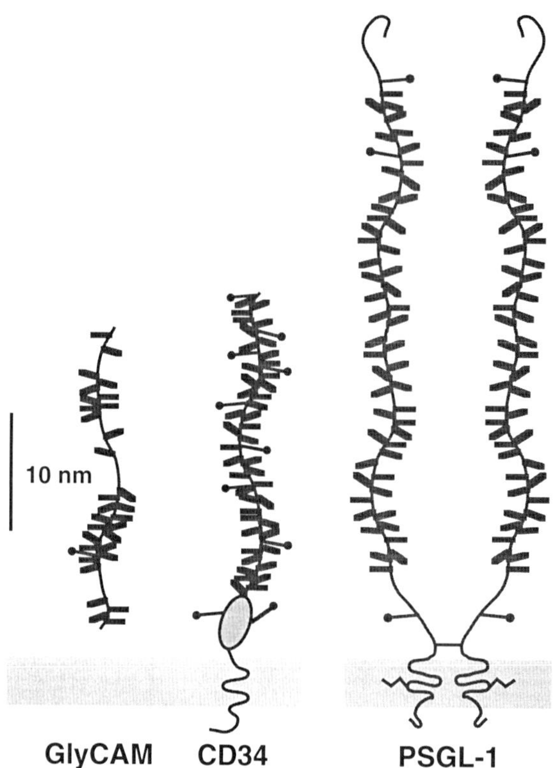

Figure 3 Mucinlike carriers of selectin ligands. The GlyCAM (37) and CD34 (50,77) molecules synthesized by peripheral lymph node HEV and MAdCAM-1 molecule synthesized by mucosal HEV (see Fig. 5) bear O-linked carbohydrates that bind to L-selectin. CD34 has a globular domain that may be Ig-like (260) and is resistant to O-glycoprotease (261). The PSGL-1 molecule on neutrophils bears O-linked carbohydrates that bind to P-selectin (55,56). A cysteine in the transmembrane region is predicted to be palmitylated. O-linked sites and N-linked sites are shown as bars and lollipops, respectively. The length of the mucinlike domains, and the percentage of serines and threonines that are O-glycosylated, are proportioned to measurements for CD43 (45 nm/224 amino acids and 75% to 90% of O-glycosylation) (72).

rolled in response to fluid drag forces. In other studies, intravascular infusion of a soluble L-selectin/IgG chimera inhibited neutrophil rolling attachments in vivo (58), as did infusion of anti-L-selectin monoclonal antibodies (59). More recent studies have shown that neutrophils roll on E-selectin in purified form (60) or on the endothelial cell surface both in vitro (61) and in vivo (62); that monoclonal antibody (mAb) to P-selectin decrease neutrophil rolling in vivo (63); and that neutrophil rolling in the microvasculature of mice genetically deficient in P-selectin is almost completely absent (64).

P- and L-selectin may cooperate with one another, because inhibition of either almost completely inhibits neutrophil rolling in vivo (58,59,64,65). E- and L-selectin also appear to cooperate (60,66–68). A class of ligand that is closely associated with L-selectin on the neutrophil surface is required for the initial tethering during flow to E-selectin bilayers, after which another class of ligands that mediates rolling takes over (69).

Selectins can mediate tethering of a flowing cell in the span of a millisecond. Other adhesion receptors require minutes to develop similar adhesive strength, and do not mediate rolling (57,70). It has been hypothesized that selectins differ from other adhesion molecules not in affinity (K_{eq}), but in having much more rapid association (k_{on}) and dissociation (k_{off}) rate constants (57), as has recently been confirmed (Table 1). Rolling is intermittent and appears mediated by random association and dissociation of selectin-ligand bonds, a small number of which tether a leukocyte to the vessel wall

Table 1 Fast On and Off Rates of a Selectin, and Affinity Modulation of an Integrin

	k_{on} ($M^{-1}sec^{-1}$)	k_{off} (sec^{-1})	K_d (μM)
P-selectin	1.4×10^{7}[a]	1[b]	0.07[c]
LFA-1 low affinity[d]	3×10^{2}	0.03	100
LFA-1 high affinity[e]	ND[f]	ND	0.6

[a]Calculated from $k_{on} = k_{off}/K_d$.
[b]At very low P-selectin densities in lipid bilayers, neutrophils, attach transiently; i.e., they subsequently detach rather than roll. Measurements of the cellular dissociation rate suggest that the $t_{1/2}$ for dissociation of a single selectin-ligand bond is about 0.7 sec. (R. Alon and T. Springer, unpublished.)
[c]For binding of monomeric, truncated P-selection to neutrophils (47).
[d]k_{on}, k_{off}, and K_d were measured by competitive inhibition by monomeric, truncated ICAM-1 of binding of Fab to LFA-1 on resting lymphocytes (111).
[e]Same as d, but for phorbol ester-stimulated lymphocytes. Approximately 20% of the cell surface LFA-1 was in the high-affinity state (111).
[f]ND = not determined.

at any one time. A rapid association rate facilitates the initial tethering in flow. A rapid dissociation rate ensures that even with multiple selectin-ligand bonds, it will not take long before the bond that is most upstream randomly dissociates, allowing the cell to roll forward a small distance until it is held by the next most upstream bond (57,71).

The elongated molecular structure of selectins and mucins, and their segmental flexibility (47,72), are predicted to enhance their accessibility for binding to counterstructures on closely opposed cells (57). P-selectin and PSGL-1 are currently the most elongated adhesion molecules known (Figs. 4, 5) and could bridge together two cells with plasma membranes about 0.1 μm apart. Expression on cytoplasmic protrusions further enhances accessibility. L-selectin is clustered on microvilli of neutrophils (67,73), which project about 0.3 μm above the surface of a cell with a diameter of 7 μm, and contain 90% of the L-selectin (D. Bainton, D. Hammer, and T. Springer, unpublished). In keeping with this topographic distribution, rolling in vivo requires the integrity of the L-selectin cytoplasmic domain and is inhibited by cytochalasin B (74). Lymphocytes bind through microvilli to HEV (75,76). Conversely, the mucinlike CD34 molecule (77) is concentrated on filopodia of nonspecialized endothelial cells found in the microvasculature of most tissues (26). These filopodia are concentrated near junctions between endothelial cells, and electron micrographs of granulocytes binding to the microvasculature in inflammatory sites suggest that the earliest binding event is to these filopodia (78).

D. Chemoattractants

Chemoattractants are important in activation of integrin adhesiveness and in directing the migration of leukocytes. In chemotaxis, cells move in the direction of increasing concentration of a chemoattractant, which typically is a soluble molecule that can diffuse away from the site of its production, where its concentration is highest (79,80). Leukocytes, which can sense a concentration difference of 1% across their diameter, move steadily in the direction of the chemoattractant. There is much interplay between adhesion molecules and chemoattractants, because adhesion to a surface is required to provide the traction necessary for migration directed by chemoattractants, and chemoattractants can activate adhesiveness.

The alternative mechanism to chemotaxis is haptotaxis. In "haptotaxis," cells migrate to the region of highest adhesiveness (81). Thus, on a gradient of an adhesive ligand affixed to the surface of other cells or to the extracellular matrix, and in the absence of a chemotactic gradient, motile cells will tend to accumulate in the region of highest ligand density. Both chemotaxis and haptotaxis can contribute to cell localization, but haptotaxis has yet to be demonstrated in vivo.

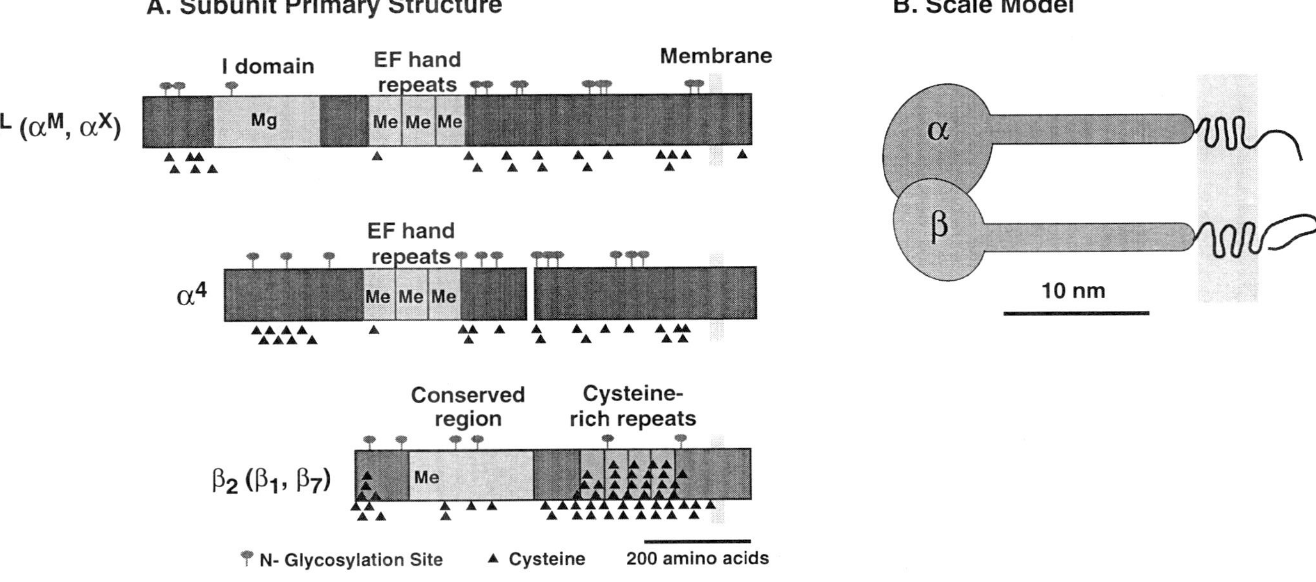

Figure 4 Integrins that bind endothelial ligands. (A) Schematics of representative integrin α and β subunits. The structures of α_L (262) and β_2 (263) integrin subunits are shown as representative of α_M and α_X or β_1 and β_7, respectively; cysteines are identical, while glycosylation sites vary but are sparse in the I domain and EF hand repeats. The EF hand repeats are divalent metal binding motifs that may bind Ca^{2+} or Mg^{2+} (labeled "Me"). A binding site for Mg^{2+} and Mn^2 but not Ca^{2+} has been identified in the I domain (110). The α_4 integrin subunit has a post translational proteolytic cleavage site (264). A putative divalent cation binding site has been defined in the conserved domain of the integrin β_3 subunit and is shown for β_2 (265). (B) Scale model of an integrin, based on electron micrographs of the integrins gpIIbIIIa (266) and VLA-05 ($\alpha_5\beta_1$) (267).

Table 2 Leukocyte Chemoattractants

Chemoattractant	Origin	Responding cells
Classical chemoattractants[a]		
N-formyl peptides	Bacterial protein processing	Monocyte, neutrophil, eosinophil, basophil
C5a	Complement activation	Monocyte, neutrophil, eosinophil, basophil
Leukotriene B4	Arachidonate metabolism	Monocyte, neutrophil
Platelet activating factor (PAF)	Phosphatidylcholine metabolism	Monocyte, neutrophil, eosinophil
CXC chemokines[b]		
IL-8/NAP-1	T lymphocyte, monocyte, endothelial cell, fibroblast, keratinocyte, chondrocyte, mesothelial cell	Neutrophil, basophil
CTAP-III/ β-thromboglobulin/ NAP-2	Successive N-terminal cleavage of platelet basic protein released from α-granules	Neutrophil, basophil, fibroblast
gro/MGSA	Fibroblast, melanomas, endothelial cell, monocyte	Neutrophil, melanomas, fibroblast
ENA-78	Epithelium	Neutrophil
CC chemokines[c]		
MCP-1	T lymphocyte, monocyte, fibroblast, endothelial cell, smooth muscle	Monocyte, T lymphocyte subpopulation, basophil
MIP-1α	Monocyte, B and T lymphocyte	Monocyte, T lymphocyte subpopulation, basophil, eosinophil
RANTES	T lymphocyte, platelets	Monocyte, T lymphocyte subpopulation, eosinophil
I-309	T lymphocyte, mast cell	Monocyte

[a](80,85).
[b](11,12,173,174,255).
[c](11,12,84,250–252,255,274–277).

whereas the receptor for IL-8 is expressed only on neutrophils (89). The receptor for MCP-1 is expressed on monocytic cells but not on neutrophils (90). Thus, the specificity of chemoattractants is regulated by the cellular distribution of their receptors.

F. Integrins

Integrins are perhaps the most versatile of the adhesion molecules. Integrin adhesiveness can be rapidly regulated by the cells on which they are expressed. Each integrin contains a noncovalently associated α and β subunit, with characteristic structural motifs (Fig. 4). Five integrins are important in the interaction of leukocytes with endothelial cells. Their cellular distribution, ligand specificity, and structure are summarized in Table 3 and Figure 4.

Table 3 Integrins in Leukocyte-Endothelial Interactions

Subunits	Names	Distribution	Ligands
Leukocyte integrins[a]			
$\alpha_L\beta_2$	LFA-1, CD11a/ CD18	B and T lymphocyte, monocyte, neutrophil	ICAM-1 ICAM-2, ICAM-3
$\alpha_M\beta_2$	Mac-1, CR3, CD11b/CD18	Monocyte, neutrophil	ICAM-1, iC3b, fibrinogen, factor X
$\alpha_x\beta_2$	p150,95, CR4, CD11c/CD18	Monocyte, neutrophil, eosinophil	iC3b, fibrinogen
α_4 *Integrins*[b]			
$\alpha_4\beta_1$	VLA-4, CD49d/ CD29	B and T lymphocyte, monocyte, neural crest-derived cells, fibroblast, muscle	VCAM-1, fibronectin
$\alpha_4\beta_7$	LPAM-1, CD49d/ CD-	B and T lymphocyte subpopulations	MAdCAM-1, VCAM-1, fibronectin

[a](36,151).
[b](36,123,124,131,132,199,278–281).

G. Activation of Integrins

The adhesiveness of LFA-1 and VLA-4 on T-lymphocytes is activated by cross-linking of the antigen receptor and other surface molecules (13,20, 36). Increased adhesiveness occurs within a few minutes, is not accompanied by any change in quantity of surface expression, and appears to result from both conformational changes that increase affinity for ligand, and altered interaction with the cytoskeleton (20,91,92). However, it is unlikely that recognition by T-cell receptors of antigen on endothelial cells (93) is a step in lymphocyte trafficking, because traffic of both lymphocytes that can and cannot recognize specific antigen is increased in antigen-induced inflammation. Although evidence has been presented that binding of neutrophils to selectins can activate adhesiveness of integrins (94), other evidence has failed to confirm this (60,95; Diacovo and Springer, unpublished).

Thus far the best candidates for activation of integrin adhesiveness within the vasculature are chemoattractants. Adhesiveness of Mac-1 and LFA-1 on neutrophils and monocytes is activated by N-formylated peptide and IL-8 (96,96–100). In contrast to LFA-1 on lymphocytes and neutrophils, Mac-1 on neutrophils is increased about 10-fold on the surface by chemoattractant-stimulated fusion of secretory granules with the plasma membrane (101); however, this is neither sufficient or necessary for increased adhesiveness (102,103). The transient nature of the activation of integrin adhesiveness (96,104) provides a mechanism for de-adhesion, and perhaps for retraction of the trailing edge of a leukocyte from the substrate during cell migration.

Conformational changes in LFA-1 and Mac-1 that are associated with increased adhesiveness are suggested by mAb and Fab that react only with these molecules after cellular activation (105-108). After chemoattractant activation of neutrophils, saturation binding shows that 10% of the surface Mac-1 molecules express an activation epitope, yet mAb to this epitope completely blocks binding to ligands such as ICAM-1 and fibrinogen. This suggests that ligand binding is mediated by a subpopulation of activated Mac-1 molecules (108). The I domain of leukocyte integrins is important in ligand binding (109,110), and expresses activation epitopes (107,108). Recent measurements of the affinity of cell surface LFA-1 for soluble, monomeric ICAM-1 (Table 1) have directly demonstrated that cellular activation increases the affinity of a subpopulation of LFA-1 molecules approximately 200-fold (111).

Surprisingly, the integrin VLA-4, by contrast to LFA-1 and Mac-1, has recently been found to be capable of supporting rolling. Lymphocytes can tether in flow and subsequently roll on VCAM-1. If activated while rolling by phorbol ester of TS2/16 mAb to the β1 subunit, the lymphocytes arrest and develop firm adhesion. Activated lymphocytes tether as efficiently as resting lymphocytes but do not roll. Fibronectin can support development of firm adhesion in static conditions but not tethering or rolling in flow. VCAM-1 is less efficient than selectins in mediating tethering and rolling (112).

H. IgSF Members on Endothelium as Integrin Ligands

In a paradigm first established with ICAM-1 binding to LFA-1, several immunoglobulin superfamily (IgSF) members, expressed on endothelium, bind to integrins expressed on leukocytes (Fig. 5). ICAM-1, ICAM-2, and ICAM-3 are products of distinct and homologous genes and were all initially identified by their ability to interact with LFA-1 (113–115). ICAM-1 has also been found to bind to Mac-1, through a distinct site in its third Ig domain (99,116,117) (Fig. 5). Induction of ICAM-1 on endothelium and other cells by inflammatory cytokines may increase cell-cell interactions and leukocyte extravasation at inflammatory sites, whereas constitutive expression of ICAM-2 may be important for leukocyte trafficking in uninflamed tissues, as in lymphocyte recirculation. ICAM-3 is restricted to leukocytes. All three of the ICAMs contribute to antigen-specific interactions, so that inhibition with mAb to all three is required to completely block LFA-1-dependent antigen-specific T cell responses (118).

Vascular cell adhesion molecule-1 (VCAM-1) is inducible by cytokines on endothelial cells, and on a more restricted subset of non-vascular cells than ICAM-1 (9). A single VCAM-1 gene gives rise through alternative splicing to a seven-domain isoform, and to a second isoform which contains either six domains or three domains and a glycosyl phosphatidyl inositol membrane anchor (119–121) (Fig. 5). VCAM-1 is a ligand for the integrin $\alpha_4\beta_1$ (VLA-4) and binds weakly to $\alpha_4\beta_7$ (122–124). In contrast to the shorter isoforms, the seven domain isoform of VCAM-1 has two binding sites for VLA-4, in highly homologous domains 1 and 4 (125–128).

An addressin for lymphocyte recirculation to mucosa is expressed on Peyer's patch HEV and on other venules (129). Now termed mucosal addressin cell adhesion molecule (MAdCAM-1), it contains three Ig-like domains and a mucinlike region interposed between domains 2 and 3 (130) (Fig. 5). MAdCAM-1 binds the integrin $\alpha_4\beta_7$ but not $\alpha_4\beta_1$ (131,132). Furthermore, carbohydrates attached to the mucinlike domain of MAdCAM-1 bind L-selectin and mediate lymphocyte rolling (133). Thus MAdCAM-1 has a dual function as an integrin and selectin ligand.

I. Other Molecules

CD31 is an IgSF member expressed on leukocytes, platelets, and at cell-cell junctions on endothelium (134–140). CD31 can bind homophilically to itself and also heterophilically to an uncharacterized counterreceptor. mAb cross-linking of CD31, similarly to many but not all other lymphocyte surface molecules, can trigger integrin adhesiveness (140). Interaction between CD31 on endothelial junctions, and CD31 on leukocytes, appears to be required for transmigration but not for integrin-mediated binding of leukocytes to endothelium (141). CD31-CD31 interaction may represent a fourth step in transendothelial migration that overlaps the integrin-mediated step and may contribute to the maintenance of the permeability barrier function of endothelia during transmigration.

CD44 is a widely distributed molecule in the body that is homologous with cartilage link protein, is extensively alternatively spliced, and can bear heparin sulfate or chondroitin sulfate side chains (142). The best-understood function of CD44 is as a major surface receptor for hyaluronate (143,144). Alternatively spliced forms of CD44 are important in tumor metastasis (145), and in localization of antibody-secreting cells (146). CD44 (H-CAM, Hermes) was at one time mistakenly thought to be the human equivalent of murine mel-14 (L-selectin). It participates in vitro in lymphocyte interaction with HEV and activated endothelium (147,148). However, lack of cell surface CD44 has no effect on lymphocyte recirculation in vivo (149).

V. TOWARD A MULTISTEP MODEL OF NEUTROPHIL EMIGRATION IN INFLAMMATION

A. Integrins and Selectins

Patients who are genetically deficient in the leukocyte integrins, owing to mutations in the common β_2 integrin CD18 subunit, provided early evidence that adhesion molecules were required for leukocyte extravasation in vivo (150,151). Leukocyte adhesion deficiency-I (LAD-I) patients have life-threatening bacterial infections, and neutrophils in these patients fail to cross the endothelium and accumulate at inflammatory sites, despite higher than normal levels of neutrophils in the circulation. In vitro, LAD-I neutrophils or normal neutrophils treated with mAb to the leukocyte integrins are deficient in binding to and migrating across resting or activated endothelial monolayers (152,153). Even though capable of binding to activated endothelium through selectins, LAD-I neutrophils fail to transmigrate (153). mAbs to the leukocyte integrin β_2 subunit, and in some cases the integrin αM subunit, have been found to have profound effects in vivo (15). These mAbs prevent the neutrophil-mediated injury that occurs when ischemic tissue is re-perfused, and thus can prevent death from shock after blood loss, limb necrosis after frostbite or after amputation and replantation, and tissue necrosis from myocardial ischemia and reperfusion. mAbs to leukocyte integrins and to ICAM-1 can also inhibit lymphocyte and monocyte-mediated antigen-specific responses in vivo, including delayed-type hypersensitivity, granuloma formation, and allograft rejection (15).

Whereas mAb to the leukocyte integrin β_2 subunit blocked accumulation of leukocytes in tissue in response to chemoattractants, and stable adhesion of leukocytes in the local vasculature, it had no effect on the number of rolling leukocytes on the vessel wall (154). Furthermore, leukocyte integrins were found to mediate binding of neutrophils to endothelial monolayers in a parallel wall flow chamber at subphysiologic, but not at physiologic shear stresses found in postcapillary venules (153,155).

Parallel studies showed that selectins were required for leukocyte accumulation in vivo and acted at an early step. Antagonists of L-selectin and E-selectin inhibit neutrophil and monocyte influx into skin, peritoneal cavity, and lung in response to inflammatory agents (40,156–159). mAb to L-selectin was shown to inhibit neutrophil accumulation on cytokine-stimulated endothelium at physiologic shear stress (100). Stimulation of neutrophils with chemoattractants results in shedding into the medium with minutes of L-selectin, with kinetics similar to upregulation of surface expression of the integrin Mac-1. Based on this and the evidence reviewed above, it was hypothesized that selectins might act at a step prior to integrins (160).

Further studies showed that selectins mediate rolling and function prior to development of firm adhesion through integrins. At sites of inflammation, leukocytes first attach to the vessel wall in a rolling interaction, the become arrested or firmly adherent at a single location on the vessel wall before diapedesis (161). This process was fully reconstituted with purified components of the endothelial surface (57). At physiologic shear stresses, neutrophils attach to and form labile rolling adhesions on phospholipid bilayers containing purified P-selectin, but not on bilayers containing ICAM-1. Chemoattractants stimulate strong, integrin-mediated adhesion to bilayers containing ICAM-1 under static conditions but not in shear flow. At physiologic shear stresses, if both P-selectin and ICAM-1 are present in the phospholipid bilayer, resting neutrophils attach and roll identically as on bilayers containing P-selectin alone. However, when chemoattractant is added to the buffer flowing through the chamber, the rolling neutrophils arrest, spread, and firmly adhere through the integrin-ICAM-1 interaction. Chemoattractant does not enhance and actually inhibits interactions of neutrophils with bilayers containing P-selectin alone. These findings show that purified adhesion molecules and chemoattractants representing the endothelial signals can reproduce the key events in leukocyte localization in vivo, and prove that the selectin-mediated step is a prerequisite for the chemoattractant and integrin-mediated steps (57). Complementary in vivo studies showed that mAb to L-selectin, or L-selectin/IgG chimeras, decreased both the number of rolling leukocytes (58,59), and the number of leukocytes that subsequently became firmly adherent, whereas mAb to the β_2-integrin subunit only decreased firm adherence of leukocytes. This suggested that L-selectin acts at a step prior to leukocyte integrins (59). In static assays, a factor derived from cytokine-stimulated endothelium induced shedding of L-selectin, and if transmigration was blocked with CD18 mAb, induced release of neutrophils from inverted endothelial monolayers. This also suggested that L-selectin acted prior to leukocyte integrin-mediated emigration (100). In elegant confirmation of a three-step model in a static assay of neutrophil adhesion to histamine-stimulated endothelium, juxtacrine cooperation between P-selectin and platelet activating factor (PAF) was found (95). P-selectin tethered neutrophils to endothelium and thereby augmented stimulation by PAF of CD18-dependent neutrophil adhesion. Stimulation of adhesiveness was by PAF and not by P-selectin, as shown with PAF receptor antagonists.

The requirement for the carbohydrate ligands of selectins for leukocyte emigration in vivo has received strong support from studies of two patients with a genetic defect in biosynthesis of fucose, and who therefore lack the ligands for E-selectin and P-selectin (162,163). The defect, designated LAD-II, has many clinical similarities to LAD-I, including strikingly depressed neutrophil emigration into inflammatory sites.

B. Chemoattractants

Chemoattractants appear to be required for transendothelial migration in vitro and in vivo, and can induce all steps required for transmigration in vivo. Injection of chemoattractants into skin or muscle leads to robust emigration of neutrophils from the vasculature and accumulation at the injection site (164). Injection of lipopolysaccharide or cytokines that induce IL-8 synthesis also elicit neutrophil emigration. Moreover, mAb to IL-8 markedly inhibits neutrophil emigration into lung and skin in several models of inflammation (165,166).

The effects of pertussis toxin provide further evidence for the importance of $G\alpha_i$ protein-coupled receptors in leukocyte emigration in vivo. Pretreatment of neutrophils with pertussis toxin inhibits emigration into inflammatory sites (167,168).

Chemoattractants impart directionality to leukocyte migration. By contrast to intradermal injection, intravascular injection of IL-8 does not lead to emigration (169). Cytokine-stimulated endothelial monolayers grown on filters secrete IL-8 into the underlying collagen layer. Neutrophils added to the apical compartment emigrate into the basilar compartment, but not when the IL-8 gradient is disrupted by addition of IL-8 to the apical compartment (82). Although IL-8 acts as an adhesion inhibitor in some assays (170), this result may be partially attributable to disruption of a gradient of IL-8 on activated endothelial monolayers when exogenous IL-8 is added on the same side as the neutrophils.

Chemoattractants act on the local tissue, as well as on leukocytes. Neutrophil chemoattractants injected into the same skin site hours apart will stimulate neutrophil accumulation the first but not the second time, whereas a second injection into a distant site will stimulate accumulation. Desensitization occurs for homologous chemoattractants only (171,172). Thus chemoattractants must act on and homologously desensitize a cell type that is localized in tissue. In some cases this localized cell may be the mast cell. Some chemoattractants stimulate the mast cell (which localizes in tissue adjacent to the vasculature) or its better studied relative the basophil, to release histamine (173,174) and TNF (175). Histamine induces P-selectin and TNF induces E-selectin on endothelium. Thus chemoattractants may indirectly increase selectin expression on endothelium, as well as directly activate integrin adhesiveness on leukocytes.

VI. THREE-STEP AREA CODE FOR SIGNALING NEUTROPHIL AND MONOCYTE TRAFFIC

The above evidence has shown that emigration from the vasculature of neutrophils and monocytes is regulated by at least three distinct molecular signals (Figs. 1, 6A). A key feature is that selectin-carbohydrate, chemoattractant-receptor, and integrin-Ig family interactions act in sequence, not in parallel. This concept has been confirmed by the observation that inhibition of any one of these steps gives essentially complete, rather than partial, inhibition of neutrophil and monocyte emigration. An important consequence of a sequence of steps, at any one of which there are choices of multiple receptors or ligands that have distinct distributions on leukocyte subpopulations or endothelium, is that it provides great combinatorial diversity for regulating the selectivity of leukocyte localization in vivo, as has been emphasized in several reviews (3,7,10,13,14).

"Area code" models for cell localization in the body (176,177) are particularly apt because it is now known that at least three sequential steps are involved. The concepts of area codes and traffic signals can be combined by thinking of how telephone traffic is routed by digital signals. Each type

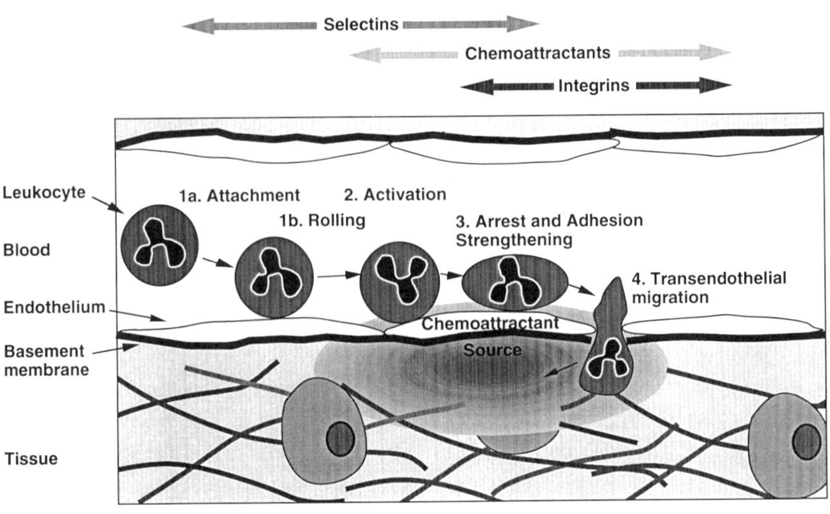

(a)

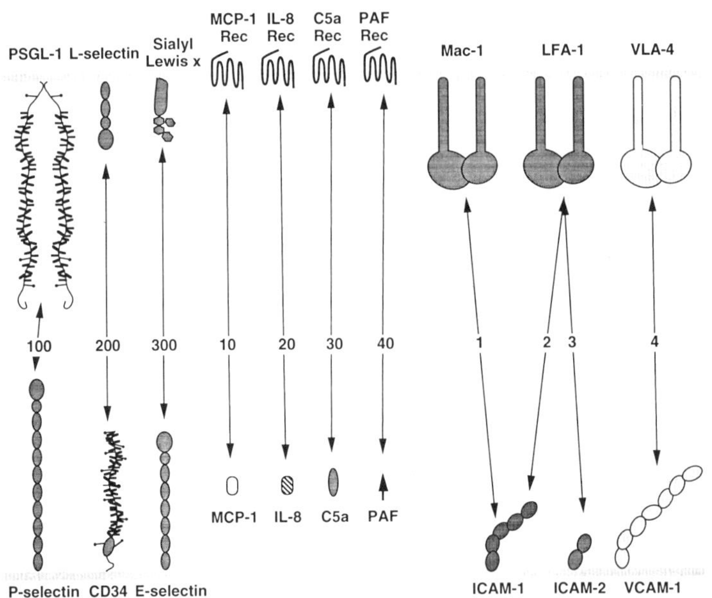

Monocyte Area Codes	111, 211, 311, 112, 212, 312, 113, 213, 313, 114, 214, 314, 134, 234, 334, 144, 244, 344
Neutrophil Area Codes	121, 221, 321, 122, 222, 322, 123, 223, 323
Monocyte and Neutrophil Area Codes	131, 231, 331, 132, 232, 332, 133, 233, 333, 141, 241, 341, 142, 242, 342, 143, 243, 343
Null Area Codes	124, 224, 324

(b)

of leukocyte responds to a particular set of area code signals. Inflammation alters the expression and location of the signals on vascular endothelium. It is as if leukocytes carry "cellular phones." An example of how this model works is shown for the two cell types for which the signals are best understood, neutrophils and monocytes (Fig. 6B). Chemoattractants provide the greatest number of molecular choices (or "digits") and the greatest cellular specificity.

Refinements to the three-step model are in order. First, selectins actually mediate two steps, initial tethering to the vessel wall and rolling (Fig. 6A), which can be distinguished for E-selectin by dependence on different classes of neutrophil ligands (69). Thus, selectins can cooperate, and some selectin-ligand combinations may be more important in tethering and others in rolling. Second, the steps are overlapping, rather than strictly sequential (Fig. 6A). Although L-selectin is shed from neutrophils soon after activation (160), the kinetics of shedding by neutrophils in whole blood (minutes) are much slower than the transition from rolling to integrin-mediated attachment (msec-sec) (59). L-selectin in shed more slowly from lymphocytes than from neutrophils (178,179). Furthermore, ligands for P-selectin (46) and E-selectin (69) remain on the neutrophil surface after activation. Thus interactions with selectins will continue after activation of integrins, probably persisting until transendothelial migration is completed. Chemoattractants are required not only for activation of integrin adhesiveness, but also for directional cues during the subsequent step of transendothelial migration. Finally, β_1-integrins that bind to extracellular matrix components are undoubtedly required during migration through the subendothelial basement membrane.

VII. LYMPHOCYTE RECIRCULATION: DISTINCT TRAFFIC PATTERNS FOR NAIVE AND MEMORY LYMPHOCYTES

Patrolling the body in search of foreign antigen, lymphocytes follow circuits through both non-lymphoid and lymphoid tissues (Fig. 7). The pe-

---

Figure 6 The three-step area code model. (a) Selectins, chemoattractants, and integrins act in sequence, with some overlap. (b) Combinatorial use of different molecules at each step can generate a large number of different area codes, and specificity for distinct leukocyte subpopulations. All of the known selectin and integrin interactions are shown in the hundreds and ones place, respectively; however, only a subset of the chemoattractants is shown in the tens place (see Table 1) owing to space limitations. The area codes symbolize how specificity for monocytes, neutrophils, or both can be generated at inflammatory sites.

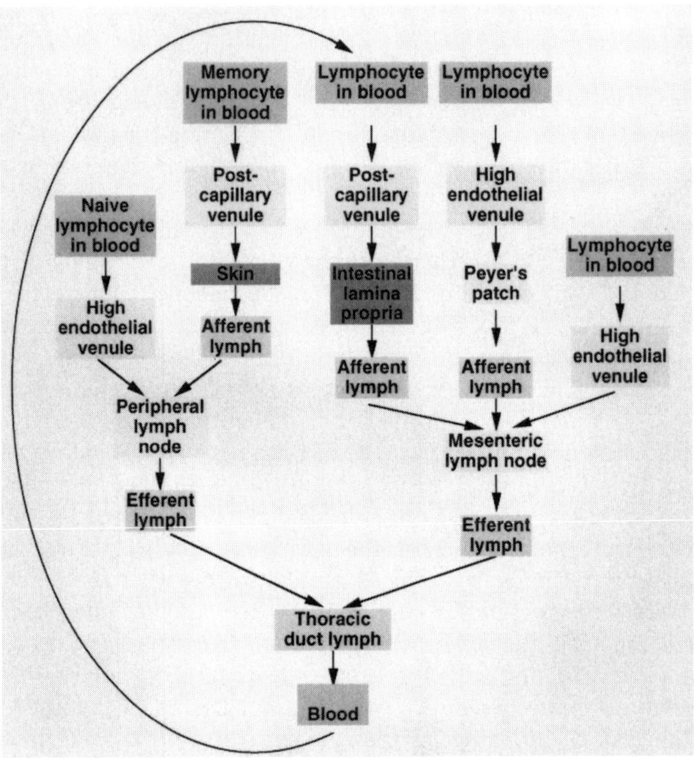

Figure 7 Lymphocyte recirculation routes.

ripheral lymph nodes draining skin and muscle, and the gut-associated lymphoid tissues such as Peyer's patch, differ in the types of antigens to which lymphocytes are exposed. When collected from lymph draining gut or skin, lymphocytes from adult animals, but not newborns, show a two-fold or higher preference to recirculate to the type of organ from which they came and to reappear in the draining lymph (1,2,17,180). This suggests that priming by specific antigen in a particular environment may induce expression of surface receptors that enable preferential recirculation to the type of secondary organ where specific antigen was first encountered. Evidence exists for separate streams of lymphocytes that recirculate through the skin, gut, and lung, and that drain into their associated lymphoid tissues (6,17).

Our understanding of the mechanisms of this selectivity has been advanced by the discovery that "naive" and "memory" lymphocytes prefer different recirculation pathways (181). When naive lymphocytes encounter antigen, those lymphocytes with receptors specific for the antigen are stimulated to clonally expand and are converted to memory lymphocytes that have altered expression of adhesion receptors and circulatory patterns. Lymphocytes that emigrate in the hind leg of a sheep through "flat" endothelium in the skin and drain through the afferent lymphatics to the popliteal lymph node are all of the memory phenotype. By contrast, lymphocytes in the efferent lymph from the popliteal lymph node, derived mostly from traffic through HEV, are predominantly of the naive phenotype. Thus, at least for peripheral tissues and lymph nodes, memory lymphocytes emigrate preferentially through tissue endothelium, whereas naive lymphocytes enter the lymph node through HEV (Fig. 7). Memory lymphocytes are more sensitive to specific antigen than naive lymphocytes, and thus better able to respond to antigen in peripheral tissues, which have fewer antigen presenting cells than lymph nodes (16).

VIII. TRAFFIC THROUGH HEV

The "high" or cuboidal-shaped endothelial cells found in HEV are specialized for emigration of lymphocytes into peripheral lymph nodes that drain skin, and the lymphoid tissues of the mucosa: Peyer's patches, tonsils, and appendix. Emigration into the spleen, by contrast, involves sinusoidal endothelia and molecular mechanisms that are distinct and not yet characterized. About 25% of lymphocytes that circulate through an HEV will bind and emigrate, a much higher percent than through nonspecialized flat venules (182,183). HEV phenotype is developmentally regulated. The carbohydrate ligands for L-selectin are absent from peripheral lymph node HEV at birth but are displayed at adult levels by 6 weeks (6). If peripheral lymph nodes are deprived of afferent lymph, the HEV convert from a high to a flat-walled endothelial morphology, lose expression of L-selectin ligands, and lose ability to support lymphocyte traffic (184,185). Introduction of antigen into the node leads to a full restoration of HEV phenotype and function. Futhermore, intense antigenic stimulation can induce formation of HEV in diverse nonlymphoid tissues (6,186).

A. Molecular Mechanisms Defined by the HEV Binding Assay

When lymphocyte suspensions are overlayed on thin sections cut from frozen lymph nodes, the lymphocytes specifically bind to the morphologically distinct HEV (187). Specific differences have been demonstrated between binding to peripheral lymph node and Peyer's patch HEV (183,188). T lymphocytes bind 1.5 fold better than B lymphocytes to peripheral lymph node HEV in vitro, and show a similar preference to recirculate to this site in vivo. B lymphocytes bind two- to threefold better to Peyer's patch than to peripheral lymph node HEV and show similar preference in recirculation in vivo. These preferences are reflected in the preponderance of T cells in peripheral lymph nodes, and the preponderance of B lymphocytes in Peyer's patch, where they are important in secretion of IgA and IgM into the mucosa (189). Certain lymphoma cells possess marked preference for binding to Peyer's patch or peripheral lymph node HEV in vitro (188), and for metastasis in vivo to mucosal or peripheral lymphoid tissue, respectively (190). Assay of lymphoma cell binding to HEV in the Stamper-Woodruff assay has led to the identification of two important adhesion pathways.

1. Molecules Involved in Binding to Peripheral Node HEV

The L-selectin molecule was initially defined in the mouse with the Mel-14 mAb as a molecule on lymphocytes required for binding to peripheral lymph node, but not Peyer's patch, HEV (38). Conversely, the MECA-79 carbohydrate antigen was defined with mAb that bound specifically to peripheral lymph node HEV and blocked lymphocyte binding. The isolated MECA-79 antigen, termed the peripheral node addressin (191), binds to L-selectin on lymphocytes (192). An L-selectin/IgG chimera was also found to specifically bind to HEV in peripheral lymph node and to block lymphocyte binding (193). The L-selectin/IgG chimera was used to isolate two distinct mucinlike ligands, GlyCAM-1, which is secreted by HEV (37), and CD34, a surface molecule on HEV (50). MECA-79 mAb recognizes a carbohydrate determinant that is expressed on multiple protein species in HEV, including GlyCAM-1 and CD34, and compared to L-selectin recognizes an overlapping but distinct set of glycoproteins (37,192). Sialylation and sulfation of the O-linked side chains of the GlyCAM-1 and CD34 molecules are required for activity in binding to L-selection (19,192,194). HEV differ from other tissues in carbohydrate processing; GlyCAM-1 and CD34 expressed in transfectants, and CD34 in other vascular endothelia do not bind L-selectin chimera under conditions in which binding to HEV is detectable (37). However, an L-selectin ligand with a presumably lower affinity is certainly present on most endothelia, as shown by L-selectin-dependent rolling in vivo and binding in vitro (58,59,68,74,100,195,196).

2. Molecules Involved in Binding to Peyer's Patch HEV

Elegant screens for mAb with specificity for Peyer's patch HEV, and ability to block lymphocyte binding to HEV, yielding mAb MECA-367 to the mucosal addressin now termed MAdCAM-1 (129). MAdCAM-1 is expressed on endothelia in mucosal tissues, not only on HEV in Peyer's patch, but also on venules in intestinal lamina propria and in the lactating mammary gland (129,197). MAdCAM-1 has both IgSF domains and a mucin-like domain (130) (Fig. 5).

Similar elegant screens for mAbs with specificity for lymphoma cells that bound to Peyer's patch HEV and with ability to block to binding to HEV in the Stamper-Woodruff assay yielded mAbs to the α_4 subunit of the Peyer's patch homing receptor (198). The α_4 subunit was found to be associated with a novel β subunit, β_p (199), which is identical to β_7 (131). The integrin $\alpha_4\beta_7$ but not $\alpha_4\beta_1$ binds to Peyer's patch HEV (131), and $\alpha_4\beta_7$ binds directly to MAdCAM-1 (132).

B. An Area Code Model for Lymphocyte Migration Through HEV

1. Peripheral Lymph Node HEV

Although the L-selectin: mucin and $\alpha_4\beta_7$: MAdCAM-1 interactions were identified in parallel assays, recent studies suggest that multiple steps are involved in lymphocyte interaction with HEV, and raise the possibility that these interactions may function in distinct, rather than parallel, steps in this process. Soon after its discovery as a lymphocyte homing receptor, L-selectin also was found to be present on neutrophils and eosinophils, and to be important in emigration of at least neutrophils (40). As expected from their strong expression of L-selectin, neutrophils and other leukocytes can bind avidly to HEV in the Stamper-Woodruff assay, yet do not normally home to peripheral lymph nodes in vivo. Injection of *E. Coli* supernatant induces acute emigration of neutrophils through HEV of the draining lymph node. Thus, signals other than those mediated by L-selectin can regulate the class of leukocyte that home into a lymph node (40). Although peripheral node HEV is far richer than any other site in the body in expression of the carbohydrate receptor for L-selectin (200), this is insufficient to explain the specificity of lymphocyte homing to this organ. The findings suggest that L-selectin is required for lymphocyte emigration through peripheral lymph node HEV, and may help regulate recirculation of the L-selectin$^+$ subset of lymphocytes; however, L-selectin is insufficient to determine the specificity of the cell types that emigrate, and other, currently undefined molecules are required to achieve specificity.

In vivo studies strongly suggest that lymphocyte emigration through HEV is a multi-step process that utilizes area code models similar to those of other leukocytes. mAb to L-selectin almost completely blocks emigration of lymphocytes from blood into peripheral lymph nodes (38,201). However, mAb to the integrin LFA-1 also markedly reduces or almost completely abolishes lymphocyte migration into peripheral lymph nodes (149,

202). Thus, molecules of step 1 and 3 are required for homing to peripheral lymph nodes in vivo. LFA-1 on blood lymphocytes requires activation for binding to its counter-structures ICAM-1 and ICAM-2 (36), which are expressed on HEV (104,203). Binding of L-selectin does not trigger activation of LFA-1 because lymphocytes attach and roll in flow on purified peripheral node addressin identically whether or not purified ICAM-1 is present on the substrate; an additional stimulus is required before lymphocytes will arrest and strengthen adhesion through LFA-1 (M. Lawrence, E. Berg, E. Butcher, and T. Springer, in preparation).

G protein-coupled receptors are required for lymphocyte recirculation and likely provide the signals required to activate the adhesiveness of LFA-1. Pertussis toxin causes lymphocytosis and profoundly depresses lymphocyte recirculation (204). Murine lymphocytes treated with pertussis toxin in vitro and reinfused fail to emigrate into either peripheral lymph nodes or Peyer's patches (205). This suggests that G protein-coupled receptors of the α_i class are required for lymphocyte emigration through HEV. Results with mice with a transgene for the ADP-ribosylating subunit of pertussis toxin selectively expressed in the T lineage suggest that $G\alpha_i$ proteins are not only required for emigration from the bloodstream, but also for emigration from the thymus (206,207). Despite lack of emigration, pertussis toxin-treated lymphocytes bind normally to lymph node HEV in vitro. These findings provided the basis for an early proposal for a two-step model, in which G protein-coupled receptors function subsequent to binding of lymphocytes to HEV (208).

Thus, emigration of lymphocytes through peripheral node HEV requires three sequential area code signals that are analogous to those involved in neutrophil emigration from the blood stream (Fig. 8). Identification of a putative lymphocyte chemoattractant secreted by peripheral lymph node HEV, and a chemoattractant receptor that is predicted to be selectively expressed on the naive subset of lymphocytes that recirculate through peripheral node HEV, will be a subject of intense research interest in coming years.

2. Peyer's Patch

mAbs to L-selectin block 50% of lymphocyte emigration from blood to Peyer's patch and to the remainder of the intestine (201,209). This is consistent with the lower level of L-selectin ligand in Peyer's patch HEV than in peripheral lymph node HEV (193,210,211). mAbs to certain epitopes on the integrin α_4 and β_7 subunits inhibit by approximately 50% recirculation of lymphocytes to Peyer's patch and intestine, but have no effect on recirculation to peripheral lymph node; furthermore, mAbs specific for the $\alpha_4\beta_7$ complex are equally as effective as mAb to α_4 (209). Moreover, recircula-

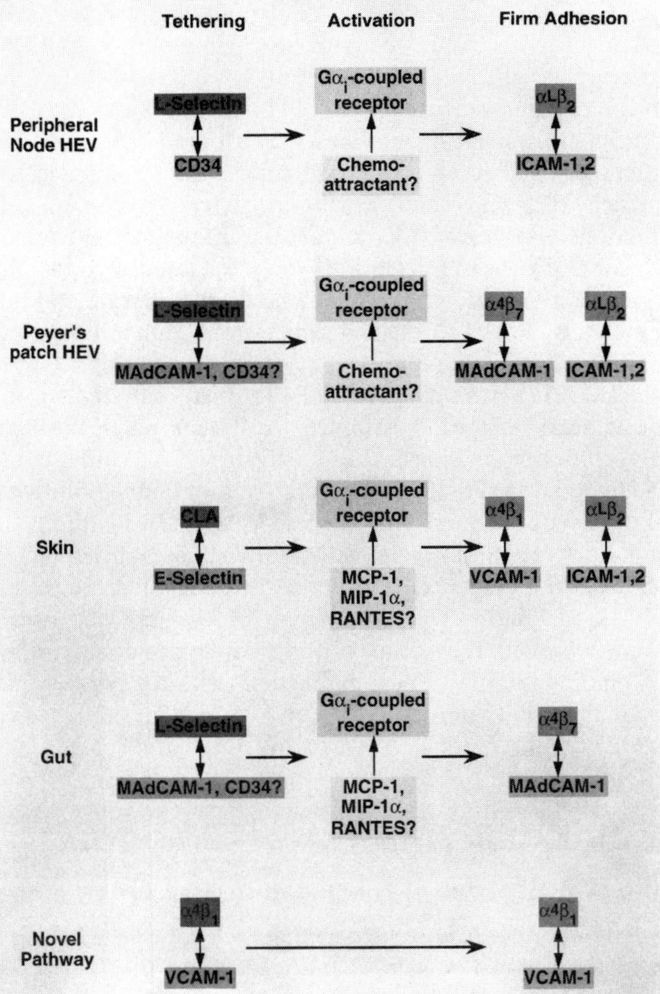

Figure 8 The three-step or four-step area code paradigm for lymphocytes. For skin and gut, the pathways shown may mediate both recirculation and increased accumulation in inflammation. The novel pathway shown at the bottom may be important when VCAM-1 expression on endothelium is induced by cytokines and may cooperate with the other illustrated pathways. For each organ, the interacting molecules are shown on the top for lymphocytes and on the bottom for endothelia. See text for support for the molecular assignments at each step, based on primarily on in vivo data.

tion is inhibited by mAb to MAdCAM-1 (129), implicating $\alpha_4\beta_7$ binding to MAdCAM-1 in recirculation to mucosal tissue. mAb to LFA-1 block recirculation to Peyer's patch by 50% to 80% but have no effect on recirculation to the remainder of the intestine (149,202). Thus, both LFA-1 and $\alpha_4\beta_7$ contribute to emigration into muscosal lymphoid tissue.

G-protein coupled receptors act subsequent to a rolling interaction in Peyer's patch HEV. In contrast with peripheral lymph nodes, Peyer's patches may be visualized by intravital microscopy (182). Normally, lymphocytes roll along Peyer's patch HEV only for a few seconds, then arrest and emigrate. However, prior treatment of lymphocytes with pertussis toxin completely blocks arrest and emigration, and prolongs the rolling indefinitely, so that the lymphocytes pass out of the Peyer's patch rather than emigrate (212). It remains to be established, but seems likely, that a chemoattractant presented or secreted by Peyer's patch binds to a $G\alpha_i$-coupled receptor on lymphocytes, and activates LFA-1 and $\alpha_4\beta_7$ to mediate arrest and emigration (Fig. 8). Lymphoma cells or lymph node lymphocytes can bind to Peyer's patch HEV or purified MAdCAM-1 without any apparent need for activation; however, activation increases the strength of binding to MAdCAM-1 (131,132). The pertussis toxin studies suggest that activation of blood lymphocytes is required for the last step of arrest and emigration (208,212). Truncation of the cytoplasmic domain of β_7 greatly decreases binding to HEV. Thus, interactions with the cytoplasmic domain can regulate the avidity of $\alpha_4\beta_7$ for MAdCAM-1 (213), similar to regulation of the avidity of LFA-1 for ICAM-1 by the β_2-integrin subunit cytoplasmic domain (214,215).

IX. RECIRCULATION OF MEMORY LYMPHOCYTES

A. Distinct Pathways Through Skin and Gut

Memory lymphocytes are imprinted so that they are more likely to return to the type of tissue, such as skin of mucosa, where they first encountered antigen (1,2,17). The surface phenotypes of gut and skin-homing memory cells are distinct (211). Furthermore, staining of lymphocytes in sections of skin and gut with mAb shows distinctive expression of adhesion molecules, which may contribute to selective extravasation in these tissues or to subsequent localization within these tissues in specific anatomic compartments (Table 4).

B. Skin Homing Lymphocytes

Lymphocytes that extravasate in the skin and appear in afferent lymph have a distinct pattern of expression of adhesion molecules (211) (Table 4). Furthermore, as shown by staining of tissue sections, T lymphocytes local-

Table 4 Naive and Memory Lymphocyte Subsets[a]

Molecule	Naive lymphocytes	Memory lymphocytes
CD45RO	negative	positive
CD45RA	high	low
CD2	low	high
LFA-3	negative	positive
L-selectin	positive	positive and negative subsets
α_4	low	high
	Memory lymphocytes subsets	
	Gut-associated	Skin-associated
CLA	negative	positive
$\alpha_E\beta_7$ (HML-1)	positive	negative
$\alpha_4\beta_7$[b]	high	low
$\alpha_4\beta_1$	low	high
α_6	low	high

[a](211,224,281).
[b]But see (282).

ized in the skin but not in the gut express a carbohydrate termed cutaneous lymphocyte associated antigen (CLA) (216). CLA is closely related to sialyl Lewis[a] and Lewis[x] (217) and is a ligand for E-selectin (218). Binding of a subpopulation of memory lymphocytes that bears CLA to E-selectin may contribute to the tropism of this subset to the skin (219–221). E-selectin is induced on dermal endothelial cells in delayed type hypersensitivity (222) and in chronically inflamed skin (20). Cloned T cells derived from challenged skin express high levels of CLA and bind to E-selectin, whereas T cell clones derived from blood lymphocytes do not (223). Both types of clones bind to P-selectin.

C. Gut Homing Lymphocytes

The most organized lymphoid structures in the wall of the gut are the Peyer's patches, which underlie follicle-associated epithelia that contain M cells, which are specialized for uptake of antigen from the gut lumen. Other lymphocytes localize more diffusely in the lamina propria underlying the digestive epithelium and in the epithelial layer. Studies on gut afferent lymph reveal the presence of both memory and naive lymphocytes (211);

whether there is differential migration of naive and memory lymphocytes through Peyer's patch HEV and lamina propria post-capillary venules, both of which contribute to gut afferent lymph (Fig. 7) remains unclear. Gut homing memory lymphocytes display a surface phenotype distinct from skin-homing lymphocytes (Table 4). When injected into the bloodstream, memory lymphocytes from gut afferent lymph display a strong preference to return to gut afferent lymph, whereas naive lymphocytes redistribute randomly (211). Gut afferent memory lymphocytes display an α_4 high, β_1-integrin low phenotype, suggesting they are $\alpha_4\beta_7{}^+$ (211) in common with a subpopulation of memory lymphocytes in blood (224). Expression of MAdCAM-1 on both Peyer's patch HEV and postcapillary venules in lamina propria (129), and 50% inhibition by mAb to α_4 and β_7 of migration into both Peyer's patch and intestine (209) suggest a role for $\alpha_4\beta_7$ interaction with MAdCAM-1 in both sites.

A subpopulation of gut lymphocytes distinct from those in lamina propria localize within the epithelium on the external surface of the basement membrane and express the human mucosal lymphocyte (HML-1) integrin $\alpha_E\beta_7$ (225–227). The α_E integrin subunit contains an I domain and a novel proteolytic cleavage site preceded by a stretch of acidic residues, just N-terminal to the I domain (228). Binding of intraepithelial lymphocytes (IEL) to epithelial cell monolayers in vitro is inhibited by mAb to α_E, suggesting that $\alpha_E\beta_7$ may help mediate localization of IEL in epithelia in vivo (229). Intraepithelial T lymphocytes may undergo thymus-independent differentiation in situ and their recirculation pattern is undefined. HML-1 is expressed on a subpopulation of 2% to 6% of blood T cells, which are in the memory subset and are CLA$^-$ and L-selectin$^-$ (230). Transforming growth factor β (TGF β) together with mitogen induces expression of HML-1 on peripheral T cells and increases expression on IEL (226,227). TGF β also induces switching of B lymphocytes to production of the IgA class of immunoglobulin (231), the predominant class secreted in the mucosa. These dual effects on differentiation of mucosal lymphocytes suggest the possibility that TGF β may be an environment-specific cytokine that imprints lymphocytes, when first exposed to antigen, to recirculate selectively to the gut.

X. ALTERATION OF LYMPHOCYTE TRAFFICKING IN INFLAMMATION

Antigen injected into the tissue of sensitized individuals induces localized accumulation of lymphocytes. These lymphocytes, and those accumulating in tissues in autoimmune disease, are almost all memory cells (232,233). The phenotype of these cells is quite similar to that of lymphocytes trafficking through these sites under basal conditions. This suggests that the signals for lymphocyte trafficking may be qualitatively the same in the basal and inflammatory states, and are upregulated in inflammation. Accumulation of lymphocytes induced by specific antigen, or by injection of IFN-γ or TNF-α, is significantly inhibited by mAb to either the LFA-1α or the integrin α_4 subunit (234–238). A combination of mAb to LFA-1 and α_4 gives almost complete inhibition of lymphocyte emigration and the resulting induration and plasma leakage (239). mAb to E-selectin and VCAM-1 also inhibit lymphocyte accumulation in delayed type hypersensitivity in skin (240). Multiple signals are thus required for augmented trafficking of lymphocytes into skin in inflammation (Fig. 8). Both antigen responsive and nonresponsive lymphocytes traffic into sites of antigenic stimulation (241). Antigen-specific lymphocytes may accumulate in the site because stimulation through their antigen receptors increases adhesiveness of integrins and causes them to be retained, whereas nonresponsive lymphocytes more rapidly enter the lymphatics and leave the site.

The interaction between VCAM-1 and VLA-4 can mediate both rolling and firm adhesion (112); thus it does not fit neatly into the three-step paradigm established for neutrophils. mAb to LFA-1 or VLA-4 alone do not completely inhibit lymphocyte accumulation in inflammation, and patients with LAD-I show delayed-type hypersensitivity reactions. This suggests that the functions of VLA-4 and LFA-1 are partially overlapping in the step of firm adhesion, but they may also act in series, as in VLA-4-mediated rolling followed by LFA-1-mediated firm adhesion. VLA-4 may act together with selectins to augment T lymphocyte tethering and rolling in the vasculature. All or most memory T lymphocytes lack L-selectin (39, 211,242,243). The CLA$^+$ subset can bind E-selectin, and T lymphocytes can also bind P-selectin (244,245). Peripheral blood T lymphocytes are substantially less efficient than neutrophils in tethering in hydrodynamic flow to E-selectin and P-selectin (T. Diacovo, R. Alon, T. Springer, unpublished); therefore, cooperation of VCAM-1 with E-selectin or P-selectin, or among all three molecules, may be important in enhancing lymphocyte accumulation in inflammation.

Inflammation also affects traffic through HEV. Antigen injected into tissue drains to the regional lymph node, and greatly increases blood flow to the node and traffic of naive lymphocytes through HEV (186). Furthermore, memory lymphocytes now appear to enter the node directly; this is associated with induction of VCAM-1 on non-HEV vascular endothelia within the node (186). Entry is inhibited by mAb to α_4, and this suggests a role for interaction of VCAM-1 with $\alpha_4\beta_1$ (186,234).

Lymphocyte chemoattractants are interesting candidates for the step 2 signal for lymphocyte accumulation at inflammatory sites. Pertussis toxin treatment inhibits lymphocyte emigration in response to antigen in delayed-type hypersensitivity (167). Identification of lymphocyte chemoattractants has been hampered by the low motility of lymphocytes compared to monocytes or neutrophils (246), and the low signal-to-background ratio, typically less than 2, in most chemotaxis assays. Recent interest has focused on chemokines (Table 2). A number of chemokines, all of which were isolated based on chemoattractive activity for neutrophils or monocytes, or by cloning genes of unknown function, have subsequently been tested and found to be chemoattractive for lymphocyte subpopulations (11,12). These include IL-8 (247) (but see (248,249), RANTES (250), MIP-1β (84), MIP-1α and β (251,252), and IP-10 (253). There are differences among reports in the subsets found to be chemoattracted, and some reports use lymphocytes pre-activated by T cell receptor cross-linking, which may be relevant to migration within inflammatory sites, but not emigration from blood. Of interest, MIP-1β can induce binding of the naive, CD8$^+$ subset to VCAM-1, either in solution or when immobilized on a substrate, mimicking presentation by an endothelial cell surface (84,254); the specific effect is modest, equal to background binding. The RANTES cytokine, by contrast to MIP-1β, selectively attracts the memory T lymphocyte subset (250).

Vascular endothelium may function to present chemoattractant to lymphocytes in a functionally relevant way, as well as to provide a permeability barrier that stabilizes the chemoattractant gradient. A transendothelial chemotaxis assay more accurately simulates lymphocyte emigration from the bloodstream than filter chemotaxis assays, and yields signals $> 10 \times$ background (255). Since lymphocytes, responding to specific antigen in tissue, signal emigration of further lymphocytes into the site, a chemoattractant was sought in material secreted by mitogen-stimulated mononuclear cells. Purification to homogeneity guided by the transendothelial lymphocyte chemotaxis assay revealed that MCP-1, previously thought to be solely a monocyte chemoattractant, is a major lymphocyte chemoattractant (255). Subsequent studies using the transendothelial chemotaxis assay have confirmed that lymphocytes respond to RANTES and MIP-1α (C-C chemokines), but do not respond to IL-8 or IP-10 (C-X-C chemokines) (256). MCP-1, RANTES, and MIP-1α all selectively attract the memory T lymphocyte subset, and both the CD4 and CD8 subsets. All also attract monocytes but not neutrophils, with MCP-1 being more potent than RANTES or MIP-1α as a monocyte chemoattractant. The physiologically relevant transendothelial assay suggests that C-C chemokines tend to attract both monocytes and lymphocytes, in agreement with the long-standing clinical observation that lymphocyte emigration into inflammatory sites is always accompanied by emigration of monocytes. The converse is not true. Monocytes sometimes emigrate in the absence of lymphocytes, correlating with activity of chemoattractants such as C5a and PAF on monocytes but not on lymphocytes. Teleologically, it is important that monocytes accompany lymphocytes into inflammatory sites in order to present antigen, and to carry out effector functions in which monocytes are activated by T lymphocytes. MCP-1 is abundantly expressed at sites of antigen challenge and autoimmune disease (12,257,258), and together with MIP-1α and RANTES, is an excellent candidate to provide the step 2 signal required to activate integrin adhesiveness and emigration of both monocytes and lymphocytes in vivo (Fig. 8).

The finding that resting T lymphocytes that tether and roll on VCAM-1 can spontaneously arrest and develop firm adhesion on VCAM-1 (112) has provocative implications for the multi-step model. It suggests that the VLA-4:VCAM-1 interaction not only can mediate the steps of rolling and firm adhesion, but may also short-circuit the step of stimulation by chemoattractants of firm adhesion through integrins. This is intriguing, since although a 2-fold stimulation of adhesiveness of VLA-4 to VCAM-1 has been demonstrated by MIP-1β in one system (84), with the chemoattractant that is most effective in eliciting transendothelial chemotaxis of T lymphocytes, MCP-1, it is difficult to detect stimulation of integrin adhesiveness on lymphocytes (Carr and Springer, unpublished). Therefore, an alternative pathway may exist in which VCAM-1 can mediate both tethering and arrest of lymphocytes, perhaps in cooperation with other endothelial molecules, prior to stimulation by chemoattractants. After arrest, chemoattractants would guide transendothelial migration, and perhaps stimulate further increases in the adhesiveness of the integrins VLA-4 and LFA-1 important in migration across the endothelium and basement membrane.

XI. CONCLUDING REMARKS

A three-step or area code model of leukocyte emigration from the bloodstream, established and validated in vitro and in vivo with neutrophils (Figs. 1, 6B), appears extendible with only slight modification to all subclasses of leukocytes including lymphocytes (Fig. 8). Multiple adhesion and chemoattractant receptors are used combinatorially in a series of steps that enable leukocytes to progress from initial tethering in flow to firm adhesion and emigration. The distinct distribution of receptors on leukocytes subsets for signals that are displayed on endothelium regulates selection of the subclasses of leukocytes that emigrate at inflammatory sites, and the distinctive recirculation behavior of lymphocyte subsets.

Many important developments await. Strong evidence suggests that G-protein coupled receptors are required for lymphocyte recirculation, but many of the putative lymphocyte chemoattractants specific to HEV, mucosa, and skin, and the receptors for these chemoattractants on lymphocytes, remain to be identified. Specific mucin-like molecules have recently emerged that present carbohydrate ligands to selectins. Are there similar mucin-like molecules on lymphocytes that present carbohydrates to P-selectin or E-selectin, and do these differ from the PSGL-1 molecule on neutrophils? It is likely that endothelial cells express molecules that retain chemoattractants on the luminal surface, preventing them from being washed away by blood flow, as already suggested for MIP-1β and IL-8. Are these molecules specifically regulated? The mucin-like ligands of selectins have many features such as extended structure, sulfation, and negative charge in common with proteoglycans, and thus might have a second function of binding chemokines through their heparin-binding sites and presenting them to leukocytes. Presenting molecules might be required not only to prevent chemoattractants from being washed away by blood flow, but also to generate maximal chemoattractant activity, analogous to proteoglycans that must bind fibroblast growth factor to enable signaling through a second receptor molecule. It will be interesting to determine whether chemoattractant receptors on leukocytes couple to distinct G-proteins and signaling effectors, allowing for selectivity in which integrins are upregulated in avidity. For example, do chemoattractants differ in ability to upregulate adhesiveness of two integrins such as LFA-1 and VLA-4 expressed on the same cell? Finally, after the area code is dialed and cells emigrate across the endothelium, much remains to be learned about the "7-digit code" that regulates leukocyte migration and localization within specific anatomic compartments.

ACKNOWLEDGMENTS

I thank the NIH for supporting most of the cited work, and Uli von Andrian for comments on the manuscript.

REFERENCES

1. Cahill RNP, Poskitt DC, Frost H, Trnka Z. Two distinct pools of recirculating T lymphocytes: migratory characteristics of nodal and intestinal T lymphocytes. J Exp Med 1977; 145:420–428.
2. Cahill RNP, Poskitt DC, Hay JB, Heron I, Trnka Z. The migration of lymphocytes in the fetal lamb. Eur J Immunol 1979; 9:251–253.
3. Springer TA. Traffic signals for lymphocyte recirculation and leukocyte emigration: the multi-step paradigm. Cell 1994; 76:301–314.
4. Carlos TM, Harlan JM. Leukocyte-endothelial adhesion molecules. Blood 1994; 84:2068–2102.
5. Granger DN, Kubes P. The microc circulation and inflammation: modulation of leukocyte-endothelial cell adhesion. J Leuk Biol 1994; 55:662–675.

6. Picker LJ, Butcher EC. Physiological and molecular mechanisms of lymphocyte homing. Annu Rev Immunol 1992; 10:561–591.
7. Lasky LA. Selectins: Interpreters of cell-specific carbohydrate information during inflammation. Science 1992; 258:964–969.
8. Bevilacqua MP, Nelson RM. Selectins. J Clin Invest 1993; 91:379–387.
9. Bevilacqua MP. Endothelial-leukocyte adhesion molecules. Annu Rev Immunol 1993; 11:767–804.
10. Butcher EC. Leukocyte-endothelial cell recognition: Three (or more) steps to specificity and diversity. Cell 1991; 67:1033–1036.
11. Baggiolini M, Dewald B, Moser B. Interleukin-8 and related chemotactic cytokines-CXC and CC chemokines. Adv Immunol 1994; 55:97–179.
12. Miller MD, Krangel MS. Biology and biochemistry of the chemokines: a family of chemotactic and inflammatory cytokines. Crit Rev Immunol 1992; 12:17–46.
13. Shimizu Y, Newman W, Tanaka Y, Shaw S. Lymphocyte interactions with endothelial cells. Immunol Today 1992; 13:106–112.
14. Zimmerman GA, Prescott SM, McIntyre TM. Endothelial cell interactions with granulocytes: tethering and signaling molecules. Immunol Today 1992; 13:93–100.
15. Harlan JM, Winn RK, Vedder NB, Doerschuk CM, Rice CL. In vivo models of leukocyte adherence to endothelium. In: Harlan JR and Liu D, eds. Adhesion: Its Role in Inflammatory Disease. New York: W.H. Freeman & Company, 1992:117–150.
16. Mackay CR. Immunological memory. Adv Immunol 1993; 53:217–265.
17. Mackay CR. Migration pathways and immunologic memory among T lymphocytes. Semin Immunol 1992; 4:51–58.
18. McEver RP. Selectins: Novel receptors that mediate leukocyte adhesion during inflammation. Thromb Haemostas, 1991; 65:223–228.
19. Rosen SD. Cell surface lectins in the immune system. Semin Immunol 1993; 5:237–247.
20. Diamond MS, Springer TA. The dynamic regulation of integrin adhesiveness. Curr Biol 1994; 4:506–517.
21. Cline MJ. The White Cell. Cambridge: Harvard University Press, 1975.
22. Issekutz AC, Issekutz TB. Quantitation and kinetics of blood monocyte migration to acute inflammatory reactions, and IL-1α, TNFα, and IFN-γ. J Immunol 1993; 151:2105–2115.
23. Ley K, Gaehtgens P. Endothelial, not hemodynamic, differences are responsible for preferential leukocyte rolling in rat mesenteric venules. Circ Res 1991; 69:1034–1041.
24. Nazziola E, House SD. Effects of hydrodynamics and leukocyte-endothelium specificity on leukocyte-endothelium interactions. Microvasc Res 1992; 44: 127–142.
25. McEver RP, Beckstead JH, Moore KL, Marshall-Carlson L, Bainton DF. GMP-140, a platelet alpha-granule membrane protein, is also synthesized by vascular endothelial cells and is localized in Weibel-Palade bodies. J Clin Invest 1989; 84(1):92–99.

26. Fina L, Molgaard HV, Robertson D, et al. Expression of the CD34 gene in vascular endothelial cells. Blood 1990; 75:2417–2426.
27. Swerlick RA, Lee KH, Wick TM, Lawley TJ. Human dermal microvascular endothelial but not human umbilical vein endothelial cells express CD36 in vivo and in vitro. J Immunol 1992; 148:78–83.
28. Zhu D, Cheng C-F, Pauli BU. Mediation of lung metastasis of murine melanomas by a lung-specific endothelial cell adhesion molecule. Proc Natl Acad Sci USA 1991; 88:9568–9572.
29. Pober JS, Cotran RS. Cytokines and endothelial cell biology. Physiol Rev 1990; 70:427–452.
30. Thronhill MH, Wellicome SM, Mahiouz DL, Lanchbury JSS, Kyan-Aung U, Haskard DO. Tumor necrosis factor combines with IL-4 or IFN-gamma to selectively enhance endothelial cell adhesiveness for T cells: the contribution of vascular cell adhesion molecule-1-dependent and -independent binding mechanisms. J Immunol 1991; 146:592–598.
31. Masinovsky B, Urdal D, Gallatin WM. IL-4 acts synergistically with IL-1 β to promote lymphocyte adhesion to microvascular endothelium by induction of vascular cell adhesion molecule-1. J Immunol 1990; 145:2886–2895.
32. Colditz IG, Watson DL. The effect of cytokines and chemotactic agonists on the migration of T lymphocytes into skin. Immunology 1992; 272:278.
33. Briscoe DM, Cotran RS, Pober JS. Effects of tumor necrosis factor, lipopolysaccharide, and IL-4 on the expression of vascular cell adhesion molecule-1 in vivo. J Immunol 1992; 149:2954–2960.
34. Issekutz TB, Stoltz JM, Meide PVD. Lymphocyte recruitment in delayed-type hypersensitivity: the role of IFN-γ. J Immunol 1988; 140:2989–2993.
35. Schmid-Schönbein GW, Usami S, Skalak R, Chien S. The interaction of leukocytes and erthrocytes in capillary and postcapillary vessels. Microvasc Res 1980; 19:45–70.
36. Springer TA. Adhesion receptors of the immune system. Nature 1990; 346: 425–433.
37. Lasky LA, Singer MS, Dowbenko D, et al. An endothelial ligand for L-selectin is a novel mucin-like molecule. Cell 1992; 69:927–938.
38. Gallatin WM, Weissman IL, Butcher EC. A cell-surface molecule involved in organ-specific homing of lymphocytes. Nature 1983; 304:30–34.
39. Kansas GS, Wood GS, Fishwild DM, Engelman EG. Functional characterization of human T lymphocyte subsets distinguished by monoclonal anti-leu-8. J Immunol 1985; 134:2995–3002.
40. Lewinsohn DM, Bargatze RF, Butcher EC. Leukocyte-endothelial cell recognition: evidence of a common molecular mechanism shared by neutrophils, lymphocytes, and other leukocytes. J Immunol 1987; 138:4313–4321.
41. Larsen E, Celi A, Gilbert GE, et al. PADGEM protein: a receptor that mediates the interaction of activated platelets with neutrophils and monocytes. Cell 1989; 59:305–312.
42. Geng J-G, Bevilacqua MP, Moore KL, et al. Rapid neutrophil adhesion to activated endothelium mediated by GMP-140. Nature 1990; 343:757–760.
43. Bevilacqua MP, Pober JS, Mendrick DL, Cotran RS, Gimbrone MA. Identi-

fication of an inducible endothelial-leukocyte adhesion molecule, E-LAM 1. Proc Natl Acad Sci USA 1987; 84:9238–9242.

44. Larsen GR, Sako D, Ahern TJ, et al. P-selectin and E-selectin: Distinct but overlapping leukocyte ligand specificities. J Biol Chem 1992; 267:11104–11110.
45. Nelson RM, Dolich S, Aruffo A, Cecconi O, Bevilacqua MP. Higher-affinity oligosaccharide ligands for E-selectin. J Clin Invest 1993; 91:1157–1166.
46. Moore KL, Stults NL, Diaz S, et al. Identification of a specific glycoprotein ligand for P-selectin (CD62) on myeloid cells. J Cell Biol 1992; 118:445–456.
47. Ushiyama S, Laue TM, Moore KL, Erickson HP, McEver RP. Structural and functional characterization of monomeric soluble P-selectin and comparison with membrane P-selectin. J Biol Chem 1993; 268:15229–15237.
48. Erbe DV, Wolitzky BA, Presta LG, et al. Identification of an E-selectin region critical for carbohydrate recognition and cell adhesion. J Cell Biol 1992; 119:215–227.
49. Graves BJ, Crowther RL, Chandran C, et al. Insight into E-selectin/ligand interaction from the crystal structure and mutagenesis of the lec/EGF domains. Nature 1994; 367:532–538.
50. Baumhueter S, Singer MS, Henzel W, et al. Binding of L-selectin to the vascular sialomucin, CD34. Science 1993; 262:436–438.
51. Foxall C, Watson SR, Dowbenko D, et al. The three members of the selectin receptor family recognize a common carbohydrate epitope, the sialyl Lewis^x oligosaccharide. J Cell Biol 1992; 117:895–902.
52. Berg EL, Magnani J, Warnock RA, Robinson MK, Butcher EC. Comparison of L-selectin and E-selectin ligand specificities: The L-selectin can bind the E-selectin ligands sialyl Le^x and sialyl Le^a. Biochem Biophys Res Commun 1992; 184:1048–1055.
53. Hemmerich S, Rosen SD. 6′-Sulfated sialyl Lewis^x is a major capping group of GlyCAM-1. Biochemistry 1994: 33:4830–4835.
54. Steininger CN, Eddy CA, Leimgruber RM, Mellors A, Welply JK. The glycoprotease of Pasteurella haemolytica A1 eliminates binding of myeloid cells to P-selectin but not to E-selectin. Biochem Biophys Res Commun 1992; 188: 760–766.
55. Norgard KE, Moore KL, Diaz S, et al. Characterization of a specific ligand for P-selectin on myeloid cells: a minor glycoprotein with sialylated O-linked oligosaccharides. J Biol Chem 1993; 268:12764–12774.
56. Sako D, Chang X-J, Barone KM, et al. Expression cloning of a functional glycoprotein ligand for P-selectin. Cell 1993; 75:1179–1186.
57. Lawrence MB, Springer TA. Leukocytes roll on a selectin at physiologic flow rates: distinction from and prerequisite for adhesion through integrins. Cell 1991; 65:859–873.
58. Ley K, Gaehtgens P, Fennie C, Singer MS, Lasky LA, Rosen SD. Lectin-like cell adhesion molecule 1 mediates leukocyte rolling in mesenteric venules in vivo. Blood 1991; 77:2553–2555.
59. Von Andrian UH, Chambers JD, McEvoy LM, Bargatze RF, Arfors KE, Butcher EC. Two-step model of leukocyte-endothelial cell interaction of in-

flammation: Distinct roles for LECAM-1 and the leukocyte β_2 integrins in vivo. Proc Natl Acad Sci USA 1991; 88:7538–7542.

60. Lawrence MB, Springer TA. Neutrophils roll on E-selectin. J Immunol 1993; 151:6338–6346.
61. Abbassi O, Kishimoto TK, McIntire LV, Anderson DC, Smith CW. E-selectin supports neutrophil rolling in vitro under conditions of flow. J Clin Invest 1993; 92:2719–2730.
62. Olofsson AM, Arfors K-E, Ramezani L, Wolitzky BA, Butcher EC, Von Andrian UH. E-selectin mediates leukocyte rolling in interleukin-1 treated rabbit mesentery venules. Blood 1994; 84:2749–2758.
63. Bienvenu K, Granger DN. Molecular determinants of shear rate-dependent leukocyte adhesion in postcapillary venules. Am J Physiol 1993; 264:H1504–H1508.
64. Mayadas TN, Johnson RC, Rayburn H, Hynes RO, Wagner DD. Leukocyte rolling and extravasation are severely compromised in P-selectin-deficient mice. Cell 1993; 74:541–554.
65. Von Andrian UH, Hansell P, Chambers JD, et al. L-selectin function is required for β_2-integrin-mediated neutrophil adhesion at physiological shear rates in vivo. Am J Physiol 1992; 263:H1034–H1044.
66. Kishimoto TK, Warnock RA, Jutila MA, et al. Antibodies against human neutrophil LECAM-1 (LAM-1/Leu-8/DREG-56 antigen) and endothelial cell ELAM-1 inhibit a common CD18-independent adhesion pathway in vitro. Blood 1991; 78:805–811.
67. Picker LJ, Warnock RA, Burns AR, Doerschuk CM, Berg EL, Butcher EC. The neutrophil selectin LECAM-1 presents carbohydrate ligands to the vascular selectins ELAM-1 and GMP-140. Cell 1991; 66:921–933.
68. Von Andrian UH, Chambers JD, Berg EL, et al. L-selectin mediates neutrophil rolling in inflamed venules through sialyl Lewis^x-dependent and -independent recognition pathways. Blood 1993; 82:182–191.
69. Lawrence MB, Bainton DF, Springer TA. Neutrophil tethering to and rolling on E-selectin are separable by requirement for L-selectin. Immunity 1994; 1: 137–145.
70. Chan P-Y, Lawrence MB, Dustin ML, Ferguson LM, Golan DE, Springer TA. Influence of receptor lateral mobility on adhesion strengthening between membranes containing LFA-3 and CD2. J Cell Biol 1991; 115:245–255.
71. Hammer DA, Apte SM. Simulation of cell rolling and adhesion on surfaces in shear flow: general results and analysis of selectin-mediated neutrophil adhesion. Biophys J 1992; 63:35–57.
72. Cyster JG, Shotton DM, Williams AF. The dimensions of the T lymphocyte glycoprotein leukosialin and identification of linear protein epitopes that can be modified by glycosylation. EMBO J 1991; 10:893–902.
73. Erlandsen SL, Hasslen SR, Nelson RD. Detection and spatial distribution of the β_2 integrin (Mac-1) and L-selectin (LECAM-1) adherence receptors on human neutrophils by high-resolution field emission SEM. J Histochem Cytochem 1993; 41:327–333.
74. Kansas GS, Ley K, Munro JM, Tedder TF. Regulation of leukocyte rolling

and adhesion to high endothelial venules through the cytoplasmic domain of L-selectin. J Exp Med 1993; 177:833–838.
75. Van Ewijk W, Brons NHC, Rozing J. Scanning electron microscopy of homing and recirculating lymphocyte populations. Cell Immunol 1975; 19:245–261.
76. Anderson AO, Anderson ND. Lymphocyte emigration from high endothelial venules in rat lymph nodes. Immunology 1976; 31:731–748.
77. Simmons DL, Satterthwaite AB, Tenen DG, Seed B. Molecular cloning of a cDNA encoding CD34, a sialomucin of human hematopoietic stem cells. J Immunol 1992; 148:267–271.
78. Cross AH, Raine CS. Central nervous system endothelial cell-polymorphonuclear cell interactions during autoimmune demyelination. Am J Pathol 1992; 139:1401–1409.
79. Wilkinson PC. Chemotaxis and Inflammation. London: Churchill Livingstone, 1982.
80. Devreotes PN, Zigmond SH. Chemotaxis in eukaryotic cells. A focus on leukocytes and *dictyostelium*. Annu Rev Cell Biol 1988; 4:649–686.
81. Carter SB. Haptotaxis and the mechanism of cell motility. Nature 1967; 213: 256–260.
82. Huber AR, Kunkel SL, Todd RF, III, Weiss SJ. Regulation of transendothelial neutrophil migration by endogenous interleukin-8. Science 1991; 254:99–102.
83. Rot A. Endothelial cell binding of NAP-1/IL-8: role in neutrophil emigration. Immunol Today 1992; 13:291–294.
84. Tanaka Y, Adams DH, Hubscher S, Hirano H, Siebenlist U, Shaw S. T-cell adhesion induced by proteoglycan-immobilized cytokine MIP-1β. Nature 1993; 361:79–82.
85. Snyderman R, Uhing RJ. Chemoattractant stimulus-response coupling. In: Gallin JI, Goldstein IM, Snyderman R, eds. Inflammation: Basic Principles and Clinical Correlates. New York: Raven Press, 1992:421–439.
86. Wu D, LaRosa GJ, Simon MI. G protein-coupled signal transduction pathways for interleukin-8. Science 1993; 261:101–103.
87. Beals CR, Wilson CB, Perlmutter RM. A small multigene family encodes G_i signal-transduction proteins. Proc Natl Acad Sci USA 1987; 84:7886–7890.
88. Amatruda TT, Gerard NP, Gerard C, Simon MI. Specific interactions of chemoattractant factor receptors with G-proteins. J Biol Chem 1993; 268: 10139–10144.
89. Murphy PM. The molecular biology of leukocyte chemoattractant receptors. Annu Rev Immunol 1994; 11:593–633.
90. Charo IF, Myers SJ, Herman A, Franci C, Connolly AJ, Coughlin AJ, Coghlin SR. Molecular cloning and functional expression of two monocyte chemoattactant protein 1 receptors reveals alternative splicing of the carboxyl-terminal tails. Proc Natl Acad Sci USA 1994; 91:2752–2756.
91. Faull RJ, Kovach NL, Harlan HM, Ginsberg MH. Stimulation of integrin-mediated adhesion of T lymphocytes and monocytes: Two mechanisms with divergent biological consequences. J Exp Med 1994; 179:1307–1316.

92. Ginsberg MH, Du X, Plow EF. Inside-out integrin signalling. Curr Opin Cell Biol 1992; 4:766–771.
93. Pober JS, Doukas J, Hughes CCW, Savage COS, Munro JM, Cotran RS. The potential roles of vascular endothelium in immune reactions. Hum Immunol 1990; 28:258–262.
94. Lo SK, Lee S, Ramos RA, et al. Endothelial-leukocyte adhesion molecule 1 stimulates the adhesive activity of leukocyte integrin CR3 (CD11b/CD18, Mac-1, $\alpha_m\beta_2$) on human neutrophils. J Exp Med 1991; 173:1493–1500.
95. Lorant DE, Patel KD, McIntyre TM, McEver RP, Prescott SM, Zimmerman GA. Coexpression of GMP-140 and PAF by endothelium stimulated by histamine or thrombin: a juxtacrine system for adhesion and activation of neutrophils. J Cell Biol 1991; 115:223–234.
96. Lo SK, Detmers PA, Levin SM, Wright SD. Transient adhesion of neutrophils to endothelium. J Exp Med 1989; 169:1779–1793.
97. Wright SD, Meyer BC. Phorbol esters cause sequential activation and deactivation of complement receptors on polymorphonuclear leukocytes. J Immunol 1986; 136:1759–1764.
98. Buyon JP, Abramson SB, Philips MR, et al. Dissociation between increased surface expression of Gp165/95 and homotypic neutrophil aggregation. J Immunol 1988; 140:3156–3160.
99. Diamond MS, Staunton DE, de Fougerolles AR, et al. ICAM-1 (CD54): A counter-receptor for Mac-1 (CD11b/CD18). J Cell Biol 1990; 111:3129–3139.
100. Smith CW, Kishimoto TK, Abbass O, et al. Chemotactic factors regulate lectin adhesion molecule 1 (LECAM-1)-dependent neutrophil adhesion to cytokine-stimulated endothelial cells in vitro. J Clin Invest 1991; 87:609–618.
101. Sengelov H, Kjeldsen L, Diamond MS, Springer TA, Borregaard N. Subcellular localization and dynamics of Mac-1 ($\alpha_m\beta_2$ in human neutrophils. J Clin Invest 1993; 92:1467–1476.
102. Philips MR, Buyon JP, Winchester R, Weissman G, Abramson SB. Upregulation of the iC3b receptor (CR3) is neither necessary nor sufficient to promote neutrophil aggregation. J Clin Invest 1988; 82:495–501.
103. Vedder NB, Harlan JM. Increases surface expression of CD11b/CD18 (Mac-1) is not required for stimulated neutrophil adherence to cultured endothelium. J Clin Invest 1988; 81:676–682.
104. Dustin ML, Springer TA. T cell receptor cross-linking transiently stimulates adhesiveness through LFA-1. Nature 1989; 341:619–624.
105. Pircher H, Groscurth P, Baumhutter S, Aguet M, Zinkernagel RM, Hengartner H. A monoclonal antibody against altered LFA-1 induces proliferation and lymphokine release of cloned T cells. Eur J Immunol 1986; 16:172–181.
106. Keizer GD, Visser W, Vliem M, Figdor CG. A monoclonal antibody (NKI-L16) directed against a unique epitope on the alpha-chain of human leukocyte function-associated antigen 1 induces homotypic cell-cell interactions. J Immunol 1988; 140:1393–1400.
107. Landis RC, Bennett RI, Hogg N. A novel LFA-1 activation epitope maps to the I domain. J Cell Biol 1993; 120:1519–1527.

108. Diamond MS, Springer TA. A subpopulation of Mac-1 (CD11b/CD18) molecules mediates neutrophil adhesion to ICAM-1 and fibrinogen. J Cell Biol 1993; 120:545–556.
109. Diamond MS, Garcia-Aguilar J, Bickford JK, Corbi AL, Springer TA. The I domain is a major recognition site on the leukocyte integrin Mac-1 (CD11b/CD18) for four distinct adhesion ligands. J Cell Biol 1993; 120:1031–1043.
110. Michishita M, Videm V, Arnaout MA. A novel divalent cation-binding site in the A domain of the $\beta 2$ integrin CR3 (CD11b/CD18) is essential for ligand binding. Cell 1993; 72:857–867.
111. Lollo BA, Chan KWH, Hanson EM, Moy VT, Brian AA. Direct evidence for two affinity states for lymphocyte function-associated antigen 1 on activated T cells. J Biol Chem 1993; 268:21693–21700.
112. Alon R, Kassner PD, Carr MW, Finger EB, Hemler ME, Springer TA. The integrin VLA-4 supports tethering and rolling in flow on VCAM-1. J Cell Biol 1995; 128:1243–1253.
113. Rothlein R, Dustin ML, Marlin SD, Springer TA. A human intercellular adhesion molecule (ICAM-1) distinct from LFA-1. J Immunol 1986; 137:1270–1274.
114. Staunton DE, Dustin ML, Springer TA. Functional cloning of ICAM-2, a cell adhesion ligand for LFA-1 homologous to ICAM-1. Nature 1989; 339:61–64.
115. De Fougerolles AR, Springer TA. Intercellular adhesion molecule 3, a third adhesion counter-receptor for lymphocyte function-associated molecule 1 on resting lymphocytes. J Exp Med 1992; 175:185–190.
116. Smith CW, Marlin SD, Rothlein R, Toman C, Anderson DC. Cooperative interactions of LFA-1 and Mac-1 with intercellular adhesion molecule-1 in facilitating adherence and transendothelial migration of human neutrophils in vitro. J Clin Invest 1989; 83:2008–2017.
117. Diamond MS, Staunton DE, Marlin SD, Springer TA. Binding of the integrin Mac-1 (CD11b/CD18) to the third Ig-like domain of ICAM-1 (CD54) and its regulation by glycosylation. Cell 1991; 65:961–971.
118. De Fougerolles AR, Qin X, Springer TA. Characterization of the function of ICAM-3 and comparison to ICAM-1 and ICAM-2 in immune responses. J Exp Med 1994; 179:619–629.
119. Moy P, Lobb R, Tizard R, Olson D, Hession C. Cloning of an inflammation-specific phosphatidyl inositol-linked form of murine vascular cell adhesion molecule-1. J Biol Chem 1993; 268:8835–8841.
120. Terry RW, Kwee L, Levine JF, Labow MA. Cytokine induction of an alternatively spliced murine vascular cell adhesion molecule (VCAM) mRNA encoding a glycosylphosphatidylinositol-anchored VCAM protein. Proc Natl Acad Sci USA 1993; 90:5919–5923.
121. Kinashi T, St. Pierre Y, Springer TA. Expression of glycophosphatidylinositol (GPI)-anchored and non-anchored isoforms of vascular cell adhesion molecule 1 in murine stromal and endothelial cells. J Leuk Biol 1995; 57:168–173.

122. Elices MJ, Osborn L, Takada Y, et al. VCAM-1 on activated endothelium interacts with the leukocyte integrin VLA-4 at a site distinct from the VLA-4/fibronectin binding site. Cell 1990; 60:577–584.
123. Rüegg C, Postigo AA, Sikorski EE, Butcher EC, Pytela R, Erle DJ. Role of integrin $\alpha4\beta7/\alpha4\beta P$ in lymphocyte adherence to fibronectin and VCAM-1 and in homotypic cell clustering. J Cell Biol 1992; 117:179–189.
124. Chan BMC, Elices MJ, Murphy E, Hemler ME. Adhesion to vascular cell adhesion molecule 1 and fibronectin: Comparison of $\alpha^4\beta_1$ (VLA-4) and $\alpha^4\beta_7$ on the human B cell line JY. J Biol Chem 1992; 267:8366–8370.
125. Osborn L, Vassallo C, Benjamin CD. Activated endothelium binds lymphocytes through a novel binding site in the alternately spliced domain of vascular cell adhesion molecule-1. J Exp Med 1992; 176:99–107.
126. Vonderheide RH, Springer TA. Lymphocyte adhesion through VLA-4: Evidence for a novel binding site in the alternatively spliced domain of VCAM-1 and an additional $\alpha4$ integrin counter-receptor on stimulated endothelium. J Exp Med 1992; 175:1433–1442.
127. Vonderheide RH, Tedder TF, Springer TA, Staunton DE. Residues within a conserved amino acid motif of domains 1 and 4 of VCAM-1 are required for binding to VLA-4. J Cell Biol 1994; 125:215–222.
128. Osborn L, Vassallo C, Browning BG, et al. Arrangement of domains, and amino acid residues required for binding of vascular cell adhesion molecule-1 to its counter-receptor VLA-4 ($\alpha4\beta1$). J Cell Biol 1994; 124 (4):601–608.
129. Streeter PR, Lakey-Berg E, Rouse BTN, Bargatze RF, Butcher EC. A tissue-specific endothelial cell molecule involved in lymphocyte homing. Nature 1988; 331:41–46.
130. Briskin MJ, McEvoy LM, Butcher EC. MAdCAM-1 has homology to immunoglobulin and mucin-like adhesion receptors and to IgA1. Nature 1993; 363: 461–464.
131. Hu MC-T, Crowe DT, Weissman IL, Holzmann B. Cloning and expression of mouse integrin $\beta_p(\beta_7)$: A functional role in Peyer's patch-specific lymphocyte homing. Proc Natl Acad Sci USA 1992; 89:8254–8258.
132. Berlin C, Berg EL, Briskin MJ, et al. $\alpha4$ $\beta7$ integrin mediates lymphocyte binding to the mucosal vascular addressin MAdCAM-1. Cell 1993; 74:185–195.
133. Berg EL, McEvoy LM, Berlin C, Bargatze RF, Butcher EC. L-selectin-mediated lymphocyte rolling on MAdCAM-1. Nature 1993; 366:695–698.
134. Muller WA, Ratti CM, McDonnell SL, Cohn ZA. A human endothelial cell-restricted, externally disposed plasmalemmal protein enriched in intercellular junctions. J Exp Med 1989; 170:399–414.
135. Albelda SM, Muller WA, Buck CA, Newman PJ. Molecular and cellular properties of PECAM-1 (endoCAM/CD31): a novel vascular cell-cell adhesion molecule. J Cell Biol 1991; 114:1059–1068.
136. Newman PJ, Berndt MC, Gorski J, et al. PECAM-1 (CD31) cloning and relation to adhesion molecules of the immunoglobulin gene superfamily. Science 1990; 247:1219–1222.
137. Simmons DL, Walker C, Power C, Pigott R. Molecular cloning of CD31, a

putative intercellular adhesion molecule closely related to carcinoembryonic antigen. J Exp Med 1990; 171:2147–2152.
138. Albelda SM, Oliver PD, Romer LH, Buck CA. EndoCAM: a novel endothelial cell-cell adhesion molecule. J Cell Biol 1990; 110:1227–1237.
139. Stockinger H, Gadd SJ, Eher R, et al. Molecular characterization and functional analysis of the leukocyte surface protein CD31. J Immunol 1990; 145: 3889–3987.
140. Tanaka Y, Albelda SM, Horgan KJ, et al. CD31 expressed on distinctive T cell subsets is a preferential amplified of $\beta 1$ integrin-mediated adhesion. J Exp Med 1992; 176:245–253.
141. Muller WA, Weigl SA, Deng X, Phillips DM. PECAM-1 is required for transendothelial migration of leukocytes. J Exp Med 1993; 178:449–460.
142. Haynes BF, Liao H-X, Patton KL. The transmembrane hyaluronate receptor (CD44): multiple functions, multiple forms. Cancer Cells 1991; 3:347–350.
143. Culty M, Miyake K, Kincade PW, Sikorski E, Butcher EC, Underhill C. The hyaluronate receptor is a member of the CD44 (H-CAM) family of cell surface glycoproteins. J Cell Biol 1990; 111:2765–2774.
144. Aruffo A, Stamenkovic I, Melnick M, Underhill CB, Seed B. CD44 is the principal cell surface receptor for hyaluronate. Cell 1990; 61:1303–1313.
145. Günthert U. CD44: a multitude of isoforms with diverse functions. Curr Top Microbiol Immunol 1993; 184:47–63.
146. Arch R, Wirth K, Hofmann M, et al. Participation in normal immune responses of a metastasis-inducing splice variant of CD44. Science 1992; 257: 682–685.
147. Jalkanen S, Bargatze RF, De los Toyos J, Butcher EC. Lymphocyte recognition of high endothelium: antibodies to distinct epitopes of an 85-95 kD glycoprotein antigen differentially inhibit lymphocyte binding to lymph node, mucosal and synovial endothelial cells. J Cell Biol 1987; 105:983–993.
148. Oppenheimer-Marks N, Davis LS, Lipsky PE. Human T lymphocyte adhesion to endothelial cells and transendothelial migration: alteration of receptor use relates to the activation status of both the T cell and the endothelial cell. J Immunol 1990; 145:140–148.
149. Camp RL, Scheynius A, Johansson C, Puré E. CD44 is necessary for optimal contact allergic responses but is not required for normal leukocyte extravasation. J Exp Med 1993; 178:497–507.
150. Anderson DC, Springer TA. Leukocyte adhesion deficiency: An inherited defect in the Mac-1, LFA-1, and p150,95 glycoproteins. Ann Rev Med 1987; 38:175–194.
151. Kishimoto TK, Larson RS, Corbi AL, Dustin ML, Staunton DE, Springer TA. The leukocyte integrins: LFA-1, Mac-1, and p150,95. Adv Immunol 1989; 46:149–182.
152. Buchanan MR, Crowley CA, Rosin RE, Gimbrone MA, Babior BM. Studies on the interaction between GP-180-deficient neutrophils and vascular endothelium. Blood 1982; 60:160–165.
153. Smith CW, Rothlein R, Hughes BJ, Mariscalco MM, Schmalstieg FC, Anderson DC. Recognition of an endothelial determinant for CD18-dependent

neutrophil adherence and transendothelial migration. J Clin Invest 1988; 82: 1746–1756.

154. Arfors KE, Lundberg C, Lindborm L, Lundberg K, Beatty PG, Harlan JM. A monoclonal antibody to the membrane glycoprotein complex CD18 inhibits polymorphonuclear leukocyte accumulation and plasma leakage in vivo. Blood 1987; 69:338–340.
155. Lawrence MB, Smith CW, Eskin SG, McIntire LV. Effect of venous shear stress on CD18-mediated neutrophil adhesion to cultured endothelium. Blood 1990; 75:227–237.
156. Jutila MA, Lewinsohn D, Berg EL, Butcher E. Homing receptors in lymphocyte, neutrophil, and monocyte interaction with endothelial cells. In: Springer TA, Anderson DC, Rosenthal AS, Rothlein R, eds. Leukocyte Adhesion Molecules. New York: Springer-Verlag, 1988:227–235.
157. Jutila MA, Rott L, Berg EL, Butcher EC. Function and regulation of the neutrophil MEL-14 antigen in vivo: Comparison with LFA-1 and MAC-1. J Immunol 1989; 143:3318–3324.
158. Watson SR, Fennie C, Lasky LA. Neutrophil influx into an inflammatory site inhibited by a soluble homing receptor-IgG chimaera. Nature 1991; 349: 164–167.
159. Mulligan MS, Varani J, Dame MK, et al. Role of endothelial-leukocyte adhesion molecule 1 (ELAM-1) in neutrophil-mediated lung injury in rats. J Clin Invest 1991; 88:1396–1406.
160. Kishimoto TK, Jutila MA, Berg EL, Butcher EC. Neutrophil Mac-1 and MEL-14 adhesion proteins inversely regulated by chemotactic factors. Science 1989; 245:1238–1241.
161. Cohnheim J. Lectures on General Pathology: A Handbook for Practitioners and Students. London: New Sydenham Society, 1889.
162. Etzioni A, Frydman M, Pollack S, et al. Recurrent severe infections caused by a novel leukocyte adhesion deficiency. N Engl J Med 1992; 327:1789–1792.
163. Von Andrian UH, Berger EM, Ramezani L, et al. In vivo behavior of neutrophils from two patients with distinct inherited leukocyte adhesion deficiency syndromes. J Clin Invest 1993; 91:2893–2897.
164. Colditz IG. Sites of antigenic stimulation: role of cytokines and chemotactic agonists in acute inflammation. In: Beh KJ, ed. Animal Health and Production in the 21st Century. Melbourne: CSIRO, 1992.
165. Mulligan MS, Jones ML, Bolanowski MA, et al. Inhibition of lung inflammatory reactions in rats by an anti-human IL-8 antibody. J Immunol 1993; 150:5585–5595.
166. Sekido N, Mukaida N, Harada A, Nakanishi I, Watanabe Y, Matsushima K. Prevention of lung reperfusion injury in rabbits by a monoclonal antibody against interleukin-8. Nature 1993; 365:654–657.
167. Spangrude GJ, Sacchi F, Hill HR, Van Epps DE, Daynes RA. Inhibition of lymphocyte and neutrophil chemotaxis by pertussis toxin. J Immunol 1985; 135:4135–4143.

168. Nourshargh S, Williams TJ. Evidence that a receptor-operated event on the neutrophil mediates neutrophil accumulation in vivo. J Immunol 1990; 145: 2633–2638.
169. Hechtman DH, Cybulsky MI, Fuchs HJ, Baker JB, Gimbrone MA Jr. Intravascular IL-8: Inhibitor of polymorphonuclear leukocyte accumulation at sites of acute inflammation. J Immunol 1991; 147:883–892.
170. Gimbrone MA, Obin MS, Brock AF, et al. Endothelial interleukin-8: a novel inhibitor of leukocyte-endothelial interactions. Science 1989; 246:1601–1603.
171. Colditz IG, Movat HZ. Desensitization of acute inflammatory lesions to chemotaxins and endotoxin. J Immunol 1984; 133:2163–2168.
172. Colditz IG. Desensitisation mechanisms regulating plasma leakage and neutrophil emigration. In: Gordon JL, ed. Vascular Endothelium: Interactions with Circulating Cells. New York: Elsevier, 1991:175–187.
173. Kuna P, Reddigari SR, Schall TJ, Rucinski D, Sadick M, Kaplan AP. Characterization of the human basophil response to cytokines, growth factors, and histamine releasing factors of the intercrine/chemokine family. J Immunol 1993; 150:1932–1943.
174. Bischoff SC, Krieger M, Brunner T, et al. RANTES and related chemokines activate human basophil granulocytes through different G protein-coupled receptors. Eur J Immunol 1993; 23:761–767.
175. Walsh LJ, Lavker RM, Murphy GF. Biology of disease. Determinants of immune cell trafficking in the skin. Lab Invest 1990; 63:592–600.
176. Hood L, Huang HV, Dreyer WJ. The area-code hypothesis: the immune system provides clues to understanding the genetic and molecular basis of cell recognition during development. J Supramolec Struct 1987; 7:531–559.
177. Springer TA. Area code molecules of lymphocytes. In: Burger MM, Sordat B, Zinkernagel RM, eds. Cell to Cell Interaction: a Karger Symposium. Basel: S. Karger AG, 1990:16–39.
178. Jung TM, Gallatin WM, Weissman IL, Dailey MO. Down-regulation of homing receptors after T cell activation. J Immunol 1988; 141:4110–4117.
179. Spertini O, Kansas GS, Munro JM, Griffin JD, Tedder TF. Regulation of leukocyte migration by activation of the leukocyte adhesion molecule (LAM-1) selectin. Nature 1991; 349:691–694.
180. Issekutz TB, Chin W, Hay JB. The characterization of lymphocytes migrating through chronically inflamed tissues. Immunology 1982; 46:59–66.
181. Mackay CR, Marston WL, Dudler L. Naive and memory T cells show distinct pathways of lymphocyte recirculation. J Exp Med 1990; 171:801–817.
182. Bjerknes M, Cheng H, Ottaway CA. Dynamics of lymphocyte-endothelial interactions in vivo. Science 1986; 231:402–405.
183. Woodruff JJ, Clarke LM, Chin YH. Specific cell-adhesion mechanisms determining migration pathways of recirculating lymphocytes. Annu Rev Immunol 1987; 5:201–222.
184. Mebius RE, Streeter, PR, Breve J, Duijvestijn AM, Kraal G. The influence of afferent lymphatic vessel interruption on vascular addressin expression. J Cell Biol 1991; 115:85–95.

185. Mebius RE, Dowbenko D, Williams A, Fennie C, Lasky LA, Watson SR. Expression of GlyCAM-1, an endothelial ligand for L-selectin, is affected by afferent lymphatic flow. J Immunol 1993; 151:6769–6776.
186. Mackay CR, Marston W, Dudler L. Altered patterns of T cell migration through lymph nodes and skin following antigen challenge. Eur J Immunol 1992; 22:2205–2210.
187. Stamper HB Jr, Woodruff JJ. Lymphocyte homing into lymph nodes: In vitro demonstration of the selective affinity of recirculating lymphocytes for high-endothelial venules. J Exp Med 1976; 144:828.
188. Butcher EC, Scollay RG, Weissman IL. Organ specificity of lymphocyte migration: mediation by highly selective lymphocyte interaction with organ-specific determinants on high endothelial venules. Eur J Immunol 1980; 10: 556–561.
189. Stevens SK, Weissman IL, Butcher EC. Differences in the migration of B and T lymphocytes: organ-selective localization in vivo and the role of lymphocyte-endothelial cell recognition. J Immunol 1982; 2:844–851.
190. Bargatze RF, Wu NW, Weissman IL, Butcher EC. High endothelial venule binding as a predictor of the dissemination of passaged murine lymphomas. J Exp Med 1987; 166:1125–1131.
191. Streeter PR, Rouse BTN, Butcher EC. Immunohistologic and functional characterization of a vascular addressin involved in lymphocyte homing into peripheral lymph nodes. J Cell Biol 1988; 107:1853–1862.
192. Berg EL, Robinson MK, Warnock RA, Butcher EC. The human peripheral lymph node vascular addressin is a ligand for LECAM-1, the peripheral lymph node homing receptor. J Cell Biol 1991; 114:343–349.
193. Watson S, Imai Y, Fennie C, Geoffroy JS, Rosen SD, Lasky LA. A homing receptor-IgG chimera as a probe for adhesive ligands of lymph node high endothelial venules. J Cell Biol 1990; 110:2221–2229.
194. Imai Y, Lasky LA, Rosen SD. Sulphation requirement for GlyCAM-1, an endothelial ligand for L-selectin. Nature 1993; 361:555–557.
195. Spertini O, Luscinskas FW, Kansas GS, et al. Leukocyte adhesion molecule-1 (LAM-1, L-selectin) interacts with an inducible endothelial cell ligand to support leukocyte adhesion. J Immunol 1991; 147:2565–2573.
196. Spertini O, Luscinskas FW, Gimbrone MA Jr, Tedder TF. Monocyte attachment to activated human vascular endothelium in vitro is mediated by leukocyte adhesion molecule-1 (L-selectin) under nonstatic conditions. J Exp Med 1992; 175:1789–1792.
197. San Gabriel-Masson C. Adhesion of Lymphocytes to the Lactating Mammary Gland in the Mouse. Ph.D. thesis, Pennsylvania State University, 1992.
198. Holzmann B, McIntyre BW, Weissman IL. Identification of a murine Peyer's patch-specific lymphocyte homing receptor as an integrin molecule with an alpha chain homologous to human VLA-4 alpha. Cell 1989; 56:37–46.
199. Holzmann B, Weissman IL. Peyer's patch-specific lymphocyte homing receptors consist of a VLA-4-like α chain associated with either of two integrin β chains, one of which is novel. EMBO J 1989; 8:1735–1741.
200. Imai Y, Singer MS, Fennie C, Lasky LA, Rosen SD. Identification of a

carbohydrate based endothelial ligand for a lymphocyte homing receptor. J Cell Biol 1991; 113:1213–1221.

201. Hamann A, Jablonski-Westrich D, Jonas P, Thiele H-G. Homing receptors reexamined: mouse LECAM-1 (MEL-14 antigen) is involved in lymphocyte migration into gut-associated lymphoid tissue. Eur J Immunol 1991; 21:2925–2929.
202. Hamann A, Westrich DJ, Duijevstijn A, et al. Evidence for an accessory role of LFA-1 in lymphocyte-high endothelium interaction during homing. J Immunol 1988; 140:693–699.
203. De Fougerolles AR, Stacker SA, Schwarting R, Springer TA. Characterization of ICAM-2 and evidence for a third counter-receptor for LFA-1. J Exp Med 1991; 174:253–267.
204. Wardlaw AC, Parton R. *Bordetella pertussis* toxins. Pharmacol Ther 1983; 19:1–53.
205. Morse SI, Barron BA. Studies on the leukocytosis and lymphocytosis induced by bordetella pertussis. III. The distribution of transfused lymphocytes in pertussis-treated and normal mice. J Exp Med 1970; 132:663–672.
206. Chaffin KE, Beals CR, Wilkie TM, Forbush KA, Simon MI, Perlmutter RM. Dissection of thymocyte signaling pathways by in vivo expression of pertussis toxin ADP-ribosyltransferase. EMBO J 1990; 9:3821–3829.
207. Chaffin KE, Perlmutter RM. A pertussis toxin-sensitive process controls thymocyte emigration. Eur J Immunol 1991; 21:2565–2573.
208. Spangrude GJ, Braaten BA, Daynes RA. Molecular mechanisms of lymphocyte extravasation. I. Studies of two selective inhibitors of lymphocyte recirculation. J Immunol 1984; 132:354–362.
209. Dunlevy JR, Couchman JR. Controlled induction of focal adhesion disassembly and migration in primary fibroblasts. J Cell Sci 1993; 105:489–500.
210. Bargatze RF, Streeter PR, Butcher EC. Expression of low levels of peripheral lymph node-associated vascular addressin in mucosal lymphoid tissues: possible relevance to the dissemination of passaged akr lymphomas. J Cell Biochem 1990; 42:219–227.
211. Mackay CR, Marston WL, Dudler L, Spertini O, Tedder TF, Hein WR. Tissue-specific migration pathways by phenotypically distinct subpopulations of memory T cells. Eur J Immunol 1992; 22:887–895.
212. Bargatze RF, Butcher EC. Rapid G protein-regulated activation event involved in lymphocyte binding to high endothelial venules. J Exp Med 1993; 178:367–372.
213. Crowe DT, Chiu H, Fong S, Weissman IL. Regulation of the avidity of integrin $\alpha_4\beta_7$ by the $\beta_7$cystoplasmic domain. J Biol Chem 1994; 269:14411–14418.
214. Hibbs ML, Xu H, Stacker SA, Springer TA. Regulation of adhesion to ICAM-1 by the cytoplasmic domain of LFA-1 integrin beta subunit. Science 1991; 251:1611–1613.
215. Hibbs ML, Jakes S, Stacker SA, Wallace RW, Springer TA. The cytoplasmic domain of the integrin lymphocyte function-associated antigen 1 β subunit: sites required for binding to intercellular adhesion molecule 1 and the phorbol ester-stimulated phosphorylation site. J Exp Med 1991; 174:1227–1238.

216. Picker LJ, Michie SA, Rott LS, Butcher EC. A unique phenotype of skin-associated lymphocytes in humans: preferential expression of the HECA-452 epitope by benign and malignant T cells at cutaneous sites. Am J Pathol 1990; 136:1053–1068.
217. Berg EL, Robinson MK, Mansson O, Butcher EC, Magnani JL. A carbohydrate domain common to both sialyl Le^a and sialyl Le^x is recognized by the endothelial cell leukocyte adhesion molecule ELAM-1. J Biol Chem 1991; 266:14869–14872.
218. Berg EL, Yoshino T, Rott LS, et al. The cutaneous lymphocyte antigen is a skin lymphocyte homing receptor for the vascular lectin endothelial cell-leuckocyte adhesion molecule 1. J Exp Med 1991; 174:1461–1466.
219. Graber N, Gopal TV, Wilson D, Beall LD, Polte T, Newman W. T cells bind to cytokine-activated endothelial cells via a novel, inducible sialoglycoprotein and endothelial leukocyte adhesion molecule-1. J Immunol 1990; 145:819–830.
220. Picker LJ, Kishimoto TK, Smith CW, Warnock RA, Butcher EC. ELAM-1 is an adhesion molecule for skin-homing T cells. Nature 1991; 349:796–798.
221. Shimizu Y, Shaw S, Graber N, et al. Activation-independent binding of human memory T cells to adhesion molecule ELAM-1. Nature 1991; 349: 799–802.
222. Cotran RS, Gimbrone MA Jr, Bevilacqua MP, Mendrick DL, Pober JS. Induction and detection of a human endothelial activation antigen in vivo. J Exp Med 1986; 164:661–666.
223. Alon R, Rossiter H, Wang X, Springer TA, Kupper TS. Distinct cell surface ligands mediate T lymphocyte attachment and rolling on P-and E-selectin under physiological flow. J Cell Biol 1994; 127:1485–1495.
224. Schweighoffer T, Tanaka Y, Tidswell M, et al. Selective expression of integrin $\alpha4\beta7$ on a subset of human $CD4^+$ memory T cells with hallmarks of gut-trophism. J Immunol 1993; 151:717–729.
225. Cerf-Bensussan N, Jarry A, Brousse N, Lisowska-Grospierre B, Guy-Grand D, Griscelli C. A monoclonal antibody (HML-1) defining a novel membrane molecule present on human intestinal lymphocytes. Eur J Immunol 1987; 17: 1279–1285.
226. Kilshaw PJ, Murant SJ. A new surface antigen on intraepithelial lymphocytes in the intestine. Eur J Immunol 1990; 20:2201–2207.
227. Parker CM, Cepek KL, Russell GJ, et al. A family of $\beta7$ integrins on human mucosal lymphocytes. Proc Natl Acad Sci USA 1992; 89:1924–1928.
228. Shaw SK, Cepek KL, Murphy EA, Russell GJ, Brenner MB, Parker CM. Molecular cloning of the human mucosal lymphocyte integrin αE subunit. J Biol Chem 1994; 269:6016–6025.
229. Cepek KL, Parker CM, Madara JL, Brenner MB. Integrin $\alpha E\beta_7$ mediates adhesion of T lymphocytes to epithelial cells. J Immunol 1993; 150:3459–3470.
230. Picker LJ, Terstappen LWMM, Rott LS, Streeter PR, Stein H, Butcher EC. Differential expression of homing-associated adhesion molecules by T cell subsets in man. J Immunol 1990; 145:3247–3255.

231. Coffman RL, Lebman DA, Shrader B. Transforming growth factor β specifically enhances IgA production by lipopolysaccharide-stimulated murine B lymphocytes. J Exp Med 1989; 170:1039–1044.
232. Pitzalis C, Kingsley G, Haskard D, Panayi G. The preferential accumulation of helper-inducer T lymphocytes in inflammatory lesions: evidence for regulation by selective endothelial and homotypic adhesion. Eur J Immunol 1988; 18:1397–1404.
233. Janossy G, Bofill M, Rowe D, Muir J, Beverley PC. The tissue distribution of T lymphocytes expressing different CD45 polypeptides. Immunology 1989; 66:517–525.
234. Issekutz TB. Inhibition of in vivo lymphocyte migration to inflammation and homing to lymphoid tissues by the TA-2 monoclonal antibody: A likely role for VLA-4 in vivo. J Immunol 1991; 147:4178–4184.
235. Issekutz TB. Inhibition of lymphocyte endothelial adhesion and in vivo lymphocyte migration to cutaneous inflammation by TA-3, a new monoclonal antibody to rat LFA-1. Immunol 1992; 149:3394–3402.
236. Chisholm PL, Williams CA, Lobb RR. Monoclonal antibodies to the integrin α-4 subunit inhibit the murine contact hypersensitivity response. Eur J Immunol 1993; 23:682–688.
237. Yednock TA, Cannon C, Fritz LC, Sanchez-Madrid F, Steinman L, Karin N. Prevention of experimental autoimmune encephalomyelitis by antibodies against $\alpha 4\beta 1$ integrin. Nature 1992; 356:63–66.
238. Scheynius A, Camp RL, Puré E. Reduced contact sensitivity reactions in mice treated with monoclonal antibodies to leukocyte function-associated molecule-1 and intercellular adhesion molecule-1. J Immunol 1993; 150:655–663.
239. Issekutz TB. Dual inhibition of VLA-4 and LFA-1 maximally inhibits cutaneous delayed type hypersensitivity-induced inflammation. Am J Pathol 1993; 143:1286–1293.
240. Silber A, Newman W, Sasseville VG, et al. Recruitment of lymphocytes during cutaneous delayed hypersensitivity in nonhuman primates is dependent on E-selectin and VCAM-1. J Clin Invest 1994; 93:1554–1563. In press.
241. McCluskey RT, Benacerraf B, McClusky JW. Studies on the specificity of the cellular infiltrate in delayed type hypersensitivity reactions. J Immunol 1963; 90:466.
242. Tedder TF, Matsuyama T, Rothstein D, Schlossman SF, Morimoto C. Human antigen-specific memory T cells express the homing receptor (LAM-1) necessary for lymphocyte recirculation. Eur J Immunol 1990; 20:1351–1355.
243. Bradley LM, Atkins GG, Swain SL. Long-term $CD4^+$ memory T cells from the spleen lack MEL-14, the lymph node homing receptor. J Immunol 1992; 148:324–331.
244. Moore KL, Thompson LF. P-Selectin (CD62) binds to subpopulations of human memory T lymphocytes and natural killer cells. Biochem Biophys Res Commun 1992; 186:173–181.
245. Damle NK, Klussman K, Dietsch MT, Mohagheghpour N, Aruffo A. GMP-

140 (P-selectin/CD62) binds to chronically stimulated but not resting CD4$^+$ T lymphocytes and regulates their production of proinflammatory cytokines. Eur J Immunol 1992; 22:1789–1793.

246. Parrott DMV, Wilkinson PC. Lymphocyte locomotion and migration. Prog Allergy 1981; 28:193–284.

24. Larsen CG, Anderson AO, Appella E, Oppenheim JJ, Matsushima K. The neutrophil-activating protein (NAP-1) is also chemotactic for T lymphocytes. Science 1989; 241:1464–1466.

248. Kudo C, Araki A, Matsushima K, Sendo F. Inhibition of IL-8-induced W3/25$^+$ (CD4$^+$) T lymphocyte recruitment into subcutaneous tissues of rats by selective depletion of in vivo neutrophils with a monoclonal antibody. J Immunol 1991; 174:2196–2201.

249. Leonard EJ, Yoshimura T, Tanaka S, Raffeld M. Neutrophil recruitment by intradermally injected neutrophil attractant/activation protein-1. J Invest Dermatol 1991; 96:690–694.

250. Schall TJ, Bacon K, Toy KJ, Goeddel DV. Selective attraction of monocytes and T lymphocytes of the memory phenotype by cytokine RANTES. Nature 1990; 347:669–671.

251. Taub DD, Conlon K, Lloyd AR, Oppenheim JJ, Kelvin DJ. Preferential migration of activated CD4$^+$ and CD8$^+$ T cells in response to MIP-1α and MIP-1β. Science 1993; 260:355–358.

252. Schall TJ, Bacon K, Camp RDR, Kaspari JW, Goeddel DV. Human macrophage inflammatory protein α (MIP-1α) and MIP-1β chemokines attract distinct populations of lymphocytes. J Exp Med 1993; 177:1821–1825.

253. Taub DD, Lloyd AR, Conlon K, et al. Recombinant human interferon-inducible protein 10 is a chemoattractant for human monocytes and T lymphocytes and promotes T cell adhesion to endothelial cells. J Exp Med 1993; 177:1809–1814.

254. Adams DH, Harvath L, Bottaro DP, et al. Hepatocyte growth factor and macrophage inflammatory protein 1β: structurally distinct cytokines that induce rapid cytoskeletal changes and subset-preferential migration in T cells. Proc Natl Acad Sci USA 1994; 91:7144–7148.

255. Carr MW, Roth SJ, Luther E, Rose SS, Springer TA. Monocyte chemoattractant protein-1 is a major T lymphocyte chemoattractant. Proc Natl Acad Sci USA 1994; 91:3652–3656.

256. Roth SJ, Carr MW, Rose SS, Springer TA. Characterization of transendothelial chemotaxis of T lymphocytes. J Immunol Methods 1995. In press.

257. Leonard EJ, Yoshimura T. Human monocyte chemoattractant protein-1 (MCP-1). Immunol Today 1990; 11:97–101.

258. Villiger PM, Terkeltaub R, Lotz M. Production of monocyte chemoattractant protein-1 by inflamed synovial tissue and cultured synoviocytes. J Immunol 1992; 149:722–727.

259. Fujimoto T, Stroud E, Whatley RE, et al. P-selectin is acylated with palmitic acid and stearic acid at cysteine 766 through a thioester linkage. J Biol Chem 1993; 268:11394–11400.

260. Barclay AN, Birkeland ML, Brown MH, et al. The Leucocyte Antigen Facts Book. London: Academic Press, 1993.
261. Sutherland DR, Marsh JCW, Davidson J, Baker MA, Keating A, Mellors A. Differential sensitivity of CD34 epitopes to cleavage by Pasteurella haemolytica glycoprotease: implications for purification of CD34-positive progenitor cells. Exp Hematol 1992; 20:590–599.
262. Larson RS, Corbi AL, Berman L, Springer TA. Primary structure of the LFA-1 alpha subunit: an integrin with an embedded domain defining a protein superfamily. J Cell Biol 1989; 108:703–712.
263. Kishimoto TK, O'Connor K, Lee A, Roberts TM, Springer TA. Cloning of the beta subunit of the leukocyte adhesion proteins: Homology to an extracellular matrix receptor defines a novel supergene family. Cell 1987; 48: 681–690.
264. Takada Y, Elices MJ, Crouse C, Hemler ME. The primary structure of a α-4 subunit of VLA-4: homology to other integrins and possible cell-cell adhesion function. EMBO J 1989; 8:1361–1368.
265. Loftus JC, O'Toole TE, Plow EF, Glass A, Frelinger AL, III, Ginsberg MH. A β_3 integrin mutation abolishes ligand binding and alters divalent cation-dependent conformation. Science 1990; 249:915–918.
266. Carrell NA, Fitzgerald LA, Steiner B, Erickson HP, Phillips DR. Structure of human platelet membrane glycoproteins IIb and IIIa as determined by electron microscopy. J Biol Chem 1985; 260:1743–1749.
267. Nermut MV, Green NM, Eason P, Yamada SS, Yamada KM. Electron microscopy and structural model of human fibronectin receptor. EMBO J 1988; 7:4093–4099.
268. Staunton DE, Dustin ML, Erickson HP, Springer TA. The arrangement of the immunoglobulin-like domains of ICAM-1 and the binding sites for LFA-1 and rhinovirus. Cell 1990; 61:243–254.
269. Kirchhausen T, Staunton DE, Springer TA. Location of the domains of ICAM-1 by immunolabeling and single-molecule electron microscopy. J Leukocyte Biol 1993; 53:342–346.
270. Simmons D, Makgoba MW, Seed B. ICAM, an adhesion ligand of LFA-1, is homologous to the neural cell adhesion molecule NCAM. Nature 1988; 331: 624–627.
271. Staunton DE, Marlin SD, Stratowa C, Dustin ML, Springer TA. Primary structure of intercellular adhesion molecule 1 (ICAM-1) demonstrates interaction between members of the immunoglobulin and integrin supergene families. Cell 1988; 52:925–933.
272. Osborn L, Hession C, Tizard R, et al. Direct cloning of vascular cell adhesion molecule 1 (VCAM-1), a cytokine-induced endothelial protein that binds to lymphocytes. Cell 1989; 59:1203–1211.
273. Polte T, Newman W, Gopal TV. Full length vascular cell adhesion molecule 1 (VCAM-1). Nucleic Acids Res 1990; 18:5901.
274. Schall TJ. Biology of the RANTES/SIS cytokine family. Cytokine 1991; 3: 165–183.

275. Rot A, Krieger M, Brunner T, Bischoff SC, Schall TJ, Dahinden CA. RANTES and macrophage inflammatory protein 1α induce the migration and activation of normal human eosinophil granulocytes. J Exp Med 1992; 176:1489–1495.
276. Alam R, Forsythe PA, Stafford S, Lett-Brown MA, Grant JA. Macrophage inflammatory protein-1α activates basophils and mast cells. J Exp Med 1992; 176:781–786.
277. Kameyoshi Y, Dörschner A, Mallet AI, Christophers E, Schröder J-M. Cytokine RANTES released by thrombin-stimulated platelets is a potent attractant for human eosinophils. J Exp Med 1992; 176:587–592.
278. Hynes RO. Integrins: versatility, modulation, and signaling in cell adhesion. Cell 1992; 69:11–25.
279. Hemler ME. VLA proteins in the integrin family: structures, functions, and their role on leukocytes. Annu Rev Immunol 1990; 8:365–400.
280. Bochner BS, Luscinskas FW, Gimbrone MA Jr, et al. Adhesion of human basophils, eosinophils, and neutrophils to interleukin 1-activated human vascular endothelial cells: contributions of endothelial cell adhesion molecules. J Exp Med 1991; 173:1553–1556.
281. Horgan KJ, Luce GEG, Tanaka Y, et al. Differential expression of VLA-α4 and VLA-β1 discriminates multiple subsets of $CD4^+CD45RO^+$ "memory" T cells. J Immunol 1992; 149:4082–4087.
282. Kilshaw PJ, Murant SJ. Expression and regulation of $\beta_7(\beta_p)$ integrins on mouse lymphocytes: relevance to the mucosal immune system. Eur J Immunol 1991; 21:2591–2597.

2

Strategies to Inhibit Cellular Adhesion Molecules

Martin R. Weiser, Simon A. L. Gibbs, and Herbert B. Hechtman
Department of Surgery, Brigham and Women's Hospital and Harvard Medical School, Boston, Massachusetts

I. NEUTROPHIL-MEDIATED INJURY

Neutrophils are a key component of the defense against infection and are recruited locally into sites of injury by various chemotactic agents including lipopolysaccharide derived from bacterial cell walls, cytokines, and eicosanoids produced by local tissue monocytes and endothelial cells, and complement-derived anaphylotoxins such as C3a and C5a. Once the neutrophils have migrated across the endothelial barrier into the site of injury, additional chemoactivators stimulate neutrophils to produce peroxides and proteases which are designed to destroy the offending organism.

During severe injury, infection, or ischemia and reperfusion damage, spillover of these activators into the systemic circulation results in cellular activation, leading to indiscriminate neutrophil-endothelial adhesion and the release of injurious agents which damage host tissues. For example, during abdominal aortic aneurysm repair, release of the aortic clamp results in a reperfusion injury of the lower extremities which is mediated in part by neutrophils. The sequence of events involves free-radical and eicosanoid production locally in the reperfused tissue and in the circulating neutrophils. Cytokine synthesis is slower since transcription is required. Finally, complement activation with release of an array of chemoactivators into the systemic circulation occurs over the first 1 to 2 hours. Systemic neutrophil and endothelial activation results in remote lung injury, characterized by an altered microvascular barrier function, with albumen leaking into the interstitial and alveolar space, the prodromes of the acute respiratory distress syndrome (ARDS). Administration of mannitol before reperfusion scavenges free radicals, limits eicosanoid synthesis, prevents neutrophil and endothelial activation, and prevents pulmonary edema (1,2).

Neutrophils have been implicaed as mediators of many types of inflammation including the local and remote effects of ischemia and reperfusion, atelectasis, acid aspiration, acute renal failure, and pancreatitis (3–9). Depleting circulating neutrophils using antineutrophil antibodies will significantly limit these injuries.

II. NEUTROPHIL-ENDOTHELIAL ADHESION

The sequence of events leading to neutrophil sequestration at sites of injury is well described (10–14). In this multistep process, circulating neutrophils are tethered to activated endothelium, become firmly attached, and eventually diapedese along chemotactic gradients through interendothelial junctions (Fig. 1).

A. Selectins

The selectin family of adhesion molecules is responsible for slowing down neutrophil movement in postcapillary venules at sites of endothelial activa-

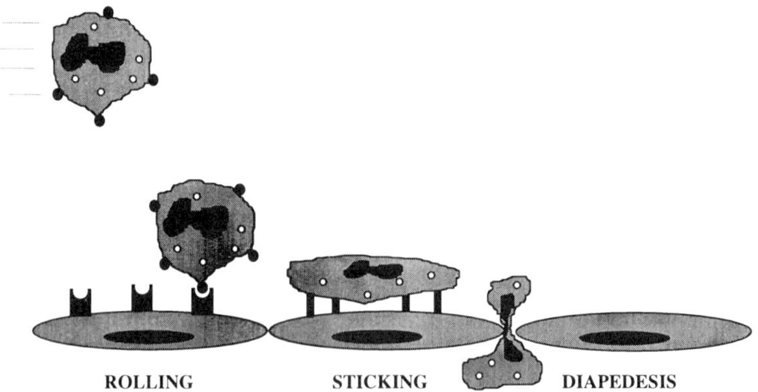

Figure 1 Neutrophils flowing in a vessels slow down by interacting with selectins expressed by activated endothelium. Subsequently they firmly adhere and stick to endothelium via the integrins, and then diapedese through gap junctions along chemotactic gradients assisted by platelet-endothelial cell adhesion molecule.

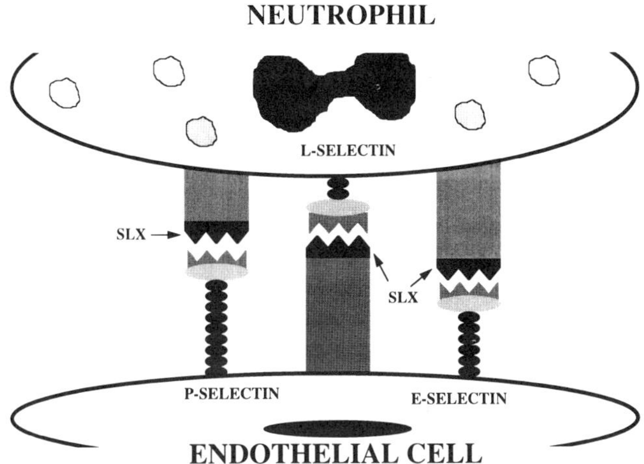

Figure 2 The selectin family of adhesion molecules is involved in the initial steps of neutrophil sequestration at sites of injury. There are three members, P-, L-, and E-selectin, and they all bind to counterreceptors containing the oligosaccharide sialyl-Lewis X.

tion. This family consists of three structurally related molecules (L-, E-, and P-selectin; Fig. 2), which contain a lectin domain that binds to oligosaccharide counterreceptors, an epidermal growth factor motif, a variable number of short consensus repeats, a transmembrane domain, and an intracytoplasmic domain (Fig. 3). All selectins recognize glycoproteins containing sialyl-Lewis X (SLX), a terminal carbohydrate containing sialic acid and fucose.

L-selectin is constitutively present on the neutrophil. It is shed from the activated cell surface (15), presumably coincident with the engagement of the leukocyte integrin adhesion molecules. E-selectin is expressed on activated endothelial cells after de novo synthesis. It is maximally up-regulated by cytokines such as interleukin-1 (IL-1) and tumor necrosis factor (TNF) after 4 to 8 hours and returns to baseline by 24 to 48 hours (16). P-selectin is located in Weibel Palade bodies of endothelial cells and alpha granules of platelets; within 2 to 5 minutes of stimulation by thrombin, histamine, eicosanoids, or C5a (17–19), it is translocated to the plasma membrane and then rapidly reinternalized. Elevated serum levels of P-selectin, measured by ELISA using antibodies against the lectin domain, are found after neutrophil or platelet activation, indicating that one or more epitopes of P-selectin are shed from the membrane surface.

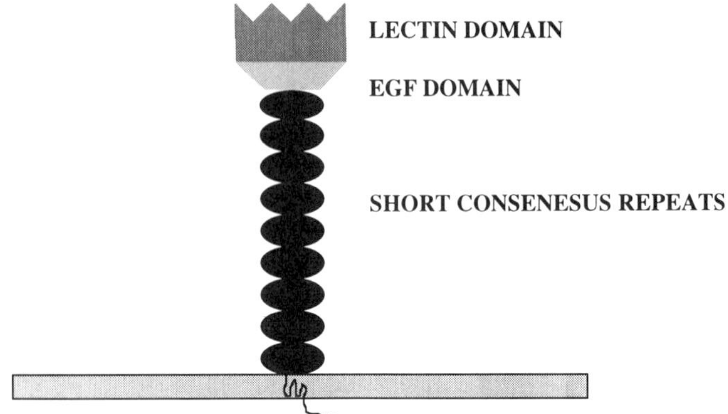

Figure 3 P-selectin, like all selectins, is composed of a lectin domain which binds to sialyl–Lewis X containing counterreceptors, an EGF motif, short consensus repeats, a transmembrane, and cytoplasmic domain.

B. Integrins

Once neutrophils are encouraged to roll and stick to endothelial cells by the selectins, they come into contact with endothelial cell derived chemoactivators such as platelet activating factor (PAF) (19) or interleukin-8 (20,21), or they bind to factors processed on the cell surface but not derived from endothelium, such as the third component of complement (iC3b) (12). Activated neutrophils rapidly become polarized and securely attached to the endothelial cell via the functionally and numerically upregulated $\beta 2$ integrin family of adhesion molecules (22) which bind to the intercellular adhesion molecule(ICAM)-1 and -2, members of the immunoglobulin superfamily (12).

The integrins are heterodimeric cell surface proteins composed of two noncovalently linked polypeptide chains, alpha and beta. The latter subunit distinguishes three subfamilies: beta-1 (CD 29), beta-2 (CD 18), and beta-3 (CD 61). The alpha and beta chains consist of transmembrane and cytoplasmic segments. The extracellular domain of the two chains bind to various ligands such as extracellular matrix glycoproteins, complement proteins, and other cell surface proteins. The cytoplasmic tail of the integrin interacts with cytoskeletal components which signal shape change, motility, and phagocytic responses.

The beta-2 family is also known as the leukocyte integrins because they are confined to white blood cells and consist of two members that are both involved in neutrophil-endothelial interactions (Fig. 4). The first, CD 11a/CD 18 (LFA-1), is constitutively expressed on neutrophil plasma membranes, is upregulated by gene transcription, and binds to ICAM-1 and -2. The second, CD 11b/CD 18 (MAC-1), binds to iC3b and ICAM-1. A second member of this family, ICAM-2, is constitutively expressed on endothelial cells whereas ICAM-1 expression is increased with cellular activation (12). CD 11b/CD 18 is present on the cell surface and in subcellular granules that are rapidly fused with the plasma membrane which permits receptor translocation upon activation. The CD 11b/CD 18 molecule also undergoes a conformational change upon activation with complement fragments, PAF, cytokines, and the eicosands (12,23,24), which functionally upregulates this adhesion molecule (22).

In addition to ICAM-1 and -2, platelet-endothelial cell adhesion molecule-1 (PECAM-1) is a member of the immunoglobulin superfamily that is localized to the endothelial cell intercellular junction. PECAM-1 is involved in neutrophil transendothelial migration, and may also act as a neutrophil activator (14). Activated neutrophils produce free radicals and proteolytic enzymes which cause cell injury by lipid peroxidation and protein damage before and after transendothelial migration (Fig. 4) (25).

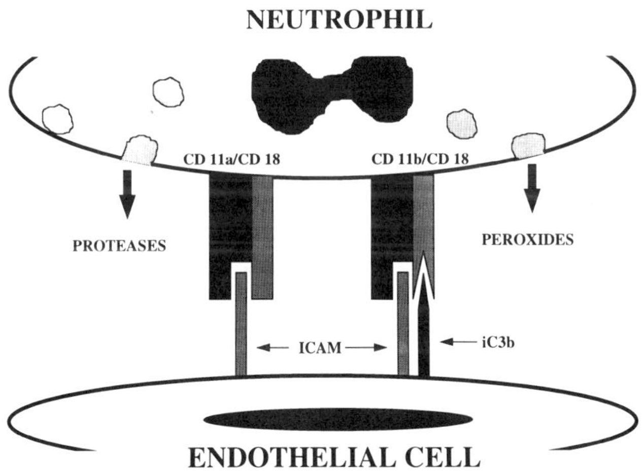

Figure 4 The integrin family of adhesion molecules is involved in firm neutrophil binding to endothelial ICAM-1 and -2. The integrins interact with various cell-bound ligands which can activate the cell, such as iC3b. Neutrophil activation can lead to release of peroxides and proteases which damage the vascular barrier.

III. STRATEGIES TO INHIBIT NEUTROPHIL BINDING TO ENDOTHELIUM

The role of neutrophils in injury has been well documented in animal models of inflammation where tissue damage can be reduced by depleting these cells. In man, removing neutrophils from the circulation is not practical. Since binding of neutrophils to endothelium is a prerequisite for injury, an alternative therapeutic goal is to temporarily prevent binding. Strategies include prevention of neutrophil and endothelial cell activation and expression of adhesion receptors as well as antagonism of the adhesion molecules themselves.

For purposes of this discussion, membrane-bound adhesion molecules are referred to as receptors. They bind to other cell-bound molecules called counterreceptors. Ligand is reserved for circulating molecules that bind to receptors, such as the cytokines (Fig. 5).

A. Antibodies

1. Structure and Function

Antibodies can be used to target cell adhesion molecules and cytokines. An antibody molecule is a glycoprotein composed of four polypeptide chains, two identical heavy chains and two identical light chains, held together by interchain disulfide bonds (Fig. 6). Based on the degree of amino acid sequence homology, the antibody chains are divided into constant, variable, and hypervariable regions. The antigen binding site is located within

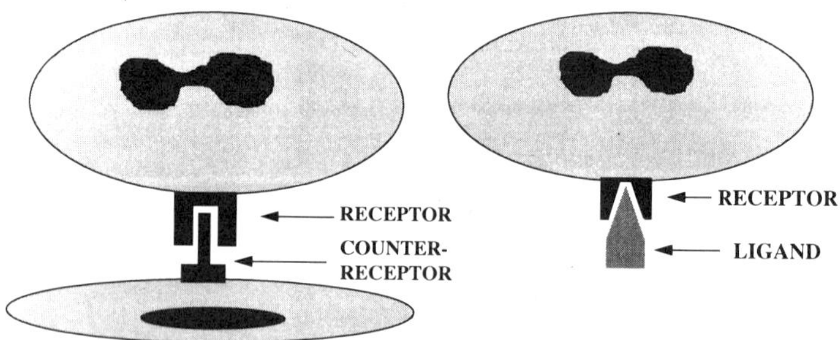

Figure 5 Cell-bound adhesion molecules are referred to as receptors, and they bind to cell-bound counterreceptors. For example, P-selectin, an endothelial adhesion receptor, binds to P-selectin glycoprotein ligand-1 (PSGL-1), its counterreceptor on neutrophils. The term ligand is reserved for soluble molecules that interact with cell-bound receptors. For example, the cell-bound IL-1 receptor binds to its soluble ligand, IL-1.

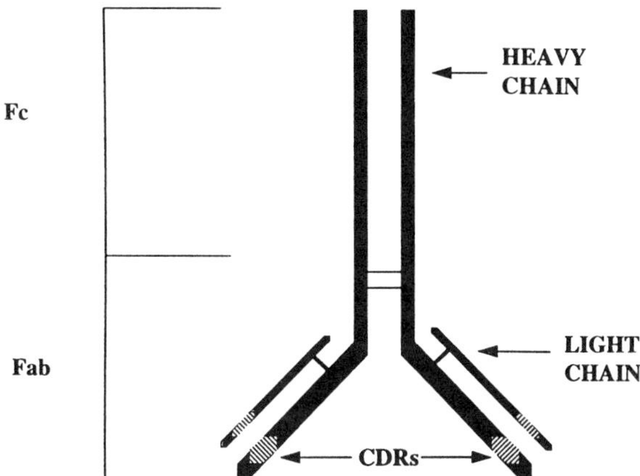

Figure 6 The immunoglobulin is composed of two heavy and two light chains and can be functionally divided into the Fab (antigen binding site) and Fc (effector function) regions. The CDRs are the hypervariable regions which bind to an epitope on the antigen.

the hypervariable portion of the antibody, referred to as the complementary determinant regions (CDRs), since this portion of the immunoglobulin is thought to form a surface that is complementary to the three-dimensional surface of a bound antigen (26). An antibody can also be functionally divided into the Fc (fraction crystallization) region and the Fab (fraction antigen binding) region (Fig. 6). The Fc region is responsible for the effector functions, or biological properties, of the antibody and is divided into domains (C_H1, C_H2, etc.). Effector functions include complement fixation, binding to leukocytes, and transfer of immunoglobulin—for example, across mucosal surfaces into the intestine or placental tissue into the fetal circulation (26–29).

The region of the antigen that interacts with CDRs of the antibody is defined as an epitope. Each antibody contains two antigen binding sites and therefore is bivalent. Cross-reactivity refers to antibody binding to an antigen other than the immunogen (28) and is the result of similarities in molecular structure of epitopes. However, similar epitopes do not necessarily imply similar functioning molecules. Conversely, similar functioning molecules from different species may have slight changes in molecular sequences, rendering the antigen unrecognizable by a particular antibody.

Immunologlobulins are described by the structure of the constant region of their heavy chain. There are five major classes (29). IgG is the principal antibody synthesized during a secondary immune response to an antigen.

The IgG serum concentrations range from 8 to 16 mg/ml and account for 80% of total immunoglobulin concentration (26). By employing complement and leukocytes, IgG is responsible for the majority of neutralization of invading organisms and their toxins. IgG binds to bacteria and forms immune complexes which activate the classical complement pathway via the C_H2 domain of the Fc region of the immunoglobulin, thereby producing direct cell lysis. Complement byproducts (C3a and C5a) at as chemoattractants and lead to additional neutrophil recruitment to the area of microbial invasion. Immune complexes and complement components on bacterial membranes also act as opsonins and enhance neutrophil attachment to bacteria through immunoglobulin Fc receptors (C_H3 domain) and iC3b receptors (CD 11b/CD 18). This facilitates phagocytosis of the organism. Antibody-coated bacteria ae also recognized and destroyed directly by natural killer (NK) lymphocytes via Fc receptors. Thus these effector functions of IgG are Fc dependent. IgG is further subclassified into IgG1-4, based on effector functions such as the affinity for Fc receptors on leukocytes and complement.

The other important class of immunoglobulin is IgM which is found in a concentration of 0.5 to 2 mg/ml in serum and comprises approximately 5% of the total immunoglobulin (26). IgM antibodies circulate as pentamers with individual molecules attached to each other by the joining (J) chain. Although individual IgM molecules have low relative binding affinities, the circulating IgM pentamers have a combined valency of 10, which substantially increases their avidity for antigen. Complement activation requires antibody cross-linking by multiple bound molecules to a surface such as a microorganism; thus IgM has the greatest complement activating ability. Many "natural" antibodies are in the IgM family. These include isohemagglutinins and antimicrobials which are found in serum and increase early in the immune response to infection.

2. Polyclonal vs Monoclonal Antibodies

Polyclonal antisera are produced by immunizing a host (e.g., rabbit or goat) with an antigen, followed by collection and purification of the immunoglobulin component from the host serum. Since a host produces antibodies against multiple epitopes of the immunogen, a heterogeneous population of antibodies is produced. IgG is used because it has the highest serum concentration and exhibits greater antigen binding affinity than monomeric IgM (22). The combination of different antigenic motifs recognized on the target molecule by polyclonal antisera results in a high binding activity compared with monoclonal antibodies that bind to a single epitope (28,29). Individual antibody-antigen affinities are increased when multiple epitopes on a target antigen are bound (26,27). Due to the heterogeneity of antibody clones in polyclonal antisera directed against different epitopes, there is a substantial amount of cross-reactivity with polyclonal antisera.

The advent of monoclonal antibody production can now provide a large supply of a homogeneous and standardized antogonist with known specificity and affinity (Fig. 7) (30,31). First described in 1975, clinical therapeutic trials with monoclonal antibodies begin in 1981 to treat allograft rejection (30). Antibody-secreting hybridomas we produced by fusing immortal myeloma cells with B cells from immunized mice. Selection of a hybridoma producing an antibody directed against a single epitope allowed for clonal expansion and harvest of the desired murine antibody.

3. Limitations of Antibody Therapy

There are limitations to administering murine antibodies to humans. These include endotoxin contamination, unpredictable clinical effects, and sensitization. To date, no episodes of viral contamination have been described with the administration of monoclonal antibodies, and the risk of infection and contamination with endotoxin may be reduced by use of totally recombinant material (30). With regard to sensitization, much smaller amounts of protein are administered with monoclonal antibodies, which reduce the risk of serum sickness and anaphylaxis (30).

Other unpredictable clinical manifestations may be due to the effector function of the immunoglobulin molecule—that is, the Fc portion. For example, neutropenia may result from the administration of an anti-L-selectin antibody (32), likely due to sequestration of the antibody-coated neutrophils in the reticuloendothelial system. Similarly, in the rat immune complex model of lung injury, administration of an anti-E-selectin anti-

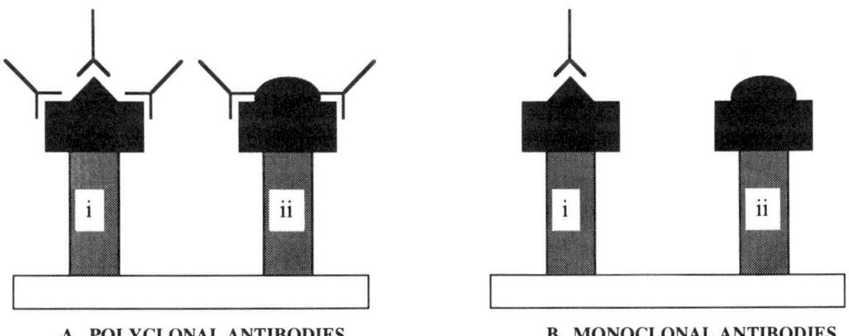

Figure 7 (A) High-affinity polyclonal antibodies are directed against many epitopes of an antigen (i), but also nonspecifically cross-react with other molecules (ii). (B) Monoclonal antibodies are directed against a single epitope of an antigen and therefore have less cross-reactivity.

body exacerbates the tissue damage. Two mechanisms could explain this observation. First, antibody-coated endothelium will attract and bind neutrophils via the Fc receptors, resulting in additional leukocyte-mediated injury. Secondly, antibody-coated endothelial cells can activate complement, also through the Fc region of the antibody molecule, leading to direct cell injury and the release of anaphylatoxins (C5a and C3a), which attract and activate additional neutrophils to the site of injury (33). Methods to reduce these adverse effects include removal of the Fc portion of the antibody molecule using pepsin digestion (Fig. 8). The resultant $F(ab')_2$ fragments are devoid of the Fc region but maintain their bivalency and their ability to block the targeted antigen (31). The drawback to treatment with $F(ab')_2$ fragments is that their increased rate of clearance requires more frequent dosing regimens (33).

Other unpredictable effects include cellular activation by the binding of the CDRs of the antibody to the receptor region of a cell adhesion molecule, leading to activation of a signaling pathology. The CDRs can either block the native receptor or counterreceptor and prevent their activation or, by interacting in another subtle manner, lead to signaling and indeed activation. Unwanted activation has been described in transplant patients receiving anti-T-cell antibody therapy where the administration of the monoclonal antibody OKT 3 resulted in a large cytokine release (34).

Another potential problem with antibody therapy is sensitization to the administered murine protein. Two types of anti-immunoglobulin antibody responses have been described: anti-isotype and anti-idiotype antibodies (31).

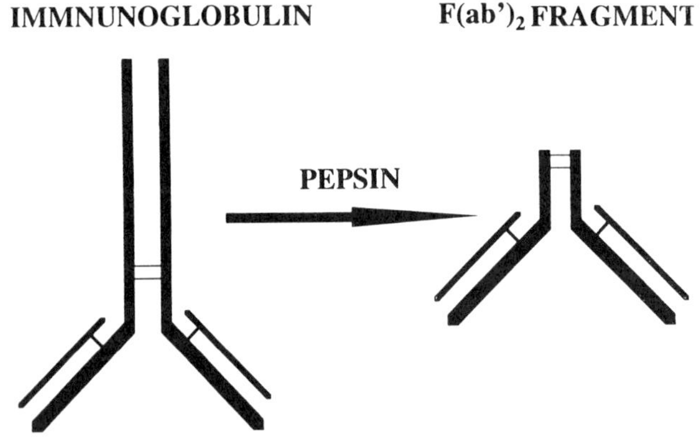

Figure 8 Immunoglobulins are treated with pepsin to remove the Fc portion of the molecule to produce $F(ab')_2$ fragments which can bind to antigens but have no effector function capability (e.g., complement activation or cell binding).

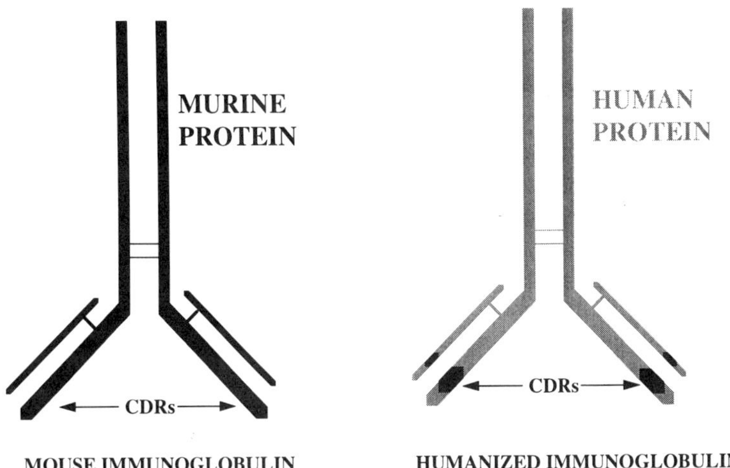

Figure 9 Humanized immunoglobulins are produced by inserting murine CDR genes into a human immunoglobulin gene. This likely reduces the immune response to the administered protein.

Anti-isotypic antibodies are antibodies against the Fc portion of the immunoglobulin molecule. In theory, this could result in serum sickness or anaphylaxis if the mouse protein is administered repeatedly. However, no cases of serum sickness have been reported, and there has only been one case of IgE-mediated anaphylaxis (31). The anti-isotope antibodies can also interfere in the immunoassays used to measure serum antibody concentrations. Antiidiotypic antibodies bind to the antigen-combining site, i.e., the CDR, and thereby reduce the therapeutic efficiency of the antibody (31).

In an attempt to decrease the adverse-effects profile of immunoglobulin administration, the amount of mouse protein in the antibody preparation can be minimized. Human-mouse chimeric antibodies have been produced by rearranging immunoglobulin DNA and fusing mouse variable regions with human constant regions, which reduces the immunogenicity of the antibody (30). A more refined approach involves incorporating the CDR genes of the murine antibody into human antibody genes. These humanized antibodies have only a small fraction of the mouse amino acid sequence and are consequently less immunogenic (Fig. 9) (30). It remains likely, however, that they will stimulate the formation of anti-idiotypic antibodies.

Although they have limitations, monoclonal antibodies directed against adhesion molecules and their counterreceptors are the gold standard for the study of cell-cell interactions. Since most antibodies are of murine origin and are directed against human epitopes, their cross-reactivity with other species must be proven before animal studies can be undertaken. Since the CDRs bind to a relatively small area on the antigen, minor changes in the epitope's amino acid sequence or configuration can reduce or abolish binding affinities (26). Proof of binding in an experimental animal is usually done by immunohistochemistry. However, antibody binding by immnohistochemical criteria does not prove neutralizing activity in functional assays. There are a number of functional assays such as blocking the binding of neutrophils to immobilized selectins or inhibition of neutrophil-neutrophil, platelet-neutrophil, or neutrophil-endothelial interactions using cells from the species of interest (35).

Thus some antibodies will block function while others bind to noncritical regions of the adhesion molecule and can only be used to stain the receptor in tissue sections. For example, G1 is a monoclonal antibody directed against P-selectin and will inhibit neutrophil binding to P-selectin incorporated into planar phospholipid membranes or to P-selectin expressed on activated endothelium (19). However, the antibody S12 will bind to P-selectin incorporated into phospholipid membranes, but will not block neutrophil interaction with the membrane bound P-selectin.

4. Antibodies Against Selectins

The early role of the selectins in neutrophil-endothelial adhesion makes them an attractive target for antiadhesion therapy, and many studies have shown efficacy of antiselectin antibody in reducing neutrophil-mediated injury. Antibodies that bind near or at the lectin domain will inhibit receptor and counterreceptor interactions (Figs. 10, 11). PB 1.3 is a blocking monoclonal antibody directed against human P-selectin which cross-reacts weakly with rabbit and rat P-selectin. When administered before reperfusion, PB 1.3 reduces neutrophil sequestration and local injury in a rabbit ear rendered ischemic for 6 hours (36). Following 1 hour of intestinal ischemia, PB 1.3 reduces neutrophil rolling in the splanchnic circulation, leukosequestration in the ileum, and tissue injury in a feline ischemia reperfusion model (37). Similarly, in a cobra venom factor infusion model of the adult respiratory distress syndrome in the rat, PB 1.3 reduces neutrophil sequestration and injury (35).

Inhibition of L-selectin with antibodies also reduces neutrophil-dependent injury. HL-60 cells are an immortal cell line with surface characteristics of neutrophils. Their rolling along endothelium is inhibited by two of three monoclonal antibodies directed against the lectin domain of L-selectin, LAM1-3 or LAM1-6, but not by the nonblocking monoclonal antibody LAM1-11 (38). In studies utilizing the rabbit ear, LAM1-3 prevented ear necrosis and reduced neutrophil sequestration after 6 hours of

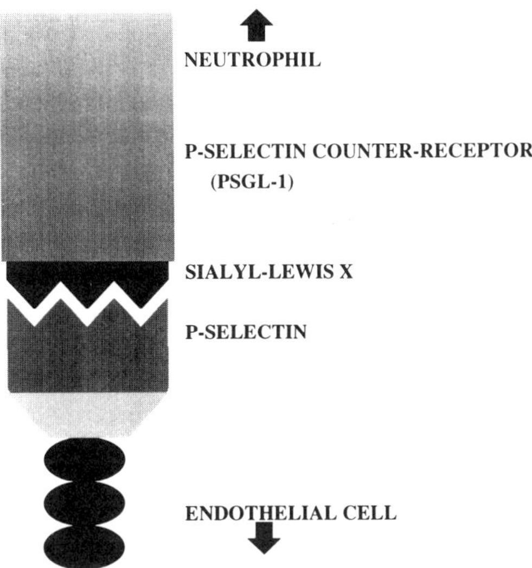

Figure 10 The sialyl–Lewis X oligosaccharide of the selectin counterreceptor binds to the lectin domain of the selectin. A magnified cartoon shows endothelial-bound P-selectin (truncated after the third short consensus repeat) binding to P-selectin glycoprotein ligand-1 (PSGL-1; also truncated in this figure) on neutrophils. PSGL-1 is a large molecule containing multiple sialylated sugars.

ischemia. The nonblocking antibody was used as the control with no amelioration of injury (32). In rat hindlimb ischemia and reperfusion, a hamster monoclonal antibody against L-selectin was infused at the time of reperfusion and reduced neutrophil sequestration as well as local muscle and remote lung injury (39). $F(ab')_2$ monoclonal antibody fragments were necessary since neutropenia was produced when the unmodified antibody molecule was used, demonstrating the difficulty of using immunoglobulin molecules with Fc fragments in antiadhesion therapy.

Inhibition of E-selectin function with antibodies is also effective in reducing neutrophil-mediated injury. A monoclonal antibody against E-selectin (CL-3) reduced peritoneal neutrophil exudates in rats produced by the intraperitoneal injection of glycogen. Similarly, in the immune complex induced lung injury model, this antibody reduced neutrophil influx and lung injury (33). The nonblocking but staining antibody CL-37 wa used as a control. Interestingly, $F(ab')_2$ fragments had to be employed since injury was increased when the whole immunoglobulin was administered. The investigators suggest that endothelial cells expressing E-selectin bound the antibody resulting in neutrophil recruitment by means of complement acti-

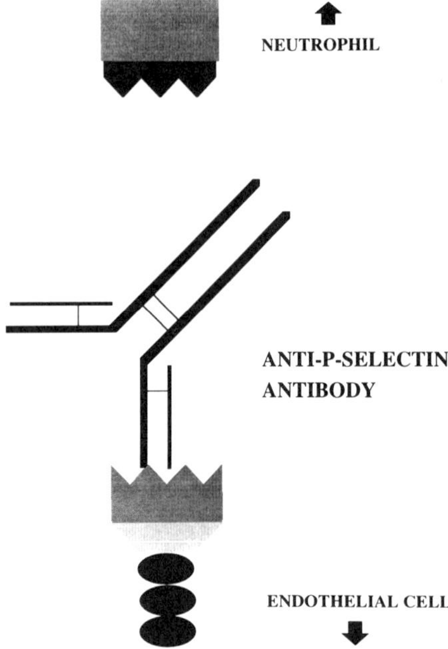

Figure 11 Antibodies directed against the lectin domain of P-selectin will inhibit binding to its counterreceptor.

vation as well as via the Fc portion of the immunoglobulin molecule. In hindlimb ischemia and reperfusion, the anti-E-selectin monoclonal antibody CL-3 (Fab′)$_2$ preparation reduced neutrophil sequestration and lung injury but did not modify local neutrophil influx or muscle injury (33).

5. Antibodies Against the Integrins

The leukocyte integrins consist of CD 11a/CD 18 (LFA-1) and CD 11b/CD 18 (MAC-1) and are a logical site for blockade since they are responsible not only for attachment but also for neutrophil activation. Thus, inhibition of all β2-integrins using an antibody against CD 18, 60.3, reduced the reperfusion injury following 10 hours of ear ischemia in the rabbit (40). Furthermore, in complete intestinal ischemia and reperfusion, blockade of CD 18 with the IgG antibody R15.7 inhibited all leukocyte integrins and reduced leukosequestration and remote lung injury (41). On the other hand, treatment with R17, an anti-CD 11b IgM which only blocks MAC-1 and does not affect LFA-1, reduced lung injury but not leukosequestration (42). These results provide evidence for a dissociation of adhesion and activation by the different integrins. MAC-1 is responsible not only for adhesion but also activation while LFA-1 only mediates adhesion. Further evidence to support this hypothesis is provided by the administration of soluble complement receptor-1 (sCR1), an anticomplement agent which prevents iC3b formation and deposition on endothelium. Administration of this agent in intestinal ischemia prevents remote lung injury but not neutrophil leukosequestration (43). This provides indirect evidence to support the thesis that iC3b, a counterreceptor for MAC-1, is the activating signal to the neutrophil and, secondly, that sequestration can occur without activation.

Other antibodies have been developed against an epitope which appears only upon activation of the CD 11b/CD 18 complex. Thus, CBRM 1/5 is directed against the CD 11b/CD 18 complex which has undergone a conformational change that functionally upregulates the receptor and exposes a new epitope (22). This type of antibody strategy may provide more specific antineutrophil therapy restricted to activated cells.

Other studies using anti-ICAM monoclonal antibodies have demonstrated the effectiveness of blocking integrin-mediated inflammatory injury. Reexpansion of atelectatic lung segments in the rabbit is associated with leukosequestration and increased vascular permeability. The vascular injury can be ameliorated by neutrophil depletion. Further, both neutrophil sequestration and the vascular injury can be reduced with the anti-CD 18 antibody R15.7 or the anti-ICAM-1 antibody RR1/1 (8). Similarly, ischemic canine gracilis muscle had improved microvascular patency when the animal received treatment with either anti-ICAM-1 antibody CL18/6C7 or R6.5. This therapy reduces neutrophil binding to endothelium and subsequent vascular occlusion, referred to as "no-reflow" (44).

6. Antibodies Against Cytokines

The cytokines tumor necrosis factor (TNF) and interleukin-1 (IL-1) play key roles in inflammatory diseases by activating cells. By inducing expression of adhesion molecules, the cytokines set the stage for neutrophil adhesion to endothelium and tissue injury. Inteleukin-8 (IL-8) has an additional important action by virtue of its function as a chemotactic peptide, attracting leukocytes to sites of injury and stimulating their release of proteases and peroxides (20). The proinflammatory effects of TNF, IL-1, and IL-8 can be antagonized with monoclonal antibodies. Anti-TNF antibodies administered to rabbits prior to the injection of endotoxin protects the animals from the development of hypotension, fibrin deposition, and death (45). In baboons, neutralizing monoclonal anti-TNF $F(ab')_2$ fragments, administered 2 hours before bacterial challenge, provided complete protection against shock and organ dysfunction, while administration of the antibody of the antibody 1 hour before protected against shock but not vital organ failure (46). Similarly, blockade of IL-1 with a monoclonal antibody directed against the IL-1 receptor provided some protection against the inflammatory injury associated with a turpentine-induced sterile abscess in the mouse (47).

Interleukin-8 is a chemotactic cytokine produced by a wide variety of cells including monocytes, endothelial cells, neutrophils, and fibroblasts (48). It contributes to the initiation of the local inflammatory response (49). For example, hypoxic endothelial cells in culture (21) produce IL-8, which induces neutrophil chemotaxis, shape change, upregulation of adhesion molecules, and a respirator burst (48). Thus, a skin burn leads to increased IL-8 levels, but not IL-6 or TNF, and is associated with neutrophil infiltration and subsequent vascular injury with protein leak (50,51). IL-8 is also produced locally in ischemic and reperfused sheep lungs (52). In ischemia and reperfused rabbit lungs the neutrophil accumulation and tissue injury can be prevented by treating with an anti-IL-8 antibody (53).

B. Soluble Receptors

1. Cell Adhesion Molecules

Recent studies have shown that soluble isoforms of adhesion molecules are found in the circulation, including the three selectins and members of the immunoglobulin superfamily (VCAM-1 and ICAM-1) (54). Increased serum levels of these molecules are seen in inflammatory disease states—e.g., sepsis and chronic infections such as extensive hydradinitis. Either they may be shed from the cell, as with L-selectin, or a secreted form may be produced by differential mRNA splicing which excludes the transmembrane portion of the receptor, as is the case with P-selectin (55). The role of these soluble receptors in the control of inflammation is unclear, but since they may serve as receptor antagonists they could modify cellular adhesion. Indeed, in the case of P-selectin, soluble circulating forms have been shown in vivo to prevent activated neutrophil adhesion to endothelium (56). It has been suggested that soluble P-selectin limits inflammatory reactions by maintaining the nonadhesiveness of neutrophils.

To take theoretical advantage of the possible antiinflammatory effects of circulating adhesion molecules, recombinant selectin chimeric molecules have been engineered, linking the extracellular domains of L-, P-, and E-selectin to the constant regions of human IgG1. The IgG domain is responsible for increasing the half-life of the circulating selectin molecule. This chimeric construct binds to the selectin counterreceptor, thereby inhibiting its interaction with cell-bound selectin receptors (Fig. 12). Thus, an L-selectin chimera successfully prevented neutrophil migration into a thio-

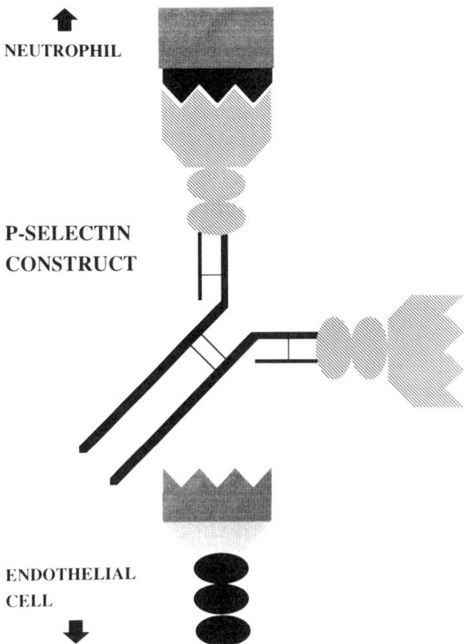

Figure 12 A P-selectin construct molecule, containing the lectin domain, EGF motif, and two SCRs regions fused to an immunoglobulin molecule, binds to the selectin counterreceptor on a neutrophil (PSGL-1), thereby inhibiting interaction of endothelial-bound P-selectin.

glycollate irritated peritoneum of the mouse (57) and prevented neutrophil rolling on rat mesenteric venules, as assessed by intravital microscopy (38). Treatment with P- or L-selectin Ig chimeras ameliorates cobra venom factor-induced, P-selectin-mediated, acute lung injury in the rat (58).

2. Cytokines

In a fashion similar to the selectin-immunoglobulin constructs, administration of soluble forms of cytokine receptors (the extracellular domain) can bind the cytokines in serum and prevent their interaction with cell-bound receptors and subsequent effector functions. Thus TNF receptors administered to nonhuman primates with lethal bacteremia attenuated the hemodynamic collapse and limited further cytokine synthesis (59). Administration of soluble IL-1 type 1 receptor increased the survival of mice undergoing allograft heart transplantation (60). These soluble cytokine receptors may act in part by preventing TNF and IL-1-induced adhesion molecule upregulation on neutrophil and endothelial cells.

Finally, chimeras produced by fusing the extracellular domain of the TNF receptor to the hinge region of IgG3 produced a slowly cleared molecule capable of reducing the lethality of endotoxemia in mice. In this model of sepsis, D-galactosamine is injected intraperitoneally and sensitizes the animal to the simultaneous injection of *E. coli* LPS (61). Again the chimeric molecule presumably affords protection by preventing neutrophil and endothelial activation and noncomitant cell adhesion molecule expression.

C. Soluble Counterreceptors

1. Adhesion Molecule Counterreceptors

The selectins bind to counterreceptors containing the oligosaccharide sialyl-Lewis X (SLX) (11). Infusion of a soluble form of SLX will bind to the receptors on endothelial cells (E- and P-selectin) or neutrophils (L-selectin), thereby blocking the attachment to their natural, cell-bound counterreceptors (Fig. 13). The result is inhibition of the initial rolling and tethering of neutrophils to endothelium at sites of inflammation. Antiadhesion therapy with SLX is advantageous since the sugar has low immunogenicity (27), it can be produced in mass quantities, and it offers reduced systemic toxicity compared to the possibilities of serum sickness associated with immunoglobulin therapy. In rodent hindlimb ischemia, skeletal muscle permeability

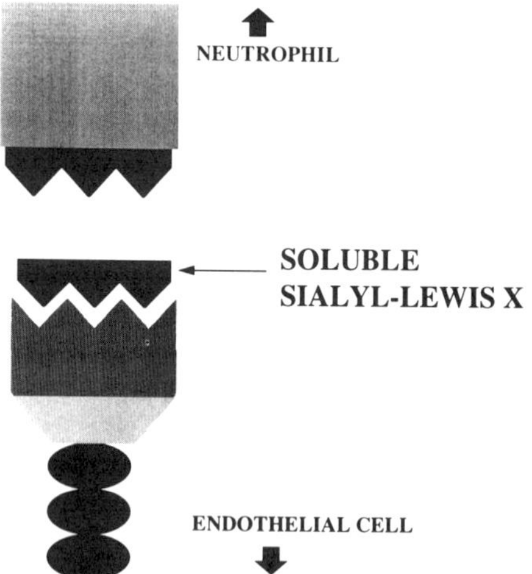

Figure 13 A soluble form of sialyl-Lewis-X binds to the lectin domain of P-selectin, thereby preventing selectin binding to its natural, cell-bound counterreceptor.

was reduced 36% with infusion of 10 mg/kg of soluble SLX before reperfusion. Treatment was associated with a 46% reduction in tissue myeloperoxidase, a neutrophil marker (3). Further evidence supporting the hypothesis that soluble SLX prevents the initial neutrophil-endothelial interaction is that neutrophil depletion resulted in a similar 36% reduction in muscle injury.

In other models of selectin-dependent inflammation, the soluble counterreceptor SLX was also beneficial. Cobra venom factor infusion into a rat resulted in a P-selectin-mediated, neutrophil-dependent acute lung injury. This was characterized by neutrophil sequestration which could be minimized by treatment with soluble SLX oligosaccharides (62). In the model of immune complex deposition in rat lung, leukosequestration is dependent on E-selectin. The neutrophil sequestration and injury could be reduced by treatment with soluble SLX oligosaccharides (63).

Recent studies have isolated the DNA for a P-selectin counterreceptor, P-selectin glycoprotein ligand-1. This mucinlike transmembrane glycoprotein consists of multiple sialylated Lewis X molecules which are available to P-selectin. Theoretically this natural counterreceptor for P-selectin will have higher binding affinity and therefore increased antiinflammatory activity in its soluble form (64).

2. Cytokines

The function of IL-1 can be blocked by IL-1 receptor antagonist (IL-1Ra), a naturally occurring substance that competes with the binding of IL-1 to its receptor. The administration of IL-1Ra reduced the lethality of endotoxin-induced shock in rabbits (65). Although initial studies found a survival benefit in patients with the systemic inflammatory response syndrome, IL-1Ra did not significantly reduce patient mortality in a recent phase III study (20,66).

D. Peptides Homologous to Receptors

1. Natural Peptides

Several peptides have been discovered that bind to adhesion receptors and prevent the natural counterreceptor from interacting with its receptor. These substances have been isolated from invading organisms which utilize them to evade host defense by blocking neutrophil adhesion and activation at sites of infection. One such agent is neutrophil inhibitory factor (NIF), a 41-kd glycoprotein from the canine hookworm *Ancylostomata caninum*. A recombinant form of this glycoprotein has been produced, and by binding to the CD 11b portion of the CD 11b/CD 18 adhesion molecule, it can inhibit neutrophil adhesion to endothelial cells and H_2O_2 production (67).

Another agent is the filamentous hemagglutinin protein of *Bordetella pertussis* which also binds to CD 11b (68) and inhibits leukocyte infiltration in a rabbit model of bacterial meningitis (69).

2. Synthesized Peptides

Synthetic peptides engineered to be homologous to receptor binding sites will interact with counterreceptors and inhibit neutrophil-endothelial interactions (Fig. 14). One area of active research is the selectin family of adhesion molecules (70). The extracellular structure of the selectins is similar in that almost all selectins contain multiple short consensus repeats, an epidermal growth factor motif, and a lectin domain. The lectin domain binds with calcium dependency to counterreceptors containing sialyl–Lewis X oligosaccharides. The amino acid sequence of the lectin domain of P-selectin has been deciphered, and overlapping peptides (10-30 amino acid sequences) of the region have been synthesized (Fig. 15). Neutrophils were treated with these peptides and assayed for adhesion to immobilized P-selectin on plastic. Peptides, homologous to regions of the lectin domain involved in neutrophil adhesion, bind to the P-selectin counterrecptor on the neutrophil and prevent their interaction with immobilized P-selectin (71). These studies have provided information on the binding region of the lectin domain and yield a new therapeutic tool. The peptides are easily

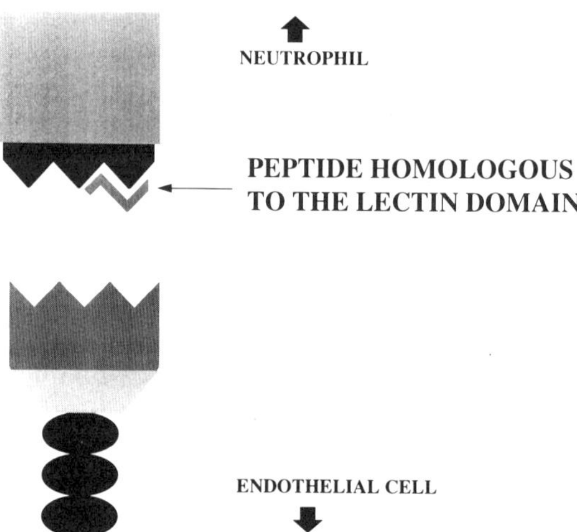

Figure 14 A peptide with sequence homology to the lectin domain of the selectins binds to the counterreceptor and inhibits its interaction with natural, cell-bound P-selectin.

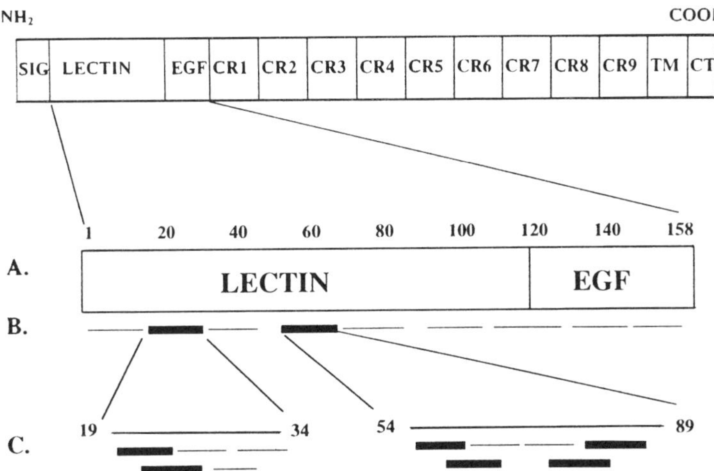

Figure 15 (A) The lectin and epidermal growth factor domains of P-selectin were sequenced, and (B) homologous, 20-30 amino acid peptides were synthesized. Peptides that bind to the P-selectin counterreceptor and inhibit neutrophil interaction with activated endothelium are depicted by thick lines. (C) These inhibitory peptides were further analyzed to determine 8-12 amino acid sequences that could inhibit cell adhesion (noted by thick lines). (Figure was modified from that published by Geng JG, Heavner GA, McEver RP. J Biol Chem 1992; 267:19846–19853.)

produced and theoretically have fewer side effects than antibodies. The potential problems are unpredicatable interactions with other epitopes, facilitated by their small size and reduced steric hindrance. These interactions may yield surprising effects such as neutrophil activation.

E. Antisense Oligonucleotides to Adhesion Molecules

An interesting new mechanism of reducing cell adhesion molecule expression is the use of antisense oligonucleotides. Oligonucleotides can be synthesized that hybridize selectively to mRNA coding for an adhesion molecule, thereby preventing new expression of that protein. The mechanism by which the antisense oligonucleotides inhibit protein synthesis varies, depending on where they bind to the native mRNA. In some instances mRNA levels in cells are reduced, indicating increased mRNA degradation, while other oligonucleotides reduce the amount of cell adhesion protein synthesized without affecting mRNA levels, indicating impaired translation. Initial studies have employed oligonucleotides directed against ICAM-1 and

E-selectin. When these oligonucleotides are added to endothelial cells before cytokine stimulation, the number of HL 60 cells adhering to the endothelial monolayer is reduced (72). As expected, there is less effect on cell binding with the anti-ICAM-1 therapy since this adhesion molecule is expressed on nonstimulated endothelial cells. This is a significant limitation of oligonucleotide therapy since it will not affect the constitutive expression of adhesion molecules.

F. Inhibitors of Cellular Activation

Neutrophil-endothelial interaction requires cellular activation. As noted above, during abdominal aortic aneurysm repair, the indiscriminate systemic release of chemoactivators upregulates adhesion molecules on neutrophils and endothelial cells remote from the site of ischemia, laying the foundation for subsequent neutrophil-endothelial interaction and tissue injury. Mechanisms that could block the upregulation of adhesion molecules include inhibitors of the eicosanoids, platelet activating factor, cytokines, and complement. Anticytokine therapy has already been discussed.

1. Eicosanoids

Leukocytes generate eicosanoids in response to injury. These agents in turn act as autocoids upregulating cell adhesion molecules and cell adhesion. Thromboxane A_2 (TXA_2) and leukotriene B_4 (LTB_4) are synthesized by platelets and leukocytes in response to stimulation by agents such as reactive oxygen metabolites that are formed after ischemia (73,74). During reperfusion of ischemic kidneys (75) or hindlimbs (74), serum levels of TXB_2 (a stable hydrolysis product of TXA_2) and LTB_4 rise within minutes followed in hours by acute tubular necrosis or in the case of hindlimbs and the lungs, leukosequestration, and permeability. Neutrophil depletion reduces the increase in eicosanoids and limits the renal and respiratory failure. Similarly, pretreatment with OKY 046 or diethylcarbamazine (a thromboxane synthetase and lipoxygenase inhibitor, respectively) reduces the elevation of TXB_2 and LTB_4 in serum and subsequent tubular necrosis and pulmonary edema. It is hypothesized that local extravascular production of TXA_2 and LTB_4 by monocytes and mast cells in the ischemic organ attracts neutrophils (75). In addition to this chemotactic effect, these eicosanoids active neutrophils to produce additional eicosanoids and free radicals as well as to upregulate CD 18 (23).

That eicosanoids moderate neutrophil chemotaxis in ischemia can be illustrated in other experimental methods. When serum from animals undergoing hindlimb ischemia is introduced into dermabrasion chambers on normal animals, neutrophils diapedese into the chamber, coincident with a rise in chamber levels of TXB_2. If the normal animals were treated with the thromboxane synthetase inhibitor OKY 046, the rise in TXB_2 and subsequent neutrophil diapedesis was limited. Treatment with the thromboxane receptor antagonist SQ 29548 did not lower TXB_2 levels but did reduce diapedesis (23).

By unknown mechanisms eicosanoids also mediate remote neutrophil sequestration in the lungs after hindlimb ischemia. Thus, bronchoalveolar lavage levels of LTB_4 rise after hindlimb ischemia. Segmental lavage of the lung with a lipoxygenase antagonist in low concentration will prevent LTB_4 synthesis by the lung and neutrophil sequestration, while elevated systemic LTB_4 levels are unaffected (76). In a rat model of acid aspiration, installation of 0.1 N HCl into a lobe of the lung results in a rise in serum TXB_2 and LTB_4, pulmonary edema, and neutrophil diapedesis into the alveolar space. Pretreatment with intravenous diethylcarbamazine, FPL-55712 (leukotriene receptor antagonist), OKY 046, or SQ 29548 limited neutrophil sequestration into the alveolar space and lung injury measured by leakage of protein (77). Also, local lavage of the affected lung segment with OKY 046 before acid instillation reduced bronchoalveolar lavage levels of TXB_2, pulmonary edema, and neutrophil accumulation in the lung (78). In addition to the importance of remote eicosanoid synthesis in leukosequestration, the data suggest a synergism between TXA_2 and LTB_4.

In a pig model of ARDS induced by lipopolysaccharide, the LTB_4 receptor antagonist LY255283 limited neutrophil activation and lung injury (79). In a rat model of intestinal ischemia, the LTB_4 receptor antagonist given before reperfusion reduced the rise in intestinal neutrophil sequestration (80). Although these data appear to offer promising therapeutic results, clinical trials with antagonists have been disappointing. These include inhibitors of thromboxane as well as ibuprofen a cyclooxygenase antagonist (81,82).

2. Platelet Activating Factor

Platelet activating factor is a phospholipid mediator of inflammation which stimulates neutrophils. Following cardiac ischemia, it is hypothesized that PAF produced by cardiac myocytes or endothelium during reperfusion promotes neutrophil adherence to endothelium.

Treatment with the PAF receptor antagonists WEB 2170 or SDZ 63-675 before reperfusion reduce neutrophil sequestration into damaged myocardium, size of the infarct, and hemodynamic alterations (83). As expected, the receptor antagonists did no affect the rise in serum concentration of PAF following reperfusion. The authors hypothesize that blocking the PAF receptor interferes with neutrophil-endothelial interaction and the subsequent inflammatory response.

3. Complement

Complement plays a key role in inflammatory injury since breakdown fragments can act both as chemoattractants and chemoactivators. Complement inhibition has been shown to modify a number of inflammatory events. One complement antagonist is the soluble form of complement receptor type-1 (sCR1), which binds C3b and C4b and accelerates the decay of the C3 and C5 convertases (84,85). Studies in rat intestinal ischemia have shown that treatment with sCR1 markedly reduce local injury but also reduces remote, neutrophil-dependent injury without affecting pulmonary leukosequestration (43). The proposed mechanism is that sCR1 inhibits systemic complement activation and indiscriminate deposition of complement fragments on the pulmonary vasculature, such as the inactivated form of C3b (iC3b) which is one of the activating ligands for CR3 (CD 11b/CD 18). Similarly, intravital microscopy studies in mice have shown that complement inhibition with sCR1 reduces neutrophil adherence to ischemic skeletal muscle (86). Possible mechanisms include reduced anaphylatoxin (C3a and C5a) production locally and reduced CD 11b/CD 18 ligand (iC3b) deposition on the endothelium.

IV. EFFECTS OF ANTI-ADHESION THERAPY ON HOST DEFENSE

The deleterious effect of neutrophils in the uncontrolled upregulation of inflammatory events following severe trauma or sepsis has been described. However, since the neutrophil plays a key role in normal homeostasis, the possibility exists that infectious consequences might follow antiadhesion therapy (87). The relevance of adhesion molecules in normal host defense is illustrated by the inherited leukocyte adherence deficiency (LAD) syndromes where these molecules are inadequate. In LAD type 1, patients lack the β-subunit of the integrin, CD 18, and suffer from recurrent bacterial infections, impaired formation of pus, and early death (88). This syndrome is particularly lethal when CD 18 expression is only 1% or 2% of normal whereas most patients with a 15% expression survive. In LAD type 2, patients have a biochemical inability to fucosylate. The resultant deficit in sialyl–Lewis X leads to a loss of selectin-mediated adhesion (88). Similar to LAD type 1, these patients suffer from recurrent, nonsuppurative bacterial infections. Because incorporation of fucose is vital to other organ systems, LAD type-2 patients exhibit other genetic defects.

Considering the above syndromes, several authors have asked whether short-term blockade of the adhesion molecules would lead to increased infection. Blockade of P-selectin with the antibody PB 1.3 does not adversely affect neutrophil accumulation and abscess formation in rabbits 7 hours after subcutaneous inoculation of *S. aureus*. Similarly, treatment with PB 1.3 does not affect neutrophil emigration into the rabbit peritoneum 4 hours after its inoculation with *E. coli* (87).

Similar studies were performed with integrin blockade in rabbits given subcutaneous injections of *P. aeruginosa* or *S. aureus*. In contrast to the use of antiselectins, the systemic administration of antibodies directed against CD 18, R 15.7 (89), leads to increased abscess formation, compared to saline controls. Use of an antibody directed against ICAM (R6.5), the CD 11/CD 18 endothelial counterreceptor, did not alter abscess formation. The effect on abscess formation with anti-CD 18 but not antiselectin or anti-ICAM therapy is hypothesized to result from the ability of the anti-CD 18 to inhibit both adhesion and neutrophil bactericidal function, the latter mediated by CR3 (CD 11b) (89,90). Not all studies were uniform in documenting adverse effects. Thus in an infection and sepsis model where rabbits underwent appendiceal devascularization, treatment with anti-CD 18 antibody 60.3 reduced neutrophil emigration into the peritoneum, but there was no change in morbidity or mortality (90). It remains likely that toxicity of antiadhesion therapy will depend on the completeness of blockade and the magnitude of the infectious challenge.

Not surprisingly, reports have indicated that some therapies worsen clinical sepsis. In a canine septic shock model, administration of the anti-CD 18 antibody R15.7 reduced the neutropenia but worsened serum endotoxemia, acidosis, and cardiovascular function (92). The authors hypothesize that the antibody impairs normal leukocyte function. Similarly, in some models anti-TNF therapy has adverse consequences. In cecal ligation and puncture in the mouse, anti-TNF antibodies (93,94) increased lethality, which could be reversed with administration of recombinant TNF to the animals (94). These data suggest that TNF may play a key homeostatic role in host defense.

V. TIMING OF ANTI-ADHESION THERAPY

The timing of therapy is critical for its applicability to human disease. In some controlled situations such as aneurysm repair, coronary bypass or transplantation, therapy can be instituted prior to ischemia or to reperfusion. In many diseases, however, the pathologic event has occurred before the patient can be evaluated and the illness identified. This is exemplified by the ischemia which attends symptoms of cerebral stroke, myocardial infarction, and intestinal or peripheral arterial embolization. In the case of atelectasis or traumatic hemorrhage, organ ischemia due to hypoperfusion is usually ongoing as the patient presents for medical therapy. The same is true for acid aspiration where the diagnosis is customarily made after the event.

In most animal models of inflammation, treatment is given before the event, such as with cobra venom infusion or immune complex-induced lung injury (58). The same is true in studies of ischemia, where treatment is given prior to reperfusion when the major injury occurs. Clinically, ischemia is a more favorable setting for treatment since antiadhesion therapy can often be given before or during reinstitution of blood flow to the injured organ. In settings of recurrent ischemia or hypotension secondary to sepsis or other low flow states, where there may be episodic release of chemoactivators, the question is unanswered whether antiadhesion treatment could blunt these recurrent stimuli to the neutrophil and thereby modify the injury.

Some studies have specifically examined the timing of therapy. In rabbit ear ischemia, treatment with anti-CD 18 antibody at different times during the 7 days of reperfusion resulted in a graded degree of injury. Treatment starting 1 hour after reperfusion was as effective as therapy begun at reperfusion; however, treatment at 4 hours of reperfusion was less effective, and therapy started at 12 hours of reperfusion showed no reduction in injury (95). Similarly, in acid aspiration in rodents, therapy aimed at reducing free radical formation or thromboxane synthesis was effective 1 hour postaspiration (9). Studies such as these ae needed to determine the relevance of antiadhesion drugs for treatment of human diseases.

REFERENCES

1. Klausner JM, Paterson IS, Mannick JA, Valeri R, Shepro D, Hechtman HB. Reperfusion pulmonary edema. J Am Med Assoc 1989; 261:1030–1035.
2. Paterson IS, Klausner JM, Goldman G, et al. Pulmonary edema after aneurysm surgery is modified by mannitol. Ann Surg 1989; 210:796–801.
3. Weiser MR, Gibbs SAL, Valeri R, Shepro D, Hechtman HB. Sialyl–Lewis X attenuates local muscle injury after hindlimb ischemia. Submitted for publication.
4. Simpson R, Alon R, Kobzik L, Valeri R, Shepro D, Hechtman HB. Neutrophil and non neutrophil-mediated injury in intestinal ischemia-reperfusion. Ann Surg 1993; 218:444–454.
5. Klausner JM, Paterson IS, Goldman G, et al. Postischemic renal injury is mediated by neutrophils and leukotrienes. Am J Physiol 1989; 256:F794–F802.
6. Guice KS, Oldham KT, Caty MG, Johnson KJ, Ward PA. Neutrophil-dependent, oxygen-radical mediated lung injury associated with acute pancreatitis. Ann Surg 1989; 210:740–747.
7. Litt MR, Jeremy RW, Weisman HF, Winkelstein JA, Becker LC. Neutrophil depletion limited to reperfusion reduces myocardial infarct size after 90 min-

utes of ischemia. Evidence for neutrophil-mediated reperfusion injury. Circulation 1989; 80:1816–1827.
8. Goldman G, Welbourn R, Rothlein R, et al. Adherent neutrophils mediate permeability after atelectasis. Ann Surg 1992; 216:372–380.
9. Goldman G, Welbourn R, Kobzik L, Valeri CR, Shepro D, Hechtman HB. Reactive oxygen species and elastase mediate lung permeability after acid aspiration. J Appl Physiol 1992; 73:571–575.
10. Springer TA. Adhesion receptors of the immune system. Nature 1990; 346: 425–434.
11. Springer TA. Traffic signals for lymphocyte recirculation and leukocyte emigration: the multistep paradigm. Cell 1994; 76:301–314.
12. Albelda SM, Smith CW, Ward PA. Adhesion molecules and inflammatory injury. FASEB J 1994; 8:504–512.
13. Butcher EC. Leukocyte-endothelial cell recognition: three (or more) steps to specificity and diversity. Cell 1991; 67:1033–1036.
14. Carlos TM, Harlan JM. Leukocyte-endothelial adhesion molecules. Blood 1994; 84:2068–2101.
15. Kishimoto TK, Jutila MA, Berg EL, Butcher EC. Neutrophil Mac-1 and MEL-14 adhesion proteins inversely regulated by chemotactic factors. Science 1989; 245:1238–1241.
16. Luscinskas FW, Brock AF, Arnaout MA, Gimbrone MA. Endothelial-leukocyte adhesion molecule-1-dependent and leukocyte (CD11/CD18)-dependent mechanisms contribute to polymorphonuclear leukocyte adhesion to cytokine-activated human vascular endothelium. J Immunol 1989; 142:2257–2263.
17. Hattori R, Hamilton KK, Fugate RD, McEver RP, Sims PJ. Stimulated secretion of endothelial von Willebrand factor is accompanied by rapid redistribution to the cell surface of the intracellular ganular membrane protein GMP-140. J Biol Chem 1989; 264:7768–7771.
18. Foreman KE, Vaporciyan H, Bonish BK, et al. C5a-induced expression of P-selectin in endothelial cells. J Clin Invest 1994; 94:1147–1155.
19. Lorant DE, Topham MK, Whatley RE, et al. Inflammatory roles of P-selectin. J Clin Invest 1993; 92:559–570.
20. Dinarello CA, Gelfand JA, Wolff SM. Anticytokine strategies in the treatment of the systemic inflammatory response syndrome. JAMA 1993; 269:1829–1835.
21. Karakurum M, Shreeniwas R, Chen J, et al. Hypoxic induction of interleukin-8 gene expression in human endothelial cells. J Clin Invest 1994; 93:1564–1570.
22. Diamond MS, Springer TS. A subpopulation of Mac-1 (CD11b/CD18) molecules mediates neutrophil adhesion to ICAM and fibrinogen. J Cell Biol 1993; 120:545–556.
23. Goldman G, Welbourn R, Klausner JM, Valeri CR, Shepro D, Hechtman HB. Thromboxane mediates diapedesis after ischemia by activation of neutrophil adhesion receptors interacting with basally expressed intercellular adhesion molecule-1. Circ Res 1991; 68:1013–1019.

24. Marks RM, Todd RF III, Ward PA. Rapid induction of neutrophil-endothelial adhesion by endothelial complement fixation. Nature 1989; 339:314–317.
25. Welbourn CRB, Goldman G, Paterson IS, Valeri CR, Shepro D, Hechtman HB. Pathophysiology of ischaemia reperfusion injury: central role of the neutrophil. Br J Surg 1991; 78:651–655.
26. Abbas AK, Lichtman AH, Pober JS. Antibodies and antigens. In: Cellular and Molecular Immunology. Philadelphia. W. B. Saunders, 1991:38–68.
27. Harlow E, Lane D. Antibodies: a laboratory manual. Cold Spring Harbor, NY: Cold Spring Harbor Laboratory, 1988.
28. Berzofsky JA, Berkower IJ, Epstein SL. Antigen-antibody interaction and monoclonal antibodies. In: Paul WE, ed. Fundamental Immunology. 3rd ed. New York: Raven Press, 1993:421–465.
29. Jefferis R, Pound JD. Immunoglobulins. In: Gallin JI, Goldstein IM, Snyderman R, eds. Inflammation: Basic Principles and Clinical Correlates. 2d ed. New York: Raven Press, 1992:11–32.
30. Bach JF, Fracchia GN, Chatenoud L. Safety and efficacy of therapeutic monoclonal antibodies in clinical therapy. Immunol Today 1993; 14:421–425.
31. Rodwell JD. Engineering monoclonal antibodies. Nature 1989; 342:99–100.
32. Mihelcic D, Schleiffenbaum B, Tedder TF, Sharar SR, Harlan JM, Winn RK. Inhibition of leukocyte L-selectin function with a monoclonal antibody attenuates reperfusion injury to the rabbit ear. Blood 1994; 84:2322–2328.
33. Mulligan MS, Varani J, Dame MK, et al. Role of endothelial-leukocyte adhesion molecules 1 (ELAM-1) in neutrophil-mediated lung injury in rats. J Clin Invest 1991; 88:1396–1406.
34. Chatenoud L, Legendre C, Ferran C, Bach JF, Kreis H. Corticosteroid inhibition of the OKT3-induced cytokine-related syndrome—dosage and kinetics prerequisites. Transplantation 1991; 51:334–338.
35. Mulligan MS, Polley MJ, Bayer RJ, Nunn MF, Paulson JC, Ward PA. Neutrophil-dependent acute lung injury, requirement for P-selectin (GMP-140). J Clin Invest 1992; 90:1600–1607.
36. Winn RK, Liggitt D, Vedder NB, Paulson JC, Harlan JM. Anti-P-selectin monoclonal antibody attenuates reperfusion injury to the rabbit ear. J Clin Invest 1993; 92:2042–2047.
37. Davenpeck KL, Gauthier TW, Albertine KH, Lefer AM. Role of P-selectin in microvascular leukocyte-endothelial interaction in splanchnic ischemia-reperfusion. Am J Physiol 1994; 267:H622–H630.
38. Ley K, Tedder TF, Kansas GS. L-selectin can mediate leukocyte rolling in untreated mesenteric venules in vivo independent of E- or P-selectin. Blood 1993; 82:1632–1638.
39. Seekamp A, Till GO, Mulligan MS, et al. Role of selectins in local and remote tissue injury following ischemia and reperfusion. Am J Pathol 1994; 144:592–598.
40. Vedder NB, Winn RK, Rice CL, Chi EY, Arfors KE, Harlan JM. Inhibition of leukocyte adherence by anti-CD18 monoclonal antibody attenuates reperfusion injury in the rabbit ear. Proc Natl Acad Sci USA 1990; 87:2643–2646.

41. Hill J, Lindsay T, Valeri CR, Shepro D, Hechtman HB. A CD18 antibody prevents lung injury but not hypotension following intestinal ischemia-reperfusion. J Appl Physiol 1992; 216:677–683.
42. Hill J, Lindsay T, Rusche J, Valeri CR, Shepro D, Hechtman HB. A Mac-1 antibody reduces liver and lung injury but not neutrophil sequestration after intestinal ischemia-reperfusion. Surgery 1992; 112:166–172.
43. Hill J, Lindsey TF, Ortiz F, Yeh CG, Hechtman HB, Moore FD Jr. Soluble complement receptor type 1 ameliorates the local and remote organ injury following intestinal ischemia-reperfusion in the rat. J Immunol 1992; 149: 1723–1728.
44. Jerome SN, Dore M, Paulson JC, Smith CW, Korthuis RJ. P-selectin and ICAM-1-dependent adherence reactions: role in the genesis of postischemic no-reflow. Am J Physiol 1994; 266:H1316–H1321.
45. Mathison JC, Wolfson E, Ulevitch RJ. Participation of tumor necrosis factor in the mediation of gram negative bacterial lipopolysaccharide-induced injury in rabbits. J Clin Invest 1988; 81:1925–1937.
46. Tracey KJ, Fong Y, Hesse DG, et al. Anti-cachectin/TNF monoclonal antibodies prevent septic shock during lethal bacteraemia. Nature 1987; 330:662–664.
47. Gershenwald JE, Fong YM, Fahey TJ 3d, et al. Interleukin 1 receptor blockade attenuates the host inflammatory response. Proc Natl Acad Sci USA 1990; 87:4966–4970.
48. Baggiolini M, Moser M, Clark-Lewis I. Interleukin-8 and related chemotactic cytokines. The Giles Filley Lecture. Chest 1994; 105:95S–98S.
49. Beaubien BC, Collins PD, Jose PJ, et al. A novel neutrophil chemoattractant generated during an inflammatory reaction in the rabbit peritoneal cavity in vivo. Purification, partial amino acid sequence and structural relationship to interleukin-8. Biochem J 1990; 271:797–801.
50. Rampart M, Van Damme J, Zonnekeyn L, Herman AG. Granulocyte chemotactic protein/interleukin-8 induces plasma leakage and neutrophil accumulation in rabbit skin. Am J Pathol 1989; 135:21–25.
51. Garner WL, Rodriguez JL, Miller CG, et al. Acute skin injury releases neutrophil chemoattractants. Surgery 1994; 116:42–48.
52. Wickersham NE, Loyd JE, Johnson JE, McCain RW, Christman JW. Acute inflammation in a sheep model of unilateral lung ischemia: the role of interleukin-8 recruitment of polymorphonuclear leukocytes. Am J Respir Cell Mol Biol 1993; 9:199–204.
53. Sekido N, Mukaida N, Harada A, et al. Prevention of lung reperfusion injury in rabbits by a monoclonal antibody against interleukin-8. Nature 1993; 365: 654–657.
54. Gearing AJ, Newman W. Circulating adhesion molecules in disease. Immunol Today 1993; 14:506–512.
55. Johnston GI, Cook RG, McEver RP. Cloning of GMP140, a granule membrane protein of platelets and endothelium; sequence similarity to proteins involved in cell adhesion and inflammation. Cell 1989; 56:1033–1044.

56. Gamble JR, Skinner MP, Berndt MC, Vadas MA. Prevention of activated neutrophil adhesion to endothelium by soluble adhesion protein GMP140. Science 1990; 249:414–417.
57. Watson SR, Fennie C, Lasky LA. Neutrophil influx into an inflammatory site inhibited by a soluble homing receptor-IgG chimera. Nature 1991; 349:164–167.
58. Mulligan MS, Watson SR, Fennie C, Ward PA. Protective effects of selectin chimeras in neutrophil-mediated lung injury. J Immunol 1993; 151:6410–6417.
59. Van Zee KJ, Kohno T, Fischer E, et al. Tumor necrosis factor soluble receptors circulate during experimental and clinical inflammation and can protect against excessive tumor necrosis factor alpha in vitro and in vivo. Proc Natl Acad Sci USA 1992; 89:4845–4849.
60. Fanslow WC, Sims JE, Sassenfeld H, et al. Regulation of alloreactivity in vivo by a soluble form of the interleukin-1 receptor. Science 1990; 248:739–742.
61. Lesslauer W, Tobuchi H, Gentz R, et al. Recombinant soluble tumor necrosis factor receptor proteins protect mice from lipopolysaccharide-induced lethality. Eur J Immunol 1991; 21:2883–2886.
62. Mulligan MS, Paulson JC, DeFrees S, Zheng Z, Lowe JB, Ward PA. Protective effects of oligosaccharides in P-selectin-dependent lung injury. Nature 1993; 364:149–151.
63. Mulligan MS, Lowe JB, Larsen RD, et al. Protective effects of sialylated oligosaccharides in immune complex-induced acute lung injury. J Exp Med 1993; 178:623–631.
64. Sako D, Chang XJ, Barone KM, et al. Expression cloning of a functional glycoprotein ligand for P-selectin. Cell 1993; 75:1179–1186.
65. Ohlsson K, Bjork P, Bergenfeldt M, Hageman R, Thompson RC. Interleukin-1 receptor antagonist reduces mortality from endotoxin shock. Nature 1990; 348:550–552.
66. Fischer CJ Jr, Dhalnaut JF, Pribble JP, Knaus WA. The Receptor Antagonist Study Group. A study evaluating the safety and efficacy of human recombinant interleukin-1 receptor antagonist in the treatment of patients with sepsis syndrome. Presented at the 13th International Symposium on Intensive Care and Emergency Medicine; March 23, 1993; Brussels, Belgium.
67. Moyle M, Foster DL, McGrath DE, et al. A hookworm glycoprotein that inhibits neutrophil function is a ligand of the integrin CD11b/CD18. J Bio Chem 1994; 269:10008–10015.
68. Relman D, Tuomanen E, Falkow S, Golenbock DT, Saukkonen K, Wright SD. Recognition of a bacterial adhesin by an integrin: macrophage CR3 ($\eta_M\beta_2$ CD11b/CD18) binds filamentous hemagglutinin of *Bordetella pertussis*. Cell 1990; 61:1375–1382.
69. Tuomanen EI, Prasad SM, George JS, et al. Reversible opening of the blood-brain barrier by anti-bacterial antibodies. Proc Natl Acad Sci USA 1993; 90: 7824–7828.
70. Geng JG, Heavner GA, McEver RP. Lectin domain peptides from selectins interact with both cell surface ligands and Ca^{2+} ions. J Biol Chem 1992; 267: 19846–19853.

71. Heavner GA, Falcone M, Kruszynski M, et al. Peptides from multiple regions of the lectin domain of P-selectin inhibiting neutrophil adhesion. Int J Peptide Protein Res 1993; 42:484–489.
72. Bennett CF, Condon TP, Grimm S, Chan H, Chiang MY. Inhibition of endothelial cell adhesion molecule expression with antisense oligonucleotides. J Immunol 1994; 152:3530–3540.
73. Klausner JM, Paterson IS, Valeri CR, Shepro D, Hechtman HB. Limb ischemia-induced increase in permeability is mediated by leukocytes and leukotrienes. Ann Surg 1988; 208:755–760.
74. Goldman G, Welbourn R, Klausner JM, Valeri CR, Shepro D, Hechtman HB. Oxygen free radicals are required for ischemia-induced leukotriene B_4 synthesis and diapedesis. Surgery 1992; 111:287–293.
75. Klausner JM, Paterson IS, Goldman G, et al. Postischemic renal injury is mediated by neutrophils and leukotrienes. Am J Physiol 1989; 256:F794–F802.
76. Goldman G, Welbourn R, Klausner JM, et al. Mast cells and leukotrienes mediate neutrophil sequestration and lung edema after remote ischemia in rodents. Surgery 1992; 112:578–586.
77. Goldman G, Welbourn R, Kobzik L, Valeri CR, Shepro D, Hechtman HB. Synergism between leukotriene B_4 and thromboxane A_2 in mediating acid-aspiration injury. Surgery 1992; 111:55–61.
78. Goldman G, Welbourn R, Klausner JM, et al. Neutrophil accumulations due to pulmonary thromboxane synthesis mediate acid aspiration injury. J Appl Physiol 1991; 70:1511–1517.
79. Wollert PS, Menconi MJ, O'Sullivan BP, Wang H, Larkin V, Fink MP. LY255283, a novel leukotriene B4 receptor antagonist, limits activation of neutrophils and prevents acute lung injury induced by endotoxin in pigs. Surgery 1993; 114:191–198.
80. Karasawa A, Guo JP, Ma XL, Tsao PS, Lefer AM. Protective actions of leukotriene B_4 antagonist in splanchnic ischemia and reperfusion in rats. Am J Physiol 261; 261:G191–G198.
81. Bernard GR, Reines HD, Halushka PV, et al. Prostacyclin and thromboxane A2 formation is increased in humam sepsis syndrome: effects of cyclooxygenase inhibition. Am Rev Respir Dis 1991; 144:1095–1101.
82. Haupt MT, Jastremski MS, Clemmer TP, Metz CA, Goris GB. Effects of ibuprofen in patients with severe sepsis: a randomized, double-blind, multicenter study. Crit Care Med 1991; 19:1339–1347.
83. Montrucchio G, Alloatti G, Mariano F, et al. Role of platelet-activating factor in polymorphonuclear neutrophil recruitment in reperfused ischemic rabbit heart. Am J Pathol 1993; 142:471–480.
84. Weisman HF, Bartow T, Leppo MK, et al. Soluble human complement receptor 1: in vivo inhibitor of complement suppressing post-ischemic myocardial inflammation and necrosis. Science 1990; 249:146–151.
85. Fearon DT. Regulation of the amplification of C3 convertase of human complement by an inhibitory protein isolated from human erythrocyte membrane. Proc Natl Acad Sci USA 1979; 76:5867–5871.
86. Pemberton M, Anderson G, Vetvicka V, Justus DE, Ross GD. Microvascular

effects of complement blockade with soluble recombinant CR1 on ischemia and reperfusion injury of skeletal muscle. J Immunol 1993; 150:5104–5113.
87. Sharar SR, Sasaki SS, Flaherty LC, Paulson JC, Harlan JM, Winn RK. P-selectin blockage does not impair leukocyte host defense against bacterial peritonitis and soft tissue infection in rabbits. J Immunol 1993; 151:4982–4988.
88. Etzioni A, Frydman M, Pollack S, et al. Brief report: recurrent severe infections caused by a novel leukocyte adhesion deficiency. N Engl J Med 1992; 327:1789–1792.
89. Sharar SR, Winn RK, Murry CE, Harlan JM, Rice CL. A CD18 monoclonal antibody increases the incidence and severity of subcutaneous abscess formation after high-dose *Staphylococcus aureus* injection in rabbits. Surgery 1991; 110:213–220.
90. Mileski WJ, Winn RK, Harlan JM, Rice CL. Transient inhibition of neutrophil adherence with the anti-CD18 monoclonal antibody 60.3 does not increase mortality rates in abdominal sepsis. Surgery 1991; 109:497–501.
91. Mileski WJ, Sikes P, Atiles L, Lightfoot E, Lipsky P, Baxter C. Inhibition of leukocyte adherence and susceptibility to infection. J Surg Res 1993; 54:349–354.
92. Eichacker PQ, Hoffman WD, Farese A, et al. Leukocyte CD 18 monoclonal antibody worsens endotoxemia and cardiovascular injury in canines with septic shock. J Appl Physiol 1993; 74:1885–1892.
93. Eskandari MK, Bolgos G, Miller C, Nguyen DeForge LE, Remick DG. Anti-tumor necrosis factor antibody therapy fails to prevent lethality after cecal ligation and puncture or endotoxemia. J Immunol 1992; 148:2724–2730.
94. Echtenacher B, Falk W, Mannel DN, Krammer PH. Requirement of endogenous tumor necrosis factor/cachectin for recovery from experimental peritonitis. J Immunol 1990; 145:3762–3766.
95. Sharar SR, Michelcic DD, Han KT, Harlan JM, Winn RK. Ischemia reperfusion injury in the rabbit ear is reduced by both immediate and delayed CD18 leukocyte adherence blockage. J Immunol 1994; 153:2234–2238.

3

Measurement of Soluble Adhesion Molecules in Biological Fluids

Y. Lebranchu, J. F. Valentin, and M. Büchler
Unité d'Immunologie Clinique et de Transplantation et Groupe Interactions Hôte-greffon, Hôpital Bretonneau, Tours, France

Adhesion molecules expressed on vascular endothelium and circulating leukocytes play a key role in inflammatory responses by regulating adhesion of leukocytes to endothelium and transendothelial migration. They are also essential in the immune response, stabilizing the cell-cell interactions and the specific binding of T lymphocyte receptors to antigens (1–3). Adhesion molecules belong to one of three families: 1. The selectins, a family of single-chain glycoproteins consisting of L-selectin, E-selectin, and P-selectin. 2. The integrins, a large family of heterodimeric glycoproteins consisting of an α chain (120 to 180 kDa) and a noncovalently associated β chain (90 to 110 kDa). Integrins of particular importance in leukocyte-EC interactions are those incorporating the α4 chain, such as VLA-4, or the β2 chain, such as LFA-1 and Mac-1. 3. The immunoglobulin superfamily, a large group of molecules with structural homology to antibodies. Within this group are a number of molecules which have a role in leukocyte-endothelial cell adhesion such as ICAM-1, ICAM-2, VCAM-1, and LFA-3.

Functionally active, soluble forms of many adhesion molecules have been described and can be detected in the plasma and other body fluids (4–7). Soluble molecules may derive from alternatively spliced transcripts that lack a transmembrane domain or from proteolytic cleavage of the membrane anchored forms. Shedding of adhesion molecules has been induced in vitro following activation of endothelial cells by cytokines (8,9). The increased levels of circulating adhesion molecules observed in some diseases may thus reflect an increased local expression of adhesion molecules in an immunopathological process, or a relatively rapid membrane turnover rate of these molecules.

As the endothelial cell surface expression of adhesion molecules is controlled independently (10), immunohistological studies have identified different patterns of tissue expression in inflammatory diseases, dependent on the nature of the stimulus (11). These different patterns of expression of membrane adhesion molecules could result in different patterns of soluble adhesion molecules in body fluids. Because it is relatively easy to analyze blood samples, there has been considerable interest in the measurement of soluble adhesion molecules as a guide either to diagnosis or to monitoring of immune diseases.

I. ICAM-1

ICAM-1 (CD54), a 90-to 115-kDa glycoprotein, has five Ig-like domains and functions as a ligand for the leukocyte integrin adhesion receptors LFA-1 (CD11a/CD18) and Mac-1 (CD 11b/CD18) (1). ICAM-1 is basally expressed in significant amounts on a limited number of cell types, including monocytes and endothelial cells. It is widely inducible, or upregulated, on many cells including lymphocytes, monocytes, dendritic cells, endothelial cells, fibroblasts, epithelial cells, and mucosal cells. Release of LPS and cytokines such as TNFα, IL-1, and IFN-γ results in increased expression of ICAM-1. Shedding of ICAM-1 has been induced in vitro following treatment of cells with cytokines (8,9). Levels of circulating forms of ICAM-1 have therefore been studied. Estimates of the mean level of ICAM-1 range from 102 to 450 ng/ml in normal individuals (12). Increased levels have been reported in allergy, inflammation, infection, cancers, and organ transplantation (Table 1).

A. Allergic Diseases

Allergic diseases are histopathologically characterized by accumulation of circulating inflammatory cells in the capillaries that migrate through the vascular endothelium to the submucosa at the sites of the allergic reaction. In bronchial asthma there is increased expression of ICAM-1 on vascular endothelium and bronchial epithelium (13,15). Serum levels of ICAM-1 obtained during an attack of bronchial asthma are higher than those obtained in clinically disease-free conditions, regardless of atopic status (16). There is a correlation between changes in serum TNF-α levels and the changes in serum ICAM-1 levels, suggesting that higher levels of ICAM-1 during asthma attacks reflect an upregulation of ICAM-1 expression in

Table 1 Increased ICAM-1 Levels in Disease by Reference

Condition	Reference
Allergic diseases	
Asthma	16, 20
Atomic dermatitis	22, 23
Nasal allergy	21
Infectious diseases	
Tuberculosis	55, 56
Neonatal sepsis	57
HIV	64
Plasmodium falciparum	59, 63
Infectious mononucleosis	58
Transplantation	
Renal graft rejection	85, 86
Liver graft rejection	79, 81, 82
Cardiac allograft	83, 84
Inflammatory and autoimmune diseases	
Rheumatoid arthritis	4, 6, 11, 24, 27, 28
SLE not increased	6, 11, 28, 29, 34
Systemic sclerosis	28
Scleroderma	30
Vasculitis	
Active vasculitis	31
Kawasaki disease	32
Wegener's disease	33, 34, 35
Miscellaneous	
IDDM	49, 50
Sarcoidosis	44, 45, 46
Idiopathic pulmonary fibrosis	47
Multiple sclerosis	37–40
Viral encephalitis	40
Inflammatory disease of CNS	39
Psoriasis	53
Inflammatory bowel disease	41, 42
Erythrodermic skin disease	54
Autoimmune liver disease	43
Graves' ophthalmolopathy and thyroid disease	51, 52
Uveitis	48
Malignant diseases	
Breast cancer	69
Hodgkin's disease	71, 72
Human malignancies	68
Bladder cancer (BCG therapy)	75
Ovarian carcinoma	74
Malignant melanoma	66, 67, 73
Metastatic cancer	4

allergic inflammation and that the soluble form of ICAM-1 may be a marker for the presence of inflammation. Similarly increased levels of ICAM-1 have been reported in the serum or sputum of patients with asthma (17–20) and/or allergy (21). Serum ICAM-1 levels are slightly higher in patients with severe atopic dermatitis than in blood transfusion donors, and dropped significantly after improvement of the skin conditions (22,23).

B. Autoimmune Diseases

1. Connective Tissue Diseases

a. Rheumatoid Arthritis. Increased levels of ICAM-1 have been reported in patients with rheumatoid arthritis (4,11,23–27). Levels of ICAM-1 were not correlated with erythrocyte sedimentation rate and C-reactive protein, and there was no relationship between increased ICAM-1 and increased VCAM-1 levels (25). Rheumatoid arthritis patients with vasculitis and/or pneumonitis showed significantly higher levels of ICAM-1 than patients without these manifestations (26). As many of the patients with very high levels of ICAM-1 had severe destructive arthritis, it is tempting to speculate that levels of ICAM-1 may reflect ongoing joint erosion. Comparison between rheumatoid arthritis patients at radiological stages I and II and between stage I and other stages has shown significantly higher levels of ICAM-1 in the sera of patients at the later stages (26). Cush reported that serum levels of ICAM-1 demonstrated a weak but significant correlation with the joint score (24). Synovial fluid ICAM-1 levels were increased but were consistently lower than serum levels, suggesting that ICAM-1 does not originate in the synovium. Because the production of ICAM-1 can be increased by cytokines (IL-1, TNF-α), elevated levels of ICAM-1 in rheumatoid arthritis may reflect systemic exposure to elevated cytokine levels (24).

b. Systemic Lupus Erythematosus. Increased expression of ICAM-1 is observed in the glomerular mesangium and endothelium, in skeletal muscle biopsies, and on endothelial cells in biopsy specimens from nonlesioned skin in patients with active systemic lupus erythematosus. Although increased levels of ICAM-1 in systemic lupus erythematosus have been described (28), other studies found levels comparable to controls, in contrast to levels of VCAM-1 which are increased (25,27,29). This emphasizes that levels of ICAM-1 and VCAM-1 may reflect different disease processes.

c. Scleroderma. Serum levels of ICAM-1, VCAM-1, and P-selectin are increased in scleroderma and correlate with their in-situ expression as well as with clinical disease activity (30).

2. Vasculitis

Release of soluble ICAM-1 from EC surfaces can be significantly enhanced by TNF-specific fixation and signaling and, prospectively, should be a sensitive indicator of intravascular inflammation in acute endothelium injury (31).

a. Kawasaki Disease. Patients with Kawasaki disease, but not those with anaphylactoid purpura or measles, have increased serum levels of ICAM-1 in the acute stage of the disease (32). Patients with active coronary artery lesions had higher levels of ICAM-1 than those without coronary artery lesions. There was a correlation between serum levels of ICAM-1 and levels of TNF-α during acute disease, suggesting that the serum ICAM-1 level is an important predictor for the severity of vascular damage in acute Kawasaki disease.

b. Wegener's Granulomatosis. Serum levels of ICAM-1 are higher in newly diagnosed, untreated patients with Wegener's granulomatosis than with controls, although patients with minor disease activity did not differ from controls (33). Levels of ICAM-1 correlated with disease activity score and with C-relative protein levels. Unfortunately, serial measurements of ICAM-1 are not likely to be helpful for monitoring disease activity since they lack both sensitivity and specificity for prediction of an impending relapse. Nevertheless, active Wegener's granulomatosis with cellular crescent formation has been associated with significantly elevated ICAM-1 levels, which could be used to predict renal outcome (34,35).

3. Inflammatory Diseases of the Central Nervous System

In the human brain, ICAM-1 is mainly expressed on the endothelial cells of small vessels, including the subependymal vessels of the choroid plexus. Cerebral endothelial cells are therefore a likely source of ICAM-1 in the cerebrospinal fluid (36).

Serum ICAM-1 levels are significantly more often elevated in patients with multiple sclerosis than in patients with other neurological and inflammatory diseases or controls (37,38) and are correlated with cerebrospinal fluid pleocytosis. ICAM-1 levels were also elevated in cerebospinal fluid in one study (37) but not in another (38). There were significant differences in serum ICAM-1 levels between clinically active patients with relapse or MRI evidence of disease progression and patients with stable multiple sclerosis; a significant increase in ICAM-1 levels was present at the time of the relapse (39).

Increased ICAM-1 levels have also been reported in other inflammatory diseases of the central nervous system (39) and in viral encephalitis (40).

4. Miscellaneous Autoimmune Diseases

Increased levels of ICAM-1 have been reported in Crohn's disease, coeliac disease (41,42), autoimmune liver diseases (43), sarcoidosis (44–46), idiopathic pulmonary fibrosis (47), uveitis (48), insulin-dependent diabetes mellitus (49,50), Grave's ophthalmopathy and thyroid diseases (51,52) psoriasis (53), and erythrodermic skin disease (54). In some diseases, such as Grave's ophthalmopathy, changes in ICAM-1 serum levels during corticosteroid therapy closely parallel changes in the degree of inflammation. However, serum levels of ICAM-1 are of no differential diagnostic or prognostic use in these diseases.

C. Infectious Diseases

1. Tuberculosis

ICAM-1 levels are significantly elevated in patients with active tuberculosis compared to those with treated disease or normal subjects (55,56). Taking the mean serum level ±2 SD in healthy controls as the upper limit of normal, ICAM-1 levels are a good marker of disease activity as the levels are elevated in about 80% of patients with active tuberculosis, 6.7% of subjects with treated disease, and 3.7% of normal subjects.

2. Neonatal Sepsis

Increased serum ICAM-1 levels might be a useful indicator for the early detection of neonatal sepsis because ICAM-1 rises more frequently and significantly earlier than currently used paramaters such as C-reactive protein (57).

3. Acute Infectious Mononucleosis

The elevated serum ICAM-1 levels in infectious mononucleosis are related to the increased counts of activated peripheral blood mononuclear cells (58).

4. Malaria

Plasmodium falciparum can bind to several adhesion molecules—e.g., CD36, ICAM-1, VCAM-1, E-selectin. Adhesion of infected erythrocytes to endothelial cells is believed to contribute to the sequestration of parasites and the obstruction of brain capillaries in brain malaria (59). Thus the higher ICAM-1 plasma or serum levels in patients with severe *Plasmodium falciparum* malaria compared with patients with mild malaria may reflect inflammatory endothelial reactions, and these reactions may be harmful for humans infected with malaria parasites (59–63).

5. HIV and Hepatitis Infection

Increased levels of ICAM-1 have been described in HIV infection related to immune activation (64) as well as in chronic hepatitis B and C (65).

D. Malignant Diseases

Increased levels of serum ICAM-1 have been reported in several malignancies such as melanoma (66,67), bladder (68), breast (68,69), gastrointestinal (68,70), ovarian (68), and renal (68) cancers; Hodgkin's disease (68,71,72), non-Hodgkin's lymphoma (68),and myeloma (68). Higher levels in more advanced or more active disease suggest that serum ICAM-1 may be a marker of disease progression in some malignancies (71,73). ICAM-1 has been found on normal and malignant tissues including melanoma, renal and intestinal cell lines (68), and ovarian carcinoma (74). Thus, at least some of ICAM-1 present in the sera of cancer patients is very likely derived from tumor tissue.

The presence of increased levels of ICAM-1 in sera of patients with malignancies may have implications for tumor metastatis because shedding of ICAM-1 by tumor cells may allow their escape from the immune response by cytotoxic T cells or NK cells. Alternatively, shedding of ICAM-1 by activated endothelial cells may saturate their counterreceptor ligands on circulating tumor cells and prevent their adhesion to endothelial cells, and thus decrease metastasis.

ICAM-1 is found in the urine of patients with bladder cancer during BCG therapy (75). As ICAM-1 expression may predispose tumor cells to cell-mediated cytotoxicity, the presence of increased shedding of ICAM-1 from tumor cells may predict a favorable clinical outcome.

E. Transplantation

Lymphocyte adhesion to and migration through the vascular endothelium of the graft are critical steps in cellular rejection and depends on cytokine-induced expression of adhesion molecules. Lymphocytes bind to vascular endothelium by E-selectin-, ICAM-1-, and VCAM-1-dependent pathways (76). During acute cellular rejection, increased or induced expression of ICAM-1 has been described on vascular endothelium and tubular epithelium of the kidney (77,78), on portal or sinusoidal endothelium and bile ducts of hepatic grafts (79), and on vessels of cardiac grafts (80).

Elevated levels of circulating ICAM-1 have been found during liver transplant rejection, although similar increases have been seen with various inflammatory but nonrejection complications, including cholangitis and hepatitis (79,81,82). Biliary ICAM-1 levels, however, are specifically elevated during acute rejection and not during infection, nor when no rejection was apparent (82). Serum levels of ICAM-1 decrease rapidly following successful treatment for rejection, whereas elevated levels persist or increase in ongoing rejection.

Increased levels of serum ICAM-1 levels have been found in cardiac rejection (80). Nevertheless, Ballantyne failed to find a correlation between ICAM-1 levels and histologial grade of the rejection (83), and Grant reported only a slight difference in ICAM-1 levels in patients with a grade 3a and grade 0 rejection (84). This modest difference, although statistically significant, is too small and too inconsistent to be useful in the diagnosis of rejection. Finally, Tanio failed to find a correlation between ICAM-1 levels and the presence or absence of cardiac graft rejection (80).

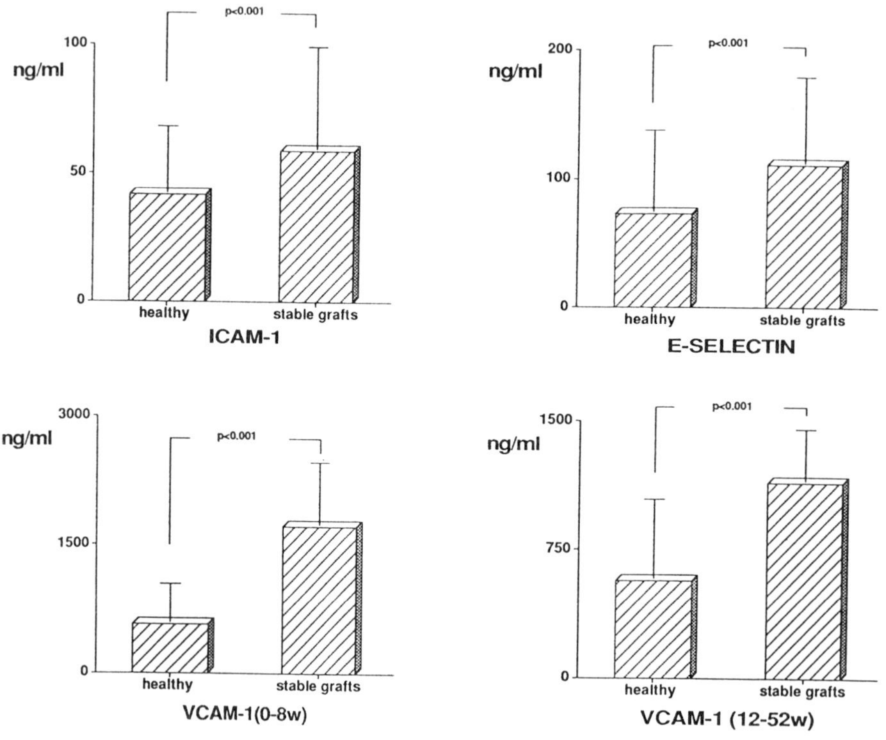

Figure 1 Soluble adhesion molecules ICAM-1, VCAM-1, and E-selectin in kidney transplant recipients with stable graft function. Eighty-nine serum samples from 10 patients were longitudinally collected during the first year after transplantation. Levels were compared to those of 101 samples of healthy controls. Histograms show the mean value, horizontal lines + 1 SD. Levels of all three adhesion molecules were significantly elevated in kidney transplant recipients (88).

Stockenhuber (85) and Kanagawa (86) demonstrated an association between serum levels of ICAM-1 and acute renal allograft rejection, although the sample size was fairly small. Such an association was not found by John (35), Bechtel (87), or ourselves (88). We studied 89 serum samples from 10 selected renal transplant patients with stable graft function and compared them with 101 normal patients. ICAM-1 was measured by ELISA as described (6). Serum levels of ICAM-1 were increased at the time of transplantation compared with healthy controls (Fig. 1) and remained elevated in stable patients during the first 12 months after transplantation (Fig. 2). As shown in Figure 3, there was no difference in the levels of ICAM-1 between 11 samples from patients with one episode of acute cellular rejection and the 89 samples from patients with stable graft function. It would be perhaps more useful to monitor ICAM-1 levels in urine because Bechtel reported an association of increased levels with acute steroid-resistant rejection (87).

II. ICAM-3

ICAM-3 (CD50) contains five Ig-like domains highly homologous to those found in ICAM-1 (CD54) and ICAM-2 (CD102) and is expressed almost exclusively on hematopoietic cells (89,90). It is the main LFA-1 ligand on

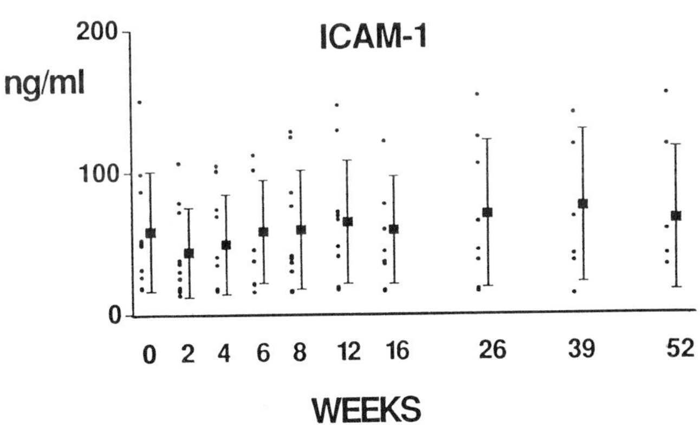

Figure 2 Longitudinal survey of serum levels of ICAM-1 in kidney transplant patients with stable graft function. Serum levels of ICAM-1 were determined in 10 patients grafting and at different times during the first year after transplantation (horizontal lines denote 1 SD; filled squares represent mean values).

Table 2 Increased VCAM-1 Levels in Disease by Reference

Condition	Reference
Inflammatory and autoimmune diseases	
Rheumatoid arthritis	6, 11
Systemic lupus erythematosus	6, 29, 34
Wegener's granulomatosis	33, 34
Inflammatory diseases of central nervous system	36, 37
Graves' disease	52
Infectious diseases	
Tuberculosis	55
Malaria	59, 98
HIV	96
Malignancies	68
Transplantation	
increased	12, 82
not increased	84, 88
increased in CMV disease	107
Miscellaneous	
IDDM	49
Renal insufficiency	49
Preeclampsia	108

resting lymphocytes, and it plays a pivotal role in cell-cell adhesion and in signal transduction (91). Soluble forms of ICAM-3 with a molecular weight of 95 kDa, which is smaller than the membrane form, have recently been described in the supernatant of stimulated peripheral blood mononuclear cells, suggesting that it can be cleaved from the cell membrane. ICAM-3 is detectable in normal human sera, with a mean concentration of 133 ± 91 ng/ml (92). Levels of ICAM-3 are significantly higher in patients with systemic lupus erythematosus, especially during active disease phases (92).

III. VCAM-1

VCAM-1 (CD106) is a single-chain glycoprotein belonging to the Ig superfamily with a molecular weight of approximately 95 to 110 kDa. VCAM-1 is expressed on cytokine-activated endothelial cells, dendritic cells, macrophages, and epithelium. In contrast to ICAM-1 and ICAM-2, VCAM-1 is not significantly expressed on unstimulated endothelial cells. The counterreceptor for VCAM-1 is VLA-4, a molecule of the integrin family. The inter-

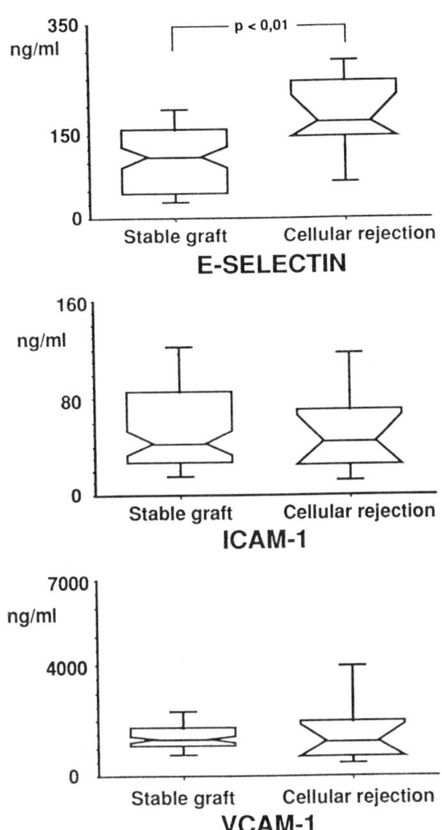

Figure 3 Soluble adhesion molecules ICAM-1, VCAM-1, and E-selectin in patients with biopsy-proven acute cellular rejection. Serum samples were collected just before treatment for rejection and levels were compared to those of 89 longitudinally collected samples of 10 patients with stable graft function. Notched boxes show the 10th, 26th, 50th (median), 75th, and 90th percentiles. Levels of E-selectin were significantly higher ($P < .008$ using the Mann-Whitney U test), whereas levels of ICAM-1 and VCAM-1 were comparable to controls with stable graft function.

acting cells are lymphocytes, monocytes, basophils, and eosinophils (1,11,93–95). Soluble VCAM-1 is released from activated endothelial cells in culture (9). There is little information on levels of VCAM-1 in the blood, but mean levels in normal individuals range from 431 to 504 ng/ml (12). Wellicome has reported a molecular weight of 85 to 90 kDa for circulating VCAM-1 molecules (6). It is now clear that release of VCAM-1 requires specific hydrolysis from the membrane-anchored protein and involves a metalloprotease (96). Levels of VCAM-1 have been measured in several diseases (Table 2).

A. Allergic Diseases

No significant modifications of VCAM-1 levels in allergic diseases have been reported.

B. Autoimmune Diseases

1. Connective Tissue Diseases

a. Rheumatoid Arthritis. Levels of VCAM-1 are significantly elevated in patients with rheumatoid arthritis (6), and these levels are correlated with disease activity as well as other markers such as C-reactive protein and erythrocyte sedimentation rate (25). In contrast, no such correlation was found with ICAM-1.

b. Systemic Lupus Erythematosus. Increases in VCAM-1 levels have been found in systemic lupus erythematosus (6,29,34); the levels were related to clinical disease activity, levels of anti-dsDNA antibodies, and complement levels. Spronk et al. (29) showed that levels of VCAM-1 were higher in patients with renal involvement than in those without. This contrasts with the lack of changes in ICAM-1 and E-selectin levels with disease activity.

c. Scleroderma. Increased VCAM-1 levels have been found in patients with scleroderma (30).

2. Vasculitis

a. Wegener's Granulomatosis. In Wegener's granulomatosis, VCAM-1 and ICAM-1 levels, but not E-selectin levels, are significantly elevated and correlate with disease activity (34). However, only VCAM-1 is significantly elevated at the time of relapse (33). The clinical relevance of elevated VCAM-1 levels is limited due to the lack of sensitivity and specificity for disease activity.

3. Inflammatory Diseases of the Central Nervous System

Increases in VCAM-1 levels have been found only in cerebrospinal fluid and not in serum samples of patients with multiple sclerosis (37). The results support a local production of this molecule within the cerebrospinal fluid (36).

4. Miscellaneous Autoimmune Diseases

a. Grave's Disease. Significantly elevated levels of VCAM-1 have been observed in patients with Grave's disease. Interestingly, the levels of

VCAM-1 normalized within 8 weeks after therapy was initiated. In patients with iodine-deficient goiter, levels of VCAM-1 were in the same range as in the control group. Serum levels of VCAM-1 correlated with the serum concentrations of antithyroid receptor antibodies (61). Dendritic-like cells present within lymphocytic infiltrates in the thyroid glands of patients with Grave's disease are positive for VCAM-1 (97). This could implicate thyroid hormones in the activation of endothelial cells, and VCAM-1 could be a useful clinical marker for disease activity in addition to thyroid hormones and autoantibodies.

C. Infectious Diseases

1. Tuberculosis

Lai et al. (55) observed that serum levels of VCAM-1 were increased in patients with active tuberculosis, particularly those with pulmonary disease. Interestingly, VCAM-1 remained elevated in subjects who had recently completed antituberculosis treatment, although ICAM-1 and E-selectin concentrations were normal. Conversely, VCAM-1 was not increased in patients with tuberculosis lymphadenitis.

2. Malaria

Increased serum concentrations of VCAM-1 have been found in patients with malaria, and their levels correlated with disease severity. VCAM-1 bind malaria-parasitized erythrocytes, and their upregulation by inflammatory cytokines may increase sequestration of parasites to endothelium in the brain, leading to cerebral malaria (59,98).

3. HIV Infection

Encephalitic brains of macaques infected by simian immunodeficiency virus were found to express abundant VCAM-1 protein on cerebral endothelium. Moreover, VCAM-1 concentrations in the cerebrospinal fluid from these animals were increased 20-fold above those from animals without AIDS encephalitis. Expression of other endothelial-related adhesion molecules (E-selectin, P-selectin, ICAM-1) was not uniformly associated with AIDS encephalitis (99).

D. Malignancies

Banks reported elevation of VCAM-1 levels in malignant diseases (68), but the relevance of these results is difficult to evaluate because of the heterogenous patient and tumor population, as well as the different forms of treatment given to these patients.

E. Transplantation

Immunohistochemical studies of cardiac, pancreatic, and hepatic allografts with rejection have demonstrated that VCAM-1 is induced on capillary endothelium in areas of mononuclear infiltration (82,100–102). In kidney transplants, VCAM-1 is also induced on the tubular epithelium during acute rejection (78,88,103–105). VCAM-1 levels are elevated in the serum of liver transplant patients with rejection in the first 21 days after transplantation compared with sera from patients with no rejection (82) (2780 ± 136 ng/ml vs 1335 ± 48). Such an increase has not been found in patients with heart graft rejection (84) (median 922 ng/ml, range 576–2505 in grade 3 versus median 970 ng/ml, range 552–912 in grade 0). Gearing and Newman found increased serum levels of VCAM-1 in kidney allograft rejection (12), but Bechtel et al. found only increased levels in the urine of patients with acute steroid-resistant rejection (87).

We studied serum VCAM-1 levels longitudinally before and during the first 12 months after transplantation and before treatment of rejection (88); VCAM-1 levels were increased in patients with stable graft function compared with healthy controls (Fig. 1), although levels decreased after 4 months (Fig. 4). There was no difference in levels between patients with acute rejection and patients with stable graft function (Fig. 3).

As CMV infects cells in vivo and induces a perivascular inflammatory cell response, induction of VCAM-1 has been studied and was shown to be

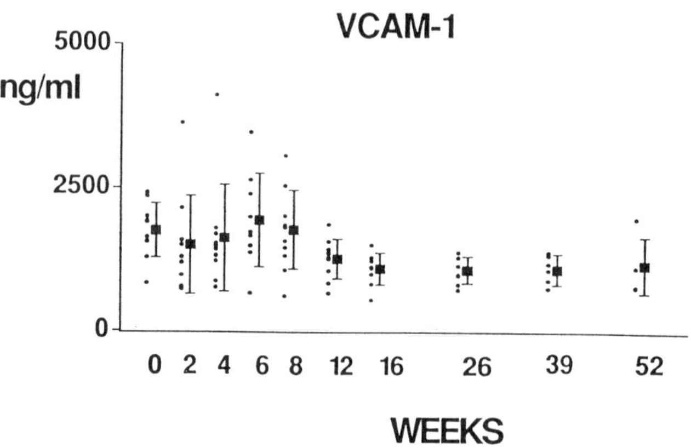

Figure 4 Longitudinal survey of serum levels of VCAM-1 in kidney transplant patients with stable graft function. Serum levels of VCAM-1 were determined in 10 patients before and at different times in the first year after transplantation (horizontal lines denote 1 SD; filled squares represent mean values).

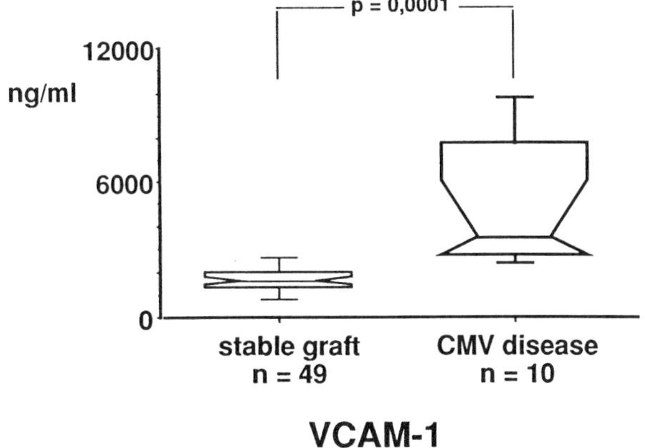

Figure 5 VCAM-1 in kidney transplant recipients with CMV disease. Levels of VCAM-1 were determined in serum samples of 10 patients with CMV disease collected before treatment and compared to those in 49 samples of 10 patients with stable graft function without CMV disease collected longitudinally during the first 12 weeks after transplantation, which is the period of occurrence of CMV disease. Notched boxes show the 10th, 25th, 50th (median), 75th, and 90th percentiles. Levels of VCAM-1 were significantly elevated in patients with CMV disease ($P = .0001$ using the Mann-Whitney U test) (107).

strongly associated with onset of CMV antigenemia in human heart transplant patients (106). We investigated whether serum VCAM-1 levels increased in kidney transplant patients with CMV disease (107). In a group of 10 patients with CMV disease we observed a significantly increased mean VCAM-1 serum concentration before starting treatment for CMV disease compared with a group with stable graft function and no CMV disease during the first 3 post-transplant months (4971 ± 2992 ng/ml vs 1711 ± 753,$P < .0001$) (Fig. 5). Moreover, a strong relationship was observed between CMV disease and a level of serum VCAM-1 above 2500 ng/ml ($P < .001$). These VCAM-1 levels dramatically decreased during ganciclovir therapy but increased again in some cases after the treatment was discontinued (Fig. 6).

F. Miscellaneous

Several diseases such as diabetes mellitus, impaired renal function, and preeclampsia have increased levels of VCAM-1 (49,108,109). However, more prospective studies are needed to establish the diagnostic, prognostic, or therapeutic utility of this finding.

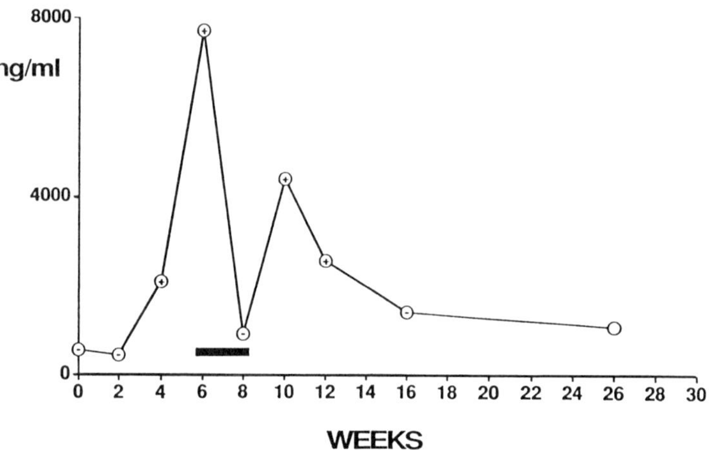

Figure 6 Changes in serum VCAM-1 levels of a typical case of a kidney transplant recipient with CMV disease during treatment with ganciclovir. Key: +, positive blood culture for CMV; −, negative blood culture for CMV; ▬, treatment with ganciclovir.

IV. E-SELECTIN (CD62E)

Circulating E-selectin is considered to be a specific marker of activated endothelium as it originates exclusively from endothelial cells. This is very different from other adhesion molecules which are expressed on endothelial and other cells and for which the origin of the circulating molecule is difficult to establish. Soluble E-selectin has been demonstrated in vitro and in vivo in a quantitative sandwich ELISA assay (110). Mean levels of E-selectin range from 16 ng/ml to 48 ng/ml in normal individuals (12). In vitro, the maximum rise of E-selectin concentration appeared in supernatants of activated endothelial cells 24 to 48 h after stimulation, while the maximal expression of the membrane form of E-selectin appeared within 4 hours of stimulation. Two forms of soluble E-selectin have been described: an 85-kDa form and a 105-kDa form. The 105-kDa form might be shed from endothelial cells since the cytoplasmic tail of the molecule is absent. It has not yet been determined whether the 85-kDa form is a degradation product of the 105-kDa molecule or whether it is a secreted form of E-selectin.

A. Allergic Diseases

Bronchial asthma, either atopic or not, is associated with an increase in plasma E-selectin levels (16). Thus the rise in E-selectin is not due to an allergic phenomenon but reflects the inflammatory component of asthma. Furthermore, it was noted that E-selectin levels remained high for some weeks after an acute attack (17).

B. Autoimmune Diseases

1. Connective Tissue Diseases

a. Systemic Lupus Erythematosus. Studies regarding circulating E-selectin levels in systemic lupus erythematosus have yielded controversial results. Although an increase in E-selectin has been observed in sera from patients with either active or inactive disease (111), decreased E-selectin levels have been reported in 22 patients with maximal activity compared to 57 healthy controls (29). Moreover, follow-up of plasma levels of E-selectin from 6 months to 1 month before maximal disease activity did not show any variation. These results were not influenced by the use of azathioprine or corticosteroids. The decrease in E-selectin levels in patients with systemic lupus erythematosus may be due to binding of the functionally active soluble molecules to their ligands on activated leukocytes. The absence of an increase in E-selectin levels in lupus nephritis was also observed in a recent study (34).

b. Sclerodermia. No significant elevation of E-selectin was observed in five patients in the early stage of systemic sclerosis compared to a control group (30).

2. Vasculitis

a. Kawasaki Disease. A 10-fold rise in E-selectin levels was found in 24 patients with Kawasaki disease compared to healthy controls (27.5 ± 10.9 ng/ml vs 3.1 ± 10. ng/ml) (112) The three patients with coronary aneurysms at the time of diagnosis had E-selectin levels of approximately 40 ng/ml. The authors observed a correlation between circulating IL-6 and TNF-α levels on the one hand and E-selectin levels on the other. Conversely, E-selectin levels were not correlated with common markers of inflammation such as C-reactive protein or erythrocyte sedimentation rate. The very low level of E-selectin in healthy controls is somewhat surprising. Results of this study are awaiting confirmation.

b. Wegener's Granulomatosis. E-selectin was measured in 22 patients with Wegener's granulomatosis and compared to 57 healthy controls (33), and the E-selectin levels were similar in both groups. Nevertheless, the authors observed a progressive increase in E-selectin before relapse. The absence of significantly elevated E-selectin levels contrasts with the increased serum levels of ICAM-1 and VCAM-1 in this disease (34) and may reflect a differential endothelial upregulation of ICAM-1, VCAM-1, and E-selectin.

c. Other. Increased levels of circulating E-selectin levels have been found in patients with polyarteritis nodosum and giant cell arteritis (111).

3. Inflammatory Diseases of the Central Nervous System

a. Multiple Sclerosis (MS). E-selectin levels are elevated in the serum and the cerebrospinal fluid of patients with multiple sclerosis (37). Particularly high levels of E-selectin were found in patients with chronic progressive MS (113).

b. Guillain-Barré Syndrome. Blood levels of E-selectin were increased in the acute stage of Guillain-Barré syndrome (40,114). The twofold increase in the adhesion molecule compared to control subjects probably reflects endothelial activation and breakdown of the blood-nerve barrier.

4. Miscellaneous Autoimmune Diseases

Elevated levels of E-selectin have been demonstrated in patients with Graves' disease (61), sarcoidosis (46), diabetes, and renal insufficiency (49).

C. Infectious Diseases

1. Tuberculosis

High levels of E-selectin are observed in patients with pulmonary tuberculosis (55). In tuberculosis with lymph node disease, these levels tended to be lower but were still significantly higher compared with control subjects. Although elevated levels of soluble ICAM-1 are the most discriminative marker of tuberculosis infection, E-selectin might be an interesting molecule reflecting disease activity.

2. Sepsis

In patients with proven bacteremia, circulating E-selectin levels were found to be elevated only if hypotension occurred. E-selectin plasma levels in patients with septic shock were elevated compared with patients who had bacteriemia without hypotension (23.3 ± 15.9 vs 1.1 ± 9 ng/ml; healthy controls: 0.9 ± 0.7 ng/ml) (110). These results indicate that high plasma E-selectin levels arise in conjunction with hemodynamic manifestations of advanced septicemia and that release of E-selectin may reflect the degree of endothelial injury or activation. High plasma concentrations of E-selectin are closely associated with multiple organ dysfunction and death (115). These studies suggest that measurement of E-selectin may be useful in the management of patients with sepsis.

3. Malaria

As E-selectin may act as a receptor for red blood cells infected with *P. falciparum* (98), a possible relationship between disease activity and E-selectin levels has been suspected. Elevated E-selectin levels are observed in patients with malaria (61). Serum levels are markedly increased on the day of diagnosis and decline rapidly after treatment. Serum levels of E-selectin are related to parasitemia, and E-selectin levels are significantly higher in patients with severe disease with cerebral involvement than in patients with mild malaria (59,63).

D. Malignancies

E-selectin is elevated in some human malignancies. Patients with ovarian, breast, and gastrointestinal cancers may have significantly higher levels of E-selectin than healthy controls (68). Large prospective studies, however, are needed to define the value of E-selectin quantification in cancers.

E. Transplantation

Graft E-selectin levels are upregulated in cardiac allografts 1 or 2 weeks before rejection (116), and it has been proposed that upregulated expression may facilitate the onset of cellular infiltration within the graft (117). As E-selectin sheds into the supernatant of cytokine-stimulated endothelial cells in vitro, we investigated whether soluble E-selectin levels increased in sera of patients with renal allograft rejection (89). The levels were elevated at the time of transplantation compared with healthy control patients (Fig. 1) and, as shown in Figure 7, remained elevated in patients with stable graft function during the first 12 months after transplantation. The absence of significant differences between the different times after transplantation allowed us to pool the different values to obtain a mean level of 110.7 ng/ml ($\pm$ 68.3) in patients with stable graft function. To assess whether changes in E-selectin levels were associated with acute rejection episodes, serum samples were collected before treatment for rejection in 11 patients. As shown in Figure 3, the serum levels were increased in patients with acute renal graft rejection compared to those with stable graft function (177.5 $\pm$ 77.7 ng/ml vs 110.7 $\pm$ 68.3 ng/ml, $P < .01$), and a relationship was found between acute cellular rejection and an E-selectin level above 140 ng/ml ($\chi^2 = 8.094$, $P < .01$). Moreover, E-selectin levels increased during OKT3 therapy and returned to prerejection values in about 10 days (Fig. 8). This

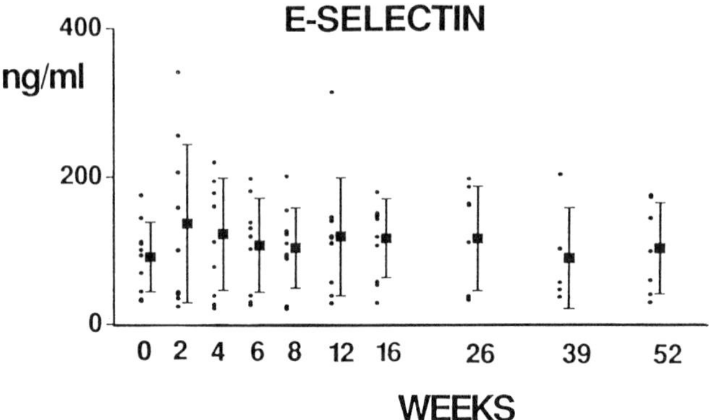

Figure 7 Longitudinal survey of serum levels of E-selectin in kidney transplant patients with stable graft function. Serum levels of E-selectin were determined in 10 patients before grafting and at different times during the first year after grafting (horizontal lines denote 1 SD; filled squares represent mean values).

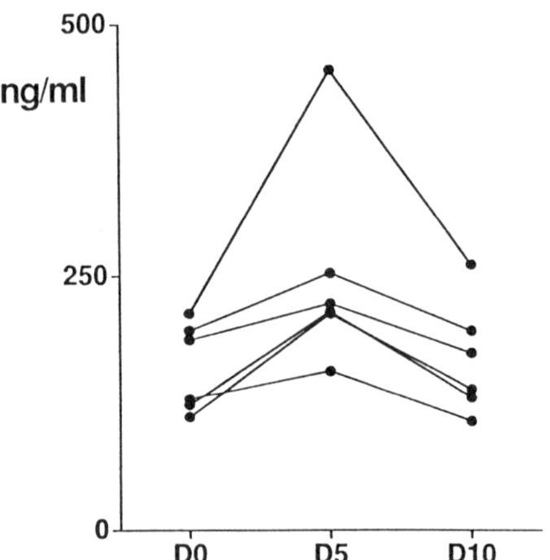

Figure 8 E-selectin during treatment of cellular rejection with OKT3. Serum samples were longitudinally collected before treatment (D0), on day 5 (D5), and on day 10 (D10) after beginning of treatment in six patients with biopsy-proven acute cellular rejection. E-selectin levels were significantly higher on D5 ($P = .027$ using Wilcoxon's test) than on D0, whereas levels at D10 were comparable ($P =$ NS) to those on D0.

rise in E-selectin levels, probably due to the release of OKT3-induced cytokines, was also observed by Weston, who, in contrast to our data, did not find any difference between blood levels in renal transplant patients with and without rejection (118), although the number of patients studied was very small. Another study examined the levels of E-selectin in a small number of cardiac transplant patients with cardiac allografts (84) and found slightly, but not significantly, elevated levels in patients with cellular rejection. The levels of E-selectin were correlated with levels of ICAM-1, which were significantly increased in patients with grade 3a rejection.

F. Miscellaneous

1. Hypertension

E-selectin levels are elevated in patients with noncontrolled hypertension compared with subjects without hypertension or controlled hypertension. Soluble E-selectin levels correlated with diastolic but not with systolic blood pressure. Interestingly, E-selectin levels did not correlate with Von Willebrand factor levels, reflecting another activation pathway of the vascular endothelium (119).

2. Atherosclerosis

A weak but significant increase in E-selectin has been reported inpatients with peripheral vascular diseases but not in ischemic heart disease (120).

V. P-SELECTIN

P-selectin (CD62P) is an adhesion receptor on activated platelets and endothelial cells and is stored in platelet alpha granules and endothelial Weibel Palade bodies. Its synthesis is increased in response to cytokines (121). Soluble forms of P-selectin have been demonstrated in human serum (122,123). Two circulating forms of P-selectin have been described: a truncated form of the membrane molecule, and an alternate spliced molecule without the transmembrane exon. Plasma concentrations in normal controls varies from 36 ng/ml to 250 ng/ml (124,125), as measured using an ELISA. Increased plasma levels have been described in patients with thrombotic thrombocytopenic purpura (TTP), hemolytic uremic syndrome (HUS) (125), malaria (126), and connective tissue diseases (127).

In patients with TTP and HUS, plasma levels were lower in patients in remission compared to the levels in patients before or during treatment, suggesting that elevated plasma P-selectin levels correlate with the microangiopathic status (125).

Mean P-selectin levels are elevated in patients with mixed connective tissue disease (1048 ng/ml) and rheumatoid arthritis (844 ng/ml), but only slightly increased in patients with systemic lupus erythematosus (306 ng/ml) compared to controls (220 ng/ml). The levels increased in active disease and were not correlated with any of the currently available laboratory parameters (WBC, CRP, ESR, antinuclear antibody, RF, anticardiolipids), except the number of platelets (127). Such increased levels have also been reported in scleroderma (30).

VI. L-SELECTIN

L-selectin (CD62L) is a cell surface receptor on granulocytes, lymphocytes, and monocytes which is responsible for the initial contact of leukocytes with the endothelium (128). The extracellular domain of L-selectin is proteolytically shed from leukocytes following cellular activation in vitro (129). The shed form is functionally active and, at high concentrations, can inhibit leukocyte attachment to the endothelium (130). Patients with sepsis and HIV infection showed markedly elevated L-selectin levels in their serum (7). Increased levels of L-selectin have been observed in CSF of patients with meningeal leukemia (131), suggesting that CSF L-selectin may be a useful marker in the detection of meningeal involvement by blast cells in patients with L-selectin + leukemia. L-selectin can be also detected in synovial fluid (132), and serum levels correlate with leukocyte count in chronic myeloid and lymphocytic leukemia and during bone marrow transplantation (133).

In contrast to these diseases with increased levels, patients who progressed to the adult respiratory distress syndrome (ARDS) had significantly lower plasma L-selectin levels on admission than those who did not and compared to normals (range 0.37–6.55) μg/ml) (134). Moreover, a correlation was found between low values of L-selectin and indices of subsequent lung injury and mortality. Thus determination of L-selectin levels may be of prognostic value in ARDS. These results may reflect the sequestration of L-selectin by widespread binding to activated endothelium in microvascular beds.

VII. DISCUSSION

This review shows that raised levels of adhesion molecules may be a useful marker for the diagnosis and management of patients with certain inflammatory diseases or organ transplants. The pathophysiologic significance of raised levels of circulating adhesion molecules in serum or plasma is unclear. Blood adhesion molecules are essentially derived from leukocytes and endothelium, where they control leukocyte recruitment into inflammatory sites. Their expression on the cell surface is highly regulated, with both induction and down-regulation. The expression of adhesion molecules on endothelial cells, therefore, needs a relatively rapid membrane turnover rate with their disappearance from the cell surface by proteolytic cleavage or internalization. Thus, raised levels could reflect differences in the degree of shedding and/or internalization (135). Nevertheless, although the amount of soluble adhesion molecules released in vitro in the supernatants of activated endothelial cells is directly correlated with their levels of surface expression, this is not directly related to a strong reduction in membrane expression (8). This demonstrates that the precise mechanisms of the regulation of the shedding are largely unknown, and that elevated levels of adhesion molecules in the blood could result from an overexpression on the cell surface and/or an increase in shedding. It is also possible that some adhesion molecules are released without previous expression on the cell surface membrane.

A significant proportion of shed molecules may escape detection because they may interact with their counterreceptors or to soluble ligands (12). Thus, circulating ICAM-1 retains the ability to bind in vitro to leukocyte LFA-1 (5), and circulating VCAM-1 molecules are capable of adhering to the Jurkatt T-cell line (6), although there is very little evidence that such a mechanism occurs in vivo. Nevertheless, L-selectin has been shown to bind to the luminal surface of endothelium in human tissues, and low levels of L-selectin have been observed in patients with the adult respiratory distress syndrome (134).

The question whether elevated levels of adhesion molecules in the blood are due to an increase in production and/or to a reduction in elimination is at present unanswered. The mechanisms of clearance of these molecules are largely unknown. The observation that the levels of ICAM-1, VCAM-1, and E-selectin are elevated in patients with renal insufficiency or chronic hemodialysis (34,49) suggests retention of adhesion molecules as a consequence of the reduction of renal function. However, we did not observe significant correlations between the serum levels of ICAM-1, VCAM-1, and E-selectin and creatinine levels in kidney transplant patients (88). Furthermore, some of the diseases leading to renal insufficiency and chronic hemodialysis are known to be associated with endothelial activation and may result in increased production of circulating adhesion molecules.

The absence of elevated blood levels of adhesion molecules in patients with inflammatory diseases could have several explanations. First, determination in the blood may not be as meaningful as within a rejecting or inflamed organ (12). Levels of ICAM-1 and VCAM-1 are increased in the urine but not the blood of renal transplant patients with acute steroid-resistant rejection (87), and biliary ICAM-1 levels are more relevant than serum levels to diagnose acute liver allograft rejection (79,82). It would therefore be of interest to determine the local release of adhesion molecules in tissues. Secondly, lower-molecular-weight components of soluble forms of cell surface receptors have been reported (136), and the blood level may thus reflect the intact extracellular component of the adhesion molecules or as a proteolyzed smaller form of the molecule. The smaller forms could have lost the epitopes recognized by antibodies and are potentially not detectable in some ELISA tests, which could explain why different ELISA kits with monoclonal antibodies with different epitope specificities give different results on the same samples.

Despite similarities in the stimuli that induce adhesion molecules such as ICAM-1, VCAM-1, and E-selectin on endothelial cells in vitro, a surprising differential increase in one of these molecules can be seen in some diseases. Both ICAM-1 and VCAM-1, but not E-selectin levels, are increased in active Wegener's granulomatosis, and ICAM-1 but not VCAM-1 levels differ significantly between active and inactive Wegener's disease. In contrast, VCAM-1 but not ICAM-1 was significantly higher in Wegener patients on hemodialysis (34). Only VCAM-1 is significantly elevated in active or inactive systemic lupus erythematosus (6,29,34). We observed a similar differential increase in kidney transplant patients with an increase in serum E-selectin levels during acute cellular rejection (88) and an increase in serum VCAM-1 during CMV disease (107). Such discrepancies could reflect the origin of the elevated soluble adhesion molecules. It is thought that ICAM-1 is mainly released from activated peripheral blood lymphocytes as a consequence of inflammation or tissue damage (5), although in vitro studies have demonstrated its release from endothelial (8,9), melanoma and ovarian carcinoma cells. In contrast, E-selectin expression is restricted to activated endothelial cells (3), and increased blood levels of E-selectin could reflect endothelial cell activation. The conditions that result in the release of VCAM-1 in vivo are still unknown; VCAM-1 is less widely distributed than ICAM-1, although it is expressed by a variety of cells outside blood vessels. Whether the elevation of VCAM-1 is the result of vascular activation with enhanced VCAM-1 expression and VCAM-1 shedding remains to be determined. Another explanation could be a differential activation of cells that express adhesion molecules in certain pathological processes. It is, for example, well established that TNF-α and IL-4 have different actions on endothelial cells in vitro than in vivo (76,94,137); both TNF-α and IL-4 induce VCAM-1 expression, while TNF-α, but not IL-4, increases or induces expression of ICAM-1 and E-selectin. An interesting possibility is that VCAM-1 is expressed by endothelial cells in situ in the presence of a T lymphocyte-mediated inflammatory response (11). Agents that selectively inhibit expression of adhesion molecules could provide new therapeutic approaches (138–143).

It is unclear whether adhesion molecules in body fluids have physiological effects, either by competing in cell-cell adhesion or by triggering a response in a ligand-bearing cell. Soluble P-selectin has been reported to inhibit neutrophil adhesion to endothelial cells (144) and neutrophil activation (145,146), and circulating L-selectin and ICAM-1 have been reported to inhibit lymphocyte functions in vitro (147,148). On the other hand, it has been reported that E-selectin can upregulate neutrophil CD 11b integrin function (149), and there is evidence for participation of the adhesion molecules in transendothelial migration of leukocytes and modulation of lymphocyte activation (150). Therefore, the release of adhesion molecules could be essential for the development of certain diseases.

REFERENCES

1. Springer TA. Adhesion receptors of the immune system. Nature 1990; 346: 425–434.
2. Pober JS, Cotran RS. The role of endothelial cells in inflammation. Transplantation 1990; 50:537–544.
3. Bevilacqua MP, Pober JS, Mendrick DL, Cotran RS, Gimbrone MA. Identification of an inducible endothelial-leukocyte adhesion molecule. Proc Natl Acad Sci USA. 1987; 84:9238–9242.
4. Seth R, Raymond FD, Makgoba MW. Circulating ICAM-1 isoforms: diagnostic prospects for inflammatory and immune disorders. Lancet 1991; 338: 83–84.
5. Rothlein R, Mainolfi AE, Czajkowski M, Marlin DS. A form of circulating ICAM-1 in human serum. J Immunol 1991; 147:3788–3793.
6. Wellicome SM, Kapahi P, Mason JC, Lebranchu Y, Yarwood H, Haskard DO. Detection of a circulating form of vascular cell adhesion molecule-1: raised levels in rheumatoid arthritis and systemic lupus erythematosus. Clin Exp Immunol 1993; 92:412–418.
7. Spertini O, Schleiffenbaum B, Whitewen C, Ruiz P, Tedder T. ELISA for quantitation of L-selectin shed from leucocytes in vivo. J Immunol Methods 1992; 156:115–123.
8. Leeuwenberg J, Smeets E, Neefjes J, et al. E-selectin and intercellular adhesion molecule-1 are released by activated human endothelial cells in vitro. Immunology 1992; 77:543–549.
9. Pigott R, Dillon LP, Hemingway IH, Gearing AJH. Soluble forms of E-selectin, ICAM-1, and VCAM-1 are present in the supernatants of cytokine activated cultured endothelial cells. Biochem Biophys Res Commun 1992; 187:584–589.
10. Munro JM, Pober JS, Cotran RS. Tumor necrosis factor and interferon-γ

induce different patterns of endothelial activation and associated leukocyte accumulation skin of *Papio anubis*. Am J Pathol 1989; 135:121–133.
11. Mason JC, Haskard DO. The clinical importance of leucocyte and endothelial cell adhesion molecules in inflammation. Vasc Med Rev 1994; 5:249–275.
12. Gearing J, Newman W. Circulating adhesion molecules in disease. Immunol Today 1993; 14:506–512.
13. Leung DYM, Pober JS, Cotran RS. Expression of endothelial-leukocyte adhesion molecule-1 in elicited late phase allergic reactions. J Clin Invest 1991; 146:1805–1809.
14. Kyan-Aung U, Haskard DO, Poston RN, Thornhill MH, Lee TH. Endothelial leukocyte adhesion molecule-1 and intellecular adhesion molecule-1 mediate the adhesion of eosinophils to endothelial cells in vitro and are expressed by endothelium in allergic cutaneous inflammation in vivo. J Immunol 1991; 146:521–528.
15. Wegner CD, Gundel RH, Peilly P, Haynes N, Letts LG, Rothlein R. Intercellular adhesion molecule-1 (ICAM-1) in the pathogenesis of asthma. Science 1990; 247:456–459.
16. Kobayashi T, Hashimoto S, Imai K, et al. Elevation of serum soluble intercellular adhesion molecule-1 (sICAM-1) and sE-selectin levels in bronchial asthma. Clin Exp Immunol 1994; 96:110–115.
17. Montefort S, Lai CK, Kapahi P, et al. Circulating adhesion molecules in asthma. Am J Respir Crit Care Med 1994; 149:1149–1152.
18. Hashimoto S, Imai K, Kobayashi T, et al. Elevated levels of soluble ICAM-1 in sera from patients with bronchial asthma. Allergy 1993; 48:370–372.
19. Schmitt M. Lymphocyte subsets, sIL2-R and sICAM-1 in blood during allergen challenge tests inasthmatic children. Pediatr Allergy Immunol. 1993; 4: 208–213.
20. Chihara J, Yamamoto T, Kurachi D, Nakajima S. Soluble ICAM-1 in sputum of patients with bronchial asthma. (Letter.) Lancet 1994; 343:1108.
21. Terada N, Konno A, Yamashita T, et al. Serum level of soluble ICAM-1 in subjects with nasal allergy and ICAM-1 mRNA expression in nasal mucosa. Arerugi 1993; 42:87–93.
22. Wuthrich B, Jollerjemelka H, Kagi MK. Levels of soluble ICAM-1 in atopic dermatitis: a new marker for monitoring the clinical activity? Allergy 1995; 50:88–89.
23. Kojima T, Ono A, Aoki T, Kameda-Hayaski N, Kobayashi Y. Circulating ICAM-1 levels in children with atopic dermatitis. Ann Allergy 1994; 73:351–355.
24. Cush JJ, Rothlein R, Lindsley HB, Mainolfi EA, Lipsky PE. Increased levels of circulating intercellular adhesion molecule 1 in the sera of patients with rheumatoid arthritis. Arthritis Rheum 1993; 36:1098–1102.
25. Mason JC, Kapahi P, Haskard DO. Detection of raised levels of circulating ICAM-1 in some patients with rheumatoid arthritis but not in systemic lupus erythematosus: lack of correlation with circulating VCAM-1. Arthritis Rheum 1993; 36:519–527.

26. Aoki S, Imai K, Yachi A. Soluble intercellular adhesion molecule-1 (ICAM-1) antigen in patients with rheumatoid arthritis. Scand J Immunol 1993; 38:485–90.
27. Machold KP, Kiener HP, Graninger W, Graninger WB. Soluble intercellular adhesion molecule-1 (ICAM-1) in patients with rheumatoid arthritis and systemic lupus erythematosus. Clin Immunol Immunopathol 1993; 68:74–78.
28. Sfikakis PP, Charalambopoulos D, Vayiopoulos G, Oglesby R, Sfikakis P, Tsokos GC. Increased levels of intercellular adhesion molecule-1 in the serum of patients with systemic lupus erythematosus. Clin Exp Rheumatol 1994; 12: 5–9.
29. Spronk PE, Bootsma H, Huitema MG, Limburg PC, Kallenberg CG. Levels of soluble VCAM-1, soluble ICAM-1, and soluble E-selectin during disease exacerbations in patients with systemic lupus erythematosus (SLE); a long term prospective study. Clin Exp Immunol 1994; 97:439–444.
30. Gruschwitz MS, Horstein OP, Von den Driesch P. Correlation of soluble adhesion molecules in peripheral blood of scleroderma patients with their in situ expression and with disease activity. Arthritis Rheum 1995; 38:184–189.
31. Sano Y, Hirai S, Katayama M, Kato I. Immunoenzymometric analysis for expression and shedding of intercellular adhesion molecule-1 on human endothelial cells stimulated with cytokines or lipopolysaccharide. Mol Cell Biochem 1994; 139:123–130.
32. Furukawa S, Imai K, Matsubara T, et al. Increased levels of circulating intercellular adhesion molecule-1 in Kawasaki disease. Arthritis Rheum 1992; 35:672–677.
33. Stegeman CA, Tervaert JWC, Huitema MG, De Jong PE, Kallenberg CGM. Serum levels of soluble adhesion molecules: intercellular adhesion molecule 1, vascular cell adhesion molecule 1 and E-selectin in patients with Wegener's granulomatosis. Arthritis Rheum 1994; 37:1228–1235.
34. Mrowka C, Sieberth HG. Detection of circulating adhesion molecules ICAM-1, VCAM-1 and E-selectin in Wegener's granulomatosis, systemic lupus erythematosus and chronic renal failure. Clin Nephrol 1995; 43:288–296.
35. John S, Neumayer HH, Weber M. Serum circulating ICAM-1 levels are not useful to indicate active vasculitis or early renal allograft rejection. Clin Nephrol 1994; 42:369–375.
36. Rieckmann P, Michel U, Albrecht M, Bruck W, Wockel L, Felgenhauser K. Cerebral endothelial cells are a major source for soluble intercellular adhesion molecule-1 in the human central nervous system. Neurosci Lett 1995; 186:61–64.
37. Sharief MK., Noori MA, Ciardi M, Cirelli A, Thomspon EJ. Increased levels of circulating ICAM-1 in serum and cerebrospinal fluid of patients with active multiple sclerosis: correlation with TNF-α and blood-brain barrier damage. J Neuroimmunol 1993; 43:15–21.
38. Dore-Duffy P, Newman W, Balabonov R, et al. Circulating soluble proteins in cerebrospinal fluid and serum of patients with multiple sclerosis: correlation with clinical activity. Ann Neurol 1995; 37:55–62.

39. Rieckmann P, Martin S, Weichselbraun I, et al. Serial analysis of circulating adhesion molecules and TNF receptor in serum from patients with multiple sclerosis: cICAM-1 is an indicator for relapse. Neurology 1994; 44:2367–2372.
40. Hartung HP, Michels M, Reiners K, Seeldrayers P, Archelos JJ, Toyka KV. Soluble ICAM-1 serum levels in multiple sclerosis and viral encephalitis. Neurology 1993; 43:2331–2335.
41. Scrivastava MD, Rossi TM, Lebenthal E. Serum soluble interleukin-2 receptor, soluble intercellular adhesion molecule-1 levels in Crohn's disease, celiac disease and systemic lupus erythematosus. Res Com Mol Pathol Pharmacol 1995; 87:21–26.
42. Nielsen OH, Langholz E, Hendel J, Brynskov J. Circulating soluble intercellular adhesion molecule-1 (sICAM-1) in active inflammatory bowel disease. Digest Dis Sci 1994; 39:1918–1923.
43. Thomson AW, Satoh S, Nussler AK, et al. Circulating intercellular adhesion molecule-1 (ICAM-1) in autoimmune liver disease and evidence for the production of ICAM-1 by cytokine-stimulated human hepatocytes. Clin Exp Immunol 1994; 95:83–90.
44. Shijubo N, Imai K, Shigehara K, et al. Soluble intercellular adhesion molecule-1 (ICAM-1) in sera and bronchoalveolar lavage fluid of patients with idiopathic pulmonary fibrosis and pulmonary sarcoidosis. Clin Exp Immunol 1994; 95:156–161.
45. Dalhoff K, Bohnet S, Braun J, Kreft B, Wiessman KJ. Intercellular adhesion molecule 1 (ICAM-1) in the pathogenesis of mononuclear cell alveolitis in pulmonary sarcoidosis. Thorax 1993; 48:1140–1144.
46. Hamblin S, Shakoor Z, Kapahi P, Haskard DO. Circulating adhesion molecules in sarcoidosis. Clin Exp Immunol 1994; 96:335–338.
47. Shijubo N, Imai K, Aoki S, et al. Circulating intercellular adhesion molecule-1 (ICAM-1) antigen in sera of patients with idiopathic pulmonary fibrosis. Clin Exp Immunol 1992; 89:58–62.
48. Zaman AG, Edelsten C, Stanford MR, et al. Soluble ICAM-1 as a marker of disease relapse in idiopathic uveoretinitis. Clin Exp Immunol. 1994; 95:60–65.
49. Gearing AJH, Hemingway I, Pigott, Hughes J, Rees AJ, Cashman SJ. Soluble forms of vascular adhesion molecules, E-selectin, ICAM-1, and VCAM-1: pathological significance. Ann NY Acad Sci 1992; 667:324–331.
50. Roep BO, Heidenthal E, De Vries RR, Kolb H, Martin S. Soluble forms of intercellular adhesion molecule-1 in insulin-dependent diabetes mellitus. Lancet 1994; 343:1590–1593.
51. Heufelder AE, Bahn RS. Soluble intercellular adhesion molecule-1 (sICAM-1) in sera of patients with Graves' ophtalmopathy and thyroid disease. Clin Exp Immunol 1993; 92:296–302.
52. Wenisch C, Myskiw D, Parschalk B, Hartmann T, Dam K, Graninger W. Soluble endothelium-associated adhesion molecules in patients with Graves' disease. Clin Exp Immunol 1994; 98:240–244.
53. Kowalzick L, Bildau H, Neuber K, Kohler I, Ring J. Clinical improvement in psoriasis during dithranol/uvb therapy does not correspond with a decrease

in elevated serum soluble ICAM-1 levels. Arch Dermatol Res 1993; 285:233-235.
54. Groves R, Kapahi P, Barker J, Haskard DO, Mcdonald D. Detection of circulating adhesion molecules in erythrodermic skin disease. J Am Acad Dermatol 1995; 32:32-36.
55. Lai CKW, Wong KC, Chan CHS, Ho SY, Chung SY, Haskard DO, Lai KN. Circulating adhesion moleculesin tuberculosis. Clin Exp Immunol 1993; 94: 522-526.
56. Shijubo N, Imai K, Nakanishi F, Yachi A, Abe S. Elevated concentrations of circulating ICAM-1 in far advanced and miliary tuberculosis. Am Rev Respir Dis 1993; 148:1298-1301.
57. Kuster H, Degitz K. Circulating ICAM-1 in neonatal sepsis. Lancet 1993; 341:506.
58. Furukawa S, Motohashi T, Matsubara T, Imai K, Okumura K, Yabuta K. Soluble ICAM-1 levels in serum during acute infectious mononucleosis. Scand J Infect Dis 1993; 25:249-252.
59. Jakobsen PH, Morris-Jones S, Ronn A, et al. Increased plasma concentrations of sICAM-1, sVCAM-1 and sELAM-1 in patients with *Plasmodium falciparum* or *P. vivax* malaria and association with disease severity. Immunology 1994; 83:655-669.
60. Graninger W, Prada J, Neifer S, Zotter G, Thalhammer F, Kremsner PG. Upregulation of ICAM-1 by *Plasmodium falciparum*: in vitro and in vivo studies. J Clin Pathol 1994; 47:653-656.
61. Wenisch C, Varijamonta S, Looareesuwan S, Graninger W, Pichler R, Wernsdorfer W. Soluble intercellular adhesion molecule-1 (ICAM-1) endothelial leukocyte adhesion molecule-1 (ELAM-1), and tumor necrosis factor receptor (55kDa TNF-R) in patients with acute *Plasmodium falciparum* malaria. Clin Immunol Immunopathol 1994; 71:344-348.
62. Hiviid L, Theander TG, Elhassan IM, Jensen JB. Increased plasma levels of soluble ICAM-1 and ELAM-1 (E-selectin) during acute *Plasmodium falciparum* malaria. Immunol Lett 1993; 36:51-58.
63. Deloron P, Dumont N, Nyongabo T, et al. Immunologic and biochemical alterations in severe falciparum malaria: relation to neurological symptoms and outcome. Clin Infect Dis 1994; 19:480-485.
64. Diez-Ruiz A, Kaiser G, Jager H, et al. Increased levels of serum intercellular adhesion molecule 1 in HIV infection are related to immune activation. Int Arch Allergy Immunol 1993; 102:56-60.
65. Horiike N. Soluble ICAM-1 in serum in chronic hepatitis B and C. Gastroenterology 1994; 29:455-459.
66. Becker JC, Dummer R, Hartmann AA, Burg G, Schmidt RE. Shedding of ICAM-1 from human melanoma cell lines induced by IFN and tumor necrosis factor. J Immunol 1991; 147:4398-4401.
67. Harning R, Mainolfi E, Bystryn JC, Henn M, Merluzzi VJ, Rothlein R. Serum levels of circulating intercellular adhesion molecule 1 in human malignant melanoma. Cancer Res 1991; 51:5003-5005.
68. Banks RE, Gearing AJH, Hemingway IK, Norfolk DR, Perren TJ, Selby

PJ. Circulating intercellular adhesion molecule-1 (ICAM-1), E-selectin and vascular cell adhesion molecule-1 (VCAM-1) in human malignancies. Br J Cancer 1993; 68:122–124.

69. Liang JT, Wang CR, Chang KJ, Chuang CY. Circulating intercellular adhesion molecule-1 and lymphocyte subsets involved in immune response of breast cancer. Chung Hua Min Kuo Wei Sheng Wu Chi Mien I Hsueh Tsa Chih 1993; 26:1–5.
70. Tsujisaki M, Imai K, Hirata H, et al. Detection of circulating intercellular adhesion molecule-1 antigen in malignant diseases. Clin Exp Immunol 1991; 85:3–8.
71. Pizzolo G, Vinante F, Nadali G, Chilosi M, Semenzato G. Circulating soluble ICAM-1 in patients with Hodgkin's disease. (Letter.) Immunol Today 1994; 15:140–141.
72. Gruss HJ, Dolken G, Brach MA, Mertelsmann R, Hermann F. Serum levels of circulating ICAM-1 are increased in Hodgkin's disease. Leukemia 1993; 7: 1245–129.
73. Almonte M, Colizzi F, Esposito G, Maio M. Circulating intercellular adhesion molecule 1 as a marker of disease progression in cutaneous melanoma. N Engl J Med 1990; 959–327(13).
74. Giavazzi R, Nicoletti MI, Chirivi RGS, et al. Soluble intercellular adhesion molecule-1 (ICAM-1) is released into the serum and ascites of human ovarian carcinoma patients and in nude mice bearing tumour xenografts. Eur J Cancer 1994; 30A:1865–1870.
75. Jackson AM, Alexandroff AB, Kelly RW, et al. Changes in urinary cytokines and soluble intercellular adhesion molecule-1 (ICAM-1) in bladder cancer patients after bacilluls Calmette-Guerin (BCG). Clin Exp Immunol 1995; 99: 369–375.
76. Galea P, Lebranchu Y, Thibault G, Bardos P. Interleukin 4 and tumor necrosis factor a induce different pathways in endothelial cells for the binding of peripheral blood lymphocytes. Scand J Immunol 1992; 36:575–585.
77. Faull RJ, Russ GR. Tubular expression of intercellular adhesion molecule-1 during renal allograft rejection. Transplantation 1989; 48:226–230.
78. Briscoe DM, Pober JS, Harmon WE, Cotran RS. Expression of vascular cell adhesion molecule-1 in human renal allografts. J Am Soc Nephrol 1992; 3: 1180–1185.
79. Adams DH, Mainolfi E, Elias E, Neuberger JM, Rothlein R. Detection of circulating intercellular adhesion molecule-1 after liver transplantation-evidence of local release. Within the liver during graft rejection. Transplantation 1993; 55:83–87.
80. Tanio JW, Basu CB, Albelda SM, Eisen HJ. Differential expression of the cell adhesion molecules ICAM-1, VCAM-1, and E-selectin in normal and posttransplantation myocardium. Cell adhesion molecules expression in human cardiac allografts. Circulation 1994; 89:1760–1768.
81. Martinez OM, Villanueva JC, Quinn M, Krams SM. Soluble ICAM-1 and soluble HLA class I as indicators of immune activation in liver allograft recipients. Hepatology 1993; 18:19–24.

82. Lang T, Krams SM, Villanueva JC, Cox K, So S, Martinez OM. Differential patterns of circulating intercellular adhesion molecule-1 (cICAM-1) and vascular cell adhesion molecule-1 (cVCAM-1) during liver allograft rejection. Transplantation 1995; 59:584–589.
83. Ballantyne CM, Minfoli EA, Young JB, et al. Relationship of increased levels of circulating ICAM-1 after heart transplant to rejection, HLA match, and survival. Circulation 1991; 84:490 (abstract).
84. Grant SCD, Lamb WR, Hutchinson IV, Brencheley PEC. Serum soluble adhesion molecules and cytokines in cardiac allograft rejection. Transplant Immunol 1994; 2:321–325.
85. Stockenhuber F, Kramer G, Schenn G, et al. Circulating ICAM-1: novel parameter of renal graft rejection. Transplant Proc 1993; 25:919–920.
86. Kanagawa K, Seki T, Nishigaki F, et al. Measurement of soluble ICAM-1 after renal transplantation. Transplant Proc 1994; 26:2103–2105.
87. Bechtel U, Scheuer R, Landgraf R, Konig A, Feucht HE. Assessment of soluble adhesion molecules (sICAM-1, sVCAM-1, sELAM-1) and complement cleavage products (sC4d, sC5b-9) in urine. Transplantation 1994; 58: 905–911.
88. Lebranchu Y, Kapahi P, Al-Najjar A, et al. Soluble E-selectin, ICAM-1 and VCAM-1 levels in renal allograft recipients. Transplant Proc 1994; 26:1873–1874.
89. De Fougerolles AR, Springer TA. Intercellular adhesion molecule 3, a third adhesion counter-receptor for lymphocyte function-associated molecule 1 on resting lymphocytes. J Exp Med 1992; 175.
90. Vazeux R, Hoffman PA, Tomita JK, et al. Cloning and characterization of a new intercellular adhesion molecule ICAM-R. Nature 1992; 360:485.
91. De Fougerolles AR, Klickstein LB, Springer TA. Cloning and expression of intercellular adhesion molecule 3 reveals strong homology to other immunoglobulin family counter-receptors for lymphocyte function-associated antigen-1. J Exp Med 1993:1187.
92. Pino-Otin MR, Vinas O, De la Fuente MA, et al. Existence of a soluble form of CD50 (intercellular adhesion molecule-3) produced upon human lymphocyte activation. Present in normal human serum and levels are increased in the serum of systemic lupus erythematosus patients. J Immunol 1995; 154:3015–3024.
93. Vonderheide RH, Springer TA. Lymphocyte adhesion through very late antigen 4: evidence for a novel binding site in the alternatively spliced domain of vascular cell adhesion molecule-1 and an additional a4 integrin counter-receptor on stimulated endothelium. J Exp Med 1992; 175:1433–1442.
94. Thornhill M, Wellicome SM, Mahiouz DL, Lanchbury JSS, Kyan-Aung U, Haskard DO. Tumor necrosis factor combines with IL-4 or IFN-γ to selectively enhance endothelial cell adhesiveness for T cells: The contribution of VCAM-1 dependent and independent binding mechanisms. J Immunol 1991; 146:592–598.
95. Rice E, Munro M, Corles C, Bevilacqua M. Vascular and nonvascular expression of INCAM-110. Am J Pathol 1991; 138:385–393.

96. Leca G, Mansur SE, Bensussan A. Expression of VCAM-1 (CD106) by a subset of TCR gamma delta-bearing lymphocyte clone. Involvement of a metalloprotease in the specific hydrolytic release of the soluble isoform. J Immunol 1995; 154:1069–1077.
97. Miyazaki A, Mirakian R, Bottazzo GF. Adhesion molecules expansion in Graves' thyroid glands; potential relevances of granule membrane protein (GMP-140) and intercellular adhesion molecule-1 (ICAM-1) in the homing and antigen presentation processes. Clin Exp Immunol 1992; 89:52–57.
98. Ockenhouse CF, Tegoshi T, Maeno Y, et al. Human vascular endothelial cell adhesion receptors for *Plasmodium falciparum*-infected erythrocytes: roles for endothelial leukocyte adhesion molecule-1 and vascular cell adhesion molecule-1. J Exp Med 1992; 176:1183–1189.
99. Sasseville VG, Newman A, Lackner AA, Smith MO, Lausen NCG, Beall D, Ringler DJ. Elevated vascular cell adhesion molecule-1 in AIDS encephalitis induced by simian immunodeficiency virus. Am J Pathol 1992; 141:1021–1030.
100. Carlos TM, Gordon D, Fishbein D, et al. Vascular cell adhesion molecule-1 is induced on endothelium during acute rejection in human cardiac allografts. J Heart Lung Transplant 1992; 11:1103–1109.
101. Bacchi CE, Marsh CL, Perkins JD, et al. Expression of vascular cell adhesion molecule (VCAM-1) in liver and pancreas allograft rejection. Am J Pathol 1993; 142:579–591.
102. Steinhoff G, Behrend M, Schrader B, Duijvestijin AM, Wonigeit K. Expression patterns of leukocyte adhesion ligand molecules on human liver endothelia. Am J Pathol 1993; 142:481–488.
103. Brockmeyer C, Ulbrecht M, Schendel DJ, et al. Distribution of cell adhesion molecules (ICAM-1, VCAM-1, ELAM-1) in renal tissue during allograft rejection. Transplantation 1993; 55:610–615.
104. Mampaso F, Sanchez-Madrid F, Marcen R, et al. Expression of adhesion molecules in allograft renal dysfunction. Transplantation 1993; 56:687–691.
105. Fuggle SV, Sanderson JB, Gray DWR, Richardson A, Morris PJ. Variation in expression of endothelial adhesion molecules in pretransplant and transplanted kidneys—correlation with intragraft events. Transplantation 1993; 55:117–123.
106. Koskinen PK. The association of the induction of vascular cell adhesion molecule-1 with cytomegalovirus antgenemia in human hart allograft. Transplantation 1993; 56:1103–1108.
107. Lebranchu Y, Al najjar A, Kapahi P, et al. The association of increased soluble VCAM-1 levels with CMV disease in human kidney allograft recipients. Transplant Proc 1995; 27:960.
108. Lyall F, Greer IA, Boswell F, Macara LM, Walker JJ, Kingdom JC. The cell adhesion molecule, VCAM-1, is selectively elevated in serum in pre-eclampsia: does this indicate the mechanism of leukocyte activation? Br J Obstet Gynaecol 1994; 101:485–487.
109. Adams DH. Endothelial activation and circulating VCAMS in alcoholic liver disease. Hepatology 1994; 19:588–594.

110. Newman W, Dawson-Beall L, Carson CW, et al. Soluble E-selectin is found in the supernatants of activated endothelial cells and is elevated in the serum of patients with septic shock. J Immunol 1993; 150:644–654.
111. Carson C, Beall L, Hunder G, Johnson C, Newman W. Serum ELAM-1 is increased in vasculitis, scleroderma, and systemic lupus erythematosus. J Rheumatol 1993; 20:809–814.
112. Kim D, Lee K. Serum soluble E-selectin levels in Kawasaki disease. Scand J Rheumatol 1994; 23:283–286.
113. Hartung H, Reiners K, Michels M, et al. Serum levels of soluble E-selectin (Elam-1) in immune-mediated neuropathies. Neurology 1994; 44:1153–1158.
114. Oka N, Akiguchi I, Kawasaki T, Ohnishi K, Kimura J. Elevated serum levels of endothelial leucocyte adhesion molecules in Guillain-Barré syndrome and chronic inflammatory demyelinating polyneuropathy. Ann Neurol 1994; 35: 621–624.
115. Cowley HC, Heney D, Gearing AJH, Hemingway I, Webster NR. Increased circulating adhesion molecule concentrations in patients with systemic inflammatory response syndrome: a prospective cohort study. Crit Care Med 1994; 22:651–657.
116. Ferran C, Peuchmaur M, Desruennes M, et al. Implications of the novo ELAM-1 and VCAM-1 expression in human cardiac allograft rejection. Transplantation 1993; 55:605–609.
117. Morgan JDT, Lycett A, Horsburgh, Nicholson ML, Veitch PS, Bell PRF. The importance of E-selectin as a marker for renal transplant rejection. Transplant Immunol 1994; 2:326–330.
118. Weston SD, Lycett AE, Edwards C, et al. Lack of correlation of soluble E-selectin level with renal transplant rejection. Transplant Immunol 1995; 3: 50–54.
119. Blann A, Tse W, Maxwell S, Waite M. Increased levels of the soluble adhesion molecule E-selectin in essential hypertension. J Hypertens 1994; 12:925–928.
120. Blann A, McCollum C. Circulating endothelial cell-leukocyte adhesion molecules in atherosclerosis. Thromb Haemostas 1994; 72:151–154.
121. Mc Ever RP. Leukocyte interactions mediated by selectins. Thromb Haemostas 1991; 66:80–88.
122. Johnston GI, Cook RG, Mc Ever RP. Cloning of GMP-140, a granule membrane protein of platelets and endothelium: sequence similarity to proteins involved in cell adhesion and inflammation. Cell 1989; 56:1033–1044.
123. Dunlop L, Skinner M, Bendall L, et al. Characterization of GMP-140 (P-selectin) as a circulating plasma protein. J Exp Med 1992; 175:1147–1150.
124. Ushiyama S, Laue T, Moore K, Erickson H, Mcevert R. Structural and functional characterization of monomeric soluble P-selectin and comparison with membrane P-selectin. J Biol Chem 1993; 268:15229–15237.
125. Katayama M, Handa M, Araki Y, et al. Soluble P-selectin is present in normal circulation and its plasma level is elevated in patients with thrombotic thrombocytopenic purpura and haemolytic uraemic syndrome. Br J Haematol 1993; 84:702–710.

126. Facer C, Theodoridou A. Elevated plasma levels of p-selectin (GMP-140/CD62p) in patients with *Plasmodium falciparum* malaria. Microbiol Immunol 1994; 38:727–731.
127. Takeda I, Kaise S, Nishimaki T, Kasukawa R. Soluble P-selectin in the plasma of patients with connective tissue disease. Int Arch Allergy Immunol 1994; 105:128–134.
128. Spertini O, Luscinskas FW, Kansas GS, et al. Leukocyte adhesion molecule-1 (LAM-1, L-selectin) interacts with an inducible endothelial cell ligand to support leukocyte adhesion and transmigration. J Immunol 1991; 147:2565–2573.
129. Kishimoto TK, Jutila MA, Berg EL, Butcher EC. Neutrophil Mac-1 and MEL-14 adhesion proteins inversely regulated by chemotactic factors. Science 1989; 245:1238–1241.
130. Schleiffenbaum B, Spertini O, Tedder TF. Soluble L-selectin is found in human plasma at high levels and retains functional activity. J Cell Biol 1992; 119:229–238.
131. Stucki A, Cordey A, Monai N, Deflaugergues J, Schapira H, Spertini O. Cleaved L-selectin concentrations in meningeal leukaemia. Lancet 1995; 345: 286–289.
132. Humbria A, Diaz-Gonzalez F, Campanero MR, et al. Expression of L-selectin, CD43 and CD44 in synovial fluid neutrophils from patients with inflammatory joint disease. Evidence for a soluble form of L-selectin in synovial fluid. Arthritis Rheum 1994; 37:342–348.
133. Zetterberg E, Richter J. Correlation between serum level of soluble L-selectin and leukocyte count in chronic myeloid and lymphocytic leukemia and during bone marrow transplantation. Eur J Haematol 1993; 51:113–119.
134. Donnelly S, Haslett C, Dransfield I, et al. Role of selectins in development of adult respiratory distress syndrome. Lancet 1994; 344:215–218.
135. Von Asmuth EJU, Smeets EF, Ginsel LA, Onderwater JJM, Leeuwenberg JFM, Buurman WA. Evidence for endocytosis of E-selectin in human endothelial cells. Eur J Immunol 1992; 22:2519–2526.
136. Ling NR. Pitfalls in the measurement of soluble forms of cell surface receptors. Clin Exp Immunol 1993; 93:139–141.
137. Galea P, Thibault G, Lacord M, Bardos P, Lebranchu Y. Il-4 but not tumor necrosis factor a increases endothelial cell adhesiveness for lymphocytes by activating a cAMP-dependant pathway. J Immunol 1993; 151:588–596.
138. Cosimi AB, Conti D, Delmonico FL, et al. In vivo effects of monoclonal antibody to ICAM-1 (CD54) in nonhuman primates with renal allografts. J Immunol 1990; 144:4604–4612.
139. Isobe M, Yagita H, Okumura K, Ihara A. Specific acceptance of cardiac allograft after treatment with antibodies to ICAM-1 and LFA-1. Science 1992; 255:1125–1127.
140. Orosz CG, Ohye RG, Pelletier RP, et al. Treatment with anti-vascular cell adhesion molecule 1 monoclonal antibody induces long-term murine cardiac allograft acceptance. Transplantation 1993; 56:453–460.
141. Issekutz TB. Inhibition of lymphocyte endothelial adhesion and in vivo lym-

phocyte migration to cutaneous inflammation by TA-3, a new monoclonal antibody to rat LFA-1. J Immunol 1992; 149:3394–3402.
142. Nishikawa K, Guo YJ, Miyasaka M, et al. Antibodies to intercellular adhesion molecule 1/lymphocyte function-associated antigen 1 prevent crescent formation in rat autoimmune glomerulonephritis. J Exp Med 1993; 177:667–677.
143. Kawasaki K, Yaoita E, Yamamoto T, Tamatani T, Miyasaka M, Kihara I. Antibodies against intecellular adhesion molecule-1 and lymphocyte function-associated antigen-1 prevent glomerular injury in rat experimental crescentic glomerulonephritis. J Immunol 1993; 150:1074–1083.
144. Gamble JR, Skinner MP, Berendt MC, Vadas MA. Prevention of activated neutrophil adhesion to endothelium by soluble adhesion protein GMP-140. Science 1990; 249:414–416.
145. Wong CS, Gamble JR, Skinner MP, Lucas CM, Berndt MC, Vadas MA. Adhesion protein GMP-140 superoxide anion release by human neutrophils. Proc Natl Acad Sci USA 1991; 88.
146. May GL, Dunlop LC, Sztelma K, Berndt MC, Sorrel TC. GMP-140 (P-selectin) inhibits human neutrophil activation by liopolysaccharide: analysis by proton magnetic resonance spectroscopy. Biochem Biophys Res Commun 1992; 183:1062–1069.
147. Schleiffenbaum B, Spertini O, Tedder TF. Soluble L-selectin is present in human plasma at high levels and retains functional activity. J Cell Biol 1992; 119:229–238.
148. Becker JC, Dummer R, Hartmann AA, Burg G, Schmidt RE. Shedding of ICAM-1 from human melanoma cell lines induced by IFN-gamma and tumor necrosis factor-alpha: functional consequences on cell-mediated cytotoxicity. J Immunol 1991; 147:4398–4401.
149. Lo SK, Lee S, Ramos RA, Rosa M, Chi-Rosso G, Wright S. Endothelial-leukocyte adhesion molecule-1 stimulates the adhesive activity of leukocyte integrin CR3 (CD11b/CD18, Mac-1, alpha mb2) on neutrophils. J Exp Med 1991; 173:1493–1500.
150. Postigo AA, Garcinavicuna R, Diaz-Gonzalez F, et al. Increased binding of synovial T lymphocytes from rheumatoid arthritis to endothelial leukocyte adhesion molecule-1 (ELAM-1) and vascular adhesion molecule-1 (VCAM-1). J Clin Invest 1992; 89:1445–1452.

4

Circulating Adhesion Receptors in Health and Disease

Oliver Spertini
Division of Hematology, University of Lausanne, Lausanne, Switzerland

I. INTRODUCTION

The constant circulation of leukocytes throughout the body is a critical feature of the immune system. Adhesive interactions between circulating leukocytes and vascular endothelium regulate leukocyte extravasation from blood into tissues (1–6). Selectins and their various ligands initiate leukocyte attachment to activated endothelium whereas integrin and immunoglobulinlike adhesion molecules have an important role in subsequent steps of leukocyte migration. In vivo and in vitro experiments have indicated that selectins mediate leukocyte rolling along endothelium of postcapillary venules at sites of inflammation. During this reversible interaction, leukocytes are exposed to local interleukins (IL), cytokines, or chemokines. These soluble factors rapidly and transiently modulate the affinity of activation-dependent adhesion receptors resulting in stable adhesion (6–17). This firm adhesion step, resistant to physiologic shear stress, is mainly mediated by the interaction of the integrins $\alpha L\beta 2$, $\alpha M\beta 2$ and $\alpha 4\beta 1$, $\alpha 7\beta 1$ with their endothelial ligands ICAM-1, ICAM-2, VCAM-1, and MadCAM (6). Finally, CD31 (PECAM-1) was shown to play a major role in the regulation of leukocyte diapedesis between endothelial cells (18–20) (Table 1; Fig. 1).

Selectins interact with glycoconjugates expressed on endothelium or leukocytes. Most of them have a lactosamine backbone, carry sialylated, sul-

Table 1 Adhesion Molecules Involved in Leukocyte-Endothelial Interactions

Molecule	Expression[a]	Ligands
Selectins		
L-selectin (CD62L)	Lymphocyte, monocyte, neutrophil, eosinophil, basophil, hematopoietic progenitors	CD34, GlyCAM-1, 200 kD glycoprotein, heparins, sLe^x various fucosylated and sulfate carbohydrates and other, yet unidentified glycoconjugates
P-selectin (CD62P)	Platelet and endothelium	PSGL-1, sLe^x and other glycoconjugates
E-selectin (CD62E)	Endothelium	ESL-1, sLe^x and other glycoconjugates
Integrins		
CD11a/CD18 ($\alpha L\beta 2$)	Most lymphocytes, monocyte, neutrophil	ICAM-1, ICAM-2, ICAM-3
CD11b/CD18 ($\alpha M\beta 2$)	Monocyte, neutrophil	ICAM-1, iC3b, fibrinogen, bacterial LPS
CD11c/CD18 ($\alpha X\beta 2$)	Monocyte, neutrophil	iC3b, fibrinogen
CD49d/Cd29 ($\alpha 4\beta 1$)	Lymphocyte, monocyte, eosinophil	VCAM-1, fibronectin
Cd49d/ CD-(LPAM-1) ($\alpha 4\beta 7$)	B and T lymphocyte subpopulation	VCAM-1, MadCAM-1, fibronectin
Immunoglobulinlike proteins		
ICAM-1 (CD54)	Lymphocyte, endothelium, macrophage	CD11a, CD11b/CD18
ICAM-2 (CD102)	Lymphocyte, endothelium	CD11a
ICAM-3 (CD50)	Lymphocyte, neutrophil	CD11a
VCAM-1	Endothelium	CD49d/CD29, LPAM-1
PECAM-1 (CD31)	Lymphocyte, platelet, endothelium, neutrophil	PECAM-1 (CD31)

[a]ICAM-1 and VCAM-1 are also expressed on fibroblast, dendritic cell. Epithelial cell can also be ICAM-1^+. VCAM-1 is also present on smooth muscle cells, synovium, kidney proximal tubule, and cytokine-activated neural cell.

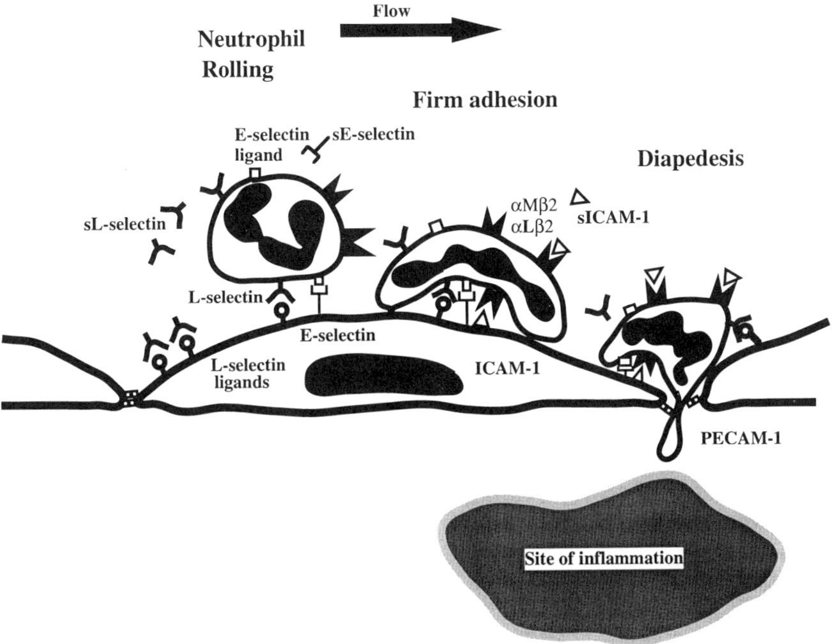

Figure 1 Soluble adhesion receptors may participate in the sequential regulation of neutrophil migration into tissues.

fated, and/or fucosylated sequences normally found at the termini of N-linked or O-linked oligosaccharides, or on glycosphingolipids (21). Some carbohydrate structures such as the tetrasaccharides sialyl Lewis^x (sLe^x) are ligands for the three selectins (22), whereas other glycoproteins such as CD34 could more specifically interact with L-, P-, or E-selectin (23–26). Two ligands for L-selectin (e.g., GlycCAM-1 and CD34) were identified on murine peripheral lymph node high endothelial venules (PLN-HEV) (23,27,28). These glycoproteins are heavily sialylated, sulfated, and fucosylated mucinlike proteins (23,27–29). Interestingly, GlyCAM-1 is not anchored into the endothelial membrane and may serve to modulate lymphocyte attachment to PLN-HEV, whereas CD34 has a transmembrane and a cytoplasmic domain and functions as an addressin for L-selectin on PLN-HEV. A third ligand of 200 kDa, which has not yet been cloned, was recently identified on PLN-HEV (29). The presence of an inducible ligand for L-selectin was previously reported on cytokine-activated human umbilical vein endothelial cells (HUVEC) (30–32). However, the structure of these ligands for L-selectin have yet not been identified. A ligand of P-selectin was identified by expression cloning. The predicted amino acid sequence revealed a novel mucinlike molecule that was called PSGL-1 (24,25). More recently, the amino acid sequence analysis of a ligand for E-selectin demonstrated that it presents a high degree of homology with the receptor for fibroblast growth factor (33).

Additional information on the understanding of the extraordinary complexity of the cell adhesion cascade has been obtained by the recent demonstration in blood of the presence of soluble forms of selectins and members of the immunoglobulin superfamily. Soluble receptors that retain functional activity may modulate leukocyte adhesion and function at sites of inflammation. Interestingly, increased concentrations of these molecules have been reported in various diseases. We will successively describe in this chapter the structure and function of soluble L-, P-, E-selectin, ICAM-1, and VCAM-1 and discuss the possibility of using them as markers of disease activity. Other soluble adhesion molecules have been reported, including CD31, CD43, CD44, and CD50 (ICAM-3) (34,35). However, their discussion is out of the scope of this review.

II. SOLUBLE L-SELECTIN (sL-SELECTIN)

L-selectin contains an amino-terminal lectin domain, an epidermallike growth factor domain, two short consensus repeats similar to those found in complement regulatory proteins, a membrane-spanning region, and a short cytoplasmic tail (36–40). It is expressed at the surface of most circulating normal and some malignant leukocytes (37,41–45). A characteristic feature of L-selectin is its rapid cleavage from the leukocyte surface after exposure to phorbol 12-myristate 13-acetate (PMA) (42–44,46–50). A large fragment of L-selectin (Mr 69,000) can be immunoprecipitated from the supernatant of activated lymphocytes or neutrophils, demonstrating that the major portion of the extracellular domain of the molecule is released from the cell surface (44,47). Although the mechanism of shedding is not known, it may result from the specific activation of a membrane-bound protease (43,51). The ability of all L-selectin$^+$ leukocytes and of most cell lines transfected with L-selectin cDNA to shed L-selectin following activation by phorbol ester suggests that the enzyme involved in L-selectin cleavage is a ubiquitous protease (44).

L-selectin shedding is dependent on protein kinase C since inhibitors of this kinase abrogate L-selectin release after leukocyte activation with PMA (44,48). Stimuli other than PMA can regulate L-selectin expression. Thus, neutrophil activation by GM-CSF, TNF-α, bacterial lipopolysaccharides, interleukin-8, or various chemoattractants induce L-selectin shedding, whereas these agents have no effect on L-selectin expressed by lymphocytes

(43). The decrease in L-selectin expression on neutrophils is accompanied by a severalfold increase in CD11b surface expression (43,47,49). The inversely regulated expression of L-selectin and CD11b/CD18 complex is consistent with the involvement of L-selectin in the initial adhesion of leukocytes on activated endothelium and suggests an early model of leukocyte-endothelial interaction (47). The time course of L-selectin shedding is much slower on activated T lymphocytes than on neutrophils (52–54). Thus, L-selectin expression decreases several hours to several days after T lymphocyte activation, whereas it disappears within minutes of neutrophil exposure to PMA, TNF-α, or other activating stimuli.

A fragment of 62,000 is released from lymphocytes whereas fragments with M_rs ranging from 75,000 to 100,000 are shed from activated neutrophils (55). Immunoprecipitation experiments suggested that the cleaved form of L-selectin is $\approx$5 kDa smaller than transmembrane molecule expressed at the cell surface (44). More recently, a peptide of $\approx$6 kDa was precipitated with a polyclonal antibody directed against the cytoplasmic domain of the molecule (51). The size of this molecule was consistent with the predicted molecular mass of the transmembrane and intracytoplasmic domain of the molecule. Although the protease involved in the shedding of L-selectin has not yet been identified, the membrane proximal cleavage site of the molecule was recently located between Lys 321 and Ser 322 (51). The cleavage of L-selectin is not influenced by the presence of inhibitors of serine proteases, metalloproteases, aspartic proteases, and thiol proteases (51,56,57).

The events that induce the cleavage of L-selectin have not been identified. The binding of L-selectin ligand to the lectin domain of the molecule could induce a conformational change of the molecule and expose a cleavage site for the involved protease. This possibility is consistent with the shedding of L-selectin induced by cross-linking with monoclonal antibodies (58). Other mechanisms may be involved in the regulation of L-selectin shedding. The activation of leukocytes by various stimuli may generate cytoplasmic signals that activate the cytoplasmic domain of the molecule, inducing conformational change of the receptor and subsequently the cleavage of the receptor.

We recently described a sandwich enzyme-linked immunosorbent assay designed to measure L-selectin shed from leukocytes in vivo. Plasma sL-selectin levels ranging from 0.65 to 3.6 μg/ml (mean $\pm$ 1SD, 2.1 $\pm$ 0.7 μg/ml) were observed among 100 healthy blood donors (45,55,59). Information on the structure of sL-selectin was obtained by epitope mapping, sL-selectin exhibited the sites identified on the lectin domain by the anti-LAM1-1, -4, -6, -7, -10, -11 mAbs. The EGF-like domain of the cleaved molecule contained the epitopes defined by the anti-LAM1-5 and -15 mAbs whereas the binding site of the anti-LAM1-1 mAb was no more expressed suggesting that the cleavage of L-selectin induces a change in L-selectin conformation (45,59). At physiological concentrations (≈ 1.5 μg/ml), sL-selectin partially inhibited lymphocyte attachment to activated human endothelium. It was proposed that sL-selectin might serve as a biological adhesion buffer system to inhibit leukocyte adhesion at vascular sites where low levels of L-selectin ligands are expressed (59). Since the number of sL-selectin molecules is much higher in blood than at leukocyte surface, it is likely that the affinity of the cleaved form of L-selectin for its endothelial ligand is lower than the affinity of the transmembrane molecule. In addition, it is possible that the rapid increase in L-selectin affinity for its ligand that follows leukocyte activation or the local redistribution of L-selectin at cell surface may allow a competitive displacement of sL-selectin bound to endothelial cell surface (14).

Much higher levels ($>$2SD above the mean) of functional sL-selectin were measured in 60% of patients with acute leukemia (sL-selectin$^+$ leukemia) (45). Plasma sL-selectin levels ranged from 0.7 to 68 μg/ml (median 4.1 μg/ml) in 93 patients with acute myelogenous leukemia (AML) and from 0.1 to 98 μg/ml in 58 patients with acute lymphoblastic leukemia (ALL). In 41 patients with AML, sL-selectin levels was related to the clinical characteristics at diagnosis. Higher sL-selectin levels were detected in patients with an enlarged spleen compared with patients without splenomegaly (45).

The effect of sL-selectin on blast cell attachment to cytokine-activated endothelium was evaluated using an in vitro attachment assay (45). sL-selectin at concentrations observed in acute leukemia completely inhibited the L-selectin-mediated adhesion of blast cells to activated endothelium. The capacity of sL-selectin to inhibit the attachment of leukemic cells to vascular endothelium suggested that this circulating adhesion receptor may contribute in vivo to maintain leukemic cells in the bloodstream.

The clinical course of patients with a sL-selectin$^+$ acute leukemia correlated perfectly with plasma sL-selectin levels. Normal levels were observed in 16 of 16 patients in complete remission whereas increased levels were detected in 8 of 8 patients with a relapsing leukemia or a therapy-resistant disease. The absence of normalization of sL-selectin after treatment was associated with a failure to achieve complete remission. It must be emphasized that this marker is not specific of acute leukemia, as increased plasma sL-selectin levels were also detected in chronic leukemia (60) and other malignant hemopathies (O. Spertini, unpublished observations).

We recently reported that cerebrospinal fluid (CSF) sL-selectin levels could be a useful marker of meningeal leukemia (61). The involvement of the central nervous system is a frequent complication of ALL and acute monoblastic leukemia, which requires intensified therapy. The identification of patients with meningeal leukemia by cytological examination of the CSF is often difficult when the blast cell numbers are low and the morphology of leukemic cells is atypical. Since the detection of cellular receptors shed from the surface of blast cells could facilitate the diagnosis of meningeal leukemia, we investigated whether measurements of sL-selectin in CSF could improve the ability to detect such involvement. CSF sL-selectin ranged from 34 to 150 ng/ml (median 60 ng/ml) in patients with sL-selectin$^+$ meningeal leukemia and were significantly higher than in 20 patients with acute sL-selectin$^+$ leukemia without meningeal involvement (1 to 39 ng/ml, median 12 ng/ml) or in 88 control patients (0–37 ng/ml, median 14 ng/ml).

Serial measurements of sL-selectin indicated that sL-selectin levels in CSF correlate with the clinical course of patients with meningeal leukemia. Increased sL-selectin levels were detected in all four patients with meningeal leukemia refractory to treatment whereas normal levels were present in all nine patients in remission after intrathecal therapy. In two patients, increased sL-selectin levels were detected several weeks before blast cells could be identified by cytological examination of CSF. These results suggested that serial assay of CSF sL-selectin may improve the management of patients with sL-selectin$^+$ leukemia and meningeal leukemia.

sL-selectin is probably a more helpful marker of malignant hemopathies than of inflammatory diseases. Much higher sL-selectin plasma levels are detected in leukemia and lymphoma than in inflammation. Smaller increases in plasma sL-selectin were reported in patients with various inflammatory diseases and in relatives from patients with diabetes (59,62) (O. Spertini, unpublished observations). Since the number of L-selectin molecules expressed at the cell surface is much lower than the number of circulating sL-selectin molecules present in the same volume of blood, the shedding of L-selectin from large numbers of leukocyte is required to induce a noticeable increase in sL-selectin plasma concentration. Accordingly, it is not surprising that only weak increases in sL-selectin are observed in some acute inflammatory diseases. Further studies will be necessary to determine whether sL-selectin could facilitate the clinical evaluation of patients with inflammatory diseases. Interestingly, a decrease in sL-selectin plasma level was recently reported in patients who develop an acute respiratory distress syndrome (ARDS) (63). The decrease in sL-selectin plasma concentration that precedes the progression to ARDS was attributed to a sequestration of the circulating adhesion receptor by widespread binding to activated endothelium in microvascular beds. It could be related to a marked activation of pulmonary vascular endothelium, which is a central feature in the pathogenesis of ARDS. These results suggested that the measurement of sL-selectin could be helpful in identifying patients at risk of developing ARDS (63).

III. SOLUBLE P-SELECTIN (sP-SELECTIN)

P-selectin (CD62P) was initially identified in platelets and was called PADGEM (platelet activation-dependent granule-external membrane protein) (64) or GMP-140 (granule membrane protein-140) (65). This 140-kDa glycoprotein is present in cytoplasmic α-granules and is rapidly translocated to platelet surface after platelet activation with various stimuli like thrombin, histamine, or activated complement C5b-9 (66,67). Subsequently, P-selectin was demonstrated in Weibel-Palade bodies of endothelial cells and to be rapidly expressed at the cell surface after cell activation (66,68). More recently, in vitro and in vivo experiments demonstrated that P-selectin mediates neutrophil and lymphocyte rolling at endothelial cell surface (11,25, 69–71).

Membrane-bound P-selectin expresses a lectin domain, an EGF-like domain, nine short consensus repeats, a transmembrane domain, and a cytoplasmic domain (72). The genomic structure of this protein suggested that two variant forms of P-selectin could arise by alternative splicing of mRNA (73). In one variant, the seventh short consensus repeat is deleted; the other variant is devoid of the transmembrane domain. This putative soluble isoform of P-selectin is translated from splice mRNA lacking the 14th exon that encodes the transmembrane domain of the intact molecule. Additional studies demonstrated that this predicted soluble isoform of P-selectin is actually synthesized in vivo and is detectable as a circulating molecule (74–76). Thus, two distinct isoforms of 135 kDa were purified from human plasma. Western blot analysis demonstrated a strong reactivity of plasma P-selectin with a polyclonal antiserum raised against a peptide corresponding to residues 762–774 located near the amino terminal portion of the cytoplasmic domain, including a region encoded by the entire exon 15. On the other hand, only a weak reactivity was observed with a polyclonal antiserum raised against the amino acid 741–760 that correspond to the carboxy-terminal halves of both the putative transmembrane domain and a region encoded by exon 14. In contrast, platelet P-selectin reacted similarly with both antibodies (74). Chong et al. (76) and Dunlop et al. (75) reported that platelet membrane P-selectin differed from plasma sP-selectin in that its molecular mass is ≈ 3 kDa lower under reducing conditions. This difference of size could correspond to the alternatively spliced sequence of the 40 amino acids of exon 14.

The spliced form of P-selectin was detectable in washed platelets, suggesting that this isoform is synthesized in megakaryocytes and/or platelets and then stored in the cells. sP-selectin is the first adhesion molecule that has been readily isolated from plasma as a soluble form synthesized from alternatively spliced mRNA in vivo. Although the major portion of plasma sP-selectin is the alternatively spliced form lacking the transmembrane domain, it is not ruled out that a smaller portion could also result from proteolytic cleavage of the membrane-bound form (74). In addition, membrane-bound forms associated to circulating microparticles were also demonstrated in plasma (74). However, it remains to be elucidated whether this soluble form of P-selectin is an in vivo product or whether it is derived from platelets present as a contaminant in the material during the purification steps.

Plasma sP-selectin levels were measured in 37 healthy controls and found to range between 0.009 and 0.53 μg/ml (median ≈0.2 μg/ml) (76–78). Plasma sP-selectin levels were not sensitive to the collection procedure. Thus, customarily, measures taken to prevent platelet in vitro activation and sample contamination by platelets/platelet dust did not influence the results of sP-selectin assay. The addition of inhibitors of platelet activation, the maintenance of the sample temperature between 0°C and 4°C, and the ultracentrifugation of samples were not required. In contrast, the omission of one of these measures resulted in increased levels of β-thromboglobulin, another marker of platelet activation (76,78).

sP-selectin was measured in various disease conditions associated with platelet activation and/or endothelial cell damage such as disseminated intravascular coagulation, thrombotic thrombocytopenic purpura, hemolytic uremic syndrome, and heparin-induced thrombocytopenia. Substantially higher plasma sP-selectin levels were found in these various diseases than in controls (76,78). In addition, increased sP-selectin levels were detected in patients with acute myocardial infarction or cerebral thrombosis (79). These results suggest that sP-selectin could serve as a marker of thrombotic diseases, particularly in those with platelet/endothelial cell activation (76,78,79). Further evidence that this soluble receptor is a marker of platelet and endothelial activation was provided by the study of patients with vasculitis. Takeda et al. reported increased plasma sP-selectin levels in patients with systemic lupus erythematosus (303 ± 120 ng/ml), mixed connective disease (1048 ± 694 ng/ml) or rheumatoid arthritis (844 ± 785 ng/ml). sP-selectin was significantly higher in patients with vascular or renal involvement (802 ± 717 ng/ml) than in uninvolved patients (480 ± 503 ng/ml) (77), suggesting that the increase in plasma sP-selectin levels observed in vasculitis could reflect vascular damage and the local platelet activation.

IV. SOLUBLE E-SELECTIN (sE-SELECTIN)

The expression of E-selectin is induced on activated endothelial cells by various cytokines such as TNF-α, IL-1, or endotoxin. Its expression requires de novo mRNA and protein synthesis. CD62E was initially isolated from human umbilical vein endothelium as a protein of 115 kDa (80). Subsequent in vitro experiments demonstrated that E-selectin was involved, like P- and L-selectin, in mediating neutrophil rolling along activated endothelium (81,82). In addition, CD62E was reported to interact through an as yet uncharacterized ligand with CD4 positive memory T cells (83). E-selectin interacts preferentially with T lymphocytes that home into the skin (84). Several in vivo animal models of antiadhesion therapy indicated that E-selectin could have a major role in the recruitment of neutrophils into inflamed tissues such as pulmonary tissues injured by immune complexes (85) or in antigen-induced acute airway inflammation and late-phase airway obstruction in monkeys (86).

In vitro studies have indicated that E-selectin is maximally expressed by human endothelial cells after 4 to 6 hours of activation with IL-1. Subsequently, E-selectin expression declines and is no more detectable by 24 to 48 hours (80). Using a quantitative ELISA, the assay of E-selectin confirmed that the in vitro expression of this adhesion receptor by IL-1-activated endothelium peaks at 4 hours (87). At that time, only a very small amount of the sE-selectin was detectable in the supernatant of the activated cells. By 24 hours, the content of E-selectin in the cell fraction progressively decreased whereas its soluble form increased in the cell supernatant. To further demonstrate that E-selectin is shed from the cell surface, 4-hour-activated endothelial cells were iodinated on the cell surface and cultured for 24 hours. The immunoprecipitation of sE-selectin from the cell supernatant showed that its molecular mass was ≈ 5 kDa smaller than the transmembrane cellular form (87).

Mean plasma sE-selectin levels detected in normal blood ranged from 1 to 50 ng/ml (87–93). The absence of reactivity of plasma sE-selectin with a polyclonal antibody directed against the cytoplasmic domain of this molecule confirmed that sE-selectin is generated by the proteolytic cleavage of the transmembrane molecule (87).

The restriction of E-selectin expression to activated endothelium could make this adhesion receptor an attractive marker of endothelial cell activation. The marked elevations ($\sim$20-fold over the normal range) of sE-selectin levels in plasma from patients with septic shock is an interesting example in which this marker could serve as a sensitive parameter of widespread endothelial cell injury. In contrast, sE-selectin was in the control range in patients with uncomplicated bacteremia (87). The measurement of sE-selectin in various clinical disorders associated with vascular injury further confirmed that sE-selectin is an early marker of endothelial cell activation (88,94,95).

Cerebral malaria is one of the most severe complications of *Plasmodium falciparum* infection. Infected erythrocytes and *Plasmodium falciparum* can bind to adhesion molecules expressed by endothelium such as VCAM-1, ICAM-1, and E-selectin (96,97). The expression of these adhesion molecules as well as the elevated concentrations of cytokines such as TNF-α (98) play an important role in the pathogenesis of cerebral malaria. TNF-α can upregulate the expression of most human vascular adhesion receptors and thus increase the number of parasites sequestered in the brain (99). Increased levels of sE-selectin, soluble VCAM-1, and soluble ICAM-1 were detected in patients with *Plasmodium falciparum* infections (88,100,101). The highest levels of sICAM-1, sVCAM-1, and sE-selectin were found in patients with severe cerebral involvement (88). The elevated levels of these adhesion receptors probably reflect the extensive endothelial inflammation found in *Plasmodium falciparium* infections complicated with cerebral manifestations. Patients with *Plasmodium vivax* infections have also increased plasma levels of sVCAM-1, sICAM-1, and sE-selectin. However *Plasmodium vivax* malaria is not complicated by cerebral malaria, suggesting that the expression of adhesion receptors is not sufficient for the development of cerebral complications and that the sequestration of parasite-infected erythrocytes at sites of activated endothelium is required for the development of severe disease.

Increased expression of ICAM-1 and E-selectin have been observed on vascular endothelium and of ICAM-1 on bronchial epithelium at sites of allergic reactions (102–106). In animal models, the injection of mAb against functional epitopes of adhesion molecules attenuated the airway hyperresponsiveness and the infiltration of the mucosa by inflammatory cells (104–106), suggesting that adhesion receptors may play a critical role in the pathogenesis of allergic inflammation. Serum sICAM-1 and sE-selectin were measured in patients with bronchial asthma during attacks and in stable condition (95,107,108). The highest serum sE-selectin and sICAM-1 levels were detected during bronchial asthma attacks (95,107), but the levels of these soluble receptors did not differ between atopic and nonatopic patients. Elevated concentrations of sICAM-1 and sE-selectin were detected for at least 28 days after an asthma attack, most likely reflecting the persistence of inflammatory activity in the airways (107). Of interest, higher serum TNF-α concentrations were also detected during asthma attacks. However, although the expression of ICAM-1 and E-selectin is upregulated by TNF-α, no correlation was found between serum levels of this cytokine and sE-selectin or sICAM-1, suggesting that other mechanisms contribute to the induction of these adhesion receptors (95).

Additional studies brought further evidence that sE-selectin is a marker of endothelial cell activation (93,109,110). sE-selectin concentrations were determined in synovial fluid from patients with inflammatory arthritis (90). One of the earliest lesions of rheumatoid arthritis is the proliferation of small synovial membrane blood vessels with perivascular collection of inflammatory cells. The expression of E-selectin and VCAM-1 was demonstrated in the rheumatoid synovium (111). Higher sE-selectin levels were detected in the synovial fluid of patients with rheumatoid or psoriasic arthritis compared with patients with osteoarthritis, gout, or calcium pyrophosphate dihydrate crystal deposition disease (90). These results suggest that the assay of synovial fluid sE-selectin may help to distinguish inflammatory from noninflammatory synovial fluid (90).

sE-selectin does not seem to be a good marker of vasculitis activity. Increased levels of sE-selectin were observed of patients with various vasculitis (112), but no correlation was found with disease activity while there was only a weak correlation with the degree of organ involvement. Other authors have not observed increased levels of sE-selectin in patients with active systemic lupus erythematosus or other vasculitides (113,114), whereas they did find that changes in sVCAM-1 levels were parallel to disease activity (113). Interestingly, elevated levels of sE-selectin were reported in essential hypertension (93). The absence of a correlation between sE-selectin levels and Von Willebrand factor, a marker of endothelial cell injury, suggests that elevated levels of sE-selectin in hypertension does not reflect endothelial cell damage but rather some form of altered endothelial cell activity. Other studies also reported increased sE-selectin in alcoholic hepatitis (115), after coronary angioplasty (116), and in acute kidney graft rejection (109).

At high concentration, sE-selectin can inhibit neutrophil adhesion (117) or activate CD11b integrin (118). Although sE-selectin levels detected in serum are too low for these effects, it may be biologically active at sites of inflammation and modulate leukocyte-endothelial interaction, thereby constituting an autoprotection that prevents leukocyte adhesion at sites of weak activation/inflammation.

V. SOLUBLE ICAM-1 (sICAM-1)

ICAM-1 is a 90-kDa molecule composed of five Ig-like extracellular domains, a transmembrane domain, and a short cytoplasmic tail (119). This cytokine-inducible adhesion molecule is expressed by multiple lineages such as tissue macrophages, mitogen-stimulated T lymphocytes, germinal center B cells, dendritic cells, endothelial cells, thymic epithelial cells, fibroblasts, and various malignant cells. It is involved in mediating granulocyte emigration into tissues, in non-MHC restricted lymphocyte-mediated cytotoxicity, and in the development of immunologic responses, and it has been identified as the receptor for rhinoviruses (6,41,117,120–126). Several mediators of inflammation such as IL-1, TNF-α, interferon-γ, and endotoxin are potent inducers of ICAM-1 expression (127,128). The first three Ig-like extracellular domains can interact with the leukocyte integrins $\alpha L\beta 2$ (CD11a/CD18) or $\alpha M\beta 2$ (CD11b/CD18) (6). This interaction not only supports cell-cell binding but can promote transmembrane signaling to the lymphocyte by engaging $\alpha L\beta 2$ (129).

A circulating form of ICAM-1 was described in sera from healthy blood donors (130–132). Serum electrophoresis on gradient gels under nondenaturing conditions indicated that sICAM-1 circulates in at least three molecular forms of about 240 kDa, 430 kDa, and more than 500 kDa (132), whereas a single molecule of 80 kDa was detected under denaturing conditions (130,132). The various circulating molecular forms may result from the formation of complexes of sICAM-1, either with itself or with other circulating molecules (132). mAb epitope mapping demonstrated that sICAM-1 contains most of the extracellular region of cell-bound ICAM-1 (130), suggesting that sICAM-1 is most likely generated by proteolytic cleavage of cell surface ICAM-1. At the present time, there is no evidence of an alternatively spliced form of ICAM-1 mRNA that would lead to the production of a soluble form of this receptor.

sICAM-1 retains functional activity, as it has the capacity to bind aLβ2 (CD11a) and support the adhesion of lymphoblastic cell lines (130). In addition, it can inhibit the intercellular adhesion dependent on the interaction of ICAM-1 with CD11a. Furthermore, it can prevent non-MHC-restricted cellular cytotoxicity and thus could represent a mechanism by which malignant cells can escape from immune surveillance (131).

sICAM-1 has been detected in the culture supernatants of lymphocytes, endothelial cells, and various malignant cells such as melanoma cells, colon carcinoma cells, and gastric, pancreatic, and ovarian carcinoma cells, suggesting that circulating ICAM can have various cellular origins (130,131, 133–137). Several mechanisms can promote the shedding of this receptor, which makes increased levels of sICAM-1 difficult to interpret.

Increased sICAM-1 levels have been detected in sera from patients with leukocyte adhesion deficiency, vascular diseases, inflammation, cancers, or infections. In the various studies, mean sICAM-1 levels detected in healthy blood donors ranged from 100 to 450 ng/ml (88,113,130,138–141).

In malignant diseases, increased sICAM-1 levels (~2-fold elevation) were reported in sera from patients with ALL or Hodgkin's disease (139, 142). In the latter disease, sICAM-1 was significantly higher in patients with disseminated disease (stage III to IV) and in patients who were symptomatic (B symptoms) (139,142). Interestingly, relapse was significantly more frequent in patients with elevated sICAM-1 levels, suggesting that increased levels could serve as a marker of poor prognosis (142). However, large prospective studies with multivariate analysis are needed to investigate this possibility. The mechanisms involved in the increase of sICAM-1 are complex and could result from an immune response against the malignant cells, a reaction of stromal cells to malignant cells, and/or the tumor burden. The production of cytokines by Reed-Sternberg cells and by cytotoxic lymphocytes may induce ICAM-1 shedding and result in increased sICAM-1 levels (139,142). Increased sICAM-1 levels found in these patients could be one of the mechanisms that lead to immunodeficiency as often found in Hodgkin's disease. Other studies reported increased sICAM-1 in sera from patients with pancreatic, gastric, or gallbladder cancer. High levels of circulating ICAM-1 were detected in patients with liver metastasis and could reflect the increased ICAM-1 expression found on malignant cells and the adjacent stromal cells (136).

Increased serum levels of sICAM-1 have also been reported in malaria (88,96,97,100,101) and in asthma (95,107,108), as discussed above. Elevated serum sICAM-1 levels were also detected in sepsis (143), in the cerebrospinal fluid from patients with inflammatory neurological diseases (91,144), in episodes of graft rejection (145–149), in sera from patients with idiopathic pulmonary fibrosis (150), and in synovial fluid from patients with rheumatoid arthritis (151). sICAM-1 and sVCAM-1 levels were higher in synovial fluid from patients with rheumatoid arthritis than in serum, suggesting that the circulating molecules may derive, at least in part, from the synovium. Further prospective longitudinal studies will be necessary to determine if the detection of elevated sICAM-1 levels reflects the degree of severity and activity of the disease. The detection of the highest levels of sICAM-1 in patients with severe erosive disease and the absence of increase of sICAM-1 in synovial fluid from patients with systemic lupus erythematosus and nonerosive arthritis supports this possibility (151).

In several studies, sICAM-1 or sVCAM-1 levels were measured in samplings obtained from the site of inflammation (61,140,144,146,152–154). In one of them, elevated sICAM-1 levels were found in tracheal aspirates from infants with bronchopulmonary dysplasia (BPD), a progressive inflammatory pulmonary syndrome affecting both the parenchyma and the airways, while normal levels were found in serum (153). Measurements of sICAM-1 at 6–7 and 12–14 days of age suggested that the detection of increased sICAM-1 concentrations in tracheal aspirates could be predictive of the development of BPD (153). This example and the others referenced above illustrate that the measurement of local levels of soluble adhesion receptors may give more information than the assays performed in blood samples.

In vascular diseases, increased sICAM-1 levels were detected in patients with atherosclerosis and peripheral vascular disease or ischemic heart disease (141). In this study, Von Willebrand factor and sICAM-1 levels were correlated, suggesting that sICAM-1 could be an index of endothelial cell activation in patients with advanced atherosclerosis (141).

In vasculitis, increased sICAM-1 and sVCAM-1 levels were measured in patients with active Wegener's granulomatosis (89,155). sICAM-1 levels, like the index of activity of the disease, were predictive of the renal outcome, and higher sICAM-1 levels were detected in hemodialyzed patients compared with patients who did not require this treatment. sICAM-1 and sVCAM-1 levels followed the clinical course of the disease; the concentration of these receptors decreased in patients in clinical remission. Interestingly, in lupus nephritis, only serum sVCAM-1 level was increased; serum sICAM-1 level was in the control range. In addition, in systemic lupus erythematosus, sVCAM-1 levels were not correlated with disease activity (155). These results suggest that the increased levels of soluble adhesion molecules reflect different pathological processes and that sICAM-1 and sVCAM-1 could be helpful markers of disease activity in Wegener's granulomatosis. sICAM-1 seems also to be a helpful marker in Kawasaki disease for determining the severity of vascular damage when the disease is active (156).

VI. SOLUBLE VCAM-1 (sVCAM-1)

VCAM-1 is a member of the immunoglobulin superfamily which is expressed on activated endothelium, macrophage-like cells in normal or inflamed tissues, bone marrow fibroblasts, smooth muscle cells, synovium, kidney proximal tubules, and cytokine-activated neural cells (157–165). It supports the adhesion of lymphocytes, monocytes, basophils, eosinophils, and hematopoietic progenitors by interacting with the integrins $\alpha4\beta1$ or $\alpha4\beta7$ (69,166–169). Cell-bound VCAM-1 is a 105- to 110-kDa protein that is composed of seven immunoglobulin-like domains, a transmembrane region, and a cytoplasmic tail (170–172). Alternate splicing can generate two variant forms of VCAM-1: in one of them the fourth immunoglobulin domain is lacking. The other variant is encoded by an alternatively spliced murine mRNA containing only the first three immunoglobulin domains (170,171). The translated protein is a glycosylphosphatidylinositol-anchored form of VCAM-1 which is cleaved from the cell surface by phosphatidylinositol-specific phospholipase C (173,174). Like the seven-immunoglobulin-domains form, these two variant forms are able to support the $\alpha4\beta1$-dependent adhesion of lymphocytes to VCAM-1. VCAM-1 can also mediate signal transduction and act as a T-cell costimulatory molecule (175–179).

A soluble form of VCAM-1 was detected in the supernatant of cytokine-activated endothelial cells and in normal plasma (135,180). VCAM-1 was isolated from plasma by immunoaffinity and size-exclusion chromatography (180). sVCAM-1 migrates as a Mr 85–90,000 protein on SDS-polyacrylamide gel (138,180). The soluble form of VCAM-1, immunoprecipitated from the supernatant of cytokine-activated endothelial cells, migrates as protein of Mr 100,000. The molecular mass of these molecules is 5 kDa lower than the transmembrane protein and corresponds to the predicted molecular mass of the cleaved VCAM-1, assuming cleavage at or close to the membrane (135,180). Immobilized plasma sVCAM-1 retains the capacity to support the $\alpha4\beta1$-dependent adhesion of lymphoid cells. In vivo, it may function as a competitive inhibitor of cell bound VCAM-1, as a signaling molecule, and/or a costimulatory molecule for T lymphocyte proliferation.

Mean sVCAM-1 concentration in normal plasma was reported to range from 400 to 625 ng/ml (113,138,141,155). Increased levels have been detected in inflammation, infectious diseases, and vasculitis. As described above, elevated sVCAM-1 levels were detected in patients with rheumatoid arthritis or systemic lupus erythematous (113,151,180), in patients with Wegener's granulomatosis (155), and in malaria (88). Elevated sVCAM-1 levels have also been reported in IgA nephropathy (181) and alcoholic cirrhosis (115). Obviously, the detection of increased sVCAM-1 levels in serum does not indicate the cellular source of the shed adhesion receptor. As discussed above for sICAM-1, sampling at site of inflammation may be more rewarding. In one study, sVCAM-1 was assayed simultaneously in plasma and in synovial fluid from patients with rheumatoid arthritis and joint effusion (151). Higher levels were found in synovial fluid, suggesting that sVCAM-1 is generated, at least in part, from the inflamed synovium. The expression of high VCAM-1 levels by synovial lining cells found in the superficial synovial lining layer supports the possibility that these cells are a major source of sVCAM-1 in inflamed synovium (165,182).

A rise in the concentration of plasma sVCAM-1 was reported in serum from some patients at the time of renal transplant rejection. The elevation of this receptor was coincident with, or slightly before, the rise in serum creatinine (138). However, different results were found in other longitudinal studies (109,154). Lebranchu et al. reported increased levels of soluble VCAM-1, ICAM-1, and E-selectin at the time of transplantation, and their concentration remained elevated for several months in stable patients. Patients with acute rejection experienced an increase in serum sE-selectin levels compared with those with stable graft function. In contrast, sVCAM-1 and sICAM-1 levels did not differ, suggesting that monitoring of serum sE-selectin, but not sVCAM-1 or sICAM-1 levels, could facilitate the diagnosis of acute rejection (109). Measurements sVCAM-1 and sICAM-1 in urine could also provide important information regarding the severity and the type of rejection (154), as higher concentrations were detected in urine from patients with acute steroid-resistant rejection compared with patients with acute steroid-sensitive rejection, chronic rejection, or stable graft function.

VII. DISCUSSION

During the past decade, the identification of adhesion receptors involved in leukocyte-endothelial cell interactions and of the mechanisms that regulate their function and their expression, has improved the understanding of the regulation of leukocyte migration into tissues. The complexity of this multistep process has been increased by the detection of circulating forms of adhesion receptors. In vitro studies have shown that circulating selectins and members of the superfamily of immunoglobulins retain functional activity, suggesting that they could modulate, in vivo, leukocyte adhesion and function as well as the immune surveillance. The detection of increased levels of these soluble molecules in some clinical disorders has led to the hypothesis that they could have a role in pathogenic procession and serve as markers of disease activity (Table 2).

The comparison of soluble adhesion molecule levels reported in several studies is difficult because of the lack of standardization of the assays used to measure them in plasma or serum. Thus, among the various studies reviewed in this chapter, mean levels of soluble adhesion receptors reported in healthy blood donors vary widely. The use of commercial kits and of common "international" standards could improve this situation (138).

Critically important will be the identification of soluble adhesion receptors that are helpful to monitor disease activity. The determination of the most suitable sites of sampling in various clinical disorders is also strongly needed. Thus, the assays performed in samples obtained at sites of local inflammatory or neoplasic involvement can be of more interest than measurements done in samples of peripheral blood (61,140,144,146,152–154). This could be particularly relevant for diseases that are not widely disseminated. The expression of adhesion molecules like ICAM-1 or VCAM-1 by multiple cell lineages often makes it difficult to interpret the significance of elevated levels. This can be true in inflammatory disease where cells of various origins are involved. In contrast, L-selectin expression is restricted to cells of hematopoietic origin; thus, increased levels of sL-selectin reflect

Table 2 Soluble Adhesion Molecule Concentrations in Various Diseases

Molecule	Measured in	Diseases	Concentrations	Controls	Ref
sL-selectin	Plasma	Acute myeloid leukemia	0.7–68 μg/ml	0.6–3.6 μg/ml	45
		Acute lymphoblastic leukemia	0.1–98 μg/ml	(n = 100)	
	CSF serum	Acute leukemia	34–150 ng/ml	0–37 ng/ml (n = 88)	61
		Sepsis	up to 3.0 μg/ml	0.5–3.5 μg/ml (n = 18)	59
		HIV infection	up to 6.0 μg/ml		
sP-selectin	Plasma	Heparin-induced thrombocytopenia	0.4–1.7 μg/ml	0.009–0.40 μg/ml (n = 37)	76
		Disseminated intravascular coagulation	0.3–1.0 μg/ml		
		Hemolytic uremic syndrome, TTP	0.2–0.7 μg/ml		
	Plasma	Mixed connective tissue disease	0.2–2 μg/ml	0.1–0.3 μg/ml (n = 12)	77
		Rheumatoid arthritis	0.2–1.8 μg/ml		
		Systemic erythematous lupus	0.2–0.6 μg/ml		
sE-selectin	Plasma	*P. falciparum* malaria	110–280 ng/ml	10–200 ng/ml (n = 9)	88
		N. meningitis infection	180–450 ng/ml		
	Serum	Bacteremia	1.1 ± 0.9 ng/ml	0.9 ± 0.6 ng/ml (n = 71)	87
		Septic shock	23.3 ± 15.9 ng/ml		
	Synovial fluid	Rheumatoid arthritis	0.18–3.9 ng/ml		90
		Psoriasic arthritis	0.88–2.31 ng/ml		
		Osteoarthritis	0.0–1.83 ng/ml		
	Serum	Rheumatoid arthritis	0.17–3.48 ng/ml	0.13–2.84 ng/ml (n = 71)	90
		Psoriasic arthritis	1.31–1.73 ng/ml		
		Osteoarthritis	0.36–3.63 ng/ml		
sICAM-1	Serum	Hodgkin's disease	277–1115 ng/ml	144–595 ng/ml (n = 10)	142
	Serum	Multiple sclerosis (active disease)	502 ± 218 ng/ml	218 ± 68 ng/ml (n = 10)	91
	Serum	Attack of bronchial asthma	27–122 ng/ml	18–48 ng/ml	95
	Serum	Wegener's granulomatosis	143–912 ng/ml	151–321 ng/ml (n = 10)	155
	Bile	Liver graft rejection	0–1850 ng/ml	0 ng/ml (n = 10)	148
sVCAM-1	Serum	Systemic erythematous lupus	446–2592 ng/ml	200–1000 ng/ml (n = 57)	113
	Urine	Acute kidney graft rejection	302 ± 39 ng/ml	26.9 ± 4.5 ng/ml (n = 30)	154

leukocyte activation or the presence of a large burden of malignant leukemic or lymphoma cells (45,59,61–63). The restriction of E-selectin expression to activated endothelium could make of sE-selectin a marker specific of endothelial cell activation (88,94,95,100,101). As expected, elevated sP-selectin levels have been reported in diseases associated with platelet and endothelial cell activation (76–78). Although the pathophysiology of some diseases suggests that elevated concentration of certain adhesion molecules should be present in the patients affected by these disorders, it should be emphasized that the presence of increased concentrations of adhesion receptors is not disease-specific.

Several reports have suggested that the levels of soluble adhesion receptors could correlate with disease activity or prognosis. Thus, relapses were more often observed in patients with Hodgkin's disease and high serum sICAM-1 levels, suggesting that this receptor could be a marker of poor outcome (142). In rheumatoid arthritis, the severity of arthritis seemed to be correlated with levels of sICAM-1 detected in synovial fluid (151). In other clinical disorders, a decrease in soluble adhesion receptor concentration correlated with a poor prognosis. Thus, a decrease in serum sL-selectin levels was associated with a high risk of developing ARDS (63). However, multivariate analysis of data collected during large prospective studies will be required to better establish the prognostic value of adhesion molecules and their role in monitoring the treatment of various diseases.

In vitro experiments have demonstrated that circulating adhesion receptors are functional (45,55,130,131,180). Purified circulating L-selectin can inhibit leukocyte adhesion to cytokine-activated endothelial cells, whereas immobilized plasma sICAM-1 and sVCAM-1 have the capacity to support leukocyte adhesion by interacting with $\beta 2$ and $\beta 1$ integrin. High levels of circulating adhesion receptors, as detected in some diseases, could have a functional role; they could inhibit leukocyte adhesion or they could activate transmembrane signal pathways. Thus, sICAM-1 could inhibit leukocyte-endothelial interaction, non-MHC-restricted cytotoxicity, and the development of lymphocyte responses requiring intercellular interactions. Several authors have proposed that the presence of elevated concentrations of shed receptors in malignant diseases such as Hodgkin's disease or melanoma could favor the escape of tumor cells from immune surveillance and promote tumor progression (131,139,142).

In health, the functional role of soluble adhesion receptors remains speculative. They could: 1. constitute a "biological buffer" that down-modulate leukocyte-endothelial interaction and/or immune responses at sites of weak inflammation; 2. represent an additional mechanism controlling the multiple steps that lead to leukocyte extravasation (soluble adhesion molecules could facilitate leukocyte deadhesion and thus their motility along endothelial cell surfaces and diapedesis) (Fig. 1); 3. generate intracellular signals that modulate integrin activity and their ligands (118) or more generally cell function; 4. transport soluble ligands (e.g., heparins or other carbohydrates that react with some selectins or GlyCAM-1, a circulating ligand for L-selectin present in normal murine blood) (21,183). In vivo, the functional effects of soluble receptors will depend on multiple parameters such as their affinity, their local concentration, their valency, and, finally, their capacity to compete with transmembrane forms for ligand binding. The importance of these various parameters remains to be determined to understand their physiological role.

The demonstration in blood of functional circulating adhesion receptors provides additional mechanisms that could modulate leukocyte adhesion and, more generally, immune function. However, further studies are needed to determine their role in vivo and how their function and release are regulated. Longitudinal studies could be of considerable help in understanding the role of these receptors in various disease syndromes. Some clinical reports already suggest that monitoring of circulating adhesion receptors could be helpful to follow patients; however, additional studies are required to better establish their role in various disease processes.

ACKNOWLEDGMENTS

We thank Prof. M. Shapira for critical comments and the Swiss National Foundation (grant 2031-43235-95) for supporting this work.

REFERENCES

1. Stoolman LM. Adhesion molecules controlling lymphocyte migration. Cell 1989; 56:907–910.
2. Yednock TA, Rosen SD. Lymphocyte homing. Adv Immunol 1989; 44:313–378.
3. Duijvestijn A, Hamann A. Mechanisms and regulation of lymphocyte migration. Immun Today 1989; 10:23–28.
4. Springer TA, Lasky LA. Sticky sugars for selectins. Nature 1991; 349:196–197.
5. Adams DH. Leucocyte-endothelial interactions and regulation of leukocyte migration. Lancet 1994; 343:831–836.
6. Springer TA. Traffic signals for lymphocyte recirculation and leukocyte emigration: the multistep paradigm. Cell 1994; 76:301–314.
7. Butcher EC. Leukocyte-endothelial cell recognition: three (or more) steps to specificity and diversity. Cell 1991; 67:1033–1036.
8. Ley K, Gaehtgens P, Fennie C, Singer MS, Lasky LA, Rosen SD. Lectin-like

cell adhesion molecule 1 mediates leukocyte rolling in mesenteric venules in vivo. Blood 1991; 77:2553–2555.
9. Von Andrian UH, Chambers JD, Berg EL, et al. L-selectin mediates neutrophil rolling in inflamed venules through sialyl Lewis X-dependent and -independent recognition pathways. Blood 1993; 82:182–191.
10. Von Andrian UH, Chambers JD, Mc Evoy LM, Bargatze RF, Arfors K-E, Butcher EC. Two step model of leukocyte-endothelial cell interaction in inflammation: distinct roles for LECAM-1 and the leukocyte β2 integrins in-vivo. Proc Natl Acad Sci 1991; 88:7538–7542.
11. Ley K, Bullard DC, Arbonés ML, et al. Sequential contribution of L- and P-selectin to leukocyte rolling in-vivo. J Exp Med 1995; 181:669–675.
12. Mayadas TN, Johnson RC, Rayburn H, Hynes RO, Wagner DD. Lymphocyte homing and leukocyte rolling and extravasation are severely compromised in P-selectin-deficient mice. Cell 1994; 74:541–544.
13. Arbonés ML. Ord DC, Ley K, et al. Lymphocyte homing and leukocyte rolling and migration are impaired in L-selectin (CD62L) deficient mice. Immunity 1994; 1:247–260.
14. Spertini O, Kansas GS, Munro JM, Griffin JD, Tedder TF. Regulation of leukocyte migration by activation of the leukocyte adhesion molecule-1 (LAM-1) selectin. Nature 1991; 349:691–693.
15. Dustin ML, Springer TA. T-cell receptor cross-linking transiently stimulates adhesiveness through LFA-1. Nature 1989; 341:619–624.
16. Shimizu Y, Van Seventer GA, Horgan KJ, Shaw S. Regulated expression and binding of three VLA(Beta 1) integrin receptors on T cells. Nature 1990; 345: 250–255.
17. Laemmli UK. Cleavage of structural proteins during the assembly of the head of bacteriophage T4. Nature 1970; 227:680–683.
18. Vaporciyan AA, DeLisser HM, Yan H-C, et al. Involvement of platelet-endothelial cell adhesion molecule-1 in neutrophil recruitment in-vivo. Science 1993; 262:1580–1582.
19. Bogen S, Pak J, Garifallou M, Deng X, Muller W. Monoclonal antibody to murine PECAM-1 (CD31) blocks acute inflammation in-vivo. J Exp Med 1994; 179:1059–1064.
20. Muller W, Weigl S, Deng X, Phillips D. PECAM-1 is required for transendothelial migration of leukocytes. J Exp Med 1993; 178:449–460.
21. Varki A. Selectin ligands. Proc Natl Acad Sci 1994; 91:7390–7397.
22. Foxall C, Watson SR, Dowbenko D, et al. The three members of the selectin receptor family recognize a common carbohydrate epitope, the sialyl Lewis x oligosaccharide. J Cell Biol 1992; 117:895–902.
23. Baumhueter S, Singer MS, Henzel W, et al. Binding of L-selectin to the vascular sialomucin CD34. Science 1993; 262:436–438.
24. Sako D, Chang X-J, Barone KM, et al. Expression cloning of a functional glycoprotein ligand for P-selectin. Cell 1993; 75:1179–1186.
25. Moore KL, Stults NL, Diaz S, et al. Identification of a specific glycoprotein ligand for P-selectin (CD62) on myeloid cells. J Cell Biol 1995; 118:445–456.

26. Lenter M, Levinovitz A, Isenmann S, Vestweber D. Monospecific and common glycoprotein ligands for E- and P-selectin on myeloid cells. J Cell Biol 1994; 125:471–481.
27. Lasky LA, Singer MS, Dowbenko D, et al. An endothelial ligand for L-selectin is a novel mucin-like molecule. Cell 1992; 69:927–938.
28. Imai Y, Singer MS, Fennie C, Lasky LA, Rosen SD. Identification of a carbohydrate-based endothelial ligand for a lymphocyte homing receptor. J Cell Biol 1991; 113:1213–1221.
29. Hemmerich S, Butcher EC, Rosen SD. Sulfation-dependent recognition of high endothelial venules (HEV)-ligands by L-selectin and MECA 79, an adhesion-blocking monoclonal antibody. J Exp Med 1994; 180:2219–2226.
30. Spertini O, Luscinskas FW, Munro JM, et al. Leukocyte adhesion molecule-1 (LAM-1, L-selectin) interacts with an inducible endothelial cell ligand to support leukocyte adhesion. J Immunol 1991; 147:2565–2573.
31. Spertini O, Luscinskas FW, Gimbrone MA Jr, Tedder TF. Monocyte attachment to activated human vascular endothelium in-vitro is mediated by leukocyte adhesion molecule-1 (L-selectin) under non-static conditions. J Exp Med 1992; 175:1789–1792.
32. Smith CW, Kishimoto TK, Abbass O, et al. Chemotactic factors regulate lectin adhesion molecule 1 (LECAM-1)-dependent neutrophil adhesion to cytokine-stimulated endothelial cells in-vitro. J Clin Invest 1991; 87:609–618.
33. Stegmaier M, Levinovitz A, Isenmann S, et al. The E-selectin ligand is a variant of a receptor for fibroblast growth factor. Nature 1995; 373:615–620.
34. Del Pozo MA, Pulido R, Munoz C, et al. Regulation of ICAM-3 (CD50) membrane expression on human neutrophils through a proteolytic shedding mechanism. Eur J Immunol 1994; 24:2586–2594.
35. Bazil V. Physiological enzymatic cleavage of leukocyte membrane molecules. Immunol Today 1995; 16:135–140.
36. Bevilacqua M, Butcher E, Furie B, et al. Selectins: a family of adhesion receptors. Cell 1991; 67:233.
37. Lasky LA. Selectins: interpreters of cell-specific carbohydrate information during inflammation. Science 1992; 258:964–968.
38. Tedder TF, Ernst TJ, Demetri GD, et al. Isolation and chromosomal localization of cDNAs encoding a novel human lymphocyte cell-surface molecule, LAM1: Homology with the mouse lymphocyte homing receptor and other human adhesion proteins. J Exp Med 1989; 170:123–133.
39. Lasky LA, Singer MS, Yednock TA, et al. Cloning of a lymphocyte homing receptor reveals a lectin domain. Cell 1989; 56:1045–1055.
40. Siegelman MH, Weissman IL. Human homologue of mouse lymph node homing receptor: evolutionary conservation at tandem cell interaction domains. Proc Natl Acad Sci USA 1989; 86:5562–5566.
41. Springer TA. Adhesion receptors of the immune system. Nature 1990; 346: 425–434.
42. Tedder TF, Penta AC, Levine HB, Freedman AS. Expression of the human leukocyte adhesion molecule, LAM1. Identity with the TQ1 and Leu-8 differentiation antigens. J Immunol 1990; 144:532–540.

43. Griffin JD, Spertini O, Ernst TJ, et al. GM-CSF and other cytokines regulate surface expression of the leukocyte adhesion molecule-1 on human neutrophils, monocytes, and their precursors. J Immunol 1990; 145:576–584.
44. Spertini O, Freedman AS, Belvin MP, Penta AC, Griffin JD, Tedder TF. Regulation of leukocyte adhesion molecule-1 (TQ1, Leu-8) expression and shedding by normal and malignant cells. Leukemia 1991; 5:300–308.
45. Spertini O, Callegari P, Cordey A-S, Hauert J, Joggi J, Von Fliedner V, Schapira M. High levels of the shed form of L-selectin (sL-selectin) are present in patients with acute leukemia and inhibit blast cell adhesion to activated endothelium. Blood 1994; 84:1249–1256.
46. Nakache M, Berg EL, Streeter PR, Butcher EC. The mucosal vascular addressin is a tissue-specific endothelial cell adhesion molecule for circulating lymphocytes. Nature 1989; 337:179–181.
47. Kishimoto TK, Julita MA, Berg EL, Butcher EC. Neutrophil Mac-1 and MEL-14 adhesion proteins inversely regulated by chemotactic factors. Science 1989; 245:1238–1241.
48. Jung TM, Dailey MO. Rapid modulation of homing receptors ($gp90^{Mel-14}$) induced by activators of protein kinase C. Receptor shedding due to accelerated proteolytic cleavage at the cell surface. J Immunol 1990; 144:3130–3136.
49. Jutila MA, Kishimoto TK, Butcher EC. Regulation and lectin activity of the human neutrophil lymph node homing receptor. Blood 1990; 76:178–183.
50. Kishimoto TK, Jutila MA, Butcher EC. Identification of a human peripheral lymph node homing receptor: a rapidly down-regulated adhesion molecule. Proc Natl Acad Sci USA 1990; 87:2244–2248.
51. Kahn J, Ingraham RH, Shirley F, Mikagi GI, Kishimoto TK. Membrane proximal cleavage of L-selectin: identification of the cleavage site and a 6-kD transmembrane fragment of L-selectin. J Cell Biol 1994; 125:461–470.
52. Chin YH, Rasmussen RA, Woodruff JJ, Easton TG. Monoclonal anti-HEBFpp antibody with specificity for lymphocyte surface molecules mediating adhesion to Peyer's patch high endothelium of rat. J Immunol 1986; 136: 2556.
53. Holzmann B, McIntyre BW, Weissman IL. Identification of a murine Peyer's patch-specific lymphocyte homing receptor as an integrin molecule with an alpha chain homologous to human VLA-4 alpha. Cell 1989; 56:37–46.
54. Kanof ME, James SP. Leu-8 antigen expression is diminished during cell activation but does not correlate with effector function of activated T lymphocytes. J Immunol 1988; 140:3701–3706.
55. Schleiffenbaum B, Spertini O, Tedder TF. Soluble L-selectin is present in human plasma at high levels and retains functional activity. J Cell Biol 1992; 119:229–238.
56. Shipp MA, Stefano GB, Switzer SN, Griffin JD, Reinherz EL. CD10 (CALLA)/neural endoprotease 24.11 modulates inflammatory peptide-induced changes in neutrophil morphology, migration, and adhesion proteins and is itself regulated by neutrophil activation. Blood 1991; 78:1834–1841.
57. Campanero MR, Pulido R, Alonso JL, et al. Downregulation of tumornecrosis factor-alpha of neutrophil cell surface expression of the sialophorin CD43

and the hyaluronate receptor CD44 through a proteolytic mechanism. Eur J Immunol 1991; 21:3045–3048.

58. Palecanda A, Walcheck B, Bishop DK. Rapid activation-independent shedding of leukocyte L-selectin induced by cross-linking of the surface antigen. Eur J Immunol 1992; 22:1279–1286.
59. Spertini O, Schleiffenbaum B, White-Owen C, Ruiz P Jr, Tedder TF. ELISA for quantitation of L-selectin shed from leukocytes in vivo. J Immunol Methods 1992; 156:115–123.
60. Zetterberg E, Richter J. Correlation between serum levels of soluble L-selectin and leukocyte count un chronic and lymphocytic leukemia and during bone marrow transplantation. Eur J Haematol 1993; 51:113–119.
61. Stucki A, Cordey A-S, Monai N, de Flaugergues J-C, Schapira M, Spertini O. Cleaved L-selectin in meningeal leukemia. Lancet 1995; 345:286–289.
62. Lampeter ER, Kishimoto TK, Rothlein R, et al. Elevated levels of circulating adhesion molecules in IDDM patients and in subjects at risk for IDDM. Diabetes 1992; 41:1668–1671.
63. Donelly SC, Haslett C, Dransfield I, et al. Role of selectins in development of adult respiratory distress syndrome. Lancet 1994; 344:215–219.
64. Hsu-Lin S-C, Berman CL, Furie BC, August D, Furie B. A platelet membrane protein expressed during platelet activation and secretion. J Biol Chem 1984; 259:9121–9126.
65. McEver RP, Martins MN. A monoclonal antibody to a membrane glycoprotein binds only to activated platelets. J Biol Chem 1984; 259:9799–9804.
66. McEver RP, Beckstead JH, Moore KL, Marshall-Carlson L, Bainton DF. GP-140, a platelet alpha-granule membrane protein, is also synthesized by vascular endothelial cells and is localized in Weibel-Palade bodies. J Clin Invest 1989; 84:92–99.
67. Lorant DE, Patel KD, McIntyre TM, McEver RP, Prescott SM, Zimmerman GA. Coexpression of GMP-140 and PAF by endothelium stimulated by histamine or thrombin: a juxtacrine system for adhesion and activation of neutrophils. J Cell Biol 1991; 115:223–234.
68. Bonfanti R, Furie BC, Furie B, Wagner DD. PADGEM (GMP-140) is a component of Weibel-Palade bodies of human endothelial cells. Blood 1989; 73:1109–1112.
69. Luscinskas FW, Ding H, Lichtman AH. P-selectin and vascular cell adhesion molecule 1 mediate rolling and arrest, respectively, of CD4+ T lymphocytes on tumor necrosis factor alpha-activated vascular endothelium under flow. J Exp Med 1995; 181:1179–1186.
70. Mayadas TN, Johnson RC, Rayburn H, Hynes RO, Wagner DD. Leukocyte rolling and extravasation are severely compromised in P-selectin-deficient mice. Cell 1993; 74:541–554.
71. Lawrence MB, Springer TA. Leukocytes roll on a selectin at physiologic flow rates: distinction from and prerequisite for adhesion through integrins. Cell 1991; 65:859–873.
72. Johnston GI, Cook RG, McEver RP. Cloning of GMP-140, a granule membrane protein of platelets and endothelium. Cell 1989; 56:1033–1044.

73. Johnston GI, Bliss GA, Newman PJ, McEver RP. Genomic structure of GMP-140, a member of the selectin family of adhesion receptors for leukocytes. J Biol Chem 1990; 34:21381–21385.
74. Ishiwata N, Takios K, Katayama M, et al. Alternative spliced isoform of P-selectin is present in-vivo as a soluble molecule. J Biol Chem 1994; 269: 23708–23715.
75. Dunlop LC, Skinner MP, Bendall LJ, et al. Characterization of GMP-140 (P-selectin) as a circulating plasma protein. J Exp Med 1992; 175:1147–1150.
76. Chong BH, Murray B, Berndt MC, et al. Plasma P-selectin is increased in thrombotic consumptive platelet disorders. Blood 1994; 83:1535–1541.
77. Takeda I, Kaise S, Nishimaki T, Kasukawa R. Soluble P-selectin in the plasma of patients with connective tissue diseases. Int Arch Allergy Immunol 1994; 105:128–134.
78. Katayama M, Handa M, Araki Y, Ambo H, Kawai Y. Soluble P-selectin is present in normal circulation and its plasma level is elevated in patients with thrombotic thrombocytopenic purpura and haemolytic uraemic syndrome. Br J Haematol 1993; 84:702–710.
79. Wu G, Li F, Li P, Ruan C. Detection of plasma alpha-granule membrane protein GMP-140 using radiolabeled monoclonal antibodies in thrombotic diseases. Haemostasis 1993; 23:121–128.
80. Bevilacqua MP, Pober JS, Mendrick DL, Cotran RS, Gimbrone MA Jr. Identification of an inducible endothelial-leukocyte adhesion molecule. Proc Natl Acad Sci USA 1987; 84:9238–9243.
81. Lawrence MB, Springer TA, Neutrophils roll on E-selectin. J Immunol 1993; 151:6338–6346.
82. Abbassi O, Kishimoto TK, McIntire LV, Anderson DC, Smith CW. E-selectin supports neutrophil rolling in-vitro under conditions of flow. J Clin Invest 1993; 92:2719–2730.
83. Shimizu Y, Shaw S, Graber N, et al. Activation-independent binding of human memory T cells to adhesion molecule ELAM-1. Nature 1991; 349:799–802.
84. Picker LJ, Kishimoto TK, Smith CW, Warnock RA, Butcher EC. ELAM-1 is an adhesion molecule for skin-homing T cells. Nature 1991; 349:796–799.
85. Mulligan MS, Warren JS, Smith CW, et al. Lung injury after deposition of IgA immune complexes: requirement for CD18 and L-arginine. J Immunol 1992; 148:3086–3092.
86. Gundel RH, Wegner CD, Torcellini CA, et al. Endothelial leukocyte adhesion molecule-1 mediates antigen-induced acute airway inflammation and late-phase airway obstruction in monkeys. J Clin Invest 1991; 88:1407–1411.
87. Newman W, Beall LD, Carson CW, et al. Soluble E-selectin is found in supernatant of activated endothelial cells and is elevated in the serum of patients with septic shock. J Immunol 1993; 150:644–654.
88. Jakobsen PH, Morris-Jones S, Rønn A, et al. Increased plasma concentrations of sICAM-1, sVCAM-1 and sELAM-1 in patients with *Plasmodium falciparum* or *P. vivax* malaria and association with disease severity. Immunology 1994; 83:665–669.
89. Stegeman CA, Tervaert JWC, Huitema MG, de Jong PE, Kallenberg CGM.

Serum levels of soluble adhesion molecules intercellular adhesion molecule 1, vascular cell adhesion molecule 1, and E-selectin in patients with Wegener's granulomatosis. Arthritis Rheum 1994; 37:1228–1235.

90. Carson CW, Beall LD, Hunder GG, Johnson CM, Newman W. Soluble E-selectin is increased in inflammatory synovial fluid. J Rheumatol 1994; 21: 605–611.
91. Rieckmann P, Martin S, Weichselbraun I, et al. Serial analysis of circulating adhesion molecules and TNF receptor in serum from patients with multiple sclerosis: cICAM-1 is an indicator for relapse. Neurology 1994; 44:2367–2372.
92. Picker LJ, Treer JR, Ferguson-Darnell PA, Collins PA, Bergstresser PR, Terstapen LW. Control of lymphocyte circulation in man. II. Differential regulation of the cutaneous lymphocyte-associated antigen, a tissue-selective homing receptor for skin-homing T cells. J Immunol 1993; 150:1122–1136.
93. Blann A, Tse W, Maxwell SJR, Waite MA. Increased levels of the soluble adhesion molecule E-selectin in essential hypertension. J Hypertens 1994; 12: 925–928.
94. Wenisch C, Varijanonta S, Looareesuwan S, Graninger W, Pichler R, Wernsdorfer W. Soluble intercellular adhesion molecule-1 (ICAM-1), endothelial leukocyte adhesion molecule-1 (ELAM-1), and tumor necrosis factor receptor (55kDa TNF-R) in patients with acute *Plasmodium falciparum* malaria. Clin Immunol Immunopathol 1994; 71:344–348.
95. Kobayashi T, Hashimoto S, Imai K, et al. Elevation of serum soluble intercellular adhesion molecule-1 (sICAM-1) and sE-selectin levels in bronchial asthma. Clin Exp Immunol 1994; 96:110–115.
96. Berendt AR, Simmons DL, Tansey J, Newbold CI, Marsh K. Intercellular adhesion molecule-1 is an endothelial cell adhesion receptor for *Plasmodium falciparum*. Nature 1989; 341:57–59.
97. Ockenhouse CF, Tegoshi T, Maeno Y, et al. Human vascular endothelial cell adhesion receptors for *Plasmodium falciparum*-infected erythrocytes: roles for endothelial leukocyte adhesion molecule-1 and vascular cell adhesion molecule-1. J Exp Med 1992; 176:1183–1189.
98. Grau GE, Taylor TE, Molyneux ME, et al. Tumor necrosis factor and disease severity in children with falciparum malaria. N Engl J Med 1989; 320:1586–1591.
99. Kwiatkowski D. Malaria: becoming more specific about non-specific immunity. Curr Opin Immunol 1992; 4:425–431.
100. Hviid L, Theander TG, Elhassan IM, Jensen JB. Increased plasma levels of soluble ICAM-1 and ELAM-1 (E-selectin) during acute *Plasmodium falciparum* malaria. Immunol Lett 1993; 36:51–58.
101. Elhassan IM, Hviid L, Satti G. Evidence of endothelial inflammation and T cell reallocation in uncomplicated *Plasmodium falciparum* malaria. Am J Trop Med Hyg 1994; 51:372–379.
102. Leung DYM, Pober JS, Corton RS. Expression of endothelial-leukocyte adhesion molecule-1 in elicited late phase allergic reactions. J Clin Invest 1991; 87:1805–1809.

103. Kyang-Aung U, Haskard DO, Poston RN, et al. Endothelial-leukocyte adhesion molecule-1 and intercellular adhesion molecule-1 mediate the adhesion of eosinophils to endothelial cells in-vitro and are expressed by endothelium in allergic cutaneous inflammation in-vivo. 1991; 146:521–528.
104. Wegner CD, Gundel RH, Peilly P, et al. Intercellular adhesion molecule-1 (ICAM-1) in the pathogenesis of asthma. Science 1990; 247:1407–1411.
105. Gundel RH, Wegner CD, Torcellini CA, et al. Endothelial leukocyte adhesion molecule-1 mediate antigen-induced acute airway inflammation and late-phase airway obstruction in monkeys. J Clin Invest 1991; 88:1407–1411.
106. Gundel RH, Wegner CD, Torcellini CA, Letts LG. The role of intracellular adhesion molecule-1 in chronic airway inflammation. Clin Exp Allergy 1992; 22:569–575.
107. Montefort S, Lai CKW, Kapahi P, et al. Circulating adhesion molecules in asthma. Am J Respir Crit Care Med 1994; 149:1149–1152.
108. Hashimoto S, Imai K, Kobayashi T, et al. Elevated levels of soluble ICAM-1 in sera from patients with bronchial asthma. Allergy 1993; 48:370–372.
109. Lebranchu Y, Kapahi P, Al Najjar A, et al. Soluble E-selectin, ICAM-1, and VCAM-1 levels in renal allograft recipients. Transplant Proc 1994; 26:1873–1874.
110. Hamblin AS, Shakoor Z, Kapahi P, Haskard D. Circulating adhesion molecules in sarcoidosis. Clin Exp Immunol 1994; 96:335–338.
111. Koch AE, Burrows JC, Haines GK, Carlos TM, Harlan JM, Leibovich SJ. Immunolocalization of endothelial and leukocyte adhesion molecules in human rheumatoid and osteoarthritic synovial tissues. Lab Invest 1991; 64:313–320.
112. Carson CW, Beall LD, Hunder GG, Johnson CM, Newman W. Serum ELAM-1 is increased in vasculitis, scleroderma, and systemic lupus erythematosus. J Rheumatol 1993; 20:809–814.
113. Spronk PE, Bootsma H, Huitema MG, Limburg PC, Kallenberg CGM. Levels of soluble VCAM-1, soluble ICAM-1, and soluble E-selectin during disease exacerbations in patients with systemic lupus erythematous (SLE): a long term prospective study. Clin Exp Immunol 1994; 97:439–444.
114. Stegeman CA, Cohen Tervaert JW, Huitema MG, de Jong PE, Kallenberg CGM. Serum levels of soluble adhesion molecules sICAM-1, sVCAM-1, and sE-selectin in patients with Wegener's granulomatosis. Arthritis Rheum 1994; 37:1228–1235.
115. Adams DH, Burra P, Hubscher SG, Elias E, Newman W. Endothelial activation and circulating vascular adhesion molecules in alcoholic liver disease. Hepatology 1994; 19:588–594.
116. Kurz RW, Graf B, Gremmel F, Wurnig C. Increased serum concentrations of adhesion molecules after coronary angioplasty. Clin Sci 1994; 87:627–633.
117. Mentzer SJ, Rothlein R, Springer TA, Faller DV. Intercellular adhesion molecule-1 (ICAM-1) is involved in the cytolytic T lymphocyte interaction with human synovial cells. J Cell Physiol 1988; 137:173–178.
118. Lo SK, van Seventer GA, Levin SM, Wright SD. Two leukocyte receptors (CD11a/CD18 and CD11b/CD18) mediate transient adhesion to endothelium by binding to different ligands. J Immunol 1989; 143:3325–3329.

119. Marlin SD, Springer TA. Purified intercellular adhesion molecule-1 (ICAM-1) is a ligand for lymphocyte function-associated antigen-1 (LFA-1). Cell 1987; 68:805–811.
120. Carlos TC, Harlan JM. Leukocyte-endothelial adhesion molecules. Blood 1994; 84:2068–2101.
121. Smith CW, Rothlein R, Hughes BJ, et al. Recognition of an endothelial determinant for CD18-dependent human neutrophil adherence and transendothelial migration. J Clin Invest 1988; 82:1746–1756.
122. Smith CW, Marlin SD, Rothlein RC, Toman C, Anderson DC. Cooperative interactions of LFA-1 and Mac-1 with intercellular adhesion molecule-1 facilitating adherence and transendothelial migration of neutrophils in-vitro. J Clin Invest 1989; 83:2008–2017.
123. Makgoba MW, Sanders ME, Ginther Luce GE, et al. Functional evidence that intercellular adhesion molecule-1 (ICAM-1) is a ligand for LFA-1 in cytotoxic T cell recognition. Eur J Immunol 1988; 18:637–640.
124. Larson RS, Springer TA. Structure and function of leukocyte integrins. Immunol Rev 1990; 114:181–217.
125. Boyd AW, Dunn SM, Fecondo JV, et al. Regulation of expression of a human intercellular adhesion molecule (ICAM-1) during lymphohematopoietic differentiation. Blood 1989; 73:1896–1903.
126. Staunton DE, Merluzzi VJ, Rothlein R, Barton R, Marlin SD, Springer TA. A cell adhesion molecule, ICAM-1, is the major surface receptor for rhinoviruses. Cell 1989; 56:849–853.
127. Dustin ML, Rothlein R, Bhan AK, Dinarello CA, Springer TA. Induction by IL 1 and interferon-gamma: tissue distribution, biochemistry, and function of a natural adherence molecule (ICAM-1). J Immunol 1986; 137:245–253.
128. Pober JS, Gimbrone MA, Lapierre LA, et al. Overlapping patterns of activation of human endothelial cells by interleukin-1, tumor necrosis factor and immune interferon. J Immunol 1986; 137:1893–1896.
129. Van Seventer GA, Shimizu Y, Horgan KJ, Shaw S. The LFA-1 ligand ICAM-1 provides an important costimulatory signal for T cell receptor-mediated activation of resting T cells. J Immunol 1990; 144:4579–4586.
130. Sadler JE. Biosynthesis of glycoproteins: formation of O-linked oligosaccharides. In: Ginsburg V, Robbins PW, eds. Biology of Carbohydrates. New York: John Wiley and Sons, 1984:199–288.
131. Becker JC, Dummer R, Hartmann AA, Burg G, Schmidt RE. Shedding of ICAM-1 from human melanoma cell lines induced by IFN-gamma and tumor necrosis factor-alpha. J Immunol 1991; 147:4398–4401.
132. Seth R, Raymond FD, Makgoba MW. Circulating ICAM-1 isoforms: diagnostic prospects for inflammatory and immune disorders. Lancet 1991; 338:83–84.
133. Giavazzi R, Chirivi RGS, Garofalo A, et al. Soluble intercellular adhesion molecule 1 is released by human melanoma cells and is associated with tumor growth in nude mice. Cancer Res 1992; 52:2628–2630.
134. Leeuwenberg JFM, Smeets EF, Neefjes JJ, et al. E-selectin and intercellular adhesion molecule-1 are released by activated human endothelial cells in-vitro. Immunology 1992; 77:543–549.

135. Pigott R, Dillon LP, Hemingway IK, Gearing AJH. Soluble forms of E-selectin, ICAM-1 and VCAM-1 are present in the supernatants of cytokine activated cultured endothelial cells. Biochem Biophys Res Commun 1992; 187:584–589.
136. Tsujisaki M, Imai K, Hirata H, et al. Detection of circulating intercellular adhesion molecule-1 antigen in malignant disease. Clin Exp Immunol 1991; 85:3–8.
137. Banks RE, Gearing AJH, Hemingway IK. Circulating intercellular adhesion molecule-1 (ICAM-1), E-selectin and vascular cell adhesion molecule-1 (VCAM-1) in human malignancies. Br J Cancer 1993; 68:122–124.
138. Gearing JH, Newman W. Circulating adhesion molecules in disease. Immunol Today 1993; 14:506–512.
139. Gruss H-J, Dölken G, Brach MA, Mertelsmann R, Herrmann F. Serum levels of circulating ICAM-1 are increased in Hodgkin's disease. Leukemia 1993; 7: 1245–1249.
140. Takahashi N, Liu MC, Proud D, Yu X-Y, Hasegawa S, Spannhake EW. Soluble intercellular adhesion molecule 1 in bronchoalveolar lavage fluid of allergic subjects following segmental antigen challenge. Am J Respir Crit Care Med 1994; 150:704–709.
141. Blann AD, McCollum CN. Circulating endothelial cell/leukocyte adhesion molecules in atherosclerosis. Thromb Haemostas 1994; 72:151–154.
142. Pui C-H, Luo X, Evans W, et al. Serum intercellular adhesion molecule-1 in childhood malignancy. Blood 1993; 82:895–898.
143. Küster H, Degitz K. Circulating ICAM-1 in neonatal sepsis. Lancet 1993; 341:506.
144. Jander S, Heidenrich F, Stoll G. Serum and CSF levels of soluble intercellular adhesion molecule-1 (ICAM-1) in inflammatory neurologic diseases. Neurology 1993; 43:1809–1813.
145. Norgard-Sumnicht KE, Varki NM, Varki A. Calcium-dependent heparin-like ligands for L-selectin in nonlymphoid endothelial cells. Science 1993; 261:480–483.
146. Lang T, Krams SM, Villanueva JC, Cox K, So S, Martinez OM. Differential patterns of circulating intercellular adhesion molecule-1 (CICAM-1) and vascular cell adhesion molecule-1 (CVCAM-1) during liver allograft rejection. Transplantation 1995; 59:584–589.
147. Stockenhuber F, Kramer G, Schenn G, et al. Circulating ICAM-1: a novel parameter of renal graft rejection. Transplant Proc 1993; 25:919–920.
148. Adams DH, Mainolfi E, Elias E, Neuberger JM, Rothlein R. Detection of circulating intercellular adhesion molecule-1 after liver transplantation—evidence of local release within the liver during graft rejection. Transplantation 1993; 55:919–920.
149. Ballantyne C, Mainolfi EA, Young JB, et al. Relationship of increased levels of circulating ICAM-1 after heart transplant to rejection, HLA match, and survival. Circulation 1991; 84 (suppl II):II-490. Abstract.
150. Shijubo N, Imai K, Aoki S, et al. Circulating intercellular adhesion molecule-1 (ICAM-1) antigen in sera of patients with idiopathic pulmonary fibrosis. Clin Exp Immunol 1992; 89:58–62.

151. Mason JC, Kapahi P, Haskard DO. Detection of increased levels of circulating intercellular adhesion molecule 1 in some patients with rheumatoid arthritis but not in patients with systemic lupus erythematous. Arthritis Rheum 1993; 36:519–527.
152. Sasseville VG, Newman WA, Lackner AA, et al. Elevated vascular cell adhesion molecule-1 in AIDS encephalitis induced by simian immunodeficiency virus. Am J Pathol 1992; 141:1021–1030.
153. Takatsugu K, Sasai M, Kobayashi Y. Increased soluble ICAM-1 in tracheal aspirates of infants with bronchopulmonary dysplasia. Lancet 1993; 342: 1023–1024.
154. Bechtel U, Scheuer R, Landgraf R, König A, Feucht HE. Assessment of soluble adhesion molecules (sICAM-1, sVCAM-1, sELAM-1) and complement cleavage products (sC4d, sC5b-9) in urine. Clinical monitoring of renal allograft recipients. Transplantation 1994; 58:905–911.
155. Mrowka C, Sieberth HG. Circulating adhesion molecules ICAM-1, VCAM-1 and E-selectin in systemic vasculitis: marked differences between Wegener's granulomatosis and systemic lupus erythematous. Clin Invest 1994; 72:762–768.
156. Furukawa S, Imai K, Matsubara T, et al. Increased levels of circulating intercellular adhesion molecule 1 in Kawasaki disease. Arthritis Rheum 1992; 35:672–677.
157. Osborn L, Hession C, Tizard R, et al. Direct expression cloning of vascular cell adhesion molecule-1, a cytokine-induced endothelial protein that binds to lymphocytes. Cell 1989; 59:1203–1211.
158. Rice GE, Bevilacqua MP. An inducible endothelial cell surface glycoprotein mediates melanoma adhesion. Science 1989; 246:1303–1306.
159. Rice GE, Munro JM, Corless C, Bevilacqua MP. Vascular and nonvascular expression of INCAM-110: a target for mononuclear leukocyte adhesion in normal and inflamed human tissues. Am J Pathol 1991; 138:385–393.
160. Rice GE, Munro JM, Bevilacqua MP. Inducible cell adhesion molecule 110 (INCAM-110) is an endothelial receptor for lymphocytes. A CD11/CD18-independent adhesion mechanism. J Exp Med 1990; 171:1369–1374.
161. Miyake K, Medina K, Ishihara M, Kimoto R, Auerbach R, Kincade P. A VCAM-like adhesion molecule on murine bone marrow stromal cells mediates binding of lymphocyte precursors in culture. J Cell Biol 1991; 114:557–565.
162. Miyake K, Weissman I, Greenberg J, Kincade P. Evidence for a role of the integrin VLA-4 in lympho-hemopoiesis. J Exp Med 1991; 173:599–607.
163. Ryan DH, Nuccie BL, Abboud C, Winslow JM. Vascular cell adhesion molecule-1 and the integrin VLA-4 mediate adhesion of human B cell precursors to cultured bone marrow adherent cells. J Clin Invest 1991; 88:995–1004.
164. Birdsall H, Lane C, Ramser M, Anderson D. Induction of VCAM-1 and ICAM-1 on human neural cells and mechanisms of mononuclear leukocyte adherence. J Immunol 1992; 148:2717–2723.
165. Marlor CW, Webb DL, Bombara MP, Greeve JM, Blue M-L. Expression of

vascular cell adhesion molecule-1 in fibroblastlike synoviocytes after stimulation with tumor necrosis factor. Am J Pathol 1992; 140:1055–1060.

166. Elices MJ, Osborn L, Takada Y, et al. VCAM-1 on activated endothelium interacts with the leukocyte integrin VLA-4 at a site distinct from the VLA-4/fibronectin binding site. Cell 1990; 60:577–584.

167. Meerschaert J, Furie MB. Monocyte use either CD11/CD18 or VLA-4 to migrate across human endothelium in-vitro. J Immunol 1994; 135:1915–1926.

168. Williams DA, Rios M, Stephens C, Patel VP. Fibronectin and VLA-4 in haematopoietic stem cell-microenvironment interactions. Nature 1991; 352: 438–441.

169. Ruegg C, Postigo AA, Sikorski EE, Butcher EC, Pytela R, Erle DJ. Role of integrin alpha 4 beta 7/alpha 4 beta P in lymphocyte adherence to fibronectin and VCAM-1 and in homotypic cell clustering. 1992; 117:179–189.

170. Hession C, Tizard R, Vassalo C, et al. Cloning of an alternate form of vascular cell adhesion molecule-1 (VCAM1). J Biol Chem 1991; 266:6682–6685.

171. Cybulsky MI, Fries JWU, Williams AJ, et al. Alternative splicing of human VCAM-1 in activated vascular endothelium. Am J Pathol 1991; 138: 815–820.

172. Polte T, Newman W, Raghunathan G, Gopal TV. Structural and functional studies of full-length vascular cell adhesion molecule-1: internal duplication and homology to several adhesion proteins. DNA Cell Biol 1991; 10:349–357.

173. Moy P, Lobb R, Tizard R, Olson D, Hession C. Cloning of an inflammation-specific phosphatidyl inositol-linked form of murine vascular adhesion molecule-1. J Biol Chem 1993; 268:8835–8841.

174. Terry RW, Kwee L, Levine JF, Labow MA. Cytokine induction of an alternatively spliced murine vascular cell adhesion molecule (VCAM) mRNA encoding a glycosylphosphatidylinositol-anchored VCAM protein. Proc Natl Acad Sci 1993; 90:5919–5923.

175. Nojima Y, Humphries MJ, Mould AP, et al. VLA-4 mediates CD3-dependent CD4+ T cell activation via the CS1 alternatively spliced domain of fibronectin. J Exp Med 1990; 172:1185–1192.

176. Nojima Y, Rothstein DM, Sugita K, Schlossman SF, Morimoto C. Ligation of VLA-4 on T cells stimulates tyrosine phosphorylation of a 105-kD protein. J Exp Med 1992; 175:1045–1053.

177. Freedman AS, Rhynhart K, Nojima Y, et al. Stimulation of protein tyrosine phosphorylation in human B cells after ligation of the $\beta 1$ integrin VLA-4. J Immunol 1993; 150:1645–1652.

178. Bednarczyk JL, McIntyre BW. A monoclonal antibody to VLA-4 a-chain (CDw49d) induces homotypic leukocyte aggregation. J Immunol 1990; 144: 777–784.

179. Damle NK, Aruffo A. Vascular cell adhesion molecule 1 induces T-cell antigen receptor-dependent activation of CD4+T lymphocytes. Proc Natl Acad Sci 1991; 88:6403–6407.

180. Wellicome SM, Kapahi P, Mason JC, et al. Detection of a circulating form

of vascular cell adhesion molecule-1: raised levels in rheumatoid arthritis and systemic lupus erythematous. Clin Exp Immunol 1993; 92:412–418.
181. Lai KN, Wong KC, Li PKT, et al. Circulating leukocyte-endothelial adhesion molecules in IgA nephropathy. Nephron 1994; 68:294–300.
182. Wilkinson LS, Poston RN, Edwards J, Haskard DO. Expression of vascular cell adhesion molecule-1 (VCAM-1) in normal and inflamed synovium. Lab Invest 1993; 68:82–88.
183. Brustein M, Kraal G, Mebius RE, Watson SR. Identification of a soluble form of a ligand for the lymphocyte homing receptor. J Exp Med 1992; 176: 1415–1419.

5

Adhesion Molecules as Signal Transduction Molecules

Andrew D. Yurochko
Lineberger Comprehensive Cancer Center,
University of North Carolina at Chapel Hill, Chapel Hill, North Carolina

I. INTRODUCTION

Much of the work regarding the role of adhesion receptors in cellular regulation has focused on the immune system. The immune system contains a complex collection of circulating leukocytes that survey the body for possible invaders or malignancies and can migrate to any site in the body. During an immune response or an inflammatory challenge, extravasating leukocytes must adhere to and pass through the intercellular junctions of capillary and postcapillary endothelial cells, the extracellular matrix (ECM; such as collagen, fibronectin, and laminin), and the basement membrane (reviewed in 1–7). This adherence of extravasating cells to endothelial cells and to extracellular and basement membrane components is mediated by cell surface receptors and serves as an essential step in the signaling of these leukocytes during an immune or an inflammatory response. Adhesion receptors are expressed on both immune and nonimmune cells. The function of these adhesion receptors and their ligands is to regulate cell migration, cellular differentiation and development, and tissue repair. A significant function recently attributed to adhesion receptors is their role in signal transduction and cell activation. This latter function will be the focus of this chapter. It is hoped that the reader will gain an increased appreciation for the function of adhesion receptors as a member of the signaling cascade.

There are now several families of adhesion receptors known to be important in the regulation of adhesion including: the integrins, the immunoglobulin (Ig) superfamily, the selectins, the cadherins, and the addressins (reviewed in 3,6,8). The former three families have been shown to be associated with cell signaling in the immune system. These families of adhesion receptors, although quite different in structure from each other, allow the adhesion of a cell to a ligand on another cell, the ECM, or some extracellular ligand, and they link the outside of the cell to its inside. It is this link between the outer membrane and the inner membrane that is hypothesized to serve as the mechanism for the start of the adhesion-induced cellular signaling. Adhesion receptors, upon engagement with a ligand, are thought to provide the initial signal for the signaling cascade. Depending on the cell involved and the type of adherence receptor engaged, the signal generated can either alone be sufficient to stimulate the cell or act in concert with an additional signal, such as the engagement of another receptor to generate the proper response. Adhesion receptors, however, do not contain a kinase domain as other receptor families have. Thus, to pass on the signal in an orderly fashion, the adhesion receptor must in some way interact with other members of the signal transduction pathway such as various kinase families, which in turn must activate different transcription factors that ultimately control gene induction and subsequent cellular activation.

II. ADHESION RECEPTOR FAMILIES AND SIGNAL TRANSDUCTION

A. The Integrins

1. Background

Integrins are a large family of glycoproteins which are generally responsible for cell/cell, cell/ECM, and cell/soluble ligand interaction (3,6,8). This family of heterodimeric proteins are made up of an α and a β subunit, noncovalently linked together. They contain a large extracellular domain, a membrane spanning domain, and a short cytoplasmic tail. There are at least eight different subfamilies (β1–β8), based on the unique structure of the different β chains with varying specificity for different cellular, ECM, and soluble protein components (3,8,9). These eight different β chains can combine with sixteen different α chains to produce at least 22 different integrin receptors (10–12).

The integrin receptors commonly associated with the immune system are: 1. the β1-or very late antigen (VLA-1-6 or CD49a-f/CD29) subfamily, which is important in cell/ECM (collagen, fibronectin, and laminin) and cell/cell interactions (1,3,6,8); 2. the β2- or Leu-CAM (CD11a,b,c/CD18–

lymphocyte function-associated antigen-1 (LFA-1), Mac-1, and gp150/95, respectively) subfamily, which is primarily associated with cell/cell interactions, but can also bind several soluble proteins (complement components, factor X, and fibrinogen) (3,6,8,13,14); and 3. the β7/CD103 subfamily which has recently been reported to bind the mucosal addressin (MAdCAM-1), fibronectin, and vascular cell adhesion molecule (VCAM-1/CD106) (6,12,15,16). The integrins function along with the selectin family in a multistep adhesion pathway involving the sequential interaction of different receptor/ligand interactions. This pathway usually involves many low-affinity interactions to capture and slow down the cell, followed by specific high-affinity interactions to strengthen the binding and allow extravasation (6,17–25). These steps apparently vary depending on the type of cell involved. In lymphocytes there is probably first a role for the selectins and then the α4 integrins (α4β7 and α4β1), although the α4 integrins have been shown to be sufficient for mediating adhesion under physiologic flow in the absence of selectin involvement (25). In contrast, monocytes show a sequential action of L-selectins, β1-integrins, and then β2-integrins (22), while in neutrophils there is first L-selectin and then β2-integrin involvement (17,18,21). The different subfamilies of integrins can generally be subdivided depending on the cell types on which they are expressed and the types of interactions they mediate (3,6,8). This rough subdivision is not absolute, but it does hint that they may serve differential signaling functions in the different cell types, and this is particularly important if one considers their potential role in the signaling cascade.

The short cytoplasmic portion of the integrin receptor is associated with the cytoskeletal framework of the cell through several accessory proteins including talin, vinculin, α-actinin, and possibly others (8,26–28). Because integrins associate with the cytoskeleton and serve as a link between the exterior of the cell and the interior, they are possible candidates for mediating the signaling that occurs following adhesion/integrin engagement. There is no intrinsic kinase activity present on the integrin receptor itself; however, there is a potential phosphorylation site on the cytoplasmic tail (29), and some integrins are phosphorylated upon stimulation (30–32), suggesting that phosphorylation could be a regulatory event in the signaling cascade. Furthermore, with the identification of a novel adhesion/integrin-associated focal adhesion kinase, $pp125^{FAK}$ (33–35), it was hypothesized that integrins played a prominent role in the signal transduction pathway (7,10,36).

2. Signaling

Integrin receptor-mediated signaling is a complex process that shares some characteristics with that of the signaling pathway described for various growth factor receptors (reviewed in 36–41). The problem presently with the knowledge of integrin-mediated signaling is that it is fragmented and the known information spans many different cell types. Generally, though, engagement of growth factor receptors results in an organized cascade of signaling, starting with the initial receptor engagement, followed by the sequential activation of various tyrosine kinases, which ultimately, along with increases in important second messengers such as Ca^{2+}, results in upregulation of transcription factors, gene induction, and cell activation. One of the first responses documented to occur following adhesion or integrin engagement is also tyrosine phosphorylation. In monocytes, adherence or treatment of cells with antibodies against the β1-integrin but not the β2-integrin subunit directly induces rapid increases in tyrosine phosphorylation of proteins of unknown function (42). In T cells, in contrast, engagement of various adhesion receptors (β1 and β2) only provides a costimulatory signal with T cell receptor (TCR)/CD3 or CD2 engagement for increased tyrosine phosphorylation (43–45). And in neutrophils, adhesion to the ECM or cellular substrates along with tumor necrosis factor (TNF) could trigger tyrosine phosphorylation (46,47) and a possible association of the β2-mediated signaling pathway with the nonreceptor *src* family member, $pp58^{FGR}$ (46).

Adhesion of fibroblasts has also been shown to upregulate tyrosine phosphorylation (33,48–54), activate mitogen-activated protein kinases (MAPK) (53,54), and promote the association of members of the signal transduction pathway (49,52,54). In addition, there have been reports of integrin-dependent activation of tyrosine phosphorylation in platelets (55–59). Cross-linking of the β1-integrin receptors (60,61) or adhesion to fibronectin stimulates tyrosine phosphorylation (61) in epidermal carcinoma cells as well.

From the work discussed above, it is clear that there is a role for adhesion receptors in the regulation of tyrosine phosphorylation, but how does adhesion actually regulate tyrosine phosphorylation and what is the mechanism for the downstream signaling? Answers to this question have been facilitated with the recent identification and cloning of a focal adhesion kinase termed, $pp125^{FAK}$ (33–35). $pp125^{FAK}$ is a novel cytoplasmic nonreceptor tyrosine kinase that was found to be localized in focal adhesions (33,34,49,61) and to be phosphorylated in response to adhesion and integrin cross-linking with antibodies (34,48,49,55–57,59–62). These results for the first time directly linked adhesion to a phosphorylation event. Other kinases have now been shown to be associated with $pp125^{FAK}$, including phosphatidylinositol 3-kinase (52) and various *src* family members ($pp60^{V\text{-}SRC}$, $pp60^{C\text{-}SRC}$, and $pp59^{FYN}$) (54,63,64) through a possible *src* homology 2 domain, suggesting that, upon integrin engagement, there is an association of different members of the signaling pathway and an ordered signaling cascade is initiated.

Activation of other kinases has also been reported following cell adhesion including other *src* family members ($pp58^{FGR}$) (46) and MAPK (53, 54). The report of the adhesion-induced activation of MAPK (53,54) along with the association of the GRB2 adapter protein to $pp60^{C\text{-}SRC}$ and $pp125^{FAK}$ suggests a link between adhesion and Ras/MAPK signaling (54). It is important to note that different cell types apparently use different kinase members in mediating the adhesion-induced signaling. For example, in fibroblasts, platelets and epidermal carcinoma cells, $pp125^{FAK}$ is present and phosphorylated/activated upon adhesion/integrin engagement (34,48,49, 55–57,59–62) while in immune cells there is no apparent use of $pp125^{FAK}$, at least under the conditions examined (42,46,65).

In lymphocytes several kinases have been documented to be involved in the signaling cascade including the *src*-related kinases, $pp56^{LCK}$ and $pp59^{FYN}$ (66–69), fakB, a $pp125^{FAK}$ homologue (65), and Zap-70, a novel tyrosine kinase (69,70). In neutrophils the *src* family member $pp58^{FGR}$ is activated by β2-mediated adhesion events (46), while in monocytes, no tyrosine kinase has yet been identified to be associated with integrin-mediated signaling, although tyrosine phosphorylation is induced by adhesion (42). These results suggest that the type of kinase activated and the subsequent signal generated may regulate the specificity of the cellular activation. The role of the adhesion receptor itself in this pathway is not well understood, but in the integrin family, the short cytoplasmic tail of the β subunit is necessary for the increase in $pp125^{FAK}$ phosphorylation (48,51) and has been shown to be phosphorylated by cellular activation, providing supporting for its role in adhesion-mediated signaling (29–32). It is not known whether the β subunit interacts directly with any of the known kinases or interacts indirectly through some integrin-associated protein (71,72). Combined, this body of recent evidence suggests that there is indeed a kinase signaling pathway that becomes activated upon adhesion, involving phosphorylation by many different kinases, and that the adhesion receptor itself is necessary in this pathway.

Additionally, the cytoskeleton is thought to play a major role because it links the outside of the cell to the inside of the cell (8,26–28,73,74). Furthermore, the cytoskeleton is found at focal contacts in association with adhesion receptors and may be a mechanism for localizing some of the important factors in the signal transduction cascade. Cytoskeletal-associated proteins are phosphorylated upon adhesion (47,49,50), suggesting that they are regulated or activated during adhesion and that this plays a role in the cytoskeletal organization. Evidence for the vital role the cytoskeleton plays in signaling comes from studies showing that cytoskeletal inhibitors block many aspects of adhesion-mediated signaling (32,50,53,55–57,59,75–78).

The downstream adhesion-mediated signal transduction pathway (following phosphorylation and kinase activation) includes the regulation of important second-messenger pathways such as the activation of phospholipase C-γ1, which results in the hydrolysis of phosphatidylinositol, the generation of active second messengers, and the subsequent mobilization of intracellular Ca^{2+}. These events have been shown to mediate the intracellular signaling in other systems (37,40,41,79). In T cells, integrin engagement can provide a costimulatory signal for the activation of phospholipase C-γ1 (44,45), inositol phospholipid hydrolysis (80,81), and increased Ca^{2+} levels (44,45,80–82). Most of the signaling described above comes in the form of a costimulatory signal along with CD3/TCR engagement; however, it has been reported that cross-linking of the β2-integrin, LFA-1, with antibodies could directly augment Ca^{2+} mobilization (44,80), suggesting that engagement of some integrin receptors can provide a direct signal for certain aspects of T cell activation. β2 integrin-dependent activation of neutrophils has been reported to regulate cytosolic Ca^{2+} (83–85), and in platelets there is an integrin-dependent activation of the Na^+/H^+ antiporter, calpain, and a rise in Ca^{2+} (58,86,87). Additionally, adhesion of fibroblasts elevated intracellular pH (88), while engagement of the fibronectin receptor could activate the Na^+/H^+ antiporter (89). These results suggest that various adhesion/integrin-mediated events can regulate many of the same second messengers that are important in other signaling pathways (36–41).

One of the end points for the signals generated by the different second messengers and phosphorylation is at the level of transcription factor regulation. This holds true in adhesion-mediated signaling as well. Recent work has shown that adhesion or receptor cross-linking with antibodies can induce the activity of transcription factors such as NF-κB and inactivate its cytosolic inhibitor, IκBα (78,90–93). The mechanism for regulating these transcription factors upon adhesion is not known, but there have been reports implicating several kinases (reviewed in 36,39,94–96), and it is interesting to speculate that these may be the same kinases activated by adhesion. Furthermore, because the *rel* family of transcription factors contains ankyrin repeats and these repeats may play a role in protein-cytoskeletal interactions (73,94,95,97,98), it is possible that adhesion, kinase activation, and cytoskeletal rearrangement are the events that lead up to the regulation of transcription factor activity and subsequent cellular activation.

a. Signaling in Monocytes. Control of cellular activation is the ultimate goal of the signaling pathways initiated with adhesion or integrin engagement. There has been an enormous amount of work documenting the many ways that different cells are activated upon adhesion. Depending on the cell, this ranges from gene induction and subsequent protein secretion to cellular proliferation. In monocytes it has been shown that the initial adherence step modulates the expression of key inflammatory mediator genes, including interleukin (IL)-1β, TNF-α, colony stimulating factor-1 (CSF-1), the proto-oncogenes, *c-fos* and *c-jun*, and several transcriptional regulators (e.g., NF-κB), suggesting that adherence serves as a primary immune regulatory stimulus for those cells (78,97,99–104). In fact, adherence is apparently obligatory for the rapid and transient induction of these cytokine messages.

Furthermore, the type of substrate the monocyte adheres to can differentially regulate message expression (78,102). The direct role for adhesion receptors in cytokine regulation was recently described using antibodies directed against various receptors (42,93,105,106). Antibodies against the β1 integrins induced monocyte inflammatory gene expression (93,106) and regulated protein secretion (106) and protein degradation and synthesis (93), suggesting that adhesion/integrin engagement could regulate transcriptional and translational processes. Surprisingly, antibodies directed against the β2-integrins could not induce these responses (42,93,106). Antibodies to various adhesion receptors have also been shown to induce homotypic aggregation (increased cell/cell binding due to a receptor induced increase in the affinity of other adhesion receptors) (74,75) or adhesion to cellular substrates (107) in monocytic cell lines.

b. Signaling in T Cells. Recently, there have been a number of reports that the engagement of T cell adhesion receptors through binding to the ECM via the VLA integrins (108–111), through binding to the purified ligands ICAM-1 and VCAM-1 (112–116), or through cross-linking of the various adhesion receptors with antibodies (82,117–119) provides a costimulatory signal for CD3/TCR-dependent T cell proliferation. Additionally, engagement of the various integrins along with coengagement of CD3/TCR could regulate IL-2, IL-4, GM-CSF, and IL-2 receptor synthesis (82,114–116,119–121). Costimulation of β1 or β2 integrins through their ligands, ICAM-1 and VCAM-1, is important in mediating activation-dependent death in antigen-primed T lymphocytes (122). Antibodies against adhesion receptors (including the various α-chains of the VLA integrins; the β-chain of the β1, β2, and β7 integrins; and LFA-1) could induce homotypic aggregation in various T and B lymphocytic cell lines or peripheral blood lymphocytes (32,75,77,123–132). These results document the role for integrin engagement in the regulation of lymphocyte function and demonstrate adhesion as a primary stimulus for lymphocytes.

c. Signaling in Neutrophils. Neutrophils are also activated through their various cell surface receptors. Adhesion of neutrophils to ECM and cellular substrates along with TNF triggers a respiratory burst that results in the production of reactive oxygen intermediates (133). Adhesion also provides a costimulatory signal, along with other activating agents such as LPS, for triggering of the respiratory burst (134). Adhesion alone to fibrinogen results in the release of the soluble forms of the TNF receptors (135). Most of the work in neutrophil adhesion-mediated signaling was discussed above in relation to tyrosine phosphorylation and Ca^{2+} regulation. Much of this adhesion-dependent signaling is through the β2 integrins (46,47,136,137), suggesting that, as in the other immune cells, integrin engagement is a primary signaling pathway responsible for controlling cellular activation. The β2 integrin-dependent activation of neutrophils is also important in the stimulation of the respiratory burst (76) and degranulation (83). Engagement of β1 or β2 integrins with antibodies could also induce a respiratory burst in another polymorphonuclear leukocyte, the eosinophil (138).

d. Signaling in Nonimmune Cells. Cellular regulation via adhesion at the level of gene induction or cellular proliferation has also been reported in nonimmune cells. Adhesion of suspension-arrested fibroblasts to fibronectin can induce expression of the proto-oncogenes *c-fos* and *c-myc* (139). Additionally, engagement of the fibronectin receptor on fibroblasts with antibodies induced the expression of the collagenase and stromelysin genes (140). Adhesion of keratinocytes to fibronectin or treatment with antifibronectin receptor antibodies inhibited terminal differentiation of these cells (141), and stimulation of the β1 integrins in osteosarcoma cells can serve as an intermediate signal for the induction of alkaline phosphatase activity (142). A hypothesized signaling pathway compiling the data from the many experiments in different cell types examining adhesion/integrin-mediated signaling is shown in Figure 1.

B. The Ig Superfamily

1. Background

The Ig superfamily encompasses a wide range of cellular adhesion molecules, including those intimately associated with the immune system and antigen recognition, the major histocompatibility complex (MHC) Class I and Class II molecules, the MHC counterreceptors CD4 and CD8, and the TCR and CD3 complex, the ligands for the integrins, intracellular adhesion molecule (ICAM)-1/CD54, ICAM-2/CD102, ICAM-3/CD50, and VCAM-1/CD106, another T cell-cell surface receptor CD2/LFA-2 and its counterreceptor, LFA-3/CD58, the platelet-endothelial cell adhesion molecule (PECAM)-1/CD31 and MAdCAM-1, and other cellular adhesion mol-

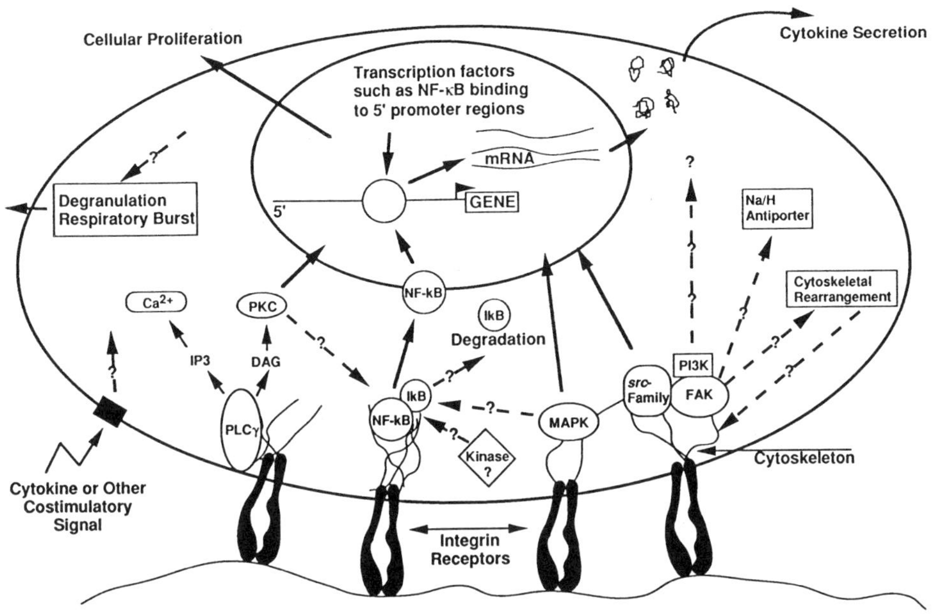

Figure 1 A model of the hypothesized signaling cascade following integrin engagement. This figure is a composite figure of the many signals that have been documented to occur following adhesion and integrin engagement in different cell types. As mentioned in the text, there are many documented differences between cell types. The reason for this is presently unclear, but it probably has to do with the nature of the signal generated and the different functions of a particular cell. Many aspects of the pathway are still unknown, so some pathways are speculative. Solid arrows denote what is thought to be the pathway of signaling, and dashed arrows denote hypothesized pathways. Abbreviations used in the figure are: PKC = protein kinase C; IP3 = inositol trisphosphate; DAG = diacylgylcerol; NF-κB = the transcription NF-κB; IκB = the cytosolic inhibitor of NF-κB; PLCγ = phospholipase Cγ; MAPK = mitogen-activated protein kinase; FAK = $pp125^{FAK}$, focal adhesion kinase; and PI3K = phosphatidylinositol-3-kinase.

ecules (3,6,8,143,144). This superfamily of receptors contains an Ig structural domain and plays a significant role in the immune signaling cascade that occurs following an immunological or an inflammatory challenge. Like the integrins, the Ig superfamily serves as a communication link between the outside and the inside of the cell. Furthermore, there is an association between nonreceptor kinases and some Ig family members (65–70,145–

151), suggesting that this family of receptors also functions as part of the adhesion-mediated signal transduction pathway.

2. Signaling

The most common signaling studied has been that generated through the direct engagement of the TCR and the CD4 or CD8 coreceptors (68). TCR stimulation by antigen/MHC on antigen presenting cells (APC) or antibody receptor cross-linking can induce a rapid intracellular signaling cascade leading to the induction of phosphorylation, Ca^{2+} mobilization, gene induction, and T cell proliferation (reviewed in (68,70,79,152–156)). Basically, the signaling pathway involves many of the same processes as described above for the integrins. These points will not be covered in any detail here, as the TCR-antigen/MHC-mediated events have been reviewed extensively elsewhere (see (68,70,79,152–161)). Nevertheless, it is important to understand that these receptor ligand interactions are mediated by the Ig family of adhesion receptors and serve as a primary signaling stimulus for both the T cell (68,70,79,152–156) and the APC (monocyte/macrophage or B cell) (162–167). This receptor-mediated activation results in the association of members of the signaling cascade (65–68,70,153). These studies showed that there is an association of *src* family members ($pp56^{LCK}$ and $pp59^{FYN}$) (66–69), fakB (65), and Zap-70 (69,70) with the TCR complex (CD3, CD4, CD8 or the ζ chain).

There is also a role for the other Ig family members in adhesion-mediated signaling. Cross-linking of ICAM-3 on T cells results in tyrosine phosphorylation through an association with $pp56^{LCK}$ and $pp59^{FYN}$ and Ca^{2+} mobilization (168). CD2 and CD28 have also been implicated in mediating tyrosine phosphorylation and T cell activation (79,116,169–172). In B cells, receptor/ligand interactions of the membrane-bound Ig antigen receptor can induce tyrosine phosphorylation, mobilization of Ca^{2+}, and an association of several *src* family members, $pp53/56^{LYN}$, $pp55^{BLK}$, $pp56^{LCK}$, and $pp59^{FYN}$, as well as the non-*src* kinase $pp72^{SYK}$, with the membrane-bound antigen receptor complex (145–151). In addition, ICAM-1 can play a role in B cell expression of the IL-2 receptor (164).

It was also shown that LFA-3 could trigger TNF-α and IL-1β protein release in monocytes (105) and that cross-linking of ICAM-1 could provide a costimulatory signal to adherent monocytes along with activating agents to induce an oxidative burst (173). Combined, these results show that Ig family-mediated engagement can regulate a number of cellular events in a variety of cells, as discussed above, resulting in transcriptional regulation, cytokine secretion, or cellular proliferation. A possible signaling pathway for the cellular activation mediated by Ig family-mediated adhesion is presented in Figure 2. The figure represents the possible signaling pathways in

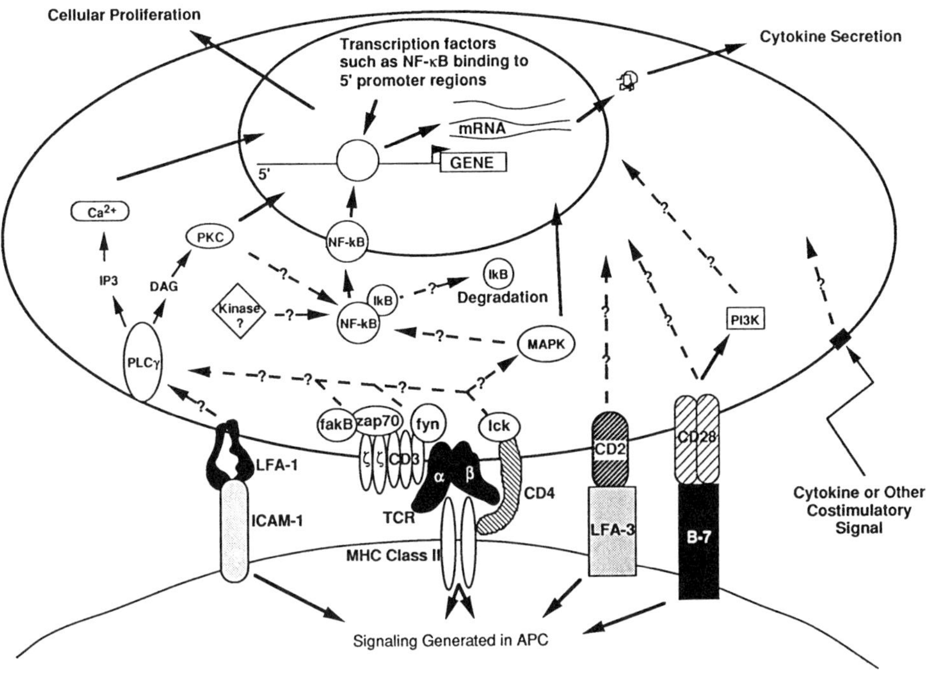

Figure 2 A model of the proposed signaling cascade following Ig-mediated adhesion in T cells. The signaling in T cells was chosen because of the abundant evidence from these cells. It is important to understand that although less is understood about the signaling pathways in the other immune cells upon Ig adhesion family-mediated adhesion, signaling does occur. In fact, the interaction of the T cell with an APC (monocyte/macrophage or B cell) clearly activates the T cell as well as the APC. Solid arrows denote what is known to be the pathway of signaling, and dashed arrows denote hypothesized pathways. Abbreviations used in the figure are: PKC = protein kinase C; IP3 = inositol trisphosphate; DAG = diacylgylcerol; NF-κB = the transcription factor NF-κB; IκB = the cytosolic inhibitor of NF-κB; PLCγ = phospholipase Cγ; MAPK = mitogen-activated protein kinase; FAK = $pp125^{FAK}$, focal adhesion kinase; fyn = the *src* family member, $pp59^{FYN}$; lck = the *src* family member, $pp56^{LCK}$; zap70 = a novel tyrosine kinase; fakB = a $pp125^{FAK}$ homologue; and PI3K = phosphatidylinositol-3-kinase.

T cells, but it is important to remember that there is a signal generated via the MHC molecule in monocytes/macrophages and via the MHC molecule and Ig receptor complex in B cells that most likely parallels the signal seen in other systems.

C. The Selectins

1. Background

The selectins are a class of adhesion receptors characterized by Ca^{2+}-dependent NH_2-terminal lectinlike domains (6,19,174–176). They also have a short cytoplasmic tail and apparently no inherent kinase activity. There are now three members in this family: E (endothelial)-selectin/CD62E or ELAM-1; L (lymphocyte)-selectin/CD62L (MEL-14, $gp90^{MEL}$, or LAM-1); and P (platelet)-selectin/CD62P (PADGEM or GMP-140). Selectins play diverse roles in adhesion-mediated events and are found on both leukocytes and endothelial cells, hinting at their vital role in leukocyte migration. The lectinlike domains found on the NH_2-terminal region of selectins suggested that the ligand for these family of receptors contains carbohydrate moieties (174–177). Recently, several carbohydrate or carbohydrate-modified proteins have been identified as counterreceptors for the selectins (reviewed in (6,177)). Selectins are rapidly upregulated (within hours after stimulation) on endothelial cells and leukocytes in response to various cytokines or inflammatory agents (174–177). They are responsible for the initial interaction between leukocytes and endothelial cells and, in most cases, are thought to serve as a mechanism to slow down and capture the leukocytes. Subsequently, the integrins strengthen the bond allowing cellular activation and extravasation (3,6,17–25). Thus, selectins were originally thought to be nonsignaling molecules, but they, like the other adhesion receptor families discussed, have a role in the adhesion-dependent signal transduction pathway.

2. Signaling

The signaling capacity of the selectins is just beginning to be realized. The primary cell investigated for selectin-mediated signaling is the neutrophil, probably because of the primary role the selectins play in mediating neutrophil adherence to the vascular endothelium. Engagement of L-selectin on neutrophils with antibodies could provide a costimulatory signal with activating agents such as TNF-α or formyl-Met-Leu-Phe for mobilization of the second messenger Ca^{2+} and for production of reactive oxygen intermediates (178). Selectin engagement could also directly regulate mobilization of Ca^{2+} (179). Antibodies against the selectins induced homotypic aggregation in neutrophils (180) and lymphocytes (181). This induction of homotypic aggregation in neutrophils was due to an increase in $\beta 2$ integrin affinity, suggesting that selectin engagement modulates the high affinity state of integrin receptors, promoting extravasation and/or activation. Triggering of L-selectin in T cells provided a costimulatory signal for proliferation (182), and in neutrophils L-selectin engagement was shown to regulate the p55 and p75 TNF receptors (183) and to induce TNF-α and IL-8 transcripts (179). This suggests that there is a whole signaling cascade that can be regulated initially by the engagement of the selectin adhesion receptor. Nothing is known about the intermediate mechanisms responsible for passing the signal from the selectin molecule to the end function of transcription and/or cellular activation. It is probable that the pathway parallels the integrin-and Ig-mediated signaling cascade.

Table 1 Summary of the Cellular Responses Mediated by the Different Adhesion Receptor Families. Similarities Between Adhesion Receptors and Receptor Tyrosine Kinases

	Cellular Response			
Adhesion receptor	Integrin[a]	Ig	Selectin	RTK[b]
Signal transduction molecule	yes	yes	yes	yes
Tyrosine phosphorylation	yes	yes	yes	yes
PLCγ activation[c]	yes	yes	?	yes
MAPK[d]	yes	yes	?	yes
Ca^{2+} mobilization	yes	yes	yes	yes
Cytoskeletal involvement	yes	?	?	yes
Transcription factor induction	yes	yes	?	yes
Gene induction	yes	yes	yes	yes
Cell proliferation	yes	yes	?	yes

[a]Note: The signals described are a compilation of the many players in the integrin-mediated signal transduction pathway. There are many differences and still many unknowns between the different cell types described in the text.
[b]RTK = receptor tyrosine kinase; see text for references.
[c]PLCγ = phospholipase Cγ.
[d]MAPK = mitogen-activated protein kinase.

III. CONCLUDING REMARKS

In conclusion, adhesion or adhesion receptor engagement is a dynamic process that results in the induction of a variety of cellular signals following an organized cascade of events from the initial receptor engagement, through phosphorylation and association of the kinase cascade, generation of important second messengers, gene induction, and finally cellular activation. The pathways for the integrin and the Ig families are only partially understood and seem to follow a mechanism similar to that responsible for signaling in receptor tyrosine kinase families, suggesting that there are many aspects in common between the different intracellular signaling pathways. A comparison of the signaling events regulated by different adhesion receptor families is shown in Table 1. The interesting question then is, if common signaling pathways exist, what mediates the specificity of the end signal? The reason for this conundrum is not understood. It is probable that the specificity starts with the unique interactions that take place when an individual receptor is initially engaged and this is followed by the cell type-specific interaction of different members of the signaling cascade. This hypothesis has some merit since different kinase members are activated by adhesion in different cells, suggesting that there are unique signaling cascades responsible for the specificity of the response, in addition to the many common pathways utilized. Nevertheless, it is clear that adhesion molecules are more than receptors simply responsible for attachment, but are instead active signal transduction molecules. Taken together, these results point to the direct role for adherence/adhesion receptor engagement in the signal transduction pathway and ultimately to the vital control of the immune or inflammatory process.

REFERENCES

1. Hemler ME. VLA proteins in the integrin family: structures, functions, and their role on leukocytes. Annu Rev Immunol 1990; 8:365–400.
2. Osborn L. Leukocyte adhesion to endothelium in inflammation. Cell 1990; 62:3–6.
3. Springer TA. Adhesion receptors of the immune system. Nature 1990; 346: 425–434.
4. Albelda SM, Smith CW, Ward PA. Adhesion molecules and inflammatory injury. FASEB J 1994; 8:504–512.
5. Beekhuizen H, Van Furth R. Monocyte adhesion to human vascular endothelium. J Leukoc Biol 1993; 54:363–378.
6. Carlos TM, Harlan JM. Leukocyte-endothelial adhesion molecules. Blood 1994; 84:2068–2101.

7. Rosales C, Juliano RL. Signal transduction by cell adhesion receptors in leukocytes. J Leukoc Biol 1995; 57:189–198.
8. Albelda SM, Buck CB. Integrins and other cell adhesion molecule. FASEB J 1990; 4:2868–2880.
9. Moyle M, Napier MA, McLean JW. Cloning and expression of a divergent integrin subunit β8. J Biol Chem 1991; 266:19650–19658.
10. Hynes RO. Integrins: versatility, modulation, and signaling in cell adhesion. Cell 1992; 69:11–25.
11. Smyth SS, Joneckis CC, Parise LV. Regulation of vascular integrins. Blood 1993; 81:2827–2843.
12. Cepek KL, Shaw SK, Parker CM, et al. Adhesion between epithelial cells and T lymphoctyes mediated by E-cadherin and the αEβ7 integrin. Nature 1994; 372:190–193.
13. Kishimoto TK, Larson RS, Corbi AL, Dustin ML, Staunton DE, Springer TA. The leukocyte integrins. Adv Immunol 1989; 46:149–182.
14. Arnaout MA. Structure function of the leukocyte adhesion molecule CD11/CD18. Blood 1990; 75:1037–1050.
15. Erle DJ, Ruegg C, Sheppard D, Pytela R. Complete amino acid sequence of an integrin β subunit (β7) identified in leukocytes. J Biol Chem 1991; 266: 11009–11016.
16. Andrew DP, Berlin C, Honda S, et al. Distinct but overlapping epitopes are involved in α4β7-mediated adhesion to vascular cell adhesion molecule-1, mucosal addressin-1, fibronectin, and lymphocyte aggregation. J Immunol 1994; 153:3847–3861.
17. Butcher EC. Leukocyte-endothelial cell recognition: three (or more) steps to specificity and diversity. Cell 1991; 67:1033–1036.
18. Von Andrian UH, Chambers JD, McEvoy LM, Bargatze RF, Arfors KE, Butcher EC. Two-step model of leukocyte-endothelial cell interaction in inflammation: distinct roles of LECAM-1 and the leuckocyte β2-integrins in vivo. Proc Natl Acad Sci USA 1991; 88:7538–7542.
19. Paulson JC. Selectin/carbohydrate-mediated adhesion of leukocytes. In: Harlan JM, Liu DY, eds. Adhesion: Its Role in Inflammatory Disease. New York: Freeman, 1992:19–42.
20. Picker LJ, Butcher EC. Physiological and molecular mechanisms of lymphocyte homing. Annu Rev Immunol 1992; 10:561–591.
21. Von Andrian UH, Hansell P, Chambers JD, et al. L-selectin function is required for β2-integrin-mediated neutrophil adhesion at physiological shear rates in vivo. Am J Physiol 1992; 263:1034–1044.
22. Abassi O, Kishimoto TK, McIntire LV, Smith CW. Neutrophil adhesion to endothelial cells. Blood Cells 1993; 19:245–259.
23. Luscinskas FW, Kansas GS, Ding H, et al. Monocyte rolling, arrest and spreading on IL-4-activated vascular endothelium under flow is mediated via sequential action of L-selectin, β1-integrins, and β2-integrins. J Cell Biol 1994; 125:1417–1427.
24. Springer TA. Traffic signals for lymphocyte recirculation and leukocyte emigration: the multistep paradigm. Cell 1994; 76:301–314.

25. Berlin C, Bargatze RF, Campbell JJ, et al. $\alpha 4$ integrins mediate lymphocyte attachment and rolling under physiologic flow. Cell 1995; 80:413–422.
26. Buck CA, Shea E, Duggan K, Horowitz AF. Integrin (the CSAT antigen): functionality requires oligomeric integrity. J Cell Biol 1986; 103:2421–2428.
27. Burridge K, Fath K, Kelly T, Nuckolis B, Turner C. Focal adhesions: transmembrane junctions between the extracellular matrix and the cytoskeleton. Annu Rev Cell Biol 1988; 4:487–525.
28. Otey CA, Pavalko FM, Burridge K. An interaction between α-actinin and the $\beta 1$ integrin subunit in vitro. J Cell Biol 1990; 111:721–729.
29. Tamkun JW, DeSimone DW, Fonda D, et al. Structure of the cysteine-rich subunit of integrin, the fibronectin receptor. Cell 1989; 46:271–282.
30. Chatila TA, Geha RS, Arnaout MA. Constitutive and stimulus-induced phosphorylation of CD11/CD18 leukocyte adhesion molecules. J Cell Biol 1989; 109:3435–3444.
31. Shaw LM, Messier JM, Mercurio AM. The activation dependent adhesion of macrophages to laminin involves cytoskeletal anchoring and phosphorylation of the $\alpha 6\beta 1$ integrin. J Cell Biol 1990; 110:2167–2174.
32. Pavlovic MD, Colic M, Pejnovic N, Tamatani T, Miyasaka M, Dujic A. A novel anti-rat CD18 monoclonal antibody triggers lymphocyte homotypic aggregation and granulocyte adhesion to plastic: different intracellular signaling pathways in resting versus activated thymoctes. Eur J Immunol 1994; 24: 1640–1648.
33. Hanks SK, Calalb MB, Harper MC, Patel SK. Focal adhesion protein-tyrosine kinase phosphorylated in response to cell attachment to fibronectin. Proc Natl Acad Sci USA 1992; 89:8487–8491.
34. Schaller MD, Borgman CA, Cobb BS, Vines RR, Reynolds AB, Parsons JT. pp125^{FAK} a structurally distinctive protein-tyrosine kinase associated with focal adhesions. Proc Natl Acad Sci USA 1992; 89:5192–5196.
35. Whitney GS, Chan P-Y, Blake J, et al. Human T and B lymphocytes express a structurally conserved focal adhesion kinase, pp125^{FAK}. DNA Cell Biol 1993; 12:823–830.
36. Juliano RL, Haskill S. Signal transduction from the extracellular matrix. J Cell Biol 1993; 120:577–585.
37. Clapham DE. Calcium signaling. Cell 1995; 80:259–268.
38. Herskowitz I. MAP kinase pathways in yeast: for mating and more. Cell 1995; 80:187–198.
39. Hill CS, Treisman R. Transcriptional regulation by extracellular signal: mechanisms and specificity. Cell 1995; 80:199–212.
40. Hunter T. Protein kinases and phosphatases: the yin and yang of protein phosphorylation and signaling. Cell 1995; 80:225–236.
41. Marshall CJ. Specificity of receptor tyrosine kinase signaling: transient versus sustained extracellular signal-regulated kinase activation. Cell 1995; 80:179–186.
42. Lin TH, Yurochko A, Kornberg L, et al. The role of protein tyrosine phosphorylation in integrin-mediated gene induction in monocytes. J Cell Biol 1994; 126:1585–1593.

43. Nojima Y, Rothstein DM, Sugita K, Schlossman SF, Morimoto C. Ligation of VLA-4 on T cells stimulates tyrosine phosphorylation of a 105-kD protein. J Exp Med 1992; 175:1045–1053.
44. Kanner SB, Grosmaire LS, Ledbetter JA, Damle NK. β2-integrin LFA-1 signaling through phospholipase C-γ1 activation. Proc Natl Acad Sci USA 1993; 90:7099–7103.
45. Dietsch MT, Chan PY, Kanner SB, et al. Coengagement of CD2 with LFA-1 or VLA-4 by bispecific ligand fusion proteins primes T cells to respond more effectively to T cell receptor-dependent signals. J Leukoc Biol 1994; 56:444–452.
46. Berton G, Fumagalli L, Laudanna C, Sorio C. β2 integrin-dependent protein tyrosine phosphorylation and activation of the FGR protein tyrosine kinase in human neutrophils. J Cell Biol 1994; 126:1111–1121.
47. Fuortes M, Jin WW, Nathan C. β2 integrin dependent tyrosine phosphorylation of paxillin in human neutrophils treated with tumor necrosis factor. J Cell Biol 1994; 127:1477–1483.
48. Guan JL, Trevithick JE, Hynes RO. Fibronectin/integrin interaction induces tyrosine phosphorylation of a 120-kDa protein. Cell Regul 1991; 2:951–964.
49. Burridge K, Turner CE, Romer LE. Tyrosine phosphorylation of paxillin and pp125^{FAK} accompanies cell adhesion to extracellular matrix: a role for cytoskeletal assembly. J Cell Biol 1992; 119:893–903.
50. Bockholt SM, Burridge K. Cell spreading on extracellular matrix proteins induces tyrosine phosphorylation of tensin. J Biol Chem 1993; 268:14565–14567.
51. Akiyama SK, Yamada SS, Yamada KM, LaFlamme SE. Transmembrane signal transduction by integrin cytoplasmic domains expressed in single-subunit chimeras. J Biol Chem 1994; 269:15961–15964.
52. Chen HC, Guan JL. Association of focal adhesion kinase with its potential substrate phosphatidylinositol 3-kinase. Proc Natl Acad Sci USA 1994; 91: 10148–10152.
53. Chen Q, Kinch MS, Lin TH, Burridge K, Juliano RL. Integrin-mediated cell adhesion activates mitogen-activated protein kinases. J Biol Chem 1994; 269: 26602–26605.
54. Schlaepfer DD, Hanks SK, Hunter T, Van der Geer P. Integrin-mediated signal transduction linked to Ras pathway by GRB2 binding to focal adhesion kinase. Nature 1994; 372:786–791.
55. Lipfert L, Haimovich B, Schaller MD, Cobb BS, Parsons JT, Brugge JS. Integrin-dependent phosphorylation and activation of the protein tyrosine kinase pp125^{FAK} in platelets. J Cell Biol 1992; 119:905–912.
56. Haimovich B, Lipfert L, Brugge JS, Shattil SJ. Tyrosine phosphorylation and cytoskeletal reorganization in platelets are triggered by integrin receptors with their immobilized ligands. J Biol Chem 1993; 268:15868–15877.
57. Huang MM, Lipfert L, Cunningham M, Brugge JS, Ginsberg MH, Shattil SJ. Adhesive ligand binding to integrin αIIb β3 stimulates tyrosine phosphorylation of novel protein substrates before phosphorylation of pp125^{FAK}. J Cell Biol 1993; 122:473–483.

58. Fox JE. Transmembrane signaling across the platelet integrin glycoprotein IIb-IIIa. Ann NY Acad Sci 1994; 18:75–87.
59. Shattil SJ, Haimovich B, Cunningham M, et al. Tyrosine phosphorylation of pp125^{FAK} in platelets requires coordinated signaling through integrin and agonist receptors. J Biol Chem 1994; 269:14738–14745.
60. Kornberg LJ, Earp HS, Turner CE, Prockop C, Juliano RL. Signal transduction by integrins: increased protein tyrosine phosphorylation caused by clustering of β1 integrins. Proc Natl Acad Sci USA 1991; 88:8392–8396.
61. Kornberg L, Earp HS, Parsons JT, Schaller M, Juliano RL. Cell adhesion or integrin clustering increases phosphorylation of a focal adhesion-associated tyrosine kinase. J Biol Chem 1992; 267:23439–23442.
62. Guan JL, Shalloway D. Regulation of focal adhesion-associated protein tyrosine kinase by both cellular adhesion and oncogenic transformation. Nature 1992; 538:690–692.
63. Cobb BS, Schaller MD, Leu TH, Parsons JT. Stable association of pp60^{src} and pp59^{fyn} with the focal adhesion-associated protein tyrosine kinase, pp125^{FAK}. Mol Cell Biol 1994; 14:147–155.
64. Xing Z, Chen HC, Nowlen JK, Taylor SJ, Shalloway D, Guan JL. Direct interaction of v-Src with the focal adhesion kinase mediated by the Src SH2 domain. Mol Biol Cell 1994; 5:413–421.
65. Kanner SB, Aruffo A, Chan PY. Lymphocyte antigen receptor activation of a focal adhesion kinase-related tyrosine kinase substrate. Proc Natl Acad Sci USA 1994; 91:10484–10487.
66. Shaw AS, Amrein KE, Hammond C, Stern DF, Sefton BM, Rose JK. The cytoplasmic domain of CD4 interacts with the tyrosine protein kinase, p56^{lck}, through it unique amino-terminal domain. Cell 1989; 59:627–636.
67. Samelson LE, Phillips AF, Luong ET, Klauser RD. Association of the fyn protein-tyrosine kinase with the T-cell antigen receptor. Proc Natl Acad Sci USA 1990; 87:4358–4362.
68. Janeway CA Jr. The T cell receptor as a multicomponent signalling machine: CD4/CD8 coreceptors and CD45 in T cell activation. Annu Rev Immunol 1992; 10:645–674.
69. Chan AC, Desai DM, Weiss A. The role of protein tyrosine kinases and protein tyrosine phosphatases in T cell antigen receptor signal transduction. Annu Rev Immunol 1994; 12:555–592.
70. Weiss A. T cell antigen receptor signal transduction: a tale of tails and cytoplasmic protein-tyrosine kinases. Cell 1993; 73:209–212.
71. Bartfeld NS, Pasquale EB, Geltosky JE, Languino LR. The $\alpha v\beta 3$ integrin associates with a 190-kDa protein that is phosphorylated on tyrosine in response to platelet-derived growth factor. J Biol Chem 1993; 268:17270–17276.
72. Lindberg FP, Gresham HD, Schwartz E, Brown EJ. Molecular cloning of an integrin associated protein: an immunoglobulin family member with multiple membrane-spanning domains implicated in $\alpha v\beta 3$-dependent ligand binding. J Cell Biol 1993; 123:485–496.

73. Luna EJ, Hitt AL. Cytoskeleton-plasma membrane interactions. Science 1992; 258:955–963.
74. Sanchez-Mateos P, Campanero MR, Balboa MA, Sanchez-Madrid F. Co-clustering of $\beta 1$ integrins, cytoskeletal proteins, and tyrosine-phosphorylated substrates during integrin-mediated leukocyte aggregation. J Immunol 1993; 151:3817–3828.
75. Campanero MR, Pulido R, Ursa MA, et al. An alternative leukocyte homotypic adhesion mechanism, LFA-1/ICAM-1-independent, triggered through the human VLA-4 integrin. J Cell Biol 1990; 110:2157–2165.
76. Berton G, Laudanna C, Sorio C, Rossi F. Generation of signals activating neutrophil functions by leukocyte integrins: LFA-1 and gp150/95, but not CR3, are able to stimulate the respiratory burst of human neutrophils. J Cell Biol 1992; 116:1007–1017.
77. Benmerah A, Badrichani A, Ngohou K, Megarbane B, Begue B, Cerf-Bensussan N. Homotypic aggregation of CD103 ($\alpha E \beta 7$)+ lymphocytes by an anti-CD103 antibody, HML-4. Eur J Immunol 1994; 24:2243–2249.
78. Yurochko AD, Huang E-S. Unpublished results.
79. Mondino A, Jenkins MK. Surface proteins involved in T cell costimulation. J Leukoc Biol 1994; 55:805–815.
80. Pardi R, Bender JR, Dettori C, Giannazza E, Engleman EG. Heterogeneous distribution and transmembrane signaling properties of lymphocyte function-associated antigen (LFA-1) in human lymphocyte subsets. J Immunol 1989; 143:3157–3166.
81. Van Seventer GA, Bonvini E, Yamada H, et al. Costimulation of T cell receptor/CD3-mediated activation of resting human CD4+ T cells by leukocyte function-associated antigen-1 ligand intercellular cell adhesion molecule-1 involves prolonged inositol phospholipid hydrolysis and sustained increase of intracellular Ca^{2+} levels. J Immunol 1992; 149:3872–3880.
82. Wacholtz MC, Patel SS, Lipsky PE. Leukocyte function-associated antigen 1 is an activation molecule for human T cells. J Exp Med 1989; 170:431–448.
83. Richter J, Ng-Sikorski J, Olsson I, Andersson T. Tumor necrosis factor-induced degranulation in adherent human neutrophils is dependent on CD11b/CD18-integrin triggered oscillations of cytosolic free Ca^{2+}. Proc Natl Acad Sci USA 1990; 87:9472–9476.
84. Jaconi ME, Theler JM, Schlegel W, Appel RD, Wright SD, Lew PD. Multiple elevations of cytosolic-free Ca^{2+} in human neutrophils: initiation of adherence receptors of the integrin family. J Cell Biol 1991; 112:1249–1257.
85. Ng-Sikorski J, Andersson R, Patarroyo M, Andersson T. Calcium signaling capacity of the CD11b/CD18 integrin on human neutrophils. Exp Cell Res 1991; 195:504–508.
86. Yamaguchi A, Tanoue K, Yamazaki H. Secondary signals mediated by GPIIb/IIIa in thrombin-activated platelets. Biochem Biophys Acta 1990; 1054:8–13.
87. Pelletier AJ, Bodary SC, Levinson AD. Signal transduction by the platelet integrin αIIb $\beta 3$: induction of calcium oscillations required for protein-

tyrosine phosphorylation and ligand-induced spreading of stably transfected cells. Mol Biol Cell 1992; 3:989–998.

88. Schwartz MA, Ingber DE, Lawrence M, Springer TA, Lechene C. Multiple integrins share the ability to induce intracellular pH. Exp Cell Res 1991; 195: 533–535.
89. Schwartz MA, Lechne C, Ingber DE. Insoluble fibronectin activates the Na/H antiporter by clustering and immobilizing integrin $\alpha5\beta1$, independent of cell shape. Proc Natl Acad Sci USA 1991; 88:7849–7853.
90. Bressler P, Pantaleo G, Demaria A, Fauci AS. Anti-CD2 receptor antibodies activate the HIV long terminal repeat in T lymphocytes. J Immunol 1991; 147:2290–2294.
91. Costello R, Lipcey C, Algarte M, et al. Activation of primary human T-lymphocytes through CD2 plus CD28 adhesion molecules induces long-term nuclear expression of NF-κB. Cell Growth Differ 1993; 4:329–339.
92. Perez JR, Higgins-Sochaski KA, Maltese JY, Narayanan R. Regulation of adhesion and growth of fibrosarcoma cells by NF-κB RelA involves transforming growth factor beta. Mol Cell Biol 1994; 14:5326–5332.
93. Lofquist AK, Mondal K, Morris JS, Haskill JS. Transcription-independent turnover of IκBα during monocyte adherence: implications for a translational component regulating IκBα/MAD-3 mRNA levels. Mol Cell Biol 1995; 15: 1737–1746.
94. Baeuerle PA, Henkel T. Function and activation of NF-κB in the immune system. Annu Rev Immunol 1994; 12:141–179.
95. Lenardo M, Siebenlist U. Bcl-3-mediated nuclear regulation of the NF-κB trans-activating factor. Immunol Today 1994; 15:145–147.
96. Siebenlist U, Franzoso G, Brown K. Structure, regulation and function of NF-κB. Annu Rev Cell Biol 1994; 10:405–455.
97. Haskill S, Beg AA, Tompkins SM, et al. Characterization of an immediate-early gene induced in adherent monocytes that encodes IκB-like activity. Cell 1991; 65:1281–1289.
98. Liou HC, Baltimore D. Regulation of the NF-κB/rel transcription factor and IκB inhibitor system. Curr Opin Cell Biol 1993; 5:477–487.
99. Fulhbrigge RC, Chaplin DD, Kiely JM, Unanue ER. Regulation of interleukin 1 gene expression by adherence and lipopolysaccharide. J Immunol 1987; 138:3799–802.
100. Thorens B, Mermod JJ, Vassalli P. Phagocytosis and inflammatory stimuli induce GM-CSF mRNA in macrophages through posttranscriptional regulation. Cell 1987; 48:671–679.
101. Haskill S, Johnson C, Eierman D, Becker S, Warren K. Adherence induces selective mRNA expression of monocyte mediators and proto-oncogenes. J Immunol 1988; 140:1690–1694.
102. Eierman DF, Johnson CE, Haskill JS. Human monocyte inflammatory mediator gene expression is selectively regulated by adherence substrates. J Immunol 1989; 140:1690–1694.
103. Shaw RJ, Doherty DE, Ritter AG, Benedict SH, Clark RAF. Adherence-

dependent increase in human monocyte PDGF(B) mRNA is associated with increases in *c-fos*, *c-jun* and EFR2 mRNA. J Cell Biol 1990; 111:2139–2148.
104. Sporn SA, Eierman DF, Johnson CE, et al. Monocyte adherence results in selective induction of novel genes sharing homology with mediators of inflammation and tissue repair. J Immunol 1990; 144:4434–4441.
105. Webb DSA, Shimizu Y, Van Seventer GA, Shaw S, Gerrard TL. LFA-3, CD44, and CD45: physiologic triggers of human monocyte TNF and IL-1 release. Science 1990; 249:1295–1297.
106. Yurochko AD, Liu DY, Eierman D, Haskill S. Integrins as a primary signal transduction molecule regulating monocyte immediate-early gene induction. Proc Natl Acad Sci USA 1992; 89:9034–9038.
107. Kovach NL, Carlos TM, Yee E, Harlan JM. A monoclonal antibody to β1 integrin (CD29) stimulates VLA-dependent adherence of leukocytes to human umbilical vein endothelial cells and matrix components. J Cell Biol 1992; 116:499–509.
108. Matsuyama T, Yamada A, Kay J, et al. Activation of CD4 cells by fibronectin and anti-CD3 antibody. A synergistic effect mediated by the VLA-5 fibronectin receptor complex. J Exp Med 1989; 170:1133–1148.
109. Davis LS, Oppenheimmer-Marks N, Bednarczyk JL, McIntyre BW, Lipsky PE. Fibronectin promotes proliferation of naive and memory T cells by signaling through both the VLA-4 and VLA-5 integrin molecules. J Immunol 1990; 145:785–793.
110. Nojima Y, Humphries MJ, Mould AP, et al. VLA-4 mediates CD3-dependent CD4+ T cell activation via the CS1 alternatively spliced domain of fibronectin. J Exp Med 1990; 172:1185–1192.
111. Shimizu Y, Van Seventer GA, Horgan KJ, Shaw S. Costimulation of proliferative responses of resting CD4+ T cells by the interaction of VLA-4 and VLA-5 with fibronectin or VLA-6 with laminin. J Immunol 1990; 145:59–67.
112. Van Seventer GA, Shimizu Y, Horgan KJ, Shaw S. The LFA-1 ligand ICAM-1 provides an important costimulatory signal for T cell receptor-mediated activation of resting T cells. J Immunol 1990; 144:4579–4586.
113. Burkly LC, Jakubowski A, Newman BM, Rosa MD, Chi-Rosso G, Lobb RR. Signaling by vascular cell adhesion molecule-1 (VCAM-1) through VLA-4 promotes CD3-dependent T cell proliferation. Eur J Immunol 1991; 21:2871–2875.
114. Damle NK, Aruffo A. Vascular cell adhesion molecule 1 induces T-cell antigen receptor-dependent activation of CD4+ T lymphocytes. Proc Natl Acad Sci USA 1991; 88:6403–6407.
115. Van Seventer GA, Newman W, Shimizu Y, et al. Analysis of T cell stimulation by superantigen plus major histocompatibility complex class II molecules or by CD3 monoclonal antibody: costimulation by purified adhesion ligands VCAM-1, ICAM-1, but not ELAM-1. J Exp Med 1991; 174:901–913.
116. Damle NK, Klussman K, Linsley PS, Aruffo A. Differential costimulatory effects of adhesion molecules B7, ICAM-1, LFA-3, and VCAM-1 on resting and antigen primed CD4+ T lymphocytes. J Immunol 1992; 148:1985–1992.

117. Van Noesel C, Miedema F, Brouwer M, de Rie MA, Aarden LA, van Lier RAW. Regulatory properties of LFA-1 α and β chains in human T-lymphocyte activation. Nature 1988; 333:850–852.
118. Bednarczyk JL, Teague TK, Wygant JN, Davis LS, Lipsky PE, McIntyre BW. Regulation of T cell proliferation by anti-CD49d and anti-CD29 monoclonal antibodies. J Leukoc Biol 1992; 52:456–462.
119. Udagawa T, McIntyre BW. A VLA-4 α-chain specific monoclonal antibody enhances CD3-induced IL-2/IL-2 receptor-dependent T-cell proliferation. Lymphokine Cytokine Res 1992; 1992; 11:193–199.
120. Damle NK, Klussman K, Aruffo A. Intracellular adhesion molecule-2, a second counter receptor for CD11a/CD18 (leukocyte function associated antigen-1), provides a costimulatory signal for T-cell receptor-initiated activation of human T-cells. J Immunol 1992; 148:665–671.
121. Reiser H, Freeman GJ, Razi-Wolf Z, Gimmi CD, Benacerraf B, Nadler LM. Murine B7 antigen provides an efficient costimulatory signal for activation of murine T lymphocytes. Proc Natl Acad Sci USA 1992; 89:271–275.
122. Damle NK, Klussman K, Leytze G, Aruffo A, Linsley PS, Ledbetter JA. Costimulation with integrin ligands intercellular adhesion molecule-1 or vascular cell adhesion molecule-1 augments activation-induced death of antigen-specific CD4+ T lymphocytes. J Immunol 1993; 151:2368–2379.
123. Keizer GD, Visser W, Vliem M, Figdor G. A monoclonal antibody (NKI-L16) directed against a unique epitope on the α-chain of human leukocyte function-associated antigen 1 induces homotypic cell-cell interactions. J Immunol 1988; 141:1393–1400.
124. Bednarczyk JL, McIntyre BW. A monoclonal antibody to VLA-4 α-chain (CDw49d) induces homotypic lymphocyte aggregation. J Immunol 1990; 144:777–784.
125. Kansas GS, Tedder TF. Transmembrane signals generated through MHC class II, CD19, CD20, CD39, and CD40 antigens induce LFA-1-dependent and independent adhesion in human B cells through a tyrosine kinase-dependent pathway. J Immunol 1991; 147:4094–4102.
126. Pulido R, Elices MJ, Campanero MR, et al. Functional evidence for three distinct and independently inhibitable adhesion activities mediated by the human integrin VLA-4. Correlation with distinct α4 epitopes. J Biol Chem 1991; 266:10241–10245.
127. Shen CX, Stewart S, Wayner E, Carter W, Wilkins J. Antibodies to different members of the β1 (CD29) integrins induce homotypic and heterotypic cellular aggregation. Cell Immunol 1991; 138:216–228.
128. Odum N, Yoshizumi H, Okamoto Y, et al. Signal transduction by HLA class II molecules in human T cells: induction of LFA-1-dependent and independent adhesion. Hum Immunol 1992; 35:71–84.
129. Robinson MK, Andrew D, Rosen H, et al. Antibody against the Leu-CAM β-chain (CD18) promotes both LFA-1-and CR3-dependent adhesion events. J Immunol 1992; 148:1080–1085.
130. Wuthrich RP. Monoclonal antibodies to the murine VLA-6 α-chain trigger homotypic lymphocyte aggregation. Immunology 1992; 77:214–218.

131. Bednarczyk JL, Wygant JN, Szabo MC, et al. Homotypic aggregation triggered by monoclonal antibody specific for a novel epitope expressed by the integrin $\beta 1$ subunit: conversion of nonresponsive cells by transfecting human integrin $\alpha 4$ subunit cDNA. J Cell Biochem 1993; 51:465–478.
132. Campanero MR, Del-Pozo MA, Arroyo AG, et al. ICAM-3 interacts with LFA-1 and regulates the LFA-1/ICAM-1 cell adhesion pathway. J Cell Biol 1993; 123:1007–1016.
133. Nathan CF. Neutrophil activation on biological surfaces. Massive secretion of hydrogen peroxide in response to products of macrophages and lymphocytes. J Clin Invest 1987; 80:1550–1560.
134. Aida Y, Pabst MJ. Neutrophil responses to lipopolysaccharide. Effect of adherence on triggering and priming of the respiratory burst. J Immunol 1991; 146:1271–1276.
135. Lantz M, Bjornberg F, Olsson I, Richter J. Adherence of neutrophils induces release of soluble tumor necrosis factor receptor forms. J Immunol 1994; 152:1362–1369.
136. Nathan C, Srimal S, Farber C, et al. Cytokine-induced respiratory burst of human neutrophils: dependence on extracellular matrix proteins and CD11/CD18 integrins. J Cell Biol 1989; 109:1341–1349.
137. Shappell SB, Toman C, Anderson DC, Taylor AA, Entman ML, Smith CW. Mac-1 (CD11b/CD18) mediates adherence-dependent hydrogen peroxide production by human and canine neutrophils. J Immunol 1990; 144:2702–2711.
138. Laudanna C, Melotti P, Bonizzato C, et al. Ligation of members of the $\beta 1$ or the $\beta 2$ subfamilies of integrins by antibodies triggers eosinophil respiratory burst and spreading. Immunology 1993; 80:273–280.
139. Dike LE, Farmer SR. Cell adhesion induces expression of growth-associated genes in suspension-arrested fibroblasts. Proc Natl Acad Sci USA 1988; 85: 6792–6796.
140. Werb Z, Tremble PM, Behrendtsen O, Crowley E, Damsky CH. Signal transduction through the fibronectin receptor induces collagenase and stromelysin gene expression. J Cell Biol 1989; 109:877–889.
141. Adams JC, Watt FM. Fibronectin inhibits the terminal differentiation of human keratinocytes. Nature 1989; 340:307–309.
142. Dedhar S. Signal transduction via the $\beta 1$ integrins is a required intermediate in interleukin-1β induction of alkaline phosphatase activity in human osteosarcoma cells. Exp Cell Res 1989; 183:207–214.
143. Freeman GJ, Freedman AS, Segil JM, Lee G, Whitman JF, Nadler LM. B7, a new member of the Ig superfamily with unique expression on activated and neoplastic B cells. J Immunol 1989; 143:2714–2722.
144. Buck CA. Immunoglobulin superfamily: structure, function and relationship to other receptor molecules. Semin Cell Biol 1992; 3:179–188.
145. Dymecki SM, Zwollo P, Zeller K, Kuhajda FP, Desiderio SV. Structure and developmental regulation of the B-lymphoid tyrosine kinase gene blk. J Biol Chem 1992; 267:4815–4823.
146. Perlmutter RM, Levin SD, Appleby MW, Anderson SJ, Alberola-Ila J. Reg-

ulation of lymphocyte function by protein phosphorylation. Annu Rev Immunol 1993; 11:451–499.
147. Aoki Y, Isselbacher KJ, Cherayil BJ, Pillai S. Tyrosine phosphorylation of Blk and Fyn Src homology 2 domain-binding proteins occurs in response to antigen-receptor ligation in B cells and constitutively in pre-B cells. Proc Natl Acad Sci USA 1994; 91:4204–4208.
148. Hata A, Sabe H, Kurosaki T, Takata M, Hanafusa H. Functional analysis of Csk in signal transduction through the B-cell antigen receptor. Mol Cell Biol 1994; 14:7306–7313.
149. Kurosaki T, Takata M, Yamanashi Y, et al. Syk activation by the Src-family tyrosine kinase in the B cell receptor signaling. J Exp Med 1994; 179:1725–1729.
150. Takata M, Sabe H, Hata A, et al. Tyrosine kinases Lyn and Syk regulate B cell receptor-coupled Ca^{2+} mobilization through distinct pathways. EMBO J 1994; 13:1341–1349.
151. Zwollo P, Desiderio S. Specific recognition of the blk promoter by the B-lymphoid transcription factor B-cell-specific activator protein. J Biol Chem 1994; 269:15310–15317.
152. Bierer BE, Sleckman BP, Ratnofsky SE, Burakoff SJ. The role of CD2, CD4, and CD8 in T-cell activation. Annu Rev Immunol 1989; 7:579–599.
153. Fraser JD, Strauss D, Weiss A. Signal transduction events leading to T-cell lymphokine gene expression. Immunol Today 1993; 14:357–361.
154. Julius M, Maroun CR, Haughn L. Distinct roles for CD4 and CD8 as co-receptors in antigen receptor signaling. Immunol Today 1993; 14:177–183.
155. Perlmutter RM, Levin SD, Applby MK, Anderson SJ, Alberola-Ila J. Regulation of lymphocyte function by protein phosphorylation. Annu Rev Immunol 1993; 11:451–499.
156. Rudd CE, Janssen O, Prasad KV, et al. *src*-related protein tyrosine kinases and their surface receptors. Biochim Biophys Acta 1993; 1155:239–266.
157. Samelson LE, Egerton M, Thomas PM, Wange RL. The T cell antigen receptor tyrosine kinase pathway. Adv Exp Med Biol 1992; 323:9–16.
158. Ledbetter JA, Deans JP, Aruffo A, et al. CD4, CD8 and the role of CD45 in T-cell activation. Curr Opin Immunol 1993; 5:334–340.
159. Allison JP. CD28-B7 interactions in T-cell activation. Curr Opin Immunol 1994; 6:414–419.
160. Isakov N, Wange RL, Samelson LE. The role of tyrosine kinases and phosphotyrosine-containing recognition motifs in regulation of the T cell-antigen receptor-mediated signal transduction pathway. J Leukoc Biol 1994; 55:265–271.
161. Webb SR, Gascoigne NR. T-cell activation by superantigens. Curr Opin Immunol 1994; 6:467–475.
162. Noelle RJ, Marshall L, Roy M, et al. Role of contact and soluble factors in the growth and differentiation of B cells by helper T cells. Adv Exp Med Biol 1992; 323:131–138.
163. Andre P, Cambier JC, Wade TK, Raetz T, Wade WF. Distinct structural

compartmentalization of the signal transducing functions of major histocompatibility complex class II (Ia) molecules. J Exp Med 1994; 179:763–768.

164. Poudrier J, Owens T. CD54/intercellular adhesion molecule 1 and major histocompatibility complex II signaling induces B cells to express interleukin 2 receptors and complements help provided through CD40 ligation. J Exp Med 1994; 179:1417–1427.

165. Scholl PR, Geha RS. MHC class II signaling in B-cell activation. Immunol Today 1994; 15:418–422.

166. Tscherning T, Claesson MH. Signal transduction via MHC class-I molecules in T cells. Scand J Immunol 1994; 39:117–121.

167. Zembala M, Siedlar M, Ruggiero I, et al. The MHC class-II and CD44 molecules are involved in the induction of tumour necrosis factor (TNF) gene expression by human monocytes stimulated with tumour cells. Int J Cancer 1994; 56:269–274.

168. Juan M, Vinas O, Pino-Otin MR, et al. CD50 (intercellular adhesion molecule 3) stimulation induces calcium mobilization and tyrosine phosphorylation through $p59^{fyn}$ and $p56^{lck}$ in Jurkat T cell line. J Exp Med 1994; 179: 1747–1756.

169. Bierer BE, Peterson A, Gorga JC, Herrmann SH, Burakoff SJ. Synergistic T cell activation via the physiological ligands for CD2 and the T cell receptor. J Exp Med 1988; 168:1145–1156.

170. Moingeon P, Chang H-C, Wallner BP, Stebbins C, Frey AZ, Reinherz EL. CD2-mediated adhesion facilitates T lymphocyte antigen recognition function. Nature 1989; 339:312–314.

171. Vandenberghe P, Freeman GJ, Nadler LM, et al. Antibody and B7/BB1-mediated ligation of the CD28 receptor induces tyrosine phosphorylation in human T cells. J Exp Med 1992; 175:951–960.

172. Rudd CE, Janssen O, Cai YC, da-Silva AJ, Raab M, Prasad KV. Two-step TCR ζ/CD3-CD4 and CD28 signaling in T cells: SH2/SH3 domains, protein-tyrosine and lipid kinases. Immunol Today 1994; 15:225–234.

173. Rothlein R, Kishimoto TK, Mainolfi E. Cross-linking of ICAM-1 induces co-signaling of an oxidative burst from mononuclear leukocytes. J Immunol 1994; 152:2488–2495.

174. Lasky LA. Selectins: interpreters of cell-specific carbohydrate information during inflammation. Science 1992; 258:964–969.

175. Bevilacqua MP, Nelson RM. Selectins. J Clin Invest 1993; 91:379–387.

176. Rosen SD. Cell surface lectins in the immune system. Semin Immunol 1993; 5:237–247.

177. Varki A. Selectin ligands. Proc Natl Acad Sci USA 1994; 91:7390–7397.

178. Waddell TK, Fialkow L, Chan CK, Kishimoto TK, Downey GP. Potentiation of the oxidative burst of human neutrophils. A signaling role for L-selectin. J Biol Chem 1994; 269:18485–18491.

179. Laudanna C, Constantin G, Baron P, et al. Sulfatides trigger increase of cytosolic free calcium and enhanced expression of tumor necrosis factor-α and interleukin-8 mRNA in human neutrophils. Evidence for a role of L-selectin as a signaling molecule. J Biol Chem 1994; 269:4021–4026.

180. Stockl J, Majdic O, Rosenkranz A, et al. Monoclonal antibodies to the carbohydrate structure Lewis X stimulate the adhesive activity of leukocyte integrin CD11b/CD18 (CR3, Mac-1, $\alpha m\beta 2$) on human granulocytes. J Leukoc Biol 1993; 53:541–549.
181. Strauch UG, Holzmann B. Triggering of L-selectin (gp90MEL-14) induces homotypic lymphocyte adhesion by a mechanism independent of LFA-1. Int Immunol 1993; 5:393–398.
182. Murakawa Y, Minami Y, Strober W, James SP. Association of human lymph node homing receptor (Leu 8) with the TCR/CD3 complex. J Immunol 1992; 148:1771–1776.
183. Richter J, Zetterberg E. L-Selectin mediates downregulation of neutrophil TNF receptors. J Leukoc Biol 1994; 56:525–527.

6

Adhesion Molecules as Mechanoreceptors

Gregory J. Fulton, Mark G. Davies, and Per-Otto Hagen
Vascular Biology and Atherosclerosis Research Laboratory, Department of Surgery, Duke University Medical Center, Durham, North Carolina

I. INTRODUCTION

Cells are subject to a variety of stimuli originating from differing energy sources (e.g., pressure, temperature, light, and sound). Receptors are transducers that convert these various forms of energy into a signal that the cells can act upon. Most receptor types respond more readily to one form of energy than to others; however, virtually all receptors can be activated by several different forms of energy if the intensity of the stimulus is sufficient. Cells differentiate among stimuli by presenting different receptor types, by quantifying the intensity of individual stimuli, and by selectively activating secondary messenger pathways. A mechanoreceptor responds to changes in forces applied to the cell. These may be externally applied or originate as the result of changes in intracellular dynamics. Mechanoreceptors reflect changes in applied force through attachment with the cytoskeleton at the cell periphery or through direct activation of second-messenger systems, thereby converting mechanical energy into the cellular language of biochemical messengers. Changes in mechanical force are reflected in alterations at cellular, molecular, and genetic levels and are produced through both rapid response pathways and chronic adaptive changes.

Mechanical forces acting on a cell vary from direct action on a cell's external structure, friction shear at a cell's surface, endogenous outward tensile forces exerted by the cytoskeleton of the cell, torsional stresses exerted across the tissue plane in which a cell rests and external hydrostatic forces (1). These mechanical forces are imposed on a preexisting force equilibrium generated by the cell's own cytoskeletal tension and are distributed through the cell by its cytoskeleton. These intracellular tensile stresses can also be transmitted to adjacent cells and to the underlying extracellular matrix, via focal adhesions on the cell surface. Endothelial cells are subject to the various physical and chemical influences experienced by other cell types; however, the shear stresses they experience from their exposure to high blood flow rates are much greater than those experienced by other cells. This makes them an excellent model for considering how mechanical forces may influence a cell. The various families of adhesion molecules form an important part of the endothelial cell responses to these influences and their interactions with adjacent cell types. This chapter will deal with the role that adhesion molecules play in response to mechanical forces and will examine the hypothesis that they may be considered as mechanoreceptors.

Two major types of mechanical forces act on the vessel wall. The first are the hemodynamic forces exerted by the blood on the endothelium; the second are those forces causing deformation within the vessel wall itself—so-called biosolid mechanical forces (2). The hemodynamic forces consist of the blood pressure, varying through the cardiac cycle, and the shear stresses that result from the flow of blood on the endothelial cell-blood interface. Shear stress ($t_w = 32\mu Q/\pi D^3$: where μ = viscosity, Q = flow and D = vessel diameter) is a measure of force per unit area and acts parallel to the surface of the arterial wall. This definition assumes laminar flow and Newtonian fluid characteristics, such as exist in larger vessels. In most larger vessels, the diameter appears to have adapted to maintain shear stress at a relatively constant value of 15 dynes/cm^2. Flow in arteries tends to be laminar, however, with the changes in direction of the vessel path and the branching off of smaller-caliber vessels, turbulence and steep velocity gradients exist at the vessel wall resulting in low flow and low shear stresses. This is most noticeable at the bifurcation of large arteries where high shears exist at the apex of the bifurcation, whereas low shear stresses exist at the outer walls.

There is a significant tendency for atherosclerotic plaques to form in such areas. Blood pressure is almost 100× greater in magnitude than shear stress, with a value approximating 1300 dynes/cm^2. It is not clear whether endothelial cells respond to constant pressure, but the rhythmic changes in pressure in the vessel wall that occur with the cardiac cycle result in a fluctuating normal stress (perpendicular to direction of flow). Within the vessel wall, this causes tension and a deformation of the smooth muscle cells, the degree of deformity being a function of the vessels' own elasticity.

The product of stress and the reciprocal of the modulus of elasticity is called strain and is a measure of displacement of the arterial wall. Experimental models of shear stress and "cyclic strain" generated by the pulsatile arterial flow have been established, permitting investigation of the cellular events resulting from changes in these variables (2).

II. MECHANICAL STRESS MECHANISMS AND THE CELL

Force transduction in endothelial cells is a combination of force transmitted through the cell's cytoskeleton and transduction of this mechanical energy to biochemical signals at specific mechanotransducer receptor sites. The microfilament network confers tension to a cell which, with microtubular rigidity, determines a cell's shape. Physical deformation of a membrane protein or cytoskeletal element produces conformational changes within an integral membrane protein, within proteins associated with the membrane or within intracellular cytoskeleton proteins to produce force transmission. Figures 1 and 2 illustrate the likely sites of transduction and the principal pathways linking stress transmission and transduction (1). Although the stress transmission pathway is common, different responses may be determined by the nature of the association between the cytoskeleton and mechanotransducers at the different locations in the cell and possibly by transducers that are independent of the cytoskeleton.

Mechanical deformation of endothelial cells has been shown to produce a rapid and large increase in intracellular calcium (3). Diamond and colleagues (4) have shown that this mechanically induced calcium mobilization is dependent on actin filaments and phospholipase in bovine aortic endothelial cells. Mechanically activated ion channels provide a rapid transduction mechanism for the conversion of stress to a biochemical response (Fig. 3). Stretch-activated, stretch-inactivated, and shear stress-activated ion channels have been identified (5). Cytoskeletal elements are necessary for stretch-inactivated and shear stress-activated ion channels. Influx of extracellular calcium via these channels is a requirement for the mobilization of intracellular calcium stores (3,6). Calcium changes attributed to increases in blood flow are less clear. Studies in both cultured cells and whole vessel preparations of shear stress induced calcium mobilization have shown inconsistent results (7–12). Nonetheless, rises in intracellular calcium are implied by the production of nitric oxide, prostacyclin, and inositol phosphate, all of which occur in endothelial cells in response to increased flow (13–15).

Phosphatidylinositol metabolism is an important aspect of signal response pathways within cells. Phospholipase activation occurs in response

Table 1 Responses of Endothelial Cells to Mechanical Forces

Force	Effect	Response Time
Shear stress	K^+ channel activation	msec
	Membrane hyperpolarization	seconds
	Biphasic increase in IP_3 concentration	<1 and 5 min
	Rise in intracellular calcium	seconds
	Increased prostacyclin release	seconds
	Increased nitric oxide release	seconds
	Increased endothelin release	minutes to hours
	Realignment of focal adhesion sites	minutes to hours
	Induction of *c-myc, c-fos*, and *c-jun* expression	minutes to hours
	Increased PDGF-A and -B mRNA expression	hours
	Increased tPA and PAI-1 expression and secretion	hours
	Cytoskeletal and glycoprotein rearrangement	hours
	Alterations in thrombomodulin expression	hours
	Decreased fibronectin synthesis	hours to days
	Increased LDL metabolism	hours to days
Stretch	Cation channel activation	msec
	Activation of adenylate cyclase	minutes
Flow	Release of ATP, substance P, and acetylcholine	seconds
	Cell proliferation	hours
	Increased cell turnover	hours
Cyclic strain	Transient rise in IP_3 concentration	msec
	Alteration in G-protein immunoreactivity	minutes to hours
	Cell and cytoskeletal realignment	minutes to hours
	Decreased collagen synthesis	hours

to changes in shear stress and stretch (14). Phospholipases generate inositol triphosphate and diacylglycerol producing acute (Ca^{2+} influx and mobilization), rapid (Ca^{2+} acting as a second messenger for enzymes such as cNOS (16)), and delayed (Ca^{2+} activation of protein kinases within subsequent changes in gene expression (17)) responses. The presence of phospholipases within the cellular membrane allows an association between these enzymes and membrane channels (Fig. 4). In addition, it has recently been demonstrated that specific phosphoinositides in association with actin binding proteins (e.g., profilin) inhibit phospholipase C.

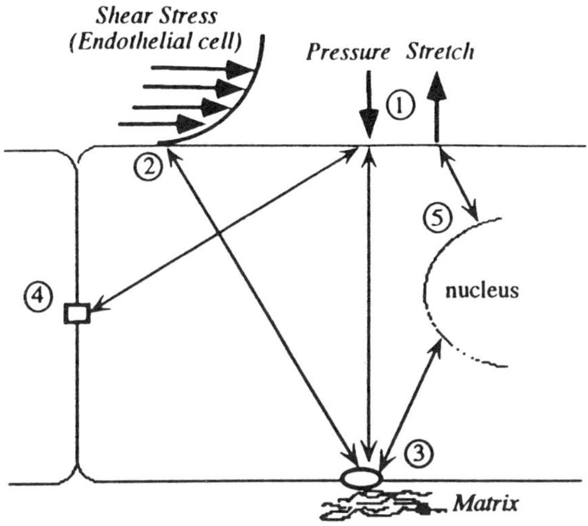

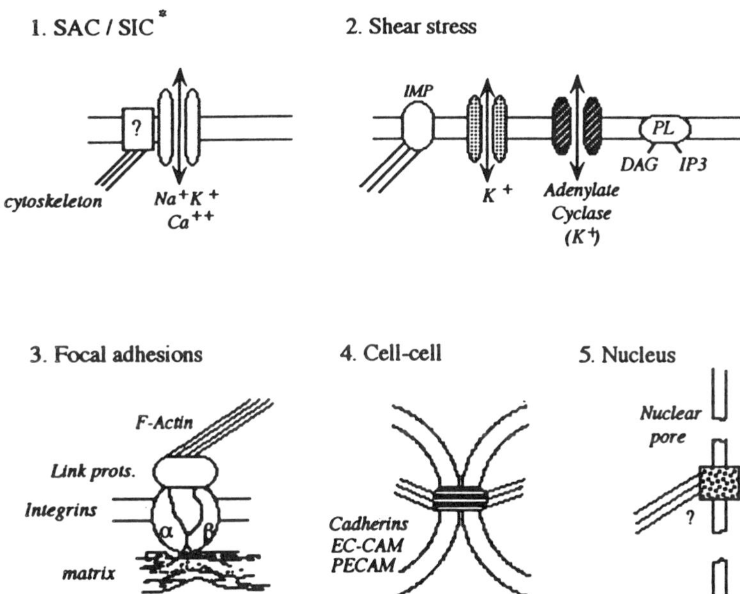

Figure 1 These diagrams show the likely sites of mechanical stress transduction in cells with interaction with the cytoskeleton. 1. SAC Stretch-activated ion channel; SIC stretch-inactivated ion channel. 2. Shear stress activation of ion channels and inositol phosphate (IP_3) formation. 3. Integrin-mediated focal adhesions. 4. Cadherin-mediated cell-cell adhesions. 5. Cytoskeletal-mediated interaction with the nucleus. (From Davies PF, Tripathi SC, Circ Res 1993; 72:239–245; by copyright permission of the American Heart Association.)

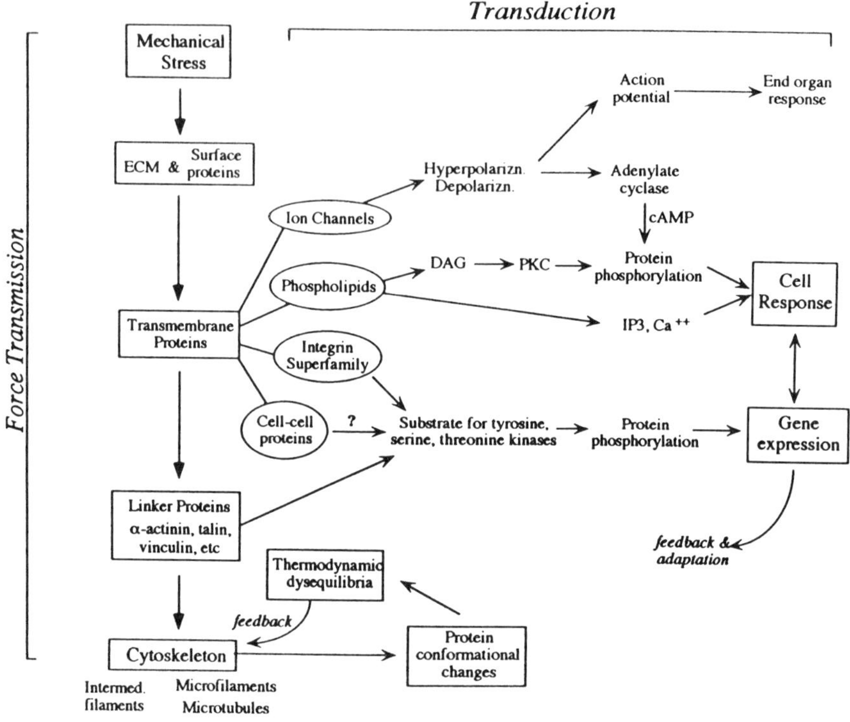

Figure 2 Diagram showing mechanisms of mechanical stress transmission and transduction in cells. ECM—extracellular matrix; DAG—diacylglycerol; PKC—protein kinase C; IP_3—inositol 1,4,5-trisphosphate. (From Davies PF, Tripathi SC, Circ Res 1993; 72:239–245; by copyright permission of the American Heart Association.)

Within any group of cells, adhesion molecules are responsible for cell-cell adhesion and the transmission of signals through interaction with the cytoskeleton, extracellular matrix, and intracellular messengers. Interaction exists not only between phenotypically identical neighbors such as adjacent vascular endothelial cells, but also between differing cell types such as vascular smooth muscle cells and fibroblasts. In the case of the endothelial cell, interaction with circulating leukocytes and platelets is governed by adhesion molecules and is subject to changes in flow and the presence of other substances such as inflammatory mediators (18–20). This has its greatest significance in leukocyte attachment and margination during the

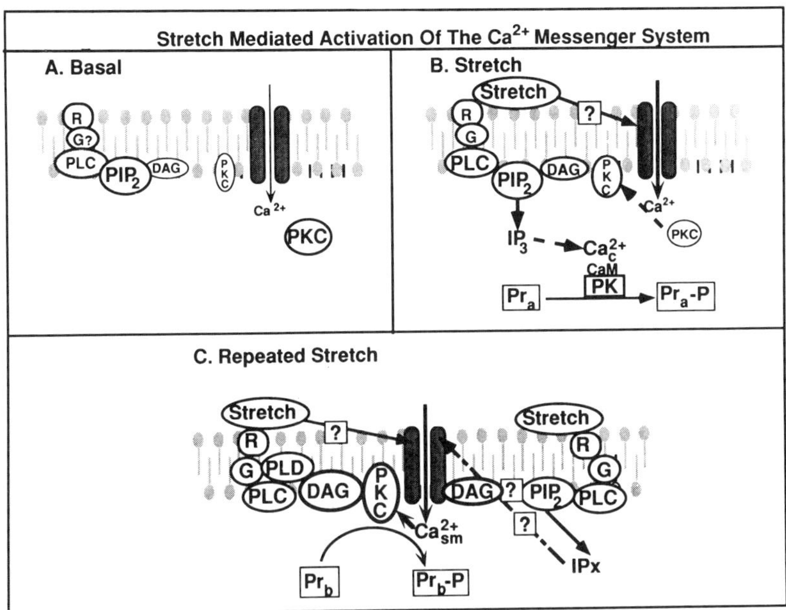

Figure 3 Model for stretch-mediated activation of the calcium messenger system. (A) A resting endothelial celi is shown. Under these conditions there is little extracellular calcium influx, only small amounts of protein kinase C (PKC) attached to the membrane, most of it being found in the cytosol and little or no phosphatidylinositol 4,5 biphosphate (PIP_2) breakdown. (B) Upon initiation of stretch, there is activation of a phosphoinositol-specific phospholipase C (Pl-PLC) leading to the generation of at least two second messengers: inositol 1,4,5-trisphosphate (IP_3) and diacylglycerol (DAG). IP_3 is water-soluble and liberates calcium from a nonmitochondrial pool. This rise in cytosolic calcium likely activates a calmodulin-dependent kinase and, in addition, helps translocate the cytosolic PKC to the plasma membrane. During the initial stretch, calcium influx is also increased, although it is unclear whether this is from direct activation of a stretch channel or a second messenger or receptor-operated calcium channel. (C) Upon repeated stretch, there is continued Pl-PLC activation, but in addition phospholipase D (PLD) is now activated and serves as an additional source of DAG. Continued phosphoinositol (Pl) turnover increases inositol tetrakisphosphate (IP_4) levels activating a calcium permeable channel, which in conjunction with the stretch-activated calcium channel leads to a sustained rise in calcium influx. In the presence of DAG and calcium influx, PKC is activated which leads to a variety of endothelial cell responses. (From Isales C, Rosales O, Sumpio BE. Chapter 6, Hemodynamic Forces and Vascular Cell biology; by copyright permission of R. G. Landes, Austin, Tex.)

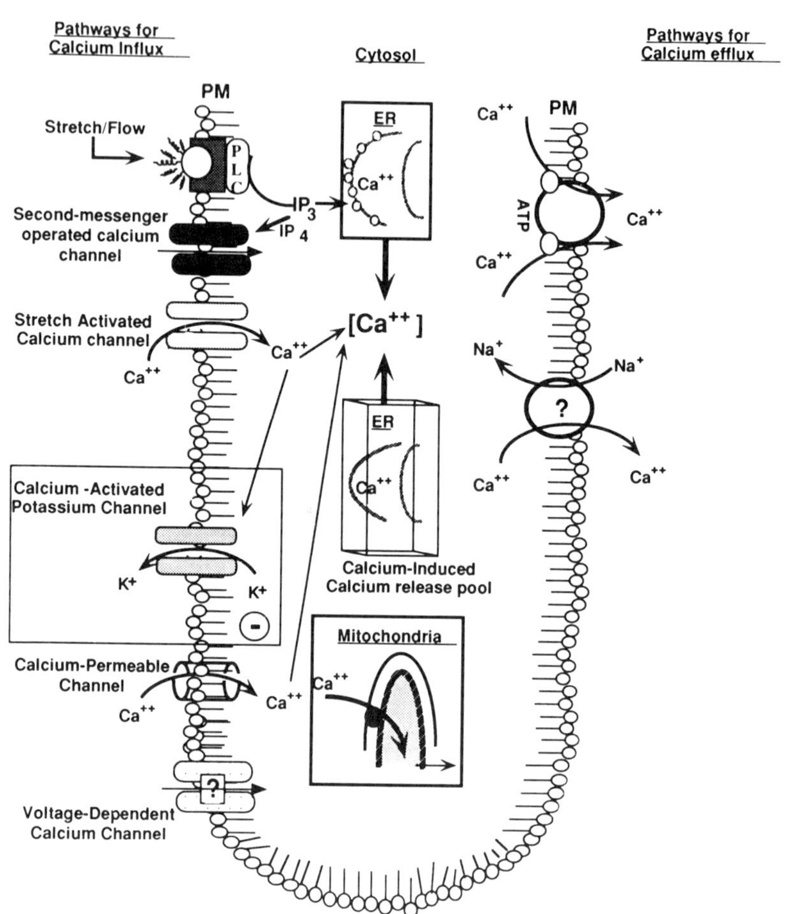

Figure 4 Cytosolic calcium concentration is the result of a balance between calcium influx, intracellular calcium redistribution, and calcium efflux. There are multiple pathways for calcium influx in endothelial cells (EC), some or all of which may or may not be present depending on the type of endothelial cell studied. Possible pathways for influx include: 1. stretch- or flow-activated increases in inositol phosphates. Inositol 1,3,4,5-tetrakisphosphate may then open a calcium permeable channel; 2. there may be a stretch-activated calcium channel; 3. a rise in Ca^{2+} through mobilization of intracellular stores may then activate a calcium-dependent potassium channel. By hyperpolarizing the EC, it favors calcium entry through a calcium permeable channel down its electrogenic gradient; 4. voltage-activated calcium channel, though present in many excitable cells, does not appear to be present in EC. There seem to be at least two intracellular calcium pools, an inositol 1,4,5-trisphosphate-mobilizable pool, and a calcium-inducible pool. The efflux pathways

inflammatory response. Expression of adhesion molecules appears to be regulated, at least in part, by mechanical forces.

The main families of adhesion molecules reviewed in this chapter are the selectins, the integrins, members of the immunoglobulin superfamily, and the cadherins.

III. ADHESION MOLECULES AND HEMODYNAMICS

The selectins are cell surface transmembrane glycoproteins with an N-terminal lectinlike domain, an epidermal growth factor (EGF) repeat, and a variable number of 60 amino acid modules similar to those found in some complement-binding proteins (21,22). It appears that the lectin and EGF domains are responsible for the binding properties while the repeating modules stereotactically enhance the presentation of these binding regions. The lectin domains render the selectins capable of binding the carbohydrate moieties, which mediate selectin attachment to other cells (21,22). Selectins are named in accordance with the cell type in which they were initially identified. They mediate the initial binding of leukocytes, particularly neutrophils, to activated endothelium.

Neutrophil adhesion is one of the first events in the acute inflammatory response, and monocyte infiltration has been implicated as an early event in the formation of atherosclerotic plaques (23). At the low shear stresses which exist in postcapillary venules (approx. 1.5 dynes/cm^2), the presence of acute inflammatory mediators such as histamine, thrombin, and hydrogen peroxide cause release of P-selectin (platelet originated) from Weibel-Palade bodies in endothelial surface within minutes and is able to adhere to circulating leukocytes. This adherence is not so strong as to lead to complete arrest of the leukocytes but rather tethers them allowing them to roll along the endothelial surface (26,27). E-selectin (endothelial derived) is not constitutively stored and is expressed several hours after the inflammatory response is initiated (22). This is generally in response to cytokines such as IL-1, TNF, and bacterial endotoxin. It too produces tethering and rolling of neutrophils on the endothelial surface (28), but also produces a similar

---

for calcium efflux are not well characterized, but calcium can be pumped out thorough a calcium ATPase pump or sodium-calcium exchanger. The latter has been shown to be involved in calcium influx in certain cells, although this does not seem to be the case for EC. (From Isales C, Rosales O, Sumpio BE. Chapter 6, Hemodynamic Forces and Vascular Cell Biology; by copyright permission of R. G. Landes, Austin, Texas.)

effect with other leukocyte subtypes such as monocytes, eosinophils, and some memory T cells (29–32).

L-selectin (leukocyte derived) is expressed on the leukocyte surface, most noticeably on the tips of membranous projections (33). This distribution is probably to facilitate leukocyte-endothelial binding. L-selectin can bind to the endothelium through a variety of ligands (34). These tend to be mainly carbohydrate ligands which bind to the lectin domain, such as the oligosaccharide sialyl–Lewis x, which is recognized by those selectins expressed on the endothelial surface (35,36). While binding between L-selectin and its endothelial-based family members is not the sole binding mechanism for either individual component monoclonal antibodies that block L-selectin have shown that this molecule accounts for over 65% of the total E-selectin binding capability at low shear stresses (28,34). In addition, anti-L-selectin antibodies also significantly inhibit neutrophil transendothelial migration (37).

L-selectin is the only member of the family that is constitutively expressed on the cell surface. It is rapidly shed by leukocytes upon activation, suggesting that selectin activity is dependent on surface expression (38). These L-selectin effects are reduced at higher shear stresses.

Integrins are cell surface heterodimeric membrane glycoproteins, which, unlike selectins, are expressed by virtually all human tissue types (39). The term integrin was designated to highlight their integration of extracellular matrix activities with the cytoskeleton; i.e., they connect the cell with its surrounding extracellular environment. Integrins are composed of α and β subunits associated through noncovalent bonds. The cell can alter how it interacts with the extracellular matrix by modifying the component subunits, thus creating a spectrum of interactions (39,40). Certain integrins play a vital role in the endothelial cell-leukocyte interactions which occur in the low-shear environment of the postcapillary venule (41). Neutrophils adhere to activated endothelium through the selectin family, causing them to roll along the endothelial surface; however, they cannot marginate beneath the endothelium unless completely arrested (18).

CD11/CD18 is a leucocyte-specific integrin which adheres to activated endothelial cells, by binding to members of the immunoglobulin supergene family, a third group of adhesion molecules (42). This glycoprotein complex consists of three heterodimers in which a common β_2 subunit (CD18) is noncovalently linked with one of three variable component α subunits (CD11). Studies with monoclonal antibodies have shown that these integrins do not bind activated endothelium unless shear stresses are low (<0.5 dynes/cm^2) (41). Under static conditions, anti-CD18 antibody decreases leucocyte binding to stimulated endothelial cells; however, there is no such effect when flow rates producing venous shear are introduced (41).

These results suggest that leukocyte velocity must be reduced to an appropriate level before adhesion to the endothelium can occur. By causing tethering and subsequent rolling of leukocytes, the selectins reduce their velocity, enabling CD11/CD18 binding. Such binding is also necessary for leukocyte margination beneath the endothelium (18). Integrin-leukocyte binding requires the presence of activating factors which increase the avidity of integrin binding potential. These include platelet activating factor (PAF), leukotriene B_4, platelet-derived growth factor (PDGF), and components of the complement system (39). CD11/CD18 members bind to members of the immunoglobulin supergene family acting as adhesion molecules, on the endothelial surface. Expression of these immunoglobulins is regulated by, among other things, the shear stresses that are exerted on the vessel wall (43–46).

Intercellular adhesion molecules (ICAMs) are members of the immunoglobulin supergene family which act as adhesion molecules. They are transmembrane glycoproteins containing five (ICAM-1) or two (ICAM-2) extracellular immunoglobulin domains (47,48). They are expressed on endothelial cells, on various leukocyte subtypes, and on a variety of other selected cell types such as fibroblasts, chondrocytes, and epithelial cells. The two domains of ICAM-2 are identical to the two N-terminal domains of ICAM-1 (48). ICAM-1 distribution on endothelial cells is similar to that of E-selectin; however, it is constitutively expressed in small amounts on the surface of unstimulated cells, and has a slower (12–24 hours), more prolonged response to cytokines. Unlike P-selectin, ICAM-1 is induced by α-interferon (αIF) (19).

ICAM-2 is constitutively expressed but is not regulated by cytokines (49). Because of the rapid initiation of CD11/CD18-mediated leukocyte adhesion following endothelial stimulation, it is constitutive rather than inducible ICAM-1 that appears to be responsible for this effect. ICAM-1 is expressed not only on the apices of endothelial cells like P- and E-selectins, but is also distributed in the basal area of the cells. This widespread expression enables ICAM-1 to act as a mediator of leukocyte migration by providing a pathway through the endothelial cell layer. Recent studies have shown that ICAM-1 expression is elevated with increases in shear stress in some tissues (43,50).

Walpola and colleagues have shown in studies on rabbit carotid artery that ICAM-1 upregulation associated with increases in shear stress is not associated with detectable leukocyte binding. While this may be due to increased mechanical forces, it may also be due to ICAM-1 being expressed in regions inaccessible to leukocytes. They demonstrated that ICAM-1 staining is most predominant at endothelial cell junctions. The reason for this is unclear; it may be due to ICAM-1 involvement in endothelial cell-cell adhesions. Subsequent upregulation of ICAM expression in response to increases in shear stress may then occur in order to protect cell junctions. Alternatively, these intercellular junctions may form a junctional ICAM-1 pool which can be rapidly mobilized upon cell activation (51).

Vascular cell adhesion molecules (VCAM) are also expressed on the surface of endothelial cells, particularly in lymph nodes. They bind monocytes and lymphocytes through the integrin VLA-4, but do not bind neutrophils. Like the ICAMs, VCAMs allow leukocyte arrest and margination. VCAMs are down-regulated by increases in shear stress—an effect that is not only rapid (within 1 hour) but also reversible (within 72 hours) (46). This response appears to be regulated at the level of gene expression as there is a decrease in the detectable amount of VCAM mRNA. This corresponds to the reduction in VCAM expression on the cell surface (45). These shear stress-mediated effects appear to be independent of other reduced flow-related changes, such as increased prostacyclin and nitric oxide production. VCAM is upregulated in low shear stress and exhibits a heterogenous distribution pattern among the endothelial cells (52). This has been shown to coincide with the adhesion of monocytes, a coincidence that occurs in experimental models of atherogenesis (53).

Other mediators of monocyte adherence may be implicated, however, as one-third of monocytes bind to areas where VCAM immunostaining is absent and there are areas of high VCAM staining where no monocyte binding is seen (50). The disparity in the response of VCAM and ICAM to shear stress may lie in the presence or absence of a promoter sequence which appears to be conserved among those proteins which are upregulated by increases in shear stress, such as platelet-derived growth factor (PDGF), tPA, and TGF-β_1 (54,55).

Resnick et al. (44) identified a cis-acting element in the PDGF-B promoter gene that is required for shear responsiveness (shear-stress-responsive element; SSRE). This core binding sequence (GAGACC) appears to bind nuclear transcription factors. The same sequence is present in the promotors for other genes that have been shown to be regulated by shear stress in the vascular endothelium. These include human, murine, and rodent tPA, human and murine TGF-β_1, and human ICAM-1 (44). The SSRE sequence is even more remarkable in that it is present in an area of the genome where there is usually poor cross-species conservation (44). Both E-selectin and VCAM are down-regulated by increases in shear stress, and neither has this core binding sequence in its promoter regions (44).

The profile of VCAM expression as well as its cell binding characteristics is different from ICAM-1. ICAM-1, unlike VCAM, is present not only at the cell apices but also in the basal layer, indicating that it may have a role in leukocyte migration (18). ICAM-1 can mediate neutrophil binding, but

VCAM cannot, as neutrophils do not express the VCAM binding ligand very-late-antigen-4 (VLA-4) (56). The distribution of ICAM-1 and VCAM tissue expression is different. VCAM expression in lymph node endothelium is high, whereas it is low in human vascular endothelial cells (52), unlike ICAM-1. This variable pattern of expression may explain the different responses observed to changes in shear stress between studies (43,45).

IV. ADHESION MOLECULES AND THE CYTOSKELETON

Regulation of cytoskeletal filament synthesis and assembly, and the maintenance of cytoskeletal organization affect fundamental processes of a cell, such as shape, locomotion, mitosis, adhesion, polarity, and contractility. Mechanical forces are distributed internally throughout the cell and externally to adjacent cells. This external communication occurs through points of contact with the extracellular matrix and other cells. Cytoskeleton filament assembly is essential for fundamental cell processes and the maintenance of cell integrity and requires attachment to the cell membrane. Both microfilaments and intermediate filaments form attachments to the cell membrane, with the F-actin fibers forming the major components of cytoskeletal attachment. The integrin and cadherin families of adhesion molecules from part of these assemblies.

Cell junctions are divided into junctional and nonjunctional adhesions. Integrins and cadherins are involved in both types (Fig. 5). Junctional adhesions can be subdivided into cell-cell and cell-substratum interactions. There is overlap between cadherins and integrins, however; cadherins form part of the intercellular adhesion complexes, while integrins mediate heterophilic binding in cellular adherence to the extracellular matrix. Cadherins tend to bind through formation of homophilic bonds, whereas integrins bind through formation of heterophilic bonds.

Integrins are expressed on almost all cells and are a major and important component of these anchoring complexes (39,57). On the ventral plasma membrane, these include sites of firm attachment called focal adhesions (15 nm or less), which, owing to their ease of study in cell culture, are one of the most extensively investigated mechanisms of cell adhesion (58). Integrins possess short, highly conserved cytoplasmic domains which appear to mediate cytoskeletal binding (57). There is a significant body of evidence that suggests that these domains bind with actin fibers through association with other cytoskeleton related proteins such as talin, vinculin, and α-actinin forming a protein complex at the cell membrane (59,60). The relative contribution of the different integrin subunits remains unclear. It appears that the more highly conserved β subunits are responsible for binding at these focal adhesions, whereas the role of the more variable α subunits

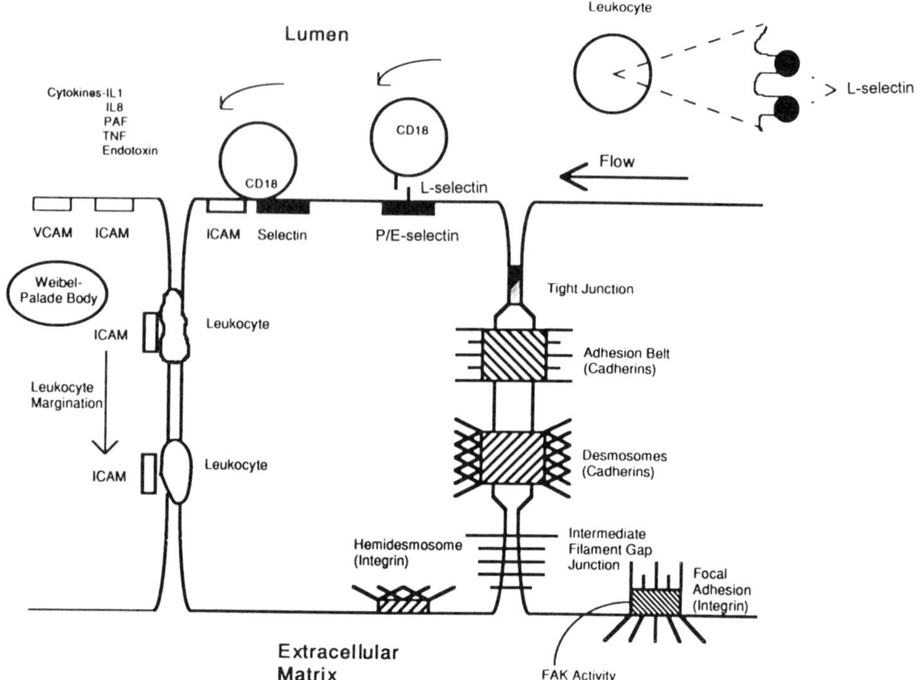

Figure 5 This shows the arrangement of adhesion molecules when expressed within the endothelial cell. Those molecules mediating endothelial-leukocyte adhesion and subendothelial margination are shown in the upper part of the figure. The mechanism of selectin-mediated leukocyte tethering, shown at the luminal surface, is followed by the binding of leukocyte integrins (noteably CD11/CD18), with subsequent leukocyte arrest and margination beneath the endothelium (left). The remainder of the figure illustrates the role of integrins and cadherins in forming cell-cell and cell-substratum junctions, and their cytoskeletal attachments.

may be in the regulation of the integrin with the other members of these protein complexes (57). Talin and α-actinin bind directly to the cytoplasmic domains of integrins, and both proteins are capable of binding F-actin (59,60). In addition, both proteins bind to vinculin which has the ability to bind F-actin. Vinculin does not appear to bind integrins directly (58,61,62).

Alpha-actinin binds integrin and F-actin through separate binding sites, which have been characterized using proteolytic enzymes (63). The integrin binding region appears to form a high-affinity bond at focal adhesions with the β-subunit, as evidenced by α-actinin binding to the β_2 subunit on the cytoplasmic domain of the leukocyte-specific LFA-1 integrin of activated neutrophils (64). Talin-integrin binding in vitro is of low affinity but is compensated for by high concentrations of these proteins at focal adhesions. Anti-talin antibodies block formation of focal adhesions but do not disrupt established adhesions, suggesting that talin has a role in the formation of focal adhesions, but not in their maintenance (65).

A number of other proteins are constituents of these integrin containing complexes. Zyxin is a constitutent protein found in such complexes. It is present in low abundance, which suggests that it may have a regulatory role. It binds to α-actinin and also to cysteine-rich protein (CRP), which possesses certain domains identified in other proteins known to regulate gene expression (66).

Some focal adhesion proteins that may interact with integrins have been implicated in the signal transduction mechanisms governing cytoskeletal organization. Both protein kinase C isoforms and various tyrosine kinases have been localized to focal adhesions (67,68). The latter group of enzymes have been implicated in integrin-mediated cell signal transduction (69). Phosphorylated tyrosine residues were first localized to the focal adhesions of normal cells in 1985 (70). The tyrosine kinase $pp60^{v\text{-}src}$ has been identified in focal adhesions of virally transformed sarcoma cells (71). In platelet studies, the phosphorylation that accompanies aggregation is dependent on integrin interaction (72,73). Other studies have shown that platelet aggregation is accompanied by $pp60^{v\text{-}src}$ association with the cytoskeleton (74).

Romer and Burridge (69) identified the focal adhesion constituents $pp125^{FAK}$ (focal adhesion kinase) and paxillin as substrates for tyrosine kinase phosphorylation during integrin-mediated adhesion in rat embryo fibroblasts. Inhibition with tyrosine kinase inhibitor herbimycin A not only blocked protein phosphorylation, but also interfered with focal adhesion formation and stress fiber assembly (69). $pp125^{FAK}$ is a tyrosine kinase in its own right, phosphorylation of which has been associated with integrin clustering and cell adhesion. It may be that the adhesion-associated increase in tyrosine phosphorylation is the result of integrin-mediated autophosphorylation of $pp125^{FAK}$ with subsequent phosphorylation of paxillin and other proteins. Romer and his colleagues (69) suggest that paxillin may be a tyrosine phosphorylated substrate with a role in cytoskeletal organization. Some cells express a shorter version of $pp125^{FAK}$ which is homologous to the C-terminal domain of its longer relative, termed FAK-related nonkinase. This enzyme localizes to focal adhesions, suggesting that it is this portion of the $pp125^{FAK}$ that binds to focal adhesions. $pp125^{FAK}$ has been shown to be autophosphorylated upon integrin clustering; however, it is not clear whether it binds integrins directly or indirectly through other proteins.

Wang et al. (75) have provided strong evidence supporting the concept that integrins act as mechanoreceptors, linking changes in cell appearance resulting from externally applied forces with integrin binding. Ferromagnetic beads were coated with antibody to β_1 integrin subunit or synthetic RGD peptide (Fig. 6). The latter represents the amino acid sequence arginine-glycine-aspartate, which is often an integrin binding sequence on their extracellular matrix (ECM) ligand. These beads were allowed to attach to capillary endothelial cells on culture plates. The beads were then magnetized in one direction with a strong, short-acting magnetic field. A second, weaker but prolonged magnetic field was then applied perpendicular to the first, thus producing a twisting action on the beads. The degree of twist was then quantified. When uncoated beads were used they rotated through 90°; however, when the beads were coated with the integrin binding compo-

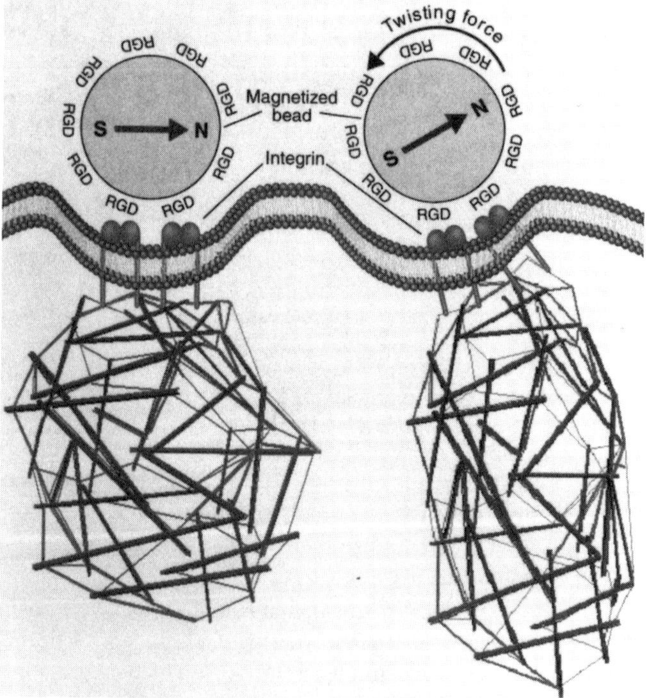

Figure 6 Magnetic beads attached to integrins are restrained by the cytoskeleton in their twisting response to a magnetic force field. The relationship between twisting force and the extent of bead twisting suggests tensegrity structure in the cytoskeleton. (From Heidemann SR. Science 260:1080–1081; by copyright permission of American Association for the Advancement of Science.)

nents, this degree of twist was reduced to less than 30°. The constraining mechanism was apparently the result of peptide-integrin-cytoskeleton binding. The results of this study support the theory that interacting molecules related to cell structure and its environment are organized according to tensional integrity; so-called "tensegrity" (76).

ICAMs also appear to interact with the cytoskeleton. It has been shown that α-actinin colocalizes with a peptide encompassing the predicted cytoplasmic domain and that this area can be mapped to the area close to the membrane spanning region. In addition, it has also been shown recently that ICAM-1 induces phosphorylation of the cytoskeleton-associated protein cortactin in studies of brain microvessel endothelial cells (77).

The cadherins share striking similarities with the integrins. While integrins for the most part mediate cell-substratum adhesion, the cadherins appear primarily to be involved in the maintenance of cell-cell adhesion (78,79). Cadherins are transmembrane glycoproteins and as with integrin adhesion, cadherin binding is Ca^{2+}-dependent. These molecules are 700 to 750 amino acid residues long with a molecular mass of 120 to 140 kDa (79). They have a large extracellular domain consisting of five homologous domains of 110 residues. The adhesive portion of the molecule is primarily localized in the first amino-terminal domain of this extracellular region, while the Ca^{2+} binding properties reside between domains 1 and 2 (80,81). The N-terminus is also responsible for specificity. Ca^{2+} binding to cadherins induces a conformational change which imparts adhesive properties (80,82). Absence of Ca^{2+} renders the cadherins susceptible to degradation by proteolytic enzymes. The remainder of the cadherin molecule consists of a single hydrophobic transmembrane region and a short, highly conserved cytoplasmic domain which can bind actin filaments through interacting with another family of proteins, the catenins (83,84).

Cadherins are classified in accordance with the tissue in which they were initially recognized. The first three cadherins identified were E-cadherin (epithelial cells), N-cadherin (nerve and muscle), and P-cadherin (placenta and epidermis); however, to date over 20 different subtypes have now been defined (79). Cadherins tend to bind to one another through homophilic bonds, therefore tending to bind to other cadherins of the same type. They can, however, form heterophilic bonds with other cadherin subtypes (79). E-cadherin is the most understood of the cadherin family and is found in the adhesion belts of epithelial cells, an adherens-type junction (79). It is the first cadherin expressed in mammalian development, causing blastomer attachment, so-called compaction in the eight-cell embryo (79). The homophilic binding property is believed to contribute to cells recognizing other cells of the same type in the developing embryo.

Cadherins are linked to the cytoskeleton at their C-terminus with bonds of high avidity. The bond strength is reflected in the failure of nonionic detergents to disturb junctional cadherins despite having the ability to extract most membrane proteins. The cytoplasmic domains are required for adhesion to the cytoskeleton (85). Fujimori and Takeichi (86) have shown that mutated N-cadherin cDNA lacking sections of the cytoplasmic domains caused disruption of cell-cell adhesions. Cadherins bind the cytoskeleton through interaction with the catenin protein family (79). This family of proteins consist of polypeptides each with a molecular mass of about 100 kDa. Three different catenins have been indentified—α, β, γ (84). Beta catenin binds most avidily to the cadherin, while the α-catenin binds to the actin filaments. Alpha-catenin is the best characterized of the three, and it is noted to have considerable homology to vinculin, which is part of the integrin-cytoskeleton protein links (87). Whether cadherins bind to other cytoskeletal proteins is unclear; however, it has been suggested that they can form complexes with ankyrin and fodrin, both of which are components of the microfilament assembly (78).

The cadherins appear to perform a similiar role to that of integrins; however, their main function is to mediate intercellular adhesion as opposed to cell-substratum adhesion. While they are not as well characterized as integrins, their link to the cytoskeleton is proven and it would therefore appear to possess the potential to respond to the mechanical forces applied to the cell.

V. CONCLUSION

Mechanical forces when applied to the cell cause changes in cell configuration with associated changes in function and protein expression. These forces are transmitted to other cells and to the extracellular matrix. Adhesion molecules link a cell with its neighboring cells and to its extracellular surroundings. Transmembrane topography and direct cytoskeletal binding enable the families of adhesion molecules to transmit changes in mechanical force equilibrium to the cell and its surroundings. The resultant alterations in cell function show that these molecules can act as mechanoreceptors. Adhesion molecules exhibit varying degrees of expression in response to different physical and chemical stimuli. This is consistent with the role of regulatory proteins, a necessary property for mechanoreceptors subject to a range of forces. In the case of ICAM expression on endothelial cells, this may be the result of possessing SSRE, a promoter sequence common to many proteins induced following increases in shear stress. Changes in adhesion molecule expression have been reported in many different disease processes (20). Current interest is focused on the influence mechanical forces have on expression of adhesion molecules, with the hope of defining means to modulate their expression.

REFERENCES

1. Davies PF, Tripathi SC. Mechanical stress mechanisms and the cell: an endothelial paradigm. Circ Res 1993; 72:239–245.
2. Sumpio BE, ed. Hemodynamic Forces and Vascular Cell Biology. Austin, Tex: R. G. Landes, 1993:1–116.
3. Sigurdson WL, Sachs F, Diamond SL. Mechanical perturbation of cultured human endothelial cells causes rapid increases of intracellular calcium. Am J Physiol 1993; 264:H1745–H1752.
4. Diamond SL, Sachs F, Sigurdson WL. Mechanically induced calcium mobilization in cultured endothelial cells is dependent on actin and phospholipase. Arterioscler Thromb 1994; 14:2000–2006.
5. Watson PA. Function follows form: generation of intracellular signals by cell deformation. FASEB J 1991; 5:2013–2019.
6. Demer LL, Wortham CM, Dirksen ER, Sanderson MJ. Mechanical stimulation induces intercellular calcium signaling in bovine aortic endothelial cells. Am J Physiol 1993; 264:H2094–H2102.
7. Curry FE, He P. Shear stress does not increase cytoplasmic calcium concentration in individually perfused microvessels. Biomedical Engineering Society. Salt Lake City, Utah. 1992.
8. Geiger RV, Berk BC, Alexander RW, Nerem RM. Flow-induced calcium transients in single endothelial cells: spatial and temporal analysis. Am J Physiol 1992; 262:C1411–C1417.
9. Shen J, Luscinska FW, Connolly A, Dewey CF, Gimbrone MA. Fluid shear stress modulates cytosolic free calcium in vascular endothelial cells. Am J Physiol 1992; 262:C1411–1417.
10. Mo M, Eskin SG, Schilling WP. Flow-induced changes in calcium signaling of vascular endothelial cells: effect of shear stress and ATP. Am J Physiol 1991; 260:H1698–H1707.
11. Dull RO, Davies PF. Flow mudulation of agonist (ATP)-response (Ca^{2+}) coupling in vascular endothelial cells. Am J Physiol 1991; 261:H149–H154.
12. Falcone JC, Kuo L, Meininger GA. Endothelial cell calcium increases during flow-induced dilation in isolated arteries. Am J Physiol 1993; 264:H1745–H1752.
13. Holtz J, Forstermann U, Pohl U, Giesler M, Bassenge E. Flow-dependent, endothelium-mediated dilation of epicardial arteries in conscious dogs: effects of cyclooxygenase inhibition. J Cardiovasc Pharmacol 1984; 6:1161–1169.
14. Nollert MU, Eskin SG, McIntire LV. Shear stress increases inositol triphosphate levels in human endothelial cells. Biochem Biophys Res Commun 1990; 170:281–287.
15. Frangos JA, Eskin SG, McIntire LV, Ives CL. Flow effects on prostacyclin production by cultured human endothelial cells. Science 1985; 227:1477–1479.

16. Sheng M, McFadden G, Greenberg ME. Membrane depolarization and calcium induce c-fos transcription via phosphorylation of transcription factor CREB. Neuron 1990; 4:571–582.
17. Yamazaki T, Tobe K, Hoh E, et al. Mechanical loading stimulates cell hypertrophy and specific gene expression in cultured rat cardiac myocytes: possible role of protein kinase C. J Biol Chem 1991; 266:1265–1268.
18. Smith CW. Leukocyte-endothelial cell interactions. Semin Hematol 1993; 30(suppl 4):45–55.
19. Bevilacqua MP. Endothelial-leukocyte adhesion molecules. Annu Rev Immunol 1993; 11:767–804.
20. Bevilacqua MP, Nelson RM, Mannori G, Cecconi O. Endothelial-leukocyte adhesion molecules in human disease. Annu Rev Med 1994; 45:361–378.
21. Bevilacqua MP, Butcher E, Furie B, et al. Selectins: a family of adhesion molecules. Cell 1991; 67:233.
22. Bevilacqua MP, Nelson RM. Selectins. J Clin Invest 1993; 91:379–387.
23. Ross R. The pathogenesis of atherosclerosis: a perspective for the 1990's. Nature (Lond) 1993; 362:801–809.
24. Bonfanti R, Furie BC, Furie B, Wagner DD. PADGEM (GMP 140) is a component of Weibel-Palade bodies of human endothelial cells. Blood 1989; 73:1109–1112.
25. McEver RP, Beckstead JH, Moore KL, Marshall-Carlson L, Bainton DF. GMP-140, a platelet alpha-granule membrane protein, is also synthesized by the vascular endothelial cells and is localized in Weibel-Palade bodies. J Clin Invest 1989; 84:92–99.
26. Jones DA, Abbassi O, McIntire LV, McEver RP, Smith CW. P-selectin mediates neutrophil rolling on histamine-stimulated endothelial cells. Biophys J 1993; 65:1560–1569.
27. Lawrence MB, Springer TA. Leukocytes roll on a selectin at physiologic flow rates: distinction from and prerequisite for adhesion through integrins. Cell 1991; 65:859–873.
28. Abbassi O, Kishimoto TK, McIntire LV, Anderson DC, Smith CW. E-selectin supports neutrophil rolling in vitro under conditions of flow. J Clin Invest 1993; 92:2719–2730.
29. Carlos T, Kovach N, Schwartz B, et al. Human monocytes bind to two cytokine-induced adhesive ligands on cultured human endothelial cells: endothelial-leukocyte adhesion molecule-1 and vascular cell adhesion molecule-1. Blood 1991; 77:2266–2271.
30. Weller PF, Rand TH, Goelz SE, Chi-Rosso G, Lobb RR. Human eosinophil adherence to vascular endothelium mediated by binding to vascular cell adhesion molecule-1 and endothelial leukocyte adhesion molecule-1. Proc Natl Acad Sci USA 1991; 88:7430–7433.
31. Shimizu Y, Shaw S, Graber N, et al. Activation-independent binding of human memory T-cells to adhesion molecule ELAM-1. Nature (Lond) 1991; 349:799–802.
32. Shimizu Y, Newman W, Gopal TV, et al. Four molecular pathways of T-cell adhesion to endothelial cells: roles of LFA-1, VCAM-1, and ELAM-1 and

changes in pathway hierarchy under different activation conditions. J Cell Biol 1991; 113:1203–1212.
33. Picker LJ, Warnock RA, Burns AR, Doerschuk CM, Bergs EL, Butcher EC. The neutrophil selectin LECAM-1 presents carbohydrate ligands to the vascular selectin ELAM-1 and GMP-140. Cell 1991; 66:921–933.
34. Smith CW, Kishimoto TK, Abbassi O, et al. Chemotactic factors regulate lectin adhesion molecule-1 (LECAM-1)-dependent neutrophil adhesion to cytokine-stimulated endothelial cells in vitro. J Clin Invest 1991; 87:609–618.
35. Philips ML, Nudelman E, Gaeta FC, et al. ELAM-1 mediates cell adhesion by recognition of a carbohydrate ligand, sialyl-Le^x. Science 1990; 250:1130–1132.
36. Walz G, Aruffo A, Kolanus W, Bevilacqua M, Seed B. Recognition by ELAM-1 of the sialyl-Le^x determinant on myeloid and tumor cells. Science 1990; 250:1132–1135.
37. Abbassi O, Lane CL, Krater S, et al. Canine neutrophil margination mediated by lectin adhesion molecule-1 in vitro. J Immunol 1991; 147:2107–2115.
38. Tedder TF. Cell-surface receptor shedding: a means of regulating function. Am J Respir Cell Mol Biol 1991; 5:305–306.
39. Hynes RO. Integrins: versatility, modulation, and signaling in cell adhesion. Cell 1992; 69:11–25.
40. Luscinskas FW, Lawler J. Integrins as dynamic regulators of vascular function. FASEB J 1994; 8:929–938.
41. Lawrence MB, Smith CW, Eskin SG, McIntire LV. Effects of venous shear stress on CD18-mediated neutrophil adhesion to cultured endothelium. Blood 1990; 75:227–237.
42. Bienvenu K, Granger DN. Molecular determinants of shear rate-dependent leukocyte adhesion in postcapillary venules. Am J Physiol 1993; 264:H1504–H1508.
43. Nagel T, Resnick N, Atkinson WJ, Dewey CFJ, Gimbrone MAJ. Shear stress selectively upregulates intracellular adhesion molecule-1 expression in cultured human endothelial cells. J Clin Invest 1994; 94:885–891.
44. Resnick N, Collins T, Atkinson W, Bonthron DT, Dewey CF, Gimbrone MA. Platelet-derived growth factor beta chain promotor contains a cis-acting fluid shear-stress-responsive element. Proc Natl Acad Sci USA 1993; 90:4591–4595.
45. Ando J, Tsuboi H, Korenaga R, et al. Shear stress adhesion of cultured mouse endothelial cells to lymphocytes by down regulating VCAM-1 expression. Am J Physiol 1994; 267:C679–C687.
46. Ohtsuka A, Ando J, Korenago R, Kamiya A, Sormachi NT, Miyasaka M. The effect of flow on the expression of vascular adhesion molecule-1 by cultured mouse endothelial cells. Biochem Biophys Res Commun 1993; 193:303–310.
47. Simmons D, Makgoba MW, Seed B. ICAM, an adhesion ligand of LFA-1, is homologous to the neural cell adhesion molecule NCAM. Nature (Lond) 1988; 331:624–626.
48. Staunton DE, Dustin ML, Springer TA. Functional cloning of ICAM-2, a cell adhesion ligand for LFA-1 homologous to ICAM-1. Nature (Lond) 1989; 339: 61–64.

49. Nortamo P, Li R, Renkonen R, et al. The expression of human intercellular adhesion molecule-2 is refractory to inflammatory cytokines. Eur J Immunol 1991; 21:2629–2632.
50. Walpola PL, Gotlieb AI, Cybulsky MI, Langille BL. Expression of ICAM-1 and VCAM-I and monocyte adherence in arteries exposed to altered shear stress. Arterioscler Thromb Vasc Biol 1995; 15:2–10.
51. Sugama Y, Tiruppathi C, Janekidevi K, Andersen TT, Fenton JW, Malik AB. Thrombin-induced expression of endothelial P-selectin and intercellular adhesion molecule-1: a mechanism for stabilizing neutrophil adhesion. J Cell Biol 1992; 119:935–944.
52. Wood KM, Cadogan MD, Ramshaw AL, Parums DV. The distribution of adhesion molecules in human atherosclerosis. Histopathology 1993; 22(437–444).
53. Cybulsky MI, Gimbrone MA. Endothelial expression of a mononuclear leukocyte adhesion molecule during atherogenesis. Science 1991; 251:788–791.
54. Hsieh H-J, Li N-Q, Frangos JA. Shear stress increases endothelial platelet-derived growth factor mRNA levels. Am J Physiol 1991; 260:H642–H626.
55. Diamond SL, Sharefkin JB, Diffenbach C, Scott KF, McIntire LV, Eskin SG. Tissue plasminogen activator messenger RNA levels increase in cultured human endothelial cells exposed to laminar stress. J Cell Physiol 1990; 143:364–371.
56. Elices MJ, Osborn L, Takada Y. VCAM-1 on activated endothelium interacts with the leucocyte integrin VLA-4 at a site distinct from the VLA-4/fibronectin binding site. Cell 1990; 60:577–584.
57. Pavalko FM, Otey CA. Role of adhesion molecule cytoplasmic domains in mediating interactions with the cytoskeleton. PSEBM 1994; 205:282–293.
58. Burridge K, Fath K, Kelly T, Nuckolls G, Turner C. Focal adhesions: transmembrane junctions between the extracellular matrix and the cytoskeleton. Annu Rev Cell Biol 1988; 4:487–525.
59. Muguruma M, Matsumura S, Fukazawa T. Direct interactions between talin and actin. Biochem Biophys Res Commun 1990; 171:1217–1223.
60. Goldman WH, Isenber G. Kinetic determination of talin-actin binding. Biochem Biophys Res Commun 1991; 178:718–723.
61. Burridge K, Nuckolls G, Otey C, Pavalko F, Simon K, Turner C. Actin-membrane interaction in focal adhesions. Cell Diff Dev 1990; 32:337–342.
62. Turner CE, Burridge K. Transmembrane molecule assemblies in cell-extracellular matrix interactions. Curr Opin Cell Biol 1991; 3:849–853.
63. Otey CA, Pavalko FM, Burridge K. An interaction between α-actinin and the β_1 integrin subunit in vitro. J Cell Biol 1990; 111:721–729.
64. Pavalko FM, LaRoche SM. Activation of human neutrophils induces an interaction between the integrin β_2 subunit (CD18) and the actin binding protein α-actinin. J Immunol 1993; 151:3795–3807.
65. Nuckolls GH, Romer LH, Burridge K. Microinjection of antibodies against talin inhibits the spreading and migration of fibroblasts. J Cell Sci 1992; 102:753–762.
66. Sadler I, Crawford AW, Michelsen JW, Beckerle MC. Zyxin and cCRP: two

interactive LIM domain proteins associated with the cytoskeleton. J Cell Biol 1992; 119:1573–1587.
67. Jaken S, Leach K, Klauck T. Association of type 3 protein kinase C with focal contacts in rat embryo fibroblasts. J Cell Biol 1989; 109:697–704.
68. Schaller MD, Borgman CA, Cobb BS, Vines RR, Reynolds AB, Parsons JT. pp125^{FAK}, a structurally distinctive protein-tyrosine kinase associated with focal adhesions. Proc Natl Acad Sci USA 1992; 89:5192–5196.
69. Romer LH, Burridge K, Turner CE. Signalling between the extracellular matrix and the cytoskeleton: tyrosine phosphorylation and focal adhesion assembly. Cold Spring Harbor Symp Quan Biol 1992; LVII:193–202.
70. Maher PA, Pasquale EB, Wang JYJ, Singer SJ. Phosphotyrosine-containing proteins are concentrated in focal adhesions and intercellular junctions in normal cells. Proc Natl Acad Sci USA 1985; 82:6576–6580.
71. Rohrschneider LR. Adhesion plaques of Rous sarcoma virus-transformed cells contain the src gene product. Proc Natl Acad Sci USA 1980; 77:3514–3518.
72. Golden A, Brugge JS. Thrombin treatment induces rapid changes in tyrosine phosphorylation in platelets. Proc Natl Acad Sci USA 1989; 86:901–905.
73. Golden A, Brugge JS, Shattil SJ. Role of platelet membrane glycoprotein IIb-IIIa in agonist-induced tyrosine phosphorylation. J Cell Biol 1990; 111: 3117–3127.
74. Horvath AR, Muszbek L, Kellie S. Translocation of pp60v^{src} to the cytoskeleton during platelet aggregation. EMBO J 1992; 11:885–61.
75. Wang N, Butler JP, Ingber DE. Mechanotransduction across the cell surface and through the cytoskeleton. Science 1993; 160:1124–1127.
76. Heidemann SR. A new twist on integrins and the cytoskeleton. Science 1993; 260:1080–1081.
77. Durieu-Trautmann O, Chaverot N, Cazaubon S, Strosberg AD, Couraud P-O. Intercellular adhesion molecule-1 activation induces tyrosine phosphorylation of the cytoskeleton-associated protein cortactin in brain microvessel endothelial cells. J Biol Chem 1994; 269(17):12536–12540.
78. Magee AI, Buxton RS. Transmembrane molecule assemblies regulated by the greater cadherin family. Curr Opin Cell Biol 1991; 3:854–861.
79. Geiger B, Ayalon O. Cadherins. Annu Rev Cell Biol 1992; 8:307–332.
80. Overduin M, Harvey TS, Bagby S, et al. Solution structure of the epithelial cadherin domain responsible for selective cell adhesion. Science 1995; 267: 386–389.
81. Wagner G. E-cadherin: a distant member of the immunoglobulin superfamily. Nature (Lond) 1995; 267:342.
82. Takeichi M. Cadherins: a molecular family important in selective cell-cell adhesion. Annu Rev Biochem 1990; 59:237–252.
83. Ozawa M, Ringwald M, Kemler R. Uvomorulin-catenin complex formation is regulated by a specific domain in the cytoplasmic region of the cell adhesion molecule. Proc Natl Acad Sci USA 1990; 87:4246–4250.
84. Ozawa M, Baribault H, Kemler R. The cytoplasmic domain of the cell adhesion molecule uvomorulin associates with three independent proteins structurally related in different species. EMBO J 1989; 8:1711–1717.

85. Nagafuchi A, Takeichi M. Cell binding function of E-cadherin is regulated by the cytoplasmic domain. EMBO J 1988; 7:3679–3684.
86. Fujimori T, Takeichi M. Distruption of epithelial cell-cell adhesion by exogenous expression of mutated nonfunctional N-cadherin. Mol Biol Cell 1993; 4: 37–47.
87. Nagafuchi A, Takeichi M, Tsukita S. The 102kd cadherin-associated protein: similarity to vinculin and posttranslational regulation of expression. Cell 1991; 65:849–857.

7

Morphogenesis of Epithelial Cells

Carmen Birchmeier and Dieter Riethmacher
Department of Medical Genetics, Max-Delbrueck Center for Molecular Medicine, Berlin, Germany

Volker Brinkmann
Department of Cell Biology, Max-Delbrueck Center for Molecular Medicine, Berlin, Germany

Morphogenesis, like differentiation, is an essential process in development, in which distinct cellular populations and structures arise from initially uniform cells. In this review, we will concentrate on morphogenesis of epithelia, which form an astounding array of different structures. From initially simple ectodermal or endodermal cells, different epithelial cell types with distinct shapes and functions emerge during development. Examples of such differentiated epithelial cells that form intricate structures are the branched tubules found in the lung or salivary gland, the one-layered villus epithelium in the small intestine with perpetually regenerating cells that differentiate continuously, or the plates of hepatocytes arranged in three dimensions that are found in the liver. Morphogenesis depends crucially on the ability of the individual epithelial cells to interact via cell adhesion molecules and on their ability to break or to modify adhesive interactions. The adhesive capacity of cells is not only dependent on a given set of adhesion molecules expressed on the cell surface but can be modified by soluble factors, for instance by scatter factor/hepatocyte growth factor (SF/HGF) which elicits cellular responses via the c-met tyrosine kinase receptor.

Epithelial cells from different organs retain, when taken into culture, differentiation markers characteristic of their origin. They "remember" their morphogenic capacity and can form in vitro, under appropriate conditions, three-dimensional structures that resemble the epithelial structures of the organ they originate from. We will discuss here selected aspects of morphogenesis of epithelia as well as selected molecules that control morphogenic processes in vitro and during embryonic development.

I. MOLECULAR AND STRUCTURAL CHARACTERISTICS OF EPITHELIA

Epithelia form continuous sheets of tightly adhesive cells that are often cuboidal in shape. In cell culture, epithelial cells often grow in aggregates and display little motility compared to other cell types. Characteristic for epithelia are specialized organelles, tight junctions, adherens junctions, and desmosomes that are responsible for intercellular contacts. Hemidesmosomes form contacts to the acellular basement membranes. An additional characteristic of epithelial cells is polarization; that is the distinct morphological appearance of basal, lateral, and apical surfaces. A consequence of the laterally located junctional complexes is the inhibition of free diffusion in the cellular membrane, and therefore adhesion molecules or receptors can be exclusively or predominantly expressed on the basolateral surface of epithelial cells (1–3).

In most organs mesenchymal cells are found below the epithelia. The mesenchymal cells are morphologically distinct, nonpolarized, and loosely associated, and they express a characteristic set of genes termed the mesenchymal program (4). A basement membrane separates the epithelial from the mesenchymal cell compartment. The basement membrane is synthesized by both epithelial and mesenchymal cells, and contains extracellular matrix molecules like laminin, collagen IV, nidogen/entactin, and basement membrane proteoglycans (5).

Adherens junctions of epithelial cells are specialized structures containing the transmembrane cell adhesion molecule E-cadherin. In a Ca^{2+}-dependent manner, E-cadherin molecules expressed on neighboring cells bind to each other in a homophilic manner. E-cadherin is the prototype of a large family of Ca^{2+}-dependent cell adhesion molecules that is still increasing in number. Other cadherins are N-cadherin (expressed in nerve cells), P-cadherin (placenta and epithelia), M-cadherin (muscle), OB cadherin (osteoblasts), LI cadherin (liver, intestine), the desmosomal proteins desmoglein and desmocollin, and others (6–12). The cytoplasmic portion of a highly conserved subgroup, the "classical" cadherins (E-, N-, P-cadherin), interacts with catenins, α- β- and γ-catenin (13). Desmosomal cadherins interact with γ-catenin (plakoglobin) and also with plakophillin and desmoplakin (14). α-Catenin shows structural similarities to vinculin located in focal contacts of fibroblasts and epithelial cells, but also in adherens junctions of epithelial cells (15,16). β-Catenin is homologous to the *Drosophilia* gene product armadillo, which has been implicated to function not only in adhesion but also in intracellular signaling (17–19). β-Catenin also interacts with the adenomatous polyposis coli (APC) tumor suppressor gene product that is frequently mutated in sporadic and inherited colon cancer (20,21). γ-Catenin (plakoglobin) has structural similarities to β-catenin and is found not only in the adherens junction, but in desmosomes as well (14,22,23). Transient expression of cDNA encoding variant β-catenin which lack sequences of different subdomains has elucidated the interactions of the molecules in the adherens junction complex (24–26). E-cadherin binds directly to β-catenin, which interacts with α-catenin, which in turn is linked directly or indirectly to the cytoskeleton. γ-Catenin interacts in a similar manner as β-catenin with E-cadherin and α-catenin. In addition, APC competes with E-cadherin for β-catenin binding (24).

In addition, various receptors with tyrosine kinase activity exist that are expressed exclusively or predominantly on epithelial cell, such as c-ros, c-met, c-erbB2, c-ret, or the receptor for keratinocyte growth factor, KGF-receptor (27–32). Since most of these receptors were identified as oncogenes by ectopic expression of mutant molecules in NIH3T3 fibroblasts, they can clearly transmit mitogenic signals (33). However, experiments with epithelial cells in culture have demonstrated that these receptors do not only regulate growth but can also influence motility, differentiation, and morphogenesis (cf. 34 for a review). Recent genetic evidence demonstrates a pivotal role for receptor tyrosine kinases in the regulation of normal epithelial physiology and development.

II. THE ROLE OF E-CADHERIN IN THE MAINTENANCE OF EPITHELIAL MORPHOLOGY

The functional integrity of the adherens junctions is essential for polarization and epithelial morphology. In cell culture, transitions of epithelial cell aggregates to single cells that resemble fibroblasts can be achieved by various agents that interfere with the adherens junctions, for instance by addition of anti E-cadherin antibodies (35). The epithelial cells treated in this manner not only change their shape, but they also acquire increased motility and become invasive. When cocultured with heart tissue, the polar Madine Darby Kidney (MDCK) cells form a single layer of epithelial sheets that surround the tissue. In the presence of anti-E-cadherin antibodies, the cells become motile and invade the heart tissue. The presence of functional adherens junctions is therefore necessary for epithelial morphology; its disruption leads to an epithelial-mesenchymal transition paralleled by an increase in motility and invasiveness of the cells (35–37).

More recently, the importance of E-cadherin in the maintenance of epithelial cell morphology was also demonstrated by the use of transdominant E-cadherin mutants (38,39). Truncated variants of E-cadherin lacking large parts of the extracellular domain but with an intact transmembrane and cytoplasmic sequence are unable to bind to E-cadherin expressed on neighboring cells. However, they can interact with the cytoplasmatically located catenins. When this transdominant mutant was expressed in large amounts, it depletes the cellular pool of catenins and interferes thus with normal E-cadherin-catenin interactions; i.e., a fibroblast morphology and an increase in cellular motility was observed in the epithelial cells.

The integrity of the adherens junction also requires functional catenin molecules. Mutant cells that lack α-catenin have been described; the cadherin-mediated cell adhesion is not functional in such cells, and can be restored by the expression of α-catenin cDNA driven by an appropriate promoter (40–42). Cell lines mutant for the β-catenin gene have also been observed (S. Hirohashi, personal communication).

III. THE ROLE OF E-CADHERIN IN MORPHOGENESIS OF EPITHELIA IN THE EMBRYO

E-cadherin is not only important for the maintenance of morphology and adhesion in differentiated epithelia, but also for the acquisition of epithelial characteristics early in embryogenesis. In mammals, cells that have formed by division of the egg are initially morphologically and functionally equivalent. The first morphogenic event in development occurs when the blastomers start to adhere to each other to form epithelial-like cells (compaction), and when inner and outer cells separate in the morula. The outer cells differentiate into trophectoderm which give rise to the extraembryonal membranes, whereas the centrally located cells give rise to the inner cell mass which later differentiates into the embryo proper.

That E-cadherin plays essential roles in the early morphogenic event during development was originally suggested by experiments with anti-E-cadherin (then called anti-uvomorulin) antibodies (43) and recently also demonstrated genetically (44,45). We have introduced a targeted mutation into the E-cadherin gene of the mouse via homologous recombination and embryonic stem cell technology. This mutation removes sequences essential for Ca^{2+} binding and thus for the adhesive function of the molecule (45). Animals that carry this mutation in a heterozygous state appear normal and are fertile. However, homozygous mutant embryos do not develop normally (Fig. 1). They can reach the morula stage and also compact, but the compacted state is not sustained (Fig. 1b). The individual morula cells in the embryo lose their morphological polarization; they become rounded but continue to divide. As a consequence, the embryos appear totally distorted

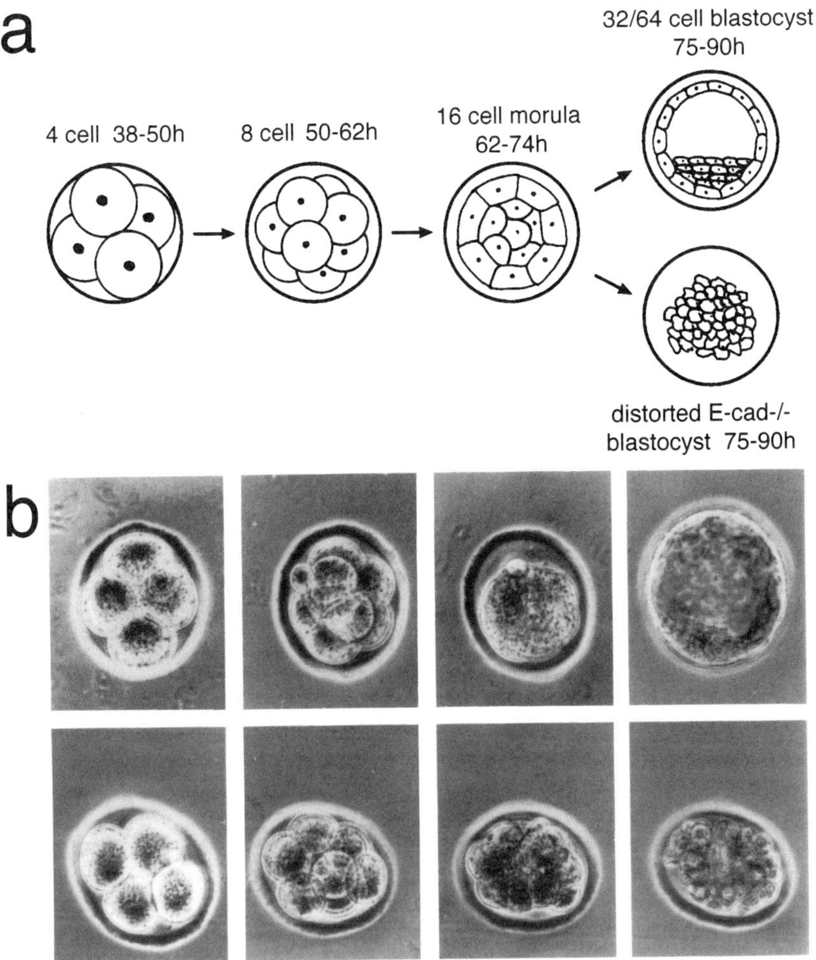

Figure 1 (a) Schematic repesentation of wild-type and E-cadherin mutant preimplantation mouse embryos at various stages (4-cell, 8-cell, 16-cell morula; 32/64 cell morula) in development. The times need to reach the depicted developmental stages are indicated. (b) Morphology of wild-type (top row) and E-cadherin mutant (bottom row) during preimplantation development. The photographs show embryos at the same stages depicted schematically in (a).

at a time when wild-type or heterozygous mutant embryos form organized blastocysts with well-formed blastocoel. Since the mutant embryos never emerge from the zona pellucida, further development, particularly implantation into the uterus, cannot take place.

It has previously been reported that removal of Ca^{2+} ions or treatment with anti-E-cadherin antibodies interferes with the compaction of the mouse morula, indicating that the initial adhesion and polarization of the epithelial-like cells already requires E-cadherin (43,46). Nevertheless, embryos that lack a functional E-cadherin gene can undergo compaction. This is due to maternally derived E-cadherin, and not to functional compensation by other cell adhesion molecules. The presence of maternal E-cadherin in mutant embryos at early stages was demonstrated by immunohistochemical techniques. It disappears when the embryos develop beyond the morula stage. In the mutant embryo, maternal E-cadherin thus suffices for the initial compaction and for formation of the first polarized, epitheliallike cells, since compaction can be prevented by anti-E-cadherin antibodies. Maternal protein does not, however, suffice for further development beyond the morula stage (45).

Electron microscopy was used to observe the cell contact sites of the distorted E-cadherin -/-embryos (Fig. 2). These were found to be severely altered in morphology. They lack adherens junctions but retain tight junctions and desmosomes. The membranes of opposing inner and outer blas-

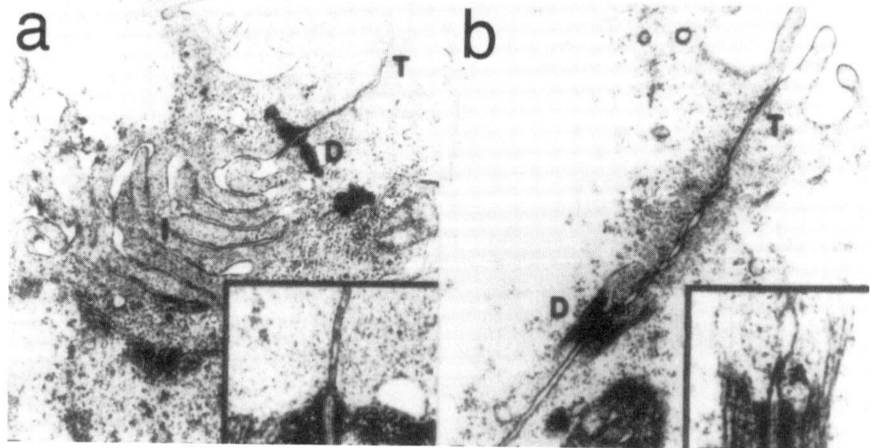

Figure 2 Cell contact sites from E-cadherin mutant (a) and wild-type (b) embryos at the same developmental stage as analyzed by electron microscopy. Desmosomes (D), tight junctions (T), and adherens junctions (A) are indicated. Bars, 100 nm.

tomers formed an irregular, interdigitating pattern that increases the area of membrane interaction (Fig. 2a). Apparently, in absence of E-cadherin-mediated cellular adhesion the afflicted embryonal cells appear to compensate by enlarging the area of interacting membranes (45).

IV. MESENCHYMAL-EPITHELIAL INTERACTIONS AS DRIVING FORCES FOR EPITHELIAL MORPHOGENESIS, GROWTH, AND DIFFERENTIATION

During organogenesis, two distinct cells types, mesenchymal and epithelial cells, interact and contribute to the formation of many different organs. Whereas the mesenchymal cells are derived from either mesoderm or neural ectoderm, the epithelial cells usually arise from the epithelial ectoderm and endoderm, and only in few exceptional cases from the mesoderm after a mesenchymal-epithelial conversion. An illustrative example of the morphogenesis of an organ is the lung, where an outgrowth of epithelial cells from the endoderm into the splanchnic mesoderm generates the anlage of the organ. All epithelia of the differentiated lung develop by further growth, branching, and differentiation of the endodermal bud. In parallel, the mesenchymal cell compartment also expands and differentiates. Growth of epithelia into mesenchymal tissues and subsequent branching and differentiation are a general theme found during development of many other organs—for example, salivary gland, pancreas, prostate, pituitary gland, kidney, and breast (3,47–50).

The anlage for the gastrointestinal tract is formed early during development. It consists initially of a poorly differentiated, simple epithelium. Later, a stratified epithelium is formed that eventually undergoes a major morphological conversion to form again a single-layered epithelium that lines the villi. At the same time—i.e., shortly before birth—terminal differentiation of the epithelium takes place. The first sign of this formation of the villus epithelium is the dissociation of cell-cell contacts and the formation of secondary lumina in the multilayered epithelium. The underlying mesenchymal cells then invade the loosened epithelium. Whereas superficial epithelial cells are exfoliated, the epithelial cells that are in direct contact with the mesenchyme proliferate (51). The outcome of this rearrangement is the one-layered villus epithelium. This is an example of a developmental process in which interconversions of different types of epithelial organizations are observed, a process guided by mesenchymal-epithelial interactions (52).

Organ development can be analyzed by the removal of the respective anlagen at the appropriate time in development and subsequent culture. For example, the murine lung anlage can be explanted on day 10 of development, after the endodermal epithelia have grown into the splanchnic mesenchyme. During the following culture, the epithelia grow and branch repeatedly (49). Alternatively, the epithelium can be dissociated from the mesenchyme by mild tryptic digestion. Cultured separately, the epithelial bud will not grow or branch; when recombined with the lung mesenchyme, growth and branching occur. Heterologous epithelia and mesenchyme can be combined in such experiments. It was thus demonstrated that the requirements of epithelia from different organs vary, since virtually every mesenchyme can support morphogenesis of epithelia from the pancreas primordia, but the ureter epithelia of the kidney are absolutely dependent on kidney mesenchyme for growth and morphogenesis (50). Thus, mesenchymal contributions to epithelial differentiation are complex, and many factors are involved.

It follows from these considerations that paracrine factors produced by the mesenchyme drive epithelial growth, morphogenesis, and differentiation during development. Three modes of interactions have been postulated to transduce the signals from the mesenchyme to the epithelial cells: 1. interactions mediated by direct cell-cell contact; 2. interactions mediated by extracellular matrix; 3. diffusion of soluble factors (cf. 3, 50, 53 for reviews). These different modes of interaction might be reflected in different types of molecules involved in signaling. For instance, signaling via wnt molecules will probably require direct cell-cell contact, since wnt is associated strongly with the cell surface. In contrast, essential signals mediated by integrins will require contact with the extracellular matrix. Signals mediated by receptor tyrosine kinases may, in principle, be triggered by soluble factors. However, many ligands of tyrosine kinase receptors are initially produced as membrane spanning molecules, or are strongly bound to heparane sulfate proteoglycans, and cell contacts might also be required for such factors.

V. THE EXTRACELLULAR MATRIX AS AN ESSENTIAL COMPONENT IN EPITHELIAL MORPHOGENESIS

The creation of a lumen in a solid epithelial structure is called cavitation, and is observed, for instance, after implantation of the mammalian embryo when the initially solid embryonic ectoderm is transformed into a hollow egg cylinder. This process is known as preamniotic cavity formation. The inner cell mass of the implanting embryo is basically a solid structure which contains as outermost layer the primitive endoderm. The inside is formed by pseudo-stratified epithelial cells, the embryonic ectoderm. A basement membrane separates the two cell types. Preamniotic cavity formation, i.e., the generation of a cavity in the center of these embryos, was recently studied using embryonic carcinoma cells (derived from teratocarcinomas) or embryonic stem cell (derived from the inner cell mass of embryos) (54). In vitro, such cells form embryoid bodies that recapitulate early steps of development under appropriate conditions. Cavity formation in embryoid bodies occurs by apoptosis and is triggered by a death signal which is given by outer endodermal cells. Only a single layer of columnar epithelia cells that line the newly formed cavity survive due to a survival signal. The signal is provided by the attachment of surviving cells to the extracellular matrix that separates endo- and ectoderm (54). This simple, two-step mechanism allows the formation of a hollow cyst. Such a mechanism might not only be used during preamniotic cavity formation, but could be fundamental to lumen formation in development. In a minor modification, this mechanism might apply wherever stratified epithelial are converted into single-layered epithelia—for instance, during rearrangement and terminal differentiation of the villi epithelium.

Preamniotic cavity formation thus provides an elegant example of a process in which extracellular matrix supplies a signal essential for epithelial rearrangement and morphogenesis. However, the contribution of the extracellular matrix as an essential component for epithelial organization has been recognized for many years. Experimental evidence for this has come mainly from reconstitution experiments, where embryonal epithelia were separated from mesenchyme and cultured in the presence of simple or complex extracellular matrix material. For such studies, purified extracellular matrix components like collagen or complex mixtures like matrigel, the extracellular matrix of Engelbreth-Holmes-Swarm sarcoma cells, not only provide mechanic support for epithelia but also allow epithelial morphogenesis to quite a remarkable degree. For instance, mouse mammary epithelia grown on collagen-coated tissue culture dishes form ductlike structures and respond to lactogenic hormones to produce milk protein (55). In collagen gels and in matrigel, three-dimensional tubular and alveolar networks are formed (56,57). However, detailed transfilter experiments have shown that, besides matrix components, morphogenesis requires soluble mesenchymal components. An example for a soluble factor that drives morphogenesis is SF/HGF (see below).

VI. EPITHELIAL RECEPTOR TYROSINE KINASES AS MODULATORS OF CELL ADHESION

Morphogenic alterations that dissociate epithelia and make them fibroblast-like in appearance can be achieved not only by interference with adherens junction but also by various soluble factors that function as ligands for tyrosine kinase receptors. Such factors, for instance SF/HGF, aFGF, EGF and others, can induce transient epithelial-mesenchymal conversion and motility in cell culture (58–60). A well-characterized motility factor is the ligand for the c-met tyrosine kinase, which has been named scatter factor (SF) because of this activity on epithelial cells in culture (58,61). In addition, the factor can induce growth of hepatocytes (hepatocyte growth factor, or HGF) and other cells (62,63). A third activity of SF/HGF is the morphogenic action that was first discovered as the ability to induce kidney epithelial cells to form tubules in vitro (64,65).

Molecules located in the adherens junction might be ultimately affected by factors that dissociate epithelial cells or induce epithelial-mesenchymal conversions. Thus, it was found that nonreceptor and receptor-type tyrosine kinases can phosphorylate β-catenin or plakoglobin on tyrosine residues (66–69). This phosphorylation correlates with dissociation and increased motility of epithelial cells. The phosphorylation by v-src or other oncogens is not temporarily regulated, and could thus cause permanent changes in cellular behavior typical for transformed epithelia, like an inability to adhere and an increase in motility and invasiveness.

SF/HGF has a unique structure since it closely resembles proteases like plasminogen (40% sequence identity) but not other ligands for tyrosine kinase receptors (62,63,70). The protein is produced as an inactive precursor molecule (90 kDa) that is cleaved outside of the producing cells into a heavy (60-kDa) and a light (30-kDa) chain (71–73). The heavy chain contains a N-terminal hairpin loop and four Kringle domains; the light chain shows extensive homologies to serine proteases. However, two of the three amino acids that form the catalytic triad of serine proteases are altered in SF/HGF, and therefore the factor has no catalytic activity.

Like many other tyrosine kinase receptors, c-met was initially identified because of its transforming activity when mutated. The met oncogene was derived from an N-methyl-N-nitro-N-nitrosoguanidine-treated osteosarcoma cell line that was used in a transfection/tumorigenicity assay (74). The gene transferred from the osteosarcoma cells was the product of a rearrangement that fused translocated promoter region (tpr) on chromosome 1 to c-met on chromosome 7 (75). The oncogenic variant of met encodes a cytoplasmatically located tyrosine kinase. In contrast, the proto oncogene product is a transmembrane glycoprotein of 190 kDa that is cleaved posttranslationally into an α- and a β-chain (76,77). A major breakthrough in the understanding of the c-met receptor was the identification of its ligand, which is SF/HGF (73,78). All the known biological activities of SF/HGF are mediated by the c-met receptor (79).

We have recently analyzed the normal physiological role of the SF/HGF and the c-met gene by the introduction of a targeted mutation in the mouse via homologous recombination and embryonic stem cell technology (80,81). Whereas animals with a heterozygous mutation in SF/HGF or c-met are normal and fertile, a homozygous mutation is not compatible with normal life. SF/HGF -/-embryos die between day 13 and day 16.5 (E13–E16.5) of development. The mutant embryos appear normal in overall morphology, but are retarded in development starting from E14, probably due to a defect in the development of the placenta. In addition, their liver is considerably reduced in size. Histological examination shows damage to the embryonic liver on E14.5 that varies in severity and is not observed on E12.5. The essential function of SF/HGF in the development of the liver is also supported by the analysis of embryonic stem cells that carry two mutant alleles in c-met (81). Such cells cannot contribute to the liver but participate in the development of a variety of other organs and cell types. Analysis of c-met -/-ES cells has revealed an additional, essential role of c-met and SF/HGF in migration of myogenic precursor cells. Myogenic precursor cells with the potential to migrate are located in the ventrolateral edge of the dermomyotome. These cells appear adhesive and have been characterized as epithelial since they are polarized and form basement membranes; they also express c-met. Upon receiving a SF/HGF signal that emanates from the target sites of migration, these cells undergo an epithelial-mesenchymal conversion. They dissociate from the dermomyotome, become fibroblastic in appearance, and also become motile. They then invade target sites, i.e., limb buds, diaphragm, and tongue, where they continue to divide and eventually differentiate to form muscle masses. Thus, SF/HGF does not only induce dissociation of epithelial cell groups, epithelial-mesenchymal conversion, and motility of cells in culture, but appears to have a similar effect on dermomyotomal cells in vivo.

VII. TYROSINE KINASE RECEPTORS AS ACTIVATORS OF MORPHOGENIC PROGRAMS OF EPITHELIAL CELLS

Among the activities of SF/HGF that were described in cell culture, the morphogenic activity is unique, and no other factor with similar properties has been described (64,65). MDCK cells, when grown in a three-dimensional collagen matrix for several days, form hollow cysts. When SF/HGF is added, individual cells dissociate and move away from the cysts. Consequently, the cells reassociate and form continuous tubules. These tubules have a lumen surrounded by well-polarized epithelial cells with a smooth basal surface in contact with the collagen matrix, and a apical surface rich in microvilli that faces the lumen. The structures formed in vitro thus resemble the tubular epithelia present in many organs (64). During development of the mammary gland, SF/HGF may play a similar role, since it can induce the branching and growth of the tubular epithelia in organ cultures (82). Organ culture experiments using anti-SF/HGF antibodies have also indicated a role of SF/HGF in the development and morphogenesis of epithelia in the kidney primordium (83,84). However, no defect in the kidney anlagen is observed in embryos that lack SF/HGF or c-met.

SF/HGF will not only induce the formation of tubular structures by MDCK cells, but can influence morphogenesis of many epithelia in reconstituted extracellular matrices (57). Thus, colon cells can be induced to rearrange to organoids with features of colonic crypts (Fig. 3), pancreas

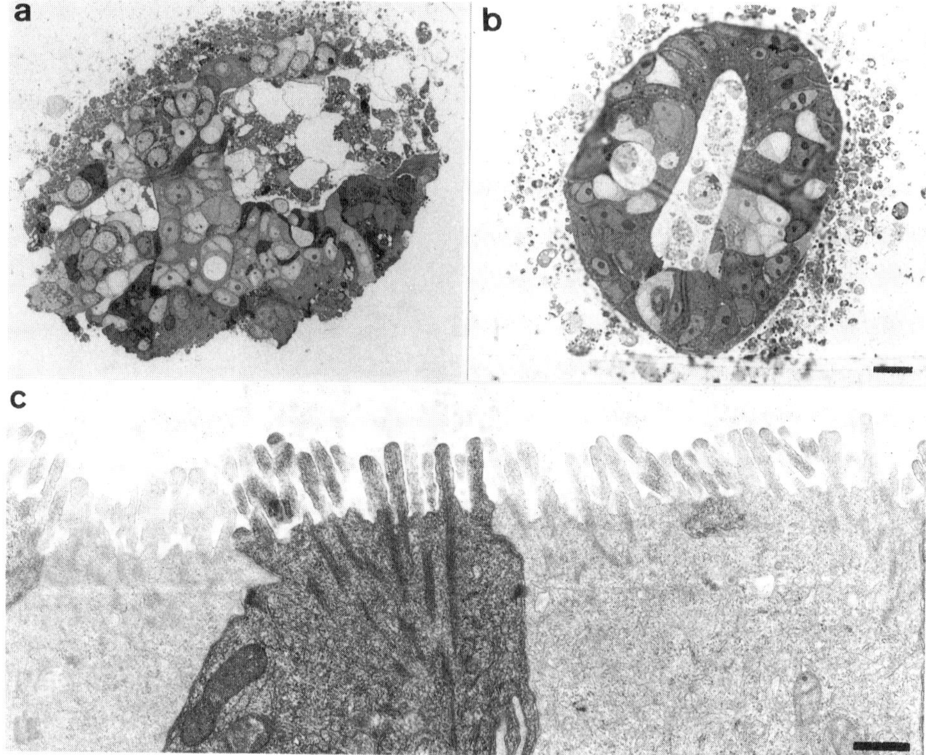

Figure 3 SW 1222 colon carcinoma cells grown in collagen gels form loose aggregates (a), and rearrange in the presence of SF/HGF to form a well organized, cryptlike structure (b, c) with microvilli that face the interior (c). Semithin sections analyzed by microscopy (a, b) and thin sections analyzed by electron microscopy (c). Bars in (b) 20 μm; in (c) 1 μm.

cells develop hollow cysts, mammary gland cells build ducts with end buds, and lung cells form alveolarlike structures in the presence of SF/HGF. The organoids induced by the factor thus resemble the organization of the epithelial cells in the respective organ of origin. Apparently, SF/HGF triggers cells to accomplish this morphogenic program, but the exact shape they assume is determined by the tissue of origin. Thus, SF/HGF does not give an instructive signal. The factor acts rather as an inducer that triggers an intrinsic morphogenic potential; i.e., it activates the respective program but does not dictate it. Instead, the intrinsic morphogenic program is specified by as yet unknown molecules.

In vitro data indicate that SF/HGF induces not only morphogenesis or motility of epithelial cells but also invasiveness into collagen matrices (61). Furthermore, it has recently been shown that SF/HGF is expressed in breast carcinomas, and that the level of expression correlates with the state of tumor progression (85). How can the same factor activate epithelial morphogenesis and invasiveness epithelial cells? Interestingly, the different morphogenic programs can only be induced by SF/HGF in cells with intact epithelial characteristics, including expression of E-cadherin and functional adherens junctions (57). On the other hand, SF/HGF was shown to induce metastasis to the lymph nodes and lung when expressed by transfection of cDNA in dedifferentiated breast carcinoma cells (S. Meiners, H. Naundorf, and W. Birchmeier, personal communications; see also 86). Thus, the differentiation state of the epithelial cells seems to determine the response toward SF/HGF: Morphogenesis requires an intact epithelial program, whereas metastatic behavior is observed in cells that have partially lost their epithelial differentiation, a consequence of their transformed state.

REFERENCES

1. Rodriguez-Boulan E, Nelson WJ. Morphogenesis of the polarized epithelial cell phenotype. Science 1989; 245:718–725.
2. Schwarz MA, Owaribe K, Kartenbeck J, Franke WW. Desmosomes and hemidesmosomes: constitutive molecular components. Annu Rev Cell Biol 1990; 6:461–491.
3. Birchmeier C, Birchmeier W. Molecular aspects of mesenchymal-epithelial interactions. Annu Rev Cell Biol 1993; 9:511–540.
4. Hay ED. Role of cell-matrix contacts in cell migration and epithelial-mesenchymal transformation. Cell Differ Dev 1990; 32:367–375.
5. Timpl R. Structure and biological activity of basement membrane proteins. Eur J Biochem 1989; 180:487–502.
6. Takeichi M. Cadherin cell adhesion receptors as a morphogenetic regulator. Science 1991; 251:1451–1455.
7. Donalies M, Cramer M, Ringwald M, Starzinski PA. Expression of M-

cadherin, a member of the cadherin multigene family, correlates with differentiation of skeletal muscle cells. Proc Natl Acad Sci USA 1991; 88:8024–8028.

8. Buxton RS, Magee AI. Structure and interactions of desmosomal and other cadherins. Semin Cell Biol 1992; 3:157–167.
9. Kemler R. Classical cadherins. Semin Cell Biol 1992; 3:149–155.
10. Sano K, Tanihara H, Heimark RL, et al. Protocadherins: a large family of cadherin-related molecules in central nervous system. EMBO J 1993; 12:2249–2256.
11. Okazaki M, Takeshita S, Kawai S, et al. Molecular cloning and characterization of OB-cadherin, a new member of cadherin family expressed in osteoblasts. J Biol Chem 1994; 269:12092–12098.
12. Berndorff D, Gessner R, Kreft B, et al. Liver-intestine cadherin: molecular cloning and characterization of a novel Ca(2+)-dependent cell adhesion molecule expressed in liver and intestine. J Cell Biol 1994; 125:1353–1369.
13. Ozawa M, Baribault H, Kemler R. The cytoplasmic domain of the cell adhesion molecule uvomorulin associates with three independent proteins structurally related in different species. EMBO J 1989; 8:1711–1717.
14. Schmidt A, Heid HW, Schafer S, Nuber UA, Zimbelmann R, Franke WW. Desmosomes and cytoskeletal architecture in epithelial differentiation: cell type-specific plaque components and intermediate filament anchorage. Eur J Cell Biol 1994; 65:229–245.
15. Nagafuchi A, Takeichi M, Tsukita S. The 102 kd cadherin-associated protein: similarity to vinculin and posttranscriptional regulation of expression. Cell 1991; 65:849–857.
16. Herrenknecht K, Ozawa M, Eckerskorn C, Lottspeich F, Lenter M, Kemler R. The uvomorulin-anchorage protein alpha catenin is a vinculin homologue. Proc Natl Acad Sci USA 1991; 88:9156–9160.
17. Peifer M, Wieschaus E. The segment polarity gene armadillo encodes a functionally modular protein that is the *Drosophila* homolog of human plakoglobin. Cell 1990; 63:1167–1176.
18. McCrea PD, Turck CW, Gumbiner B. A homolog of the armadillo protein in *Drosophila* (plakoglobin) associated with E-cadherin. Science 1991; 254:1359–1361.
19. Butz S, Stappert J, Weissig H, Kemler R. Plakoglobin and beta-catenin: distinct but closely related. (Letter.) Science 1992; 257:1142–1144.
20. Rubinfeld B, Souza B, Albert I, et al. Association of the APC gene product with beta-catenin. Science 1993; 262:1731–1734.
21. Su LK, Vogelstein B, Kinzler KW. Association of the APC tumor suppressor protein with catenins. Science 1993; 262:1734–1737.
22. Cowin P, Kapprell HP, Franke WW, Tamkun J, Hynes OR. Plakoglobin: a protein common to different kinds of intercellular adhering junctions. Cell 1986; 46:1063–1073.
23. Knudsen KA, Wheelock MJ. Plakoglobin, or an 83-kD homologue distinct from beta-catenin, interacts with E-cadherin and N-cadherin. J Cell Biol 1992; 118:671–679.
24. Huelsken J, Birchmeier W, Behrens J. E-cadherin and APC compete for the

interaction with beta-catenin and the cytoskeleton. J Cell Biol 1994; 127:2061–2069.

25. Nathke IS, Hinck L, Swedlow JR, Papkoff J, Nelson WJ. Defining interactions and distributions of cadherin and catenin complexes in polarized epithelial cells. J Cell Biol 1994; 125:1341–1352.
26. Rubinfeld, B, Souza B, Albert I, Munemitsu S, Polakis P. The APC protein and E-cadherin form similar but independent complexes with alpha-catenin, beta-catenin, and plakoglobin. J. Biol Chem 1995; 270:5549–5555.
27. Kokai Y, Cohen JA, Drebin JA, Greene MI. Stage-and tissue-specific expression of the neu oncogene in rat development. Proc Natl Acad Sci USA 1987; 84:8498–8501.
28. Quirke, P, Pickles A, Tuzi NL, Mohamdee O, Gullick WJ. Pattern of expression of c-erbB-2 oncoprotein in human fetuses. Br J Cancer 1989; 60:64–69.
29. Sonnenberg E, Godecke A, Walter B, Bladt F, Birchmeier C. Transient and locally restricted expression of the ros 1 protooncogene during mouse development. EMBO J 1991; 10:3693–3702.
30. Sonnenberg E, Meyer D, Weidner KM, Birchmeier C. Scatter factor/hepatocyte growth factor and its receptor, the c-met tyrosine kinase, can mediate a signal exchange between mesenchyme and epithelia during mouse development. J Cell Biol 1993; 123:223–235.
31. Orr-Urtreger A, Bedford MT, Burakova T, et al. Developmental localization of the splicing alternatives of fibroblast growth factor receptor-2 (FGFR2). Dev Biol 1993; 158:475–486.
32. Pachnis V, Mankoo B, Costantini F. Expression of the c-ret proto-oncogene during mouse embryogenesis. Development 1993; 119:1005–1017.
33. Aaronson SA. Growth factors and cancer. Science 1991; 254:1146–1153.
34. Birchmeier C, Sonnenberg E, Weidner KM, Walter B. Tyrosine kinase receptors in the control of epithelial growth and morphogenesis during development. Bioessays 1993; 15:185–190.
35. Behrens J, Mareel MM, Van RF, Birchmeier W. Dissecting tumor cell invasion: epithelial cell acquire invasive properties after the loss of uvomorulin-mediated cell-cell adhesion. J Cell Biol 1989; 108:2435–2447.
36. Frixen UH, Behrens J, Sachs M, et al. E-cadherin-mediated cell-cell adhesion prevents invasiveness of human carcinoma cells. J Cell Biol 1991; 113:173–185.
37. Vleminckx K, Vakaet LJ, Mareel M, Fiers W, Van RF. Genetic manipulation of E-cadherin expression by epithelial tumor cells reveals an invasion suppressor role. Cell 1991; 66:107–119.
38. Kintner C. Regulation of embryonic cell adhesion by the cadherin cytoplasmic domain. Cell 1992; 69:225–236.
39. Fujimori T, Takeichi M. Disruption of epithelial cell-cell adhesion by exogenous expression of a mutated nonfunctional N-cadherin. Mol Biol Cell 1993; 4:37–47.
40. Hirano S, Kimoto N, Shimoyama Y, Hirohashi S, Takeichi M. Identification of a neural alpha-catenin as a key regulator of cadherin function and multicellular organization. Cell 1992; 70:293–301.

41. Oda T, Kanai Y, Shimoyama Y, Nagafuchi A, Tsukita S, Hirohashi S. Cloning of the human alpha-catenin cDNA and its aberrant mRNA in a human cancer cell line. Biochem Biophys Res Commun 1993; 193:897–904.
42. Shimoyama Y, Nagafuchi A, Fujita S, et al. Cadherin dysfunction in a human cancer cell line: possible involvement of loss of alpha-catenin expression in reduced cell-cell adhesiveness. Cancer Res 1992; 52:5770–5774.
43. Hyafil F, Babinet C, Jacob F. Cell-cell interactions in early embryogenesis: a molecular approach to the role of calcium. Cell 1981; 26:447–454.
44. Larue L, Ohsugi M, Hirchenhain J, Kemler R. E-cadherin null mutant embryos fail to form a trophectoderm epithelium. Proc Natl Acad Sci USA 1994; 91:8263–8267.
45. Riethmacher D, Brinkmann V, Birchmeier C. A targeted mutation in the mouse E-cadherin gene results in defective preimplantation development. Proc Natl Acad Sci USA 1995; 92:855–859.
46. Vestweber D, Kemler R. Rabbit antiserum against a purified surface glycoprotein decompacts mouse preimplantation embryos and reacts with specific adult tissues. Exp Cell Res 1984; 152:169–178.
47. Grobstein C. Morphogenetic interaction between embryonic mouse tissues separated by a membrane filter. Nature 1953; 172:869–871.
48. Wessells NK, Cohen JH. Effects of collagenase on developing epithelia in vitro: lung, ureteric bud, and pancreas. Dev Biol 1968; 18:294–309.
49. Spooner BS, Wessells NK. Mammalian lung development: interactions in primordium formation and bronchial morphogenesis. J Exp Zool 1970; 175:445–454.
50. Saxen L. Organogenesis of the Kidney. Cambridge: Cambridge University Press, 1987.
51. Madara JL, Neutra MR, Trier JS. Junctional complexes in fetal rat small intestine during morphogenesis. Dev Biol 1981; 86:170–178.
52. Haffen K, Kedinger M, Simon AP. Mesenchyme-dependent differentiation of epithelial progenitor cells in the gut. J Pediatr Gastroenterol Nutr 1987; 6:14–23.
53. Grobstein C. Inductive tissue interaction in development. Adv. Cancer Res 1956; 4:187–236.
54. Coucouvanis E, Martin GR. Signals for death and survival: a two-step mechanism for cavitation in the vertebrate embryo. Cell 1995; 83:279–287.
55. Streuli CH, Bailey N, Bissell MJ. Control of mammary epithelial differentiation: basement membrane induces tissue-specific gene expression in the absence of cell-cell interaction and morphological polarity. J Cell Biol 1991; 115:1383–1395.
56. Soriano JV, Pepper MS, Nakamura T, Orci L, Montesano R. Hepatocyte growth factor stimulates extensive development of branching duct-like structures by cloned mammary gland epithelial cells. J Cell Sci 1995; 108:413–430.
57. Brinkmann V, Foroutan H, Sachs M, Weidner KM, Birchmeier W. Hepatocyte growth factor/scatter factor induces a variety of tissue-specific morphogenic programs in epithelial cells. J Cell Biol 1995; 131:1573–1586.

58. Stoker M, Gherardi E, Perryman M, Gray J. Scatter factor is a fibroblast-derived modulator of epithelial cell mobility. Nature 1987; 327:239–242.
59. Jouanneau J, Gavrilovic J, Caruelle D, et al. Secreted or nonsecreted forms of acidic fibroblast growth factor produced by transfected epithelial cells influence cell morphology, motility, and invasive potential. Proc Natl Acad Sci USA 1991; 88:2893–2897.
60. Manske M, Bade EG. Growth factor-induced cell migration: biology and methods of analysis. Int Rev Cytol 1994; 155:49–96.
61. Weidner KM, Behrens J, Vandekerckhove J, Birchmeier W. Scatter factor: molecular characteristics and effect on the invasiveness of epithelial cells. J Cell Biol 1990; 111:2097–2108.
62. Miyazawa K, Tsubouchi H, Naka D, et al. Molecular cloning and sequence analysis of cDNA for human hepatocyte growth factor. Biochem Biophys Res Commun 1989; 163:967–973.
63. Nakamura T, Nishizawa T, Hagiya M, et al. Molecular cloning and expression of human hepatocyte growth factor. Nature 1989; 342:440–443.
64. Montesano R, Schaller G, Orci L. Induction of epithelial tubular morphogenesis in vitro by fibroblast-derived soluble factors. Cell 1991; 66:697–711.
65. Montesano R, Matsumoto K, Nakamura T, Orci L. Identification of a fibroblast-derived epithelial morphogen as hepatocyte growth factor. Cell 1991; 67: 901–908.
66. Behrens J, Vakaet L, Friis R, et al. Loss of epithelial differentiation and gain of invasiveness correlates with tyrosine phosphorylation of the E-cadherin/beta-catenin complex in cells transformed with a temperature-sensitive v-SRC gene. J Cell Biol 1993; 120:757–766.
67. Hamaguchi M, Matsuyoshi N, Ohnishi Y, Gotoh B, Takeichi M, Nagai Y. p60v-src causes tyrosine phosphorylation and inactivation of the N-cadherin-catenin cell adhesion system. EMBO J 1993; 12:307–314.
68. Shibamoto S, Hayakawa M, Takeuchi K, et al. Tyrosine phosphorylation of beta-catenin and plakoglobin enhanced by hepatocyte growth factor and epidermal growth factor in human carcinoma cells. Cell Adhes Commun 1994; 1:295–305.
69. Hoschuetzky H, Aberle H, Kemler R. Beta-catenin mediates the interaction of the cadherin-catenin complex with epidermal growth factor receptor. J Cell Biol 1994; 127:1375–1380.
70. Weidner KM, Arakaki N, Hartmann G, et al. Evidence for the identity of human scatter factor and human hepatocyte growth factor. Proc Natl Acad Sci USA 1991; 88:7001–7005.
71. Hartmann G, Naldini L, Weidner KM, et al. A functional domain in the heavy chain of scatter factor/hepatocyte growth factor binds the c-Met receptor and induces cell dissociation but not mitogenesis. Proc Natl Acad Sci USA 1992; 89:11574–11578.
72. Lokker NA, Mark MR, Luis EA, et al. Structure-function analysis of hepatocyte growth factor: identification of variants that lack mitogenic activity yet retain high affinity receptor binding. EMBO J 1992; 11:2503–2510.

73. Naldini L, Weidner KM, Vigna E, et al. Scatter factor and hepatocyte growth factor are indistinguishable ligands for the MET receptor. EMBO J 1991; 10: 2867–2878.
74. Cooper CS, Park M, Blair DG, et al. Molecular cloning of a new transforming gene from a chemically transformed human cell line. Nature 1984; 311:29–33.
75. Park M, Dean M, Cooper CS, et al. Mechanism of met oncogene activation. Cell 1986; 45:895–904.
76. Gonzatti-Haces M, Seth A, Park M, Copeland T, Oroszlan S, Vandewoude GF. Characterization of the TPR-MET oncogene p65 and the MET protooncogene p140 protein-tyrosine kinases. Proc Natl Acad Sci USA 1988; 85:21–25.
77. Giordano S, Ponzetto C, Di RM, Cooper CS, Comoglio PM. Tyrosine kinase receptor indistinguishable from the c-met protein. Nature 1989; 339:155–156.
78. Bottaro DP, Rubin JS, Faletto DL, et al. Identification of the hepatocyte growth factor receptor as the c-met proto-oncogene product. Science 1991; 251:802–804.
79. Weidner KM, Sachs M, Birchmeier W. The Met receptor tyrosine kinase transduces motility, proliferation, and morphogenic signals of scatter factor/hepatocyte growth factor in epithelial cells. J Cell Biol 1993; 121:145–154.
80. Schmidt C, Bladt F, Goedecke S, et al. Scatter factor/hepatocyte growth factor is essential for liver development. Nature 1995; 373:699–702.
81. Bladt F, Riethmacher D, Isenmann S, Aguzzi A, Birchmeier C. Essential role for the c-met receptor in the migration of myogenic precursor cells into the limb bud. Nature 1995; 376:768–771.
82. Yang Y, Spitzer E, Meyer D, et al. Sequential requirement of scatter factor/hepatocyte growth factor (SF/HGF) and neu differentiation factor/neuregulin (NDF/HRG) in the morphogenesis and differentiation of the mammary gland. J Cell Biol 1995; 131:215–226.
83. Santos OF, Barros EJ, Yang XM, et al. Involvement of hepatocyte growth factor in kidney development. Dev Biol 1994; 163:525–529.
84. Woolf AS, Kolatsi JM, Hardman P, et al. Roles of hepatocyte growth factor/scatter factor and the met receptor in the early development of the metanephros. J Cell Biol 1995; 128:171–184.
85. Yamashita J, Ogawa M, Yamashita S, et al. Immunoreactive hepatocyte growth factor is a strong and independent predictor of recurrence and survival in human breast cancer. Cancer Res 1994; 54:1630–1633.
86. Rosen EM, Knesel J, Goldberg ID, et al. Scatter factor modulates the metastatic phenotype of the EMT6 mouse mammary tumor. Int J Cancer 1994; 57: 706–714.

8

Adhesion Molecules in Immunoregulation

Daniel R. Salomon
Departments of Molecular and Experimental Medicine, and Immunology, The Scripps Research Institute, La Jolla, California

While subsequent chapters will provide detailed information regarding the function of specific adhesion molecules in a variety of situations, the purpose of the present chapter is to explore a number of their fundamental properties which determine their role in the regulation of the cellular immune response. Specific types that will be covered include:

- Extracellular matrix and a three-dimensional view of immunity
- Integrating cell-cell vs. cell-matrix interactions
- Three basic functions of adhesion molecules in cellular immunity: traffic, directed migration, and cell activation
- Shear stress and the physics of the immune response
- Adhesion molecules in B and T cell development

I. EXTRACELLULAR MATRIX AND A THREE-DIMENSIONAL VIEW OF IMMUNITY

Any site in the body can be considered in terms of its physical or three-dimensional space. Thus, I propose that cellular immunity must ultimately be understood in the context of real intercellular distances and events that are shaped by the complex framework structures that *determine* and *maintain* the organization of normal tissues. The corollary of this three-dimensional view is that normal organ or tissue function requires a normal structure. We have all had the tendency to view immune-mediated injury in two dimensions in which activated T cells recognize antigen via the T cell receptor and engage a cascade of inflammatory cytokines and cytotoxic cells. However, to understand a pathological state we must also consider a third dimension of space and structure in which immune-mediated injury results in the disruption of normal cellular organization and function.

Extracellular matrix (ECM) proteins compose the basic structural framework for creating the body's three-dimensional structure. These proteins include fibronectin, the laminin family, vitronectin, the collagen family, osteopontin, and a number of plasma molecules such as fibrinogen. In each organ, the mix of extracellular matrix proteins may be different, and there are many examples of matrix changing at specific stages in development (1,2). In any case, the ECM proteins are critical to determining the organization and function of each tissue's microenvironment. All ECM proteins, despite their structural diversity, share three common features: 1. they are deposited by the cells in a microenvironment and may be cross-linked and organized in the extracellular space into larger fibrillar structures; 2. they contain specific protein sequences that act as ligands for different adhesion molecules (integrins); and 3. their deposition and organization are regulated during development and altered in response to injury. Two key concepts explain the link between ECM structures and tissue function.

ECM proteins are synthesized by stromal cells such as fibroblasts as well as endothelial and epithelial cells. Thus, during the development of a tissue its cells participate actively in creating the normal structure within which they will function. The health of an adult tissue also depends on the constant remodeling of the ECM structure such as occurs in the bone with osteopontin (3) or in the skin with laminin (4,5).

Stromal and epithelial cells also express an array of cell surface adhesion molecules which recognize specific binding sequences on the ECM molecules; these adhesion molecules belong to the integrin family (Chapter 1). Integrins mediate the connection between the cells and the framework structures created by ECM. By expression of integrins on specific surfaces, the cell can also determine its polarity with respect to an underlying basement membrane structure. For example, epithelial cells lining the gut or kidney tubular epithelial cells express certain integrins on their basolateral surfaces which attach them to the laminin and collagen of the basement membrane in a polar fashion.

Both these features emphasize the relationship between normal structure and normal function. Thus, tissue structure is *determined* during development by ECM and integrin adhesion molecule interactions. However, tissue structure is also *maintained* in the mature organism by ongoing ECM/integrin interactions.

The importance of integrins and ECM proteins in the immune response can now be considered in the context of a three-dimensional view in acute and chronic injury.

A. Acute Injury

Leukocytes or lymphocytes are attracted to a site of vascular injury—for example, acute rejection of a transplanted organ, wound infection, or autoimmune vasculitis. Cytokines mediate the upregulation of several adhesive ligands on the endothelial cell surfaces that mediate leukocyte and lymphocyte adhesion. One of these ligands, vascular cell adhesion molecule 1 (VCAM-1) can trigger cells attracted to this site via the integrin VLA-4 ($\alpha 4\beta 1$, CD49) to release various enzymes, including members of the metalloproteinase family. In turn, these enzymes can digest the ECM of the basement membrane and allow cells access to the interstitial tissue beneath (6). As injury evolves, the normal endothelial cell monolayer is increasingly disrupted and the underlying vascular basement membrane is exposed. In turn, this disrupts the vascular integrity of the tissue, increased downstream ischemia, and facilitates the extravasation of other cells and inflammatory mediators to fuel interstitial injury.

In parallel, the lumen of the vessel is filled with a growing mass of activated platelets, releasing a number of chemotactic cytokines and factors. Furthermore, activation of complement components occur which may act as chemokines or bind to integrin molecules such as p150/95 (CD11b) to activate leukocytes. The clotting cascade is activated by basement membrane tissue factors as well as local platelet aggregation and activation. Polymerization of fibrin in the vessel lumen and its extravasation into the interstitium expose the cells to ECM proteins trapped from the circulation including fibrin, fibronectin, and vitronectin. The immobilization of these ECM proteins at the site of acute injury uncovers integrin binding sites that are normally hidden on the proteins in the circulation. The exposure of these ECM binding sites for integrin attachment accelerates cell recruitment and directs migration of the activated cells into the underlying tissue spaces.

Thus, the sequence of events in acute injury demonstrates how the evolving disruption of tissue structures at the site of injury can accelerate the process. It also demonstrates how multiple mechanisms of cell adhesion, migration, and activation interact with multiple ligands and ECM proteins to determine the progression of the immune response. Finally, this example demonstrates how the three-dimensional space of the vascular lumen can be connected directly across the endothelial monolayer and underlying basement membrane with events taking place in the interstitium. In fact, these three spatially separated compartments (lumen, vascular wall, interstitium) each present a unique combination of ECM proteins and structural features which contribute to the immune mechanisms evolving during inflammation.

B. Chronic Tissue Injury

Chronic rejection of a kidney transplant is typically associated with progressive interstitial fibrosis and vascular narrowing, resulting in ischemia and a progressive loss of renal function. Similar lesions can be found in other organs with chronic injury. One view of chronic rejection suggests that it represents the final expression of a series of low-grade acute immune rejections which are further complicated by secondary mechanisms of progressive tissue injury. Starting with our description of acute injury, we can view the start of chronic rejection in the gross disruption of the three-dimensional tissue space.

The loss of ECM framework structure occurs as a consequence of cell- and cytokine-mediated killing of epithelial and stromal cells, destructive enzymes released by activated inflammatory cells, and tissue ischemia. In the healing phase of the injury, the interstitial space is infiltrated with a new set of cells arising from the interstitial stroma which include fibroblasts, cells that secrete ECM proteins to heal the internal wounds. Other cells, including monocytes, may migrate from the vascular space. In parallel, the vascular basement membrane is reestablished by endothelial cells migrating into the damaged area, though some vessel sites will be damaged beyond repair. In these sites, downstream ischemia will never be resolved, and whole tissue units will be lost. The tubular epithelial cells of the kidney that were damaged in the original inflammation must find an intact tubular basement membrane upon which to attach their integrins, proliferate, and organize their polarity to the tubular lumen. While they can secrete ECM proteins to aid this process, they do require some underlying structure first. If they do not find this ECM, they will undergo a form of programmed cell death or apoptosis signaled through the integrin adhesion molecules (7,8). Unfortunately, the epithelial ECM frameworks of the kidney which were created originally during fetal development cannot be repaired by the infiltrating fibroblasts and monocytes. As a result, this critical organization element of the tissue is lost. The combination of tissue disorganization and ischemia results in increased organ dysfunction, a stress that activates secondary mechanisms of chronic injury such as hypertrophy and hyperfiltration in remaining nephrons. Areas of injured blood vessels may also never heal to their original state creating altered flow, ischemia, hypertension, vessel wall stress, low-grade inflammatory cell, and platelet activation leading to progressive vascular injury.

As this example illustrates, while healing the wounds of acute injury may be successful in a limited sense, the impact on the whole tissue at the organizational or three-dimensional level is much more difficult to repair. Moreover, the resulting disruption of normal organ structure produces organ dysfunction triggering secondary mechanisms of chronic tissue injury. While the connection between fetal organ development and adult organ function may not be apparent at first, integrin and ECM protein interactions determine the successful outcome of both development and healing.

II. INTEGRATING CELL-MATRIX VS. CELL-CELL INTERACTIONS

Integrin molecules mediate two basic adhesive phenomena: adhesion of cells to extracellular matrix proteins, and adhesion of cells to other cells. In fact, this dual capability of integrins is fundamental to understanding their function. However, it is critical to understand that the *expression* of a given integrin on a cell surface does not mean that it can bind its ligand. The first level of regulating integrin adhesion is that integrins must be *activated* to bind their ligand. In most situations this really means that the cells must be activated and communicate this change to the integrin molecule on the cell surface by cytoplasmic signaling. Cytoplasmic signaling appears to induce a conformational change in the integrin allowing the ligand stable access to the binding site. As a consequence, studies that simply measure integrin expression on different cells during acute inflammation do not necessarily tell the true story; integrin activity must also be determined.

We have made the point that ECM proteins create the three-dimensional structure around which all tissues are organized. Thus, a second level of regulating cell behavior through integrin adhesion is the presentation of different ECM proteins in different tissues. For example, joints tend to express substantial amounts of the collagen family proteins while the vascular basement membrane is rich in laminin and vitronectin. Thus, the targeting of acute inflammatory cells to either site can be mediated selectively by activating different integrins recognizing either collagen or laminin. A corollary to this concept is that one strategy to understand and potentially manipulate the recruitment of activated cells during immune-mediated injury is to determine the ECM content of a given site. In this context it is also interesting that several ECM ligands can be regulated by alternative splicing of the messenger RNA transcripts to produce multiple ECM proteins isoforms capable of interacting with integrin receptors differentially. For example, the CS1 binding site on fibronectin for the integrin VLA4 can be alternatively spliced, allowing the cell to produce at least two forms of this ECM protein—one that binds VLA4, and one that cannot. In fact the potential clinical significance of alternative splicing was demonstrated in patients with rheumatoid arthritis where the expression of the CS1 form of FN was specifically segregated to the afferent arterioles where inflammatory cells accumulated and not to the interstitial areas of the joint, which were spared (9).

It is also important to remember that ECM expression is not a static phenomenon. In a site of acute injury a number of mechanisms including the release of proteolytic enzymes, ischemia, the clotting cascade, and inflammatory cytokines will expose ECM proteins to inflammatory cells and, thus, shape the evolving immune response. We also made the point in the example of acute injury that the immobilization of normally circulating plasma ECM proteins in the vascular lumen or interstitium can reveal hidden binding sites for integrin adhesion.

In parallel, integrins mediate a number of cell-cell interactions. Two of the best-characterized are LFA-1 binding to intercellular adhesion molecule-1 (ICAM-1) and VLA-4 binding to VCAM-1. Both ICAM-1 and VCAM-1 are typically presented by cytokine-activated endothelial cells at sites of acute inflammation. Leukocytes use their integrins to bind these ligands as a first stage in trafficking to a site of injury (10,11). Alternative splicing of these cellular ligands may also be involved in regulating integrin function during inflammation. VCAM-1 exists in both seven- and six-domain forms where the difference is the deletion of one of the VLA4 binding sites by alternative splicing of the six-domain version (12). The six-domain form of VCAM-1 is only produced in cytokine-activated endothelial cells.

Finally, VLA-4 binds to the CS1 site of fibronectin (13), and LFA-1 can bind fibrinogen (14). Thus, the same integrins can mediate both cell-cell and cell-matrix interactions, demonstrating that the integrin binding sequences on ECM proteins share structural homologies with the binding sites on cellular ligands like VCAM-1 and ICAM-1. This fact will be important in later chapters when strategies to inhibit integrin binding in states of acute inflammation are discussed. One of the most exciting areas of integrin research is the effort to understand the complex structure of the integrin binding site and determine the molecular rules governing the regulation of adhesive ligand recognition.

There are three basic functions of adhesion molecules in cellular immunity: traffic, directed migration, and cell activation.

A. Traffic

Integrin binding to an ECM protein or cell ligand directs three basic functions. The first consequence of adhesion to the vascular surface at a site of inflammation is the local accumulation of activated leukocytes. In retrospect, it is surprising how long it took to recognize the importance of this remarkably efficient system of recruiting activated cells to specific sites and directing the traffic of cells from peripheral lymphoid reservoirs such as the spleen and lymph nodes. Now we realize that this cellular traffic is also bidirectional such that antigen presenting cells will also move from inflammatory sites back to the lymphoid germinal centers, where antigen presentation drives immune amplification and enhanced specificity. Integrin-mediated cell traffic is also involved in moving hematopoietic cells derived from stem cells in the bone marrow to the spleen and lymph nodes, T cell progenitors to the thymus, and mature T cells from the thymus to the peripheral lymphoid organs. Thus, cell traffic is critical in both health and disease states. This is another fundamental issue to consider in the context of clinical strategies designed to manipulate integrin function with novel agents such as peptide inhibitors, engineered proteins, or antibodies.

B. Directed Migration

The second integrin-mediated mechanism is directed migration. We have described adhesion in a one dimensional way. The integrin binds a sequence on the matrix or cell ligand, conformational changes facilitate a stable complex, and the cell adheres. However, a unique feature of integrin binding is the cytoplasmic signaling that follows the first encounter of integrin with its ligand (15). A large number of cytoskeletal proteins such as vinculin and talin and signaling proteins, including multiple protein kinases, accumulate in a complex with the cytoplasmic tail of the integrin (16,17). This complex effectively links the cell's cytoskeleton through the integrin on the cell surface to the ligand binding site. Indeed, the cytoskeletal complex actually determines the shape of the cell by providing it a leading edge where the integrin is bound to the ligand. The cell can then release the integrins bound on its leading surface and recruit new integrins from the rear of the cell, resulting in a directed movement or migration in the direction of the matrix presented. One way to visualize migration is to think of the cell surface like a tank tread rolling over the matrix surface binding and releasing its integrins in sequence. Thus, adhesion leads to migration, and a one-dimensional view yields to three dimensions of space and motion.

C. Cell Activation

We have described the roles of adhesion molecules in physical events such as adhering to a fixed cell surface or migrating on a matrix; we have also made the fundamental point that integrin expression is not equal to integrin activity, as some form of cell activation is required to activate the integrin molecules. However, this cell activation is typically derived from nonintegrin signals such as inflammatory cytokines or the engagement of the T cell receptor by antigen. The third key is that integrins, when they engage their ligand, can also deliver a signal to the cell which can participate in the process of cell activation. This process of integin-mediated activation is often referred to as *costimulation* since it still requires a primary signal for complete cell activation. For example, T cells exposed to suboptimal concentrations of antigen or anti-T-cell receptor antibody will not be activated unless they are also exposed to fibronectin which delivers a costimulatory signal through the VLA4 and VLA5 integrins (18). Another example pertinent to immunoregulation is that helper T cells exposed to antigen presenting cells require a costimulatory signal delivered by LFA-1 binding to ICAM-1 expressed on the presenting cell's surface (19). In fact, immunosuppressive strategies to block integrin-mediated costimulatory signals have been proposed by several groups (20).

Finally, we must consider the implications of the concepts that some form of cell activation is required for integrin function and that integrin binding to its ligand can costimulate cell activation. This means that activation cannot be viewed as a one-dimensional event, in other words "on" vs. "off." Therefore, cells and integrins must exist in a whole series of relative activation states that determine their behavior at any given time in development or during an immune response. Moreover, if a cell enters the site of a developing immune response, it will participate in the response but it will also be directly affected by the evolution and nature of the response. Thus, a cell may be activated to bind one ligand to enter the site of inflammation at the vascular surface and upon further activation use additional ligands and/or integrins to mediate migration into the tissue or mediate an effector response such as cell killing.

In fact, the concept of *activation* is increasingly problematic. What does it mean to "activate" a cell in the context of multiple states of activation capable of regulating the cell's function? Similarly, what does it mean to say a cell or an integrin is "activated" if there are multiple levels of activation? This complexity emphasizes the importance we have attached to viewing the immune response in terms of space and time. Certainly, we will have to be careful to consider these issues in the design of trials to manipulate integrin-mediated immune phenomenon in patients.

III. SHEAR STRESS AND THE PHYSICS OF THE IMMUNE RESPONSE: A THEORY

We have described the vascular surface several times as the initial site of immune cell entry into a site of inflammation such as transplant organ rejection or autoimmune vasculitis. Now we must consider this problem with a three-dimensional view since there is no such thing as a generic "vascular surface." The vascular bed of any tissue is anatomically complex and comprises arterioles, capillaries, and postcapillary venules. In turn, these three anatomic sites differ in the nature of their endothelial surfaces and structure. Can we integrate this complexity of the vascular surface with our understanding of integrins? We can describe a theory.

The first clue was the observation that early accumulations of inflammatory cells in classic acute transplant rejection appear to begin in the postcapillary venules and only later may spread to involve the arterioles and capillary beds. This form of rejection predominantly involves interstitial accumulation of activated lymphocytes and cell-mediated cytotoxicity. Thus, it would seem to involve efficient cell adhesion and a prompt migration into the underlying tissue. In contrast, the vasculitis associated with autoimmune disease or the acute vascular rejection of an allograft typically starts on the arteriolar side of the capillary bed. Though antibody-mediated injury may play an important role in early stages, there is an active involvement of inflammatory leukocytes and T lymphocytes. This form of vasculitis is characterized by tissue ischemia, hemorrhage, activation of the clotting and complement cascades, and platelet aggregation, events suggesting a substantial degree of cell activation in the vascular lumen. Thus, the pathology of an immune response is clearly related to the anatomic site of the vascular bed.

The second clue was the discovery that the vascular ligands for integrins are differentially displayed by the vascular endothelium of the arterioles, capillaries, and venules. VCAM-1, ICAM-1, and another family of vascular surface adhesion molecules, called selectins, have been studied (11,20–23). VCAM-1 is expressed primarily on postcapillary endothelium, presumably requiring the induction of inflammatory cytokines such as IL-1 and TNFα. In contrast, ICAM-1 and the selectins are expressed constitutively on arteriolar endothelium though they are upregulated by cytokine exposure. Thus the anatomic segregation of the vascular ligands is consistent with the concept that different mechanisms of inflammatory cell recruitment might characterize the phenotype of the immune response on either side of the capillary bed.

The third clue has derived from recent studies of the role of *blood flow* in the mechanics of cell-cell interactions in the vascular lumen. It is a basic principle of physics that flowing fluid creates a shear force. Thus, shear stress is created within the vascular lumen by blood flow, and it can be described and measured in very precise physical terms. In fact it is now clear that any understanding of a cell's adhesion to the vascular surface must consider the physics of shear stress in the lumen as applied to the cell (24). As a result, we have realized that the physical complex of any adhesion molecule with its ligand has a tensile strength that must be measured under flow conditions (25). What is even more challenging is that many adhesion mechanisms described in classic, one-dimensional assays as involving one receptor/one ligand, turn out to require multiple receptor/ligand events when studied under flow conditions. Finally, it is logical that adhesion in the high shear stress of the arteriole will involve a complex of receptor ligand interactions fundamentally different from those required for adhesion in the relatively low shear conditions of the postcapillary venule.

A number of fundamental new insights about adhesion molecules have already followed these flow studies of cell adhesion. For example, to the extent that cell activation increases the activity of a selectin or integrin molecule, it will also increase its binding efficiency or the cell's ability to arrest its forward motion on an endothelial surface. The local concentration of inflammatory cytokines will also influence the state of cell activation as well as endothelial ligand expression. The activation of the clotting cascade and platelet aggregation will create disturbances in the flow of the vessel and also influence the ability of inflammatory cells to stop and participate in the evolving immune response. Thus, the accumulation of lymphocytes in classic acute cellular rejection may be favored by interactions between the integrin VLA4 and VCAM-1 presented early in the immune response by the postcapillary venules while acute vasculitis appears to favor early accumulation of inflammatory leukocytes using LFA-1/ICAM-1 and selectin adhesion to stop their motion in the much higher shear stress of the arteriole.

This theory is only a first attempt to integrate the complex nature of integrin functions within a flowing system and the different phenotypes of immune responses observed in clinical situations. Clearly, the problem of defining "activation" described above is an issue here also. However, it is reasonable to assume that shear stress is an important factor in determining the location and pattern of cell recruitment initiated by any form of vascular injury.

IV. ADHESION MOLECULES IN B AND T CELL DEVELOPMENT

To this point in the chapter we have concentrated on the role of integrins in the behavior of mature leukocytes recruited and activated in sites of acute inflammation. We have also noted the important role played by integrins of epithelial and stromal cells in wound healing and tissue organization. On the other hand, integrins are also fundamentally involved in the development of the immune system.

T cell development is a complex progression involving multiple stages of immature progenitors and takes place in the thymus. The thymus is divided into several anatomically distinct compartments: subcapsule, cortex, corticomedullary junction, and medulla. The fact that the developing T cells move from one thymic compartment to another during their development suggests two things: 1. that cell migration and adhesion may be important in positioning the thymocytes in different compartments, and 2. that each compartment must deliver a unique set of signals to the developing T cells. In fact, we have determined that thymocytes express a number of integrin molecules which function to mediate adhesion and migration on extracellular matrix proteins within the thymus (26–28). Integrin-mediated interactions with matrix proteins can also costimulate the proliferation of more mature thymocyte subsets in collaboration with activation of the T cell receptor for antigen. In contrast, the least mature T cell progenitors appear to use integrins for firm adhesion to cellular ligands expressed predominantly on cortical epithelial cells. These studies have suggested that integrins are regulated during development to play specific roles as the T cells mature and move to the different thymic compartments. Thus, integrin function can be related to the fundamental process of T cell selection and development. This process determines the antigen response repertoire of the adult immune system.

B cell development also involves integrin and ECM protein interactions. For example, the integrin VLA4 interacting with the cellular ligand VCAM-1 is required for B cell progenitor development in the bone marrow (29,30). At later stages, the B cell may use a number of integrins to migrate to the peripheral lymphoid reservoirs, spleen, and lymph nodes, where they position themselves in the germinal centers in close proximity to antigen presenting cells expressing ICAM-1 and VCAM-1.

V. CONCLUSIONS

We have described integrin and ECM interactions in the context of a three-dimensional view of immunity. A number of fundamental features of integrin function must be considered. Integrins are expressed on many kinds of cells involved in the immune response including the vascular endothelium, tissue epithelium, and stromal cells of the tissue site, and the circulating blood cells which traffic to a site of injury including platelets, lymphocytes, and leukocytes. In parallel, we must integrate the expression and function of a large number of ECM proteins and cellular ligands for adhesion which determine the structural organization of a tissue during development and maintain its normal structure in the adult.

Because normal tissue structure is required for normal function, we must also consider how immune-mediated injury alters these elements of structure and function. Integrins require "activation" for ligand binding, although our current view of integrin activation has become more complicated. A more pragmatic view of the immune response required for developing clinical applications has led us to recognize the physics of shear stress and study its significance in real situations of inflammation at the vascular surface. Finally, just as integrin/matrix interactions are fundamental to the development and function of tissue structures, it is now clear that they are also integrally involved in the development and selection of the immune system.

REFERENCES

1. Bissell MJ, Barcellos-Hoff MH. The influence of extracellular matrix on gene expression: is structure the message? J Cell Sci 1987; suppl 8:327.
2. Sorokin L, Ekblom P. Development of tubular and glomerular cells of the kidney. Kidney Int 1992; 41:657.
3. Brown LF, Berse B, Van de Water L, et al. Expression and distribution of osteopontin in human tissues: widespread association with luminal epithelial surfaces. Mol Biol Cell 1992; 3:1169.
4. Jones PH, Watt FM. Separation of human epidermal stem cells from transit amplifying cells on the basis of differences in integrin function and expression. Cell 1993; 73:713.
5. Watt FM, Jones PH. Expressions and function of the keratinocyte integrins. In: Ingham P, Brown A, Arias AM, eds. Signals, Polarity and Adhesion in Development. Dev Suppl 1993; 185–192.
6. Romanic AM, Madri JA. The induction of 72-kD gelatinase in T cells upon adhesion to endothelial cells is VCAM-1 dependent. J Cell Biol 1994; 125: 1165.
7. Frisch SM, Francis H. Disruption of epithelial cell-matrix interactions induces apoptosis. J Cell Biol 1994; 124:619.
8. Meredith JE Jr, Fazeli B, Schwartz MA. The extracellular matrix as a cell survival factor. Mol Biol Cell 1993; 4:953.
9. Elices MJ, Tsai V, Strahl D, et al. Expression and functional significance of alternatively spliced CS1 fibronectin in rheumatoid arthritis microvasculature. J Clin Invest 1994; 93:405.
10. Carlos TM, Schwartz BR, Kovach NL, et al. Vascular cell adhesion molecule-1 mediates lymphocyte adherence to cytokine-activated cultured human endothelial cells. Blood 1990; 76:965. (Published erratum appears in Blood 1990; 76(11):2420.)
11. Carlos TM, Harlan JM. Leukocyte-endothelial adhesion molecules. Blood 1994; 84:2068.
12. Cybulsky MI, Fries JW, Williams AJ, et al. Gene structure, chromosomal location, and basis for alternative mRNA splicing of the human VCAM1 gene. Proc Natl Acad Sci USA 1991; 88:7859.

13. Guan J-L, Hynes RO. Lymphoid cells recognize an alternatively spliced segment of fibronectin via the integrin receptor $\alpha_4\beta_1$. Cell 1990; 60:53.
14. Languino LR, Plescia J, Duperray A, et al. Fibrinogen mediates leukocyte adhesion to vascular endothelium through an ICAM-1-dependent pathway. Cell 1993; 73:1423.
15. Schwartz MA, Ingber DE. Integrating with integrins. Mol Biol Cell 1994; 5: 389.
16. Clark EA, Brugge JS. Integrins and signal transduction pathways: the road taken. Science 1995; 268:233.
17. Miyamoto S, Akiyama SK, Yamada KM. Synergistic roles for receptor occupancy and aggregation in integrin transmembrane function. Science 1995; 267: 883.
18. Matsuyama T, Yamada A, Kay J, et al. Activation of CD4 cells by fibronectin and anti-CD3 antibody: a synergistic effect mediated by the VLA-5 fibronectin receptor complex. J Exp Med 1989; 170:1133.
19. Geppert TD, Davis LS, Gur H, Wacholtz MC, Lipsky PE. Accessory cell signals involved in T-cell activation. In: Immunological Reviews. Copenhagen: Munksgaard, 1990:5–66.
20. Recent Developments in Transplantation Medicine. Vol II: Adhesion Molecules, Fusion, Proteins, Novel Peptides, and Monoclonal Antibodies. Glenview, IL: Physicians & Scientists Publishing Company, 1995: 1–197.
21. Mampaso F, Sanchez-Madrid F, Marcen R, et al. Expression of adhesion molecules in allograft renal dysfunction. A distinct diagnostic pattern in rejection and cyclosporine nephrotoxicity. Transplantation 1993; 56:687.
22. Shimizu Y, Newman W, Tanaka Y, Shaw S. Lymphocyte interactions with endothelial cells. Immunol Today 1992; 13:106.
23. Postigo AA, Teixido J, Sanchez-Madrid F. The alpha 4 beta 1/VCAM-1 adhesion pathway in physiology and disease. Res Immunol 1993; 144:723.
24. Dustin ML, Springer TA. Role of lymphocyte adhesion receptors in transient interactions and cell locomotion. Annu Rev Immunol 1991; 9:27.
25. Alon RA, Hammer DA, Springer TA. Lifetime of the P-selectin carbohydrate bond and its response to tensile force in hydrodynamic flow. Nature 1995; 374:539.
26. Salomon DR, Mojcik CF, Chang AC, et al. Constitutive activation of integrin $\alpha4\beta1$ defines a unique stage of human thymocyte development. J Exp Med 1994; 179:1573.
27. Chang AC, Salomon DR, Wadsworth S, et al. $\alpha3\beta1$ and $\alpha6\beta1$ integrins mediate laminin/merosin binding and function as costimulatory molecules for human thymocyte proliferation. J Immunol 1994 (in press).
28. Mojcik CF, Salomon DR, Chang AC, Shevach EM. Differential expression of integrins on human thymocyte subpopulations. Blood 1995 (in press).
29. Miyake K, Medina K, Ishihara K, Kimoto M, Auerbach R, Kincade PW. A VCAM-like adhesion molecule on murine bone marrow stromal cells mediates binding of lymphocyte precursors in culture. J Cell Biol 1991; 114:557.
30. Miyake K, Weissman IL, Greenberger JS, Kincade PW. Evidence for a role of the integrin VLA-4 in lympho-hemopoiesis. J Exp Med 1991; 173:599.

9

Regulation of Endothelial Selectins and Their Ligands

Dietmar Vestweber
Institute for Cell Biology, ZMBE, University of Munster, Munster, Germany

I. INTRODUCTION

Cell adhesion molecules on the endothelial cell surface determine where leukocytes leave the bloodstream (1,2). In inflammatory processes such adhesion molecules are induced on the plasma membrane, in response to cytokines or pro-inflammatory mediators.

Selectins are carbohydrate binding adhesion molecules which mediate the initial contacts between extravasating leukocytes and the blood vessel wall. Such contacts are at first reversible, and, due to the rapidly flowing blood, they support a rolling movement of leukocytes on the vessel wall (3–5). Other cell adhesion molecules, mainly integrins and members of the Ig superfamily, mediate a second, firmer adhesion step, which eventually allows leukocytes to migrate actively through the blood vessel wall barrier.

The selectin family is comprised of three glycoproteins designated by the prefixes L (leukocytes), E (endothelium), and P (platelets). Only L-selectin is found on leukocytes; the other two are expressed by endothelial cells. The selectins share a similar structural organization, having an N-terminal lectin domain, a single EGF-type domain, and various numbers of consensus repeats with sequences similar to those in complement regulatory proteins (6,7).

L-selectin was originally found as a lymphocyte homing receptor (8), mediating the entry of lymphocytes into peripheral lymph nodes. It was soon found that L-selectin is also expressed on most other types of leukocytes (9–11) and that it is an essential molecule for the entry of PMNs into inflamed tissues (12,13). E-selectin was identified as a cytokine inducible adhesion molecule on human umbilical vein endothelial (HUVE) cells, and supports the binding of PMNs and monocytic cells (14,15). In addition, a subset of lymphocytes, the skin-homing T cells, bind to E-selectin (16,17). P-selectin, first found as a membrane protein in platelet α-granules (18,19), was later also found in Weibel-Palade bodies of endothelial cells (20). After translocation of P-selectin to the cell surface, it mediates the binding of myeloid cells (21). In addition, P-selectin mediates the binding of certain lymphocyte subpopulations, which differ from those that bind to E-selectin (22–27).

In vivo studies have shown that the selectins are involved in various physiological and pathophysiological processes, such as lymphocyte homing, neutrophil influx into inflamed tissues, and ischemia/reperfusion damage. Several of these studies have been reviewed recently (28) and will also be covered in other chapters in this book. This review is limited to two aspects of selectin biology; it will focus on the regulation of the endothelial selectins and on the current data on the identification and characterization of their ligands.

II. REGULATION OF ENDOTHELIAL SELECTINS

The regulation of selectins is of central importance for the initiation of leukocyte entry into sites of inflammation. Since L-selectin is constitutively expressed on leukocytes, its involvement in inflammatory processes requires the inducibility of its endothelial ligand(s). Indeed there is good, albeit indirect, evidence that endothelial ligands for L-selectin exist that are cytokine-inducible (29,30). However, the molecular identity of these ligands has not yet been elucidated. The L-selectin ligands that have been identified so far are GlyCAM-1, CD34, and MAdCAM-1 (31–33). They are all restricted to lymph node endothelial venules and are probably involved in the lymphocyte homing process.

E-selectin is transcriptionally induced by cytokines such as IL-1β, TNF-α, and TNF-β and by lipopolysaccharide (LPS) (14,34). Cell surface expression of the protein on HUVE cells reaches maximal levels 4 to 6 hours after stimulation and rapidly declines again to basal levels after another 12 to 16 hours. Expression kinetics are similar on several mouse endothelioma cell lines, with maximal expression levels 3 to 4 hours after stimulation (35,36). It was soon found that NF-κ_B is one of the intracellular key regulators for transcriptional induction (37,38). Analysis of the promotor region of the human E-selectin gene has revealed four positive regulatory domains that are required for maximal expression levels of the E-selectin mRNA upon induction by TNF-α (39–41). Of these regulatory domains, one is a consensus NF-κ_B site, two are novel adjacent binding sites for NF-κ_B and one is a CRE/ATF (cAMP-responsive element/activating transcription factor) site. TNF-α-induced E-selectin transcription is decreased by c-AMP, possibly by changing the composition of the proteins that bind to the CRE/ATF site (42).

Down-regulation of E-selectin protein at the cell surface is most likely achieved by endocytosis. Von Asmuth et al. (43) showed that TNF-α-activated HUVE cells constitutively internalize E-selectin. Endocytosis of E-selectin was also observed in transfected AtT-20 cells, where it is dependent on the presence of the cytoplasmic tail of E-selectin (44).

Upregulation in vivo has been analyzed in man (45), baboon (46–48), and mouse (49). In baboon skin, E-selectin was induced within 2 hours after local injection of endotoxin, as demonstrated by immunostaining of cryostat tissue sections. E-selectin staining was virtually absent again by 9 hours (48). In another study in baboons, E-selectin was induced within 2 hours after intracutaneous injection of TNF-α; the E-selectin staining was present up to 24 hours (46). Intravenous injection of live *E. coli* bacteria (injected intravenously) induced widespread expression of E-selectin in baboon endothelium in most tissues (47); similar results have been obtained after intravenous injection of TNF-α or LPS in the mouse (49).

In chronic inflammatory lesions in human skin, E-selectin is chronically expressed. It was suggested that differences in the stability of E-selectin transcript variants might be responsible for the different half-life times of E-selectin expression (50). Because of differential usage of three polyadenylation sites in the 3′ untranslated region of the E-selectin gene, three alternatively processed E-selectin transcripts are generated in primary human endothelial cells. The shortest transcript lacks six mRNA destabilizing elements which are thought to mediate rapid degradation of the corresponding mRNA. Only this type of E-selectin transcript was found in skin biopsies of chronic inflammatory lesions.

In contrast to E-selectin, P-selectin is stored in the membrane of intracellular storage granules in platelets (α-granules) (18,19) and in endothelial cells (Weibel-Palade bodies) (20). Fusion of these granules with the plasma membrane is rapidly stimulated by pro-inflammatory mediators such as histamine and thrombin (51), thereby exposing P-selectin on the cell surface within minutes. P-selectin is rapidly down-regulated by internalization into the endothelial cell. The cytoplasmic domain of P-selectin targets newly synthesized P-selectin from the Golgi compartment into storage granules (52,53). This domain was also crucial for the rapid internalization of P-selectin (54). In contrast to E-selectin, which is targeted to lysosomes after internalization, P-selectin can recycle from endosomes into storage granules, as was found in transfected AtT 20 cells (44). In CHO cells, which lack secretory granules, transfected P-selectin was rapidly internalized and targeted into lysosomes by the cytoplasmic tail domain (54).

Rapid upregulation of P-selectin on the cell surface was also observed after treatment of endothelial cells with oxygen radicals (55). This induction mechanism did not require protein synthesis, and the cell surface expression was sustained between 1 and 4 hours. Thus, oxygen radicals most likely induce transport of stored P-selectin to the plasma membrane and prolong the half-life time of P-selectin at the cell surface.

In addition to the regulation of its transport to the cell surface, P-selectin is inducible by cytokines (TNF-α) and LPS on mouse endothelioma cells. Maximal expression levels of the P-selectin mRNA were reached 2 hours after stimulation, and maximal protein levels were reached after 3 to 4 hours (35,36). Newly synthesized P-selectin reached the cell surface with no need for activating the transport of storage granules to the plasma membrane. In fact, stored P-selectin was still present inside the cell when maximal levels of TNF-α-stimulated, newly synthesized P-selectin was present at the cell surface. Additional stimulation of P-selectin secretion at this time point could further increase the amount on the plasma membrane (36). Recently, TNF-α-induced expression of P-selectin was also demonstrated on HUVE cells at 6 hours after stimulation (56).

Cytokine-induced upregulation of P-selectin has also been found in vivo in mouse tissue, at both the RNA level (57) and the protein level (49,58). Endothelium in mouse brain and in the leptomeninges is devoid of constitutive expression and storage of P-selectin. However, P-selectin is upregulated on venules in the leptomeninges and (although much less strongly) in brain parenchyme upon systemic administration of TNF-α or LPS (49). E-selectin and VCAM-1 were also upregulated in the same vessels. However, endothelium of arterioles in the leptomeninges showed only expression of E-selectin and not of P-selectin after stimulation. Thus, although cytokines can induce the synthesis of both endothelial selectins, certain endothelia can respond selectively to such stimuli by upregulation of only one of the two selectins.

III. GLYCOPROTEIN LIGANDS FOR E-SELECTIN

The identification of a C-type lectin domain (59) in E-selectin initiated the search for carbohydrate ligands. Since PMNs were the first group of leukocytes found to bind to E-selectin, uncommon terminal carbohydrate sequences as found on glycolipids and glycoproteins of PMNs were first suspected to be ligands for E-selectin. Indeed, expression of the tetrasaccharide sialyl Lewis X (sLe^x, Neu5Acα2-3Galβ1-4[Fucα1-3]GlcNAc) on PMNs, monocytic cell lines, and other cells correlated with the binding of these cells to E-selectin (60,61). Antibodies that recognized sLe^x could inhibit cell adhesion in these assays (60). In addition, crude protein preparations from amnionic fluid containing sLe^x determinants could also block adhesion in such assays (61). Neoglycoproteins consisting of bovine serum albumin loaded with sLe^x, or its stereoisomer sLe^a, could support binding of E-selectin transfected cells (62). Soon it was found that sLe^x indeed could bind to all three selectins (63) and could even block neutrophil migration into inflamed tissue in some models in vivo (64–66). Thus, it is likely that the carbohydrate structures on the physiological selectin ligands somehow resemble the sLe^x structure. However, whether they are identical with sLe^x is not known. Several glycoproteins are known which are strongly modified with sLe^x, yet do not bind to P-selectin (67) in affinity isolation experiments. In addition HT29 cells with well detectable levels of sLe^x on the cell surface do not bind to P-selectin (67). From these data it is clear that it takes more than the mere presence of sLe^x to define a glycoprotein as a selectin ligand.

Besides neutrophils and monocytic cells, a subset of T lymphocytes that preferentially home to the skin was also found to bind to E-selectin (16,17). This subset of T-cells expresses an sLe^a-related carbohydrate epitope which is defined by the mAb HECA 452 (68,69). A series of glycoproteins purified with the help of this mAb could support the binding of E-selectin expressing cells (70).

The first glycoprotein ligands for which direct binding to a soluble form of a selectin was shown were the two L-selectin ligands GlyCAM-1 (31) and CD34 (32). They are expressed by endothelial cells in high endothelial venules (HEV) of lymph nodes, where they are thought to be involved in the lymphocyte homing process. These ligands were identified by affinity isolation using an L-selectin-immunoglobulin fusion protein as affinity probe (71). Both ligands carry densely packed O-linked carbohydrate side chains which are rich in sialic acid. This type of carbohydrate modification defines them as so-called sialomucins; it is essential for the binding to L-selectin. Also a third ligand for L-selectin, the vascular addressin MAd-CAM-1, carries a sialomucinlike domain (33,72). This protein was originally found as an addressin on HEV of Peyer's patches, where it mediates the binding to the integrin $\alpha_4\beta_7$, a lymphocyte homing receptor on lymphocytes which home to Peyer's patches (73). A subpopulation of the MAd-CAM-1 molecules carry carbohydrate modifications, possibly in the mucin type domain, which enable them to bind to L-selectin.

The same approach that led to the identification of high-affinity ligands for L-selectin was used to search for glycoprotein ligands for E-selectin.

With the help of a fusion protein containing the first four domains of mouse E-selectin and the Fc region of human IgG1, a single 150-kDa glycoprotein was affinity isolated from a metabolically labeled mouse neutrophilic cell line (32DcI3) and mouse neutrophils (74). Binding of this E-selectin ligand (named ESL-1) to E-selectin was Ca^{2+}-dependent and required the presence of sialic acid on the ligand. Although low-affinity carbohydrate ligands such as sLe^x are able to bind to all three selectins, ESL-1 was not recognized by an analogous mouse P-selectin Ig fusion protein (75). Thus, high-affinity binding motifs on selectin ligands can be specific for one selectin and irrelevant for another.

In contrast to the other high-affinity selectin ligands, ESL-1 is not a sialomucin and requires N-linked (instead of O-linked) carbohydrates for binding. Transfection experiments with the ESL-1 cDNA revealed that the ESL-1 glycoprotein could only be affinity-isolated from transfected CHO cells when coexpressed with a fucosyltransferase (76). Thus, modification with fucose is imperative for binding to E-selectin. The amino acid sequence of ESL-1 is 94% identical (over 1078 amino acids) to a novel chicken cysteine-rich fibroblast growth factor receptor (CFR) (77), except for a unique 70-amino acid amino-terminal domain of mature ESL-1. Whether ESL-1 is a splicing variant of a putative mouse equivalent of the chicken CFR is not known. The strong structural homology to the chicken FGF-receptor allows one to speculate that ESL-1 may have some signaling function. Since E-selectin mediates the initial cell contact between leukocytes and endothelial cells, a step that is followed by the activation of leukocyte integrins, it is intriguing to speculate that ESL-1 may be involved in triggering signals which lead to integrin activation. Indeed, evidence has been reported that soluble E-selectin can induce activation of PMNs (78). However, so far no evidence has been presented that ESL-1 or the chicken CFR are able to mediate signal transduction.

While the receptorlike function of ESL-1 is still speculative, its function as a cell adhesion molecule has been well established (76). An immunoglobulin fusion protein of ESL-1 containing the complete extracellular part of ESL-1 and modified by fucose was able to support the binding of E-selectin-transfected CHO cells. No binding was seen to a fucose-containing L-selectin Ig fusion protein. Furthermore, affinity-purified polyclonal antibodies against ESL-1 blocked the binding of the neutrophilic cell line 32DcI3 to immobilized E-selectin IgG. Similarly, the binding of these cells and of mouse PMNs to cytokine-induced mouse endothelioma cells could be blocked by the anti-ESL-1 antibodies. The cytokine-induced endothelioma cells expressed E-and P-selectin. Since the binding of PMNs was predominantly mediated by P-selectin in these assays, the inhibitory effect of the anti-ESL-1 antibodies was only detectable when P-selectin was blocked simultaneously with a mAb against P-selectin. This is additional proof that ESL-1 is a ligand specific for E-selectin and not for P-selectin (Fig. 1).

ESL-1 was not the only glycoprotein ligand that could be affinity-isolated with E-selectin Ig from metabolically labeled mouse PMNs. In addition, a protein running as a sharp band of 250 kDa apparent MW (230 kDa nonreducing) in SDS-polyacrylamide gel electrophoresis (SDS-PAGE) was seen, which was not detectable on any tested myeloid cell line (74,75). Like ESL-1, this protein did not bind to P-selectin Ig. As a much weaker signal, a pair of proteins running at 230 and 130 kDa in SDS-PAGE was isolated (75). These proteins were detected with similar efficiency in affinity isolation experiments with P-selectin Ig. While the 230-kDa protein could not be shifted in MW under normal reducing conditions, raising the concentration of the reducing reagent permitted a shift of the apparent MW to 130 kDa, indicating that the 230-kDa form was a disulfide-linked dimeric

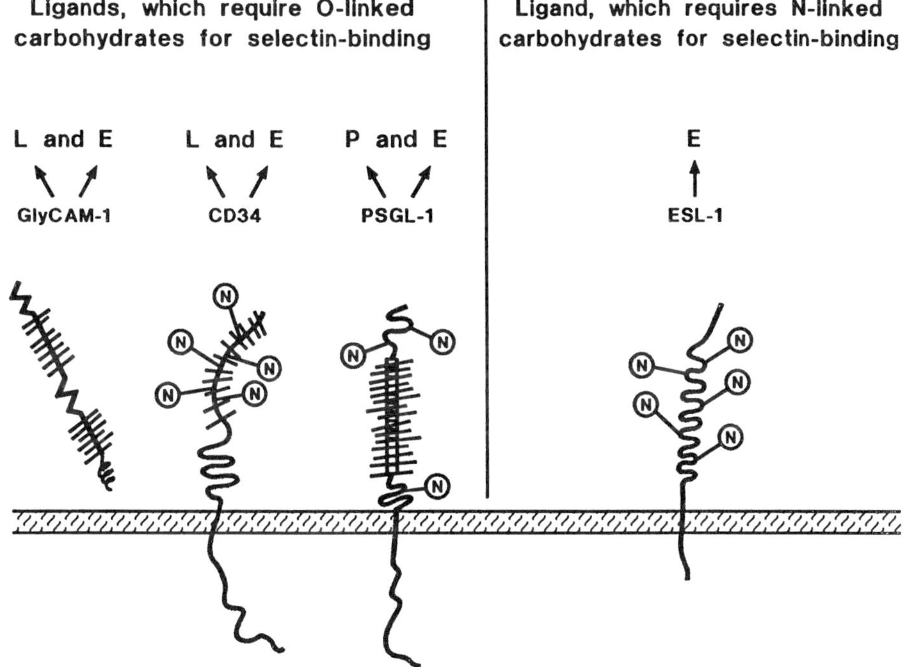

Figure 1 Cloned selectin ligands, which can be affinity isolated with immobilized selectins as affinity probes. The sialomucin ligands bind to more than one selectin, whereas ESL-1, which requires N-linked carbohydrates for binding, is monospecific for E-selectin.

structure (Borges and Vestweber, unpublished). Since this was reminiscent of the recently identified P-selectin ligand PSGL-1 (79) (see below), we raised an antiserum against PSGL-1. Indeed, the 230/130-kDa ligand was recognized by this antiserum (Borges and Vestweber, unpublished).

In an elegant approach, a 250-kDa ligand (280 kDa under reducing conditions) was isolated from bovine γ/δ T lymphocytes, using as affinity probe complete human E-selectin purified from transfected L cells and immobilized with the help of a non-adhesion-blocking anti E-selectin mAb (80). Like the selectin ligands in the mouse, this glycoprotein ligand is not recognized by either the anti sLe^x antibody CSLEX-1 or the anti sLe^a antibody HECA 452.

Several other glycoproteins were defined as ligands for E-selectin, based either on observations that antibodies against them blocked leukocyte binding to E-selectin or that the immobilized glycoprotein could support the binding of E-selectin-expressing cells. However, none of these ligands has been tested yet to see whether it can bind to E-selectin with sufficient affinity to allow affinity isolation with E-selectin, as described above for the "high-affinity" ligands. Whether "high affinity" or "low affinity" is necessary for a selectin ligand to be of physiological relevance is being debated (81,82).

The most prominent of these ligands is L-selectin, which was suggested to serve as a carbohydrate-presenting ligand for E-selectin and P-selectin (83,84). Only L-selectin from human neutrophils, but not from human lymphocytes, was able to support binding of E-selectin transfected cells. In addition, antibodies against L-selectin blocked binding of PMNs to cells expressing E-selectin or P-selectin. Immunogold labeling revealed that L-selectin is concentrated on the tips of PMN pseudopods, which allows a prominent exposure of L-selectin to endothelial cells. This is a very valuable observation, underlining the importance of L-selectin for the initial interactions of PMNs with the endothelium, whether it functions as a carbohydrate-presenting molecule or as a lectin. L-selectin on PMNs is indeed involved in neutrophil binding to cytokine-induced endothelial cells, and indirect evidence has been presented that cytokine-inducible endothelial binding partners for L-selectin exist that are independent of the endothelial selectins (29,30).

Other suggested E-selectin ligands include members of the NCA family (nonspecific crossreactive antigens) present on human neutrophils (85), a subpopulation of the $\beta 2$ integrins that carry sLe^x (86), and the heavily sLe^x-modified lysosomal membrane protein lamp-1 (87). Indeed, increasing the cell surface expression of lamp-1 on transfected cells correlated with an increase in the binding of these cells to E-selectin. It is conceivable that various colon carcinoma cells that display increased levels of lamp-1 at the cell surface may bind via lamp-1 to E-selectin-expressing endothelium.

IV. GLYCOPROTEIN LIGANDS FOR P-SELECTIN

The first P-selectin ligand was found by using the complete P-selectin membrane protein, purified from human platelets, as an affinity probe (67). A 250-kDa disulfide-linked dimeric glycoprotein (reduced monomeric form: 120 kDa) was detected in neutrophil extracts by immunoblot-like experiments using ^{125}I-labeled P-selectin as a probe. In addition, this protein could be isolated from ^{3}H-glucosamine-labeled monocytic cells (HL60) using immobilized P-selectin as an affinity matrix. This ligand required O-linked carbohydrates for binding to P-selectin (88,89). In a separate, independent experimental approach, the same protein was identified as a ligand for P-selectin by expression cloning (79). An expression library of HL60 cells was transfected into COS-7 cells which were cotransfected with fucosyltransferase III. COS cell clones were selected for their ability to bind to P-selectin Ig fusion protein immobilized on plastic. The identified ligand was reported to migrate on SDS-PAGE at 110 kDa under reducing conditions (220 kDa nonreducing) and was named PSGL-1. Fucosylation of PSGL-1 was imperative for binding to P-selectin.

An immobilized fusion protein of PSGL-1 supported binding of P-selectin-as well as of E-selectin-expressing cells (79). This is in agreement with studies on the 230/130-kDa ligand (described above) which Lenter et al. (75) found on mouse neutrophils and which is most likely identical with PSGL-1. This protein was isolated by mouse E-selectin IgG as efficiently as by mouse P-selectin IgG. In careful titration experiments, Moore et al. (89) showed that ^{125}I-labeled PSGL-1 binds with higher affinity to immobilized truncated human P-selectin than to an analogous form of human E-selectin.

Recently, the mAbs PL1 and PL2 were raised against human PSGL-1 (90). Only PL1 blocks adhesion of neutrophils to immobilized P-selectin under static conditions. This antibody also inhibited rolling of neutrophils on immobilized P-selectin, establishing PSGL-1 as an essential P-selectin ligand for rolling of neutrophils.

Beside PSGL-1, a protein of 160 kDa was found on metabolically labeled mouse and human myeloid cells using mouse P-selectin Ig as affinity matrix (75). Although this protein was undetectable in SDS-PAGE when reduced by boiling with β-mercaptoethanol, it was shifted to 80 kDa when reduced under mild conditions at room temperature. This protein only bound to P-selectin but not to E-selectin. Like ESL-1, this ligand was shown to require N-linked carbohydrates for binding. A ligand of 160 kDa was also found for human P-selectin, although it was more difficult to detect than PSGL-1 (67). This protein was not shifted in size by reduction.

The strongly glycosylated, heat-stable antigen (HSA) from mouse neutrophils and some lymphoid cells was found to bind to mouse P-selectin (91). When HSA from different types of leukocytes was immobilized on plastic, P-selectin Ig but not E-selectin Ig bound in an ELISA-like assay. No binding was observed to HSA isolated from erythrocytes. As analyzed by lectin-binding studies, HSA from different types of leukocytes displays extensive heterogeneity in carbohydrate composition. Binding to P-selectin correlated with the presence of the L2/HNK-1 epitope, although antibodies against this epitope did not interfere with the binding to P-selectin.

The mAb HNK-1 also defines an unusual class of sulfated glycosphingolipids, the sulfoglucuronyl-containing neolactosylceramides (SGNL-lipids), which bind to P-and L-selectin but not to E-selectin (92). In contrast, sulfated Le^x or Le^a and sulfatides (galactosyl ceramides) were found to bind to all three selectins (92–95).

V. SIGNAL TRANSDUCTION VIA THE SELECTINS AND THEIR LIGANDS

Transmembrane signaling by cell adhesion molecules has been documented for integrins (96), cadherins, and members of the Ig superfamily (97). Only a few reports document signal transduction functions for the selectins and their ligands. The selectin-mediated initial step of leukocyte adhesion to endothelium is followed by triggering the activation of leukocyte integrins. It has been well documented that this activation process can be triggered by chemokines or the phospholipid PAF (platelet activating factor) which are presented on the surface of endothelial cells and bind to receptors on the leukocyte surface (98–100). In addition, the endothelial selectins may directly trigger integrin activation by binding to their ligands on the leukocyte surface. The first evidence toward this concept was presented by Lo et al. (78), who showed that incubation of human PMNs with cytokine-activated monolayers of endothelial cells activated Mac-1 on the neutrophils. This effect was blocked by anti-E-selectin antibody. Activation of Mac-1 could also be induced by incubation of PMNs with a recombinant soluble form of E-selectin immobilized on plastic surfaces.

Similar effects were reported for P-selectin (101). Binding of human PMNs to purified recombinant P-selectin enhanced the phagocytosis of unopsonized zymosan particles, a process that is mediated by β_2-integrins. In addition, P-selectin caused increased binding of the mAb 24 to PMNs;

this antibody recognizes an activation epitope on α-chains of β2-integrins. However, other activation phenomena can be blocked by P-selectin. A soluble form of P-selectin can inhibit CD18 (β_2-integrin)-dependent adhesion of TNF-stimulated neutrophils to endothelium (102) and can inhibit superoxide generation by neutrophils (103). In yet another series of experiments, Lorant et al. (98,99) showed that P-selectin did not stimulate neutrophil function directly. Instead, P-selectin had only a function as an anchoring molecule; signaling was mediated by PAF. The identification of PSGL-1 and possibly further ligands should prove helpful in clarifying the issue.

Another recent report presents clear evidence that P-selectin can indeed stimulate leukocyte effector functions. Incubation of monocytes with soluble, purified P-selectin stimulated tissue factor expression in monocytes (104). In addition, binding of monocytes to P-selectin-transfected CHO cells increased expression of tissue factor on the mRNA as well as the protein level. This effect could be blocked by antibodies against P-selectin and was not observed with E-selectin-transfected or untransfected CHO cells. Interestingly, activated platelets with surface expression of P-selectin induced tissue factor expression much more efficiently than P-selectin-expressing CHO cells or purified P-selectin, arguing for additional factors (e.g., PAF) which amplify the signaling process significantly.

Information about a potential function of selectins as signal transducing receptors is now starting to accumulate. Sulfatides that are able to bind L-selectin as well as antibodies against L-selectin can trigger increases in cytosolic Ca^{2+} in human neutrophils (105). Waddell et al. confirmed this effect of anti-L-selectin antibodies on intracellular Ca^{2+} levels (106). However, L-selectin cross-linking did not trigger production of H_2O_2 by itself but significantly enhanced the subsequent response to the soluble activating agents formyl-Met-Leu-Phe and TNF. Antibody-mediated cross-linking of L-selectin on the surface of neutrophils can trigger the release of the 55-kDa TNF receptor (107). Recently, all three selectins were reported to be involved in shape control of cells. The shape change which was induced in lymphocytes upon binding to high endothelial cells was inhibited by anti-L-selectin antibodies (108). Binding of antibodies against E- or P-selectin to HUVE cells induced a cell shape described as "rounding up" of the IL-1 or thrombin-activated cells (109). No such effect was observed with antibodies against ICAM-1 or Von Willebrand factor.

VI. CONCLUDING REMARKS

Like most cell adhesion molecules, each of the selectins bind to more than one ligand. The function of each ligand has been demonstrated in in vitro cell adhesion assays. For only four ligands (GlyCAM-1, CD34, PSGL-1, and ESL-1) has direct biochemical evidence for selectin-binding been demonstrated. For none of the ligands known today has its physiological role as a selectin ligand been demonstrated in vivo.

It is quite possible that some of the different ligands that bind to the same selectin serve different molecular functions. The secreted L-selectin ligand GlyCAM-1 almost certainly has a different function than the transmembrane ligand CD34. L-selectin itself, as a carbohydrate-presenting ligand for E-selectin, was recently demonstrated to mediate tethering of human neutrophils on E-selectin, while rolling movement, which immediately follows, was independent of L-selectin (110). Possibly ESL-1 or PSGL1 or one of the other suggested ligands may be responsible for the rolling process. Yet another function of a ligand may be signal transduction. The fact that some ligands (e.g., PSGL-1) bind to both endothelial selectins and some (e.g., ESL-1) to only one is a further indication that different ligands may serve different physiological functions.

It has almost become a rule that expression of the protein scaffold of a selectin ligand is not restricted to the cell types that use this ligand for binding to a selectin. Thus, the modifying enzymes that generate the correct posttranslational modifications necessary for selectin binding are of central importance for the regulation of cell-type-specific expression of the selectin ligands. For all selectin ligands modification by a fucosyltransferase seems essential. In addition, the L-selectin ligands GlyCAM-1 and CD34 require modification by a sulfotransferase. While no candidate for a sulfotransferase has yet been identified, it is possible that some of the five cloned human fucosyltransferases (111) may be involved in the generation of selectin ligands. The recent identification of high-affinity ligands for each of the selectins now permits analysis of the binding motifs recognized by the selectins. If these motifs are indeed formed exclusively by carbohydrate structures, it will be interesting to learn how they determine the binding specificity for the selectins. At present the most favored hypothesis for the sialomucin type of selectin ligands is that "the unique clustering of relatively common oligosaccharides" explains the specificity of recognition (82). For a ligand like ESL-1, which requires some of the five potential N-linked carbohydrate side chains for binding, the situation is different. Besides the possibility of a direct participation of the amino acid backbone in the binding to E-selectin, which can not be excluded at present, glycosyltransferases may be able to generate specific carbohydrate structures selectively on only a few (or even only one) protein scaffold. Whether this is the case and how this is controlled is still unknown.

REFERENCES

1. Springer TA. Traffic signals for lymphocyte recirculation and leukocyte emigration: the multistep paradigm. Cell 1994; 76:301–314.
2. Carlos TM, Harlan JM. Leukocyte-endothelial adhesion molecules. Blood 1994; 84:2068–2101.
3. Lawrence MB, Springer TA. Leukocytes roll on selectin at physiologic flow rates: distinction from and prerequisite for adhesion through integrins. Cell 1991; 65:859–873.
4. Ley K, Gaehtgens P, Fennie C, Singer MS, Lasky LA, Rosen SD. Lectin-like cells adhesion molecule 1 mediates leukocyte rolling in mesenteric venules in vivo. Blood 1991; 12:2553–2555.
5. Von Andrian UH, Chambers JD, McEvoy LM, Bargatze RF, Arfors KE, Butcher EC. Two-step model of leukocyte-endothelial cell interaction in inflammation: distinct roles for LECAM-1 and the leukocyte b2 integrins in vivo. Proc Natl Acad Sci USA 1991; 88:7538–7542.
6. Lasky LA. Selectins: interpreters of cell-specific carbohydrate information during inflammation. Science 1992; 258:964–969.
7. Vestweber D. Selectins: cell surface lectins which mediate the binding of leukocytes to endothelial cells. Sem Cell Biol 1992; 3:211–220.
8. Gallatin WM, Weissman IL, Butcher EC. A cell-surface molecule involved in organ-specific homing of lymphocytes. Nature 1983; 304:30–34.
9. Camerini D, James SP, Stamenkovic I, Seed B. Leu8/TQ1 is the human equivalent of the MEL-14 lymph node homing receptor. Nature 1989; 342: 78–82.
10. Griffin JD, Spertini O, Ernst TJ, et al. GM-CSF and other cytokines regulate surface expression of the leukocyte adhesion molecule-1 on human neutrophils, monocytes and their precursors. J Immunol 1991; 145:576–584.
11. Tedder TF, Penta AC, Levine HB, Freedman AS. Expression of the human leukocyte adhesion molecule, LAM1. J Immunol 1990; 144:532–540.
12. Lewinsohn DM, Bargatze RF, Butcher EC. Leukocyte-endothelial cell recognition: evidence of a common molecular mechanism shared by neutrophils, lymphocytes, and other leukocytes. J Immunol 1987; 138:4313–4321.
13. Watson SR, Fennie C, Lasky LA. Neutrophil influx into an inflammatory site inhibited by a soluble homing receptor-IgG chimaera. Nature 1991; 349: 164–167.
14. Bevilacqua MP, Pober JS, Mendrick DL, Cotran RS, Gimbrone MA. Identification of an inducible endothelial-leukocyte adhesion molecule. Proc Natl Acad Sci USA 1987; 84:9238–9242.
15. Bevilacqua MP, Stengelin S, Gimbrone MA, Seed B. Endothelial leukocyte adhesion molecule 1: an inducible receptor for neutrophils related to complement regulatory proteins and lectins. Science 1989; 243:1160–1165.
16. Picker LJ, Kishimoto TK, Smith CW, Warnock RA, Butcher EC. ELAM-1 is an adhesion molecule for skin-homing T cells. Nature 1991; 349:796–799.
17. Shimizu Y, Shaw S, Graber N, et al. Activation-independent binding of

human memory T cells to adhesion molecule ELAM-1. Nature 1991; 349: 799–802.
18. Hsu-Lin SC, Berman CL, Furie BC, August D, Furie B. A platelet membrane protein expressed during platelet activation and secretion. J Biol Chem 1984; 259:9121–9126.
19. McEver RP, Martin MN. A monoclonal antibody to a membrane glycoprotein binds only to activated platelets. J Biol Chem 1984; 259:9799–9804.
20. McEver RP, Beckstead JH, Moore KL, Marshall-Carlson L, Bainton DF. GMP-140, a platelet alpha-granule membrane protein, is also synthesized by vascular endothelial cells and is localized in Weibel-Palade bodies. J Clin Invest 1989; 84:92–99.
21. Geng J-G, Bevilacqua MP, Moore KL, et al. Rapid neutrophil adhesion to activated endothelium mediated by GMP-140. Nature 1990; 343:757–760.
22. Damle NK, Klussman K, Dietsch MT, Mohagheghpour N, Aruffo A. GMP-140 (P-selectin/CD62) binds to chronically stimulated but not resting CD4+ T lymphocytes and regulates their production of proinflammatory cytokines. Eur J Immunol 1992; 22:1789–1793.
23. Moore KL, Thompson LF. P-selectin (CD62) binds to subpopulations of human memory T lymphocytes and natural killer cells. Biochem Biophys Res Commun 1992; 186:173–181.
24. Kunzendorf U, Notter M, Hock H, Distler A, Diamantstein T, Walz G. T cells bind to the endothelial adhesion molecule GMP-140 (P-selectin). Transplantation 1993; 56:1213–1217.
25. Rossiter H, van Reijsen F, Mudde GC, et al. Skin disease-related T cells bind to endothelial selectins: expression of cutaneous lymphocyte antigen (CLA) predicts E-selectin but not P-selectin binding. Eur J Immunol 1994; 24:205–210.
26. Alon R, Rossiter H, Wang X, Springer TA, Kupper TS. Distinct cell surface ligands mediate T lymphocyte attachment and rolling on P and E selectin under physiological flow. J Cell Biol 1994; 127:1485–1495.
27. Postigo AA, Marazueal M, Sánchez-Madrid F, de Landázuri MO. B-lymphocyte binding to E-and P-selectins is mediated through the de novo expression of carbohydrates on in vitro and in vivo activated human B cells. J Clin Invest 1994; 94:1585–1596.
28. Parekh RB, Edge CJ. Selectins—glycoprotein targets for therapeutic interaction in inflammation. Trends Biotechnol 1994; 12:339–345.
29. Spertini O, Luscinskas FW, Kansas VX, et al. Leukocyte adhesion molecule-1 (LAM-1, L-selectin) interacts with an inducible endothelial cell ligand to support leukocyte adhesion. J Immunol 1991; 147:2565–2573.
30. Brady HR, Spertini O, Jimenez W, Brenner BM, Marsden PA, Tedder TF. Neutrophils, monocytes, and lymphocytes bind to cytokine-activated kidney glomerular endothelial cells through L-selectin (Lam-1) in vitro. J Immunol 1992; 149:2437–2444.
31. Lasky LA, Singer MS, Dowbenko D, et al. An endothelial ligand for L-selectin is a novel mucin-like molecule. Cell 1992; 69:927–938.

32. Baumheuter S, Singer MS, Henzel W, et al. Binding of L-selectin to the vascular sialomucin CD34. Science 1993; 262:436–438.
33. Berg EL, McEvoy LM, Berlin C, Bargatze RF, Butcher EC. L-selectin-mediated lymphocyte rolling on MAdCAM-1. Nature 1993; 366:695–698.
34. Pober JS, Lapierre LA, Stolpen AH, et al. Activation of cultured human endothelial cells by recombinant lymphotoxin: comparison with tumor necrosis factor and interleukin 1 species. J Immunol 1987; 138:3319–3324.
35. Weller A, Isenmann S, Vestweber D. Cloning of the mouse endothelial selectins: expression of both E-and P-selectin is inducible by tumor necrosis factor-α. J Biol Chem 1992; 267:15176–15183.
36. Hahne M, Jäger U, Isenmann S, Hallmann R, Vestweber D. Five TNF-inducible cell adhesion mechanisms on the surface of mouse endothelioma cells mediate the binding of leukocytes. J Cell Biol 1993; 121:655–664.
37. Montgomery KF, Osborn L, Hession C, et al. Activation of endothelial-leukocyte adhesion molecule 1 (ELAM-1) gene transcription. Proc Natl Acad Sci USA 1991; 88:6523–6527.
38. Whelan J, Ghersa P, Von Huijsduijene RH, et al. An NFk_B-like factor is essential but not sufficient for cytokine induction of endothelial leukocyte adhesion molecule 1 (ELAM-1) gene transcription. Nucl Acids Res 1991; 19: 2645–2653.
39. Whitley MZ, Thanos D, Read MA, Maniatis T, Collins T. A striking similarity in the organization of the E-selectin and beta interferon gene promoters. Mol Cell Biol 1994; 14:6464–6475.
40. Schindler U, Baichwal VR. Three-NF-κB binding sites in the human E-selectin gene required for maximal tumor necrosis factor alpha-induced expression. Mol Cell Biol 1994; 14:5820–5831.
41. Lewis H, Kaszubska W, DeLamarter JF, Whelan J. Cooperativity between two NF-κB complexes, mediated by high-mobility-group protein I(Y), is essential for cytokine induced expression of the E-selectin promoter. Mol Cell Biol 1994; 14:5701–5709.
42. De Lucas LG, Johnson DR, Whitley MZ, Collins T, Pober JS. cAMP and tumor necrosis factor competitively regulate transcriptional activation through and nuclear factor binding to the cAMP-responsive element/activating transcription factor element of the endothelial leukocyte adhesion molecule-1 (E-selectin) promoter. J Biol Chem 1994; 269:19193–19196.
43. Von Asmuth EJU, Smeets EF, Ginsel LA, Onderwater JJM, Leeuwenberg JFM, Buurman WA. Evidence for endocytosis of E-selectin in human endothelial cells. Eur J Immunol 1992; 22:2519–2526.
44. Subramaniam M, Koedam JA, Wagner DD. Divergent fates of P-and E-selectins after their expression on the plasma membrane. Mol Biol Cell 1993; 4:791–801.
45. Cotran RS, Gimbrone MA Jr, Bevilacqua MP, Mendrick DL, Pober JS. Induction and detection of a human endothelial activation antigen in vivo. J Exp Med 1986; 164:661–666.
46. Munro JM, Pober JS, Cotran RS. Tumor necrosis factor and interferon-γ

induce distinct patterns of endothelial activation and associated leukocyte accumulation in skin of *Papio anubis*. Am J Pathol 1989; 135:121–133.
47. Redl H, Dinges HP, Buurman WA, et al. Expression of endothelial leukocyte adhesion molecule-1 in septic but not traumatic/hypovolemic shock in the baboon. Am J Pathol 1991; 139:461–466.
48. Munro JM, Pober JS, Cotran RS. Recruitment of neutrophils in the local expression response: association with de novo endothelial expression of endothelial leukocyte adhesion molecule-1. Lab Invest 1991; 64:295–299.
49. Gotsch U, Jäger U, Dominis M, Vestweber D. Expression of P-selectin on endothelial cells is upregulated by LPS and TNF-a in vivo. Cell Adhesion Commun 1994; 2:7–14.
50. Chu W, Presky DH, Swerlick RA, Burns DK. Alternatively processed human E-selectin transcripts linked to chronic expression of E-selectin in vivo. J Immunol 1994; 153:4179–4189.
51. Stenberg PE, McEver RP, Shuman MA, Jacques YV, Bainton DF. A platelet alpha-granule membrane protein (GMP-140) is expressed on the plasma membrane after activation. J Cell Biol 1985; 101:880–886.
52. Disdier M, Morrissey JH, Fugate RD, Bainton DF, McEver RP. Cytoplasmic domain of P-selectin (CD62) contains the signal for sorting into the regulated secretory pathway. Mol Biol Cell 1992; 3:309–321.
53. Koedam JA, Cramer EM, Briend E, Furie B, Furie BC, Wagner DD. P-selectin, a granule membrane protein of platelets and endothelial cells, follows the regulated secretory pathway in AtT-20 cells. J Cell Biol 1992; 116: 617–625.
54. Greem SA, Setiadi H, McEver RP, Kelly RB. The cytoplasmic domain of P-selectin contains a sorting determinant that mediates rapid degradation in lysosomes. J Cell Biol 1994; 124:435–448.
55. Patel KD, Zimmerman GA, Prescott SM, McEver RP, McIntyre TM. Oxygen radicals induce human endothelial cells to express GMP-140 and bind neutrophils. J Cell Biol 1991; 112:749–759.
56. Luscinskas FW, Ding H, Tedder TF, Cumming D, Gerritsen ME. L-selectin and P-selectin preferentially mediate monocyte rolling and adhesion to TNF-α activated vascular endothelium under flow. Meeting abstract, Eighth International Symposium on the Biology of Vascular Cells, Heidelberg, September 1994.
57. Sanders WE, Wilson RW, Ballantyne CM, Beaudet AL. Molecular cloning and analysis of in vivo expression of murine P-selectin. Blood 1992; 80:795–800.
58. Lobow MA, Norton CR, Rumberger JM, et al. Characterization of E-selectin-deficient mice: demonstration of overlapping function of the endothelial selectins. Immunity 1994; 1:709–720.
59. Drickamer K. Two distinct classes of carbohydrate-recognition domains in animal lectins. J Biol Chem 1988; 263:9557–9560.
60. Phillips ML, Nudelman E, Gaeta FCA, et al. ELAM-1 mediates cell adhesion by recognition of a carbohydrate ligand, sialyl-Le^x. Science 1990; 250:1130–1132.

61. Walz G, Aruffo A, Kolanus W, Bevilacqua M, Seed B. Recognition by ELAM-1 of the sialyl-Le^x determinant on myeloid and tumor cells. Science 1990; 250:1132–1135.
62. Berg EL, Robinson MK, Mansson O, Butcher EC, Magnani JL. A carbohydrate domain common to both sialyl le^a and sialyl le^x is recognized by the endothelial cell leukocyte adhesion molecule ELAM-1. J Biol Chem 1991; 266:14869–14872.
63. Foxall C, Watson SR, Dowbenko D, et al. The three members of the selectin receptor family recognize a common carbohydrate epitope, the sialyl Lewis^x oligosaccharide. J Cell Biol 1992; 117:895–902.
64. Mulligan MS, Paulson JC, De Frees S, Zheng ZL, Lowe JB, Ward PA. Protective effects of oligosaccharides in P-selectin-dependent lung injury. Nature 1993; 364:149–151.
65. Mulligan MS, Lowe JB, Larsen RD, et al. Protective effects of sialylated oligosaccharides in immune complex-induced acute lung injury. J Exp Med 1993; 178:623–631.
66. Buerke M, Weyrich AS, Zheng ZL, Gaeta FA, Forrest MJ, Lefer AM. Sialyl lewis^x-containing oligosaccharide attenuates myocardial reperfusion injury in cats. J Clin Invest 1994; 93:1140–1148.
67. Moore KL, Stults NL, Diaz S, et al. Identification of a specific glycoprotein ligand for P-selectin (CD62) on myeloid cells. J Cell Biol 1992; 118:445–456.
68. Picker LJ, Terstappen LWMM, Rott LS, Streeter PR, Stein H, Butcher EC. Differential expression of homing-associated adhesion molecules by T cell subsets in man. J Immunol 1990; 145:3247–3255.
69. Picker LJ, Michie SA, Rott LS, Butcher EC. A unique phenotype of skin-associated lymphocytes in man: preferential expression of the HECA-452 epitope by benign and malignant T-cells at cutaneous sites. Am J Pathol 1990; 136:1053–1068.
70. Berg EL, Yoshino T, Rott LS, et al. The cutaneous lymphocyte antigen is a skin lymphocyte homing receptor for the vascular lectin endothelial cell-leukocyte adhesion molecule 1. J Exp Med 1991; 174:1461–1466.
71. Imai Y, Singer MS, Fennie C, Lasky LA, Rosen SD. Identification of a carbohydrate-based endothelial ligand for a lymphocyte homing receptor. J Cell Biol 1991; 113:1213–1221.
72. Briskin MJ, McEvoy LM, Butcher EC. MAdCAM-1 has homology to immunoglobulin and mucin-like adhesion receptors and to IgA1. Nature 1993; 363: 461–464.
73. Berlin C, Berg EL, Briskin MJ, et al. $\alpha 4\beta 7$ integrin mediates lymphocyte binding to the mucosal vascular addressin MAdCAM-1. Cell 1993; 185–195.
74. Levinovitz A, Mühlhoff J, Isenmann S, Vestweber D. Identification of a glycoprotein ligand for E-selectin on mouse myeloid cells. J Cell Biol 1993; 121:449–459.
75. Lenter M, Levinovitz A, Isenmann S, Vestweber D. Monospecific and common glycoprotein ligands for E-and P-selectin on myeloid cells. J Cell Biol 1994; 125:471–481.
76. Steegmaier M, Levinovitz A, Isenmann S, et al. The E-selectin-ligand ESL-1

is a variant of a receptor for fibroblast growth factor. Nature 1995; 373:615–620.

77. Burrus LW, Zuber ME, Lueddecke BA, Olwin BB. Identification of a cysteine-rich receptor for fibroblast growth factors. Mol Cell Biol 1992; 12:5600–5609.
78. Lo SK, Lee S, Ramos RA, et al. Endothelial-leukocyte adhesion molecule 1 stimulates the adhesive activity of leukocyte integrin CR3 (CD11b/CD18, Mac-1 $\alpha_M\beta_2$) on human neutrophils. J Exp Med 1991; 173:1493–1500.
79. Sako D, Chang XJ, Barone KM, et al. Expression cloning of a functional glycoprotein ligand for P-selectin. Cell 1993; 75:1179–1186.
80. Walcheck B, Watts G, Jutila MJ. Bovine γ/δ T cells bind E-selectin via a novel glycoprotein receptor: first characterization of a lymphocyte/E-selectin interaction in an animal model. J Exp Med 1993; 178:853–863.
81. Van der Merwe PA, Barclay AN. Transient intercellular adhesion: the importance of weak protein-protein interactions. TIBS 1994; 19:354–358.
82. Varki A. Selectin ligands. Proc Natl Acad Sci USA 1994; 91:7390–7397.
83. Picker LJ, Warnock RA, Burns AR, Doerschuk CM, Berg EL, Butcher EC. The neutrophil selectin LECAM-1 presents carbohydrate ligands to the vascular selectins ELAM-1 and GMP-140. Cell 1991; 66:921–933.
84. Kishimoto TK, Warnock RA, Jutila MA, et al. Antibodies against human neutrophil LECAM-1 (LAM-1/LEU-8/DREG-56 antigen) and endothelial cell ELAM-1 inhibit a common CD18-independent adhesion pathway in vitro. Blood 1991; 78:805–811.
85. Kuijpers TW, Hoogerwerf M, Van der Laan LCW, et al. CD66 nonspecific cross-reacting antigens are involved in neutrophil adherence to cytokine-activated endothelial cells. J Cell Biol 1992; 118:457–466.
86. Kotovuori P, Tontti E, Pigott R, et al. The vascular E-selectin binds to the leukocyte integrins CD11/CD18. Glycobiol 1993; 3:131–136.
87. Sawada R, Lowe JB, Fukuda M. E-selectin-dependent adhesion efficiency of colonic carcinoma cells is increased by genetic manipulation of their cell surface lysosomal membrane glycoprotein-1 expression levels. J Biol Chem 1993; 268:12675–12681.
88. Norgard KE, Moore KL, Diaz S, et al. Characterization of a specific ligand for P-selectin on myeloid cells. J Biol Chem 1993; 268:12764–12774.
89. Moore KL, Eaton SF, Lyons DE, Lichenstein HS, Cummings RD, McEver RP. The P-selectin glycoprotein ligand from human neutrophils displays sialylated, fucosylated, O-linked poly-N-acetyllactosamine. J Biol Chem 1994; 269:23318–23327.
90. Moore KL, Patel KD, Bruehl RE, et al. P-selectin glycoprotein ligand-1 mediates rolling of human neutrophils on P-selectin. J Cell Biol 1995; 128:661–671.
91. Sammar M, Aigner S, Hubbe M, et al. Heat-stable antigen (CD24) as ligand for mouse P-selectin. Intern Immunol 1994; 6:1027–1036.
92. Needham LK, Schnaar RL. The HNK-1 reactive sulfoglucuronyl glycolipids are ligands for L-selectin and P-selectin but not E-selectin. Proc Natl Acad Sci USA 1993; 90:1359–1363.

93. Aruffo A, Kolanus W, Walz G, Freedman P, Seed B. CD62/P-selectin recognition of myeloid and tumor cell sulfatides. Cell 1991; 67:35–44.
94. Green PJ, Tamatani T, Watanabe T, et al. High affinity binding of the leukocyte adhesion molecule L-selectin to 3′-sulphated-Le^a and -Le^x oligosaccharides and the predominance of sulphate in this interaction demonstrated by binding studies with a series of lipid-linked oligosaccharides. Biochem Biophys Res Commun 1992; 188:244–251.
95. Yuen CT, Lowson AM, Chai W, et al. Novel sulfated ligands for the cell adhesion molecule E-selectin revealed by the neoglycolipid technology among O-linked oligosaccharides on an ovarian cystadenoma glycoprotein. Biochem 1992; 31:9126–9131.
96. Schwartz MA. Transmembrane signalling by integrins. Trends Cell Biol 1992; 2:304–307.
97. Mason I. Do adhesion molecules signal via FGF receptors? Curr Biol 1994; 4:1158–1161.
98. Lorant DE, Patel KD, McIntyre TM, McEver RP, Prescott SM, Zimmerman GA. Coexpression of GMP-140 and PAF by endothelium stimulated by histamine or thrombin: a juxtacrine system for adhesion and activation of neutrophils. J Cell Biol 1991; 115:223–224.
99. Lorant DE, Topham MK, Whatley RE, et al. Inflammatory roles of P-selectin. J Clin Invest 1993; 92:559–570.
100. Tanaka Y, Adams DH, Hubscher S, Hirano H, Siebenlist U, Shaw S. T-cell adhesion induced by proteoglycan-immobilized cytokin MIP-1b. Nature 1993; 361:79–82.
101. Cooper D, Butcher CM, Berndt MC, Vadas MA. P-selectin interacts with b2-integrin to enhance phagocytosis. J Immunol 1994; 153:3199–3209.
102. Gamble JR, Skinner MP, Berndt MC, Vadas MA. Prevention of activated neutrophil adhesion to endothelium by soluble adhesion protein GMP140. Science 1990; 249:414–417.
103. Wong CS, Gamble JR, Skinner MP, Lucas CM, Berndt MC, Vadas MA. Adhesion protein GMP-140 inhibits superoxide anion release by human neutrophils. Proc Natl Acad Sci USA 1991; 88:2397–2401.
104. Celi A, Pellegrini G, Lorenzet R, et al. P-selectin induces the expression of tissue factor on monocytes. Proc Natl Acad Sci USA 1994; 91:8767–8771.
105. Laudanna C, Constantin G, Baron P, et al. Sulfatides trigger increase of cytosolic free calcium and enhanced expression of tumor necrosis factor-a and interleukin-8 mRNA in human neutrophils. J Biol Chem 1994; 269:4021–4026.
106. Wadell TK, Fialkow L, Chan CK, Kishimoto TK, Downey GP. Potentiation of the oxidative burst of human neutrophils. A signaling role for L-selectin. J Biol Chem 1994; 269:18485–18491.
107. Richter J, Zetterberg E. L-selectin mediates downregulation of neutrophil TNF receptors. J Leukoc Biol 1994; 56:525–527.
108. Harris H, Miyasaka M. Reversible stimulation of lymphocyte motility by cultured high endothelial cells: mediation by L-selectin. Immunol 1995; 85: 47–54.

109. Kaplanski G, Farnarier C, Benoliel A-M, Foa C, Kaplanski S, Bongrand P. A novel role for E-and P-selectins: shape control of endothelial cell monolayers. J Cell Sci 1994; 107:2449–1457.
110. Lawrence MB, Bainton DF, Springer TA. Neutrophil tethering to and rolling on E-selectin are separable by requirement for L-selectin. Immunity 1994; 1: 137–145.
111. Natsuka S, Gersten KM, Zenita K, Kannagi R, Lowe JB. Molecular cloning of a cDNA encoding a novel human leukocyte α-1,3-fucosyltransferase capable of synthesizing the sialyl Lewis x determinant. J Biol Chem 1994; 269: 16789–16794.

10

Role of Chemoattractants in Neutrophil Recruitment to Sites of Inflammation

Peter Tan and Shervanthi Homer-Vanniasinkam
Vascular Surgical Research Unit, The General Infirmary at Leeds, Leeds, England

Michael A. Gimbrone, Jr. and Francis W. Luscinskas
Department of Pathology, Brigham and Women's Hospital and Harvard Medical School, Boston, Massachusetts

I. INTRODUCTION

The histological hallmark of acute inflammation is a localized leukocytic infiltrate that consists predominantly of neutrophils. In response to an injurious stimulus, the local vascular endothelium is activated either directly or by the generation of inflammatory cytokines, thus inducing the expression of endothelial-leukocyte adhesion molecules and the secretion of chemotactic factors. Peripheral blood neutrophils then adhere to and emigrate across the activated vascular endothelium via a "cascade" of adhesion molecules and infiltrate the extravascular tissue spaces. Based on in vitro and in vivo studies, these processes appear to involve the sequential and reciprocal engagement of multiple pairs of cell adhesion molecules in a so-called adhesion cascade. Chemotactic substances generated at sites of injury and inflammation also appear to play an integral part in this neutrophil recruitment process (1,2).

Chemoattractants are a chemically diverse class of potent bioeffector molecules (peptides and lipids) produced either by peripheral blood leukocytes, tissue macrophages and mast cells, vascular endothelium, and connective tissue cells, or generated as cleavage products of plasma proteins, and therefore may be released into either intravascular or extravascular compartments. A correlation between the degree of neutrophil infiltration at an extravascular inflammatory site and the local concentration of one or more chemoattractants has been shown in a number of clinical studies and animal models (3–6). Paradoxically, however, chemotactic factors present in the intravascular compartment have also been shown to down-regulate neutrophil recruitment to inflammatory sites (7–9).

The ability of neutrophils to extravasate and localize at sites of infection is crucial for host defense and survival as demonstrated by patients with the rare autosomal recessively inherited disease, leukocyte adhesion deficiency type I (LAD I) (10). Patients with this syndrome have molecular defects in the leukocyte β_2-integrins (CD11/CD18) that render their neutrophils unable to phagocytose microorganisms, or to extravasate normally, if at all. Conversely, excessive neutrophil recruitment may result in inflammatory diseases with ensuing neutrophil-mediated tissue damage such as in active rheumatoid arthritis and idiopathic lung fibrosis. Thus, it appears that in adaptive pathophysiological processes, neutrophil recruitment is a carefully regulated process.

The literature on neutrophil recruitment during inflammation has been extensively reviewed in recent years (11–16). In this chapter, we will first provide a brief summary of neutrophil and endothelial adhesion molecules and their functions in neutrophil recruitment to sites of inflammation. We will then examine the role of chemoattractants per se in the inflammatory response. In particular, we will consider data indicating that chemoattractants can act to either promote or attenuate neutrophil recruitment at inflammatory sites, depending on their relative concentrations in extravascular or intravascular compartments, and possibly inherent differences in their receptor-mediated effects on the neutrophil.

II. NEUTROPHIL ADHESION MOLECULES

Neutrophils normally comprise around 55% of the circulating mass of human peripheral blood leukocytes and can undergo dramatic increases in number in the face of infection. They constitute major effector cells in natural immunity by accumulating at sites of infection, injury, and inflammation, and eliminating foreign matter and dead, damaged, or abnormal cells and debris. To enable them to play their role in host defense, neutrophils are equipped with a number of cell surface, plasma membrane, and cytoplasmic structures (see Fig. 1):

1. Cell adhesion molecules such as β_2-integrins, L-selectin, and selectin ligands
2. Specific chemotactic receptors and receptors for other inflammatory mediators

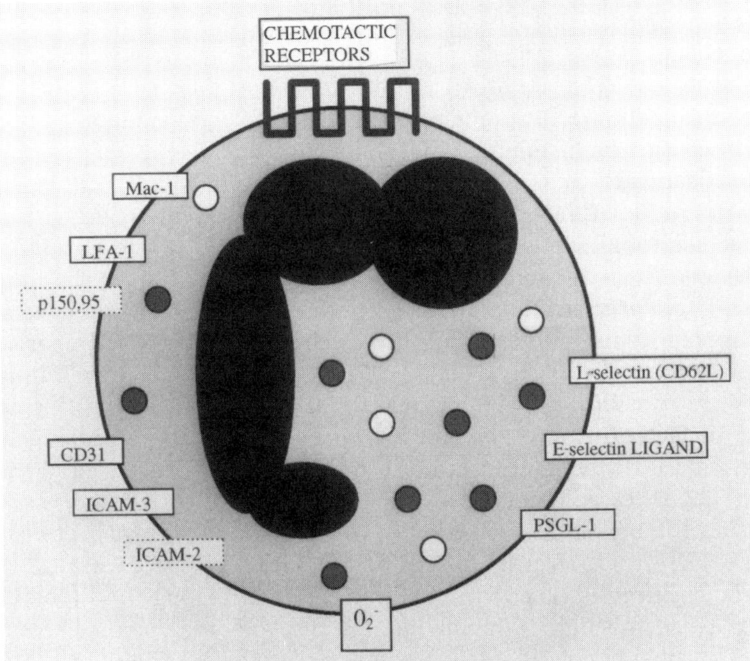

Figure 1 Defense arsenal utilized by blood neutrophils. This figure depicts the major neutrophil cell surface molecules that function during recruitment to sites of inflammation.

3. Characteristic cytoplasmic granules containing a variety of proteases, as well as enzymes located within the plasma membrane and cytosol that are involved in the generation of the respiratory burst necessary for intracellular killing of bacteria

A detailed discussion of all the above is beyond the scope of this chapter, and the reader is referred to recent reviews on the subject (11–21). In this section, we will briefly summarize the cell adhesion receptor-ligand mechanisms employed by peripheral blood neutrophils to extravasate, interact with extracellular matrix proteins, and migrate efficiently toward sites of infection or inflammation.

A. Neutrophil β_2-Integrins

The integrins are a family of transmembrane heterodimeric glycoproteins that are widely expressed on many cell types and that function in a broad sense to integrate and modulate a given cell's interactions with its extracellular environment (12,13,19–21, and references therein). Integrins, being components of both outside-in and inside-out signal transduction, are involved in a wide spectrum of cellular functions. Integrins important for neutrophil adhesive functions comprise the β_2-integrin subfamily and consist of a common β_2 subunit (CD18) noncovalently associated with one of three α subunits, α_L (CD11a, LFA-1), α_M (CD11b, Mac-1, CR3), and α_X (CD11c, p150, CR4) (13). Each $\alpha\beta$ heterodimer is expressed at high density on the neutrophil surface, and additional CD11b/CD18 and CD11c/CD18 molecules are sequestered in intracellular granules. β_2-integrin adhesive function requires divalent cations (i.e., Ca^{2+}, Mg^{2+}). As a consequence of neutrophil activation by stimuli such as chemoattractants and phorbol esters, the intracellular reservoir of CD11b/CD18 and CD11a/CD18 molecules are released onto the leukocyte plasma membrane, thus increasing their surface expression by severalfold. However, this increased surface expression is neither sufficient nor even necessary for enhanced ligand affinity (22,23) or adhesion to endothelium (24) since recent studies suggest that integrin affinity is a function of conformational changes. The regulation of CD11a/CD18 affinity for ligand(s) is less well characterized.

The β_2-integrins function in neutrophils by mediating firm adhesion to vascular endothelium through interactions with their counterreceptors, ICAM-1 and possibly ICAM-2, and by mediating adhesion to extracellular matrix proteins such as laminin and fibronectin. In addition, certain of the neutrophil surface β_2-integrins serve as binding sites for the activated complement fragment 3bi (C3bi), C4, LPS, and fibrinogen (12). Neutrophil adherence via β_2-integrins leads to enhanced cellular functions (e.g., respiratory burst, F-actin increase) (25). A critical role for β_2-integrins in host defense is demonstrated by the inherited disease leukocyte adhesion deficiency syndrome type I (LAD I), in which patient's phagocytic cells have defective or absent β_2-integrins resulting in severe adhesion and motility defects (10). A lack of pus formation at sites of infection indicates failure of neutrophil extravasation in these patients, who typically suffer from recurrent life-threatening bacterial infections (10,13 for review of LAD I).

B. Selectin Family of Adhesion Molecules

Peripheral blood neutrophils constitutively express high levels of L-selectin (CD62L; formerly known as LAM-1, TQ-1, Leu 8, murine MEL-14), which is a member of the selectin gene family (26,27). Two other selectins also have been characterized—E-selectin (CD62E; formerly known as ELAM-1) (28), and P-selectin (CD62P; formerly known as GMP-140, PADGEM) (29)—neither of which is expressed on leukocytes (reviewed in 30,31). E-selectin was originally described as an inducible endothelial-specific adhesion molecule for leukocytes (32). The E-selectin gene is silent in resting endothelial cells, but transcription and expression of E-selectin protein may be rapidly and transiently induced by inflammatory stimuli such as endotoxin (LPS), or inflammatory cytokines, (IL-1 and TNF-α) (28,33). mRNA for E-selectin is detectable in cultured human umbilical vein endothelial cells within 1 hour after treatment with one of these inflammatory stimuli, with peak surface protein expression around 4 hours followed by its disappearance within 24 hours (28). E-selectin is uniquely expressed on activated endothelial cells; i.e., it is endothelial cell-specific. P-selectin is sequestered in Weibel-Palade bodies in endothelium and in α-granules in platelets (34). P-selectin molecules are rapidly mobilized to the plasma membrane surface by secretagogues such as histamine, thrombin, and leukotrienes C and D, and its gene can also be transcriptionally activated by TNF-α or LPS. For a more detailed discussion of E- and P-selectins and their leukocyte ligands, the reader is referred to Chapter 9 in this book.

L-selectin surface expression may be modulated by soluble agonists (for a detailed discussion, see 35,36). For example, neutrophil activation induces shedding (i.e., loss, not internalization) of surface expressed L-selectin concomitant with increased surface expression of β_2-integrins (36,37). This design for regulation of L-selectin and β_2-integrins may have important implications during neutrophil adhesion to endothelium at sites of inflammation (see below, section III). L-selectin appears to recognize specific carbohydrate determinants (ligands) carried on molecules such as GLYcam I (38), MAdCAM-1 (39), and CD34 (40) found in the specialized high endothelial venules (HEV) characteristic of lymphoid tissues, as well as less well defined ligand(s) expressed on activated peripheral vascular endothelium (41–43).

Neutrophils also express complex cell surface ligands for E- and P-selectin (30,44,45). Structurally, these selectin ligands contain sialylated polylactosamines such as sialylated Lewis^x and sialylated Lewis^a determinants (30,31). Currently, the efforts of many investigators are focused on determining the exact structures that "present" these carbohydrate determinants to E-, P-, and L-selectin (46).

The importance of selectins in leukocyte recruitment is revealed by a second, rare heritable disease, termed LAD type II, in which a genetic defect in fucose metabolism results in failure to synthesize selectin ligands such as sialyl Lewis^x and related carbohydrate determinants (47,48). LAD type II patients also suffer from severe, recurrent bacterial infections, and exhibit reduced neutrophil mobility and lack of pus formation at sites of inflammation. Deficient selectin ligand expression presumably interferes with efficient neutrophil attachment and subsequent adhesive interactions with activated vascular endothelium in vivo (see below), and is manifested in vitro as a failure in E-selectin-dependent neutrophil adhesion to IL-1β-activated human umbilical vein endothelium (HUVEC) (47). These data further indicate the importance for selectins during leukocyte initial adhesive interactions with vascular endothelium under fluid shear conditions (see Fig. 2, step I).

C. Immunoglobulin Family of Adhesion Molecules

The immunoglobulin (IgG) gene family of adhesion molecules, all of which contain varying numbers of characteristic IgG domains, includes ICAM-1 (CD54), ICAM-2 (CD102), ICAM-3 (CD50), VCAM-1 (CD106), and PECAM-1 (CD31). ICAM-1 is expressed by a wide variety of cell types including vascular endothelium, fibroblasts, epithelial cells, and some leukocytes (but not neutrophils) (49). ICAM-1 is constitutively expressed on vascular endothelium, and its expression can be markedly increased by inflammatory cytokines such as IL-1, TNF-α, interferon-γ, and bacterial endotoxin. ICAM-2 is constitutively expressed on endothelium, but its expression levels are not regulated by inflammatory stimuli (50,51). Unlike ICAM-1 and ICAM-2, ICAM-3 is highly expressed on all leukocytes, but not endothelium (52). To date the roles of ICAM-2 and ICAM-3 in neutrophil function have not been extensively explored. On the leukocyte surface, the best-characterized counterreceptor for ICAM-1 and ICAM-2 is CD11a (LFA-1) (53,54). CD11b/CD18 (Mac-1) also appears to interact with ICAM-1, albeit with significantly lower avidity (55), but not with ICAM-2 (56).

VCAM-1 is expressed at low or barely detectable levels by unactivated vascular endothelium, but can be focally upregulated by inflammatory and atherogenic stimuli (57–61). It interacts with VLA-4 ($\alpha_4\beta_1$, CD49d/CD29) (62), which is present on lymphocytes, monocytes, eosinophils, and basophils, as well as $\alpha_4\beta_7$, which is present on basophils, B lymphocytes and certain T lymphocyte subsets (63). The reader is referred to Chapter 11 for a detailed discussion of the role of VCAM-1 in lymphocyte recruitment. Unactivated neutrophils do not express $\alpha_4\beta_7$ or $\alpha_4\beta_1$ and thus do not directly interact with VCAM-1 during their adhesion to activated endothelium (64).

CD31, or platelet-endothelial cell adhesion molecule-1 (PECAM-1), was originally described as an "endothelial cell-restricted, externally disposed surface molecule" (65). Once molecularly cloned, it was found to be a member of the IgG gene family. PECAM-1 is constitutively expressed on neutrophils, monocytes, certain lymphocytes, and platelets. PECAM-1 has recently been demonstrated to be important in both neutrophil and mono-

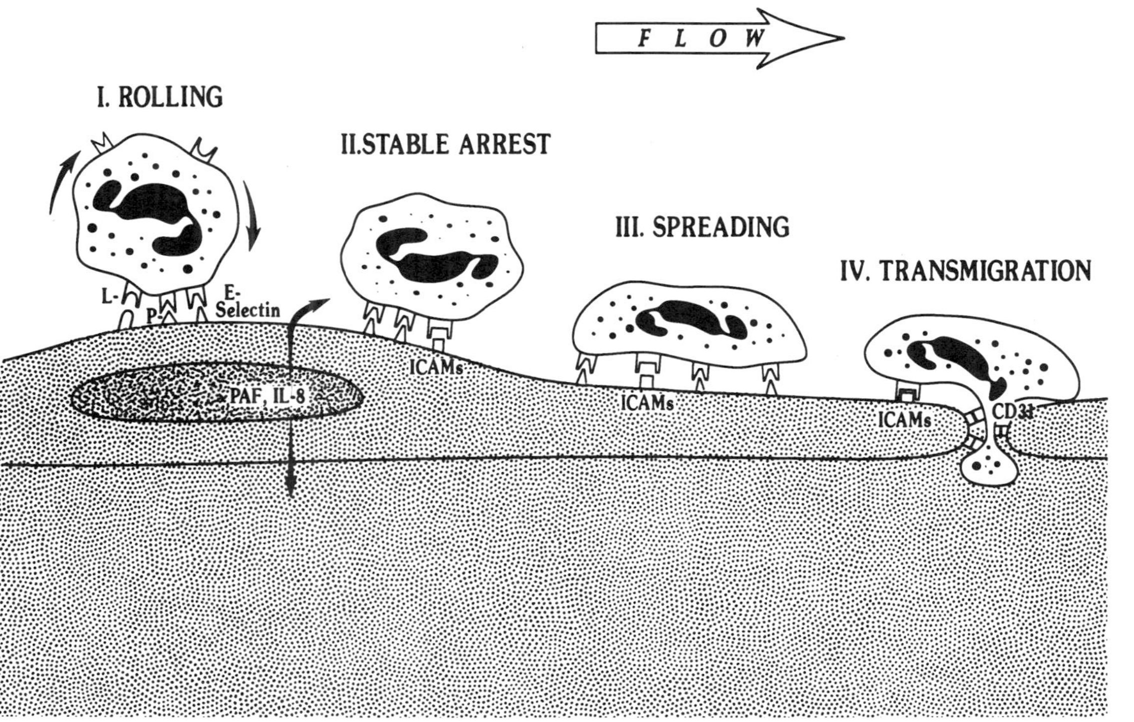

Figure 2 Multistep adhesion molecule cascade for neutrophil attachment and transmigration across endothelium under flow. This figure depicts the four sequential and overlapping cellular events (steps I–IV) observed during neutrophil adhesive interactions with activated vascular endothelial monolayers under defined laminar flow in vitro. L-, P-, and E-selectin can mediate neutrophil initial attachment and rolling (step I) on the endothelial monolayer and also facilitate β_2-integrin-dependent arrest (step II) and spreading (step III). β_2-integrins are involved in neutrophil spreading and movement to intercellular borders and, in conjuction with CD31, mediate neutrophil transmigration (step IV) under flow.

cyte transmigration across normal and cytokine-activated vascular endothelium. Muller and colleagues have proposed that PECAM-1 mediates adhesive interactions that are distal to firm adhesion (e.g., β_2-integrin-dependent) in the adhesion cascade and crucial for transendothelial migration of neutrophils (see Fig. 2, step IV) (66). In this regard, both in vivo (67,68) and in vitro (66) studies have implicated PECAM-1 in leukocyte transmigration and recruitment. However, PECAM-1 expression does not appear to be necessary for transmigration, since not all leukocytes express PECAM-1 (e.g., memory T cells; 69).

III. CHEMOATTRACTANTS

A. The Process of Chemotaxis

"Chemotaxis" may be defined as the unidirectional locomotion of a cell along a concentration gradient of a chemoattractant substance, which is referred to as a "chemotaxin" (70). For over a century, it has been recognized that cells such as leukocytes exhibit directional movement toward chemical substances or microorganisms located at extravascular sites, e.g., the anterior ocular chamber (71) or the peritoneal cavity (72), where they accumulate as part of the acute inflammatory response. It is now clear that leukocytes possess specific cell surface receptors by which they "sense" even small differences in concentration of a chemotaxin across their diameter and move "up" this concentration gradient (73). However, leukocytes placed in a solution of a chemotaxin where its concentration is uniform will undergo enhanced random (directionless) movements called "chemokinesis" (74).

1. In Vitro Chemotaxis Assays

A "chemotaxin," to be defined as such, must be shown to induce directional movement of leukocytes. Traditionally, chemotaxis has been assessed in vitro using bioassays such as the Boyden chamber (75) or the agarose gel migration assay (76). In the Boyden chamber, leukocytes are separated from a putative chemotaxin by a nitrocellulose filter with pores of such size that leukocytes cannot "fall through" passively, but have to "squeeze through" in order to traverse the filter. In the agarose gel assay, leukocytes migrate from a reservoir under a layer of agarose in response to a chemotaxin which is placed in another reservoir at a fixed distance. In both assays, a diffusion barrier located between the leukocytes and a reservoir of chemotaxin establishes a concentration gradient of chemoattractant.

Much of our present understanding of the action of chemotactic substances is derived from such in vitro systems. Clearly, their roles in in vivo pathophysiological settings may be more complex. Indeed, most of the mediators empirically classified as chemotaxins, and listed below, have wide-ranging biological activities on leukocytes apart from chemotaxis such as priming, (77), shape change, degranulation, induction of the respiratory burst, F-actin polymerization, and protein phosphorylation, and additionally may have effects on other nonleukocyte cell types. For example, interleukin-8 (IL-8) has been reported to induce angiogenesis (78), and PAF can cause vasodilation and bronchoconstriction (reviewed in 79).

2. Haptotaxis

Haptotaxis may be defined as the movement of cells on an immobilized gradient of adhesive or chemotactic molecules in the direction of increasing concentration. Several tumor cell lines have been shown to exhibit directional movement on immobilized gradients of extracellular matrix proteins or adhesive peptides (80–82). Haptotaxis has been proposed as an alternative mechanism involved in leukocyte recruitment in inflammation (83). In support of this notion, IL-8 has been reported to bind to the surface of endothelial cells (84) and dermal cells (85) in vitro. However, a clear demonstration of the existence of haptotactic gradients and leukocyte haptotaxis in vivo is lacking at present.

B. Neutrophil Chemoattractants

1. "Classical" Chemoattractants

A variety of bioactive peptides and lipids of both endogenous and exogenous origin are known to be potent neutrophil chemotaxins. C5a anaphylatoxin, a product of complement activation, as well as $C5a_{des\ Arg}$, its cleavage product, were shown to exhibit in vitro chemoattractant activity for neutrophils (86), with the former being about 10-fold more potent. fMLP, a formylated bacterial oligopeptide has also been shown to induce neutrophil and macrophage chemoattraction (87). This fundamental difference in protein metabolism between prokaryotic organisms such as bacteria that commence protein synthesis by N-formylmethionine and eukaryotes such as humans may be an important means by which human neutrophils recognize the presence of bacterial infection. Activated neutrophils produce leukotriene B_4 (LTB_4), an oxidized phospholipid, as well as platelet-activating factor (PAF), an ether phospholipid, both of which are chemotactic for neutrophils in vitro (88,89). PAF-like molecules are also secreted by activated endothelium in a membrane-bound form (90–92). It should be noted that these "classical" chemoattractants do not exhibit specificity of action for any subset of leukocytes.

2. Interleukin-8

The chemotactic polypeptide now known as interleukin-8 (IL-8; a potent and specific neutrophil chemoattractant with leukocyte adhesion inhibitory properties) was first isolated from psoriatic scales (93) and activated monocytes (94) as a soluble factor which was chemotactic for neutrophils but distinct from IL-1. It was initially called neutrophil-activating peptide or protein (NAP) or monocyte-derived neutrophil chemotactic factor (MD-NCF). It was molecularly cloned in 1988 (95,96). It is now clear that IL-8 can be secreted by a wide variety of cells in response to a wide range of stimuli such as inflammatory cytokines (96), bacterial endotoxin (94), viruses, and uric acid crystals. Most cells with receptors for IL-1β or TNFα appear to secrete IL-8 in response to these cytokines.

IL-8 is a member of a superfamily of chemotactic cytokines known as the intercrine or chemokine family (97). These low-molecular-weight cytokines have been further subclassified into two groups on the basis of their peptide sequence and the major leukocyte subset on which they exert their chemotactic activity. The C-X-C or α chemokines, of which IL-8 is a member, have an amino acid intervening between the first two of the four conserved cysteines at the N terminal. The C-C or β chemokines lack an intervening amino acid at this position. While the C-X-C chemokines predominantly chemoattract neutrophils, the C-C chemokines exert their effects primarily on monocytes and lymphocytes.

In 1988, cultured human endothelial cells stimulated with cytokines or endotoxin were reported to secrete a soluble factor which significantly inhibited adhesion of neutrophils to activated but not unactivated endothelium (98). This factor was called "leukocyte adhesion inhibitor" (LAI) and subsequently identified as an N-terminally extended 77 amino acid form of IL-8 secreted by activated vascular endothelium (99). Both the 72-amino acid (predominant form secreted by leukocytes) and the 77-amino acid (predominant endothelial product) forms of IL-8 have been shown to reduce neutrophil-endothelial adhesion in an in vitro assay system (98,100–102), and the 77-amino acid form is thought to be cleaved by thrombin in vivo into the 72-amino acid form, which functions as a more potent LAI (100).

3. In Vitro Studies of IL-8

A number of studies have demonstrated a role for IL-8 in neutrophil transmigration across cytokine-activated endothelium (103–105). In these studies, addition of either IL-8 or anti-IL-8 antiserum to the upper compartment of an in vitro chemotaxis assay reduced neutrophil transmigration. In one of the above in vitro studies, IL-8 secreted by the cytokine-activated endothelium appeared to be immunolocalized to both the endothelial monolayer and the underlying interstitium (103). Based on these studies, it would appear that IL-8 could play a stimulatory role in neutrophil transmigration at sites of inflammation.

4. Clinical Studies of IL-8

IL-8 has been implicated in a wide variety of diseases in which neutrophils are believed to mediate much of the tissue damage. In many of these diseases, the levels of IL-8 found in various extravascular tissue spaces correlate with the degree of neutrophil infiltration and disease severity. For example, in patients with the adult respiratory distress syndrome, high levels of IL-8 have been found in their bronchoalveolar lavage fluid and its concentration correlated with clinical outcome (106). Similarly, high levels of IL-8 as well as leukocyte infiltrates rich in neutrophils have been found in pleural effusions of patients with empyema (3). The clinicopathology of IL-8 has already been well reviewed (100,107) and will not be discussed at length here. However, it should be noted that in these diseases, the high levels of IL-8 were primarily sequestered in extravascular tissue compartments and have been assumed to exert a specific and potent chemoattraction for neutrophils.

IV. INTEGRATION OF ADHESION MOLECULES AND CHEMOATTRACTANTS IN NEUTROPHIL EMIGRATION

Neutrophil emigration from the intravascular into the extravascular compartment is a complex process involving the following steps:

1. Close apposition of the neutrophil to the vascular endothelial surface and stable arrest under physiological conditions of blood flow, an interaction that is mediated via adhesion molecules
2. Engagement of sensory (e.g., chemotactic) receptors, which directs leukocyte migration out of the blood vessel and then, within the extravascular tissues, toward a site of infection or inflammation
3. Neutrophil binding to extracellular matrix proteins and interstitial tissues such as fibroblasts, interactions that are also mediated by adhesion molecules
4. Cytoskeletal elements that integrate chemotactic and other sensory stimuli with neutrophil motor functions, thus enabling the cell to deform and change shape, move purposefully through tissues, and then phagocytose invading microorganisms

Below are summarized important experimental and conceptual advances in this area that are relevant to understanding the roles of adhesion molecules and chemoattractants in leukocyte recruitment. For further information on mononuclear leukocyte-endothelial interactions, please see Chapters 1 and 11 in this book.

Early in vitro studies of leukocyte adhesion to cultured human umbilical vein endothelial cells were typically performed under static conditions and indicated that basal adhesion of blood neutrophils was low compared to monocytes (108). Activation of the endothelium with cytokines TNFα and IL-1β or endotoxin (LPS) resulted in a striking (10- to 50-fold) increase in neutrophil adhesion (109). This enhanced attachment reflects the induction of multiple adhesion molecules on the endothelial cell surface (for review, see 12,35). In addition, activated endothelial cells also synthesize leukocyte chemoattractants, including IL-8, PAF, and MCP-1 (see below and 90,92,97,99,110; for review see 12). That multiple adhesion molecules and leukocyte chemoattractants are induced suggests redundancy or overlap in their function. However, recent experiments performed in vivo, or in vitro under defined laminar flow conditions, have suggested that multiple receptor-ligand pairs actually function in a sequential and orchestrated fashion to mediate leukocyte attachment (35,41,111–114). The following sections will refer to the video frames shown in Figure 3 that are from an in vitro experiment performed in our laboratory that examined neutrophil adhesive interactions with TNF-α-activated endothelium in a flow chamber (41,112, 115,116). The cascade of molecular and cellular processes during neutrophil-endothelial interactions under flow is detailed in Figure 2.

A. Multistep Cascade of Leukocyte Adhesion

1. Selectin-Mediated Initial Attachment

In the molecular models that have been proposed, principally by Butcher and co-workers (35,111,117), the initial attachment of peripheral blood neutrophils to vascular endothelium, particularly under physiological conditions of blood flow, is mediated by members of the selectin gene family as depicted in Figure 2. L-selectin, P-selectin, and E-selectin interact with their carbohydrate ligands to mediate initial attachments such as "rolling," as well as "tethering," adhesive interactions between leukocytes and the vascular endothelium. Recent studies have suggested that the ability of selectin molecules to mediate the experimentally observed rolling and tethering adhesive interactions, which require fast "on-off" kinetics, is related to their rates of bond formation and breakage (reviewed in 118,119). Under conditions of fluid flow in which the wall shear stress exceeds 0.5 dynes/cm^2, neutrophils appear to require E- and/or L-selectin-mediated adhesion to decelerate and roll on activated vascular endothelium (120,121) (Fig. 2, step I). In vitro, neutrophils exhibit transient adhesion to endothelium (<5 sec) followed by detachment and release, or initiation of slow rolling (~ 10

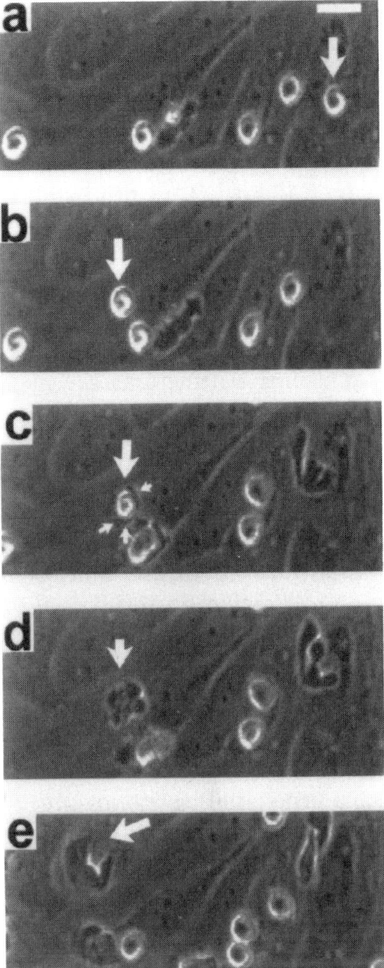

Figure 3 Neutrophil rolling, arrest, spreading, and transmigration across TNF-α-activated vascular endothelium under defined laminar flow at 1.8 dynes/cm^2. Sequential video frames from a flow experiment illustrating neutrophil rolling, arrest, spreading, and transmigration were digitized, and specific regions of interest from each video frame were combined using commercial software (Adobe photoshop, ver 2.5) to create the composite image (a–e). This composite photomicrograph depicts the various stages of neutrophil adhesion to 6-hour TNF-activated human umbilical vein endothelium under flow (direction of flow is from right to left). (a) Arrow indicates neutrophil that has just attached to the apical endothelial surface (bar = 20 μm). (b) The same neutrophil has rolled downstream and stopped. (c) Neutrophil has stably arrested, spread, and projected pseudopods (small arrows) into the border between two endothelial cells. Pseudopods extended down (perpendicular to plane of diagram) into the subendothelial space are not visible. (d) The neutrophil has completed transendothelial migration across the activated endothelial monolayer, exhibiting a flattened phase-dark appearance. (e) Once transmigrated, the neutrophil is able to move in the subendothelial space.

μm/sec) on endothelial cells downstream (120). These initial rolling or tethering interactions are reversible unless leukocytes are activated to undergo firm adhesion (arrest). In Figure 3a, a neutrophil (identified by arrow) initially attaches to 6-hr TNF-α-activated endothelium and then rolls downstream at 10 μm/sec before stably arresting on the apical surface of the endothelial cell under flow at 1.8 dynes/cm^2 (panel b; flow is from right to left).

2. β_2-Integrin Activation-Dependent Stable Arrest

It is at this juncture that chemoattractants are hypothesized to critically effect neutrophil-endothelial adhesion, namely, that β_2-integrin activation is triggered by locally derived endothelial chemoattractants, IL-8 and PAF (well characterized as mediators of leukocyte activation) or GM-CSF, or possibly via ligation of E-selectin (122) or CD31 (123) (Fig. 2, step II). Activation dramatically increases the adhesiveness of leukocytes for endothelium primarily by upregulated surface expression and affinity of β_2-integrins for its ligand. Neutrophil β_2-integrin engagement of its endothelial ligand, ICAM-1 (and possibly ICAM-2), results in stable arrest. This may occur in vitro at a wall shear stress of 1.8 dynes/cm^2 and is shown in Figure 3b (stably arrested neutrophil is identified by large arrow).

3. Spreading of the Neutrophil on the Endothelial Surface

Once it has arrested, the neutrophil begins to spread, and pseudopods are projected into the intercellular borders between endothelial cells (Fig. 3, panel c; pseudopods identified by small arrows), again via β_2-integrins engaging their ligand(s), presumably ICAM-1 or ICAM-2. This step most likely requires activation of integrins, as well as coordination of sensory receptors, adhesion molecules, and cytoskeletal elements. Since activated endothelium produces IL-8 and PAF, there is much circumstantial evidence that β_2-integrin activation is triggered via these endothelial-derived chemoattractants (Fig. 3, step III). The adhesion bonds formed during stable arrest and spreading presumably have to be closely regulated such that the adherent neutrophil can resist fluid shear forces but can also adhere and de-adhere to allow migration across the luminal endothelial surface prior to transendothelial migration. The exact role of individual alpha subunits remains unresolved, although all of them appear to be involved (124,125).

4. β_2-Integrin and CD31-Mediated Transmigration

As shown in Figure 3d, once stably arrested and flattened, neutrophils migrate to an adjacent intercellular junction and diapedese between endothelial cells to the abluminal surface (large arrow). This latter process involves homotypic adhesion of PECAM-1 (CD31) expressed on both leukocytes and endothelium as well as β_2-integrins (Fig. 2, step IV). The importance of β_2-integrins in migration is clearly demonstrated by the heritable disease termed LAD I (see above section, Integrins). The molecular structure of endothelial cell junctions also has been studied recently, and intracellular proteins such as plakoglobin, α-, and β-catenins appear to associate with transmembrane proteins such as VE cadherin (126). The role of these molecules in neutrophil transendothelial migration remains to be further defined.

5. Chemoattractant-Directed Infiltration of Tissues

It is likely that chemoattractants like IL-8, which may be secreted and sequestered in extravascular tissues, may be important in directing neutrophil movement following transmigration into the interstitial space. At sites of infection, chemoattractant-mediated activation of neutrophils promotes phagocytosis and other functions necessary for efficient killing of microorganisms. Neutrophil binding of various extracellular matrix proteins and extravascular tissues by both β_2-integrin-dependent and -independent (127) adhesive interactions may further activate and prime the neutrophil and increase its bactericidal capabilities (25).

B. Can Chemoattractants Down-Regulate Neutrophil Recruitment?

Under certain pathological circumstances such as overwhelming sepsis, high concentrations of circulating IL-8 may be found in association with defective neutrophil function. Additionally, in vivo and in vitro experimental studies have documented the ability of IL-8 and certain other chemoattractants to down-regulate neutrophil adhesion. This has been referred to as the "leukocyte adhesion inhibitory," or LAI effect, of IL-8.

V. LAI EFFECT

A. Experimental and Clinical Data

1. In Vitro Studies

In static adhesion assays in vitro, IL-8, C5a, and fMLP inhibit neutrophil adhesion to activated HUVEC monolayers in a dose-dependent manner, while the chemotactic lipids PAF and LTB_4 do not have this effect (102). This inhibitory effect appears to be a direct effect on the neutrophil, as pretreatment of the endothelium with the above chemoattractants had no effect. However, this leukocyte adhesion inhibitory effect is probably not due to L-selectin shedding since all of the chemoattractants tested (IL-8,

C5a, fMLP, PAF, and LTB_4) induced shedding of L-selectin and upregulation of CD11b/CD18 in a similar fashion, whereas only IL-8, C5a, and fMLP demonstrated the LAI effect. Pretreatment of neutrophils with cytochalasin B abolished the LAI effect of IL-8, C5a, and fMLP, suggesting an actin microfilament-dependent mechanism (102).

2. In Vivo Studies

Consistent with these in vitro studies, IL-8 introduced into the intravascular compartment appears to down-regulate neutrophil recruitment at sites of inflammation in vivo. In a rabbit model of acute inflammation induced by the intradermal injections of equipotent doses of IL-1, fMLP, C5a, LTB_4, or IL-8, neutrophil accumulation at each of these sites was comparably reduced by an intravascular injection of IL-8 (9). This result was not related to the transient neutropenia observed following injection of IL-8. Similarly, in an intravital microscopy study of leukocyte rolling and chemoattractant induced emigration in mesenteric venules in rabbits, intravascular injection of IL-8 again was found to reduce neutrophil emigration. Of interest, this was not due to loss of neutrophil L-selectin surface expression or changes in rolling leukocyte flux, both of which remained within normal limits (8). Recently, transgenic mice genetically engineered to secrete and maintain high plasma levels of human IL-8 were found to have severely impaired neutrophil migration into inflamed peritoneal cavities induced by injection of thioglycollate or IL-8, despite having markedly increased circulating leukocyte counts (7).

3. Clinical Studies

High circulating levels of IL-8 have been reported in patients with overwhelming sepsis (128,129). Additionally, in experimental human endotoxemia, neutrophils from volunteers injected with bacterial endotoxin are defective in chemotaxis as assayed in a Boyden chamber (130). While none of these studies conclusively demonstrate a causal relationship between high circulating levels of IL-8 and deficient neutrophil function, it is conceivable that the high intravascular concentration of IL-8 which can down-regulate neutrophil transmigration, may have had an adverse effect on the ability of these patients with severe sepsis to overcome their infection.

B. Possible Mechanisms of the LAI Effect of IL-8

1. Collapse of an Endogenous IL-8 Gradient

The ability of IL-8 in the intravascular compartment to reduce neutrophil adhesion to the luminal endothelial surface and subsequent transmigration has been attributed to a “collapse” of a preexistent endogenous gradient of

IL-8, presumably generated by activated endothelium and other cells in the inflamed area. The available data do not fully support this hypothesis because other chemoattractants apart from IL-8 (namely, C5a and fMLP) also appear able to mediate the LAI effect in vitro studies when presented in the apical compartment (101,102). Similarly, regardless of whether IL-8, C5a, or fMLP was introduced at the intradermal injection site, intravenous IL-8 reduced neutrophil recruitment in the in vivo rabbit dermal inflammation model (9). Furthermore, as PAF and LTB_4, both potent chemoattractants, do not exhibit the LAI effect when present in the "apical" compartment (101), the LAI effect cannot be explained as merely a reversal of chemotaxin gradients.

2. Effect of Tissue Compartments and Chemoattractant Concentration

Based on our knowledge of chemotaxis in vitro, it appears that for a chemotaxin to function as such it has to be initially separated from the leukocytes by a diffusion barrier which acts to establish the chemical gradient to which the neutrophils can then respond. In the Boyden chamber, the nitrocellulose membrane acts as this barrier, while in vivo the vascular endothelium and vessel wall may serve this purpose by initially separating the circulating peripheral blood neutrophils from an adjacent extravascular tissue compartment where the chemotaxin may be present at higher concentration. The absence of this diffusion barrier would theoretically result in neutrophils being exposed to uniform concentrations of chemotaxin which may induce enhanced but directionless motion akin to chemokinesis.

IL-8 is secreted by a wide variety of cell types upon their activation by an inflammatory or injurious stimulus. Therefore, depending on the pathology, high levels of IL-8 may accumulate either in intravascular or extravascular tissue spaces. Under certain pathological circumstances such as septicemia, abnormally high concentrations of IL-8 may occur in the circulation (128,129) despite the fact that circulating red blood cells provide a large sink of IL-8 receptors (131). The available data from both in vitro and in vivo animal studies (discussed above) suggest that the presence of high levels of IL-8 in the intravascular compartment in these conditions may reduce neutrophil transendothelial migration. In contrast, in other situations, e.g., infections in soft tissues, the presence of a high concentration of IL-8 in an extravascular tissue compartment may be sensed by marginated peripheral blood neutrophils. Having transmigrated across the endothelium, these neutrophils would then follow the chemotactic gradient toward the tissue space where the concentration of IL-8 was highest. Thus, IL-8 and other chemoattractants may play an important role following the transmigration event, in directing neutrophil movement.

3. Effect of Differences in Signal Transduction Between LAI and Non-LAI Chemoattractants

While the existence of chemotaxin gradients and differential concentrations in various tissue compartments may partly explain how IL-8 and other peptide chemotaxins mediate the LAI effect, they do not account for why PAF and LTB_4 lack this inhibitory effect. IL-8, C5a, fMLP, and PAF engage specific and distinct receptors on the neutrophil surface that belong to a family of seven transmembrane G-protein coupled chemotactic receptors (12, and references therein). The LTB_4 receptor has not been cloned, although available data suggest that it is a G-protein coupled receptor (C. N. Serhan, personal communication). However, there do appear to be differences in the postreceptor signaling events between LAI agonists and non-LAI agonists. One such identified difference is the pattern of actin polymerization-depolymerization that these two groups of chemoattractants induce in neutrophils. IL-8, C5a, and fMLP induce rapid and extensive actin polymerization, followed by gradual depolymerization at a rate inversely proportional to the chemoattractant concentration, whereas PAF and LTB_4 cause rapid actin polymerization followed by an almost equally rapid depolymerization (102). Elucidation of any other differences in signal transduction between LAI and non-LAI chemoattractants awaits further studies.

4. Desensitization of Neutrophils by Receptor Down-Regulation

Exposure of neutrophils to IL-8 results in rapid down-regulation of its surface receptors by internalization (132), and reexpression by recycling within 10 minutes of removal of the agonist. This process is known as desensitization and has been proposed as being a possible mechanism by which IL-8 exerts its LAI effect. However, the non-LAI agonist PAF exhibits a similar phenomenon (79), thus suggesting that desensitization is not the primary mechanism by which certain chemoattractants exert their LAI effect.

VI. ANTI-INFLAMMATORY THERAPEUTIC STRATEGIES

While neutrophil infiltration at sites of injury and inflammation has been observed for at least 100 years (72), it is only recently that tremendous strides have been made in understanding the cellular and molecular mechanisms of neutrophil recruitment. This knowledge has afforded the possibility of pharmacologically modulating the inflammatory response in a controlled fashion. Although normal neutrophil function is crucial to maintaining host defenses, there exists a long list of inflammatory diseases, affecting various organs, in which unchecked neutrophil activity is responsible for extensive tissue damage. The ability to abrogate neutrophil-mediated tissue damage without immunocompromise would offer hope for reducing both the morbidity of chronic inflammatory disorders such as rheumatoid arthritis and inflammatory bowel disease, and the high mortality associated with adult respiratory distress syndrome or myocardial ischemia-reperfusion injury. For a detailed discussion of adhesion molecules in various disease/injury situations, the reader is referred to Chapters 14 through 25 of this book.

If the major cause of tissue damage in a disease is indeed neutrophil-mediated, then by reducing neutrophil recruitment to the affected organ, a significant reduction in morbidity and mortality, as well as measurable benefit in histological and functional terms should be the result. Tabulated in Table 1 are a number of potential molecular targeting sites to achieve the desired reduction or ablation in neutrophil recruitment. In the following sections we will briefly discuss some of these strategies.

A. Cell Adhesion Molecule Blockade

1. Selectins

The importance of selectins in neutrophil recruitment in vivo has recently been vividly demonstrated in mice rendered genetically deficient in P-selectin (133) and L-selectin (134,135). Leukocyte recruitment to experimen-

Table 1 Potential Therapeutic Targets Against Neutrophil Recruitment

Target	Function	Pharmacological agent
Selectins	Initial attachment and rolling	Monoclonal antibodies Soluble selectins Carbohydrate selectin ligand analogs
β_2-integrins and Ig	Stable arrest, spreading, and transmigration	Monoclonal antibodies Peptide inhibitors
Chemoattractants	Directional endothelial transmigration and migration into tissue Activation of β_2-integrins	Neutralizing monoclonal antibodies
Gene transcription of selectin, IgG, integrin, and chemoattractant	mRNA synthesis Protein synthesis	Proteosome/serine protease inhibitors Antisense oligonucleotides

tally induced sites of inflammation was significantly reduced in these selectin knockout animals. Antiselectin monoclonal antibodies (mAb) have also been shown to significantly reduce leukocyte accumulation in a number of animal models: anti-L-selectin in peritoneal inflammation (136); anti-P-selectin in ischemia-reperfusion injury of the gut (137) and myocardium (138); and anti-L-selectin in neutrophil-mediated lung injury (139,140).

There are a number of important considerations raised by these studies regarding the potential translation of antiselectin mAb therapy into the clinical setting for use in human subjects. With regard to efficacy, mAb reduced leukocyte recruitment by 50% to 90% as compared to sham-treated controls, but it was never completely ablated. Therefore, there remained 10% to 50% leukocytes which were able to infiltrate the tissues in mAb-treated subjects. There are some in vivo data (P. Kubes, University of Calgary, personal communication) indicating that, unless the efficiency of selectin blockade exceeds 95%, an overall biologically significant reduction in neutrophil recruitment and tissue damage may not be achieved. As the temporal and spatial expression of selectins in vivo is not clearly known, particularly in the various disease settings in humans, specific and timely intervention to block a relevant adhesion molecule remains problematic. Other important considerations include the pharmacokinetics and safety of the formulation of the mAb. Alternatives to the use of mAb for selectin blockade might be either soluble forms of recombinant selectin molecules or small synthetic molecules which can compete with the native molecule for ligand binding.

2. β_2-Integrins and the Ig Gene Superfamily

There have been numerous animal models of hypovolemic shock and fluid resuscitation (141), dermal inflammation (142), myocardial ischemia reperfusion injury (143), joint inflammation (144), lethal endotoxin challenge (145), and other models of neutrophil-mediated diseases in which blockade of CD11a, CD11b, CD18, or ICAM-1 reduced or inhibited neutrophil accumulation. A notable exception to these findings was in models of systemic shock or sepsis, in which treatment with anti-CD18 mAb did not significantly reduce lung injury. This may reflect the existence of CD18-independent mechanisms of leukocyte recruitment in the pulmonary circulation (92,141), thus highlighting the importance of appropriately targeting adhesion molecules with respect to their organ distribution.

Another field in which the use of mAb blockade of cell adhesion molecules may be beneficial is organ transplantation. It was observed that acutely rejected renal and liver allografts expressed heightened levels of ICAM-1. Additionally, blockade of ICAM-1 may also ameliorate any ischemia-reperfusion injury of the allograft which occurs during transplantation. Anti-ICAM-1 therapy in patients receiving renal allografts appears to be well tolerated (146).

A major concern with blockade of CD18-dependent adhesion pathways, in particular, has been the possibility of sepsis as a result of interference with the ability of neutrophils to extravasate and kill bacteria at sites of infection. While a brief period of treatment with anti-CD18 mAb did not increase mortality in a rabbit model of abdominal sepsis (147), in a canine septic shock model, anti-CD18 mAb worsened endotoxemia and myocardial dysfunction (148). Depending on the size of the bacterial inoculum, treatment with an anti-CD18 mAb either did not affect or increased the size of skin abscesses (149), although it should be noted that clinically relevant inoculations did not worsen the soft tissue sepsis. However, the risk of sepsis as a complication during treatment with anti-CD18 mAb remains a concern.

B. Blockade of Chemoattractants

1. IL-8

The major role of IL-8 in neutrophil recruitment in certain disease states has been shown in several experimental models of inflammation. In a rabbit endotoxin-induced pleurisy model, mAb against IL-8 reduced neutrophil migration by 70% to 80% (4). Interestingly, desensitization of the neutrophils with recombinant rabbit IL-8 also reduced neutrophil recruitment by 72% (150). Similar results were obtained in endotoxin-induced dermatitis (151) and arthritis (152) in rabbits treated with anti IL-8 mAb. Treatment with anti-IL-8 also prevented pulmonary injury in rabbits in an acid aspiration model (153) and in a reperfusion injury model (154). Treatment with anti-IL-8 mAb does not appear to be associated with an increased risk of sepsis in these animals as assessed by bacterial inoculation of dermal sites (C. Hebert, personal communication).

Intravenous infusion of IL-8 caused a transient pulmonary sequestration followed by a granulocytosis which persists with elevated plasma IL-8 levels but without causing any adverse hemodynamic changes in primates (155). The LAI effect seen in vivo (9) with intravenously administered IL-8 begs the question of whether administration of recombinant IL-8 may be an efficacious and safe means of down-regulating neutrophil recruitment at sites of inflammation in certain clinical settings. Conversely, whether the reported defect in neutrophil function seen in patients with overwhelming sepsis and high circulating levels of IL-8 (128) may be safely corrected by anti-IL-8 mAb, remains to be tested.

2. Other Chemoattractants

Blockade of C5a-mediated neutrophil recruitment with a recombinant soluble form of CR1 (156), and of LTB_4 with a receptor antagonist (157), in rat models of ischemia-reperfusion injury of the myocardium and skeletal muscle, respectively, reduced muscle damage significantly in both studies. LTB_4 also has been implicated in myocardial ischemia-reperfusion injury, but studies with receptor antagonists in animal models have produced equivocal results (158,159). Thus, blockade of chemoattractants per se may effectively reduce neutrophil-mediated damage.

VII. CONCLUSION

Neutrophil accumulation at sites of inflammation is a precisely regulated process requiring complex interactions between leukocytes, vascular endothelium, and extravascular connective tissues, all of which may secrete chemotactic cytokines and express cell adhesion molecules. The relative roles of various cell adhesion receptors and chemoattractants in mediating the neutrophil-endothelial adhesion cascade in inflammation is in turn dependent upon other factors, such as a given organ's microvascular bed and its particular pathology. Development of effective and safe clinical therapies that target the neutrophil recruitment process demands a precise knowledge of these mechanisms and modifying factors.

ACKNOWLEDGMENTS

The authors wish to thank Drs. Caroline Hebert, Myron Cybulsky, and Charles N. Serhan for preprints and helpful discussions; Dr. Han Ding for excellent technical assistance in the experimental results presented; and the members of the Vascular Research Division, Brigham and Women's Hospital, for helpful discussions. Funding support for this work was provided by National Institutes of Health grants PO1-HL36028 and HL47646.

REFERENCES

1. Issekutz TB, Movat HZ. The in vivo quantitation of and kinetics of rabbit neutrophil leukocyte accumulation in the skin in reponse to chemotactic agents and *Escherichia coli*. Lab Invest 1980; 42:310–317.
2. Cybulsky MI, McComb DJ, Movat HZ. Protein synthesis dependent and independent mechanisms of neutrophil emigration. Am J Pathol 1989; 135: 227–237.
3. Broaddus VC, Hebert CA, Vitangcol RV, Hoeffel JM, Bernstein MS, Boylan AM. Interleukin-8 is a major neutrophil chemotactic factor in pleural liquid of patients with emphysema. Am Rev Respir Dis 1992; 146:825–830.

4. Broaddus VC, Boylan AM, Hoeffel JM, et al. Neutralization of IL-8 inhibits neutrophil influx in a rabbit model of endotoxin-induced pleurisy. J Immunol 1994; 152:2960–2967.
5. Izzo RS, Witkon K, Chen AI, Hadjiyane C, Weinstein MI, Pellechia C. Interleukin-8 and neutrophil markers in colonic mucosa from patients with ulcerative colitis. Am J Gastroenterol 1992; 87:1447–1452.
6. Koch AE, Kunkel SE, Burrows JC, et al. Synovial tissue macrophages as a source of the chemotactic cytokine IL-8. J Immunol 1991; 147:2187–2195.
7. Simonet WS, Hughes TM, Nguyen HQ, Trebasky LD, Danilenko DM, Medlock ES. Long-term impaired neutrophil migration in mice overexpression human interleukin-8. J Clin Invest 1994; 94:1310–1319.
8. Ley K, Baker JB, Cybulsky MI, Gimbrone MA Jr, Luscinskas FW. Intravenous IL-8 inhibits granulocyte emigration from rabbit mesenteric venules without altering L-selectin expression or leukocyte rolling. J Immunol 1993; 151:6347–6357.
9. Hechtman DH, Cybulsky MI, Fuchs HJ, Baker JB, Gimbrone MA Jr. Intravascular IL-8. Inhibitor of polymorphonuclear leukocyte accumulation at sites of acute inflammation. J Immunol 1991; 147:883–892.
10. Arnaout MA. Leukocyte adhesion molecule deficiency; its structural basis, pathophysiology and implications for modulating the inflammatory response. Immunol Rev 1990; 114:147–180.
11. Carlos TM, Harlan JM. Leukocyte-endothelial adhesion molecules. Blood 1994; 84:2068–2101.
12. Springer TA. Traffic signals for lymphocyte recirculation and leukocyte emigration: the multistep paradigm. Cell 1994; 76:301–314.
13. Arnaout MA. Structure and function of leukocyte adhesion molecules CD11/CD18. Blood 1990; 75:1037–1050.
14. Pober JS, Cotran RS. The role of endothelial cells in inflammation. Transplantation 1990; 50:537–544.
15. Faruqi R, de la Motte C, DiCorleto PE. α-Tocopherol inhibits agonist-induced monocytic cell adhesion to cultured human endothelial cells. J Clin Invest 1994; 94:592–600.
16. Granger DN, Kubes P. The microcirculation and inflammation: modulation of leukocyte-endothelial cell adhesion. J Leuko Biol 1994; 55:662–675.
17. Pober JS. Cytokine-mediated activation of vascular endothelium: physiology and pathology. Am J Pathol 1988; 133:426–433.
18. Pober JS, Cotran RS. Cytokines and endothelial cell biology. Physiol Rev 1990; 70:427–451.
19. Rosales C, Juliano RL. Signal transduction by cell adhesion receptors in leukocytes. J Leuko Biol 1995; 57:189–198.
20. Luscinskas FW, Lawler J. Integrins as dynamic regulators of vascular function. FASEB J 1994; 8:929–938.
21. Hynes RO. Integrins: Versatility, modulation, and signaling in cell adhesion. Cell 1992; 69:11–25.
22. Diamond MS, Springer TA. A subpopulation of Mac-1 (CD11b/CD18) mole-

cules mediates neutrophil adhesion to ICAM-1 and fibrinogen. J Cell Biol 1993; 120:545–556.
23. Dransfield I, Cabanas C, Craig A, Hogg N. Divalent cation regulation of the function of the leukocyte integrin LFA-1. J Cell Biol 1992; 116:219–226.
24. Vedder NB, Harlan JM. Increased surface expression of CD11b/CD18 (Mac-1) is not required for stimulated neutrophil adherence to cultured endothelium. J Clin Invest 1988; 81:676–682.
25. Nathan C, Srimal S, Farber C, et al. Cytokine-induced respiratory burst of human neutrophils: dependence on extracellular matrix proteins and CD11/CD18 integrins. J Cell Biol 1989; 109:1341–1349.
26. Tedder TF, Isaacs CM, Ernst TJ, Demetri GD, Adler DA, Disteche CM. Isolation and chromosomal localization of cDNA's encoding a novel human lymphocyte cell surface molecule, LAM-1. J Exp Med 1989; 170:123–133.
27. Tedder TF, Penta AC, Levine HB, Freedman AS. Expression of the human leukocyte adhesion molecule, LAM-1: identify with TQ1 and Leu-8 differentiation antigens. J Immunol 1990; 144:532–540.
28. Bevilacqua MP, Stengelin S, Gimbrone MA Jr, Seed B. Endothelial-leukocyte adhesion molecule-1: an inducible receptor for neutrophils related to complement regulatory proteins and lectins. Science 1989; 243:1160–1165.
29. Johnston GI, Cook RG, McEver RP. Cloning of GMP-140, a granule membrane protein of platelets and endothelium. Cell 1989; 56:1033–1044.
30. Bevilacqua MP, Nelson RM. Selectins. J Clin Invest 1993; 91:379–387.
31. Kuijpers TW. Terminal glycosyltransferase activity: a selective role in cell adhesion. Blood 1993; 81:873–882.
32. Bevilacqua MP, Pober JS, Mendrick DL, Cotran RS, Gimbrone MA Jr. Identification of an inducible endothelial-leukocyte adhesion molecule. Proc Natl Acad Sci USA 1987; 84:9238–9242.
33. Collins T, Williams A, Johnston GI, et al. Structure and chromosomal location of the gene for endothelial-leukocyte adhesion molecule-1. J Biol Chem 1991; 266:2466–2473.
34. McEver RP, Beckstead JK, Moore KL, Marshall-Carlson L, Bainton DF. GMP-140, a platelet α-granule membrane protein, is also synthesized by vascular endothelial cells and is localized in Weibel-Palade bodies. J Clin Invest 1989; 84:92–99.
35. Butcher EC. Leukocyte-endothelial cell recognition: three (or more) steps to specificity and diversity. Cell 1991; 67:1033–1036.
36. Kishimoto TO, Jutila MA, Berg EL, Butcher EC. Neutrophil Mac-1 and MEL-14 adhesion proteins inversely regulated by chemotactic factors. Science 1989; 245:1238–1241.
37. Kishimoto TO, Jutila MA, Butcher EC. Identification of a human peripheral lymph node homing receptor: a rapidly down-regulated adhesion molecule. Proc Natl Acad Sci USA 1990; 87:2244–2248.
38. Lasky LA, Singer MS, Dowbenko D, et al. An endothelial ligand for L-selectin is a novel mucin-like molecule. Cell 1992; 69:927–938.
39. Briskin MJ, McEvoy LM, Butcher EC. MAdCam-1 has homology to immu-

noglobulin and mucin-like adhesion receptors and to IgA1. Nature 1993; 363: 461–464.
40. Baumhueter S, Singer MS, Hensel WJ, et al. Binding of L-selectin to the vascular sialomucin CD34. Science 1993; 262:436–438.
41. Luscinskas FW, Kansas GS, Ding H, et al. Monocyte rolling, arrest and spreading on IL-4-activated vascular endothelium under flow is mediated via sequential action of L-selectin, β_1-integrins, and β_2-integrins. J Cell Biol 1994; 125:1417–1427.
42. Spertini O, Luscinskas FW, Gimbrone MA Jr, Tedder TF. Monocyte attachment to activated human vascular endothelium in vitro is mediated by leukocyte adhesion molecule-1 (L-selectin) under non-static conditions. J Exp Med 1992; 175:1789–1792.
43. Kuijpers TW, Hakkert BC, van Mourik JA, Roos D. Distinct adhesive properties of granulocytes and monocytes to endothelial cells under static and stirred conditions. J Immunol 1990; 145:2588.
44. Bevilacqua MP. Endothelial-leukocyte adhesion molecules. Annu Rev Immunol 1993; 11:767–804.
45. Brandley BK, Swiedler SJ, Robbins PW. Carbohydrate ligands of the LEC cell adhesion molecules. Cell 1990; 63:861–863.
46. Picker LJ, Warnock RA, Burns AR, Doerschuk CM, Berg EL, Butcher EC. The neutrophil selectin LECAM-1 presents carbohydrate ligands to the vascular selectins ELAM-1 and GMP-140. Cell 1991; 66:921–933.
47. Etzioni A, Frydman M, Pollack S, et al. Recurrent severe infections caused by a novel leukocyte adhesion deficiency. N Engl J Med 1992; 327:1789–1792.
48. Price TH, Ochs HD, Gershoni-Baruch R, Harlan JM, Etzioni A. In vivo neutrophil and lymphocyte function studies in a patient with leukocyte adhesion deficiency type II. Blood 1994; 1635–1639.
49. Rothlein R, Dustin ML, Marlin SD, Springer TA. A human intercellular adhesion molecule (ICAM-1) distinct from LFA-1. J Immunol 1986; 137: 1270–1274.
50. Gahmberg CG, Nortamo P, Zimmerman D, Rouslahti E. The human leukocyte-adhesion ligand, intercellular-adhesion molecule 2. Expression and characterization of the protein. Eur J Biochem 1991; 195:177–182.
51. de Fougerolles AR, Stacker SA, Schwarting R, Springer TA. Characterization of ICAM-2 and evidence for a third counter-receptor for LFA-1. J Exp Med 1991; 174:253–267.
52. de Fougerolles AR, Springer TA. Intercellular adhesion molecule 3, a third adhesion counter-receptor for lymphocyte function-associated molecule 1 on resting lymphocytes. J Exp Med 1992; 175:185–190.
53. Marlin SD, Springer TA. Purified intercellular adhesion molecule-1 (ICAM-1) is a ligand for lymphocyte function-associated antigen 1 (LFA-1). Cell 1987; 51:813–819.
54. Makgoba MW, Sanders ME, Luce GEG, et al. ICAM-1: a ligand for LFA-1 dependent adhesion to B, T and myeloid cells. Nature 1988; 331:86–88.

55. Groves RW, Allen MH, Barker JNWN, Haskard DO, MacDonald DM. Endothelial leucocyte adhesion molecule-1 (ELAM-1) expression in cutaneous inflammation. Br J Dermatol 1991; 124:117–123.
56. Diamond MS, Staunton DE, de Fougerolles AR, et al. ICAM-1 (CD54): a counter-receptor for Mac-1 (CD11b/CD18). J Cell Biol 1990; 111:3129–3139.
57. Osborn L, Hession C, Tizard R, et al. Direct expression of vascular cell adhesion molecule 1, a cytokine-induced endothelial protein that binds to lymphocytes. Cell 1989; 59:1203–1211.
58. Rice GE, Munro JM, Bevilacqua MP. Inducible cell adhesion molecule 110 (INCAM-110) is an endothelial receptor for lymphocytes. A CD11/CD18 independent adhesion mechanism. J Exp Med 1990; 171:1369–1374.
59. Cybulsky MI, Gimbrone MA Jr. Endothelial expression of a mononuclear leukocyte adhesion molecule during atherogenesis. Science 1991; 251:788–791.
60. Kume N, Cybulsky MI, Gimbrone MA Jr. Lysophosphatidylcholine, a component of atherogenic lipoproteins, induces mononuclear leukocyte adhesion molecules in cultured human and rabbit arterial endothelial cells. J Clin Invest 1992; 90:1138–1144.
61. Ross R. The pathogenesis of atherosclerosis: a perspective for the 1990's. Nature 1993; 362:801–809.
62. Elices MJ, Osborn L, Takada Y, et al. VCAM-1 on activated endothelium interacts with the leukocyte integrin VLA-4 at a site distinct from the VLA-4/fibronectin binding site. Cell 1990; 60:577–584.
63. Berlin C, Bargatze RF, Campbell JJ, et al. $\alpha 4$ Integrins mediate lymphocyte attachment and rolling under physiologic flow. Cell 1995; 80:413–422.
64. Hemler ME. VLA proteins in the integrin family: structures, functions, and their role on leukocytes. Annu Rev Immunol 1990; 8:365–400.
65. Muller WA, Ratti CM, McDonnell SL, Cohn ZA. A human endothelial cell-restricted, externally disposed plasmalemmal protein enriched in intercellular junctions. J Exp Med 1989; 170:399–414.
66. Muller WA, Weigl SA, Deng X, Phillips DM. PECAM-1 is required for transendothelial migration of leukocytes. J Exp Med 1993; 178:449–460.
67. Bogen S. Pak J, Garifallou M, Deng X, Muller WA. Monoclonal antibody to murine PECAM-1 (CD31) blocks acute inflammation in vivo. J Exp Med 1994; 179:1059–1064.
68. Vaporciyan AA, Delisser HM, Yan H-C, et al. Involvement of platelet-endothelial cell adhesion molecule 1 in neutrophil recruitment in vivo. Science 1993; 262:1580–1583.
69. Tanaka Y, Albelda SM, Horgan KJ, et al. CD31 expressed on distinctive T cell subsets is a preferential amplifier of β_1 integrin-mediated adhesion. J Exp Med 1992; 176:245–253.
70. Movat HZ. Chemotaxis. In: The Inflammatory Reaction. Amsterdam: Elsevier, 1985; 203–257.
71. Leber T. Die enstehung der Entzundung. Die Wirkung der entzundungserregenden Schadlichkeiten nach vorzugsweise am Auge angestellten Untersuchungen. Leipzig: Verlag von Wilhelm Engelmann, 1891.

72. Metchnikoff E. Lectures on the Comparative Pathology of Inflammation. London: Kegan, Paul, Trench, Trubner & Co. 1893.
73. Devreotes PN, Zigmond SH. Chemotaxis in eukaryotic cells: a focus on leukocytes and *Dictyostelium*. Annu Rev Cell Biol 1988; 4:649–686.
74. Keller HU, Wilkinson PC, Abercrombie M, et al. A proposal for the definition of terms related to locomotion of leukocytes and other cells. Clin Exp Immunol 1977; 27:377–380.
75. Boyden S. The chemotactic effect of mixtures of antibody and antigen on polymorphonuclear leukocytes. J Exp Med 1962; 115:453–466.
76. Cutler JE. A simple in vitro method for the study of chemotaxis. Proc Soc Exp Biol Med 1974; 147:471–474.
77. Pabst MJ. Priming of neutrophils. In: Hellewell PG, Williams T, eds. Immunopharmacology of Neutrophils. London: Academic Press, 1994:195–220.
78. Koch AE, Polverini PJ, Kunkel SL, et al. Interleukin-8 as a macrophage-derived mediator of angiogenesis. Science 1992; 258:1798–1801.
79. Evangelou AM. Platelet-activating factor (PAF): implications for coronary heart and vascular diseases. Prostaglandins Leukotrienes and Essential Fatty Acids 1994; 50:1–28.
80. Taraboletti G, Roberts DD, Liotta LA. Thrombospondin-induced tumor cell migration: haptotaxis and chemotaxis are mediated by different molecular domains. J Cell Biol 1987; 105:2409–2415.
81. Klominek J, Robert KH, Sundqvist KG. Chemotaxis and haptotaxis of human malignant mesothelioma cells: effects of fibronectin, laminin, type IV collagen, and an autocrine motility factor-like substance. Cancer Res 1993; 15:4376–4382.
82. Brandley BK, Schnaar RL. Tumour cell haptotaxis on covalently immobilized linear and exponential gradients of a cell adhesion peptide. Dev Biol 1989; 135:74–86.
83. Mansfield PJ, Suchard SJ. Thrombospondin promotes chemotaxis and haptotaxis of human peripheral blood monocytes. J Immunol 1994; 153:4219–4229.
84. Tanaka Y, Adams DH, Shaw S. Proteoglycans on endothelial cells present adhesion-inducing cytokines to leukocytes. Immunol Today 1993; 14:111–115.
85. Rot A. Biding of neutrophil attractant/activation protein-1 (interleukin-8) to resident dermal cells. Cytokine 1992; 4:347–352.
86. Ward PA, Newman LJ. A neutrophil chemotactic factor from human C′5. J Immunol 1969; 102:93–99.
87. Schiffman E, Corcoran B, Wahl SM. N-formylmethionyl peptides as chemoattractants for leukocytes. Proc Natl Acad Sci USA 1975; 72:1059–1062.
88. Goetzl EJ, Picket WC. Novel structural determinants of the human neutrophil chemotactic activity of leukotriene B_4. J Exp Med 1981; 153:482–487.
89. Czarnetzki BM, Benveniste J. Effect of 1-O-octadecyl-2-O-acetyl-sn-glycero-3-phosphocholine (PAC-aceter) on leukocytes. Analysis of the in vitro migration of human neutrophils. Chem Phys Lipids 1981; 29:317–326.
90. Prescott SM, Zimmerman GA, McIntyre TM. Human endothelial cells in

culture produce platelet-activating factor (1-alkyl-2-acetyl-sn-glycero-3-phosphocholine) when stimulated with thrombin. Proc Natl Acad Sci USA 1984; 81:3534–3538.

91. Kuijpers TW, Hakkert BC, Hart MHL, Roos D. Neutrophil migration across monolayers of cytokine-prestimulated endothelial cells: a role for platelet-activating factor and IL-8. J Cell Biol 1992; 117:565–572.
92. Hellewell PG, Henson PM. In: Gordon JL, ed. Vascular Endothelium: Interactions with Circulating Cells. Amsterdam: Elsevier Science Publishers, 1991: 143–160.
93. Schroeder JM, Mrowietz U, Morita E, Christophers E. Purifications and partial biochemical characterization of a human monocyte-derived neutrophil-activating peptide that lacks IL-1 activity. J Immunol 1987; 139:3474–3483.
94. Yoshimura TK, Matsushima K, Oppenheim JJ. Leonard EJ. Neutrophil chemotactic factor produced by lipopolysaccharide (LPS)-stimulated human blood mononuclear leukocytes: partial characterization and separation from interleukin-1 (IL-1). J Immunol 1987; 139:788–793.
95. Lindley I, Aschauer H, Seifert JM, et al. Synthesis and expression in *Escherichia coli* the gene encoding monocyte-derived neutrophil-activating factor: biological equivalence between natural and recombinant neutrophil-activating factor. Proc Natl Acad Sci USA 1988; 85:9199–9203.
96. Matsushima K, Morishita K, Yoshimura T, et al. Molecular cloning of a human monocyte-derived neutrophil chemotactic factor (MDNCF) and the induction of MDNCF mRNA by interleukin-1 and tumour necrosis factor. J Exp Med 1988; 167:1883–1893.
97. Oppenheim JJ, Zachariae COC, Mukaida N, Matsushima K. Properties of the novel proinflammatory supergene "intercrine" cytokine family. Annu Rev Immunol 1991; 9:617–648.
98. Wheeler ME, Luscinskas FW, Bevilacqua MP, Gimbrone MA Jr. Cultured human endothelial cells stimulated with cytokines or endotoxin produce an inhibitor of leukocyte adhesion. J Clin Invest 1988; 82:1211–1218.
99. Gimbrone MA Jr, Obin MS, Brock AF, et al. Endothelial interleukin-8: a novel inhibitor of leukocyte-endothelial interactions. Science 1989; 246:1601–1603.
100. Hebert CA, Baker JB. Interleukin-8: a review. Cancer Invest 1993; 11:743–750.
101. Luscinskas FW, Kiely J-M, Ding H, et al. In vitro inhibitory effect of IL-8 and other chemoattractants on neutrophil-endothelial adhesive interactions. J Immunol 1992; 149:2163–2171.
102. Westlin WF, Kiely J-M, Gimbrone MA Jr. Interleukin-8 induces changes in human neutrophil actin conformation and distribution: relationship to inhibition of adhesion to cytokine-activated endothelium. J Leuko Biol 1992; 52:43–51.
103. Huber AR, Hunkel SL, Todd RF III, Weiss SJ. Regulation of transendothelial neutrophil migration by endogenous interleukin-8. Science 1991; 254:99–102.

104. Kuijpers TW, Hakkert BC, Hart MHL, Roos D. Neutrophil migration across monolayers of cytokine-prestimulated endothelial cells: a role for platelet-activating factor and IL-8. J Cell Biol 1992; 117:565–572.
105. Smart SJ, Casale TB. TNF-alpha induced transendothelial migration is IL-8 dependent. Am J Physiol 1994; 266:L238–L245.
106. Miller EJ, Cohen AB, Nagao S, et al. Elevated levels of NAP-1/interleukin-8 are present in the airspaces of patients with the adult respiratory distress syndrome and are associated with increased mortality. Am Rev Respir Dis 1992; 146:427–432.
107. Strieter RM, Koch AE, Antony VB, Fick RB Jr, Standiford TJ, Kunkel SL. The immunopathology of chemotactic cytokines: the role of interleukin-8 and monocyte chemoattractant protein-1. J Lab Clin Med 1994; 123:183–197.
108. Pawlowski NA, Abraham EL, Pontier S, Scott WA, Cohn ZA. Human monocyte-endothelial cell interaction in vitro. Proc Natl Acad Sci USA 1985; 82:8208–8212.
109. Bevilacqua MP, Pober JS, Wheeler ME, Cotran RS, Gimbrone MA Jr. Interleukin-1 acts on cultured human vascular endothelium to increase the adhesion of polymorphonuclear leukocytes, monocytes, related leukocyte cell lines. J Clin Invest 1985; 76:2003–2011.
110. Rollins BJ, Pober JS. Interleukin-4 induces the synthesis and secretion of MCP-1/JE by human endothelial cells. Am J Pathol 1991; 138(6):1315–1319.
111. Von Andrian UH, Chambers JD, McEvoy LM, Bargatze RF, Arfors KE, Butcher EC. Two-step model of leukocyte-endothelial cell interaction in inflammation: distinct roles of LECAM-1 and the leukocyted β_2-integrins in vivo. Proc Natl Acad Sci USA 1991; 88:7538–7542.
112. Luscinskas FW, Ding H, Lichtman AH. P-selectin and VCAM-1 mediate rolling and arrest of CD4+ T-lymphocytes on TNF-α-activated vascular endothelium under flow. J Exp Med 1995; 181:1179–1186.
113. Bargatze RF, Butcher EC. Rapid G protein-regulated activation event involved in lymphocyte binding to high endothelial venules. J Exp Med 1993; 178:367–372.
114. Lawrence MB, Springer TA. Leukocytes roll on a selectin at physiologic flow rates: distinct from and prerequisite for adhesion through integrins. Cell 1991; 65:1–20.
115. Shen J, Gimbrone MA Jr, Luscinskas FW, Dewey CF Jr. Regulation of nucleotide concentration at endothelium-fluid interface by viscous shear flow. Biophys J 1992; 64:1323–1330.
116. Shen J, Luscinskas FW, Connolly A, Dewey CF Jr, Grimbrone MA Jr. Fluid shear stress modulates cytosolic free calcium in vascular endothelial cells. Am J Physiol 1992; 262:C384–C390.
117. Von Adrian UH, Hansell P, Chambers JD, et al. L-selectin function is required for β_2-integrin-mediated neutrophil adhesion at physiological shear rates in vivo. Am J Physiol 1992; 263:H1034–H1044.
118. Tozeren A, Ley K. How do selectins mediate leukocyte rolling in venules? Biophys J 1992; 63:700–709.

119. Hammer DA, Apte SM. Simulation of cell rolling and adhesion on surfaces in shear flow: general results and analysis of selectin-mediated neutrophil adhesion. Biophys J 1992; 63:35–37.
120. Abbassi O, Kishimoto TK, Mcintire LV, Anderson DC, Smith CW. E-selectin supports neutrophil rolling in vitro under conditions of flow. J Clin Invest 1993; 92:2719–2730.
121. Lawrence MB, Smith CW, Eskin SG, Mcintire LV. Effect of venous shear stress on CD18-mediated neutrophil adhesion to cultured endothelium. Blood 1990; 75:227–237.
122. Lo SK, Lee S, Ramos RA, et al. Endothelial-leukocyte adhesion molecule 1 stimulates the adhesive activity of leukocyte integrin CD3 (CD11b/CD18, Mac-1, $\alpha_m\beta_2$) on human neutrophils. J Exp Med 1991; 173:1493–1500.
123. Berman ME, Muller WA. Ligation of platelet/endothelial cell adhesion molecule 1 (PECAM-1/CD31) on monocytes and neutrophils increases biding capacity of leukocyte CR3 (CD11b/CD18). J Immunol 1995; 154:299–307.
124. Luscinskas FW, Brock AF, Arnaout MA, Gimbone MA Jr. Endothelial-leukocyte adhesion molecule-1 (ELAM-1)-dependent and leukocyte (CD11/CD18)-dependent mechanisms contribute to polymorphonuclear leukocyte adhesion to cytokine-activated human vascular endothelium. J Immunol 1989; 142:2257–2263.
125. Smith CW, Marlin SD, Rothlein R, Toman C, Anderson DC. Cooperative interactions of LFA-1 and Mac-1 with intercellular adhesion molecule-1 in facilitating adherence and transendothelial migration of human neutrophils in vitro. J Clin Invest 1989; 83:2008–2017.
126. Lampugnani MG, Corada M, Caveda L, et al. The molecular organization of endothelial cell to cell junctions: differential association of plakoglobin, β-catenin, and α-catenin with vascular endothelial cadherin (VE-cadherin). J Cell Biol 1995; 129:203–218.
127. Suchard SJ, Burton MJ, Dixit VM, Boxer LA. Human neutrophil adherence to thrombospondin occurs through a CD11/CD18-independent mechanism. J Immunol 1991; 146:3945–3952.
128. Marty C, Misset B, Tamion F, Fitting C, Carlet J, Cavaillon JM. Circulating IL-8 concentrations in patients with multiple organ failure of septic and non-septic origin. Crit Care Med 1994; 22:673–679.
129. Hack CE, Hart M, van Schijndel RJ, et al. Interleukin-8 in sepsis: relation to shock and inflammatory mediators. Infect Immun 1992; 60:2835–2842.
130. Territo MC, Golde DW. Granulocyte function in experimental endotoxemia. Blood 1976; 47:539–544.
131. Darbonne WC, Rice GC, Mohler MA, et al. Red blood cells are a sink for interleukin-8, a leukocyte chemotaxin. J Clin Invest 1991; 88:1362–1369.
132. Samanta A, Oppenheim J, Matsushima J. Interleukin-8 (MDNCF) dynamically regulates its own receptor expression on human neutrophils. J Biol Chem 1989; 265:183–189.
133. Mayadas TN, Johnson RC, Rayburn H, Hynes RO, Wagner DD. Leukocyte rolling and extravasation are severely compromised in P-selectin deficient mice. Cell 1993; 74:541–554.

134. Arbones ML, Ord DC, Ley K, et al. Lymphocyte homing and leukocyte rolling and migration are impaired in L-selectin-deficient mice. Immunity 1995; 1:247–260.
135. Tedder TF, Steeber DA, Pizcueta P. L-selectin deficient mice have impaired leukocyte recruitment into inflammatory sites. J Exp Med 1995. 181:2259–68.
136. Pizcueta P, Luscinskas FW. Monoclonal antibody blockade of L-selectin inhibits mononuclear leukocyte recruitment to inflammatory sites in vivo. Am J Pathol 1994; 145:461–469.
137. Davenpeck KL, Gauthier TW, Albertine KH, Lefer AM. Role of P-selectin in microvascular leukocyte-endothelial interaction in splanchnic ischemia-reperfusion. Am J Physiol 1994; 267:H622–H630.
138. Weyrich AS, Ma XY, Lefer DJ, Albertine KH, Lefer AM. In vivo neutralization of P-selectin protects feline heart and endothelium in myocardial ischemia and reperfusion injury. J Clin Invest 1993; 91:2620–2629.
139. Mulligan MS, Varani J, Dame MK, et al. Role of endothelial-leukocyte adhesion molecule 1 (ELAM-1) in neutrophil-mediated lung injury in rats. J Clin Invest 1991; 88:1396–1406.
140. Mulligan MS, Miyasaka M, Tamatani T, Jones ML, Ward PA. Requirements for L-selectin in neutrophil-mediated lung injury in rats. J Immunol 1994; 152:832–840.
141. Vedder NB, Winn RK, Rice CL, Chi EY, Arfors KE, Harber JM. A monoclonal antibody to the adherence promoting leukocyte glycoprotein CD18 reduces organ injury and improves survival from hemorrhagic shock and resuscitation in rabbits. J Clin Invest 1988; 81:939–944.
142. Issekutz AC, Issekutz TB. The contribution of LFA-1 (CD11a/CD18) and Mac-1 (CD11b/CD18) to the in vivo migration of polymorphonuclear leukocytes to inflammatory reactions in the rat. Immunology 1992; 76:655–661.
143. Winquist R, Frei P, Harrison P, et al. An anti CD18 mAb limits the size in primates following myocardial ischemia and reperfusion. Circulation 1990; 82:701.
144. Jasin HE, Lightfoot E, Davis LS, Rothlein R, Faanes RB, Lipsky PE. Amelioration of antigen induced arthritis in rabbits treated with monoclonal antibodies to leukocyte adhesion molecules. Arthritis Rheum 1992; 35:541–549.
145. Xu H, Gonzalo JA, St. Pierre Y, et al. Leukocytosis and resistance to septic shock in intercellular adhesion molecule 1-deficient mice. J Exp Med 1994; 180:95–109.
146. Cosimi AB, Conti D, Delmonico FI, et al. In vivo effects of monoclonal antibody to ICAM-1 (CD54) in non-human primates with renal allografts. J Immunol 1990; 144:4604–4613.
147. Mileski WJ, Winn RK, Harlan JM, Rice CM. Transient inhibition of neutrophil adherence with the anti-CD18 monoclonal antibody 60.3 does not increase mortality rates in abdominal sepsis. Surgery 1991; 109:497–501.
148. Eichacker PQ, Hoffman WD, Farese A, et al. Leukocyte CD18 monoclonal antibody worsens endotoxemia and cardiovascular injury in canines in septic shock. J Appl Physiol 1993; 74:1885–1892.

149. Sharar SR, Winn RK, Murry CE, Harlan JM, Rice CR. A CD18 monoclonal antibody increases the incidence and severity of subcutaneous abscess formation after high dose *S. aureus* injection in rabbits. Surgery 1991; 116:213–219.
150. Boylan AM, Hebert CA, Sadick M, et al. Interleukin-8 is a major component of pleural fluid chemotactic activity in a rabbit model of endotoxin pleurisy. Am J Physiol 1994; 267:L137–L144.
151. Harada A, Sekido N, Akahoshi T, Wada T, Mukaida N, Matsushima K. Essential involvement of interleukin-8 (IL-8) in acute inflammation. J Leuk Biol 1994; 56:559–564.
152. Akahoshi Y, Endo H, Kondo H, et al. Essential involvement of interleukin-8 in neutrophil recruitment in rabbits with acute experimental arthritis induced by lipopolysaccharide and interleukin-1. Lymphokine Cytokine Res 1993; 13: 113–116.
153. Folkesson HG, Matthay MA, Hebert CA, Broaddus VC. Acid aspiration induced lung injury in rabbits is mediated by interleukin-8 dependent mechanisms. J Clin Invest 1995; 96:107–116.
154. Sekido N, Mukaida N, Harada A, Nakanishi I, Watanabe Y, Matsushima K. Prevention of lung reperfusion injury in rabbits by a monoclonal antibody against interleukin-8. Nature 1993; 365:654–657.
155. Van Zee K, Fischer E, Hawes A. Effects of intravenous IL-8 administration in nonhuman primates. J Immunol 1992; 148:1746–1752.
156. Weisman HF, Barton T, Leppo MK, et al. Soluble human complement receptor type I: in vivo inhibitor of complement suppressing post-ischemia myocardial inflammation and necrosis. Science 1990; 249:146–151.
157. Homer-Vanniasinkam S, Gough MJ. Role of lipid mediators in the pathogenesis of skeletal muscle infarction and oedema during reperfusion after ischemia. Br J Surg 1994; 81:1500–1503.
158. Hahn RA, MacDonald BR, Simpson PJ, Potts BD, Paril CJ. Antagonism of leukotriene B_4 receptors does not limit canine myocardial infarct size. J Pharmacol Exp Ther 1995; 253:58–66.
159. Taylor AA, Gacic AC, Kitt TM, et al. A specific leukotriene B_4 antagonist protects against myocardial ischemia-reperfusion injury. Clin Res 1989: A528.

11

Adhesion Molecules in Lymphocyte-Mediated Inflammation

Aiyappa Palecanda and Thomas B. Issekutz
Department of Medicine and Immunology, University of Toronto, Toronto, Ontario, Canada

I. INTRODUCTION

There has been a considerable advance in the understanding of the mechanisms of lymphocyte, neutrophil and monocyte migration out of the blood into the tissue. Many of the adhesion molecules mediating the attachment of these blood leukocytes to the endothelium in the blood vessel have been identified. In addition, numerous chemotactic stimuli which promote leukocyte extravasation in inflammation have been isolated. This chapter will focus on the adhesion molecules that are involved in the migration of lymphocytes out of the blood with special emphasis on in vivo studies of T-cell recruitment to sites of inflammation.

II. MAJOR PLAYERS IN LYMPHOCYTE ENDOTHELIAL ADHESION

Studies by Gowans and Knight (1) in rats and by Hall and Morris (2) in sheep first showed that lymphocytes migrate from lymph nodes through lymphatics to the blood and back to lymph nodes. This process was referred to as recirculation to distinguish this pattern of migration from that of other leukocytes which predominantly circulated in the blood and did not reenter the blood after migrating out of the vasculature. Subsequently, Smith showed that lymphocytes also traffic through virtually all nonlymphoid tissues and entered afferent lymphatics which carried them to peripheral or, in the case of the gut, mesenteric lymph nodes (3). The recognition of two populations of lymphocytes—namely, naive lymphocytes, which were generated in the primary lymphoid tissues and had not been activated through their antigen receptor; and memory lymphocytes, which had encountered antigen, together with markers for these T-cells based on alternate splicing of CD45, led to the recognition of different recirculation pathways for antigen-experienced and -inexperienced lymphocytes (4). Naive lymphocytes predominantly recirculate from lymph nodes via efferent lymphatics to the blood and back to lymph nodes (5). Lymphocytes recirculating through nonlymphoid tissues, on the other hand, appear to be predominantly T lymphocytes of the memory phenotype under normal noninflammatory conditions (5). During inflammation, however, there is a marked change in T-cell migration through the inflamed tissue. There is an enormous increase in cells migrating out of the blood, trafficking through the inflammatory site, and on into the regional lymph node (3). The magnitude of this traffic can be huge and may include both memory and naive T lymphocytes (6,7). A key step in controlling this migration of T-cells is the adhesion and activation steps involved in the interaction of lymphocytes with the vascular endothelial cells in the inflamed tissue (8).

At least four major classes of adhesion molecules are involved in the adhesion process (Fig. 1), including the selectins, the integrins, the sialomucins, and members of the immunoglobulin supergene family (reviewed in Chapter 1 and Ref. 9). Selected members of each of these families have been shown to be involved in lymphocyte migration. These include L-selectin on lymphocytes and E-selectin on the endothelium which bind to O-linked oligosaccharides on the sialomucins, and possibly other glycoproteins. The integrin family members shown to be involved in lymphocyte migration include: LFA-1 ($\alpha_L\beta_2$ or CD11a/CD18), VLA-4 ($\alpha_4\beta_1$ or CD49d/CD29), and LPAM-1 ($\alpha_L\beta_7$). LFA-1 binds to ICAM-1 (CD54) and ICAM-2 (CD102). VLA-4 binds to VCAM-1 (CD106), the CS-1 fragment of fibronectin; and $\alpha_4\beta_7$ binds to the mucosal addressin cell adhesion molecule, MAdCAM-1, VCAM-1, and the CS-1 fragment of fibronectin. The structures of each of these molecules have been described in detail elsewhere in this book, so we will concentrate on the studies demonstrating the role of these molecules in lymphocyte migration.

In addition to these adhesion molecules, lymphocyte migration also involves chemotactic factors which contribute to the activation of the lymphocyte and may direct the movement of the cell. Members of the chemokine family of adhesion molecules appear to be particularly important in this regard (8,10,11).

The migration of lymphocytes, as well as most leukocytes, out of the

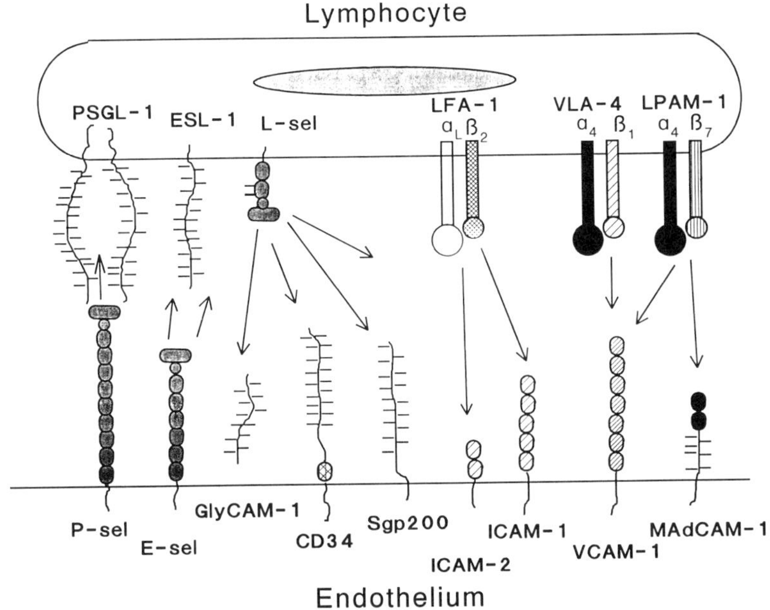

Figure 1 Schematic diagram of the major adhesion receptors involved in lymphocyte endothelial cell adhesion.

circulation occurs predominantly in the postcapillary venule. Studies using intravital microscopy have demonstrated that lymphocytes initially interact with the endothelial cells in the venule by rolling on the blood vessel wall for a short distance. This is followed by firm adhesion and transmigration. The initial rolling interaction with the vessel wall is mediated by the binding of the selectins to the sialomucins, and to some extent adhesion mediated by the α_4 integrins. Chemokine-mediated activation of the lymphocyte is next thought to increase the affinity of the lymphocyte integrins, allowing a firm adhesion and arrest of the T-cell. (8). This is followed by migration across the endothelial cells of the blood vessel into the inflammatory site. The transendothelial migration is thought to occur at the tight junctions between endothelial cells or very nearby and, for monocytes and neutrophils, has been shown to involve the platelet endothelial cell adhesion molecule-1 (CD31) or PECAM-1, which is concentrated in the region of the tight junctions (12,13).

This model of lymphocyte interaction with endothelium is thought to occur at sites of inflammation. In parallel with the identification of the various adhesion receptors, in vivo studies using blocking monoclonal antibodies (mAbs) have attempted to determine the contribution of these adhesion receptors in lymphocyte migration in inflammation. These investigations have highlighted the key role of the adhesion receptors in lymphocyte extravasation in vivo, and demonstrated both independent and overlapping functions of the adhesion pathways. In addition, these in vivo studies have shown striking organ-specific adhesion receptor interactions.

III. CONTRIBUTION OF THE SELECTINS TO LYMPHOCYTE MIGRATION IN INFLAMMATION

A. L-selectin (CD62L)

L-selectin is a 70 to 90-kDa glycoprotein expressed on virtually all naive lymphocytes and on a portion of memory T-cells (14–17). It is also expressed on neutrophils, monocytes, and eosinophils in the blood (18). It was one of the first lymphocyte receptors to be shown to mediate adhesion to the high endothelial venules of peripheral and mesenteric lymph nodes through which most recirculation by lymphocytes occurs (19). Monoclonal antibodies to L-selectin nearly abolish lymphocyte homing to peripheral lymph nodes and partially inhibit migration to mesenteric nodes (19,20). Mice rendered L-selectin deficient by gene targeting have also shown a profound inhibition of lymphocyte accumulation in peripheral lymph nodes and poor lymph node development (21). Thus, L-selectin appears essential for lymphocytes, especially naive lymphocytes, to home to peripheral lymph nodes, and L-selectin contributes to the traffic of lymphocytes to mesenteric nodes.

The role of L-selectin in lymphocyte migration to areas of inflammation is less well defined. Using the mAb MECA-79, which reacts with the carbohydrate ligand on L-selectin-binding glycoproteins, it has been shown that L-selectin ligands are present on the blood vessels in areas of inflammation (22). However, the functional importance of L-selectin ligands for lymphocyte migration into inflamed tissue are unknown. Furthermore, studies with blocking mAbs to L-selectin have been problematic because of the rapid shedding of L-selectin after it is bound by antibody (23).

L-selectin has been suggested to play a role in lymphocyte-mediated inflammatory reactions. Anti-L-selectin antibodies, low dose chymotrypsin treatment of lymphocytes, which removes L-selectin, and sialidase treatment of the tissue all inhibit the binding of lymphocytes from rats with a rejecting kidney allograft to the peritubular capillary's endothelium in an in vitro cell adhesion assay (24). Glycam-1, an L-selectin ligand produced by endothelial cells in high endothelial venules, and thought to contribute to lymphocyte migration into lymph nodes, has also been shown to be expressed in the Islets of Langerhans in nonobese diabetic mice infiltrated with lymphocytes (25). Studies of lymphocyte recruitment to inflammation in vivo have shown that treatment with an anti-L-selectin antibody reduced lymphocyte accumulation in the peritoneum by 90% in mice (26). Furthermore, L-selectin knockout mice have impaired delayed-type hypersensitivity reactions and show 70% decrease in lymphocyte migration to the thioglycolate-stimulated peritoneum (27). These studies suggest that L-selectin plays a role both in normal lymphocyte recirculation and in lymphocyte recruitment to some sites of inflammation. The extent of the contribution by L-selectin to lymphocyte migration in other inflammatory sites is not well defined, however, and its contribution relative to the other selectins and leukocyte integrins is unknown.

B. P-selectin (CD62P)

P-selectin is a 140-kDa glycoprotein expressed on activated platelets and activated endothelial cells (28–31). P-selectin is stored in the α granules of platelets and the Weibel-Palade bodies of endothelial cells, and is rapidly translocated to the cell surface following activation (28–31). Endothelial cell P-selectin expression is stimulated by numerous pro-inflammatory mediators, including histamine, thrombin, IL-1, and TNFα (30–33). P-selectin binds to the sialomucin family glycoprotein PSGL-1 on leukocytes (34,35). Neutrophils will roll on monolayers of P-selectin, and P-selectin can mediate the binding of neutrophils to activated endothelium and activated platelets (36–38). The contribution of P-selectin to T-cell migration is less clear. The finding that T-cell clones bind to P-selectin and that T cells can roll on immobilized P-selectin in flow chambers suggests that T cells express functional ligands for P-selectin (39–42). This interaction may contribute to T-cell recruitment at inflammatory sites. It has also been shown that P-selectin is expressed at sites of chronic inflammation associated with extensive lymphocyte infiltration such as in the synovium in rheumatoid arthritis (43). Mice made genetically deficient in P-selectin show an absence of initial leukocyte rolling and reduced recruitment of neutrophils into the inflamed peritoneum (44). The lymph nodes in these mice are apparently normal, but these animals have a decrease in cutaneous contact sensitivity (45). Further studies of lymphocyte recruitment into sites of inflammation in the absence of P-selectin are needed.

C. E-selectin (CD62)

E-selectin is a 115-kDa glycoprotein expressed on activated endothelial cells (46). Unlike P-selectin, it is not stored but is newly synthesized after endothelial activation with cytokines such as IL-1 and TNFα (46). E-selectin binds to the sialyl Lewis^x (sLe^x) and sLe^a carbohydrates decorating glycoproteins on neutrophils, monocytes, and a subpopulation of T lymphocytes (47–49). The specific glycoprotein ligands on leukocytes have not been fully identified, but on mouse neutrophils the protein ESL-1 and L-selectin can bind E-selectin (50–52). A subset of T lymphocytes can also bind to E-selectin in vitro (53). This subset is found particularly in association with inflammation in the skin. These T-cells express a 200-kDa glycoprotein known as the cutaneous lymphocyte antigen (CLA) (54). The CLA antigen reacts with the mAb HECA-452. This mAb recognizes a carbohydrate, most likely sLe^x or sLe^a, which is one of the ligands for E-selectin (47,54). The CLA$^+$ cells are a subset of CD4$^+$ memory lymphocytes (53). The HECA-452 mAb blocks the adhesion of CLA$^+$ T cells to E-selectin. This CLA E-selectin interaction is thought to be crucial for the accumulation of CLA$^+$ T-cells in cutaneous inflammation (53,54). HECA-452 also selectively stains cutaneous T-cell lymphomas, further supporting its role in targeting T-cells to the skin (55). In addition, T-cell clones grown from the skin of atopic subjects can bind to E-selectin in vitro (41).

Direct evidence for E-selectin mediating T-cell migration comes from antibody-blocking studies which have suggested that in tuberculin DTH reactions in some macaque monkeys, treatment with anti-E-selectin can partially inhibit the intensity of T-cell infiltrates in the skin (56). Bovine $\gamma\delta$ T-cells have also been reported to bind to E-selectin in the skin, suggesting that the $\gamma\delta$ T-cells, which are found in cutaneous tissues, may accumulate in these sites through the CLA E-selectin adhesion pathway (57). Studies in E-selectin knockout mice have also shown that contact sensitivity reactions are reduced in such animals if they are treated with an anti-P-selectin-blocking mAb (58). These reactions, which are mediated by T-cells but heavily infiltrated with neutrophils, also suggest a role for E-selectin in cutaneous inflammation induced by T lymphocytes.

IV. CONTRIBUTION OF THE INTEGRINS TO LYMPHOCYTE MIGRATION IN INFLAMMATION

A. Lymphocyte Function-Associated Antigen-1 (LFA-1, CD11a/CD18)

LFA-1 (CD11a/CD18) is a member of the CD18 (β_2) integrin family of adhesion molecules, which also includes Mac-1 (CD11b/CD18) and P150/95 (CD11c/CD18) (59). Of these three family members, all lymphocytes express LFA-1 and a small subpopulation may express Mac-1 (59,60). LFA-1 consists of an α chain of 180 kDa, which is noncovalently associated with a 95-kDa β chains (59). It is the predominant β_2 integrin mediating lymphocyte adhesion. LFA-1 is involved in multiple aspects of lymphocyte function, including cytotoxic T-cell-mediated killing, T helper cell interactions with B cells and macrophages, and lymphocyte adhesion to endothelium (61). LFA-1 binds to at least three ligands, including ICAM-1, ICAM-2, and ICAM-3 (62–64). LFA-1 was one of the first receptors shown to mediate lymphocyte adhesion to endothelial cells in vitro, and among the first molecules shown to inhibit lymphocyte migration in vivo (65,66). LFA-1 on T-cells can bind to either ICAM-1 or ICAM-2 expressed on the surface of endothelial cells. ICAM-1 is a 110-kDa protein expressed at a low level on vascular endothelium in most tissues; it is also found on some lymphocytes, monocytes, and NK cells. This expression can be markedly increased by endothelial activation with a number of cytokines including IL-1, TNFα, and IFN-γ (67,68). ICAM-2 is constitutively expressed on endothelial cells and also appears to be an important ligand for lymphocyte LFA-1, but is not upregulated following endothelial cell activation (63,69).

There are numerous studies examining the contribution of LFA-1 and ICAM-1 to lymphocyte-mediated inflammatory reactions. Many of these investigations have shown dramatic effects on inflammation as a result of inhibiting the LFA-1/ICAM-1-mediated interaction. Since the LFA-1/ICAM-1 pathway is important in the generation of immune responses through the interaction of T-cells with antigen-presenting cells, where it may provide important costimulatory signals as well as in T-cell-mediated cytotoxicity, the effect of LFA-1 and/or ICAM-1 blockade on inflammatory reactions may only partially be related to alterations in lymphocyte endothelial cell adhesion and T-cell migration. Blockade of LFA-1 inhibits the adhesion of lymphoblastoid cell lines and human and rodent blood lymphocytes to endothelial cells in vitro (65,70–72). This adhesion is enhanced by the stimulation of the lymphocytes through CD3 or a number of other costimulatory molecules such as CD2, CD28, and CD43 (73–75). In addition, T cell activation by chemokines, such as MCP-1 and RANTES, may also increase LFA-1 avidity for its ligands on the endothelium (8).

In vivo studies have shown that treatment with anti-LFA-1 mAbs reduces the migration of peripheral lymph node lymphocytes to peripheral and mesenteric lymph nodes and Peyer's patches by 40% to 60% in the mouse (66). Our laboratory has shown that LFA-1 blockade also inhibits the migration of peripheral lymph node lymphocytes, mesenteric lymph node lymphocytes, and spleen T-cells to peripheral and mesenteric lymph nodes and Peyer's patches by about 50% (72). However, the migration of lymphoblasts and antigen-primed peritoneal exudate lymphocytes to peripheral nodes appears to be less dependent on LFA-1. Studies of lymphocyte migration into inflammation have shown that LFA-1 blockade strongly inhibits lymphocyte migration to cutaneous DTH reactions, LPS-induced inflammation, and to the T-cell-recruiting cytokines IFN-γ and TNFα. The contribution of LFA-1 appears to depend on the T-cell population migrating to these inflammatory sites. T lymphocytes from the spleen of normal animals are inhibited up to 80% by anti-LFA-1, while antigen-primed lymphocytes from an inflammatory exudate are substantially more LFA-1-independent in both their adhesion to endothelium in vitro and their migration to cutaneous inflammatory sites in vivo (72,76). In mice, treatment with anti-LFA-1 markedly reduced the ear swelling following challenge with a contact-sensitizing agent (77). The inflammatory infiltrate, which includes large numbers of neutrophils, was also significantly diminished. LFA-1 blockade in rats only partially inhibited the DTH- and IFN-γ-induced inflammation in the skin and lymphocyte and monocyte migration to these reactions (76,78). Very little neutrophil accumulation occurs to these inflammatory sites in the rat. The remainder of the lymphocyte migration to these reactions appears to be mediated through the VLA-4 integrin, since blocking both LFA-1 and VLA-4 virtually abolishes T-cell accumulation to these reactions (76).

The contribution of LFA-1 to lymphocyte migration to inflammation in other tissues is different. Anti-LFA-1 treatment alone was not able to inhibit T lymphocyte accumulation in inflamed joints in adjuvant arthritis in the rat, even though it strongly inhibited lymphocyte migration to the skin (79). Anti-CD18 inhibits inflammation in an acute rabbit arthritis model (80). Anti-LFA-1 treatment also did not inhibit lymphoblast migration into the parasite-infected gut, suggesting that LFA-1 may be important in infiltration of lymphocytes across specific endothelial cell beds (81).

Numerous studies have demonstrated the importance of LFA-1 in transplantation and graft survival (82–84). Prolongation of graft survival was shown by treatment with anti-LFA-1 in a number of studies using cardiac allografts (82,84).

There has also been extensive investigation of the role of ICAM-1 in various types of inflammation. Studies of lymphocyte endothelial cell adhesion showed that ICAM-1 is an important ligand for LFA-1 on T-cells and mediates T-cell adhesion to endothelial cells and transendothelial migration (85,86). Investigations of the role of ICAM-1 in inflammation have employed blocking mAbs and, in more recent studies, disruption of the ICAM-1 gene. Treatment with anti-ICAM-1 blocking mAb significantly prolongs renal allograft survival from 9 days to 24 days in cyanomegalus monkeys (87). Many other investigations of T-cell-mediated inflammation have utilized the combination of blocking both LFA-1 and ICAM-1. Treatment with the combination of anti-LFA-1 and anti-ICAM-1 shortly after birth prevented the development of diabetes in the NOD mouse (88). Antibody blockade of ICAM-1 and LFA-1 was shown also to induce specific tolerance to peripheral nerve allografts in rats (89). Dual blockade of LFA-1 and ICAM-1 could also produce indefinite cardiac allograft survival and tolerance in mice (90). However, similar studies in rats have led to accelerated allograft rejection (91). The reason for the longterm graft survival and tolerance in the mouse with this treatment and for the accelerated graft rejection in the rat is unclear.

ICAM-1-deficient mice, generated by gene targeted disruption, have an increase in the circulating blood lymphocyte count and a 50% decrease in the intensity of contact hypersensitivity reactions (92,93). The survival of cardiac allografts in ICAM-1 knockout mice was not different from controls, and there was no evidence of decreased mononuclear infiltration in the rejecting grafts (94). These studies suggest that lymphocyte mediated immune responses and lymphocyte infiltration can occur in the absence of ICAM-1 although anti-ICAM-1 can nevertheless have an anti-inflammatory effect. The contribution of ICAM-1 to the interaction of lymphocytes with endothelial cells in vivo in lymphocyte migration is not clear. Alternative ligands for LFA-1, such as ICAM-2, and other integrins on the lymphocyte may be able to substitute partially for the LFA-1 ICAM-1 adhesion pathway.

B. Very Late Activation Antigen-4 (VLA-4, $\alpha_4\beta_1$, CD49d/CD29)

VLA-4 is a member of the β_1 integrin family of adhesion molecules, which include VLA-1 to VLA-6 (95). Each of these integrins can bind to extracellular matrix proteins, such as collagen, laminin, and fibronectin. VLA-4 consists of a 150-kDa α chain noncovalently associated with a 130-kDa β_1 chain. It is expressed on lymphocytes, monocytes, eosinophils, and mast cells, and has recently been reported to be present at low levels on neutrophils (96,97). VLA-4 binds to the vascular cell adhesion molecule-1 (VCAM-1) and the CS-1 fragment of fibronectin and to thrombosponding (96,98–100). In addition, other ligands for $\alpha_4\beta_1$ include the bacterial outer membrane protein invasin and a cellular ligand which can mediate homotypic aggregation of lymphocytes stimulated by antibodies to $\alpha_4\beta_1$ (101,102).

VLA-4 can mediate the adhesion of blood lymphocytes to endothelial cells expressing VCAM-1 (103–105). Unstimulated endothelial cells do not express VCAM-1. TNFα, IL-1, IL-4, and LPS can stimulate endothelial cells to synthesize VCAM-1 and to express it on its luminal surface (67,106–108). Endothelial cells also express fibronectin containing the CS-1 binding site of VLA-4 on the luminal surface in some areas of chronic inflammation and lymphocytes bind to this endothelium (109). The binding of lymphocytes to VCAM-1 through VLA-4 appears to require a lower state of activation of this integrin than the adhesion to fibronectin regimens (110). In addition to mediating lymphocyte adhesion to endothelium, engagement of the VLA-4 by antibody or binding to its ligands can deliver costimulatory signals leading to T-cell proliferation, cytokine production, tyrosine phosphorylation, and production of a 72-kDa gelatinase by lymphocytes (111–113).

VCAM-1 is a member of the immunoglobulin superfamily of cell adhesion molecules and exists in several forms (9). The major form expressed on activated endothelial cells has seven domains and is 110 kDa. Domains 1 and 4 can bind to VLA-4 on lymphocytes. There is also an alternately spliced 6-domain form of VCAM-1 in which domain 4 has been spliced out resulting in only the one VLA-4 binding site on domain 1. The 7- and 6-domain forms of VCAM-1 are type 1 transmembrane proteins, but in mice and rats there is also a glycosylphosphotidylinositol-linked form of 3-domain VCAM-1 expressed on the endothelium (114).

Studies regarding the role of the VLA-4 VCAM-1 adhesion pathway in lymphocyte migration have relied largely on blocking mAbs. Targetted gene disruption of VCAM-1 has shown that the VLA-4 VCAM-1 adhesion pathway also mediates adhesion involved in uterine implantation, resulting in the VCAM-1 knockout being a lethal mutation (115). Antibody-blocking studies, however, have shown that lymphocytes from several species bind to VCAM-1 on endothelial cells in vitro and that lymphocyte activation by mitogens can enhance lymphocyte endothelial cell adhesion (116,117). Furthermore, lymphocytes from peripheral lymph nodes display relatively little adhesion to cytokine stimulated endothelium, but lymphocytes obtained from peripheral lymph nodes after stimulation with antigen are enriched in lymphoblasts which bind extremely well to VCAM-1 (118). Thus, antigen exposure in peripheral lymph nodes stimulates the generation of lymphoblasts after 3 to 4 days, and these cells express an activated form of VLA-4. These lymphoblasts exit the peripheral lymph node via the efferent lymphatics and enter the blood, where they will selectively encounter VCAM-1 expressed on endothelial cells in areas of inflammation. This may be an important mechanism for targetting both antigen-specific and recent polyclonally activated T cells responding to the inflammatory antigenic stimulus from the lymph node to the inflammatory site.

Several studies in animals have shown that VLA-4 blockade can inhibit inflammation. Our laboratory showed that the migration of radiolabeled lymphocytes from the blood to several types of inflammatory stimuli, including IFN-γ, TNF-α, classic delayed-type hypersensitivity reactions, and LPS, was inhibited by the VLA-4-blocking monoclonal antibody TA-2 (119). T-cell migration was inhibited by about 60%, and this inhibition was observed with activated lymphoblasts, antigen-stimulated peritoneal exudate lymphocytes, and resting T-cells. Treatment with anti-VLA-4 was also shown to decrease T-cell migration out of the circulation, causing a prolonged half-life of these lymphocytes in the blood. In addition, treatment with antibodies to α_4 induced lymphocytosis, probably by inhibiting lymphocyte migration out of the vasculature.

Investigation of lymphocyte migration to other T-cell-mediated inflammatory reactions have also shown an important role for VLA-4. Previous studies by us demonstrated that the migration of activated T-cells to inflamed joints of rats with adjuvant arthritis was inhibited by 40% to 60% by treatment with anti-VLA-4 mAb (79). Experimental allergic encephalomyelitis (EAE) is a T-cell-dependent autoimmune disease of the central nervous system. Anti-VLA-4 mAbs inhibit the adhesion of lymphocytes to blood vessels in frozen sections of the brain from animals with EAE, and treatment with anti-VLA-4 delayed the development of neurologic injury in EAE induced by adoptive transfer of immune T cells (120). Furthermore, adoptive transfer of disease correlated with the surface expression of α_4 on T-cells and could be blocked by treatment with anti-α_4 or anti-VCAM-1 (121). Our studies have shown that a single injection of anti-VLA-4 in animals with active disease can completely inhibit T lymphocyte migration into the central nervous system (122). Thus, there is a considerable amount of direct evidence from *in vivo* studies that VLA-4 can mediate lymphocyte migration to inflammatory reactions induced by T cells, IFN-γ, and TNF-α.

Additional studies have examined the effect of blocking the VLA-4 VCAM-1 interaction in a variety of other inflammatory disease models. T-cell recruitment to allergen induced tracheal inflammation in mice is partially dependent upon this pathway (123). Treatment with anti-VLA-4 antibody inhibited infiltration of the bronchial wall by T-cells in antigen-induced hypersensitivity reactions in guinea pigs (124).

There have also been dramatic anti-inflammatory effects of treatment with anti-VLA-4 antibodies. Anti-VLA-4 mAb treatment inhibits neurological injury in EAE, development of diabetes in NOD mice, and tissue injury in nephrotoxic or mercuric chloride-induced nephritis, and VLA-4 blockade 8 delays cardiac allograft rejection (120,121,125–128). Studies have also shown marked inhibition of allergic lung inflammation in several models. Anti-VLA-4 antibody treatment inhibited airway hyperresponsiveness in both a rat and a sheep model of asthma (129,130). This effect, at least partially, appears to be independent of inhibition of T-cell or eosinophil migration.

VCAM-1, one of the main endothelial cell ligands for VLA-4 is upregulated in areas of inflammation. Its expression is increased on endothelial cells in synovium from rheumatoid arthritis and osteoarthritis patients

(131). During acute graft rejection, VCAM-1 is also expressed on renal tubular epithelium and in several autoimmune disorders including vasculitis, Grave's disease, and thyroiditis (132–134). Expression of VCAM-1 was also reported on human cardiac allograft biopsies (135). Anti-VCAM-1 treatment was also shown to reduce adoptive transfer of EAE and prolong cardiac allografts in mice (136,137).

Our laboratory has also examined the effect of blocking both the VLA-4 and LFA-1 integrins on lymphocyte migration in inflammation. Blockade of VLA-4 and LFA-1 each alone inhibited T-cell migration to cutaneous sites of IFN-γ injection and to tuberculin delayed-type hypersensitivity reactions. Blocking both these integrins simultaneously virtually abolished (>98%) T-cell migration to these inflammatory reactions in the skin (76). This suggests that both VLA-4 and LFA-1 can mediate part of the T-cell migration to cutaneous inflammation, but blockade of both pathways prevents endothelial adhesion and migration to these inflammatory stimuli in the skin.

As outlined above, lymphocyte recruitment to inflammation in other tissues, such as in the central nervous system, inflamed joints, and lungs, appears to be mediated by VLA-4 and LFA-1 to a variable extent. The effect of blocking VLA-4 together with LFA-1 has also been examined in various allograft models. Treatment with anti-VLA-4 plus anti-LFA-1 did not significantly prolong cardiac allograft survival beyond that seen with anti-LFA-1 alone, but dramatically increased islet cell allograft survival and induced a degree of tolerance (138,139). Thus, blockade of these two integrin pathways in some tissues could virtually abolish T-cell accumulation, while in others only partial inhibition was observed. It is possible that additional integrins or integrin-independent adhesion pathways may be contributing in the latter types of inflammation.

C. α_4/β_7 (LPAM-1)

The integrin α_4, in addition to associating with β_1, can also form a functional receptor with β_7, namely $\alpha_4\beta_7$ (140,141). β_7 is a 120-kDa transmembrane protein, which can also pair with another α chain α_E to give $\alpha_E\beta_7$, an integrin found in association with mucosal intraepithelial lymphocytes (142,143). $\alpha_4\beta_7$ is expressed on virtually all lymphocytes in the newborn, and is widely expressed on T and B lymphocytes from adults (144). Naive lymphocytes uniformly express this integrin, while memory T-cells vary in their expression of $\alpha_4\beta_7$. Natural killer cells and eosinophils also express this receptor. $\alpha_4\beta_7$ was initially identified as a receptor-mediating adhesion of lymphocytes to Peyer's patch high endothelial venules on frozen sections (140,145). The ligand for β_7 in this adhesion is the mucosal addressin cell adhesion molecule (MAdCAM-1) (146). MAdCAM-1 is a 66-kDa protein that contains at least two domains that belong to the immunoglobulin superfamily and a serine threonine rich region that is heavily glycosylated and is related to the sialomucin family of proteins (147). $\alpha_4\beta_7$ mediates binding to MAdCAM-1 in the region of the immunoglobulin homologous domains. In addition, lymphocytes have also been shown to be able to bind to MAdCAM-1 through the interaction of L-selectin with carbohydrates in the sialomucin portion of the molecule (148).

Lymphocyte adhesion mediated by $\alpha_4\beta_7$ to MAdCAM-1 appears to occur without the requirement for specific cell activation (149). However, activation of $\alpha_4\beta_7$ by phorbol esters or Mn^{2+} can mediate binding of these lymphocytes to VCAM-1 and to the CS-1 region of fibronectin, similar to VLA-4 (147,149,150). Thus, it appears that under unstimulated conditions $\alpha_4\beta_7$ mediates adhesion to MAdCAM-1 on endothelium in intestinal associated tissues, and it can mediate lymphocyte adhesion to VCAM-1 and fibronectin by activated lymphocytes. $\alpha_4\beta_7$ has also been shown to mediate lymphocyte interaction with VCAM-1 and MAdCAM-1 under conditions of flow, suggesting a role for this receptor in mediating the initial stage of T-cell attachment to the endothelium (151).

Initial in vivo studies of the contribution of $\alpha_4\beta_7$ to lymphocyte migration came from investigations of the effect of anti-α_4 mAbs on lymphocyte migration to intestinal lymphoid tissues. Treatment of rats with the mAb, TA-2, which can bind to both $\alpha_4\beta_1$ and $\alpha_4\beta_7$, demonstrated that blockade of α_4 integrins completely inhibited T lymphocyte migration to Peyer's patches and reduced migration to mesenteric lymph nodes by 80% (119). Further studies with anti-α_4 blockade in mice and anti-β_7 mAbs have demonstrated that $\alpha_4\beta_7$ appears to be the major receptor mediating the homing of lymphocytes to these intestinal lymphoid tissues (152).

Evidence for a role of $\alpha_4\beta_7$ or MAdCAM-1 in lymphocyte migration to inflamed tissues is less clear. A murine brain endothelial cell line can be induced to express MAdCAM-1 by activation with TNFα, and the TK1 lymphoma which expresses $\alpha_4\beta_7$ can bind to MAdCAM-1 on these cytokine-activated endothelial cells (153). Anti-MAdCAM-1 mAb was also shown to react with endothelial cells in the spinal cord of mice with chronic relapsing EAE and in the blood vessels of the islets of Langerhans in NOD mice with insulitis (22,154). There is also evidence that the synovium in patients with rheumatoid arthritis have an increased proportion of $\alpha_4\beta_7$-expressing T-cells as compared to the percent of $\alpha_4\beta_7^+$ T-cells in the blood, suggesting preferential recruitment or at least increased expression of $\alpha_4\beta_7$ on T-cells in this inflammatory tissue (155). Similarly, infection with *Borrelia burgdorferi* in SCID mice have also been reported to increase the expression of MAdCAM-1 in the synovium (156). Thus, MAdCAM-1 may contribute to

T-cell adhesion to endothelium in selected inflammatory tissues, and $\alpha_4\beta_7$ may mediate migration to these sites. Since $\alpha_4\beta_7$ can also mediate binding of T-cells to VCAM-1 and alternately spliced fibronectin, in addition to mediating T-cell migration through adhesion to MAdCAM-1, $\alpha_4\beta_7$ may contribute to T-cell migration previously thought to involve only $\alpha_4\beta_1$. Further studies utilizing antibodies and other specific antagonists of $\alpha_4\beta_7$ will be required to investigate the relative contributions of $\alpha_4\beta_1$ and $\alpha_4\beta_7$ to T lymphocyte migration.

CONCLUSION

The studies described above demonstrate the considerable complexity of lymphocyte migration through normal and inflamed tissues. Although T- and B-cells appear to utilize the same adhesion receptors as other leukocytes for interaction with endothelial cells, the extensive differentiation of lymphocytes in the secondary lymphoid tissues, the role of many of the adhesion receptors as costimulatory molecules in lymphocyte activation and the heterogeneity of lymphocytes make the analysis of the pathways of lymphocyte migration a particular challenge. Nevertheless, there continues to be rapid progress in our understanding of these cellular interactions with the identification of new receptors and the appreciation of their roles.

REFERENCES

1. Gowans JL, Knight EJ. The recirculation of lymphocytes in the rat. Proc R Soc Lond Series B 1964; 159:257–282.
2. Hall MB, Morris B. The origin of the cells in the efferent lymph from a single lymph node. J Exp Med 1965; 121:901–910.
3. Smith JB, McIntosh GH, Morris B. The traffic of cells through tissues: a study of peripheral lymph in sheep. J Anat 1970; 107:87–100.
4. MacKay CR. T-cell memory: the connection between function, phenotype and migration pathways. Immunol Today 1991; 12:189–192.
5. MacKay CR, Marston WL, Dudler L. Naive and memory T cells show distinct pathways of lymphocyte recirculation. J Exp Med 1990; 171:801–817.
6. Issekutz TB, Chin W, Hay JB. Lymphocyte traffic through granulomas: difference in the recovery of Indium^{111}-labeled lymphocytes in afferent and efferent lymph. Cell Immunol 1980; 20:79–86.
7. MacKay CR, Marston W, Dudler L. Altered patterns of T cell migration through lymph nodes and skin following antigen challenge. Eur J Immunol 1992; 22:2205–2210.
8. Springer TA. Traffic signals for lymphocyte recirculation and leukocyte emigration: the multistep paradigm. Cell 1994; 76:301–314.
9. Carlos TM, Harlan JM. Leukocyte-endothelial adhesion molecules. Blood 1994; 84:2068–2101.

10. Oppenheim JJ, Zachariae COC, Mukaida N, Matsushima K. Properties of the novel proinflammatory supergene "intercrine" cytokine family. Annu Rev Immunol 1991; 9:617–648.
11. Miller MD, Krangel MS. Biology and biochemistry of the chemokines: a family of chemotactic and inflammatory cytokines. Crit Rev Immunol 1992; 12:17–46.
12. Newman PJ, Berndt MC, Gorski J, et al. PECAM-1 (CD31) Cloning and relation to adhesion molecules of the immunoglobulin gene superfamily. Science 1990; 247:1219–1222.
13. Muller WA, Weigl SA, Deng X, Phillips DM. PECAM-1 is required for transendothelial migration of leukocytes. J Exp Med 1993; 178:449–460.
14. Butcher EC, Scollay R, Weissman IL. Organ specificity of lymphocyte migration: mediation by highly selective lymphocyte interaction with organ-specific determinants on high endothelial venules. Eur J Immunol 1980; 10:556–561.
15. Lasky LA, Singer MS, Yednock TA, et al. Cloning of a lymphocyte homing receptor reveals a lectin domain. Cell 1989; 56:1045–1055.
16. Tedder TF, Isaacs CM, Ernst TJ, Demetri GD, Adler DA. Isolation and chromosomal localization of cDNAs encoding a novel human lymphocyte cell surface molecule, LAM-1. Homology with the mouse lymphocyte homing receptor and other human adhesion proteins. J Exp Med 1989; 170:123–133.
17. Picker LJ, Terstappen LWMM, Rott LS, Streeter PR, Stein H, Butcher EC. Differential expression of homing-associated adhesion molecules by T-cell subsets in man. J Immunol 1990; 145:3247–3255.
18. Lewinsohn DM, Bargatze RF, Butcher EC. Leukocyte-endothelial cell recognition: evidence of a common molecular mechanism shared by neutrophils, lymphocytes, and other leukocytes. J Immunol 1987; 138:4313–4321.
19. Gallatin WM, Weissman IL, Butcher EC. A cell-surface molecule involved in organ-specific homing of lymphocytes. Nature 1983; 304:30–34.
20. Hamann A, Jablonski-Westrich D, Jonas P, Thiele H-G. Homing receptors reexamined: mouse LECAM-1 (MEL-14 antigen) is involved in lymphocyte migration into gut-associated lymphoid tissue. Eur J Immunol 1991; 21:2925–2929.
21. Arbones ML, Ord DC, Ley K, et al. Lymphocyte homing and leukocyte rolling and migration are impaired in L-selectin-deficient mice. Immunity 1994; 1:247–260.
22. Hánninen A, Taylor C, Streeter PR, et al. Vascular addressins are induced on islet vessels during insulitis in nonobese diabetic mice and are involved in lymphoid cell binding to islet endothelium. J Clin Invest 1993; 92:2509–2515.
23. Lepault F, Gagnerault MC, Faveeuw C, Boitard C. Recirculation, phenotype and functions of lymphocytes in mice treated with monoclonal antibody MEL-14. Eur J Immunol 1994; 24:3106–3112.
24. Turunen JP, Paavonen T, Majuri M-L, et al. Sialyl Lewis^x- and L-selectin-dependent site-specific lymphocyte extravasation into renal transplants during acute rejection. Eur J Immunol 1994; 24:1130–1136.
25. Baumhueter S, Dybdal N, Kyle C, Lasky LA. Global vascular expression of

murine CD34, a sialomucin-like endothelial ligand for L-selectin. Blood 1994; 84:2554–2565.
26. Pizcueta P, Luscinskas FW. Monoclonal antibody blockade of L-selectin inhibits mononuclear leukocyte recruitment to inflammatory sites in vivo. Am J Pathol 1994; 145:461–469.
27. Tedder TF, Steeber DA, Pizcueta P. L-selectin-deficient mice have impaired leukocyte recruitment into inflammatory sites. J Exp Med 1995; 181:2259–2263.
28. Stenberg PE, McEver RP, Shuman MA, Jacques YV, Bainton DF. A platelet alpha-granule membrane protein (GMP-140) is expressed on the plasma membrane after activation. J Cell Biol 1985; 101:880–886.
29. Berman CL, Yeo EL, Wencel-Drake JD, Furie BC, Ginsberg MH, Furie B. A platelet alpha granule protein that is associated with the plasma membrane after activation. J Clin Invest 1986; 78:130–137.
30. McEver RP, Beckstead JH, Moore KL, Marshall-Carlson L, Bainton DF. GMP-140, a platelet alpha-granule protein, is also synthesized by vascular endothelial cells and is localized in Weibel-Palade bodies. J Clin Invest 1986; 84:92–99.
31. Bonfonti R, Furie BC, Furie B, Wagner DD. PADGEM (GMP140) is a component of Weibel-Palade bodies of human endothelial cells. Blood 1989; 73:1109–1112.
32. Toothill VJ, Van Mourik JA, Niewenhuis HK, Metzelaar MJ, Pearson JD. Characterization of the enhanced adhesion of neutrophil leukocytes to thrombin-stimulated endothelial cells. J Immunol 1990; 145:283–291.
33. Gotsch U, Jager U, Dominis M, Vestweber D. Expression of P-selectin on endothelial cells is upregulated by LPS and TNF-alpha in vivo. Cell Ad Comm 1994; 2:7–14.
34. Moore KL, Stults NL, Diaz S, Smith DF, Cummings RD, Varki A, McEver RP. Identification of a specific glycoprotein ligand for P-selectin (CD62) on myeloid cells. J Cell Biol 1992; 118:445–456.
35. Sako D, Chang X-J, Barone KM, et al. Expression cloning of a functional glycoprotein ligand for P-selectin. Cell 1993; 75:1179–1186.
36. Lawrence MB, Springer TA. Leukocytes roll on a selectin at physiologic flow rates: distinction from and prerequisite for adhesion through integrins. Cell 1991; 65:859–873.
37. Geng JG, Bevilacqua MP, Moore KL, et al. Rapid neutrophil adhesion to activated endothelium mediated by GMP-140. Nature 1990; 343:757–760.
38. Hamburger SA, McEver RP. GMP-140 mediates adhesion of stimulated platelets to neutrophils. Blood 1990; 75:550–554.
39. Kunzendorf U, Notter M, Hock H, Distler A, Diamanstein T, Walz G. T cells bind to the endothelial adhesion molecule GMP-140 (P-selectin). Transplantation 1993; 556:1213–1217.
40. Alon R, Rossiter H, Wang X, Springer TA, Kupper TS. Distinct cell surface ligands mediate T lymphocyte attachment and rolling on P and E selectin under physiological flow. J Cell Biol 1994; 127:1485–1495.

41. Rossiter H, Van Reijsen F, Mudde GC, et al. Skin disease-related T cells bind to endothelial selectins: expression of cutaneous lymphocyte antigen (CLA) predicts E-selectin but not P-selectin binding. Eur J Immunol 1994; 24:205–210.
42. Damle NK, Klussman K, Dietsch MT, Mohagheghpour N, Aruffo A. GMP-140 (P-selectin/CD62) binds to chronically stimulated but not resting $CD4^+$ T lymphocytes and regulates their production of proinflammatory cytokines. Eur J Immunol 1992;22:1789–1793.
43. Grober JS, Bowen BL, Ebling H, et al. Monocyte-endothelial adhesion in chronic rheumatoid arthritis. J Clin Invest 1993; 91:2609–2619.
44. Mayadas TN, Johnson RC, Rayburn H, Hynes RO, Wagner DD. Leukocyte rolling and extravasation are severely compromised in P selectin-deficient mice. Cell 1993; 74:541–554.
45. Subramaniam M, Saffaripour S, Watson SR, Mayadas TN, Hynes RO, Wagner DD. Reduced recruitment of inflammatory cells in a contact hypersensitivity response in P-selectin-deficient mice. J Exp Med 1995; 181:2277–2282.
46. Bevilacqua MP, Pober JS, Mendrick DL, Cotran RS, Gimbrone MA Jr. Identification of an inducible endothelial-leukocyte adhesion molecule. Proc Natl Acad Sci USA 1987; 84:9238–9242.
47. Berg EL, Robinson MK, Mansson O, Butcher EC, Magnani JL. A carbohydrate domain common to both sialyl Le^a and sialyl Le^x is recognized by the endothelial cell leukocyte adhesion molecule ELAM-1. J Biol Chem 1991; 266:14869–14872.
48. Walz G, Aruffo A, Kolanus W, Bevilacqua M, Seed B. Recognition by ELAM-1 of the sialyl-Le^x determinant of myeloid and tumor cells. Science 1990; 250:1132–1135.
49. Munro JM, Lo SK, Corless C, et al. Expression of sialyl-Lewis x, an E-selectin ligand, in inflammation, immune processes, and lymphoid tissues. Am J Pathol 1992; 141:1397–1408.
50. Levinovitz A, Mühlhoff J, Isenmann I, Vestweber D. Identification of a glycoprotein ligand for E-selectin in mouse myeloid cells. J Cell Biol 1993; 121:449–459.
51. Steegmaler M, Levinovitz A, Isenmann S, et al. The E-selectin-ligand ESL-1 is a variant of a receptor for fibroblast growth factor. Nature 1995; 373:615–620.
52. Picker LJ, Warnock RA, Burns AR, Doerschuk CM, Berg EL, Butcher EC. The neutrophil selectin LECAM-1 presents carbohydrate ligands to the vascular selectins ELAM-1 and GMP-140. Cell 1991; 66:921–933.
53. Picker LJ, Kishimoto TK, Smith CW, Warnock RA, Butcher EC. ELAM-1 is an adhesion molecule for skin-homing T cells. Nature 1991; 349:796–799.
54. Berg EL, Yoshino T, Rott LS, et al. The cutaneous lymphocyte antigen is a skin lymphocyte homing receptor for the vascular lectin endothelial cell-leukocyte adhesion molecule 1. J Exp Med 1991; 174:1461–1466.
55. Picker LJ, Michie SA, Rott LS, Butcher EC. A unique phenotype of skin-

associated lymphocytes in humans: preferential expression of the HECA-452 epitope by benign and malignant T cells at cutaneous sites. Am J Pathol 1990; 136:1053–1068.

56. Silber A, Newman W, Sasseville VG, et al. Recruitment of lymphocytes during cutaneous delayed hypersensitivity in nonhuman primates is dependent on E-selectin and vascular cell adhesion molecule 1. J Clin Invest 1994; 93:1554–1563.
57. Walcheck B, Watts G, Jutila MA. Bovine gamma/delta T cells bind E-selectin via a novel glycoprotein receptor: first characterization of a lymphocyte/E-selectin interaction in an animal model. J Exp Med 1993; 178(3):853–863.
58. Labow MA, Norton CR, Rumberger JM, et al. Characterization of E-selectin deficient mice: demonstration of overlapping function of endothelial selectins. Immunity 1994; 1:709–720.
59. Arnaout MA. Structure and function of the leukocyte adhesion molecules CD11/CD18. Blood 1990; 75:1037–1050.
60. Kishimoto TK, Anderson DC. The role of integrins in inflammation. In: Gallatin JI, Goldstein IM, Snyderman R, eds. Inflammation: Basic Principles and Clinical Correlates (ed 2). New York: Raven, 1992:353–378.
61. Springer TA. Adhesion receptors of the immune system. Nature 1990; 346: 425–434.
62. Marlin D, Springer TA. Purified intercellular adhesion molecule-1 (ICAM-1) is a ligand for lymphocyte function-associated antigen 1 (LFA-1). Cell 1987; 51:813–819.
63. Staunton DE, Dustin ML, Springer TA. Functional cloning of ICAM-2, a cell adhesion ligand for LFA-1 homologous to ICAM-1. Nature 1989; 339: 61–64.
64. DeFougerolles AR, Springer TA. Intercellular adhesion molecule 3, a third adhesion counter-receptor for lymphocyte function-associated molecule 1 on resting lymphocytes. J Exp Med 1992; 175:185–190.
65. Haskard D, Cavender D, Beatty P, Springer T, Ziff M. T lymphocyte adhesion to endothelial cells: mechanisms demonstrated by anti-LFA-1 monoclonal antibodies. J Immunol 1986; 137:2901–2906.
66. Hamann A, Jablonski-Westrich D, Duijvestijn A, et al. Evidence for an accessory role of LFA-1 in lymphocyte-high endothelium interaction during homing. J Immunol 1988; 140:693–699.
67. Pober JS, Gimbrone MA Jr, Lapierre LA, et al. Overlapping patterns of activation of human endothelial cell by interleukin-1, tumor necrosis factor and immune interferon. J Immunol 1986; 137:1893–1896.
68. Dustin ML, Rothlein R, Bhan AK, Dinarello CA, Springer TA. Induction by IL-1 and interferon-g: tissue distribution, biochemistry, and function of a natural adherence molecule (ICAM-1). J Immunol 1986; 137:245–254.
69. De Fougerolles AR, Stacker SA, Schwarting R, Springer TA. Characterization of ICAM-2 and evidence for a third counter-receptor for LFA-1. J Exp Med 1991; 174:253–267.
70. Dustin ML, Springer TA. Lymphocyte associated antigen-1 (LFA-1) interac-

tion with intercellular adhesion molecule-1 (ICAM-1) is one of at least three mechanisms for lymphocyte adhesion to cultured endothelial cells. J Cell Biol 1988; 107:321-331.
71. Makgoba MW, Sanders ME, Luce GE, et al. ICAM-1 a ligand for LFA-1 dependent adhesion of B, T and myeloid cells. Nature 1988; 331:86-88.
72. Issekutz TB. Inhibition of lymphocyte endothelial adhesion and in vivo lymphocyte migration to cutaneous inflammation by TA-3, a new monoclonal anitbody to rat LFA-1. J Immunol 1992; 149:3394-3402.
73. Dustin ML, Springer TA. T cell receptor cross-linking transiently stimulates adhesiveness through LFA-1. Nature 1989; 341:619-624.
74. Van Seventer GA, Shimizu Y, Horgan KJ, Luce GEG, Webb D, Shaw S. Remote T cell co-stimulation via LFA-1/ICAM-1 and CD2/LFA-3: demonstration with immobilized ligand/mAb and implication in monocyte-mediated co-stimulation. Eur J Immunol 1991; 21:1711-1718.
75. Shimizu Y, Van Seventer GA, Ennis E, Newman W, Horgan KJ, Shaw S. Crosslinking of the T cell-specific accessory molecules CD7 and CD28 modulates T cell adhesion. J Exp Med 1992; 175:577-582.
76. Issekutz TB. Dual inhibition of VLA-4 and LFA-1 maximally inhibits cutaneous delayed-type hypersensitivity-induced inflammation. Am J Pathol 1993; 143:1286-1293.
77. Scheynius A, Camp RL, Pure E. Reduced contact sensitivity reactions in mice treated with monoclonal antibodies to leukocyte function-associated molecule-1 and intercellular adhesion molecule-1. J Immunol 1993; 150:655-663.
78. Issekutz TB. In vivo blood monocyte migration to acute inflammatory reactions, IL-1a, TNFa, IFN-t, and C5a utilizes LFA-1, Mac-1 and VLA-: The relative importance of each integrin. J Immunol 1995; 154:6533-6540.
79. Issekutz TB, Issekutz AC. T lymphocyte migration to arthritic joints and dermal inflammation in the rat: differing migration patterns and the involvement of VLA-4. Clin Immunol Immunopathol 1991; 61:436-447.
80. Jasin HE, Lightfoot E, Davis LS, Rothlein R, Faanes RB, Lipsky PE. Amelioration of antigen-induced arthritis in rabbits treated with monoclonal antibodies to leukocyte adhesion molecules. Arthritis Rheum 1992; 35:541-549.
81. Bell RG, Issekutz TB. Expression of a protective intestinal immune response can be inhibited at three distinct sites by treatment with anti-a4 integrin. J Immunol 1993; 151:4790-4802.
82. Nakakura EK, McCabe SM, Zheng B, et al. Potent and effective prolongation by anti-LFA-1 monoclonal antibody monotherapy of non-primarily vascularized heart allograft survival in mice without T cell depletion. Transplantation 1993; 55:412-417.
83. Talento A, Nguyen M, Blake T, et al. A single administration of LFA-1 antibody confers prolonged allograft survival. Transplantation 1993; 55:418-422.
84. Cvazzana-Calvo M, Sarnicki S, Haddad E, et al. Prevention of bonemarrow and cardiac graft rejection in an H-2 haplotype disparate mouse combination by an anti-LFA-1 antibody. Transplantation 1995; 59:1576-1582.

85. Oppenheimer-Marks N, Davis LS, Lipsky PE. Human T lymphocyte adhesion to endothelial cells and transendothelial migration. J Immunol 1990; 145:140–148.
86. Oppenheimer-Marks N, Davis LS, Bogue DT, Ramberg J, Lipsky PE. Differential utiliztion of ICAM-1 and VCAM-1 during the adhesion and transendothelial migration of human T lymphocytes. J Immunol 1991; 147:2913–2921.
87. Cosimi AB, Conti D, Delmonico FL, et al. In vivo effects of monoclonal antibody to ICAM-1 (CD54) in nonhuman primates with renal allografts. J Immunol 1990; 144:4604–4612.
88. Hasegawa Y, Yokono K, Taki T, et al. Prevention of autoimmune insulin-dependent diabetes in non-obese diabetic mice by anti-LFA-1 and anti-ICAM-1 mAb. Int Immunol 1994; 6:831–838.
89. Nakao Y, MacKinnon SE, Hertl MC, Miyasaka M, Hunter DA, Mohakumar T. Monoclonal antibodies against ICAM-1 and LFA-1 prolong nerve allograft survival. Musc Nerve 1995; 18:93–102.
90. Isobe M, Yagita H, Okumura K, Ihara A. Specific acceptance of cardiac allograft after treatment with antibodies to ICAM-1 and LFA-1. Science 1992; 255:1125–1127.
91. Morikawa M, Tamatani T, Miyasaka M, Uede T. Cardiac allografts in rat recipients with simultaneous use of anti-ICAM-1 and anti-LFA-1 monoclonal antibodies leads to accelerated graft loss. Immunopharmacology 1994; 28:171–182.
92. Sligh JD, Ballantyne CM, Rich SS, et al. Inflammatory and immune responses are impaired in mice deficient in intercellar adhesion molecule 1. Proc Natl Acad Sci USA 1993; 90:8529–8533.
93. Xu H, Gonzalo JA, St Pierre Y, et al. Leukocytosis and resistance to septic shock in intercellular adhesion molecule 1-deficient mice. J Exp Med 1994; 180:95–109.
94. Schowengerdt KO, Zhu JY, Stepkowski SM, Tu Y, Entman ML, Ballantyne CM. Cardiac allograft survival in mice deficient in intercellular adhesion molecule-1. Circulation 1995; 92:82–87.
95. Hynes RO. Integrins: Versatility, modulation, and signaling in cell adhesion. Cell 1992; 69:11–25.
96. Hemler ME, Elices MJ, Parker C, Takada Y. Structure of the integrin VLA-4 and its cell-cell and cell-matrix adhesion functions. Immunol Rev 1990; 114:45–65.
97. Kubes P, Niu X-F, Smith CW, Kehrli ME Jr, Reinhardt PH, Woodman RC. A novel b1-dependent adhesion pathway on neutrophils: a mechanism invoked by dihydrocytochalasin B or endothelial transmigration. FASEB J 1995; 9:1103–1111.
98. Osborn L, Vassallo C, Benjamin CD. Activated endothelium binds lymphocytes through a novel binding site in the alternately spliced domain of vascular cell adhesion molecule-1. J Exp Med 1992; 176:99–107.
99. Vonderheide RH, Springer TA. Lymphocyte adhesion through very late antigen 4: evidence for a novel binding site in the alternatively spliced domain of

vascular cell adhesion molecule 1 and an additional a4 integrin counter-receptor on stimulated endothelium. J Exp Med 1992; 175:1433–1442.
100. Yabkowitz R, Dixit VM, Guo N, Roberts DD, Shimizu Y. Activated T-cell adhesion to thrombospondin is mediated by the alpha-4/beta-1 (VLA-4) and alpha-5/beta-1 (VLA-5) integrins. J Immunol 1993; 151:149–158.
101. Ennis E, Isberg RR. Shimizu Y. Very late antigen 4-dependent adhesion and costimulation of resting human T cells by the bacterial b1 integrin ligand invasin. J Exp Med 1993; 177:207–212.
102. Takada Y, Elices MJ, Crouse C, Hemler ME. The primary structure of the a4 subunit of VLA-4: homology to other integrins and a possible cell-cell adhesion function. EMBO J 1989; 8:1361–1368.
103. Elices MJ, Osborn L, Takada Y, et al. VCAM-1 on activated endothelium interacts with the leukocyte integrin VLA-4 at a site distinct from the VLA-4/fibronectin binding site. Cel. 1990; 60:577–584.
104. Chan BMC, Elices MJ, Murphy E, Hemler ME. Adhesion to vascular cell adhesion molecule 1 and fibronectin. Comparison of $\alpha_4\beta_1$ (VLA-4) and $\alpha_4\beta_7$ on the human B cell line JY. J Biol Chem 1992; 267:8366–8370.
105. Issekutz TB, Wykretowicz A. Effect of a new monoclonal antibody, TA-2, that inhibits lymphocyte adherence to cytokine stimulated endothelium in the rat. J Immunol 1991; 147:109–116.
106. Rice GE, Munro JM, Bevilacqua MP. Inducible cell adhesion molecule 110 (INCAM-110) is an endothelial receptor for lymphocytes. A CD11/CD18-independent adhesion mechanism. J Exp Med 1990; 171:1369–1374.
107. Thornhill MH, Kyan-Aung U, Haskard DO. Il-4 increases human endothelial cell adhesiveness for T cells but not for neutrophils. J Immunol 1990; 144: 3060–3065.
108. Thornhill MH, Wellicome SM, Mahiouz DL, Lanchbury JSS, Kyan-Aung U, Haskard DO. Tumor necrosis factor combines with IL-4 or IFN-gamma to selectively enhance endothelial cell adhesiveness for T cells: the contribution of vascular cell adhesion molecule-1-dependent and -independent binding mechanisms. J Immunol 1991; 146:592–598.
109. Elices MJ, Tsai V, Strahl D, et al. Expression and functional significance of alternatively spliced CS1 fibronectin in rheumatoid arthritis microvasculature. J Clin Invest 1994; 93:405–416.
110. Masumoto A, Hemler ME. Multiple activation states of VLA-4. Mechanistic differences between adhesion to CS1/fibronectin and to vascular cell adhesion molecule-1. J Biol Chem 1993; 268:228–234.
111. Yurochko AD, Liu DY, Eierman D, Haskill S. Integrins as a primary signal transduction molecule regulating monocyte immediate-early gene induction. Proc Natl Acad Sci USA 1992; 89:9034–9038.
112. Romanic AM, Madri JA. The induction of 72-kD gelatinase in T cells upon adhesion to endothelial cells is VCAM-1 dependent. J Cell Biol 1994; 125: 1165–1178.
113. Nojima Y, Rothstein DM, Sugita K, Schlossman SF, Morimoto C. Ligation of VLA-4 on T cells stimulates tyrosine phosphorylation of a 105-kD protein. J Exp Med 1992; 175:1045–1053.

114. Moy P, Lobb R, Tizard R, Olson D, Hession C. Cloning of an inflammation-specific phosphatidyl inositol-linked form of murine vascular cell adhesion molecule-1. J Biol Chem 1993; 268:8835–8841.
115. Gurtner GC, Davis V, Li H, McCoy MJ, Sharpe A, Cybulski MI. Targeted disruption of the murine VCAM1 gene: essential role of VCAM-1 in chorioallantoic fusion and placentation. Genes Dev 1995; 9:1–14.
116. Shimizu Y, Newman W, Gopal TV, et al. Four molecular pathways of T cell adhesion to endothelial cells: roles of LFA-1, VCAM-1, and ELAM-1 and changes in pathway hierarchy under different activation conditions. J Cell Biol 1991; 113:1203–1212.
117. Shimizu Y, Van Seventer GA, Horgan KJ, Shaw S. Regulated expression and binding of three VLA (β1) integrin receptors on T cells. Nature 1990; 345: 250–253.
118. Issekutz TB. Effect of antigen challenge on lymph node lymphocyte adhesion to vascular endothelial cells and the role of VLA-4 in the rat. Cell Immunol 1991; 138:300–312.
119. Issekutz TB. Inhibition of in vivo lymphocyte migration to inflammation and homing to lymphoid tissues by the TA-2 monoclonal antibody: A likely role for VLA-4 in vivo. J Immunol 1991; 147:4178–4184.
120. Yednock TA, Cannon C, Fritz LC, Sanchez-Madrid F, Steinman L, Karin N. Prevention of experimental autoimmune encephalomyelitis by antibodies against a4b1 integrin. Nature 1992; 356:63–66.
121. Baron JL, Reich E-P, Visintin I, Janeway CA Jr. The pathogenesis of adoptive murine autoimmune diabetes requires an interaction between α4-integrins and vascular cell adhesion molecule-1. J Clin Invest 1994; 93:1700–1708.
122. Issekutz TB. Effect of anti-LFA-1 and anti-VLA-4 on T lymphocyte migration to skin, joint, and CNS inflammation and lymph nodes. Int Cong Immunol 1992; 8:288. Abstract.
123. Nakajima H, Sano H, Nashimura T, Yoshida S, Iwamoto T. Role of VCAM-1/VLA-4 and ICAM-1/LFA-1 interactions in antigen induced eosinophil and T cell recruitment into the tissue. J Exp Med 1994; 179:1145–1154.
124. Pretolani M, Ruffie C, Lapa e Silva JR, Joseph D, Lobb RR, Vargaftig BB. Antibody to very late activation antigen 4 prevents antigen-induced bronchial hyperractivity and cellular infiltration in the guinea pig. J Exp Med 1994; 180:795–805.
125. Mulligan MS, Johnson KJ, Todd RF III, et al. Requirements for leukocyte adhesion molecules in nephrotoxic nephritis. J Clin Invest 1993; 91:577–587.
126. Yang XD, Michie SA, Tisch R, Karin N, Steinman L, McDewitt HO. A predominant role of the integrin alpha-4 in the spontaneous development of autoimmune diabetes in nonobese diabetic mice. Proc Natl Acad Sci USA 1994; 91:12604–12608.
127. Molina A, Sanchez-Madrid F, Brico T, et al. Prevention of mercuric chloride-induced nephritis in the brown Norway rat by treatment with antibodies against alpha-4 integrins. J Immunol 1994; 153:2313–2320.
128. Paul LC, Davidoff A, Paul DW, Benediktsson H, Issekutz TB. Monoclonal

antibodies against LFA-1 and VLA-4 inhibit graft vasculitis in rat cardiac allografts. Transplant Proc 1993; 25:813–814.
129. Rabb HA, Olivenstein R, Issekutz TB, et al. The role of the leukocyte adhesion molecules VLA-4, LFA-1, and Mac-1 in allergic airway responses in the rat. Am J Respir Crit Care Med 1994; 149:1186–1191.
130. Abraham WM, Seilczak MW, Ahmed A, et al. $\alpha 4$ integrins mediate antigen-induced late bronchial responses and prolonged airway hyperresponsiveness in sheep. J Clin Invest 1994; 93:776–787.
131. Koch AE, Burrows JC, Haines GK, Carlos TM, Harlan JM, Leibovich SJ. Immunolocalization of endothelial and leukocyte adhesion molecules in human rheumatoid and osteoarthritic synovial tissues. Lab Invest 1991; 64:313–320.
132. Lin Y, Kirby JA, Browell DA, et al. Renal allograft rejection: expression and function of VCAM-1 on tubular epithelial cells. Clin Exp Immunol 1993; 92: 145–151.
133. Burrows MP, Molina FA, Terenghi G, et al. Comparison of cell adhesion molecule expression in cutaneous leucocytoclastic and lymphocytic vasculitis. J Clin Pathol 1994; 47:939–944.
134. Marazuela M, Postigo AA, Acevedo A, et al. Adhesion molecules from the LFA-1/ICAM-1,3 and VLA-4/VCAM-1 pathways on T lymphocytes and vascular endothelium in Graves' and Hashimoto's thyroid glands. Eur J Immunol 1994; 24:2483–2490.
135. Herskowitz A, Mayne AE, Willoughby SB, Kanter K, Ansari AA. Patterns of myocardial cell adhesion molecule expression in human endomyocardial biopsies after cardiac transplantation. Induced ICAM-1 and VCAM-1 related to implantation and rejection. Am J Pathol 1994; 145:1082–1094.
136. Baron JL, Madri JA, Ruddle NH, Hashim G, Janeway CA Jr. Surface expression of $\alpha 4$ integrin by CD4 T cells is required for their entry into brain parenchyma. J Exp Med 1993; 177:57–68.
137. Isobe M, Suzuki J, Yagita H, et al. Immunosuppression to cardiac allografts and soluble antigens by anti-vascular cell adhesion molecule-1 and anti-very late antigen-4 monoclonal antibodies. J Immunol 1994; 153:5810–5818.
138. Paul LC, Davidoff A, Issekutz TB. The efficacy of LFA-1 and VLA-4 antibody treatment in rat vascularized cardiac allograft rejection. Transplantation 1993; 55:1196–1199.
139. Yang H, Issekutz TB, Wright JR. Prolongation of rat islet allograft survival by treatment with monoclonal antibodies against VLA-4 and LFA-1. Transplantation 1995; 60:71–76.
140. Holzmann B, McIntyre BW, Weissman IL. Identification of a murine Peyer's patch-specific lymphocyte homing receptor as an integrin molecule with an a chain homologous to human VLA-4α. Cell 1989; 56:37–46.
141. Erle DJ, Ruegg C, Sheppard D, Pytela R. Complete amino acid sequence of an integrin β subunit identified in leukocytes. J Biol Chem 1991; 266:11009–11016.
142. Parker CM, Cepek KL, Russell GJ, et al. A family of β_7 integrins on human mucosal lymphocytes. Proc Natl Acad Sci USA 1992; 89:1924–1928.

143. Parker CM, Cepek KL, Russell GJ, et al. A family of $\beta 7$ integrins on human mucosal lymphocytes. Proc Natl Acad Sci USA 1992; 89:1924–1928.
144. Erle DJ, Briskin MJ, Butcher EC, Garcia-Pardo A, Lazarovits AI, Tidswell M. Expression and function of the MAdCAM-1 receptor, integrin $\alpha 4\beta 7$, on human leukocytes. J Immunol 1994; 153:517–528.
145. Hu MC-T, Crowe DT, Weissman IL, Holzmann B. Cloning and expression of mouse integrin $\beta_p(\beta_7)$: a functional role in Peyer's patch-specific lymphocyte homing. Proc Natl Acad Sci USA 1992; 89:8254–8258.
146. Berlin C, Lang EL, Briskin MJ, et al. $\alpha 4\beta 7$ integrin mediates lymphocyte binding to the mucosal vascular addressin MAdCAM-1. Cell 1993; 74:1–20.
147. Briskin MJ, McEvoy M, Butcher EC. MAdCAM-1 has homology to immunoglobulin and mucin-like adhesion receptors and to IgA1. Nature 1993; 363: 461–464.
148. Berg EL, McEvoy LM, Berlin C, Bargatze RF, Butcher EC. L-selectin-mediated lymphocyte rolling on MAdCAM-1. Nature 1993; 366:695–698.
149. Andrew DP, Berlin C, Honda S, et al. Distinct but overlapping epitopes are involved in alpha4beta7-mediated adhesion to vascular cell adhesion molecule-1, mucosal addressin-1, fibronectin, and lymphocyte aggregation. J Immunol 1994; 153:3847–3861.
150. Strauch UG, Lifka A, Gosslar U, Kilshaw PJ, Clements J, Holzmann B. Distinct binding specificities of integrins alpha-4/beta-7 (LPAM-1), alpha-4/beta-1 (VLA-4) and alpha-IEL/beta-7. Int Immunol 1994; 6:263–275.
151. Berlin C, Bargatze RF, Campbell JJ, et al. Alpha 4 integrins mediate lymphocyte attachment and rolling under physiologic flow. Cell 1995; 80:413–422.
152. Hamann A, Andrew DP, Jablonski-Westrich D, Holzmann B, Butcher EC. Role of α_4-integrins in lymphocyte homing to mucosal tissues in vivo. J Immunol 1994; 152:3282–3293.
153. Sikorski EE, Hallmann R, Berg EL, Butcher EC. The Peyer's patch high endothelial receptor for lymphocytes, the mucosal vascular addressin, is induced on a murin endothelial cell line by tumor necrosis factor-α and IL-1. J Immunol 1993; 151:5239–5250.
154. O'Neill JK, Butter C, Baker D, et al. Expression of vascular addressins and ICAM-1 by endothelial cells in the spinal cord during chronic relapsing experimental allergic encephalomyelitis in the Biozzi AB/H mouse. Immunology 1991; 72:520–525.
155. Lazarovits AI, Karsh J. Differential expression in rheumatoid synovium and synovial fluid of $\alpha 4\beta 7$ integrin. J Immunol 1993; 151:6482–6489.
156. Schaible UE, Vestweber D, Butcher EC, Stehle T, Simon MM. Expression of endothelial cell adhesion molecules in joints and heart during *Borrelia burgdorferi* infection of mice. Cell Ad Comm 1994; 2:465–479.

12

P-Selectin and Its Ligands: Structure and Function

Carol T. Mei, Colleen Sweeney Crovello, Barbara C. Furie, and Bruce Furie
New England Medical Center, Tufts University School of Medicine, Boston, Massachusetts

P-selectin was discovered as a result of efforts to identify neoantigenic determinants expressed on the activated platelet but not on the unstimulated platelet. These investigations led to the discovery in 1984 of the cell adhesion molecule now known as P-selectin (PADGEM, GMP-140, CD62). This protein is the antigenic target of monoclonal antibodies KC4 and S12 directed against an activation-dependent platelet cell surface structure (1,2). P-selectin is an integral membrane protein with a single polypeptide chain of 140 kDa and is expressed at approximately 13,000 surface molecules per platelet. Surface expression is secretion-dependent. Immunoelectron microscopy and subcellular organelle fractionation studies localized P-selectin to alpha granules of resting platelets and showed translocation to the cell surface upon thrombin activation (3,4).

Although originally discovered in platelets, P-selectin is also synthesized in endothelial cells (5,6). Immunofluorescence double staining of permeabilized human umbilical vein endothelial cells with the KC4 antibody and anti-Von Willebrand factor antibody shows colocalization to the Weibel-Palade bodies, the endothelial cell storage granules. As in platelets, P-selectin migrates to the surface on cell activation, but surface expression is 10-fold lower than in platelets.

I. ROLE AS AN ADHESION MOLECULE

When P-selectin was cloned in 1989 (7), it demonstrated a high degree of structural homology with two other proteins involved in cell adhesion,

E-selectin and L-selectin. E-selectin was known to mediate the interaction of interleukin-1 (IL-1) or TNF-treated endothelium with monocytes and neutrophils (8); L-selectin was implicated in lymphocyte adhesion to high endothelial venules of peripheral lymph nodes (9). As platelets had previously been shown to interact with monocytes and neutrophils (10), P-selectin was proposed to mediate the interaction of platelets and endothelium with monocytes and neutrophils (11). Functional homology of P-selectin with E-selectin and L-selectin as an adhesion molecule was demonstrated by Larsen and co-workers in 1989 (12) in experiments examining the adhesion of activated platelets to monocytes and neutrophils. EDTA, anti-P-selectin antibodies, or purified P-selectin significantly inhibited this interaction. In addition, purified P-selectin incorporated into lipospheres mediated specific adhesion of lipospheres to monocytes and neutrophils. Hamburger and McEver (13) confirmed the role of P-selectin in platelet adhesion to monocytes and neutrophils, and Geng and colleagues (14) demonstrated that rapid neutrophil adhesion to stimulated endothelium was also P-selectin dependent. Thus, a role for P-selectin in one or more aspects of leukocyte localization to sites of vascular injury and thrombosis was defined (6,12).

A. Leukocyte Rolling Model

Studies employing intravital microscopy, an experimental approach initially performed over 100 years ago (15), have demonstrated that leukocyte emigration from the bloodstream to tissue involves several distinct events (16–18). The leukocyte is first observed to "roll" along the endothelium, adjacent to the site of injury, in the direction of flow. This is followed by spreading and firm adhesion of the leukocyte to the vessel wall. Subsequently, the leukocyte crosses the endothelial barrier into tissue by crawling through cellular junctions, and migrates toward the inflammatory stimulus (16).

Patients with leukocyte adhesion deficiency type I, characterized by recurrent bacterial infections, provided early clues that the integrin family of cell adhesion molecules were important in leukocyte extravasation (reviewed in 19). Neutrophils from these patients are unable to firmly adhere to the vessel wall and are therefore incapable of crossing the endothelium and accumulating at sites of inflammation. This syndrome is associated with the absence of membrane glycoproteins on neutrophils (20). The genetic defect in this disease has been identified as mutations in the β_2-integrin subunit, a subunit shared by several members of the integrin superfamily, including CD11/CD18. In vitro studies employing antibodies to CD18 confirmed the importance of the β_2-integrin subunit for firm adhesion of leukocytes to the vessel wall (21). These antibodies had no effect on leukocyte rolling (22).

Lawrence and Springer (17), in experiments performed in vitro, provided evidence of P-selectin's role in leukocyte rolling. Under flow conditions at physiologic shear rates, P-selectin incorporated into phospholipid bilayers mediates neutrophil rolling while bilayers containing intercellular adhesion molecule-1 (ICAM-1), a ligand for leukocyte CD11/CD18, are incapable of doing so. Under static conditions in the presence of chemoattractants, ICAM-1 mediates firm neutrophil adhesion, while chemoattractants do not influence neutrophil rolling on P-selectin-containing bilayers. When both proteins are incorporated into the bilayer, neutrophils roll until chemoattractant is added. Neutrophils are then observed to arrest, flatten, and firmly adhere. These experiments demonstrate that rolling precedes firm adhesion and exemplify the essential role that P-selectin plays in capture of flowing leukocytes. In vivo evidence supporting the in vitro data was provided by Bienvenu and Granger (23) in experiments demonstrating the inhibition of leukocyte rolling by an anti-P-selectin antibody. Further evidence supporting P-selectin's role in leukocyte rolling in vivo comes from experiments performed on P-selectin-deficient mice (24). In these mice, leukocyte rolling in mesenteric venules is virtually absent. There is also a 2-hour lag in recruitment of neutrophils to inflammatory sites.

An elevated neutrophil count and recurrent severe infections are found in patients with leukocyte adhesion deficiency type II (25,26). Because of a genetic defect in the biosynthesis of fucose, these patients are unable to synthesize sialyl Lewis x, a component of the P-selectin ligand. Von Andrian and colleagues (27) have used intravital microscopy to demonstrate markedly impaired rolling of neutrophils obtained from a LAD II patient. Price and colleagues (28), using skin window and skin chamber techniques, have demonstrated that in LAD II patients, levels of in vivo neutrophil accumulation at sites of inflammation are less than 6% of normal.

B. Role in Inflammation

The current model of leukocyte extravasation to sites of inflammation holds that the selectins mediate both the initial capture and tethering of the flowing leukocyte to the inflamed endothelium as well as the rolling that follows (16,29). Yeo and colleagues (30) have shown in an in vitro flow chamber system that P-selectin expressed on the surface of a monolayer of activated platelets leads to capture and rolling of neutrophils, as would be expected at sites of vascular injury or thrombi. Yeo et al. suggest that activated platelets at sites of injury may provide a physiologically important adhesive surface, with a sevenfold increase in surface density of P-selectin over endothelium and a more stable surface expression pattern. P-selectin is transiently expressed on inflamed endothelium, and disappears within 30 minutes (31) while platelet P-selectin may persist for up to 6 hours (32). Leukocyte rolling is proposed to bring the cell in close proximity to chemoattractants expressed at the site of inflammation, leading to leukocyte activation via G protein-linked chemoattractant receptors and an increase in the avidity of leukocyte integrins for endothelial ligands, members of the immunoglobulin superfamily (27). This leads to firm, shear-resistant adhesion. The adherent leukocyte then migrates across the endothelium to adjacent tissue; both leukocyte CD11/CD18 (33) and PECAM (34,35) have been implicated in this process.

C. Role in Immune Response

Data from several laboratories have extended the role of P-selectin as a cell adhesion molecule to include mediating adhesion of platelets and endothelium to T cells (36–38), natural killer cells (36,39), eosinophils (39,40), basophils (41), and solid tumor cells (39,42,43). Moore and Thompson (36) demonstrated binding of purified P-selectin to subpopulations of natural killer cells as well as CD4+ and CD8+ T cells. Binding was inhibited by a blocking monoclonal antibody, EDTA, or sialidase pretreatment of lymphocytes. CD8+ T cells bound severalfold better than CD4+ T cells. Among the T cell population, memory cells bound significantly better than naive cells. T cell activation had no effect on binding. The authors propose that P-selectin may play a role in the migration of memory cells into sites of inflammation. De Bruijne-Admiraal and co-workers (41) demonstrate binding of activated platelets to eosinophils, basophils, natural killer cells, and peripheral T lymphocytes from both CD4 and CD8 classes. Binding was inhibited by a blocking monoclonal antibody to P-selectin as well as by EDTA. Damle and co-workers (44) demonstrate binding of a P-selectin immunoglobulin chimera to chronically stimulated, but not resting CD4+ T cells, in contrast to Moore and Thompson (36). T cells isolated from the synovia of rheumatoid arthritis patients also bind to this chimera. Neuraminidase pretreatment of T cells inhibited binding. Binding was also inhibited by sulfated proteoglycans. When primed T cells were cultured with immobilized combinations of P-selectin chimera and a monoclonal antibody directed against either glycoprotein CD19 or the T cell receptor, production of GM-CSF was enhanced while IL-8 production was inhibited. This regulation of the production of inflammatory cytokines may be a factor in the inflammatory response. P-selectin chimera had no effect on T cell proliferation. Kunzendorf and co-workers (37), using the same P-selectin chimera as Damle and colleagues, found binding to CD4+ and

CD8+ T cells, without prior activation. No differences were observed in the levels of CD4 or CD8 binding. Neuraminidase pretreatment of T cells inhibits binding. As Damle and colleagues found, no effects of the chimera on T cell proliferation were observed.

Most recently, Rossiter and colleagues (38) found that partially purified P-selectin or a P-selectin immunoglobulin chimera binds to skin and blood derived T cell clones from sensitized individuals. This binding can be inhibited by anti P-selectin antibodies. A T cell clone negative for sialyl Lewis x expression by immunofluorescence still bound significantly to the P-selectin chimera, raising the possibility that these cells may interact with P-selectin in a sialyl Lewis x-independent manner. Symon and co-workers (40) have demonstrated that P-selectin mediates the interaction of eosinophils with nasal polyp endothelium, but not with vessels in normal skin. They suggest that P-selectin may be involved in eosinophil infiltration into tissue in allergic disease.

II. GENE STRUCTURE

A. Genomic Organization

In 1989, Johnston and colleagues (7) isolated a cDNA clone for P-selectin from a human umbilical vein endothelial cell cDNA library. Later analysis showed that the human gene spans at least 50 kb, including 17 exons and 16 introns (45). The majority of exons are noted to encode distinct domains, suggesting evolution of the selectins through exon rearrangement and duplication. In support of this, 14 of the 16 exons contain symmetrical phase 1 intron-exon boundaries, promoting duplication and insertion into other genes.

B. Promoter

An analysis of the 5′ flanking region of the P-selectin gene (46) by primer extension, RNase protection, and anchored PCR cloning revealed that the transcriptional start sites of the P-selectin gene are heterogeneous, most likely owing to the absence of TATA and CCAAT motifs in the 5′ flanking region. Unlike other genes lacking TATA motifs, a "GC" box for transcription factor SP1 was lacking. To test the promoter activity of this region, chimeric constructs were prepared containing progressively deleted fragments fused to a promoterless luciferase gene. Measurement of luciferase expression following transfection of these constructs into bovine aortic endothelial cells showed that the region from −309 to −13 contained elements positively regulating P-selectin expression. Basal expression was observed in COS-7, 293, or HeLa cells, demonstrating tissue-specific regulation.

Further analysis unveiled three positive regulatory domains located within this region at −249 to −197, −196 to −147, and −128 to −100. The first domain contains an inverted repeat, CTTCCATGGAAG, that may form a binding site for GA binding protein (GABP), an oligomeric transcription factor homologous to the ETS family of transcription factors (47). Also within this region is an inverted repeat, GGGGTGACCCC, noted in other instances to bind members of the NF-κB/rel family (48) as well as the zinc finger proteins MBP-1 and MBP-2 (49,50). In the second domain, a putative GATA element, TTATCA, resides at position −158. Mutational analysis demonstrated that this six-base sequence is essential for high-level transcription as mutation to TTAAGA decreased luciferase expression by nearly 80%. An oligonucleotide probe encompassing this element formed several specific complexes of varying intensity when incubated with nuclear extracts from K562 cells, HEL cells, bovine aortic endothelial cells, human umbilical vein endothelial cells, or Jurkat cells. One complex, termed complex B, appeared to be due to the interaction of a GATA family member with the GATA element contained within the probe. A probe within which the core GATA sequence was mutated did not yield this complex. Nuclear extracts from COS-7 cells expressing human GATA-2 also formed a specific complex with this oligonucleotide probe, demonstrating that GATA family members can recognize this region. The third domain contains a probable ETS family site as well as a motif similar to the GT-IIC element found in the SV40 enhancer, shown to activate transcription in response to phorbol esters (51). Further analysis is needed to identify the *trans*-acting factors responsible for tissue-specific expression of the P-selectin gene.

C. Transcriptional Regulation

Weller and co-workers (52) show that the mouse P-selectin transcript in endothelial cells is dramatically induced by TNFα treatment, as has been observed for E-selectin (53), and also demonstrate a concomitant rise in P-selectin protein. The majority of newly synthesized P-selectin is transported to the cell surface rather than sequestered in granules. TNFα treatment of bovine endothelial cells likewise leads to an increase in P-selectin synthesis. Lipopolysaccharide is noted to generate a similar expression pattern to that induced by TNFα. Auchampach and co-workers (54), in studies in rat, demonstrate a potent induction of P-selectin mRNA in lung, heart, kidney, thymus, and spleen and lower levels in brain, liver, and small intestine derived from lipopolysaccharide-treated rats. These data suggest that in addition to the rapid phase of P-selectin cell surface expression following granule secretion, there is a subsequent synthetic phase originating from transcriptional induction by inflammatory mediators. This implies that P-selectin transcription may play a critical role in the inflammatory response.

D. Domain Structure

The deduced amino acid sequence predicts the presence of six distinct structural domains. The first domain, comprising 41 amino acids, is a signal peptide based on the presence of several positively charged amino acids, followed by a hydrophobic stretch and a region abundant in polar residues. The second domain, encompassing 118 amino acids, is a lectin or carbohydrate recognition domain based on homologies to sequences found in other lectins, including chicken hepatic lectin (55) and rat mannose-binding protein (56). The third domain, containing 40 amino acids, demonstrates a pattern of cysteine residues characteristic of epidermal growth factor (EGF)-like domains, first described in epidermal growth factor precursor (57,58). The fourth domain consists of nine tandem consensus repeats of 62 amino acids each, with extensions of eight and four residues at the end of the seventh and ninth repeats. This region of P-selectin is most homologous to repeats found in proteins that are known to regulate complement in plasma and on cell surfaces (59). The fifth domain comprises the 24 residue putative transmembrane domain. It is followed by the sixth domain, a cytoplasmic domain.

Among the four endothelial cell clones obtained, two different in-frame deletions were observed. One predicted a protein lacking a seventh consensus repeat while the other lacked a transmembrane domain, predicting a soluble version of P-selectin. Northern analysis detected a single transcript in human umbilical vein endothelial cells and platelets as well as in CHRF-288 and HEL cells, two human leukemia cells with features of megakaryocytes (7). Transcripts were not detected in Daudi cells, a human B cell line; or in K562 cells, a human myeloid cell line. Southern blot analysis of human genomic DNA suggested the presence of a single gene encoding P-selectin.

E. Alternative Splicing

Following up on the observation of cDNA clones predicting the presence of alternative transcripts, studies employing PCR confirmed the presence of an alternatively spliced messenger RNA for a soluble P-selectin molecule lacking the transmembrane domain encoded by exon 14 in platelets and a transcript for a P-selectin isoform lacking the seventh consensus repeat encoded by exon 11 in endothelial cells.

Recently, cDNA clones encoding P-selectin have been obtained from mouse (52,60) and rat (54). Both mouse and rat P-selectin contain only eight consensus repeats, lacking the second consensus repeat. Alternative transcripts encoding a soluble P-selectin isoform were not detected in either mouse or rat. Seven potential N-linked glycosylation sites are conserved among human, mouse, and rat, suggesting that this modification may be important for P-selectin function. The sequence of the cytoplasmic domain from these three species is also highly conserved.

F. Soluble Form

As suggested from the existence of an alternatively spliced message encoding a soluble molecule, P-selectin protein has been detected in normal plasma at levels of 36 to 250 ng/ml (61–63). In a static assay, fluid phase P-selectin inhibits TNFα-activated neutrophil adhesion to resting endothelium, a CD18-dependent process (64). The authors have speculated that soluble P-selectin plays some role in maintaining neutrophils in a nonadhesive state. Wong and co-workers (65) have shown that TNFα-activated neutrophils adhering to P-selectin-coated wells generate significantly less superoxide anion than those adhering to other surfaces. Fluid-phase P-selectin is also capable of inhibiting superoxide anion production. TNFα-activated neutrophils adhering to P-selectin-coated glass slides remain rounded while those adhering to uncoated glass slides become flattened and polarized. These data suggest that soluble P-selectin may function as an antiinflammatory agent.

In disorders including thrombotic thrombocytopenic purpura, hemolytic uremic syndrome, adult respiratory distress syndrome, disseminated intravascular coagulation, and heparin-induced thrombocytopenia, plasma P-selectin is markedly elevated (66). Working with a recombinant form of soluble P-selectin expressed and secreted from 293 cells, Ushiyama and co-workers have shown that it exists as a monomer, while full length P-selectin oligomerizes (61). In HL60 cell adhesion assays, these authors have also shown that full-length P-selectin is fivefold more efficient at mediating adhesion than the soluble isoform. Using equilibrium binding techniques, a K_d of 49 nM was obtained for the interaction between soluble P-selectin and human neutrophils. The range of 36 to 250 ng/ml corresponds to a concentration range of approximately 0.25 to 2 nM (65), well below the Kd, suggesting that small amounts of soluble P-selectin present in normal plasma may not be physiologically relevant.

A longstanding question about the origin of soluble P-selectin was answered recently by Ishiwata and colleagues (67). Theoretically, P-selectin detected in plasma could be authentic soluble protein derived from the alternatively spliced message or derived from proteolytic cleavage of the membrane-bound form just before the transmembrane domain, or it could be full length P-selectin, associated with platelet microparticles. Ultracentrifugation of plasma samples does not reduce antigen levels, arguing against microparticles (61). Proteolyzed P-selectin would lack the cytoplasmic domain while the alternatively spliced isoform would contain glutamine^{721} of the ninth consensus repeat adjacent to aspartate^{762} of the cytoplasmic domain. Purification of P-selectin from plasma followed by protein sequencing demonstrated that a major form of P-selectin present in plasma contains such a sequence. This form was also detected in washed platelets, suggesting that the soluble form may be stored in platelets. This provides the first evidence that the alternatively spliced message is translated into protein in vivo.

III. STRUCTURE-FUNCTION RELATIONSHIPS OF INDIVIDUAL DOMAINS

As the function of P-selectin as a leukocyte receptor became clarified, attention has focused on the structure-function relationships of individual domains.

A. Lectin Domain

Quantitative amino acid and carbohydrate analysis demonstrated that P-selectin has an actual molecular weight of 126,4000, of which 29% is comprised of complex N-linked oligosaccharides as deduced from tunicamycin treatment (68). The potential N- and O-linked glycosylation sites are clustered in the 118-amino acid lectin domain. The lectin domain is responsible for P-selectin binding to specific carbohydrate residues.

Domain-swapping studies have aided in identification of some of the features responsible for the binding specificities of the three selectins. Only P-selectin is able to bind to 2,6 sialylated Lewis x, albeit weakly (69). Exploiting this interaction, Erbe and co-workers (70) engineered the construct PE-1 containing an E-selectin IgG chimera in which the E-selectin lectin domain was replaced with that of P-selectin. The secondary structure did not appear disrupted as the construct bound antibody against the complement repeat domain of E-selectin. Verifying the successful domain swap, antibody to the P-selectin lectin domain but not an antibody against the E-selectin lectin domain recognized the construct. PE-1 bound to glycolipids presenting 2,3 sialylated Lewis x (sLe^x), and sulfatides, mirroring the pattern of P-selectin IgG chimera more closely than that of the E-selectin/IgG chimera, which only bound to 2,3 sLe^x. These data suggest that the lectin domain of P-selectin is sufficient for binding to these glycolipids. In binding assays, the PE-1 construct bound to P-selectin ligand on HL60 cells, but not as well as a P-selectin IgG chimera, suggesting that the EGF and/or complement repeat domain, although not strictly necessary for binding, may enhance affinity.

Deletion of the EGF-like or complement-repeat domain of L-selectin prevents recognition of ligand on high endothelial venules by a soluble form of L-selectin (71). In cell adhesion studies, when chimeras with the lectin domains of L- and P-selectin exchanged are expressed in CHO cells, the transfectants exhibit the adhesion and ligand recognition properties of the lectin domain donor. In binding inhibition studies, P-selectin lectin domain peptides, specifically those spanning amino acids 6–89 and 109–118, decrease P-selectin-mediated binding to HL60 cells or neutrophil, whereas P-selectin EGF domain peptides do not affect binding (72,73). Taken together, these studies point to the lectin domain as the main ligand recognition site of P-selectin, but suggest that the EGF domain may play a role as well.

Site-directed mutagenesis was used to demonstrate the importance of potential glycosylation sites tyrosine^{48}, tyrosine^{94}, lysine^{111}, and lysine^{113} within the lectin domain of P-selectin for sialyl Lewis x binding (70,74). Additionally P-selectin containing a mutation at tyrosine^{48}, tyrosine^{94}, or lysine^{113} did not bind to HL60 cells while P-selectin containing a mutation at lysine^{111} bound less well than wild-type P-selectin to the cells. Tyrosine^{48}, lysine^{111}, and lysine^{113} are predicted to be in the protein fold thought to be common to all C-type animal lectins based on the crystal structure of the C-type lectin mannose-binding protein complexed to an oligosaccharide ligand (75).

B. EGF Domain

Speaking for a role for the EGF-like domain in ligand recognition, a chimera coupling the L-selectin lectin domain to the P-selectin EGF domain acquired dual ligand specificity when expressed at very high levels (71). Bolstering the argument, CHO cells expressing P-selectin/L-selectin chimeras containing P-selectin lectin and EGF domains at physiologic density bound HL60 cells with greater affinity than chimeras containing just the P-selectin lectin domain (76). The feature that distinguishes these studies and gives them increased credence is the attention paid to optimization of surface density expression of P-selectin. The importance of surface density in intercellular interactions mediated by cell adhesion molecules has been highlighted by other groups as well. Dustin and Springer (77) demonstrated that adhesion of lymphocyte cell lines to ICAM-1 incorporated into a planar phospholipid bilayer varies with the density of ICAM-1 and is most sensitive in the physiologic range of ICAM-1 expression on lymphocytes. Larkin and co-workers (78) have shown that the nature of the carbohydrate component of the ligand recognized by CHO cells expressing E-selectin varies with the surface density of the E-selectin.

Possible mechanisms for the part played by P-selectin EGF domain in ligand recognition in these chimeras have been advanced. The EGF domain may alter the conformation of the selectin lectin domain, creating an optimal carbohydrate recognition site. Alternatively, the EGF domain may interact directly with the ligand, as has been described for the thrombomodulin EGF-like domain (79). Adhesion-blocking antibodies have been isolated that recognize EGF-like epitopes at L-selectin (80–82).

C. Complement Repeats

Studies of E-selectin suggest that its complement repeat domain has a role in ligand binding (83). Soluble constructs with variable numbers of complement repeats have been expressed in CHO cells. At a minimum, the lectin and EGF domains are necessary to support in vitro HL60 adhesion. However, in binding inhibition studies, the construct with all six complement repeat domains repeats (compared to one or two repeats) is most potent in blocking neutrophil or HL60 adhesion to immobilized E-selectin on microtiter plates or to cytokine (IL-1 or TNF) stimulated human umbilical vein endothelial cells. A domain swap of the complement repeat domains of P- and L-selectin resulted in no detectable effect (71). Gibson and co-workers (76) showed that in P-selectin/L-selectin chimeras containing P-selectin lectin and EGF domains, the addition of the P-selectin complement repeat domain did not further enhance binding to HL60 cells. These findings may be interpreted to show that the P-selectin complement repeat domain does not contribute to binding or, alternatively, that the L-selectin complement repeat domain is able to substitute for the P-selectin complement repeat domain. The role of P-selectin complement repeat domain remains to be determined.

D. Transmembrane Domain

A role for the transmembrane domain of P-selectin in oligomerization has been suggested by experiments performed by Ushiyama and colleagues (61). Two soluble forms of P-selectin, one lacking the transmembrane domain (asPS) and one truncated after the ninth consensus repeat (tPS), were expressed in 293 cells and purified from cell media. These recombinant proteins, as well as full-length P-selectin purified from platelets, were characterized by electron microscopy, sedimentation equilibrium, and sedimentation velocity analyses. These experiments demonstrated that the soluble forms of

P-selectin were present as monomers, while full-length P-selectin existed primarily as dimers but also as oligomers, with a stoichiometry estimated between three and six P-selectin molecules per oligomer. This multimerization was detergent-resistant, even at detergent concentrations well above the critical micelle concentration. In assays measuring HL60 cell adhesion to all three forms of immobilized P-selectin, half-maximal HL60 cell adhesion supported by the soluble forms of P-selectin required site densities four- to fivefold higher than full-length P-selectin. The increased affinity of full-length P-selectin for its ligand may be physiologically significant if multimerization occurs in the membrane of intact cells.

E. Cytoplasmic Tail

Upon platelet or endothelial cell activation by an agonist, P-selectin is translocated from the alpha granules of platelets (3,4) and the Weibel-Palade bodies of endothelial cells (5,6) with accompanying fusion of granule membrane with the plasma membrane (31). In both cell locations, granule membrane and plasma membrane, the carboxy-terminal region of P-selectin is cytoplasmic.

1. Role of the P-Selectin Cytoplasmic Tail in Granule Targeting

Eukaryotic secretory proteins can follow one of two intracellular pathways: a route that leads directly to the cell surface (constitutive secretion), or one that detours through secretory granules (regulated secretion). The latter pathway is observed in storage-competent cells such as those involved with exocrine or endocrine function and requires an as yet unidentified granule-targeting signal (84). The mouse anterior pituitary cell line AtT-20, which stores and secrets ACTH in the regulated mode of secretion, is able to correctly process and store heterologous secretory proteins (85). To examine whether the targeting of P-selectin to secretory granules is specific for platelets and endothelial cells, P-selectin cDNA was expressed in AtT-20 cells (86). P-selectin colocalized with ACTH in secretory granules. P-selectin was thus shown to contain an independent sorting signal targeting it to storage granules in a heterologous cell line.

The granule-targeting sequence was demonstrated to be contained in the cytoplasmic tail of P-selectin (79). Whereas full-length P-selectin expressed in AtT-20 cells initially localized to secretory granules, P-selectin lacking the last 23 amino acids of the cytoplasmic tail migrated directly to the plasma membrane. When the cytoplasmic tail of tissue factor, a transmembrane protein normally found in the plasma membrane of cells in which it is expressed, was replaced with the cytoplasmic domain of P-selectin, the chimera was routed to secretory granules. Thus, the cytoplasmic tail is required to target P-selectin to storage granules.

Phosphorylation of the EGF receptor and the CD4 antigen, as part of activation of their respective signaling cascades, results in translocation of these molecules from the plasma membranes to lysosomes, in what is termed receptor down-regulation. In P-selectin-transfected CHO cells, which lack storage granules, the cytoplasmic tail was found to contain a sorting signal targeting P-selectin to lysosomes (88). When lysosomal proteases were inhibited in these cells, P-selectin colocalized with lgp-B, a lysosomal membrane protein. While deletion of 10 amino acids of the cytoplasmic domain of P-selectin interfered with lysosomal targeting, attaching the cytoplasmic tail sequence of P-selectin to the LDL receptor gene accelerated turnover of the LDL receptor, consistent with targeting to lysosomes. Green and co-workers (88) proposed, based on these findings, that this sorting event, which occurs in the endosomes of transfected CHO cells, represents a constitutive equivalent of receptor down-regulation and may be a means of regulating P-selectin expression on activated endothelial cells. Troubling, though, is their inability to demonstrate colocalization of P-selectin with lgp-A, a lysosomal membrane protein, in transfected neuroendocrine PC-12 cells, which contain storage granules.

Expression of P-selectin in endothelial cells is transient, with endocytosis occurring 30 to 60 minutes after activation (31,87). The cytoplasmic tail may also be necessary for endocytosis as it mediates interaction with coated pits (88). Subramaniam and co-workers (87) traced the intracellular movement of P-selectin in IL-1-stimulated endothelial cells using immunofluorescent staining with the nonblocking anti-P-selectin antibody AC-1.2. After translocation to the cell surface, P-selectin migrated through endosomes (colocalizing with horseradish peroxidase), to the Golgi (colocalizing with wheat germ agglutinin), to the Weibel-Palade bodies (colocalizing with Von Willebrand factor). In parallel studies, E-selectin moved instead from endosome to the lysosomal compartment. P-selectin trafficking may be different in cells with storage granules, specifically platelets and endothelial cells, compared to an experimental heterologous models without a regulated pathway.

2. Phosphorylation

P-selectin is phosphorylated during cell activation (89,90). The cytoplasmic domain of P-selectin contains five potential phosphorylation sites: two serines, two threonines, and one tyrosine (7). In the first analysis, P-selectin was rapidly phosphorylated on serine, threonine, and tyrosine residues in response to platelet activation with thrombin. Phosphorylation peaked between 15 and 30 seconds following platelet activation, and declined with a half-life of approximately 60 seconds. Phosphoserine persisted, while phosphotyrosine and phosphothreonine were swiftly removed. P-selectin was also phosphorylated in response to ADP, collagen, epinephrine, and a thrombin receptor peptide capable of mimicking the effects of thrombin, SFLLRN (91). Stoichiometric analyses revealed a lower limit of 0.52 moles of phosphate per mole of P-selectin. Resting platelets displayed low levels of phosphorylated P-selectin, but platelet activation with thrombin led to a 15-fold increase in the amount of phosphorylated P-selectin. Phosphotyrosine can mediate interactions with proteins containing SH2 domains (92, 93). Modderman and colleagues (94) found that strong tyrosine phosphorylation of P-selectin could be induced in intact platelets by the phosphatase inhibitor, pervanadate. A fraction of immunoprecipitated P-selectin from resting platelets was found to exist in a disulfide-linked complex with the tyrosine kinase, c-src. P-selectin could be phosphorylated in vitro by this kinase, but, unconventionally, this requires that a disulfide bond exist between src and P-selectin. Recent studies have demonstrated that in addition to being phosphorylated on serine, threonine, and tyrosine residues, P-selectin is also inducibly phosphorylated on two histidine residues within the cytoplasmic domain (95). This finding has implications for the function of phosphohistidine in signal transduction.

In the independent studies by Fujimoto and McEver (90), P-selectin phosphorylation is observed in platelets activated either with thrombin or phorbol myristic acid (PMA) and in endothelial cells stimulated with PMA. Kinetic analyses demonstrated that P-selectin phosphorylation increased by approximately twofold after 5 minutes of thrombin stimulation of platelets and fourfold following 10 minutes of PMA treatment. Phosphoserine and a low level of phosphothreonine were observed in these experiments. Phosphotyrosine was not observed. Stoichiometric analyses revealed 0.05 moles of phosphate per mole of P-selectin in thrombin-activated platelets. Inhibitor studies suggested that protein kinase C may play some role in P-selectin phosphorylation, while protein phosphatases 1 and/or 2A may be involved in P-selectin dephosphorylation. Expression of a cytoplasmic "tail-minus" mutant of P-selectin in CHO cells suggested that phosphorylation occurs on the cytoplasmic domain. Expression of phosphorylation site mutants in these cells suggested that P-selectin was primarily phosphorylated at serine^{778}, with minor amounts of phosphate associated with the other serine and threonines. As expected, the pattern of constitutive phosphorylation of P-selectin in CHO cells is very different from the pattern observed in activated platelets (95; Crovello et al., unpublished data). The discrepancies between the two studies in platelets could be explained by the time points chosen. Since Fujimoto and McEver's (90) first time point was 1 minute, they may have examined postpeak phosphorylation. This may also explain the absence of phosphotyrosine and the lower level of phosphothreonine in the former study, since both are rapidly dephosphorylated.

3. Acylation

Fujimoto and colleagues have described the acylation of the cytoplasmic tail of P-selectin with both palmitic and stearic acid (96). In this study, platelets were labeled with ^{3}H-palmitate, and P-selectin was isolated and analyzed by autoradiography. Further experiments suggested the presence of a thioester linkage of palmitate to P-selectin, suggesting that the single cysteine, Cys^{766}, present in the cytoplasmic tail of P-selectin, was the site of modification. Combined gas chromatographic/mass spectral analysis demonstrated that purified platelet P-selectin contained both palmitate and stearate. Analysis of wild-type and mutant P-selectin expressed in COS-7 cells confirmed that Cys^{766} was the site of fatty acid modification. A functional significance of P-selectin acylation remains to be demonstrated. The authors suggest that regulation of the lateral mobility of P-selectin in the membrane may affect the rate of contact with its ligand on leukocytes.

IV. LIGANDS FOR P-SELECTIN

At the outset, the search for a functional P-selectin counterreceptor focused on carbohydrates as an antileukocyte antibody directed against a lineage-specific carbohydrate moiety blocked P-selectin interaction with monocytes and neutrophils (97). The search for a specific ligand has been impeded by the adulterous nature of P-selectin. This promiscuous tendency runs in the lectin family, as many lectins consort with more than one carbohydrate moiety. And, as has become increasingly apparent, carbohydrates are not monogamous.

Insight into the nature of the P-selectin ligand has been gained through a number of approaches. Larsen and co-workers (97) began by testing antileukocyte antibodies for ability to block bonding to neutrophils, monocytes, and leukocyte cell lines to activated platelets. Anti-CD15 antibody, which recognizes lacto-N-fucopentose III (LNFIII, Lewis x antigen), inhibited such leukocyte-platelet interactions. Binding of U937 cells to P-selectin-expressing COS cells and to P-selectin-containing phospholipid vesicles was also impaired by anti-CD15 antibody. Thus, Lewis x antigen was advanced as a potential component of the P-selectin ligand. Additional light was cast by the finding that rosetting of neutrophils or HL60 cells by activated platelets was inhibited by broad-spectrum sialidases, speaking for a requirement for sialic acid on the P-selectin ligand (98). However, Moore and co-workers (99) showed no inhibition of ^{125}I-P-selectin to neutrophils by anti-CD15 monoclonal antibody 82H5 or by the neoglycoproteins Le^{x}-BSA and sialyl Le^{x}-BSA in concentrations up to 6.5 μM. These findings suggested that neither Lewis x antigen nor sialyl Lewis x alone constitute the P-selectin ligand.

Characterization of the interaction between P-selectin and its ligand was further elucidated by disruption of the interaction through enzymatic modification of the ligand on the leukocyte surface. Using a radioligand binding assay, Moore and co-workers (99) showed that treatment of neutrophils with neuraminidase, which removes sialic acid residues, diminished binding of human neutrophils to radiolabeled purified P-selectin, whereas digestion with protease such as trypsin or elastase abolished such binding altogether. Norgard and colleagues (100) showed that O-sialoglycoprotease eliminates direct binding of P-selectin to HL60 cells as well as HL60 binding to immobilized P-selectin. Inferring from the specificity of this O-glycanase, they suggest that P-selectin displays a clustered saccharide patch of sialylated O-linked oligosaccharides and that it is the multivalent carbohydrate array that confers the specificity and high affinity for P-selectin.

In cell adhesion assays, Larsen and co-workers (101) demonstrated that fucosidase or neuraminidase treatment of HL60 myelomonocytic cells decreased binding to CHO cells expressing P-selectin. In the same system, tunicamycin treatment, which blocks N-linked glycosylation, also decreased binding of HL60 cells to P-selectin-expressing CHO cells. Parenthetically, Moore and co-workers (102) caution that tunicamycin also can arrest the migration of some proteins from the endoplasmic reticulum. With that caveat, these data imply that P-selectin binds to a cell surface glycoproteinbearing carbohydrate chains containing terminal sialic acids and pointed toward sialyl Lewis x involvement in P-selectin binding. L-selectin, which contains sialic acid, has also been proposed as a putative P-selectin ligand (103).

A. Sialyl Lewis x Antigen

The sialyl Lewis x antigen (sLe^x) is a sialylated, fucosylated tetrasaccharide. It was found to be a critical component of the P-selectin ligand through studies showing complete inhibition of P-selectin-mediated activated platelet binding to neutrophils by monoclonal antibody to sLe^x, but only partial inhibition by monoclonal antibody to Le^x (104). Synthetic sLe^x proved 30 times more potent an inhibitor than Le^x of this binding interaction. Furthermore, activated platelets bind to the glycosylation mutant CHO cell line Lec-11, which expresses only sLe^x, but not to Lec-12 cells, which express only Le^x. Activated platelet binding to Lec-11 cells was abolished following treatment with a specific sialidase converting sLe^x to Le^x. Nonsialylated Le^x (LNF III) may act as a low-affinity ligand, as concentrations as high as 200 μg/ml of an oligosaccharide containing Le^x are required to maximally inhibit binding of activated platelets to neutrophils (97). Neither sLe^x nor Le^x alone is sufficient to support P-selectin-mediated cell interaction; transfection of CHO cells with specific fucosyl transferases leading to expression of these carbohydrate moieties on the cell surface does not convert these cells to P-selectin binders (105). Moreover, soluble P-selectin binding to CHO cells expressing sLe^x is low-affinity and nonsaturable (99).

B. Sulfatides

Clouding the picture, other studies showed that a chimera of the extracellular domains of P-selectin attached to the immunoglobulin hinge region binds to purified sulfatide (galactose-4-sulfate ceramide) and to myeloid cell lines (43) with about equal affinity. Notably, half-maximal binding required 20-fold higher concentrations than previously reported for detergent-solubilized full-length P-selectin binding to myeloid cells (99,106). Sulfatides at high concentrations blocked binding of the chimera to the myeloid cell line U937 and, in the converse experiment, blocked adhesion of U937 cells to the chimera immobilized on plastic wells. Additionally, treatment of myeloid cells with sodium selenate, an inhibitor of sulfation, partially impaired ability to bind P-selectin. Aruffo and co-workers (43) speculated that, as granulocytes excrete large enough amounts of sulfatides (100 fg/24 h) to block adhesion, the role of granulocyte sulfatides may be to permit these cells to detach from their receptors in order to diapedese through the endothelial junctions, much as a salamander slides out of the clutches of its predator by detaching its tail. Interaction with sulfated groups may be responsible for the observed interference of P-selectin binding to neutrophils by heparin and other sulfated glycans (106).

There are conflicting data regarding sulfatides as a physiologic ligand for P-selectin. Arguing against a physiologic role, the interaction between P-selectin and sulfatides is not calcium-dependent. Also, since activated platelets express sulfatides and P-selectin, if sulfatides can act as a P-selectin ligand, platelets should aggregate in the absence of a soluble ligand, but this does not occur (1,2). Moreover, antisulfatide antibody does not inhibit platelet aggregation. However, LYP20 monoclonal antibody to P-selectin partially inhibits platelet aggregation without effect on ^{125}I-fibrinogen binding to thrombin-stimulated platelets. Other anti-P-selectin antibodies such as KC4 and S12 do not affect platelet aggregation (107). Notably, after thrombin stimulation of platelets, P-selectin clusters in areas of contact between platelets. It has been suggested that soluble P-selectin may not encounter the steric hindrance membrane-bound P-selectin does in binding to plasma membrane surface sulfatides (39). Also arguing against sulfatides as a physiological ligand, neuraminidase, which should not affect sulfatide expression, abolishes P-selectin binding to myeloid cells. As yet, the role of sulfatides in P-selectin binding in vivo remains open to question.

C. PSGL-1

Attempts to expression clone the P-selectin ligand were thwarted until recently due to lineage-specific differences in ability of the transfected cell to decorate the ligand protein with the appropriate carbohydrate residues. COS cells, which neither bind P-selectin nor synthesize Le^x or sLe^x, were cotransfected with an HL60 cDNA library and a plasmid containing the $\alpha(1,3/1,4)$ fucosyltransferase gene necessary for synthesis of sLe^x. This novel strategy resulted in the cloning of P-selectin glycoprotein ligand (PSGL-1) (108). PSGL-1 mRNA was found in the myelomonocytic HL60 and the monocytic THP-1 and U937 cell lines as well as in freshly isolated monocytes and neutrophils. PSGL-1 is expressed as a homodimer of 220 kDa under nonreducing conditions and as a single-chain polypeptide of 110 kDa under reducing conditions. In addition to an N-terminal signal sequence, the extracellular domain contains three potential N-glycosylation sites, three potential tyrosine sulfation sites, multiple serine, threonine, and proline sites for potential O-glycosylation, and a single cysteine. Save for a 27 amino acid stretch in the putative transmembrane domain with 48% homology to amino acid 125–151 of EV12A, a human homolog of a murine gene implicated in leukemogenesis (109), PSGL-1 does not bear significant homology to any known protein.

A soluble PSGL-1/bacteriophage T7 capsid protein chimera binds to not only P-selectin but also E-selectin in a calcium-dependent manner (108). The latter finding is likely due to E-selectin interaction with sLe^x. Treatment of the PSGL-1-expressing COS cells with neuraminidase diminishes binding to a P-selectin/F_c chimera. Further studies using the same system show greater decrement in binding upon treatment with O-glycanase than with N-glycanase, suggesting that O-linked glycosylated sites may be more important than N-linked sites for P-selectin binding. These data agree with prior experiments identifying the importance of sialylated oligosaccharides as critical moieties for P-selectin binding.

More recently, using PSGL-1 purified from human neutrophils to explore possible differences in the polypeptide backbone as well as in the sugar decoration between recombinant and native PSGL-1, Moore and colleagues (102) demonstrated that antibody to a peptide derived from the published recombinant PSGL-1 sequence recognized the neutrophil PSGL-1. However, the neutrophil-derived PSGL-1, in contrast to the recombinant PSGL-1, was less susceptible to endo-α-N-galactosaminidase, indicating fewer simple O-linked core 1 Galβ1-3Gal Nac disaccharides. They suggest the disparity in findings stems from the difference in glyosylation pattern resulting from the α(1-3,4) fucosyltransferase used in the cloning strategy from the one that modifies P-selectin in vivo. Moreover, the binding of the neutrophil-derived PSGL-1 to P-selectin is not abrogated by cleavage of at least two of three N-linked poly-N-acetyl lactosamines by peptide:N-glycosidase F treatment, nor is sLe^x expression detectably altered. They contend that these data show that neutrophil PSGL-1 displays sialylated, fucosylated O-linked poly-N-acetyl lactosamines that promote high-affinity binding to P-selectin but not to E-selectin.

The murine PSGL-1 has also been cloned and expressed in heterologous cells (110). It possesses 50% homology to human PSGL-1, although there is marked homology in the region of the cytoplasmic tail, suggesting an important conserved function for this intracellular domain.

D. Factors Affecting Ligand Binding

The complexity of the findings for selectin ligand binding is likely rooted in the different conditions employed in binding studies. The binding has been studied at artificially high density of P-selectin or sLe^x expressed in transfected cells or in high concentrations of soluble forms, primarily under nonphysiologic conditions that do not approximate the shear stress of normal blood flow. P-selectin-mediated adhesion to leukocytes is critically dependent on the P-selectin density in the cell membrane (76). Varying known concentrations of purified P-selectin were incorporated into fluorescent lipospheres, and the binding of these lipospheres to HL60 cells was analyzed. A critical P-selectin density of about 100 molecules/μm^2 is required to sustain maximal cell adhesion. In parallel studies, CHO cells expressing P-selectin at three different densities showed near maximal binding to HL60 cells when P-selectin was expressed at 200 molecules/μm^2, a surface density comparable to that estimated for activated platelets. Also important in selectin ligand binding interactions is the context in which the carbohydrate moiety is presented. The protein portion of PSGL-1 may contribute to specificity and affinity by holding the sugars in correct configuration, by presenting an array of multivalent sugars to enhance binding affinity, or through protein-protein contact (111).

V. ROLE OF P-SELECTIN IN DISEASE STATES AND POTENTIAL THERAPEUTIC APPLICATIONS

With the recognition of the role that the selectin family plays in the inflammatory response to vascular injury, many laboratories are exploring the possibilities of selectin antagonists as useful therapies in inflammatory disease.

A. ARDS

An animal model of human adult respiratory distress syndrome (ARDS) has been developed in rats, and several potential therapies have been evaluated in this model. Intravenous infusion of cobra venom factor into rats results in acute lung injury mediated by neutrophils infiltrating tissue that is shown to be P-selectin-dependent (112). The injury was inhibited by a blocking monoclonal anti-P-selectin antibody directed against human P-selectin and cross-reactive with rat P-selectin. Infusion of sialyl Lewis x prior to the cobra venom treatment dramatically reduced lung injury and tissue accumulation of neutrophils in a dose-dependent fashion. Injury-induced increase in vascular permeability, quantitated by ^{125}I-labeled bovine serum albumin in lung tissue, decreased 67% with sLe^x pentasaccharide pretreatment. Intraalveolar hemorrhage decreased 47%, and neutrophil accumulation decreased 49%. Infusion with sialyl-N-acetyllactosamine, a nonfucosylated form of sLe^x, had no effect. Protective effects of sLe^x plateaued at a dose of 200 μg, corresponding to a blood concentration of less than 1 μM (113). The protection against acute lung injury afforded by sLe^x was due to inhibition of P-selectin-, not E-selectin-, mediated cell-cell interaction. This was shown by inhibition of lung injury by pretreatment with P-selectin-IgG chimera, but not by E-selectin-IgG chimera, prior to cobra venom factor infusion (114).

More recently, selectin chimeras containing the extracellular domains of the selectins fused to the CH2 and CH3 domains of human IgG1 have been characterized in the rat model described above as well as in a lung injury model in which the lesions are induced by intrapulmonary deposition of IgG immune complexes, also known to be neutrophil-dependent (114). In the cobra venom model, administration of P- or L-selectin chimera prior to infusion of the venom significantly reduced tissue injury. The P-selectin chimera inhibited permeability, hemorrhage, and neutrophil accumulation by about 50%. Pretreatment with the E-selectin chimera had no effect, probably owing to the early time point (30 minutes) at which tissue damage was evaluated. In acute alveolitis induced by intrapulmonary IgG immune complex deposition, P-selectin chimera had no effect while treatment with either E- or L-selectin chimera significantly reduced tissue damage. The end point in this model is 4 hours, a time point when significant expression of E-selectin could be expected on inflamed endothelium. As selectin-human IgG chimeras have long half-lives in the circulation and have been shown not to induce neutropenia, they have significant potential as therapeutic agents.

B. Gram-Negative Sepsis

Coughlan and colleagues (115), using a rat model of systemic endotoxemia in response to intravenous administration of lipopolysaccharide, have investigated the role of P-selectin in neutrophil deposition within tissue during acute inflammation. They showed that lipopolysaccharide infused intravenously into rats resulted in development of neutropenia within 5 minutes. Concurrently, P-selectin expression on endothelium in kidney, liver, and lung increased. Monoclonal antibody to P-selectin blocked neutropenia from developing and prevented neutrophil accumulation in tissue but only transiently, conferring protection for 15 minutes. After 20 minutes, P-selectin antibody had no effect on blood neutrophil counts, and neutropenia ensued. Pretreatment of animals with a platelet activating factor antagonist gave results similar to animals treated with antibody. These data may explain the neutropenia that can accompany overwhelming sepsis as LPS is known to mediate gram-negative sepsis.

C. Inflammation

Ceconni and co-workers (116) have evaluated inositol polyanions, low-molecular-weight noncarbohydrate structures consisting of six carbon rings esterified with phosphate or sulfate groups, as antiinflammatory agents. Peritoneal inflammation in mice induced by thioglycollate leads to accumulation of neutrophils in the peritoneum, and the early phase of this accumulation is thought to be L- (117,118) and P-selectin-dependent (24). In an in vitro assay, both inositol hexakissulfate and inositol hexakisphosphate were able to inhibit binding of a P-selectin-IgG chimera, but not E-selectin-IgG chimera, to immobilized BSA-sLe^x conjugate. There was partial inhibition of an L-selectin-IgG chimera binding to the BSA-sLe^x conjugate. Inositol hexakisphosphate was also able to significantly inhibit adhesion of human neutrophils to COS cells transfected with P-selectin but not to COS cells transfected with E-selectin. Using this model of acute peritoneal inflammation, Ceconni et al. (116) demonstrate that inositol hexakisphosphate significantly inhibits such accumulation in vivo. Inositol hexakisphosphate is also capable of inhibiting peritoneal accumulation of neutrophils in response to zymosan. A fivefold higher amount of inositol hexakissulfate was required to give somewhat lesser inhibition. Neither agent reduced blood neutrophil counts. These experiments provide indirect evidence for a role for P-selectin and L-selectin in inflammation in vivo.

D. Allergy

Symon and co-workers (40) have demonstrated that eosinophil adhesion to nasal polyp endothelium is P-selectin-dependent. Asthma and related diseases are characterized by eosinophil infiltration into tissue, suggesting that P-selectin antagonists may provide a useful therapy. In support of this, a blocking monoclonal antibody directed against P-selectin nearly obliterated eosinophil adhesion. Antibodies to L-selectin had no effect. A chimeric molecule containing the lectin, EGF, and one consensus domain of P-selectin fused to the Fc portion of human IgG was equally effective, while the L-selectin chimera had no effect. This suggests that therapies targeting P-selectin may be useful in treating allergic disease. Grober and colleagues (119) demonstrated the constitutive expression of P-selectin on synovial vascular endothelium in rheumatoid arthritis and suggested that monocyte adhesion to synovium was largely due to interactions with P-selectin, raising the possibility that P-selectin antagonists may be useful in this scenario as well.

E. Thrombotic Platelet Disorders

Plasma P-selectin has been found to be increased in thrombotic consumptive platelet disorders such as disseminated intravascular coagulation, heparin-induced thrombocytopenia, thrombotic thrombocytopenia purpura/hemolytic uremic syndrome, but not in ITP (120,121). Under reducing conditions, the plasma P-selectin was found to migrate faster than platelet membrane P-selectin, consistent with a decreased molecular size. The authors attribute this difference in size to lack of the transmembrane domain in soluble P-selectin. However, differences in posttranslational modifications could also explain this disparity. The authors postulate that the increase in plasma P-selectin found in these platelet disorders reflects platelet and endothelial cell activation and/or damage.

F. Role in Thrombosis and Ischemia/Reperfusion Injury

The ability of P-selectin to capture neutrophil and monocytes at sites of vascular injury may be a means of enhancing wound healing. A possible role for P-selectin in thrombosis has hypothesized. In a primate model, an experimental thrombus was developed using a Dacron graft implanted within an arteriovenous shunt (122). The graft accumulated activated platelets expressing P-selectin, leukocytes, and fibrin. Blocking anti-P-selectin antibodies not only prevented adhesion of indium-labeled leukocytes at the wound, but also inhibited fibrin deposition on the graft without affecting platelet accumulation. The authors hypothesize that the presence of leukocytes in platelet-rich thrombi, mediated via interactions with P-selectin, contributed directly to fibrin deposition. More recently, Celi and colleagues (123) have demonstrated that P-selectin is not only involved in cell adhesion, but is capable of inducing tissue factor expression by monocytes, leading to fibrin production. Tissue factor production by monocytes peaks approximately 6 hours following stimulation with P-selectin. The authors propose a model in which injured endothelium flags down passing monocytes by expression of P-selectin. P-selectin-monocyte interaction results in monocyte tissue factor production, which in turn promotes fibrin deposition (124). The fibrin patch provides a substrate on which to overlay fibroblast and endothelial tissue for wound repair.

P-selectin expression occurs after focal brain ischemia and reperfusion in a primate model (125). A surgically implanted device in the middle cerebral artery was employed to examine the effect of occlusion of 2 to 4 hours followed by reperfusion for 1, 4, or 24 hours on P-selectin and ICAM-1 expression in the lenticulostriatal microvessels, primarily precapillary arterioles and postcapillary venules. Both molecules showed significantly increased surface expression, lasting at least 24 hours for P-selectin and 4 hours for ICAM-1. GPIIbIIIa antibody staining to assess the potential platelet contribution to P-selectin expression was found to be quite low at 1 hour of reperfusion, but increased at the 4-hour and 24-hour time points. These results suggest that during early ischemia and reperfusion endothelial cell expression of P-selectin and ICAM-1 may precipitate a cycle of neutrophil adhesion and platelet deposition, leading to further cytokine release and tissue injury.

In a feline model of myocardial ischemia and reperfusion (126), administration of a blocking anti-P-selectin antibody (PB1.3) 10 minutes prior to reperfusion inhibited myocardial necrosis by 58% and preserved endothelium-dependent relaxation in response to vasodilators in the left anterior descending coronary artery. In vivo administration of PB1.3 also inhibited neutrophil adherence to the isolated coronary artery. Similarly, in a rabbit ear model of ischemia and reperfusion (127), antibody PB1.3 significantly inhibited tissue necrosis.

G. Malignancy

In studies using a soluble P-selectin immunoglobulin chimera, Aruffo and co-workers demonstrated binding of this chimera to tumor cells, including those derived from breast, colon, and lung tumors in a saturable fashion (128), and sulfatides (43) were the proposed P-selectin ligand. They show that a monoclonal antibody sulph-1 highly specific for sulfatides bound to two breast adenocarcinoma cell lines, H3630 and H3396, and that P-selectin-Ig chimera also bound to these cells in a saturable fashion. Sulfatides, 3-sulfated galactosyl ceramides, are expressed on the surface of and secreted by granulocytes and some tumor cells.

P-selectin has been shown to mediate adhesion of platelets to neuroblastoma and small-cell lung cancer cells in a rosetting phase contrast cell adhesion assay (39). Inhibitory P-selectin antibodies blocked this interaction, whereas neither RGDS peptide nor antibody to GPIIbIIIa had any effect, suggesting that the interaction was not mediated by GPIIbIIIa, as is the case in some tumors. P-selectin-containing fluorescent lipospheres adhered to neuroblastoma and small-cell lung cancer cell lines and to frozen sections of the primary tumors of these cell types. Although neuraminidase or trypsin treatment of the malignant cells interfered with the binding, suggesting that the glycoprotein ligand PSGL-1 might be recognized, no Le^x or sLe^x antigen could be detected on the malignant cells using monoclonal antibodies CD15 and CSLEX1, respectively. The authors note that these data may indicate that CSLEX1 is not able to recognize the sLe^x antigen in the context within which it is displayed on the tumor cells studied or that a different carbohydrate group is sialylated. Malignant transformation upregulates glycosyltransferases, perhaps including a fucosyl or sialyl transferase that may lead to formation of the P-selectin ligand (129–131). Stone and Wagner (39) have suggested that endothelial P-selectin may be involved in the initial attachment of tumor cells to the vessel wall while P-selectin on platelets may form a protective platelet cloak to surround the arrested cancer cells, perhaps aiding in escaping recognition by the immune system.

Although anti-sLe^x antibody inhibition of the binding was not directly examined, the P-selectin binding activity of the breast carcinoma cell lines or of frozen carcinoma tissue sections was shown to be refractory to neuraminidase and incompletely sensitive to trypsin treatment (42). The binding of tumor cells to endothelial P-selectin at sites of vascular injury may have a role in the metastatic process.

VI. SUMMARY

P-selectin helped define the selectin family of adhesion molecules, with E-selectin and L-selectin. A glycoprotein whose surface expression on platelets and endothelial cells is activation-dependent, P-selectin serves as a leukocyte receptor important for the initial rolling step in the three-step model of leukocyte adhesion to endothelium. The signaling pathways resulting in mobilization of P-selectin out of storage granules to the plasma membrane remain to be mapped, as do events triggered by the engagement of leukocyte PSGL-1 by P-selectin. P-selectin and its ligand, PSGL-1, stand at the crossroads of inflammation and thrombosis and hold promise as the potential targets of therapy to retard or interrupt the cycle of vascular perturbation, inflammatory response, and thrombosis in a broad range of pathologic states from autoimmune disease to ischemia and reperfusion injury.

REFERENCES

1. Hsu-Lin SC, Berman C, Furie BC, August D, Furie B. A platelet membrane protein expressed during platelet activation and secretion. J Biol Chem 1990; 259:9121–9126.

2. McEver RP, Martin NM. A monoclonal antibody to a membrane glycoprotein binds only to activated platelets. J Biol Chem 1984; 259:9799–9804.
3. Berman CL, Yeo EL, Wencel-Drake JD, Furie BC, Ginsberg M, Furie B. A platelet alpha-granule membrane protein that is associated with the plasma membrane after action. J Clin Invest 1986; 78:130–137.
4. Stenberg PE, McEver RP, Shuman MA, Jacques YV, Bainton DF. A platelet alpha granule membrane protein (GMP140) is expressed on the plasma membrane after activation. J Cell Biol 1985; 101:880–886.
5. Bonfanti R, Furie BC, Furie B, Wagner DD. PADGEM (GMP-140) is a component of Weibel-Palade bodies of human endothelial cells. Blood 1989; 73:1109–1112.
6. McEver RP, Beckstead JH, Moore KL, Marshall-Carlson L, Bainton DF. GMP-140, a platelet alpha-granule membrane protein, is also synthesized by vascular endothelial cells and is localized in Weibel-Palade bodies. J Clin Invest 1989; 84:92–99.
7. Johnston GI, Cook RG, McEver RP. Cloning of GMP-140, a granule membrane protein of platelets and endothelium: sequence similarity to proteins involved in cell adhesion and inflammation. Cell 1989; 56:1032–1044.
8. Bevilacqua MP, Pober JS, Mendrick DL, Cotran RS, Gimbrone MA Jr. Identification of an inducible endothelial-leukocyte adhesion molecule. PNAS 1987; 84(24):9238–9242.
9. Lasky LA, Singer MS, Yednock TA, et al. Cloning of a lymphocyte homing receptor reveals a lectin domain. Cell 1989; 56:1045–1055.
10. Jungi TW, Spycher MO, Nydegger JE, Barndum S. Platelet-leukocyte interactions: selective binding of platelets to human monocytes, polymorphonuclear leukocytes, and related cell lines. Blood 1986; 64:629–636.
11. Johnston GI, Cook RG, McEver RP. Cloning of GMP-140, a granule membrane protein of platelets and endothelium: sequence similarity to proteins involved in cell adhesion and inflammation. Cell 1989; 56:1033–1044.
12. Larsen E, Celi A, Gilbert GE, et al. PADGEM protein: a receptor that mediates the interaction of activated platelets with neutrophils and monocytes. Cell 1989; 59:305–312.
13. Hamburger SA, McEver RP. GMP-140 mediates adhesion of stimulated platelets to neutrophils. Blood 1990; 75:550–556.
14. Geng J-G, Bevilacqua MP, Moore KL, et al. Rapid neutrophil adhesion to activated endothelium mediated by GMP-140. Nature 1990; 343:757–760.
15. Cohnheim J. Lectures on General Pathology: A Handbook for Practitioners and Students. London: New Sydenham Society 1889.
16. Butcher EC. Leukocyte-endothelial recognition: three (or more) steps to specificity and diversity. Cell 1991; 67:1033–1036.
17. Lawrence MB, Springer TA. Leukocytes roll on a selectin at physiologic flow rates: distinction from and prerequisite for adhesion through integrins. Cell 1991; 65:859–873.
18. Sugama Y, Tiruppathi C, Jankidevi K, Andersen TT, Fenton JW II, Malik AB. Thrombin-induced expression of endothelial P-selectin and intercellular adhesion molecule-1: a mechanism for stabilizing neutrophil adhesion. J Cell Biol 1992; 119(4):935–954.

19. Anderson DC, Springer TA. Leukocyte adhesion deficiency: an inherited defect in the Mac-1, LFA-1, and p150, 95 glycoproteins. Annu Rev Med 1987; 38:175–194.
20. Crowley CA, Curnutte JT, Rosin RE, et al. An inherited abnormality of neutrophil adhesion: its genetic transmission and its association with a missing protein. NEJM 1980; 302:1163–1168.
21. von Adrian UH, Chambers JD, McEvoy LM, Bargatze RF, Arfors KE, Butcher EC. Two-step model of leukocyte-endothelial cell interaction in inflammation: distinct roles for LECAM-1 and the leukocyte $\beta 2$ integrins in vivo. PNAS 1991; 88:7538–7542.
22. Arfors KE, Lundberg C, Lindborm L, Lundberg K, Beatty PG, Harlan JM. A monoclonal antibody to the membrane glycoprotein complex CD18 inhibits polymorphonuclear leukocyte accumulation and plasma leakage in vivo. Blood 1987; 69:338–340.
23. Bienvenu K, Granger DN. Molecular determinants of shear rate dependent leukocyte adhesion in postcapillary venules. Am J Phys 1993; 2664:H1504.
24. Mayades TN, Johnson RC, Rayburn H, Hynes RO, Wagner DD. Leukocyte rolling and extravasation are severely compromised in P-selectin-deficient mice. Cell 1993; 74:1–20.
25. Etzioni A, Frydman M, Pollack S, et al. Brief report: recurrent severe infection caused by a novel leukocyte adhesion deficiency. NEJM 1992; 327(25): 1789–1792.
26. Frydman M, Etzioni H, Eidlitz-Markus T. Ramban Hasharon syndrome of mental retardation, short-limbed dwarfism, defective neutrophil chemotaxis, and Bombay phenotype. Am J Med Genet 1992; 44(3):297–302.
27. von Andrian UH, Berger EM, Ramezani L, et al. In vivo behavior of neutrophils from two patients with distinct inherited leukocyte adhesion deficiency syndromes. J Clin Invest 1993; 91:2893–2897.
28. Price TH, Ochs HD, Gershoni-Baruch R, Harlan JM, Etzioni A. In vivo neutrophil and lymphocyte function studies in a patient with leukocyte adhesion deficiency type II. Blood 1994; 84(5):1635–1639.
29. Carlos TM, Harlan JM. Leukocyte-endothelial adhesion molecules. Blood 1994; 84(7):2068–2101.
30. Yeo EL, Sheppard J-AI, Feuerstein IA. Role of P-selectin and leukocyte activation in polymorphonuclear cell adhesion to surface adherent activated platelets under physiologic shear conditions (an injury vessel wall model). Blood 1994; 83(9):2498–2507.
31. Hattori R, Hamilton KK, Fugate RD, McEver RP, Sims PJ. Stimulated secretion of endothelial von Willebrand factor is accompanied by rapid redistribution to the cell surface of the intracellular granule membrane protein GMP-140. J Biol Chem 1989; 2264(14):7768–7771.
32. Yeo E, Gemmel C, Rand ML. In vitro stability of P-selectin expression on human platelets. Blood 1992; 80(10)(suppl 1):56a. Abstract.
33. Furie MB, Tancinco MC, Smith CW. Monoclonal antibody to leukocyte integrins CD11a/CD18 and CD11b/CD18 or intercellular adhesion molecule-1 inhibit chemoattractant-stimulated neutrophil transendothelial migration in vitro. Blood 1991; 78(8):2089–2097.

34. Muller WA, Weigel SA, Deng X, Phillips DM. PECAM-1 is required for transendothelial migration of leukocytes. J Exp Med 1993; 178(2):449–460.
35. Vaporciyan AA, Delisser HM, Yan HC, et al. Involvement of platelet-endothelial cell adhesion molecule-1 in neutrophil recruitment in vivo. Science 1993; 262(5139):1580–1582.
36. Moore KL, Thompson LF. P-selectin (CD62) binds to subpopulations of human memory T lymphocytes and natural killer cells. Biochem Biophys Res Commun 1992; 186(1):173–181.
37. Kunzendorf U, Notter M, Hock H, Distler A, Diamantstein T, Walz G. T cells bind to the endothelial adhesion molecule GMP-140 (P-selectin). Transplantation 1993; 56(5):1213–1217.
38. Rossiter H, van Reijsen F, Mudde GC, et al. Skin disease-related T cells bind to endothelial selectins: expression of cutaneous lymphocyte antigen (CLA) predicts E-selectin but not P-selectin binding. Eur J Immun 1994; 24:205–210.
39. Stone JP, Wagner DD. P-selectin mediates adhesion of platelets to neuroblastoma and small cell lung cancer. J Clin Invest 1993; 92:804–813.
40. Symon FA, Walsh GM, Watson SR, Wardlaw AS. Eosinophil adhesion to nasal polyp endothelium is P-selectin dependent. J Exp Med 1994; 180(1): 371–376.
41. de Bruijne-Admiraal LG, Modderman PW, Von dem Borne AEG, Sonnenberg A. P-selectin mediates Ca^{2+} dependent adhesion of activated platelets to many different types of leukocytes: detection by flow cytometry. Blood 1992; 80(1):134–142.
42. Aruffo A, Dietsch MT, Wan H, Hellstrom KE, Hellstrom I. Granule membrane protein 140(GMP-140) binds to carcinomas and carcinoma-derived cell lines. PNAS 1992; 89:2292–2296.
43. Aruffo A, Kolanus W, Walz G, Fredman P, Seed B. CD62/P-selectin recognition of myeloid and tumor cell sulfatides. Cell 1991; 67:35–44.
44. Damle NK, Klussman K, Dietsch MT, Mohagheghpour N, Aruffo A. GMP-140 (P-selectin/CD62) binds to chronically stimulated but not resting CD4+ T lymphocytes and regulates their production of proinflammatory cytokines. Eur J Immun 1992; 221:789–793.
45. Johnston GI, Bliss GA, Newman PJ, McEver RP. Structure of the human gene encoding granule membrane protein-140, a member of the selectin family of adhesion receptors for leukocytes. J Biol Chem 1990; 265:21381–21385.
46. Pan J, McEver RP. Characterization of the promoter for the human P-selectin gene. J Biol Chem 1993; 268(30):22600–22608.
47. Thompson CC, Brown TA, McKnight SL. Convergence of Ets- and Notch-related structural motifs in a heteromeric DNA-binding complex. Science 1991; 253:762–768.
48. Blank V, Kouritsky P, Israel A. NF-kB and related proteins: Rel/dorsal homologies meet ankyrin-like repeats. Trends Biochem Sci 1992; 17:135–140.
49. Baldwin AS Jr, LeClair KP, Singh H, Sharp PA. A large protein containing zinc finger domains binds to related sequence elements in the enhancers of the class I major histocompatibility complex and kappa immunoglobulin genes. Mol Cell Biol 1990; 10:1406–1414.

50. Fan C-M, Maniatis T. A DNA-binding protein containing two widely separated zinc finger motifs that recognize the same DNA sequence. Genes Dev 1990; 4:29–42.
51. Chiu R, Imagawa M, Imbra RJ, Bockoven JR, Karin M. Multiple *cis*- and *trans*-acting elements mediate the transcriptional response to phorbol esters. Nature 1986; 329(6140):648–651.
52. Weller A, Isenman S, Vestweber D. Cloning of the mouse endothelial selectins: expression of both E- and P-selectin is inducible by tumor necrosis factor α. J Biol Chem 1992; 267(21):15176–15183.
53. Bevilacqua MP, Pober JS, Mendrick DL, Cotran RS, Gimbrone MA Jr. Identification of an inducible endothelial-leukocyte adhesion molecule, ELAM-1. PNAS 1987; 84:9238–9242.
54. Auchampach JA, Oliver MG, Anderson DC, Manning AM. Cloning, sequence comparison, and in vivo expression of the gene encoding rat P-selectin. Gene 1994; 145(2):251–255.
55. Drickamer K. Complete amino acid sequence of a membrane receptor for glycoproteins. J Biol Chem 1981; 256:5827–5839.
56. Drickamer K, Dordal MS, Reynolds L. Mannose-binding proteins isolated from rat liver contain carbohydrate-recognition domains linked to collagenous tails. J Biol Chem 1986; 261:6878–6887.
57. Gray A, Dull TJ, Ullrich A. Nucleotide sequence of epidermal growth factor cDNA predicts a 128,000 molecular weight protein precursor. Nature 1983; 303:236–240.
58. Scott J, Urdea M, Quiroga M, et al. Structure of a mouse submaxillary messenger RNA encoding epidermal growth factor and seven related proteins. Nature 1983; 221:236–240.
59. Hourcade P, Miesner DR, Atkinson JP, Holers VM. Identification of an alternative complement receptor type 1 transcriptional unit and prediction of a secreted form of complement receptor type 1. J Exp Med 1988; 168(4): 1255–1270.
60. Sanders WE, Wilson RW, Ballantyne CM, Beaudet AL. Molecular cloning and analysis of an in vivo expression of murine P-selectin. Blood 1992; 80(3): 795–800.
61. Ushiyama S, Laue TM, Moore KL, Erickson HP, McEver RP. Structural and functional characterization of monomeric soluble P-selectin and comparison with membrane P-selectin. J Biol Chem 1993; 269(20):15229–15237.
62. Katayama M, Handa M, Ambo H, et al. A monoclonal antibody-based enzyme immunoassay for human GMP-140/P-selectin. J Immun Meth 1992; 153:41–48.
63. Dunlop LC, Skinner MP, Bendall LJ, et al. Characterization of GMP-140 (P-selectin) as a circulating plasma protein. J Exp Med 1992; 175:1147–1150.
64. Gamble JR, Skinner MP, Berndt MC, Vadas MA. Prevention of activated neutrophil adhesion to endothelium by soluble adhesion protein GMP-140. Science 1990; 249:414–417.
65. Wong CS, Gamble JR, Skinner MP, Lucas CM, Berndt MC, Vadas MA.

Adhesion protein GMP-140 inhibits superoxide anion release by human neutrophils. PNAS 1991; 88:2397–2401.
66. Chong BH, Murray, Berndt MC, Dunlop LC, Brighton T, Chestermen CN. Plasma P-selectin is increased in thrombotic consumptive platelet disorders. Blood 1994; 83(6):1535–1541.
67. Ishiwata N, Takio K, Katayama M, et al. Alternatively spliced isoform of P-selectin is present in vivo as a soluble molecule. J Biol Chem 1994; 269(38): 23708–23715.
68. Johnston GI, Kurosky A, McEver RP. Structural and biosynthetic studies of the granule membrane protein GMP-140 from human platelets and endothelial cells. J Biol Chem 1989; 264:1816–1823.
69. Foxall C, Watson SR, Dowbenko D, et al. The three members of the selectin family recognize a common carbohydrate epitope, the sialyl Lewis x oligosaccharide. J Cell Biol 1992; 117:895–902.
70. Erbe DV, Watson SR, Presta LG, et al. P- and E-selectin use common sites for carbohydrate ligand recognition and cell adhesion. J Cell Biol 1993; 120(5):1227–1235.
71. Kansas GS, Saunders KB, Ley K, et al. A role for the epidermal growth factor-like domain of P-selectin in ligand recognition and cell adhesion. J Cell Biol 1994; 124(4):609–618.
72. Freedman SJ, Furie B, Furie BC. The lectin domain but not the EGF domain is a potent inhibitor of P-selectin-mediated cellular adhesion. Blood 1993; 82(suppl 1):341A.
73. Heavner GA, Falcone M, Kruszynski M, et al. Peptides from multiple regions of the lectin domain of P-selectin inhibiting neutrophil adhesion. Int J Peptide Protein Res 1993; 42:484–489.
74. Hollenbaugh D, Bajorath J, Stenkamp R, Aruffo A. Interaction of P-selectin (CD62) and its cellular ligand: analysis of critical residues. Biochemistry 1993; 32:2960–2966.
75. Weis WI, Drickamer K, Hendrickson WA. Structure of a C-type mannose-binding-protein complexed with an oligosaccharide. Nature 1992; 360:127–134.
76. Gibson RM, Kansas GS, Tedder TF, Furie B, Furie BC. The lectin and EGF domains of P-selectin at physiologic density are the recognition unit for leukocyte binding. Blood 1995; 85:151–158.
77. Dustin ML, Springer TA. Lymphocyte function associated antigen-1 (LFA-1) interaction with intracellular adhesion molecule-1 (ICAM-1) is one of at least three mechanisms for lymphocyte adhesion to cultured endothelial cells. J Cell Biol 1988; 107:321–331.
78. Larkin M, Ahern TJ, Stoll MS, et al. Spectrum of sialylated and nonsialylated fuco-oligosaccharides bound by the endothelial-leukocyte adhesion molecule E-selectin. J Biol Chem 1992; 267:13661–13668.
79. Disdier M, Morrissey JH, Fugate RD, Bainton DF, McEver RP. Cytoplasmic domain of P-selectin (CD62) contains the signal for sorting in to the regulated secretory pathway. Mol Biol Cell 1992; 3:309–321.
80. Kansas GS, Spertini O, Stoolman LM, Tedder TF. Molecular mapping of

functional domains of the leukocyte receptor for endothelium LAM-1. J Cell Biol 1991; 114:351–358.

81. Spertini O, Kansas GS, Reimann KA, Mackay CR, Tedder TF. Functional and evolutionary conservation of distinct epitope on the leukocyte adhesion molecule LAM-1 that regulate leukocyte migration. J Immun 1991; 147:942–949.
82. Siegelman MH, Cheng IC, Weisman IL, Wakeland EC. The mouse lymph node homing receptor is identical with the lymphocyte cell surface molecule LY-22: role of the EGF domain in endothelial binding. Cell 1990; 61(4):611–622.
83. Li SH, Burns DK, Rumberger JM, et al. Consensus repeat domains of E-selectin enhance ligand binding. J Biol Chem 1994; 269(6):4431–4437.
84. Burgess TL, Kelly RB. Constitutive and regulated secretion of proteins. Annu Rev Cell Biol 1987; 3:243–293.
85. Moore HP, Walker MD, Lee F, Kelly RB. Expressing a human proinsulin cDNA in a mouse ACTH-secreting cell. Intracellular storage, proteolytic processing, and secretion on stimulation. Cell 1983; 35:531–538.
86. Koedam JA, Cramer EM, Friend E, Furie B, Furie BC, Wagner DD. P-selectin, a granule membrane protein of platelets and endothelial cells, follows the regulated secretory pathway in AtT20 cells. J Cell Biol 1992; 116(3): 617–625.
87. Subramaniam M, Koedam JA, Wagner DD. Divergent fates of P- and E-selectin after their expression on the plasma membrane. Mol Biol Cell 1993; 4:791–801.
88. Green SA, Setiadi, McEver RP, Kelly RB. The cytoplasmic domain of P-selectin contains a sorting determinant that mediates rapid degradation in lysosomes. J Cell Biol 1994; 124:435–448.
89. Crovello CS, Furie BC, Furie B. Rapid phosphorylation and selective dephosphorylation of P-selectin accompanies platelet activation. J Biol Chem 1993; 268:1–4.
90. Fujimoto T, McEver RP. The cytoplasmic domain of P-selectin is phosphorylated on serine and threonine residues. Blood 1993; 82(6):1758–1768.
91. Coughlan AF, Hau H, Dunlop LC, Berndt MC, Hancock WW. P-selectin and platelet-activating factor mediate initial endotoxin-induced neutropenia. J Exp Med 1994; 179:329–334.
92. Cantley LC, Auger KR, Carpenter C, et al. Oncogenes and signal transduction. Cell 1991; 64:281–302.
93. Koch CA, Anderson D, Moran MF, Ellis C, Pawson T. SH2 and SH3 domains: elements that control interactions of cytoplasmic signalling proteins. Science 1991; 252:668–674.
94. Modderman PW, von Dem Borne AEG, Sonnenberg A. Tyrosine phosphorylation of P-selectin in intact platelets and in a disulfide linked complex with immunoprecipitated pp60 c-src. Biochem J 1994; 299:613–621.
95. Crovello CS, Furie BC, Furie B. Histidine phosphorylation of P-selectin upon stimulation of human platelets: a novel pathway for activation-dependent signal transduction. Cell 1995; 82(2):279–86.
96. Fujimoto T, Stroud E, Whatley RE, et al. P-selectin is acylated with palmitic

acid and stearic acid at cysteine 766 through a thioester linkage. J Biol Chem 1993; 268(15):11394–11400.

97. Larsen E, Palabrica T, Sajer S, et al. PADGEM-dependent adhesion of platelets to monocytes and neutrophils is mediated by a lineage-specific carbohydrate, LNFIII (CD15). Cell 1990; 63:467–474.
98. Corral L, Singer MS, Macher BA, Rosen SD. Requirement for sialic acid on neutrophils in a GMP-140 (PADGEM) mediated interaction with activated platelets. Biochem Biophys Res Commun 1990; 172:1349–1356.
99. Moore KL, Varki A, McEver RP. GMP-140 binds to a glycoprotein receptor on human neutrophils: evidence for a lectin-like interaction. J Cell Biol 1991; 112(3):491–499.
100. Norgard KE, Moore KL, Diaz S, et al. Characterization of a specific ligand for P-selectin on myeloid cells. J Biol Chem 1993; 268(17):12764–12774.
101. Larsen GR, Sako D, Ahern TJ, et al. P-selectin and E-selectin: distinct but overlapping leukocyte ligand specificities. J Biol Chem 1992; 267(16):11104–11110.
102. Moore KL, Eaton SF, Lyons DE, Lichentstein HS, Cummings RD, McEver RP. The P-selectin glycoprotein ligand from human neutrophils displays sialylated, fucosylated, O-linked poly-N-acetyllactosamine. J Biol Chem 1994; 269(37):23318–23327.
103. Picker LJ, Warnock RA, Burns AR, Doerschuk CM, Berg EL, Butcher EC. The neutrophil selectin LECAM-1 presents carbohydrate ligands to the vascular selectins ELAM-1 and GMP-140. Cell 1991; 66:921–933.
104. Polley MJ, Phillips ML, Wayner E, et al. CD62 and endothelial cell-leukocyte adhesion molecule 1 (ELAM-1) recognize the same carbohydrate ligand, sialyl-Lewis x. PNAS 1991; 88:6224–6228.
105. Zhou Q, Moore KL, Smith DF, Varki A, McEver RP, Cummings RD. The selectin GMP-140 binds to sialylated, fucosylated lactosaminoglycans on both myeloid and nonmyeloid cells. J Cell Biol 1991; 115(2):557–564.
106. Skinner MP, Lucas CM, Burns GF, Chesterman CN, Berndt MC. GMP-140 binding to neutrophils is inhibited by sulfated glycans. J Biol Chem 1991; 266:5371–5374.
107. Parmentier S, McGregor L, Catimel B, Leung LLK, McGregor JL. Inhibition of platelet functions by a monoclonal antibody (LYP20) directed against a granule membrane glycoprotein (GMP-140/PADGEM). Blood 1991; 77(8):1734–1739.
108. Sako D, Chang X-J, Barone KM, et al. Expression cloning of a functional glycoprotein ligand for P-selectin. Cell 1993; 75:1179–1186.
109. Cawthon R, O'Connell P, Buchberg A, et al. Identification and characterization of transcripts from the neurofibromatosis 1 region: the sequence and genomic structure of EV12 and mapping of other transcripts. Genomics 1990; 7:555–565.
110. Yang J, Furie BC, Furie B. Cloning of the murine homologue of the human P-selectin glycoprotein ligand-1. Blood 1994; 84(10)(suppl 1):33a. Abstract.
111. Geng J-G, Heavner GA, McEver RP. Lectin domain peptides interact with both cell surface ligands and Ca^{2+} ions. J Biol Chem 1992; 267:19846–19853.

112. Mulligan MS, Polley MJ, Bayer RJ, Nunn MF, Paulson JC, Ward PA. Neutrophil-dependent acute lung injury. Requirement for P-selectin (GMP-140). J Clin Invest 1992; 90:1600–1607.
113. Mulligan MS, Paulson JC, De Frees S, Zheng Z-L, Lowe JB, Ward PA. Protective effects of oligosaccharides in P-selectin-dependent lung injury. Nature 1993; 364:149–151.
114. Mulligan MJ, Watson SR, Fennie C, Ward PA. Protective effects of selectin chimeras in neutrophil-mediated lung injury. J Immun 1993; 151:6410–6417.
115. Coughlan AF, Hau H, Dunlop LC, Berndt MC, Hancock WW. P-selectin and platelet-activating factor mediate initial endotoxin-induced neutropenia. J Exp Med 1994; 179:329–334.
116. Ceconni O, Nelson RM, Roberts WG, et al. Inositol polyanions—noncarbohydrate inhibitors of L- and P-selectin that block inflammation. J Biol Chem 1994; 269(21):15060–15066.
117. Lewinsohn DM, Bargatze RF, Butcher EC. Leukocyte-endothelial cell recognition: evidence of a common molecular mechanism shared by neutrophils, lymphocytes, and other leukocytes. J Immun 1987; 138:4313–4321.
118. Watson SR, Fennie C, Lasky LA. Neutrophil influx into an inflammatory site inhibited by a soluble homing receptor-IgG chimera. Nature 1991; 349: 164–167.
119. Grober JS, Bowen EL, Ebling H, et al. Monocyte-endothelial adhesion in chronic rheumatoid arthritis. In situ detection and integrin-dependent interactions. J Clin Invest 1993; 91(6):2609–2619.
120. Katayama M, Handa M, Araki Y, et al. Soluble P-selectin is present in normal circulation and its plasma level is elevated in patients with thrombotic thrombocytopenic purpura and hemolytic uremic syndrome. Br J Hematol 1993; 84(4):702–710.
121. Chong BH, Murray B, Berndt MC, Dunlop LC, Brighton T, Chesterman CN. Plasma P-selectin is increased in thrombotic consumptive platelet disorders. Blood 1994; 83(6):1535–1541.
122. Palabrica T, Lobb R, Furie BC, et al. Leukocyte accumulation promoting fibrin deposition is mediated in vivo by P-selectin on adherent platelets. Nature 1992; 359(6398):848–851.
123. Celi A, Pellegrini G, Lorenzet R, et al. P-selectin induces the expression of tissue factor on monocytes. PNAS 1994; 92:8767–8771.
124. Altieri DC, Mannucci PM, Capitanio AM. Binding of fibrinogen to human monocytes. J Clin Invest 1986; 78:968–976.
125. Okada Y, Copeland BR, Mori E, Tung M-M, Thomas WS, del Zoppo GJ. P-selectin and intercellular adhesion molecule-1 expression after focal brain ischemia and reperfusion. Stroke 1994; 25(1):202–211.
126. Weyrich AS, Ma X-L, Lefer AM. Anti-P-selectin monoclonal antibody PB 1.3 pretreatment decreased myocardial damage in a feline nyocardial ischemia/reperfusion model. J Clin Invest 1994; 91:2620–2629.
127. Winn RK, Vedder NB, Paulson JC, Harlan JM. Monoclonal antibodies to P-selectin are effective in preventing reperfusion injury to rabbit ears. J Clin Invest 1993; 92:2042–2047.

128. Aruffo A, Dietsch MT, Wan H, Hellstrom KE, Hellstrom I. Granule membrane protein 140 (GMP-140) binds to carcinomas and carcinoma-derived cell lines. PNAS 1992; 89:2292–2296.
129. Ogata S, Muramatsu T, Kobata A, New structural characteristics of the large glycopeptides from transformed cells. Nature 1976; 259:580–582.
130. Wagner DD, Ivatt R, Destree AT, Hynes RO. Similarities and differences between the fibronectins of normal and transformed hamster cells. J Biol Chem 1981; 256:11708–11715.
131. Hubbard CS. Differential effects of oncogenic transformation on N-linked oligosaccharide processing at individual glycosylation sites of viral glycoproteins. J Biol Chem 1987; 262:16403–16411.

13

Use of Host Adhesion Molecules by Infectious Agents

Anthony R. Berendt and Christopher J. McCormick
Adhesion and Infection Group, Nuffield Department of Medicine, University of Oxford, Oxford, England

I. INTRODUCTION: THE IMPORTANCE OF ADHESION IN INFECTION

Adhesion is a central event in microbial pathogenesis. Not only are adhesion-dependent host responses critical in determining the outcome of infection (see, for example, the disastrous effects of the leukocyte adhesion deficiency syndromes, 1,2), but in addition, pathogens establish infection in specific target organs or tissues in a multistep, adhesion-dependent process. This begins with early events such as colonization of epithelial surfaces, often followed by invasion leading to disease. Many pathogens have the ability to traverse epithelial barriers (either primarily or by exploiting traumatic breaches) and to invade deeper tissues, including the bloodstream, which can then act as a means for dissemination throughout the body. While certain organisms (such as *Plasmodium* spp., the causative agents of malaria) establish clinically important infection exclusively in the vascular space, others circulate only transiently in the bloodstream, exiting it to enter specific sites such as the Peyers patches in the gut in the case of *Salmonella typhi* (the cause of typhoid fever), endothelial cells in the case of measles virus, or the central nervous system for poliovirus. All of these interactions involve the recognition of, and adhesion to, specific structures in target tissues, but a full discussion of these diverse interactions is outside the scope of this chapter.

A particularly interesting subset of host-pathogen adhesive interactions concern cases where the pathogen utilizes as its adhesion receptor a structure that is also an adhesion molecule for the host. The evolutionary pressures that drive such biological behavior are unclear, but several questions spring to mind. Does it confer a particular advantage on the pathogen, perhaps directly interfering with adhesion-dependent host responses? Does it confer tissue specificity? Does it lead to signal transduction events in the host cell that benefit the pathogen? Or are host adhesion molecules simply part of a wide range of potential receptors that the host cannot afford to delete or mutate, that some pathogens have evolved to adhere to? It is likely that all these possibilities are true in different contexts.

In the rest of this review, we shall consider only those host-pathogen interactions where adhesion molecules are utilized as receptors. Because of the diversity of microorganisms involved, we shall consider them under the different families of host structures used, relating them where possible to physiological function and to the molecules used by the pathogen to interact. Finally, we shall speculate on the prospects for clinical intervention in this area.

II. DEFINING AN ADHESION RECEPTOR FOR A PATHOGEN

Certain basic criteria must be fulfilled to identify a given molecule as a receptor, and in many respects these also apply to defining it as an adhesion or invasion receptor. The receptor must be expressed on target cells for adhesion or invasion and appropriate polyclonal or monoclonal antibodies (mAbs) directed against the receptor should inhibit the interaction in question. Purified native or recombinant receptor should mediate binding if presented on a suitable surface, and this interaction should also be inhibited by appropriate antibodies. Transfection of a nonadherent cell line with complementary DNA (cDNA) leading to expression of the putative receptor should confer on the transfected cell the ability to interact with the pathogen, once again inhibited by antireceptor antibodies. Finally, soluble receptor or peptides derived from it may inhibit adhesion or invasion, although this criterion may not always be fulfilled if the affinity of the interaction is low.

In addition to these requirements for proof of receptor activity, the biological relevance of a particular adhesion pathway must be assessed if a therapeutic intervention is the long-term aim. In many cases multiple adhesion mechanisms exist, and in these situations it may be that interrupting a single pathway is of little use. Such assessments of relevance often require more complex systems than those used to identify a receptor, so that where they exist, good animal models remain important.

Finally, discussion of host receptors would be incomplete without consideration of the counterreceptors on the pathogen. In this regard three different categories of interaction have been defined (3): *masking*, where a host component binds to the pathogen surface allowing interaction with the host component receptor(s); *mimicry*, where the pathogen has evolved a ligand with an identical motif to the physiological ligand of the host receptor, hence interacting in an indistinguishable manner; and *ancillary ligand recognition*, where the host receptor is bound at a site outside the physiologic one or by a structure unique to the pathogen.

III. HOST ADHESION MOLECULES USED AS RECEPTORS

A. Immunoglobulin Superfamily

1. ICAM-1

This 90-112 kDa transmembrane glycoprotein is used as an adhesion receptor by two quite different pathogens, rhinovirus (4–6) (a causative agent of the common cold), and the malarial parasite *Plasmodium falciparum* (7). ICAM-1 is made up of five tandemly arranged immunoglobulin-like domains, a transmembrane anchor, and a short cytoplasmic domain (8,9). In the case of rhinovirus, the molecule functions as both an attachment and apparently an invasion receptor on nasal epithelium; in the case of malaria, as an attachment receptor for the adhesion of infected erythrocytes to vascular endothelial cells. The binding sites on ICAM-1 for these two interactions have been mapped, demonstrating distinctive mechanisms of adhesion when the two pathogens are compared. Much is known about the structure on the rhinovirus that mediates this interaction. By contrast, the nature of the cognate ligand on the surface of the malaria-infected red cell is less well understood, although it is almost certainly borne on the erythrocyte variant surface antigen, clones of which have recently been isolated and sequenced.

a. Rhinovirus. A number of agents can cause the infection of nasal and upper airways epithelium that gives rise to the recognizable symptoms of the common cold, but the rhinoviruses are the most common cause. There are a large number of antigenically distinct serotypes, each of which generates a specific neutralizing antibody response in vivo. This means that many infections can occur before there is a significant probability that an individual has been exposed to a given serotype in the past; hence repeated infections are the rule.

Early studies showed that rhinoviruses would adhere specifically to certain target cells in vitro; the adhesion of labeled rhinovirus could be inhibited by an excess of unlabeled virus of the same serotype. It then became clear that the vast majority of rhinovirus serotypes would compete for the binding of an individual strain, suggesting that they all bound to the same receptor. These serotypes were therefore classified as the major serogroup, and they comprise some 85 of the identified 101 serotypes (10). Two minor serogroups, each apparently recognizing a different receptor (11), comprise the remaining serotypes (12).

The identification of the major serogroup receptor became an area of considerable interest. Monoclonal antibodies (mAbs) raised against HeLa cells were identified that were capable of inhibiting adhesion of members of the major serogroup (13). Using such antibodies, the receptor was initially characterized as an 89-kDa membrane glycoprotein with a substantial proportion of its molecular mass accounted for by N-linked carbohydrate (14). Subsequently, ICAM-1 was identified as the major serogroup receptor by three different approaches. A rhinovirus research group identified ICAM-1 by screening for virus binding to a panel of mouse-human hybrid cell lines (4). At the same time, following the molecular cloning of a cDNA encoding ICAM-1, a leukocyte adhesion group screened COS cells transiently expressing ICAM-1 for the ability to bind rhinovirus. Transfectants bound virus, and this interaction was inhibited by known anti-ICAM-1 mAbs (5). Finally, a third group used their blocking mAb against the major serogroup receptor on HeLa cells to screen an expression library, and clones encoding ICAM-1 were identified (6).

The finding that ICAM-1 was the receptor for the major serogroup was of some biological interest, not only because of the central role that ICAM-1 plays in immune cell adhesion via the interaction with its physiological ligands LFA-1 and Mac-1 but also because of known features about the structure of the viral capsid. As a picornavirus, rhinovirus comprises an RNA genome and an icosahedral capsid made up of 60 copies of each of four protein subunits designated VP1, VP2, VP3, and VP4. Crystallographic studies indicate that at each of the fivefold axes of symmetry, a deep groove or "canyon" is formed by two of the subunits on each face, effectively constituting a circular depression around the fivefold axis (15). Nucleic acid sequencing studies had indicated that the residues comprising the outer, surface-exposed borders of the canyon were hypervariable, while the floor was conserved, implying that it might form the binding site on the viral capsid for the cellular receptor (16).

Further evidence for this hypothesis came from mutagenesis studies on the virus, in which key residues for binding were identified as lying predominantly on the canyon floor. The dimensions of the canyon are such that the floor is inaccessible to the paired immunoglobulin domains of an antibody combining site; neutralizing antibody must instead bind to the hypervariable residues at the rim of the canyon, and hence rhinovirus neutralization is serotype-specific (17).

ICAM-1 has a five-domain structure which appears as a rod-shaped monomer on electron microscopy of recombinant soluble protein. The monomer has appropriate dimensions to fit into the canyon (18). A molecular model of ICAM-1 has been visually "docked" into the coordinates of the canyon (19), and cryoelectron microscopy of complexes of rhinovirus with a 2-domain soluble form of ICAM-1 shows a viral capsid decorated with receptor in a manner strongly suggestive of docking with the canyon (20).

The rhinovirus binding site on ICAM-1 has been mapped using domain deletion techniques, site-directed mutagenesis, and mouse-human chimeras (homolog scanning mutagenesis). In these studies it has been established that the binding site is on the first, amino-terminal domain, as is that for LFA-1. To generate the proper conformation of domain 1, domain 2 is also needed. Blocking mAbs map to domains 1 and 2 (21); critical residues map to domain 1 at the predicted apex (the B-C and F-G loops in the immunoglobulin fold) and on the sides of the domain (18,22,23). These data are also consistent with the docking of the first domain into the rhinovirus canyon; mAbs may inhibit viral adhesion either by covering the binding site or by sterically hindering the domain as a whole from engaging with the canyon.

b. Plasmodium falciparum. An integral part of the life-cycle of the malaria parasite *Plasmodium falciparum* is the adhesion of erythrocytes infected with mature parasites to postcapillary venular endothelial cells. This results in the withdrawal of such forms from the blood (which therefore only contains red cells infected with the immature ring forms) and their sequestration in deep vascular beds in a number of organs. The localization of large numbers of infected cells in the microvasculature of vital organs is thought to be an important factor in the pathogenesis of the disease, and when this process occurs in the brain, the condition of cerebral malaria may result. This feared complication of malaria is characterized by coma, often accompanied by convulsions, and carries a high mortality.

Laboratory studies have demonstrated that the process of adhesion to endothelial cells is a specific one mediated by receptor-ligand interactions. A number of host molecules have been identified that can act as receptors in vitro; these include the platelet, monocyte, and microvascular endothelial membrane glycoprotein CD36 (24–26); the secreted glycoprotein thrombospondin (27) (for which CD36 acts as a cellular receptor, 28); ICAM-1 (7); and for a minority of strains E-selectin (29), VCAM (29), and chondroitin-

4-sulfate (30). There is considerable diversity between parasite strains, with CD36 and thrombospondin (TSP) acting as apparently universal receptors, while a subset of strains adhere to other receptors. It has recently become clear that *Plasmodium falciparum* undergoes clonal variation in the expression of both surface antigens and adhesive properties (31,32), and there is now strong evidence that the variant surface antigen is also the ligand for the host receptors.

By contrast, two studies have addressed the binding site for the infected erythrocyte on the ICAM-1 molecule (33,34). Using domain deletion constructs, point mutants, and mouse-human chimeras, both groups showed that the binding site is located in domain 1. The greatest reductions in binding were seen with mutations predicted to lie on the B strand in the C-E region, which was predicted to form an extended loop running essentially at right angles to the plane of the β-strands (34). Consistent with this, the epitopes for inhibitory monoclonal antibodies were identified as being located on closely related parts of the molecule, also mapping to the C-E loop. These findings are therefore in contrast with those for rhinovirus and indeed with previous data on the binding site for LFA-1, which binds to the opposite face of the β-barrel, in addition making critical contacts with the B-C and F-G loops. Thus certain mAbs that block infected erythrocyte adhesion fail to block LFA-1 and vice versa, with a group of mAbs blocking both interactions.

The corollary of the distinct natures of the binding sites on ICAM-1 for these two pathogens is that the structures of the cognate ligands are likely to be quite different. Although there are similar themes in the great degree of antigenic diversity displayed by rhinovirus and *P. falciparum* malaria, the localized nature of the malarial binding site suggests the lack of a canyon-type structure on the parasite-derived ligand, especially as the region of ICAM-1 in question would not be predicted to have a prominent structure that would "dock" into a canyon. The parasite-derived ICAM-1 binding moiety is almost certainly carried on the variant surface antigen, a 200- to 350-kDa protein inserted in the infected erythrocyte membrane that is encoded by a larger diverse gene family, members of which have recently been cloned (35–37). Although the structure-function relationships of this molecule have not yet been elucidated, it does contain between one and four copies of a consensus domain also found in a number of malarial proteins that are involved in red cell invasion, one of which has been shown to bind to the Duffy blood group antigen. Thus the DBL (Duffy binding-like) domains may represent receptor-recognizing modules the precise specificities of which may be determined by primary sequence differences in an analogous manner to immunoglobulin superfamily members.

Does the use of ICAM-1 as an adhesion receptor have implications for pathogenesis? In the case of rhinovirus, no particular significance beyond its suitability as a picornavirus receptor has been inferred. ICAM-1 is expressed on nasal epithelium where it may be expected to play a role in local inflammatory responses to infection, but where it confers susceptibility to viral attack. Its monomeric nature gives it the right dimensions to fit into the canyon, and its crucial role in immune responses means that residues in the first domain are probably unlikely to mutate successfully in a manner that confers resistance to rhinovirus while preserving LFA-1 binding.

In the case of malaria, however, we have argued that the fact that only a subset of parasite strains can adhere to ICAM-1, together with its cytokine-regulated expression, may explain why only a proportion of patients with *P. falciparum* malaria develop severe disease (38). Levels of TNF correlate with severity in malaria (39), and this may be matched by variable expression of ICAM-1. In the absence of a good animal model of cerebral malaria (although primate models have recently been suggested; 40), this hypothesis has had to be tested in an indirect manner. We have recently shown enhanced levels of ICAM-1 on endothelium in the brains of fatal cases of cerebral malaria and found that parasitized red cells colocalized with ICAM-1 expression and to a lesser extent with other receptors such as CD36 and E-selectin (41).

Furthermore evidence of endothelial activation is seen on skin biopsies from nonfatal malaria (G. Turner and A. Berendt, in preparation), is inferred from elevated levels of circulating adhesion molecules in the plasma of malarious patients (42–45) and has been seen in the recently-developed primate models of cerebral malaria (40,46). A comprehensive comparison of the adhesive characteristics of field isolates from Kenyan children showed that binding to ICAM-1 was greater in isolates from cases of clinical malaria than in cases of asymptomatic parasitemia, though binding to ICAM-1 was not significantly associated with disease severity (C. I Newbold et al., in preparation).

Finally, we have recently demonstrated that binding of infected erythrocytes to microvascular endothelium is greatly enhanced by ICAM-1 expression even when CD36 is expressed. Based on antibody inhibition experiments, it appears that ICAM-1 and CD36 operate synergistically when expressed together (C. J. McCormick and A. Berendt, in preparation). Interestingly, ICAM-1 has been shown to act as a rolling receptor for infected red cells (47). CD36, by contrast, is a stationary receptor, adhesion to which may be enhanced by prior rolling on ICAM-1. Thus for malaria, the case is increasingly strong that adhesion to ICAM-1 plays an important role in the pathogenesis of disease, though other receptors, notably CD36, probably also play key roles.

2. VCAM-1

This 110-kDa member of the immunoglobulin superfamily is composed of seven Ig domains, though splice variants exist. It shows inducible expression on vascular endothelium where it mediates adhesion via the integrin VLA-4 ($\alpha_4\beta_1$). It is also expressed on developing myoblasts, on stromal cells in the bone marrow, and on Bowman's capsule in the kidney. It is known to be subverted as a receptor by two pathogens; encephalomyocarditis virus (EMC virus) in mice and *Plasmodium falciparum*-infected erythrocytes.

a. EMC virus. This picornavirus is capable of causing several diseases in experimental animals, including diabetes, pancreatitis, and myocarditis. In a recent report (48), an anti-murine VCAM-1 mAb inhibited (by c. 50%) the lysis of murine cardiac endothelial cells by the EMC-D strain. CHO cells transfected with murine VCAM-1 became sensitive to viral lysis; control transfectants were minimally affected. Anti-VCAM-1 mAb inhibited lysis of transfectants by over 80%. In direct binding studies, radiolabeled virions bound to VCAM-1 transfectants but not to control transfectants and this binding was abolished by the inhibitory mAb. The study leaves some unanswered questions; for example, the inhibitory effect on endothelial cell lysis was incomplete, suggesting that other receptors might be operating. Nonetheless, this appears to be another clear example of a picornavirus binding to an Ig superfamily receptor. Although the viral structures involved have not received the same scrutiny as poliovirus or rhinovirus, the characteristics of VCAM-1 would also make it a suitable receptor for docking into a canyon.

b. Plasmodium falciparum. *Plasmodium falciparum*-infected erythrocytes adhere to VCAM-1 (29), but few details are available about this interaction. In experiments with a Thai strain, adhesion to activated endothelium was seen that was not blocked by anti-ICAM-1 mAbs and that had a time course more characteristic of VCAM-1 expression. The same parasite was shown to adhere to recombinant VCAM-1 (expressed in CHO cells) such that it could be blocked with an anti-VCAM-1 mAb. Subsequently, a parasite line was selected in vitro by panning sequentially on CD36, ICAM-1, VCAM-1, and E-selectin that had the ability to bind all four receptors. Such parasite strains would appear to be very rare in vivo; in a study of Kenyan field isolates, levels of adhesion to VCAM-1 were low and to E-selectin minimal (A. Berendt, unpublished; C. I. Newbold et al., in preparation). Further details of this interaction and its relationship to antigenic variation are unavailable. The pathological significance is unclear, but the same arguments may apply as for ICAM-1 adhesion; certainly VCAM-1 can also be detected on multiple vascular beds in the tissues of individuals who have died from malaria (29,41). Furthermore VCAM-1 may be able to synergize with CD36 just as ICAM-1 appears to, enhancing its role.

3. Other IgSF Receptors

Other members of the immunoglobulin superfamily act as pathogen receptors but are not proven to be adhesion molecules. These include the major receptor for human immunodeficiency virus (HIV), the T cell marker CD4. As for interactions of pathogens with ICAM-1, the binding site for the HIV gp120 molecule is in the first domain of CD4 (49). This is also the case for the poliovirus receptor, a monomeric 3-domain member of the IgSF which mediates viral adhesion and uptake (50). The function of the poliovirus receptor is unknown, but its expression on neural tissue is thought to be suggestive of a role in intercellular adhesion in the nervous system. Finally, murine hepatitis virus has been shown to adhere to members of the carcinoembryonic antigen (CEA) family of molecules (51).

B. Integrins

An extensive number of organisms subvert integrins as cellular receptors. We shall discuss these under the broad headings of β_1-, β_2-, and β_3-integrins.

1. β_1-Integrins as Receptors for Pathogens

a. Echovirus. Echoviruses are picornaviruses that may cause fatal disseminated infections in the newborn, meningitis, encephalitis, and a range of less severe illnesses such as fever with rash or respiratory symptoms. In view of the strong relationship between picornaviridae and Ig superfamily receptors, the fact that integrin $\alpha_2\beta_1$ (VLA-2) is the echovirus 1 receptor (52) is of considerable interest and somewhat surprising. This was demonstrated by raising mAbs against HeLa cells, which are permissive for infection, and screening for protection in a cellular assay. Two independent mAbs were obtained, both of which inhibited lysis of HeLa cells. One of these, AA10, protected at high multiplicity of infection and inhibited binding of radiolabeled virions by 90%. Each mAb immunoprecipitated a 125-kDa and a 145-kDa protein pair from lysates of surface-labeled HeLa cells. This pattern is the same as that seen on immunoprecipitating with antibodies to α_2 or β_1, and preclearing experiments demonstrated that each mAb indeed recognized one of the two subunits of VLA-2. When the α_2 subunit was transfected into RD rhabdomyosarcoma cells, which express β_1 but, in their normal state, no α_2, the transfectants acquired the ability to bind labeled virions and became susceptible to infection. These effects were inhibited with the mAb AA10. Thus VLA-2 acts as a receptor for echovirus 1 (52), breaking the paradigm of monomeric Ig-superfamily receptors for picornaviridae binding via the viral canyon.

Subsequent studies have demonstrated that the critical element of the VLA-2 structure for the interaction with echovirus 1 is the α_2 subunit (53). In common with the β_2-integrins, this has the 200-amino acid insertion similar to the A domain of Von Willebrand factor, cartilage matrix protein, and certain complement components, which is termed the I-domain. It is now clear that for the β_2-integrins CD11a/CD18 and CD11b/CD18, the I domain is a ligand binding site (54–56). This issue has been examined for echovirus 1/VLA-2 interactions. For the binding of both the physiologic ligand (collagen) and echovirus 1, the I-domain is essential and inhibitory mAbs are mapped to this region (57,58).

In addition, a recombinant soluble I-domain–glutathione-S-transferase fusion protein (I-GST) directly mediated the adhesion of virus, whereas GST alone or GST fused to the I-domain of the CD11b chain failed to mediate adhesion (59). Inhibitory mAbs both recognized the I-GST protein and blocked adhesion to it. Finally, I-GST blocked adhesion to HeLa cells and inhibited plaque formation, with 50% reduction in adhesion using 3 to 30 nM I-GST. Thus the I-domain of the α_2 subunit is necessary and sufficient for virus attachment and can act as a functional inhibitor (59).

b. Yersinia: The Invasin Paradigm. One of the most remarkable and complete accounts of a host-pathogen interaction at the molecular level has come from studies of the attachment to and invasion of epithelial cells by the bacterial pathogen *Yersinia. Y. enterocolitica* is a cause of diarrhea and abdominal pain; *Y. pseudotuberculosis* a cause of mesenteric adenitis, with enlarged lymph nodes in the abdomen. Both infections are widely distributed in animal reservoirs, and both are almost always self-limiting in humans. In experimental models, *Y. enterocolitica* causes a disseminated disease, and the first step in this is the invasion of intestinal epithelial cells. Such cells are not normally competent to take up particles or organisms, but are induced to do so by specific bacterial products. It has become clear that a key role in this phenomenon is played by the bacterial protein invasin, which uses as a cellular receptor the β_1-integrins, though two other bacterial gene products have also been identified that mediate uptake.

Invasin was identified using a genetic approach (60). The noninvasive *E. coli* strain K-12 was used to screen for genes from the invasive *Y. pseudotuberculosis* that would allow uptake into HEp2 cells. A 3.2-kb locus designated *inv* was identified that contained a gene coding for a 986-amino acid, 103-kDa outer membrane protein (61). Using a similar approach, the invasin of *Y. enterocolitica* was shown to be a 2.5-kb gene coding for an 835-amino acid polypeptide with a predicted molecular mass of approximately 91 kDa. This molecule is homologous with 85% of the *Y. pseudotuberculosis* invasin and within these regions is 73% identical at the amino acid level.

Direct proof of the role of invasin came from the demonstration that outer membrane protein extracts from *Y. pseudotuberculosis* invasin-expressing *E. coli*, immobilized on plastic surfaces, would mediate the adhesion of HEp2 cells (62). By contrast, outer membrane extracts of isogenic, invasin-negative *E. coli* could not mediate adhesion. When the outer membrane preparations were subjected to polyacrylamide gel electrophoresis and transferred to nitrocellulose, adhesion of HEp2 cells as detected using amido black staining or scanning electron microscopy occurred only to the extract of invasin-expressing *E. coli* and only to a set of protein bands unique to that strain. Immunoblotting using invasin-specific mAbs confirmed that the invasin bands comigrated with the bands mediating attachment in both transfected *E. coli* strains and in *Y. pseudotuberculosis*. Extracts of an *inv*-negative mutant strain of *Y. pseudotuberculosis* expressed no invasin as determined by immunoblotting and did not mediate adhesion of HEp2 cells to the filters. Finally, the anti-invasin mAbs were shown to inhibit adhesion of transfected *E. coli* strains.

This unequivocal identification of invasin as a protein causing attachment and entry of *Y. pseudotuberculosis* to a range of target cells was followed by the localization of the binding domain to the C-terminus (63). Using a panel of mAbs raised against invasin and a set of invasin deletion constructs expressed on cell surfaces or as fusions to maltose-binding protein, inhibitory mAbs were mapped to the C-terminal 192 amino acids (residues 795–986) (64). The C-terminal 192-amino acids were necessary and, when expressed as the fusion protein, sufficient to mediate adhesion of HEp2 cells. Furthermore, a secreted 53 kDa deletion construct of invasin containing the C-terminus not only bound HEp2 cells but also inhibited invasion by invasin-expressing bacteria. One critical feature within the 192-amino acid adhesive molecule was subsequently shown, by mutagenesis and analysis of invasion of bacteria expressing the mutant protein or of adhesion of cells to purified protein, to be a 76-amino acid disulfide-bonded loop (65). A further mutagenesis study of the C-terminal 192-amino acids has identified a critical 11-amino acid stretch from residues 903–913, containing an essential aspartate at position 911 that may function similarly to the aspartate in an RGD motif (66). Finally, the C-terminus may be of additional importance as addition of residues affects function, perhaps by altering the conformation of the nearby crucial sequence (67).

Of considerably more interest in the context of our discussion, however, is the finding that the receptor(s) for invasin are members of the β_1-integrin family. Since the secreted 53 kDa invasin construct bound in a manner indistinguishable from wild-type invasin, it was used to make an affinity column down which octylglucoside extracts of surface-biotinylated HEp2 cells were passed. The column was eluted with EDTA (the invasin-invasin receptor interaction had been found to be divalent cation-dependent) to yield two protein bands of molecular mass 130 to 135 kDa on reducing gels with other properties suggestive of an integrin. Polyclonal anti-β_1 antiserum recognized one of the proteins, and the pair was recognized by a polyclonal antiserum raised against human placenta-derived fibronectin receptor, $\alpha_5\beta_1$. Since multiple cell lines, many of which do not express $\alpha_5\beta_1$, bind invasin, extracts of other cell lines (EJ, HPB-MLT, and K562) were passed over the invasin column. In addition to obtaining $\alpha_5\beta_1$ again (from the K562 cells), the eluate yielded $\alpha_3\beta_1$ from EJ cells and $\alpha_4\beta_1$ from HPB-MLT cells. A platelet extract was invasin affinity-purified to obtain material for N-terminal sequencing, and this yielded $\alpha_6\beta_1$. This material was reconstituted into phosphatidylcholine vesicles which proved able to bind active but not inactive invasin and which also bound the 120-kDa cell attachment fragment of fibronectin (68).

Despite the diverse array of β_1-integrins competent to bind invasin, the process is specific and not simply a function either of the β_1 subunit alone or of all β_1-integrins. Thus $\alpha_2\beta_1$ was never observed to bind even when present in cells being analyzed. When incorporated into phospholipid vesicles, human platelet $\alpha_2\beta_1$, isolated by collagen-agarose affinity chromatography, was able to bind type 1 collagen but not invasin, whereas the platelet integrin isolated from the invasin affinity column bound invasin but not collagen. Further proof that the β_1-integrins are the invasin receptors was supplied by using polyclonal anti-$\alpha_5\beta_1$ to block the attachment of HEp2 cells to immobilized invasin. MAbs against $\alpha_5\beta_1$ or against the β_1 subunit (both of which block the physiological function of the relevant integrins) inhibited attachment of K562 cells to invasin, and mAbs against β_1 or α_5 blocked invasin-promoted entry of a transfected *E. coli* strain into HEp2 cells.

These findings raised a number of interesting questions, some of which have already been resolved. Where on the β_1-integrins was invasin binding? How did this binding result in cell invasion? And what was the significance of these findings for our understanding of the pathogenesis of disease due to *Yersiniae*?

Subsequent studies have shown that for the interaction with the fibronectin receptor VLA-5 ($\alpha_5\beta_1$) invasin competes for the binding of fibronectin. Furthermore, mAbs against VLA-5 either block both interactions or neither. Finally, the RGDS peptide blocks binding of invasin. On the basis of these data, although invasin does not have an RGD motif in the primary sequence, it appears to bind to a site very close, if not identical, to that for the physiological ligand (69). The affinity of this interaction is high (approx. Kd 5 nM), about 1000-fold higher than that of fibronectin. This may explain the observation that for adhesion via VLA-4 ($\alpha_4\beta_1$), binding is relatively unaffected by the integrin activation essential for binding of most integrin ligands (70). Indeed, a mutagenesis study of VLA-4, which tested the roles of the divalent cation-binding domains in integrin function, found little effect on invasin binding of mutations which affect divalent cation binding and markedly affect binding of VCAM-1 or the CS-1 fragment of fibronectin (71).

The mechanism by which β_1-integrin engagement leads to cellular uptake has recently been examined, and it appears that uptake can be triggered as a function of receptor density and occupancy (72). This occurs with invasin even though β_1-integrins are not generally recognized as receptors involved in phagocytosis. Particles coated with fibronectin attach to cells but are not internalized; yet invasin-bearing cells are. To elucidate the mechanism of this, bacteria (*Staphylococcus aureus*, which does not normally enter epithelial cells and which will bind antibodies via surface-expressed protein A) were coated with anti-integrin mAbs and then used in an invasion assay with HEp-2 cells. It was found that coated bacteria were efficiently internalized if anti-$\alpha_3\beta_1$ or anti-β_1 mAbs were used. In HEp-2 cells, anti-$\alpha_5\beta_1$ mAbs mediated poor uptake, but this could be increased by overexpression, through transfection, of $\alpha_5\beta_1$ in HEp-2 cells.

Using a panel of mAbs against $\alpha_5\beta_1$ that were shown to recognize multiple independent epitopes, it was found that the phenomenon did not depend on the blocking activity of the mAb or on the epitope recognized. There was, however, a clear correlation between the affinity of the mAb and the efficiency with which it directed internalization of coated bacteria. This was also seen with mAb-coated latex beads and in further support of a critical role for affinity in directing internalization, a mutant of invasin with low efficiency of invasion was found to have a much reduced affinity of binding. Finally, the issue of competition for receptor was assessed by assaying adherence and uptake of latex particles coated with anti-$\alpha_5\beta_1$ mAbs by cells plated on substrates coated with mAbs of varying affinity for $\alpha_5\beta_1$. It was found that high-affinity mAbs could direct invasion, effectively competing for receptors on the basolateral surfaces of the HEp-2 cells, but that this was most efficient when the substrate was a low-affinity mAb or contained none at all. By contrast, lower-affinity mAbs were able to direct internalization only if the cells were plated on low-affinity substrates (72).

These data suggest a reason for the marked difference in the properties of a fibronectin-coated particle and an invasin-coated one. Fibronectin, as a low-affinity ligand, directs attachment but not internalization because many receptors in the region of the attached particle remain unoccupied, and hence the signal for internalization is not generated. An invasin-coated particle has sufficient affinity to recruit many integrin molecules at once, including those that may have been attached to the basal extracellular matrix. This process of recruitment leads to rearrangements of the cytoskeleton and generates signals for internalization. Although the nature of this is unclear, it appears to involve protein phosphorylation as inhibitors of tyrosine kinase inhibit invasin-mediated uptake (73).

The biological significance of invasin remains to be fully determined. Although its broad receptor range confers a wide cellular host range on it, its role as a virulence factor is complex. It seems clear from studies with mutants failing to express invasin that it plays an important role in the early stages of infection via the oral route. It is not indispensable, however, and in the later stages of challenge appears to play little part. Furthermore, if a double mutant of *Y. pseudotuberculosis* (which is genetically related to the more virulent *Y. pestis*, the plague bacillus) is constructed that lacks both the *inv* gene and the gene for an outer membrane protein *yopA*, the mutant is surprisingly *more* virulent, not less (74). Since *Y. pestis* expresses neither gene but has a nonfunctional *yopA* gene, it may well have arisen by mutation from *Y. pseudotuberculosis*. Irrespective of the role of invasin in natural infection, the elucidation of the details of this host-pathogen interaction has not only told us much about a highly specialized bacterium, but has also yielded valuable insights into integrin function.

c. Trypanosoma cruzi. This protozoan parasite infects 10 million to 12 million people in South America (over half of them in Brazil) and causes Chagas' disease, a chronic illness affecting the heart and gut in particular. The parasite is introduced via abrasions in the skin that are contaminated by the feces of various species of blood-feeding insects. Once inside the human host, parasites invade a variety of cell types and multiply intracellularly before being released to disseminate via the blood and invade cells in other tissues.

Using human macrophages as a model system, a panel of anti-β_1-integrin mAbs were used to assess the role of integrins in invasion (75). It was already known that *Trypanosoma cruzi* can adhere to fibronectin and that this is mediated by the RGD motif (76). Certain anti-β_1 mAbs were found to inhibit uptake of trypanosomes and to protect cells against death; these activities correlated with the ability of the mAbs to block binding of labeled fibronectin to the macrophages. Levels of inhibition of the uptake of parasites approached 90%, and the same levels of cytoprotection were seen. Which α-subunits are involved was not resolved. Thus it seems likely that parasites are able to bind fibronectin and subsequently enter cells using a range of β_1-integrins. However, it is important to note that interactions with nonphagocytic cells (clearly relevant *in vivo* in the invasion of muscle) proceed by a quite distinct and novel mechanism involving the recruitment of lysosomes at the site of invasion. In this respect certain sulfated glycoconjugates may be important as receptors, and the integrins may be of lesser importance (see 77 for review).

d. Toxoplasma gondii. The clinical manifestations of this obligate intracellular protozoan range from asymptomatic infection through an acute glandular feverlike illness to congenital infections and in immunosuppressed individuals, an encephalitis with mass lesions in the brain. *Toxoplasma* can survive both in specialized phagocytic cells and in nonphagocytic cells. Its attachment and invasion mechanisms have been reviewed recently (78,79). These include carbohydrate-mediated ones, but of relevance to this discussion is the role of laminin and its receptors (80).

Laminin, but not fibronectin, can be demonstrated by indirect immunofluorescence and by Western blotting on the surface of parasites isolated from infected mouse tissues. Antibodies to laminin inhibit binding to CHO cells and by mAb studies, the C-terminus seems particularly important. Maximal levels of inhibition with anti-laminin antibody vary from 30% (polyclonal) to 60% (mAb). It seems that $\alpha_6\beta_1$ plays an important role as a receptor for this parasite-bound laminin. Similar maximal levels of inhibition can be achieved on both CHO cells and human foreskin fibroblasts with anti-α_6 mAb (60%) and anti-β_1 mAb (30% to 40% on human fibroblasts). Interestingly, anti-α_3 mAbs and polyclonal anti-$\alpha_3\beta_1$ consistently enhance adhesion by up to 40%. Whether this is due to altered interactions with laminin on the parasite surface is unclear. Although only 60% inhibition can be achieved with the anti-α_6 mAb (80), it seems unlikely that other laminin-binding proteins play a major role, partly because of the data with polyclonal anti-laminin antibodies and partly because of the lack of effect of antisera against the known non-integrin laminin-binding proteins.

2. β_2-Integrins: Mac 1

A wide range of intracellular pathogens use the β_2-integrin CD11b/CD18 to recognize and enter macrophages, which are then used as host cells. In almost all cases, a central role in the process is played by adsorbed complement component C3bi to direct the uptake of the pathogen. These organisms include *Rhodococcus equi* (a cause of pneumonia in immunocompromised humans and in foals) (81), *Legionella pneumophila* (Legionnaires' disease) (82), *Bordetella pertussis* (whooping cough) (83), *Mycobacterium tuberculosis* (tuberculosis) (84), *Mycobacterium leprae* (leprosy) (85,86), *Histoplasma capsulatum* (histoplasmosis), *Cryptococcus neoformans* (meningitis, usually in immunocompromised individuals), and *Leishmania major*, *Leishmania donovani*, and *Leishmania mexicana* (causes of leishmaniasis).

Why a cell as potentially hostile as the macrophage should be such a favored ecological niche for so many pathogens is an enigma; perhaps it reflects its avidly phagocytic nature, the wide variety of mechanisms by which it can take up particles, and the inherent longevity of the monocyte-macrophage lineage. Once inside the cell, there are a variety of means by which the pathogen can either prevent the formation of a typical phagolysosome or survive the conditions within it, but one mechanism appears to be to enter via the Mac-1 pathway which, in the absence of simultaneous Fc-receptor engagement, fails to trigger the respiratory burst. Of all these organisms using similar pathways, the greatest number of studies have been performed on *Leishmania*, studies that have highlighted an important potential pitfall in this area.

a. Leishmania–Mac-1 Interactions. Organisms in the genus *Leishmania* affect large numbers of individuals in the tropics. The clinical disease ranges from a chronic but ultimately self-healing skin ulcer through a progressively destructive disease of the mucosal lining of the mouth, nose, and larynx to a disseminated disease affecting liver, spleen, and bone marrow, depending in part on the species and in part on poorly defined host factors. Pathologically, all species are intracellular pathogens that inhabit macrophages. It was recognized some time ago that the third component of complement was involved both in invasion and in intracellular survival of *Leishmania*.

With attention therefore focused on macrophage complement receptors, it subsequently emerged that two further different surface structures interacted with Mac-1, apparently directly—the lipophosphoglycan (LPG) that is a major component of the cell membrane (87), and a parasite surface glycoprotein, gp63. These appeared to bind to different sites on Mac-1; it was speculated that the LPG bound at the same site as bacterial lipopolysaccharide. Further support for direct interaction between these parasite ligands and Mac-1 appeared to come when the sequence of gp63 was determined and an RGD motif was found within it; particularly as binding of gp63 to Mac-1 was inhibited by RGD-containing peptides or antibodies to them. It later transpired, however, that the ascertainment of an RGD motif in the gp63 protein was erroneous, owing to a sequencing error (88).

Fascinatingly, however, another region was identified containing the sequence SRYD which was shown to mimic an RGDS sequence in its activity (89), including blocking parasite attachment and antibody binding when contained within a larger peptide. These data appeared to demonstrate a direct interaction, but using purified Mac-1 immobilized on surfaces, three species of *Leishmania* were found to be dependent on complement for their adhesion to Mac-1 (90). It seems likely that in many experimental systems, the target macrophages synthesize and release enough complement components to opsonize the *Leishmania* for adhesion to Mac-1.

This consideration applies to other pathogens; *Rhodococcus* is absolutely dependent on complement for entry into macrophages (81), as are the mycobacteria, *Legionella* and *Cryptococcus*. *Histoplasma* may interact directly, as may *Bordetella*, though these have not been assessed on purified Mac-1. The observation did, however, demonstrate that great care is necessary to prove direct interactions between host receptor and pathogen, especially when the host receptor binds a range of ligands, any one of which might be synthesized by the target cell, adsorbed onto its surface from the culture medium, or adsorbed onto the surface of the pathogen.

3. β_3-Integrins

Five pathogens have been shown to adhere to cells via β_3-integrins. These are two more picornaviruses, coxsackie A9 and the agriculturally important foot-and-mouth disease virus (FMDV); *Neisseria meningitidis*, a major cause of bacterial meningitis; *Mycobacterium avium-intracellulare*, an opportunist pathogen in immunosuppressed, chiefly HIV-positive, patients; and finally the bacterial agent of Lyme disease, the spirochete *Borrelia burgdorferi*. In addition adenovirus uses a β_3-integrin (and a β_5) to enter host cells, though not to attach to them. Although β_3-integrins do not classically have a phagocytic role, $\alpha_v\beta_3$ has been shown to be involved in the phagocytosis of apoptotic neutrophils by macrophages (91) and must therefore be capable of interacting with the cytoskeleton in a manner that can result in internalization of particles bearing ligands.

a. Coxsackievirus A9. Coxsackieviruses cause a range of illnesses ranging from respiratory tract infections through febrile illnesses with rashes to viral meningitis, encephalitis, and myocarditis. There is a large number of serotypes, some of which use ICAM-1 as a receptor. However, coxsackievirus A9 was found to have an unusual feature on sequence analysis; an insertion of 15-amino acids at the C-terminus of the VP1 capsid subunit including an RGD sequence (92). Since RGD-containing peptides inhibit infection, it was predicted that an integrin might be the cellular receptor. Studies in GMK (green monkey kidney) cells indicated that the integrin in question is $\alpha_v\beta_3$ (93). Antisera against $\alpha_v\beta_3$, but not against $\alpha_5\beta_1$, inhibited infection. A GMK lysate purified on virus or viral-peptide columns yielded $\alpha_v\beta_3$, whereas $\alpha_5\beta_1$ was purified from the same lysate on a GRGDSP column. Proteins affinity-purified on viral columns inhibited infection. In a survey of a variety of other enteroviruses, only echovirus 22 competed for the binding of radiolabeled coxsackievirus A9 to GMK cells, indicative of binding to the same receptor or to another molecule found on the cell surface uniquely in association with $\alpha_v\beta_3$.

b. Foot-and-Mouth-Disease Virus (FMDV). This is another example of a picornavirus that contains an RGD sequence in one of the capsid subunits. A considerable amount is known about the structure of the virus and the binding region. As has been shown for the RGD sequence in the type III modules of fibronectin (94), the FMDV RGD sequence is present on an exposed loop. There are a variety of serotypes of the virus, and it appears that neutralizing antibody recognizes the flanking residues, not the RGD sequence itself. Recognition by antibody appears to be hampered by a degree of structural disorder in the loop. Mutational analysis has shown that the RGD region is essential for binding (95) and RGD-containing peptides inhibit binding. Thus it was surmised for some time, but only recently proved, that an RGD-recognizing integrin or integrins could act as a cellular receptor.

When coxsackievirus A9 (CAV-9) was compared with FMDV, it was found that both would replicate well in the rhesus monkey cell line LLC-MK2 and that in addition FMDV would replicate well in BHK-21, HeLa, and BK-LF cells, while CAV-9 would not (96). However, radiolabeled CAV-9 was able to attach efficiently to BHK-21 cells, even though it could not replicate. Binding of labeled FMDV or CAV-9 to BHK-21 cells was inhibited by excess unlabeled homologous or heterologous virus, suggesting that the two shared the same cellular receptor.

On LLC-MK2 cells the CAV-9 binding was inhibited by FMDV, but not vice versa. Scatchard plots indicated a higher affinity of interaction on LLC-MK2 cells with FMDV, and it was speculated that FMDV was able to displace CAV-9 by virtue of its higher affinity. A polyclonal antiserum directed against the vitronectin receptor (anti-$\alpha_v\beta_3/\beta_5$) inhibited FMDV adhesion to LLC-MK2 cells (by 74%), though plaque formation was only inhibited by 55%. One mAb directed against β_3 inhibited binding by 53% but did not affect plaque formation; a mAb directed against the α_v subunit inhibited binding by 39%, but inhibited plaque formation by 79% (96). These data imply distinctions between epitopes critical for binding and those for internalization, and although adhesion to purified $\alpha_v\beta_3$ has not been studied, the data do appear to support a role for this integrin in adhesion of FMDV.

Whether other RGD-recognizing integrins also play roles remains to be proved. The pathway is not the only one capable of mediating invasion; antibody-neutralized virions can also infect cells, including normally non-permissive ones, apparently via an Fc-receptor-dependent mechanism (95,97).

c. *Neisseria meningitidis.* Of the causes of bacterial meningitis, perhaps that most feared by clinicians is the gram-negative diplococcus *Neisseria meningitidis*. As well as having the ability to cause rapidly progressive meningitis, it can cause overwhelming septicemia either independently or combined with meningeal infection. A hallmark of meningococcal septicemia is the development of necrotic skin lesions containing bacteria, and it is assumed that such lesions result from interactions of bloodborne bacteria with endothelial cells and subsequent tissue damage. Furthermore, with this organism meningitis itself arises from hematogenous seeding of bacteria, an event that must at some point require transmigration of the endothelial barrier. For these reasons, the adhesion of *Neisseria meningitidis* to endothelial cells has come under scrutiny and avid adherence has been demonstrated in vitro to human umbilical vein endothelial cells (98).

Two different bacterial structures—class I pili (and in particular the PilC subunit it carries), and an outer membrane protein designated Opc—appear to contribute to adhesion. Pili project a considerable distance from the cell surface and are particularly important in adherence when a polysaccharide capsule is expressed, presumably because this is otherwise inhibitory to adhesion. In the absence of pili, an encapsulated meningococcus does not adhere to HUVEC, but if pili are expressed, such organisms will adhere. If pili are lacking but the bacterium is unencapsulated, adhesion can still occur via the Opc protein (99), suggesting that it is much closer to the membrane and unable to interact with its receptor if capsule is present.

Host receptors have not been identified for the class I pili, but for Opc, $\alpha_v\beta_3$ has been shown to be important (100). Adhesion of unencapsulated, nonpiliated, Opc^+ strains is serum-dependent, and this effect is also seen with purified vitronectin. Bacteria adhere to purified vitronectin, and these effects are blocked with a mAb against Opc. RGDS peptide, but not RGES, inhibits serum- or vitronectin-enhanced binding to HUVEC, and both interactions are blocked by mAbs directed against α_v or β_3 but not by mAbs against $\alpha_5\beta_1$. Interestingly, the importance of this receptor is most evident for established monolayers of endothelial cells; other receptors may be exposed if endothelial cells are partially detached (100). What is clear is that this represents another example of adhesion via "masking," with a crucial host component adsorbed to the bacterial surface and then interacting with its physiological receptor.

d. Mycobacterium avium-intracellulare. MAI, as this organism is commonly known to clinicians, is an important cause of late morbidity and accelerated mortality in the acquired immunodeficiency syndrome (AIDS). It causes a disseminated disease with large numbers of organisms present in many organs, the principal symptoms of which are high fever and weight loss. In common with other mycobacteria already discussed, it is predominantly an intracellular pathogen.

A single report has suggested that it uses $\alpha_v\beta_3$ as an adhesion receptor (101). Crude monocyte extract immobilized on a nitrocellose filter was shown to bind labeled bacteria, whereas in these experiments human serum and fibronectin did not (though fibronectin has previously been shown to act as a receptor when immobilized on the polystyrene; 102). An extract of the monocytic cell line THP-1 also bound, as did purified human placental $\alpha_v\beta_3$. Adhesion to the monocyte extract was qualitatively reduced by depletion with suspensions of MAI or of GRGDSPK peptide coupled to sepharose 4B beads, but not with sepharose 4B coupled to an anti-α_v mAb. However, an 8-M urea eluate of the mAb-coupled beads did show some binding, as (very weakly) did an EDTA eluate of the peptide-coupled beads or the whole organisms. This material was subjected to polyacrylamide gel electrophoresis and Western blotting, and bands of 138- and 95-kDa (reduced) or 122- and 103-kDa (nonreduced) were detected with an anti-$\alpha_v\beta_3$ polyclonal. There was some inhibition (up to 40%) in adhesion to monocyte-derived macrophages with mAb LM609, which recognizes the $\alpha_v\beta_3$ complex, and a similar level of inhibition was seen if adherent macrophages were depleted of $\alpha_v\beta_3$ by growth on mAb-coated matrix. Depletion of CD14, CD18, and $\alpha_2\beta_1$ had little effect. Thus $\alpha_v\beta_3$ appears to play a role in adhesion, but based on these data it is highly likely that other quantitatively important adhesion pathways exist.

e. Adenovirus: $\alpha_v\beta_3$- (and $\alpha_v\beta_5$)-Directed Internalization. Adenovirus is a nonenveloped DNA virus associated with childhood respiratory infection and infantile diarrhea. Structurally the virion has 12 vertices, each formed from two distinct components; a 400-kDa protein called the penton base, composed of five identical subunits, and a 30-nm-long protein, the fiber protein, composed of three identical 62-kDa subunits. The fiber protein inserts into a central cavity in the penton base.

Adenovirus-infected cells release a soluble factor which can cause tissue culture cells infected in vitro to detach from their underlying surface. This proved to be soluble penton base. As the coding sequence of this predicts an RGD sequence at residues 485–488, it was possible that detachment resulted from direct competition for RGD-dependent integrins by penton base. This hypothesis was tested by examining the adhesion of cell lines to recombinant penton base, vitronectin, laminin, or type I collagen. A number of cell lines bound to penton base in a manner indistinguishable from attachment to vitronectin—inhibited by EDTA and the GRGDSP peptide, but not by GDRESP. Adhesion to collagen or laminin was not affected by any of these conditions. GRDGSP or soluble penton base reduced infectivity, presumably by competing for receptor, and plating cells on penton base or vitronectin, expected to deplete the cell surface of receptor by recruitment to the basolateral surface, also reduced infectivity. Cells lacking the α_v integrin subunit bound to vitronectin and penton base poorly and were relatively refractory to infection; transfection with α_v led to high levels of binding to both, and increased infectivity. Both binding to penton base and infectivity were inhibited specifically by GRGDSP and by a combination of the mAbs LM609 (anti-$\alpha_v\beta_3$) and P3G2 (anti-$\alpha_v\beta_5$) though not by these mAbs individually (103).

When different components of the virus were examined it was found that the fiber protein bound with high affinity to cells. However, only penton base and the penton base-fiber protein complex were internalized significantly, even though the affinity of binding of penton base was some 30 times lower. Internalization of whole virus was shown to have similar kinetics to that of penton base, or intact penton-fiber protein complex. Using the α_v-deficient cell line and its α_v-transfected derivative, internalization of intact virus was shown to be dependent on the presence of the α_v subunit. Binding of virus was unaffected by peptides or mAbs, but internalization was inhibited by the same mAb combination (anti-$\alpha_v\beta_3$ and anti-$\alpha_v\beta_5$) and by GRGDSP (103). Thus distinct mechanisms operate for the attachment of adenovirus to cells (the fiber protein binding to an unknown receptor) and internalization (penton base binding to $\alpha_v\beta_3$ and $\alpha_v\beta_5$), and this appears to operate for multiple serotypes (104). As in all the other cases, the nature of the signals generated after integrin engagement remains unclear.

f. Borrelia burgdorferi. The Lyme disease spirochete is inoculated into the human by the bite of the *Ixodes* tick, and initially it causes a reddened skin lesion which slowly enlarges and is known as *erythema annulare*. At some point the organism is disseminated, and subsequently a number of different complications can arise including arthritis, heart conduction and rhythm abnormalities, and neurological problems.

The nature of the skin lesion and the ability to disseminate implies that adhesion to endothelium is important and indeed *Borrelia* will invade endothelial cells (105), though the mechanisms are unknown. The only molecular information available to date concerns adhesion to platelets. It was found that a low-passage, infectious strain would attach to activated, but not resting, platelets; a high-passage, noninfectious strain bound 30- to 50-fold less well (106). This binding was inhibited with EDTA, with an RGD peptide at 2 mg/ml and with the γ-peptide of fibrinogen at 1 mg/ml.

Using monoclonal antibodies to block attachment to immobilized platelets, $\alpha_{IIb}\beta_3$ was identified as the receptor; a mAb against either chain inhibited binding. Bacteria failed to bind to platelets from an individual with Glanzmann's thrombocytopenia (lacking $\alpha_{IIb}\beta_3$). Furthermore using affinity-purified $\alpha_{IIb}\beta_3$, bacteria were shown to bind directly to the integrin, and this was inhibited by the blocking mAbs, but not by an anti $\alpha_5\beta_1$ mAb (106). Thus $\alpha_{IIb}\beta_3$ is necessary and sufficient for binding to platelets. This does, however, leave unresolved the mechanism of binding to endothelial cells, which do not express $\alpha_{IIb}\beta_3$. It is not known whether a serum component plays a role in this adhesion, though the adherent strain does remain adhesive after some passages in serum-free medium.

IV. PROSPECTS FOR INTERVENTION

How does this information help us in the therapeutic sense? Just as for interventions aimed at other adhesion-dependent functions such as inflammation, there are a number of key questions to be answered. These center around efficacy, the kind of therapeutic agent to be used, toxicity, and, ultimately, the commercial market.

For an agent to be efficacious, it must successfully reduce adhesion of the pathogen to a level that makes a significant impact on clinical disease. The size of such a required reduction is unclear. Although we often do not know what the relationship is between pathogen load and disease severity, it is assumed that substantial reductions in pathogen adhesion are necessary (to prevent the next step in the disease process). In some cases, such as the invasion of cells by virus, complete inhibition may be necessary to prevent infection.

In order to achieve this, however, there are a number of requirements. First, it is essential that the inhibitor interact with a sufficiently high affinity. If the inhibitor used is a soluble receptor or a peptide derived from its primary sequence, this may not necessarily be the case, as some cell-cell interactions are of low affinity. Second, even if a soluble receptor is effective, the possibility must be considered that the pathogen might mutate to use a novel receptor, a novel site on the same receptor, or to recognize the receptor with lowered affinity such that it can still adhere to a cell by multiple receptor-ligand bonds but can no longer be blocked by a monomeric inhibitor. Finally, many pathogens have multiple receptors, and it then becomes important to identify which ones are critical in the form of disease one wants to prevent or treat and how much ability there is to switch from one receptor to another.

In addition to these considerations, issues of toxicity include the potential for the inhibitor to interfere with the physiological function of the adhesion molecule. This is particularly the case when the pathogen uses mimicry or masking to interact with the binding site for the physiological ligand or when, despite using ancillary ligand recognition, it nonetheless binds at a closely related site. The final concern is the hard reality of the commercial environment in which drug development takes place. Only those major infections that present a pressing need for new therapies are likely to be worth considering. In this context it is hard to see most of the infectious agents discussed here offering attractive markets; the most prevalent infections (such as malaria) occur in the poorest areas of the world where the problem is not affording new treatments but affording any treatment at all.

Despite this pessimism, clear proof of principle is available that numerous key host-pathogen adhesive events can be inhibited in vitro. Furthermore since there is intense interest in adhesion molecules as therapeutic targets in inflammatory disease, it may also be that drugs developed to affect their expression or their signaling functions may usefully inhibit the interaction with the pathogen in some situations. In addition, the development of adhesion inhibitors generates further information about the mechanisms and consequences of the host-pathogen interaction, which may in itself be a pointer to other therapeutic approaches.

With these points in mind we can see a wide range of situations where soluble receptor or peptide inhibits interactions. Soluble ICAM-1 (107–109) or ICAM-1-immunoglobulin Fc chimeras (110) inhibit the binding of a wide range of the serotypes in the major serogroup of rhinovirus. As expected, a dimeric or higher order multimeric construct is more efficient at performing this function as the avidity of the molecule is raised by its multivalency (110). In the case of poliovirus, it was possible readily to select for lower-affinity mutants which still bound to cell surfaces, but were not inhibited by individual molecules in solution (111). A panel of such mutants has been extensively characterized, indicating that almost all mutations are on the floor and walls of the canyon, probably at the residues that contact the poliovirus receptor (112). Such mutants have also proved possible to select with rhinovirus (113), and moderate levels of resistance persist in the absence of continued selection.

The frequency of mutation to resistance for both rhinovirus and poliovirus appears to be approximately 1 in 10^4 to 1 in 10^5, which is consistent with rates of escape from neutralizing mAbs (112,113). However, although there is a range in susceptibility of wild type rhinoviruses, *de novo* resistant clinical isolates have not been seen (114). Study of the nature of rhinovirus particles neutralized by soluble ICAM-1 has shown that the particles rapidly become "empty" with uncoating and loss of the RNA contents (115–117). It appears that the intact virions contain a hydrophobic molecule in a pocket beneath the floor of the canyon which acts to prevent uncoating, but which is displaced when the receptor binds. Soluble ICAM-1 may offer a therapeutic option in specialized situations where rhinovirus poses particular risks, particularly as no alternative receptor pathways have been described for viruses of the major serogroup.

Soluble receptor inhibition has also been described for the adhesion of malaria-infected red cells to ICAM-1 (33); again, an ICAM-1-Fc chimera was more efficient on a molar basis (118). In addition, two peptides based on the ICAM-1 primary sequence were found to inhibit binding at concentrations in the millimolar range (33). Translating these findings into clinical practice may be difficult, partly because multiple adhesion pathways are known to exist and the evidence that ICAM-1 is the critical one remains circumstantial, but partly too for commercial reasons.

Peptide analogs have been shown to inhibit in many other situations as outlined above; such experiments often form part of the proof that a host receptor has been identified. These include peptides based on fibronectin for the adhesion of *Trypanosoma cruzi* (76) and RGD-containing peptides for *Leishmania*, *Yersinia* (69), *Neisseria meningitidis* (100), and the RGD-presenting viruses (foot-and-mouth-disease virus, coxsackievirus A9, and adenovirus). Such agents might, however, have undesirable effects on host RGD-dependent functions, including coagulation.

The potential for intervention does not stop at the generation of adhesion-blocking peptides, mimeotopes, or other inhibitors. Identification of the molecular mechanism of a critical event in pathogenesis also brings with it a knowledge of the important structure on the surface of the pathogen which could form a component of a vaccine. In this regard it is of interest that immunization of mice with live *aroA*-recombinant (i.e., attenuated) *Salmonella typhimurium*-producing invasin induces anti-invasin antibodies and inhibits intestinal translocation of *Yersinia pseudotuberculosis* (119). Although this does not affect dissemination and ultimate mortality, the kinetics of the infection are very different over the first 2 or 3 days and indicate substantial inhibition of the earliest event in invasive disease. Certainly invasin could be one component of a vaccine against animal or human yersiniosis.

V. CONCLUSIONS

We have discussed the burgeoning group of organisms for which host adhesion molecules have been identified as receptors and the differing ways in which these receptors are subverted. A dominant theme is the use of integrins to direct the uptake of a wide range of invasive pathogens, presumably reflecting proven or possible physiological roles (certainly in the cases of CD11bCD18 and $\alpha_v\beta_3$). Secondary to this is the widespread use of processes of molecular mimicry (with the RGD motif particularly vulnerable to this) or of masking by host proteins such as laminin, vitronectin, and fibronectin. Considering the very large number of additional organisms known to adhere to such extracellular matrix receptors, it is highly possible that yet more pathogens enter host cells either to pass through them or to take up residence inside them by means of integrins. Finally, it remains a realistic hope that as well as enhancing our understanding of host and pathogen biology, these insights will lead to the identification of critical targets in the armamentarium of the pathogen and hence to the identification of new vaccine candidates or drug targets for therapy.

ACKNOWLEDGMENTS

We thank our partners for personal support and Dr. Paul for a degree of patience that exceeded all reasonable expectation. ARB is a Lister Institute Research Fellow.

REFERENCES

1. Anderson DC, Springer TA. Leukocyte adhesion deficiency: an inherited defect in the Mac-1, LFA-1, and p150,95 glycoproteins. Annu Rev Med 1987; 38:175–194.
2. Harlan JM. Leukocyte adhesion deficiency syndrome: insights into the molecular basis of leukocyte emigration. Clin Immunol Immunopathol 1993; 67: S16–S24.
3. Hoepelman AI, Tuomanen EI. Consequences of microbial attachment: directing host cell functions with adhesions. Infect Immun 1992; 60:1729–1733.
4. Greve JM, Davis G, Meyer AM, et al. The major human rhinovirus receptor is ICAM-1. Cell 1989; 56:839–847.
5. Staunton DE, Merluzzi VJ, Rothlein R, Barton R, Marlin SD, Springer TA. A cell adhesion molecule, ICAM-1, is the major surface receptor for rhinoviruses. Cell 1989; 56:849–853.
6. Tomassini JE, Graham D, De Witt CM, Lineberger DW, Rodkey JA, Colonno RJ. cDna cloning reveals that the major group rhinovirus receptor on HeLa cells is intercellular adhesion molecule 1. Proc Natl Acad Sci USA 1989; 86:4907–4911.
7. Berendt AR, Simmons DL, Tansey J, Newbold CI, Marsh K. Intercellular adhesion molecule-1 is an endothelial cell adhesion receptor for *Plasmodium falciparum*. Nature 1989; 341:57–59.
8. Simmons D, Makgoba MW, Seed B. ICAM, an adhesion ligand of LFA-1, is homologous to the neural cell adhesion molecule NCAM. Nature 1988; 331: 624–627.
9. Staunton DE, Marlin SD, Stratowa C, Dustin ML, Springer TA. Primary structure of ICAM-1 demonstrates interaction between members of the immunoglobulin and integrin supergene families. Cell 1988; 52:925–933.
10. Hamparian VV, Colonno RJ, Cooney MK, et al. A collaborative report: rhinoviruses—extension of the numbering system from 89 to 100. Virology 1987; 159:191–192.
11. Mischak H, Neubauer C, Kuechler E, Blaas D. Characteristics of the minor group receptor of human rhinoviruses. Virology 1988; 163:19–25.
12. Uncapher CR, De Witt CM, Colonno RJ. The major and minor group recep-

tor families contain all but one human rhinovirus serotype. Virology 1991; 180:814–817.

13. Colonno RJ, Callahan PL, Long WJ. Isolation of a monoclonal antibody that blocks attachment of the major group of human rhinoviruses. J Virol 1986; 57:7–12.
14. Tomassini JE, Maxson TR, Colonno RJ. Biochemical characterization of a glycoprotein required for rhinovirus attachment. J Biol Chem 1989; 264: 1656–1662.
15. Rossmann MG, Arnold E, Erickson JW, et al. Structure of a human common cold virus and functional relationship to other picornaviruses. Nature 1985; 317:145–153.
16. Rossmann MG, Palmenberg AC. Conservation of the putative receptor attachment site in picornaviruses. Virology 1988; 164:373–382.
17. Rossmann MG. The canyon hypothesis. Hiding the host cell receptor attachment site on a viral surface from immune surveillance. J Biol Chem 1989; 264:14587–14590.
18. Staunton DE, Dustin ML, Erickson HP, Springer TA. The arrangement of the immunoglobulin-like domains of ICAM-1 and the binding sites for LFA-1 and rhinovirus. Cell 1990; 61:243–254. (Published errata appear in Cell 1990; 61(2):1157, and 1991; 66(6):following 1311.)
19. Giranda VL, Chapman MS, Rossmann MG. Modeling of the human intercellular adhesion molecule-1, the human rhinovirus major group receptor. Proteins 1990; 7:227–233.
20. Olson NH, Kolatkar PR, Oliveira MA, et al. Structure of a human rhinovirus complexed with its receptor molecule. Proc Natl Acad Sci USA 1993; 90:507–511.
21. Lineberger DW, Graham DJ, Tomassini JE, Colonno RJ. Antibodies that block rhinovirus attachment map to domain 1 of the major group receptor. J Virol 1990; 64:2582–2587.
22. McClelland A, de Bear J, Yost SC, Meyer AM, Marlor CW, Greve JM. Identification of monoclonal antibody epitopes and critical residues for rhinovirus binding in domain 1 of intercellular adhesion molecule 1. Proc Natl Acad Sci USA 1991; 88:7993–7997.
23. Register RB, Uncapher CR, Naylor AM, Lineberger DW, Colonno RJ. Human-murine chimeras of ICAM-1 identify amino acid residues critical for rhinovirus and antibody binding. J Virol 1991; 65:6589–6596.
24. Barnwell JW, Ockenhouse CF, Knowles DM. Monoclonal antibody OKM5 inhibits the in vitro binding of *Plasmodium falciparum*-infected erythrocytes to monocytes, endothelial, and C32 melanoma cells. J Immunol 1985; 135: 3494–3497.
25. Ockenhouse CF, Chulay JD. *Plasmodium falciparum* sequestration: OKM5 antigen (CD36) mediates cytoadherence of parasitized erythrocytes to a myelomonocytic cell line. J Infect Dis 1988; 157:584–588.
26. Ockenhouse CF, Tandon NN, Magowan C, Jamieson GA, Chulay JD. Identification of platelet membrane glycoprotein as a falciparum malaria sequestration receptor. Science 1989; 243:1469–1471.

27. Roberts DD, Sherwood JA, Spitalnik SL, et al. Thrombospondin binds falciparum malaria parasitized erythrocytes and may mediate cytoadherence. Nature 1985; 318:64–66.
28. Asch AS, Barnwell J, Silverstein RL, Nachman RL. Isolation of the thrombospondin membrane receptor. J Clin Invest 1987; 79:1054–1061.
29. Ockenhouse CF, Tegoshi T, Maeno Y, et al. Human vascular endothelial cell adhesion receptors for *Plasmodium falciparum*-infected erythrocytes: roles for endothelial leukocyte adhesion molecule 1 and vascular cell adhesion molecule 1. J Exp Med 1992; 176:1183–1189.
30. Rogerson SJ, Chaiyoroj SC, Ng K, Reeder JC, Brown GV. Chondroitin sulphate A is a cell surface receptor for *Plasmodium falciparum*-infected erythrocytes. J Exp Med 1995; 182:15–20.
31. Biggs BA, Gooze L, Wycherley K, et al. Antigenic variation in *Plasmodium falciparum*. Proc Natl Acad Sci USA 1991; 88:9171–9174.
32. Roberts DJ, Craig AG, Berendt AR, et al. Rapid switching to multiple antigenic and adhesive phenotypes in malaria. Nature 1992; 357:689–692.
33. Ockenhouse CF, Betageri R, Springer TA, Staunton DE. *Plasmodium falciparum*-infected erythrocytes bind ICAM-1 at a site distinct from LFA-1, Mac-1, and human rhinovirus. Cell 1992; 68:63–69. (Published erratum appears in Cell 1992; 68(5):following 994.)
34. Berendt AR, McDowall A, Craig AG, et al. The binding site on ICAM-1 for *Plasmodium falciparum*-infected erythrocytes overlaps, but is distinct from, the LFA-1-binding site. Cell 1992; 68:71–81.
35. Baruch DI, Pasloske BL, Singh HB, et al. Cloning the *P. falciparum* gene encoding PfEMP1, a malarial variant antigen and adherence receptor on the surface of parasitized human erythrocytes. Cell 1995; 82:77–87.
36. Su X, Heatwole VM, Wertheimer SP, et al. The large diverse gene family *var* encodes proteins involved in cytoadherence and antigenic variation of *Plasmodium falciparum*-infected erythrocytes. Cell 1995; 82:89–100.
37. Smith JD, Chitnis CE, Craig AG, et al. Switches in expression of *Plasmodium falciparum var* genes correlate with changes in antigenic and cytoadherent phenotypes of infected erythrocytes. Cell 1995; 82:101–110.
38. Berendt AR, Ferguson DJ, Gardner J, et al. Molecular mechanisms of sequestration in malaria. Parasitology 1994; 108:S19–S28.
39. Kwiatkowski D, Hill AV, Sambou I, et al. TNF concentration in fatal cerebral, non-fatal cerebral, and uncomplicated *Plasmodium falciparum* malaria. Lancet 1990; 336:1201–1204.
40. Aikawa M, Brown A, Smith CD, et al. A primate model for human cerebral malaria: *Plasmodium coatneyi*-infected rhesus monkeys. Am J Trop Med Hyg 1992; 46:391–397.
41. Turner GD, Morrison H, Jones M, et al. An immunohistochemical study of the pathology of fatal malaria. Evidence for widespread endothelial activation and a potential role for intercellular adhesion molecule-1 in cerebral sequestration. Am J Pathol 1994; 145:1057–1069.
42. Hviid L, Theander TG, Elhassan IM, Jensen JB. Increased plasma levels of

soluble ICAM-1 and ELAM-1 (E-selectin) during acute *Plasmodium falciparum* malaria. Immunol Lett 1993; 36:51–58.

43. Wenisch C, Looareesuwan S, Parschalk B, Graninger W. Soluble vascular cell adhesion molecule 1 is elevated in patients with *Plasmodium falciparum* malaria. J Infect Dis 1994; 169:710–711. Letter.
44. Wenisch C, Varijanonta S, Looareesuwan S, Graninger W, Pichler R, Wernsdorfer W. Soluble intercellular adhesion molecule-1 (ICAM-1), endothelial leukocyte adhesion molecule-1 (ELAM-1), and tumor necrosis factor receptor (55 kDa TNF-R) in patients with acute *Plasmodium falciparum* malaria. Clin Immunol Immunopathol 1994; 71:344–348.
45. Jakobsen PH, Morris-Jones S, Ronn A, et al. Increased plasma concentrations of sICAM-1, sVCAM-1 and sELAM-1 in patients with *Plasmodium falciparum* or *P. vivax* malaria and association with disease severity. Immunology 1994; 83:665–669.
46. Fujioka H, Millet P, Maeno Y, et al. A nonhuman primate model for human cerebral malaria: rhesus monkeys experimentally infected with *Plasmodium fragile*. Exp Parasitol 1994; 78:371–376.
47. Cooke BM, Berendt AR, Craig AG, MacGregor J, Newbold CI, Nash GB. Rolling and stationary cytoadhesion of red blood cells parasitized by *Plasmodium falciparum*: separate roles for ICAM-1, CD36 and thrombospondin. Br J Haematol 1994; 87:162–170.
48. Huber SA. VCAM-1 is a receptor for encephalomyocarditis virus on murine vascular endothelial cells. J Virol 1994; 68:3453–3458.
49. Peterson A, Seed B. Genetic analysis and monoclonal antibody and HIV binding sites on the human lymphocyte antigen CD4. Cell 1988; 54:65–72.
50. Mendelsohn CL, Wimmer E, Racaniello VR. Cellular receptor for poliovirus: molecular cloning, nucleotide sequence, and expression of a new member of the immunoglobulin superfamily. Cell 1989; 56:855–865.
51. Yokomori K, Lai MM. Mouse hepatitis virus utilizes two carcinoembryonic antigens as alternative receptors. J Virol 1992; 66:6194–6199.
52. Bergelson JM, Shepley MP, Chan BM, Hemler ME, Finberg RW. Identification of the integrin VLA-2 as a receptor for echovirus 1. Science 1992; 255: 1718–1720.
53. Bergelson JM, St. John N, Kawaguchi S, et al. Infection by echoviruses 1 and 8 depends on the $\alpha2$ subunit of human VLA-2. J Virol 1993; 67:6847–6852.
54. Diamond MS, Garcia-Aguilar J, Bickford JK, Corbi AL, Springer TA. The I domain is a major recognition site on the leukocyte integrin Mac-1 (CD11b/CD18) for four distinct adhesion ligands. J Cell Biol 1993; 120:1031–1043.
55. Randi AM, Hogg N. I domain of $\beta2$ integrin lymphocyte function-associated antigen-1 contains a binding site for ligand intercellular adhesion molecule-1. J Biol Chem 1994; 269:12395–12398.
56. Landis RC, McDowall A, Holness CL, Littler AJ, Simmons DL, Hogg N. Involvement of the "I" domain of LFA-1 in selective binding to ligands ICAM-1 and ICAM-3. J Cell Biol 1994; 126:529–537.

57. Kamata T, Puzon W, Takada Y. Identification of putative ligand binding sites within I domain of integrin $\alpha2\beta1$ (VLA-2, CD49b/CD29). J Biol Chem 1994; 269:9659–9663.
58. Bergelson JM, St. John NF, Kawaguchi S, et al. The I domain is essential for echovirus 1 interaction with VLA-2. Cell Adhes Commun 1994; 2:455–464.
59. King SL, Cunningham JA, Finberg RW, Bergelson JM. Echovirus 1 interaction with the isolated VLA-2 I domain. J Virol 1995; 69:3237–3239.
60. Isberg RR, Falkow S. A single genetic locus encoded by *Yersinia pseudotuberculosis* permits invasion of cultured animal cells by *Escherichia coli* K-12. Nature 1985; 317:262–264.
61. Isberg RR, Voorhis DL, Falkow S. Identification of invasin: a protein that allows enteric bacteria to penetrate cultured mammalian cells. Cell 1987; 50: 769–778.
62. Isberg RR, Leong JM. Cultured mammalian cells attach to the invasin protein of *Yersinia pseudotuberculosis*. Proc Natl Acad Sci USA 1988; 85:6682–6686.
63. Leong JM, Fournier RS, Isberg RR. Identification of the integrin binding domain of the *Yersinia pseudotuberculosis* invasin protein. EMBO J 1990; 9: 1979–1989.
64. Leong JM, Fournier RS, Isberg RR. Mapping and topographic localization of epitopes of the *Yersinia pseudotuberculosis* invasin protein. Infect Immun 1991; 59:3424–3433.
65. Leong JM, Morrissey PE, Isberg RR. A 76-amino acid disulfide loop in the *Yersinia pseudotuberculosis* invasin protein is required for integrin receptor recognition. J Biol Chem 1993; 268:20524–20532.
66. Leong JM, Morrissey PE, Marra A, Isberg RR. An aspartate residue of the *Yersinia pseudotuberculosis* invasin protein that is critical for integrin binding. EMBO J 1995; 14:422–431.
67. Isberg RR, Yang Y, Voorhis DL. Residues added to the carboxyl terminus of the *Yersinia pseudotuberculosis* invasin protein interfere with recognition by integrin receptors. J Biol Chem 1993; 268:15840–15846.
68. Isberg RR, Leong JM. Multiple $\beta1$ chain integrins are receptors for invasin, a protein that promotes bacterial penetration into mammalian cells. Cell 1990; 60:861–871.
69. Van Nhieu GT, Isberg RR. The *Yersinia pseudotubersulosis* invasin protein and human fibronectin bind to mutually exclusive sites on the $\alpha5\beta1$ integrin receptor. J Biol Chem 1991; 266:24367–24375.
70. Ennis E, Isberg RR, Shimizu Y. Very late antigen 4-dependent adhesion and costimulation of resting human T cells by the bacterial $\beta1$ integrin ligand invasin. J Exp Med 1993; 177:207–212.
71. Masumoto A, Hemler ME. Mutation of putative divalent cation sites in the $\alpha4$ subunit of the integrin VLA-4: distinct effects on adhesion to CS1/fibronectin, VCAM-1 and invasin. J Cell Biol 1993; 123:245–253.
72. Tran Van Nhieu G, Isberg RR. Bacterial internalization mediated by $\beta1$ chain

integrins is determined by ligand affinity and receptor density. EMBO J 1993; 12:1887–1895.
73. Rosenshine I, Duronio V, Finlay BB. Tyrosine protein kinase inhibitors block invasin-promoted bacterial uptake by epithelial cells. Infect Immun 1992; 60: 2211–2217.
74. Rosqvist R, Skurnik M, Wolf-Watz H. Increased virulence of *Yersinia pseudotuberculosis* by two independent mutations. Nature 1988; 334:522–525.
75. Fernandez MA, Munoz-Fernandez MA, Fresno M. Involvement of $\beta 1$ integrins in the binding and entry of *Trypanosoma cruzi* into human macrophages. Eur J Immunol 1993; 23:552–557.
76. Ouaissi MA, Cornette J, Afchain D, Capron A, Gras-Masse H, Tartar A. *Trypanosoma cruzi* infection inhibited by peptides modeled from a fibronectin cell attachment domain. Science 1986; 234:603–607.
77. Andrews NA. Lysosome recruitment during host cell invasion by *Trypanosoma cruzi*. Trends Cell Biol 1995; 5:133–137.
78. Kasper LH, Mineo JR. Attachment and invasion of host cells by *Toxoplasma gondii*. Parasitol Today 1994; 10:184–188.
79. Sibley LD. Invasion of vertebrate cells by *Toxoplasma gondii*. Trends Cell Biol 1995; 5:129–132.
80. Furtado GdC, Cao Y, Joiner KA. Laminin on *Toxoplasma gondii* mediates parasite binding to the $\beta 1$ integrin receptor $\alpha 6\beta 1$ on human foreskin fibroblasts and chinese hamster ovary cells. Infect Immun 1992; 60:4925–4931.
81. Hondalus MK, Diamond MS, Rosenthal LA, Springer TA, Mosser DM. The intracellular bacterium *Rhodococcus equi* requires Mac-1 to bind to mammalian cells. Infect Immun 1993; 61:2919–2929.
82. Payne NR, Horwitz MA. Phagocytosis of *Legionella pneumophila* is mediated by human monocyte complement receptors. J Exp Med 1987; 166:1377–1389.
83. Relman D, Tuomanen E, Falkow S, Golenbock DT, Saukkonen K, Wright SD. Recognition of a bacterial adhesion by an integrin: macrophage CR3 ($\alpha_M\beta_2$, CD11b/CD18) binds filamentous hemagglutinin of *Bordetella pertussis*. Cell 1990; 61:1375–1382.
84. Schlesinger LS, Bellinger-Kawahara CG, Payne NR, Horwitz MA. Phagocytosis of *Mycobacterium tuberculosis* is mediated by human monocyte complement receptors and complement component C3. J Immunol 1990; 144:2771–2780.
85. Schlesinger LS, Horwitz MA. Phagocytosis of *Mycobacterium leprae* by human monocyte-derived macrophages is mediated by complement receptors CR1 (CD35), CR3 (CD11b/CD18), and CR4 (CD11c/CD18) and IFN-gamma activation inhibits complement receptor function and phagocytosis of this bacterium. J Immunol 1991; 147:1983–1994.
86. Schlesinger LS, Horwitz MA. Phagocytosis of leprosy bacilli is mediated by complement receptors CR1 and CR3 on human monocytes and complement component C3 in serum. J Clin Invest 1990; 85:1304–1314.
87. Talamas-Rohana P, Wright SD, Lennartz MR, Russell DG. Lipophosphogly-

can from *Leishmania mexicana* promastigotes binds to members of the CR3, p150,95 and LFA-1 family of leukocyte integrins. J Immunol 1990; 144: 4817–4824.

88. Miller RA, Reed SG, Parsons M. Leishmania gp63 molecule implicated in cellular adhesion lacks an Arg-Gly-Asp sequence. Mol Biochem Parasitol 1990; 39:267–274.
89. Soteriadou KP, Remoundos MS, Katsikas MC, et al. The Ser-Arg-Tyr-Asp region of the major surface glycoprotein of *Leishmania* mimics the Arg-Gly-Asp-Ser cell attachment region of fibronectin. J Biol Chem 1992; 267:13980–13985.
90. Mosser DM, Springer TA, Diamond MS. *Leishmania* promastigotes require opsonic complement to bind to the human leukocyte integrin Mac-1 (CD11b/CD18). J Cell Biol 1992; 116:511–520.
91. Savill J, Hogg N, Ren Y, Haslett C. Thrombospondin cooperates with CD36 and the vitronectin receptor in macrophage recognition of neutrophils undergoing apoptosis. J Clin Invest 1992; 90:1513–1522.
92. Chang KH, Auvinen P, Hyypia T, Stanway G. The nucleotide sequence of coxsackievirus A9; implications for receptor binding and enterovirus classification. J Gen Virol 1989; 70:3269–3280.
93. Roivainen M, Piirainen L, Hovi T, et al. Entry of coxsackievirus A9 into host cells: specific interactions with $\alpha v\beta 3$ integrin, the vitronectin receptor. Virology 1994; 203:357–365.
94. Main AL, Harvey TS, Baron M, Boyd J, Campbell ID. The three-dimensional structure of the tenth type III module of fibronectin: an insight into RGD-mediated interactions. Cell 1992; 71:671–678.
95. Mason PW, Rieder E, Baxt B. RGD sequence of foot-and-mouth disease virus is essential for infecting cells via the natural receptor but can be bypassed by an antibody-dependent enhancement pathway. Proc Natl Acad Sci USA 1994; 91:1932–1936.
96. Berinstein A, Roivainen M, Hovi T, Mason PW, Baxt B. Antibodies to the vitronectin receptor (integrin $\alpha v\beta 3$) inhibit binding and infection of foot-and-mouth disease virus to cultured cells. J Virol 1995; 69:2664–2666.
97. Mason PW, Baxt B, Brown F, Harber J, Murdin A, Wimmer E. Antibody-complexed foot-and-mouth disease virus, but not poliovirus, can infect normally insusceptible cells via the Fc receptor. Virology 1993; 192:568–577.
98. Virji M, Kayhty H, Ferguson DJ, Alexandrescu C, Heckels JE, Moxon ER. The role of pili in the interactions of pathogenic *Neisseria* with cultured human endothelial cells. Mol Microbiol 1991; 5:1831–1841.
99. Virji M, Makepeace K, Ferguson DJ, Achtman M, Sarkari J, Moxon ER. Expression of the Opc protein correlates with invasion of epithelial and endothelial cells by *Neisseria meningitidis*. Mol Microbiol 1992; 6:2785–2795.
100. Virji M, Makepeace K, Moxon ER. Distinct mechanisms of interactions of Opc-expressing meningococci at apical and basolateral surfaces of human endothelial cells; the role of integrins in apical interactions. Mol Microbiol 1994; 14:173–184.

101. Rao SK, Ogata K, Catanzaro A. *Mycobacterium avium–M. intracellulare* binds to the integrin receptor $\alpha v\beta 3$ on human monocytes and monocyte-derived macrophages. Infect Immun 1993; 61:663–670.
102. Rao SP, Gehlsen KR, Catanzaro A. Identification of a $\beta 1$ integrin on *Mycobacterium avium–Mycobacterium intracellulare*. Infect Immun 1992; 60: 3652–3657.
103. Wickham TJ, Mathias P, Cheresh DA, Nemerow GR. Integrins $\alpha v\beta 3$ and $\alpha v\beta 5$ promote adenovirus internalization but not virus attachment. Cell 1993; 73:309–319.
104. Mathias P, Wickham T, Moore M, Nemerow G. Multiple adenovirus serotypes use αv integrins for infection. J Virol 1994; 68:6811–6814.
105. Comstock LE, Thomas DD. Characterization of *Borrelia burgdorferi* invasion of cultured endothelial cells. Microb Pathog 1991; 10:137–148.
106. Coburn J, Leong JM, Erban JK. Integrin $\alpha IIb\beta 3$ mediates binding of the Lyme disease agent *Borrelia burgdorferi* to human platelets. Proc Natl Acad Sci USA 1993; 90:7059–7063.
107. Marlin SD, Staunton DE, Springer TA, Stratowa C, Sommergruber W, Merluzzi VJ. A soluble form of intercellular adhesion molecule-1 inhibits rhinovirus infection. Nature 1990; 344:70–72.
108. Arruda E, Crump CE, Marlin SD, Merluzzi VJ, Hayden FG. In vitro studies of the antirhinovirus activity of soluble intercellular adhesion molecule-1. Antimicrob Agents Chemother 1992; 36:1186–1191.
109. Martin S, Martin A, Staunton DE, Springer TA. Functional studies of truncated soluble intercellular adhesion molecule 1 expressed in *Escherichia coli*. Antimicrob Agents Chemother 1993; 37:1278–1284.
110. Martin S, Casasnovas JM, Staunton DE, Springer TA. Efficient neutralization and disruption of rhinovirus by chimeric ICAM-1/immunoglobulin molecules. J Virol 1993; 67:3561–3568.
111. Kaplan G, Peters D, Racaniello VR. Poliovirus mutants resistant to neutralization with soluble cell receptors. Science 1990; 250:1596–1599.
112. Colston E, Racaniello VR. Soluble receptor-resistant poliovirus mutants identify surface and internal capsid residues that control interaction with the cell receptor. EMBO J 1994; 13:5855–5862.
113. Arruda E, Crump CE, Hayden FG. In vitro selection of human rhinovirus relatively resistant to soluble intercellular adhesion molecule-1. Antimicrob Agents Chemother 1994; 38:66–70.
114. Ohlin A, Hoover-Litty H, Sanderson G, et al. Spectrum of activity of soluble intercellular adhesion molecule-1 against rhinovirus reference strains and field isolates. Antimicrob Agents Chemother 1994; 38:1413–1415.
115. Greve JM, Forte CP, Marlor CW, et al. Mechanisms of receptor-mediated rhinovirus neutralization defined by two soluble forms of ICAM-1. J Virol 1991; 65:6015–6023.
116. Hoover-Litty H, Greve JM. Formation of rhinovirus-soluble ICAM-1 complexes and conformational changes in the virion. J Virol 1993; 67:390–397.
117. Casasnovas JM, Springer TA. Pathway of rhinovirus disruption by soluble

intercellular adhesion molecule 1 (ICAM-1): an intermediate in which ICAM-1 is bound and RNA is released. J Virol 1994; 68:5882–5889.

118. Staunton DE, Ockenhouse CF, Springer TA. Soluble intercellular adhesion molecule 1-immunoglobulin G1 immunoadhesion mediates phagocytosis of malaria-infected erythrocytes. J Exp Med 1992; 176:1471–1476.

119. Simonet M, Fortineau N, Beretti JL, Berche P. Immunization with live *aroA* recombinant *Salmonella typhimurium* producing invasin inhibits intestinal translocation of *Yersinia pseudotuberculosis*. Infect Immun 1994; 62:863–867.

14

Modulation of Leukocyte Trafficking in Infectious Diseases

Elaine Tuomanen
Rockefeller University, New York, New York

The classical paradigm of the acute inflammatory reaction in response to infection has been developed using endotoxin as the noxious stimulus. As described in detail throughout this book, activated endothelial cells recruit rolling leukocytes via selectins. Leukocytes are then activated via chemokines such as platelet-activating factor (PAF) and interleukin-8 (IL-8) and finally arrest and transmigrate to the site of inflammation through integrins, intercellular adhesion molecules (ICAMs), and platelet-endothelial cell adhesion molecule (PECAM).

Dissection of these molecular events has been accomplished in impressive detail and spawned significant hope of controlling inflammation and leukocyte trafficking to the benefit of the inflamed host. However, it is well recognized that leukocytes are required for control of infection and that leukocyte dysfunction renders the host at increased risk of serious disease. How can the technology of modulating leukocyte trafficking be applied to inflammation, specifically the setting of infectious diseases?

It is recognized that a patient requires medical attention because an acute phase response is already in progress. In fact, the diagnosis of infection frequently relies on the presence of leukocytes at the site of infection. Thus, the question posed here must be carefully rephrased to ask, "Can modulation of leukocyte trafficking in a host with leukocytes present at an infected site benefit the course of recovery in conjunction with appropriate antibiotic therapy?" This broad issue is addressed in three parts.

1. Is the mechanism of leukocyte trafficking to sites of infection similar from one class of infectious agent to another? This question addresses the chances of finding a broadly active antiinflammatory agent for infections.
2. When in the course of an infectious disease is there a rationale to modulate leukocyte trafficking? Although patients present to medical attention at vastly different times in the course of infection, this question addresses the possibility that there exists a point in time where initiation of antiinflammatory therapy is appropriate for all patients.
3. Which steps in leukocyte trafficking are amenable to down-modulation of inflammation in the context of infection? This section will review the studies that have tested the effect of blockade of steps in the adhesion cascade in assays specifically using infectious stimuli in vitro or in animal models.

I. MOLECULAR EVENTS IN GRAM-POSITIVE VS. GRAM-NEGATIVE INFLAMMATION

In the 1970s and 1980s, many biotechnology ventures were formed to take advantage of an apparently clear-cut enterprise in which the mediators of the acute-phase response were believed to be largely defined and therapeutic interventions to control inflammation were only a pharmaceutical race away. Toward the end of the 1980s and into the 1990s, however, there arose considerable skepticism that all was proceeding toward a well-defined finish line. Application of what seemed to be obvious new avenues of therapy did not yield predicted benefits. Often results of animal models or clinical trials were equivocal. A major assumption underlying these early studies was that the mechanisms of inflammation in the major forms of infection were similar. This is now known to be a misconception. Reevaluation of this assumption has set the stage for cautious application of a more complete understanding of inflammation in the context of infection such that future therapeutics will yield the benefits expected of them.

Classical descriptions of inflammation document a sequence of pathological events at an infected site proceeding from the development of a serous exudate, to leakage of erythrocytes and deposition of fibrin, and finally to the arrival of leukocytes (1). The initial reversible binding between polymorphonuclear leukocytes and vascular endothelial cells is mediated through endothelial surface proteins of the selectin family (2). The expression of selectins on endothelial cells is upregulated by the effects of inflammatory cytokines. These adhesion molecules recognize carbohydrate ligands on leukocytes and result in margination of leukocytes along the vessel walls and a reversible binding to the endothelial cells detected as "leukocyte rolling." Following engagement of selectin molecules, the leukocyte becomes tethered to the endothelium by interactions involving platelet activating factor (3). This tethering then leads to upregulation of the CD18 integrins on the surface of leukocytes. The family of CD18 adhesion molecules mediates the stable association of leukocytes with endothelial cells by binding ICAMs and other unknown ligands (4). Upon subsequent ligation of PECAM at the junctions between endothelial cells, leukocytes transmigrate into the underlying tissue (5).

In the case of gram-negative bacteria, the acute-phase response occurs in response to endotoxin. However, insight into the possibility that inflammation may arise by mechanisms distinct from those generated by endotoxin has come from recent analysis of gram-positive infections. Since gram-positive bacteria account for over 50% of serious infections, a difference in pathogenesis would have a major impact on outcome of clinical trials of agents designed to interrupt inflammation strictly from the point of view of endotoxin. This is particularly true since, in the real-life clinical situation, institution of therapy most often occurs rapidly, well before differentiation of gram-negative and gram-positive infection can be made.

How does inflammation arise in the case of gram-positive bacteria that do not harbor endotoxin? Recent work has described the potent inflammatory properties of the cell wall, a single macromolecule that serves as an exoskeleton for all bacteria. This network comprises the multilayered, exposed surface of gram-positive bacteria as opposed to residing in the periplasm for gram-negative bacteria. The cell wall is a dynamic structure composed of over a dozen distinct glycopeptides that continuously shuffle in and out of the rigid matrix. The glycopeptide building blocks have several unique components not found in eukaryotic systems, most notably N-acetyl muramic acid in the glycan backbone, and D-amino acids in the peptides cross-linking adjacent glycan chains into the wall superstructure. The complete composition of the glycopeptide network is known for several gram-negative bacteria (6,7) and two gram-positive pathogens, *Staphylococcus aureus* and *Streptococcus pneumoniae* (8,9).

Teichoic acids classically decorate the glycopeptide network of gram-positive bacteria only (10). Each type of bacteria present a unique and characteristic composition of glycopeptides in its cell wall. The bioactivities of cell wall components are extraordinarily diverse and potent, and there is strong evidence that the release of these biological effectors stored in the meshwork of the wall plays a major role in the course of inflammation, especially gram-positive infection (11,12). As the bioactivities of cell wall components differ, so do the spectrum of inflammatory responses generated by the different libraries of these components on each bacterial surface (13,14).

Purified preparations of either endotoxin or cell wall components recreate all of the signs of inflammation (11,12,15,16). They can elicit separation of endothelial cells consistent with formation of edema (17). Both strongly activate procoagulant activity on the surface of endothelial cells, compatible with the deposition of fibrin (18,19). The generation of edema and fibrin can be sufficient to induce death of the experimental animal. The binding of either endotoxin or cell wall complexes to epithelia, endothelia, and macrophages induces the secretion of arachidonate metabolites and cytokines (20,21).

It is at this point that the inflammatory cascades induced by gram-positive and gram-negative pathogens may differ (Table 1). It is also at this point that pathways of inflammation diverge between different body sites, a phenomenon well recognized for the arachidonate cascade. CD14, a leukocyte surface receptor, can bind endotoxin and some cell walls (22). However, the resultant cytokine profile differs—tumor necrosis factor being a

Table 1 Comparison of LPS and Cell Wall-Induced Inflammatory Cascades

Step in cascade	Differences between LPS and cell wall
Binding properties	
Serum binding protein	LBP for LPS; ? for cell wall
Bind to CD14	Both but C3H HeJ mice resistant to LPS but sensitive to cell wall
Bind to PAF receptor	Pneumococcal cell wall only
Induction of inflammatory components	
Activation of NFκB	Both, but p50 knockout mouse hypersensitive to pneumococcus but not gram-negative
Induction of TNF	Stronger for LPS; TNF receptor knockout mouse protected from LPS but not gram-positive
Induction of IL-1	Stronger for cell wall
Induction of PAF	Stronger for cell wall
Leukocyte transmigration	
Selectins	P and E involved in both
CD18-dependent	Both
CD18-independent	Pneumococcus in the lung only
ICAM-1	Knockout mouse hypersensitive to pneumococcus but not gram-negative
Effects on endothelial cell	
Induction of cell separation	Both
Induction of tissue factor	Both

Source: Adapted from (81) with permission.

very predominant product engendered by endotoxin, but being at least 100-fold less prominent following cell walls (20,21). Furthermore, leukocytes from C3H HeJ mice bind endotoxin to CD14 but are resistant to further downstream effects, whereas gram-positive infection proceeds unaltered. Telling data emerging from studies using transgenic mice deficient in various components of the inflammatory cascade point to several instances where defects lead to protection from the effects of endotoxin but worsen gram-positive infection, such as mice lacking p50 of NFkB (23), the 55-kDa receptor for tumor necrosis factor (24), and ICAM-1 (25).

Further differences between gram-positive and gram-negative inflammation are suggested in the participation of selectins, chemokines, and integrins in the influx of leukocytes. Both classes of infection appear to induce selectin-mediated rolling, although this has not been rigorously compared. Differences in the participation of individual selectins depending on the body site or the type of inflammatory challenge have been documented for several types of sterile inflammation, but remain to be tested for infection (26).

Although PAF receptor antagonists attenuate leukocyte recruitment in gram-positive and gram-negative settings, the importance of chemokines for the tethering of leukocytes may differ significantly between the two classes of infection. PAF is particularly prominent in driving the inflammation of pneumococcal infection where it serves as an actual receptor for the invading bacteria (27,28). Firm adherence of leukocytes to endothelium is recognized to occur by two different mechanisms, one dependent and one independent of the CD18 family of leukocyte adhesion molecules (27,29, 30). The CD18-dependent pathway is well recognized and is presented in detail in other contributions to this book.

Much less is known of the CD18-independent pathway. This pathway is intriguing since it appears to be operative in infections caused by gram-positive pathogens, being particularly prominent in the case of pneumococcal infection of the lung (27,30,31). For instance, the mAb IB4 directed against the CD18 leukocyte adhesion molecules reduces completely the leukocytosis in lung fluids of rabbits challenged with gram-negative bacteria, but is only partially effective in animals challenged with *S. pneumoniae* (27). In the latter instance, addition of a platelet-activating factor (PAF) antagonist to the anti-CD18 antibody greatly inhibits leukocytosis, suggesting that PAF-dependent leukotaxis can occur independent of CD18 in the lung. A CD18-independent mechanism has been studied in vitro using stimulation of endothelial cells by thrombin (29).

Putting these findings into perspective, however, it seems reasonable to state that most leukocyte transmigration in response to either gram-positive or gram-negative pathogens into most body compartments depends heavily on CD18 integrins. The participation of endothelial ligands such as ICAM-s and PECAM remains to be documented in gram-positive infection although studies in mice lacking ICAM-1 indicate that susceptibility to endotoxin decreases while that to gram-positive bacteria increases (25). Finally, it has been suggested that the coagulation cascade could contribute to leukocyte adherence by providing a fibrinogen bridge between the leukocyte fibrinogen receptor and ICAM-1 on endothelium (32). This possibility remains to be investigated in the context of infection.

As summarized in Table 1, gram-positive and gram-negative inflammation differ at nearly every step of the acute-phase response. Which of these differences are critical to take into account when designing therapeutics to be deployed in acute care settings remains to be fully defined. The differences in the cytokine cascade are very significant. Although TNF and interleukin-1 (IL-1) are present in both classes of infection and some attenuation of an ongoing inflammatory response is likely to occur with either agent, it would be expected that agents targeted to TNF would be more effective in settings with endotoxin while those targeted to IL-1 would more prominently counteract the effects of cell walls. One step that is attractive as a common element in both classes of infection is leukocyte rolling, adherence, and transmigration. This will be discussed more fully next. One caveat to consider here, however, is the unusual mechanisms of trafficking that seem to occur in pneumococcal infection, a particularly important invasive pathogen.

II. RATIONALE FOR THE USE OF ANTI-INFLAMMATORY AGENTS DURING THE THERAPY OF INFECTIOUS DISEASES

For many sterile inflammatory states, the presence of almost any degree of inflammation is considered harmful. Such is not the case for infectious diseases. However, it has become clear that some inflammation is good but that more inflammation can be harmful. Evolution of a rationale to improve outcome of infections has become a twofold problem:

1. Define when in the course of disease the amount of inflammation has crossed the line from good to bad.
2. Given that leukocyte trafficking is a shared therapeutic target, at least in part, in both classes of infections, prove that the leukocyte mediates a large part of the tissue injury during infection.

Endotoxin and cell wall components can generate the entire picture of meningitis, pneumonia, otitis media, and arthritis. The profound attenuation of these events in infections caused by bacteria that do not release cell walls or where endotoxin is neutralized attests to the essential role of these components in inflammation (33,34). A striking corollary to the understanding of the mechanism driving inflammation has been the realization that as antibiotics kill bacteria, they trigger the release of high concentrations of bacterial walls and endotoxin as a shower of inflammatory debris (35–37). This is sensed by the host which in turn generates an equally intense but, in this context, inappropriate inflammatory response. In model systems and in patients, the leukocyte density can increase 2 orders of magnitude over the few hours after initiation of antibiotic therapy (38). This burst is sufficiently disruptive to injure host tissues, a finding particularly well documented in meningitis (for review see 16). The implication that the host's inflammatory response participates in the process of injury identifies a universal point in all infections when the inflammatory response, not just the bacteria themselves, has become an appropriate target for therapeutic agents. Attenuation of the inflammatory burst at the onset of antibiotic therapy may be a key opportunity to improve the outcome of a wide variety of serious infections.

Based on the considerable differences in the cytokine components driving inflammation in various infectious diseases, it has been of major importance to determine if leukocyte transmigration represents a common key element to injury. Furthermore, agents that prevent leukocyte adhesion are promising therapeutic tools because they can be easily administered intravenously to prevent leukocytes from exiting the vascular compartment.

Traditionally, testing of the effect of inhibition of leukocyte recruitment on the course of infection has been undertaken in models of sepsis, and the results have been conflicting and difficult to interpret. In part, this may have arisen because the anti-inflammatory agent has been given well after, as opposed to in combination with, the initiation of antibiotic therapy, and maximal potential benefit may therefore not have been achieved. This is demonstrated by the clear benefit to outcome, consistently observed in animal model systems, in which inflammation can be monitored closely over time. An example of such a system is bacterial meningitis, an infection of a fluid-filled compartment (the subarachnoid space) which can be accessed multiple times over the course of the disease in a single animal (39). Convincing evidence for a key role for the leukocyte in brain injury was obtained in studies of the treatment of animals with bacterial meningitis with antibiotics supplemented by intravenous anti-CD18 integrin antibody (36,40). Upon masking of these integrins, leukocytes are transiently unable to marginate into tissues. Intravenously administered anti-CD18 antibody prevented recruitment of leukocytes into the subarachnoid space and resulted not only in protection from blood brain barrier injury but also improved the outcome of infection. Antibody treatment enhanced survival from 25% to 100%. The anti-inflammatory effect of the antibody extended to both gram-positive and gram-negative pathogens. A beneficial effect on tissue injury has been further documented in other infected body sites such as the lung (27) and the ear (41). Consistent with the suggestion that down-modulation of the effects of initial doses of the antibiotic is important to a successful outcome, the antibody was effective only when administered at the outset of antibiotic therapy and not thereafter.

When coupled in time with initiation of antibiotic therapy, anti-inflammatory strategies, particularly those directed at the leukocyte, can benefit outcome of infectious diseases. Many infectious-diseases laboratories have confirmed this observation, and it is fair to state that it represents proof of principle that anti-inflammatory strategies are applicable to infectious diseases. Clinically, this strategy is in practice with the use of dexamethasone in the treatment of bacterial meningitis in children (38). The aim of this approach is to diminish the burst of inflammation associated with the release of bacterial components during bacterial lysis and death.

This context will be important to recall in the following sections, which discuss in more detail studies on inhibitors of specific steps in the leukocyte recruitment cascade. In fact, it may be important to revisit some of the discarded therapies that failed in animal models and clinical trials of sepsis controlling for this apparently critical linkage in time of initiation of antibiotic and anti-inflammatory agent.

III. EFFECTIVE BLOCKADE OF THE LEUKOCYTE ADHESION CASCADE IN THE CONTEXT OF INFECTION

This section will review the studies that have specifically tested the effect of blockade of steps in the adhesion cascade in assays using infectious stimuli in vitro or in animal models. For an excellent review of the use of these agents in largely noninfectious settings, the reader is referred to Albelda et al. (42). For comparison to agents that affect the generation of inflammation in infection by mechanisms other than leukocyte adhesion, the reader is referred to Lynn and Cohen (43).

A. Selectins

Upon activation by various stimuli, eukaryotic cells rapidly alter the presentation of surface receptors and thereby gain the ability to support the adherence of neutrophils. Initial leukocyte rolling involves the sequential appearance of three selectins (2,44–46). P-selectin and E-selectin appear on activated endothelial cells and promote the margination and reversible rolling of leukocytes by recognition of sialylated and nonsialylated determinants. L-selectin is present on leukocytes and mediates rolling as well as homing to lymphoid tissue.

The importance of the selectin system to protection from infection is exhibited by the clinical syndrome of leukocyte adhesion deficiency type II, in which the fucosylation of the glycoconjugates which serve as selectin ligands is absent (47). This results in an inability to initiate rolling under conditions of flow with resultant enhanced susceptibility to infection. Transgenic mice engineered for the complete absence of P-selectin demonstrate a profound deficit in leukocyte rolling and delayed recruitment of leukocytes to the peritoneal cavity in response to thioglycollate challenge (48). Antisense oligonucleotides have been used to decrease expression of E-selectin, resulting in less rolling in vitro (49).

There are a number of possible classes of agents that could interfere with leukocyte rolling in a selectin-dependent manner. These include antibodies or peptides based on selectins or competitive carbohydrates based on the selectin ligands. Studies using all these approaches have been undertaken in systems relevant to infection.

1. Antibodies

Intravenous administration of an Fab fragment of a monoclonal antibody against P-selectin inhibits the accumulation of leukocytes in the rabbit model of bacterial meningitis (50). Antibodies to E- or L-selectin have not been used in the context of infection but are likely to be of benefit based on in vivo results using sterile inflammatory challenges. Anti-E-selectin protects against leukocytosis and injury in three settings: a model of asthma in primates (51), glycogen-induced peritonitis in rat, and IgG immune complex-induced inflammation in lung (26). Anti-L-selectin Mel-14 antibody decreases leukocyte accumulation in response to intraperitoneal challenge with thioglycollate (52,53).

2. Carbohydrates

Fucosylated polylactosamines are shared ligands for the three selectins. Sialylated Lewis x administered intravenously decreased P-selectin-dependent leukocyte transmigration into lung in response to complement depletion (54). Tetra- and pentasaccharide derivatives of sialylated Lewis x reduced E-selectin-mediated leukocyte influx into bronchoalveolar lavage fluid in response to IgG immune complex challenge (55). In the context of models of genuine infection, treatment of rabbits with pneumococcal meningitis with fucoidin, a polymer of fucosylated polylactosamine that competitively inhibits selectin-leukocyte recognition, strongly decreased meningeal inflammation (56).

3. Peptides

E- and L-selectin, which undergo shedding from activated cells, can be demonstrated in increased concentration in soluble form in blood of patients with bacteremia (57). This presumably could be a natural mechanism to inhibit leukocyte rolling by competing with cell bound selectin forms. Similarly, chimeras of L-, P-, and E-selectin and IgG1 have been constructed, and when introduced intravenously they competitively decrease leukocyte accumulation in response to intraperitoneal thioglycollate (52, 53). Complement depletion leads to P- and L-selectin-dependent lung injury, which is reduced by P- and L-selectin chimeras but not by the E-selectin chimera. Conversely, IgG immune complex challenge of the lung leads to injury-dependent on L- and E-selectin but not P-selectin, as evidenced by the antiinflammatory activity of the appropriate chimeras (58).

Peptides derived from various regions of the selectins are under investigation in many pharmaceutical laboratories, but published data on these studies are as yet meager in the context of models of infection. Indication that peptides could be useful agents comes from the apparent efforts of bacteria to subvert leukocyte trafficking during the course of infections. For example, pertussis toxin, a major virulence factor of *Bordetella pertussis*, is a lectin that contains two subunits with crystal structure superimposing in large part on that of the selectins (59,60). In particular, residues 19–53 of subunits S2 and S3 demonstrate primary amino acid sequence similarity to residues 15–45 of E-, P-, and L-selectin (61). Coupled with the defect in leukocyte trafficking that is pathognomonic of whooping cough, this mimicry suggested a source of peptides that could potentially competitively inhibit selectin-mediated leukocyte adhesion events. A variety of toxin-derived peptides inhibited binding of neutrophils to endothelial cells and selectin-coated surfaces (62). When administered intravenously in a pneumococcal meningitis model, strongly active 18-, 15-, and 5-mers reduced by >70% the leukocyte accumulation in cerebrospinal fluid (63).

B. Chemokines

In the presence of cytokines or thrombin, adherence of leukocytes to endothelia increases in two waves (3). One wave is rapid, occurring within 10 minutes of the addition of thrombin and decaying by 30 minutes back to baseline. The second increase is sustained over hours after TNF or IL-1 treatment. During the first wave, PAF is expressed on the surfaces of activated endothelia and activates selectin-tethered leukocytes by binding the PAF receptor, a G-protein-coupled ligand (64). Ligation of the receptor by PAF is believed to act in an autocrine fashion to stimulate the upregulation of the CD11b/CD18 integrin on the leukocyte surface. This link between selectin engagement, PAF receptor ligation, and activation of the integrin promotes the progression from leukocyte rolling to firm leukocyte-endothelial cell attachment.

PAF receptor antagonists have been previously shown to markedly attenuate inflammation during experimental pneumonia and meningitis, particularly in the case of pneumococcal disease (27). This activity may now be appreciated to arise from interruption of the activation of integrin by the chemokine pathway. Why then is the effect so prominent specifically for pneumococcal disease? Recent evidence suggests that pneumococci attach to the PAF receptor and that specific PAF receptor antagonists completely inhibit the enhanced pneumococcal adherence to activated cells (28). Thus, the PAF receptor antagonists may act not only to prevent PAF activation of the receptor but also potentially to elute pneumococci attached to host cells. Although applicable to only this specific pathogen, this combination of activities by PAF receptor antagonists may be important to exploit, given the extent and significance of pneumococcal disease.

C. CD18 Integrins

Complete absence of the CD18 integrins occurs in the immunodeficiency disease of man known as leukocyte adhesion deficiency (LAD) type I. LAD type 1 is characterized by an inability of leukocytes to transmigrate into tissues and present early in life with multiple forms of infection (65). This phenotype is mimicked in genetically engineered mice with greatly decreased CD18 expression; these animals demonstrate a delayed leukocyte response to chemical peritonitis (66). Decreased expression of CD11b/CD18 has been described as a mechanism for the ability of leumedins to protect against endotoxin challenge in mice (67). Theoretically, therapeutic interference with CD18-dependent leukocyte adhesion could be achieved by antibodies to one or all of the three CD18 integrins or peptides based on CD11/CD18 ligands. Studies using all these approaches have been undertaken in systems relevant to infection.

1. Antibodies

The ability of anti-CD18 to protect against tissue injury was first noted in a model of sterile inflammation of the skin in which rabbits were challenged with fMLP, LTB4, C5a, or histamine (68). More recently, anti-CD18 blocked delayed type hypersensitivity reactions in rabbit skin (69). Variable efficacy has been demonstrated in models of sterile inflammation for individual CD11 antibodies. Using a model of lung injury associated with complement depletion in rats, anti-CD11a resulted in ~30% less leukocyte accumulation and less hemorrhage, anti-CD11b produced a ~50% decrease, and anti-ICAM-1 was most effective (60% to 75% decrease) (70). It should be noted that mononuclear leukocyte migration in models of more chronic infection fails to respond to anti-CD18. For instance, in animal models of encephalitis, the anti-CD49d-CD29 ($\alpha_4\beta_1$) antibody blocks monocyte influx into brain whereas anti-CD18 does not (71).

In models of challenge with nonreplicating bacterial components, anti-CD11 or anti-CD18 antibodies have proven very effective in decreasing inflammation and damage in a wide variety of settings. Anti-CD11b protects against endotoxin challenge in mice, but anti-CD11a does not (67). Anti-CD18 completely blocked the leukocyte accumulation in the rabbit meningitis model to challenges with endotoxin, cell wall, and the three common meningeal pathogens: *S. pneumoniae*, *H. influenzae*, and *N. meningitidis* (36). Anti-CD11a and -b were not effective in this setting. In a rabbit model of pneumonia, the anti-CD18 antibody blocked all leukocyte recruitment in response to endotoxin but only half of the response to cell wall or *S. pneumoniae* (27). Complete attenuation of recruitment in settings of challenge with pneumococci in various body compartments other than the brain appears to require interruption of a non-CD18-mediated recruitment mechanism, in some cases by antagonists of PAF (see above).

In models of genuine infection, distinction must be drawn between increased susceptibility to infection in animals that are rendered functionally neutropenic by antibody and then challenged with an infectious agent versus treating with antibody in conjunction with antibiotics in the setting of established infection. Obviously in the first instance, the course of infection will generally be more severe in the absence of leukocyte recruitment from the start of the challenge. For instance, in a model of *Haemophilus influenzae* otitis media, guinea pigs treated with anti-CD18 antibody before challenge develop fewer leukocytes and less edema and epithelial damage in the middle ear, but there are more bacteria (41). This is contrasted to the extremely effective use of antibody in animals with established meningitis who undergo antibiotic therapy (36,40). Animals receiving the antibody together with antibiotics are protected against the burst of leukocytosis associated with the release of inflammatory debris from dying bacteria. Concomitantly they experience less brain damage, and survival improves from 25% to 100% (72).

2. Peptides

The endothelial ligands for CD18 integrins are only partially defined. Therefore, peptides that mimic these ligands and therefore compete with CD18-dependent leukocyte migration are still undergoing formulation. ICAM-1 and -2 support adhesion mediated by CD11a/CD18 and, in part, that mediated by CD11b/CD18. One ICAM-derived peptide, KELLLPGN-NRKV (from residues 40–64), decreased adherence of Molt-4 cells to TNF-stimulated endothelial cells in culture (73).

An unanticipated source of peptides capable of modulating leukocyte adherence was found in prokaryotes. Filamentous hemagglutinin (FHA), a 220-kDa secreted protein of *B. pertussis*, inhibits leukocyte adherence to endothelia in vitro and, when administered intravenously, prevents accumulation of leukocytes in the subarachnoid space in animal models of meningitis (74). The mechanism of this effect appears to be twofold. FHA binds cell surface and purified CD11b/CD18 (75,76), most probably at the binding site for factor 10 of the coagulation cascade (74,77). This binding is most strongly attributed to a region of FHA which mimics factor 10 since the peptide ETKEVDG derived from this region exhibits potent ability to inhibit leukocyte accumulation in the model of bacterial meningitis. Secondly, FHA contains an RGD sequence that is believed to influence activation of CD11b/CD18 by binding to the leukocyte response integrin/integrin activating protein complex (78). Similarly, peptides derived from the RGD region show a dose-dependent inhibition of leukocyte accumulation in the animal model of bacterial meningitis (74).

D. ICAMs

Consistent with the ability of ICAM to serve as a cognate ligand for the CD18 integrins, mice engineered to lack ICAM-1 exhibit a delayed influx of neutrophils in response to chemical peritonitis (79). These mice demonstrate strikingly different outcomes when challenged with bacteria in a model of sepsis and meningitis. Knockout mice show improved survival as compared to controls in the setting of gram-negative challenge, but are more susceptible to gram-positive disease (25).

Anti-ICAM-1 is more strongly effective at diminishing lung injury in complement-depleted rats than either anti-CD11a or CD11b (70). However, although ICAMs-1 and -2 have been detected on inflamed cerebral microvessels, antibodies failed to inhibit monocyte accumulation in the chronic inflammation of the central nervous system (71).

E. PECAM

Published studies exploring the use of agents blocking PECAM in infectious settings are sparse. Anti-PECAM antibody reduces leukocyte influx by 75% in models of glycogen-induced peritonitis, immune complex-induced lung injury, and TNF injected into human skin grafted onto mouse (80). Given that PECAM is believed to shepherd the leukocyte through endothelial junctions during transmigration (5), this approach may yield potent agents in the future, when its ligands and mechanism of action are more completely understood.

IV. SUMMARY

Modulation of inflammation can decidedly improve the outcome of infectious diseases, especially when used in combination with antibiotics. However, the deployment of these agents has been complicated by the diversity of pro-inflammatory events incited by different infectious agents. One area that may provide grounds for common therapeutic initiatives is leukocyte adhesion. Currently, interruption of the functions of selectins, chemokines, integrins, and intercellular adhesion molecules by antibodies, carbohydrates, peptides, or other agents has become clearly feasible. It is left to future studies to apply this technology to the therapeutic setting of the critically ill patient.

REFERENCES

1. Wood B. Studies on the cellular immunology of acute bacterial infections. Harvey Lect 1952; 47:72–98.
2. Lasky LA. Selectins: interpreters of cell-specific carbohydrate information. Science 1992; 258:964–969.
3. Zimmerman G, Prescott S, McIntyre T. Endothelial cell interactions with granulocytes: tethering and signalling molecules. Immunol Today 1992; 13: 93–100.
4. Springer TA. Adhesion receptors of the immune system. Nature 1990; 346: 425–434.
5. Muller W, Ratti C, McDonnell S, Cohn Z. The role of Pecam in leukocyte transmigration across endothelia. J Exp Med 1989; 170:399–404.
6. Glauner B, Schwarz U. The analysis of murein composition with high-pressure liquid chromatography. In: Hackenbeck R, Schwarz U, eds. The Target of Penicillin. Berlin: de Gruyter, 1983; 29–34.
7. Burroughs M, Chang Y-S, Gage D, Tuomanen E. Composition of the peptidoglycan of *Haemophilus influenzae*. J Biol Chem 1993; 268:11594–11598.
8. Garcia-Bustos J, Tomasz A. A biological price of antibiotic resistance: major changes in the peptidoglycan structure of penicillin-resistant pneumococci. Proc Natl Acad Sci USA 1990; 87:5415–5419.
9. De Jonge B, Chang Y-S, Gage D, Tomasz A. Peptidoglycan composition of a highly methicillin-resistant *Staphylococcus aureus* strain. J Biol Chem 1992; 267:11248–11254.
10. Jennings H, Lugowski C, Young N. Structure of the complex polysaccharide C-substance from *Streptococcus pneumoniae*. Biochem 1980; 19:4712–4719.
11. Tuomanen E, Liu H, Hengstler B, Zak O, Tomasz A. The induction of meningeal inflammation by components of the pneumococcal cell wall. J Infect Dis 1985; 151:859–868.
12. Tuomanen EI, Tomasz A, Hengstler B, Zak O. The relative role of bacterial cell wall and capsule in the induction of inflammation in pneumococcal meningitis. J Infect Dis 1985; 151:535–540.
13. Tuomanen E. Bacterial cell walls and the mechanisms of inflammation in experimental meningitis. In: Schrinner E, Richmond MH, Seibert G, Schwarz

U, eds. Surface Structures of Microorganisms and Their Interactions with the Mammalian Host. Proceedings of the 18th Workshop Conference Hoechst, Schloss Ringberg, October 20–23, 1987. New York: VCH Publishers, 1987; 79–90.

14. Burroughs M, Rozdzinski E, Geelen S, Tuomanen E. A structure-activity relationship for induction of meningeal inflammation by muramyl peptides. J Clin Invest 1993; 92:297–302.
15. Tauber MG, Shibl AM, Hackbarth CJ, Larrick JW, Sande MA. Antibiotic therapy, endotoxin concentration in cerebrospinal fluid, and brain edema in experimental *Escherichia coli* meningitis in rabbits. J Infect Dis 1987; 156:456–462.
16. Quagliarello V, Scheld WM. Bacterial meningitis: pathogenesis, pathophysiology and progress. N Engl J Med 1992; 327:864–872.
17. Geelen S, Battacharyya C, Toumanen E. Cell wall mediates pneumococcal attachment and cytopathology to human endothelial cells. Infect Immun 1993; 61:1538–1543.
18. Naworth PP, Stern DM. Modulation of endothelial cell hemostatic properties by tumour necrosis factor. J Exp Med 1986; 163:740–745.
19. Geelen S, Bhattacharyya C, Tuomanen E. Induction of procoagulant activity on human endothelial cells by *Streptococcus pneumoniae*. Infect Immun 1992; 60:4179–4183.
20. Riesenfeld-Orn I, Wolpe S, Garcia-Bustos JF, Hoffman MK, Tuomanen E. Production of interleukin-1 but not tumour necrosis factor by human monocytes stimulated with pneumococcal cell surface components. Infect Immun 1989; 57:1890–1893.
21. Heumann D, Barras C, Severin A, Glauser M, Tomasz A. Gram positive cell walls stimulate synthesis of tumour encrosis factor alpha and interleukin-6 by human monocytes. Infect Immun 1994; 62:2715–2721.
22. Pugin J, Heumann D, Tomasz A, et al. CD14 is a pattern recognition receptor. Immunity 1994; 1:509–516.
23. Sha W, Liou H, Tuomanen E, Baltimore D. Targeted disruption of the p50 subunit of NF-kB leads to multifocal defects in immune responses. Cell 1995; 80:321–330.
24. Pfeffer K, Matsuyama T, Kundig T, Mak T. Mice deficient for the 55kD tumour necrosis factor receptor are resistant to endotoxic shock, yet succumb to *L. monocytogenese* infection. Cell 1993; 73:457–467.
25. Tan T, Smith C, Hawkins E, Mason E, Kaplan S. Hematogenous bacterial meningitis in an ICAM-1 deficient infant mouse model. J Infect Dis 1995; 171: 342–349.
26. Mulligan M, Varani J, Dame M, et al. Role of endothelial-leukocyte adhesion molecule 1 in neutrophil-mediated lung injury in rats. J Clin Invest 1991; 88: 1396–1406.
27. Cabellos C, MacIntyre DE, Forrest M, Burroughs M, Prasad S, Tuomanen E. Differing roles of platelet-activating factor during inflammation of the lung and subarachnoid space. J Clin Invest 1992; 90:612–618.
28. Cundell D, Gerard N, Gerard C, Tuomanen E. The platelet activating factor

receptor anchors *Streptococcus pneumoniae* to eukaryotic cells. Nature 1995; 377:435–438.
29. Zimmerman G, McIntyre T. Neutrophil adherence to human endothelium in vitro occurs by CDw18 glycoprotein-dependent and independent mechanisms. J Clin Invest 1988; 81:531–537.
30. Doerschuk CM, Winn RK, Coxson HO, Harlan JM. CD18-dependent and -independent mechanisms of neutrophil emigration in the pulmonary and systemic microcirculation of rabbits. J Immunol 1990; 144:2327–2333.
31. Mileski W, Harlan J, Rice C, Winn R. *Streptoccus pneumoniae*-stimulated macrophages induce neutrophils to emigrate by a CD18-independent mechanism of adherence. Circ Shock 1990; 31:259–267.
32. Languino L, Plescia J, Cuperray A, et al. Fibrinogen mediates leukocyte adhesion to vascular endothelium through an ICAM-1-dependent pathway. Cell 1993; 73:1423–1434.
33. Tuomanen E, Pollack H, Parkinson A, et al. Microbiological and clinical significance of a new property of defective lysis in clinical strains of pneumococci. J Infect Dis 1988; 158:36–43.
34. Tauber MG, Burroughs M, Niemoller UM, Kuster H, Borschberg U, Tuomanen E. Differences of pathophysiology in experimental meningitis caused by three strains of *Streptococcus pneumoniae*. J Infect Dis 1991; 163:806–811.
35. Tuomanen E, Hengstler B, Rich R, Bray M, Zak O, Tomasz A. Nonsteroidal anti-inflammatory agents in the therapy of experimental pneumococcal meningitis. J Infect Dis 1987; 155:985–990.
36. Tuomanen E, Saukkonen K, Sande S, Cioffe C, Wright SD. Reduction of inflammation, tissue damage, and mortality in bacterial meningitis in rabbits treated with monoclonal antibodies against adhesion-promoting receptors of leukocytes. J Exp Med 1989; 170:959–969.
37. Tuomanen E. Breaching the blood brain barriers. Sci Am 1993; 268:80–84.
38. Lebel MH, Freij BJ, Syrogiannopoulos GA, et al. Dexamethasone therapy for bacterial meningitis. N Engl J Med 1988; 15:964–971.
39. Dacey R, Sande MA. Effect of probenecid on cerebrospinal fluid concentrations of penicillin and cephalosporin derivatives. Antimicrob Agents Chemother 1974; 6:437–441.
40. Saez-Llorens X, Jafari H, Severien C, et al. Enhanced attenuation of meningeal inflammation and brain edema by concomitant administration of anti-CD18 monoclonal antibodies and dexamethasone in experimental *Haemophilus* meningitis. J Clin Invest 1991; 88:2003–2011.
41. Patel J, Chonmeitree R, Schmalsteig F. Effect of modulation of polymorphonuclear leukocyte migration with anti-CD18 antibody on pathogenesis of experimental otitis media in guinea pigs. Infect Immun 1993; 61:1132–1135.
42. Albelda S, Smith C, Ward P. Adhesion molecules and inflammatory injury. FASEB J 1994; 8:504–512.
43. Lynn W, Cohen J. Adjunctive therapy for septic shock: a review of experimental approaches. Clin Infect Dis 1995; 20:143–158.
44. Bevilacqua MP, Stengelin S, Gimbrone M, Seed B. Endothelial leukocyte

adhesion molecule 1: an inducible receptor for neutrophils related to complement regulatory proteins and lectins. Science 1989; 243:1160–1164.
45. Lawrence M, Springer T. Leukocytes roll on a selectin at physiologic flow rates: distinction from and prerequisite for adherence through integrins. Cell 1991; 65:859–873.
46. Springer TA, Lasky LA. Cell adhesion. Sticky sugars for selectins. Nature 1991; 349:196–197.
47. Etzioni A, Frydman M, Pollack S, et al. Brief report: recurrent severe infections caused by a novel leukocyte adhesion deficiency. N Engl J Med 1993; 327:1789–1792.
48. Mayadas T, Johnson R, Rayburn H, Hynes R, Wagner D. Leukocyte rolling and extravasation are severely compromised in P selectin-deficient mice. Cell 1993; 74:541–554.
49. Bennett C, Condon T, Grimm S, Chan H, Chiang M. Inhibition of endothelial cell adhesion molecule expression with antisense oligonucleotides. J Immunol 1994; 152:3530–3540.
50. Spellerberg B, Tuomanen E. The pathophysiology of pneumococcal meningitis. Ann Med 1994; 26:411–418.
51. Gundel R, Wegner C, Torcellini C, et al. Endothelial leukocyte adhesion molecule-1 mediates antigen-induced acute airway inflammation and late-phase airway obstruction in monkeys. J Clin Invest 1991; 88:1407–1411.
52. Watson SR, Imai Y, Fennie C, Geoffroy JS, Rosen SD, Lasky LA. A homing receptor-IgG chimera as a probe for adhesive ligands of endothelia. J Cell Biol 1990; 110:2221–2229.
53. Watson S, Fennie C, Lasky L. Neutrophil influx into an inflammatory site inhibited by a soluble homing receptor-IgG chimera. Nature 1991; 349:164–167.
54. Mulligan M, Paulson J, Ce Frees S, Zheng Z, Lowe J, Ward P. Protective effects of oligosaccharides in P-selectin-dependent lung injury. Nature 1993; 364:149–151.
55. Mulligan M, Lowe J, Larsen R, et al. Protective effects of sialylated oligosaccharides in immune complex-induced acute lung injury. J Exp Med 1993; 178: 623–631.
56. Granert C, Raud J, Xie X, Lindquist L, Lindbom L. Inhibition of leukocyte rolling with polysaccharide fucoidin prevents pleocytosis in experimental meningitis in the rabbit. J Clin Invest 1994; 93:929–936.
57. Newman W, Beall L, Carson C, et al. Soluble E-selectin is found in supernatants of activated endothelial cells and is elevated in serum of patients with septic shock. J Immunol 1993; 150:644–654.
58. Mulligan M, Watson S, Fennie C, Ward P. Protective effects of selectin chimeras in neutrophil-mediated lung injury. J Immunol 1993; 151:6410–6417.
59. Sandros J, Rozdzinski E, Zheong J, Cowburn D, Tuomanen E. Lectin domains of the adhesions of *Bordetella pertussis*: selectin mimicry linked to microbial pathogenesis. Glycoconjugates 1994; 11:1–6.
60. Stein P, Boodhoo A, Armstrong G, Cockle S, Klein M, Read R. The crystal structure of pertussis toxin. Structure 1994; 2:45–57.

61. Saukkonen K, Burnette WN, Mar V, Masure HR, Tuomanen E. Pertussis toxin has eukaryotic-like carbohydrate recognition domains. Proc Natl Acad Sci USA 1992; 89:118–122.
62. Rozdzinski E, Burnette WN, Jones T, Mar V, Tuomanen E. Prokaryotic peptides which block adherence of leukocytes to selectins. J Exp Med 1993; 178:917–924.
63. Sandros J, Rozdzinski E, Tuomanen E. Peptides from pertussis toxin interfere with neutrophil adherence in vitro and counteract inflammation in vivo. Microbial Pathogenesis 1994; 16:213–220.
64. Kunz D, Gerard N, Gerard C. The human leukocyte platelet activating factor receptor. J Biol Chem 1992; 267:9101–9106.
65. Anderson D, Schmalsteig F, Finegold M, et al. The severe and moderate phenotypes of heritable Mac-1m LFA-1 deficiency: their quantitative definition and relation to leukocyte dysfunction and clinical features. J Infect Dis 1985; 152:668–689.
66. Wilson R, Ballantyne C, Smith C, et al. Gene targeting yields a CD18-mutant mouse for study of inflammation. J Immunol 1993; 151:1571–1578.
67. Burch R, Noronha-Blob L, Bator J, Lowe V, Sullivan J. Mice treated with a leumedin or antibody to Mac-1 to inhibit leukocyte sequestration survive endotoxin challenge. J Immunol 1993; 150:3397–3403.
68. Arfors K, Lundberg C, Lindbom L, Lundberg K, Beatty P, Harlan J. Monoclonal antibody to the membrane glycoprotein complex CD18 inhibits polymorphonuclear leukocyte accumulation and plasma leakage in vivo. Blood 1986; 69:338–340.
69. Lindbom L, Lundberg C, Prieto J, et al. Rabbit leukocyte adhesion molecules CD11/CD18 and their participation in acute and delayed inflammatory responses and leukocyte distribution in vivo. Clin Immunol Immunopathol 1990; 57:105–119.
70. Mulligan M, Smith C, Anderson D, et al. Role of leukocyte adhesion molecules in complement-induced lung injury. J Immunol 1993; 150:2401–2406.
71. Yednock TA, Cannon C, Fritz L, Sanchez-Madrid F, Steinman L, Karin N. Prevention of experimental autoimmune encephalomyelitis by antibodies against alpha 4 beta 1 integrin. Nature 1992; 356:63–66.
72. Tuomanen E. Adjuncts to the therapy of bacterial meningitis. Pediatr Infect Dis J 1990; 9:782–783.
73. Ross L, Hassman F, Molony L. Inhibition of Molt-4-endothelial adherence by synthetic peptides from the sequence of ICAM-1. J Biol Chem 1992; 267: 8537–8543.
74. Rozdzinski E, Spellerberg V, van der Flier M, et al. Peptide from a prokaryotic adhesin inhibits leukocyte migration in vitro and in vivo. J Infect Dis 1995; 172:785–793.
75. Relman DA, Domenighini M, Toumanen E, Rapppuoli R, Falkow S. Filamentous hemagglutinin of *Bordetella pertussis*: nucleotide sequence and crucial role in adherence. Proc Natl Acad Sci USA 1989; 86:2637–2641.
76. Van Strijp JAG, Russell DG, Tuomanen E, Brown EJ, Wright SD. Ligand

specificity of purified complement receptor type 3: indirect effects of an Arg-Gly-Asp sequence. J Immunol 1993; 151:3324–3336.
77. Altieri DC, Morrissey JH, Edgington TS. Adhesive receptor Mac-1 coordinates the activation of factor X on stimulated cells of monocytic and myeloid differentiation: an alternative initiation of the coagulation protease cascade. Proc Natl Acad Sci USA 1988; 85:7462–7466.
78. Ishibashi Y, Claus S, Relman D. *Bordetella pertussis* filamentous hemagglutinin interacts with a leukocyte signal transduction complex and stimulates bacterial adherence to monocyte CD3. J Exp Med 1994; 180:1225–1233.
79. Sligh J, Ballantyne C, Rich S, et al. Inflammatory and immune responses are impaired in mice deficient in intercellular adhesion molecule 1. Proc Natl Acad Sci USA 1993; 90:8529–8533.
80. Vaporciyan A, DeLisser H, Horng-Chin Y, et al. Involvement of platelet-endothelial cell adhesion molecule-1 in neutrophil recruitment in vivo. Science 1993; 262:1580–1582.
81. Cundell D, Masure H, Tuomanen E. The molecular basis of pneumococcal infection: an hypothesis. Clin Infect Dis 1995; 21:12–18.

15

Adhesion Molecules in Atherosclerosis

Quirino Orlandi
Whitaker Cardiovascular Institute, Boston University School of Medicine, Boston, Massachusetts

Joseph Loscalzo
Department of Pathology, Brigham and Women's Hospital and Harvard Medical School, Boston, Massachusetts

I. ATHEROSCLEROSIS AS AN INFLAMMATORY DISEASE

Advances in processing pathological specimens (1) and the use of monoclonal antibodies to identify cell subtypes (2–4) have allowed better characterization of the cellular composition of atherosclerotic lesions. With the use of these tools, the role of inflammatory cells in atherogenesis, initially proposed in 1941 (1), has become clear. Currently, there is ample evidence that inflammatory cells not only are present in atheromata, but are also central to the development of atherosclerosis. Atherosclerosis has many features in common with inflammatory injury, viz., leukocyte accumulation, fibrosis, and angiogenesis (5), and, correspondingly, the cellular makeup of lesions includes activated macrophages and activated T cells (2–4). There is also growing evidence for the involvement of immune mechanisms in atherosclerosis (6), and chronic infection has also been implicated by some investigators (7).

Current theory holds that endothelial injury induced by factors known to promote atherosclerosis leads to a molecular and cellular response cascade that ultimately results in the formation of an atherosclerotic plaque (1,8). This premise is based on early observations that denudation of the endothelial monolayer produces vascular lesions with smooth muscle proliferation which resemble atherosclerotic plaques, and that platelet adhesion and aggregation at the site of injury induce smooth muscle proliferation (by growth factors secretion) (1). This hypothesis has been modified to include endothelial dysfunction (without morphological changes) as a sufficient component in the initiation of atherosclerosis (8,9). Furthermore, the role of inflammatory cells in promoting injury by secreting a variety of cytokines, growth factors, bioactive lipids, proteolytic enzymes, coagulation factors, and reactive oxygen species is now recognized (8,10,11).

II. ENDOTHELIAL DYSFUNCTION (Figs. 1, 2)

Under normal conditions the endothelium modulates vascular smooth muscle tone and proliferation, inhibits platelet activation and the coagulation cascade, and does not support the adhesion of inflammatory cells and platelets. Endothelial dysfunction, as induced by a number of factors, will attenuate or eliminate these protective mechanisms and lead to the initiation and propagation of atherosclerotic lesions. A key factor in endothelial dysfunction is insufficient production of endothelium-derived relaxing factor/nitric oxide (EDRF/NO), which inhibits platelet activation, reduces vascular injury from free radicals, modulates normal vessel tone (12), and inhibits leukocyte adhesion to the endothelium by inactivating superoxide (13, 14). In addition, Radomski and colleagues have shown that NO inhibits the adhesion of platelets to the vascular endothelium (15), an effect unique to nitric oxide and one which distinguishes it from prostacyclin.

In 1980 Furchgott and Zawadzki (16) showed that the endothelium is necessary for the arterial vasodilator response to muscarinic agonists. This effect occurs because the release of EDRF/NO induced by the muscarinic agonist in normal endothelium overcomes the direct vasoconstricting action of the agonist on arterial smooth muscle cells (17). Since this initial seminal observation, a number of vasodilator responses have been demonstrated to be endothelium-dependent and include response to serotonin, histamine, and, in part, reactive hyperemia (increased flow-induced vasodilation) (17,18). Impaired endothelium-dependent vasodilation, related to an insufficient release or local inactivation of EDRF/NO, or possibly to an increased release of vasoconstrictor substances, is thus a marker for endothelial dysfunction and has been extensively studied in recent years in animal models and humans. Among the many markers of dysfunctional endothelium is included the expression of adhesion molecules that facilitate interactions with circulating blood leukocytes. Importantly, blunted endothelium-dependent vasodilator responses are accompanied by enhanced adhesion molecule expression by endothelial cells in many disorders or disease states associated with atherosclerosis.

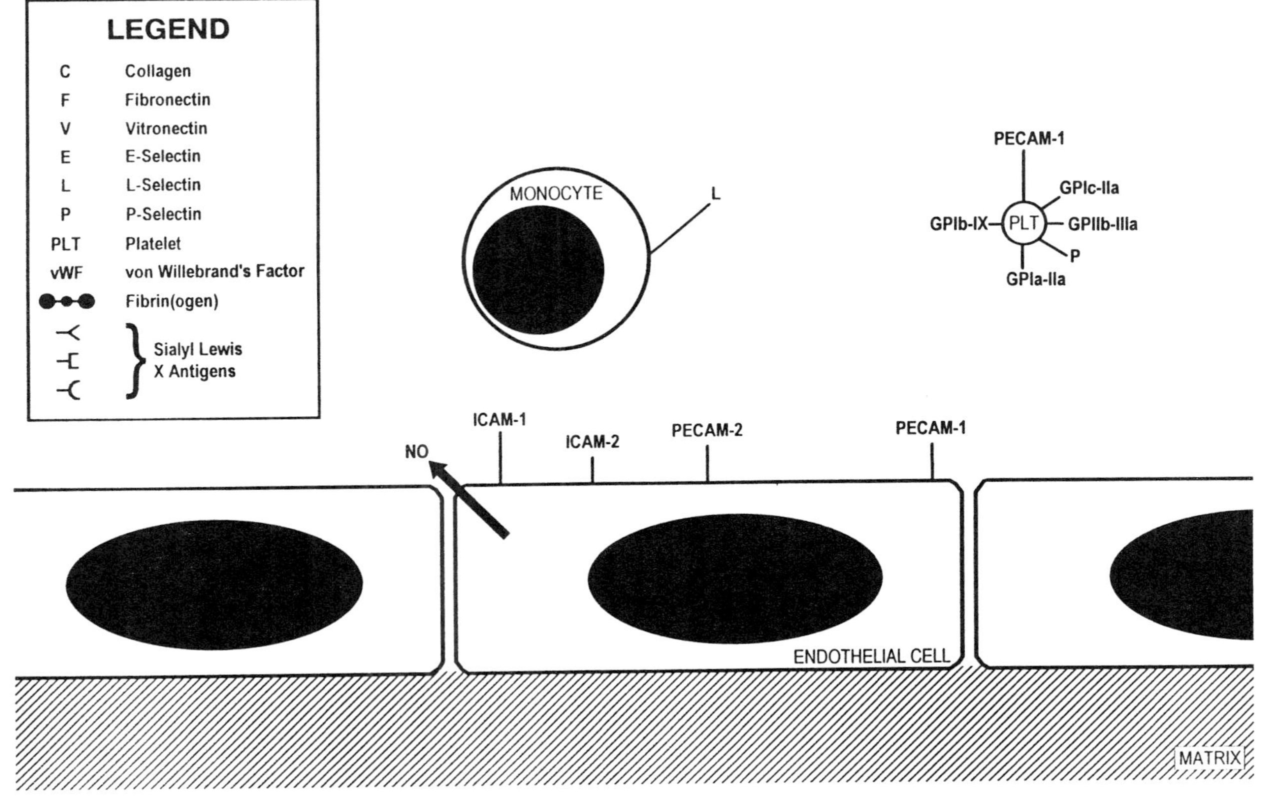

Figure 1a Normal endothelium demonstrating the constitutive expression of adhesion molecules on endothelial cells, blood monocytes, and blood platelets, with normal production of nitric oxide.

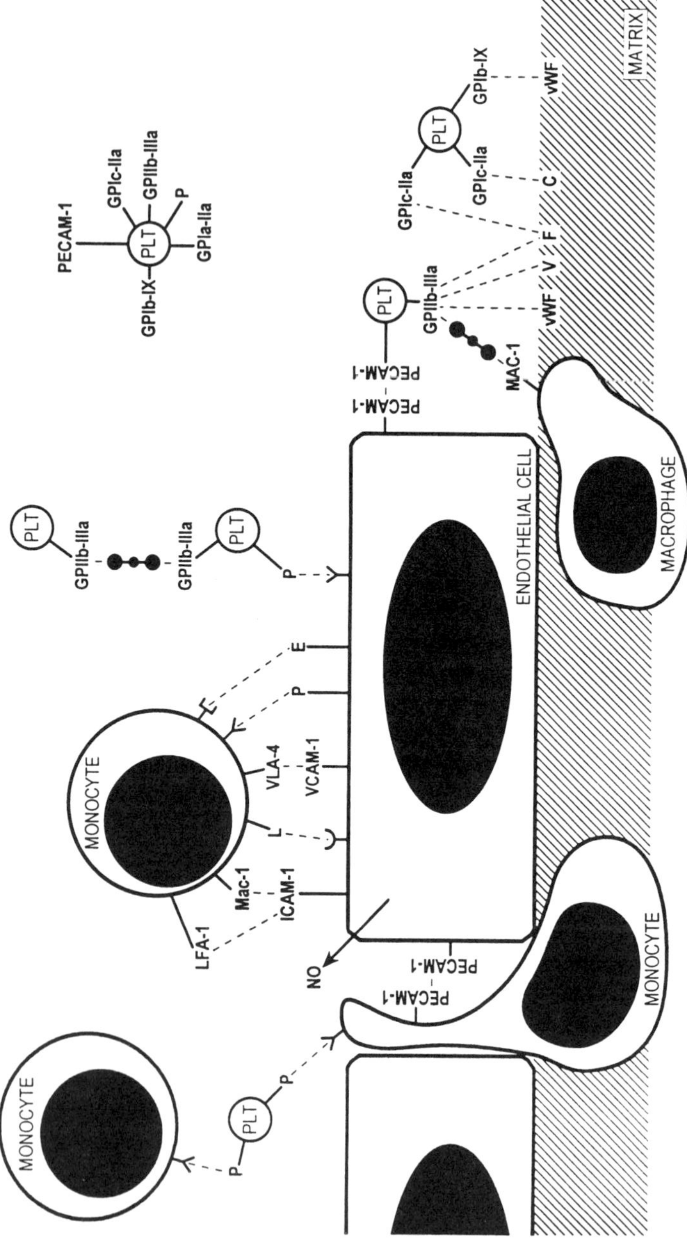

Figure 1b Dysfunctional endothelium demonstrating upregulation of constitutively expressed adhesion molecules (ICAM-1), and adhesion molecules present on activated endothelial cells and blood cells, with diminished production of nitric oxide. Dashed lines denoted ligand-counterligand interactions.

III. LEUKOCYTE-ENDOTHELIAL CELL INTERACTIONS

The initial studies of leukocyte interaction with endothelial cells focused on the vascular response to inflammatory mediators. Consequently, the mechanism by which neutrophils cross areas of inflamed endothelium has been relatively well characterized (19) by studying in vivo (20–22) models and in vitro (23–25) approximations of postcapillary venule flow. The principal adhesion molecules involved and their counterreceptors are listed in Table 1.

Under normal conditions, leukocytes do not adhere to endothelial cells; however, following activation of endothelial cells, a series of events is initiated that allows leukocytes to cross the endothelium. The process is divided into three phases: rolling (mediated by selectins), activation and firm adhesion (mediated by integrins and members of the immunoglobulin superfamily) and diapedesis (requiring PECAM-1 and a chemotactic gradient) (10,19).

A. Rolling

Leukocyte rolling is mediated by selectins, a group of proteins with lectin-like domains capable of binding to oligosaccharides related to sialyl Lewis

Table 1 Adhesion Molecules and Counterligands Implicated in Atherosclerosis

Molecule	Cell type	Ligands
Selectins		
L-selectin	M, PMN, L	CD34, GlyCAM-1
P-selectin	EC, P	PSGL-1, SLX
E-selectin	EC	SLX and others
Integrins		
$\alpha_L\beta_4$ (LFA-1)	PMN and L	ICAM-1 and ICAM-2
$\alpha_M\beta_4$ (Mac-1)	PMN	ICAM-1 and others
$\alpha_4\beta_1$ (VLA-4)	E and L	VCAM-1 and others
IG superfamily		
ICAM-1	L and EC	LFA-1, Mac-1
ICAM-2	L, EC, and P	LFA-1
VCAM-1	EC	VLA-4
PECAM-1	EC, PMN, L, P	PECAM-1 and others

Abbreviations: ICAM-1,-2: intercellular adhesion molecule-1,-2; VCAM-1: vascular cell adhesion molecule -1; PECAM-1: platelet-endothelial cell adhesion molecule-1; E: eosinophils; EC: endothelial cells; M: monocytes; PMN: polymorphonuclear neutrophils; L: lymphocytes; P: platelets.

x and A antigens; phosphorylated mono- and polysaccharides; and sulfated polysaccharides and lipids (26). E-selectin is synthesized by endothelial cells in response to IL-1 and TNFα and supports the adhesion of neutrophils, monocytes, lymphocytes, and eosinophils. P-selectin is stored within the α-granules of platelets and the Weibel-Palade bodies of endothelial cells, and is translocated to the plasma membrane within minutes following exposure of the cells to thrombin or histamine. Like E-selectin, P-selectin supports the adhesion of neutrophils, monocytes, lymphocytes, and eosinophils. L-selectin is constitutively expressed on neutrophils, monocytes, and lymphocytes, and mediates the adhesion of these leukocytes to activated endothelial cells (10). One of the functions of L-selectin may be to present sialyl Lewis x antigen molecules to P- and E-selectin (19,20). Leukocyte rolling mediated by these selectins has been demonstrated on artificial layers of E- and P-selectin and in vivo. Recent data using mice, in which the selectins are eliminated by targeted gene disruption, suggest that leukocyte rolling is depressed by P-selectin deficiency (26a), but not in all vascular beds (26b), and that a combined deficiency of P- and E-selectin is associated by a profound reduction in leukocyte rolling an enhanced susceptibility to infection (26c). Rolling is a prerequisite for firm adhesion in certain models of inflammation, but may not be necessary in all cases. Of note, deficiencies in both P-selectin and intercellular adhesion molecule-1 (ICAM-1) are accompanied by a complete absence of leukocyte rolling (26d), indicating that the physiologic correlates of selected adhesion molecules are not necessarily exclusive. Importantly, from the standpoint of atherogenesis, activated monocytes can bind to unactivated endothelial cells (which do not express selectins) in vitro (19). This adhesion event may occur through upregulation of integrin expression by monocyte chemoattractant protein-1 (MCP-1) (27) and the subsequent binding of these integrins (such as $\alpha_M\beta_4$) to constitutively expressed ICAM-1.

B. Firm Adhesion

The integrins responsible for firm adhesion of leukocytes to endothelial cells are $\alpha_L\beta_4$ (LFA-1), $\alpha_M\beta_4$ (Mac-1), and $\alpha_4\beta_1$ (VLA-4). They are composed of noncovalently bound heterodimeric α and β subunits and require activation to bind counterreceptors of the immunoglobulin superfamily on endothelial cells (ICAM-1, ICAM-2, VCAM-1). Activation can occur by translocation of one of the subunits to the cell surface or by conformational changes in surface-expressed molecules. Once activation occurs, LFA-1 binds ICAM-1 and -2, Mac-1 binds ICAM-1, and VLA-4 binds VCAM-1. Activation is accompanied by the shedding of selectins and a transient reduction in integrin activity that may facilitate diapedesis (10,19).

C. Diapedesis

Migration of leukocytes across the endothelium is not a direct consequence of adhesion and has been shown to be dependent on a chemotactic gradient and PECAM-1, a member of the immunoglobulin superfamily that can function as its own counterligand. It has been localized at the junctions between endothelial cells (28,29). More importantly, Muller and colleagues showed that antibodies to PECAM-1 block leukocyte migration across an endothelial monolayer without decreasing adhesion (30). This dependence of diapedesis on PECAM-1 has also been confirmed recently in vivo (31).

Neutrophils are the primary circulating leukocytes, and despite the fact that monocytes and neutrophils share the same adhesion molecules, atherosclerotic plaques are comprised primarily of monocytes/macrophages (and lymphocytes) with only rare neutrophil involvement. Clearly, chemoattractants need play a critical role in the specific recruitment of monocytes to areas of activated or dysfunctional endothelium prone to support the development of atheromata. The signaling between monocytes and endothelial cells may occur via soluble mediators; however, for two principal reasons, this interaction is more likely to occur after or coincident with the adhesion of leukocytes to the endothelium. Firstly, mediators released into the arterial circulation would be diluted and quickly carried downstream. Secondly, since many of the monocyte chemoattractants also cause shedding of selectins and can thereby inhibit leukocyte adhesion, it seems likely that they would have an effect on recruiting leukocytes only after the initiation of leukocyte adhesion to the endothelium.

D. Monocyte Chemoattractant Protein-1

A number of molecules can specifically attract monocytes. Of particular interest is monocyte chemoattractant protein-1 (MCP-1), also known as smooth muscle cell-derived chemotactic factor, endothelium-derived chemotactic factor, lymphocyte-derived chemotactic factor, and JE gene product (11). It is a potent chemoattractant for monocytes and binds to high-affinity receptors found only on monocytes (32,33). MCP-1 is produced by endothelial cells, vascular smooth muscle cells, and macrophages (11); in endothelial cell cultures it is constitutively expressed (although established endothelial cell cultures may represent a state of injury or dysfunction with upregulation of expression). Nevertheless, cytokines and inflammatory mediators are capable of upregulating mRNA for MCP-1 and increasing protein expression in cell culture, and the induction of IL-1 and IL-6 production in monocytes by MCP-1 (27) provides a mechanism for autoamplification of cytokines responsible for monocyte recruitment. Cushing and colleagues (34) have shown that minimally modified low-density lipoprotein (LDL) increases expression of MCP-1 in human endothelial and smooth muscle cell cultures. MCP-1 increases expression of both the α and β subunits of the integrins MAC-1 and p150,95 by monocytes and increases monocyte adhesion to endothelial cells (35), providing yet another mechanism by which hyperlipidemia promotes monocyte adhesion.

MCP-1, thus, completes a relatively well-defined sequence by which hyperlipidemia, after inducing expression of adhesion molecules and promoting leukocyte adhesion, can also induce monocyte transmigration into the subendothelial space to foster foam cell formation. Foam cells are resident vascular macrophages containing cholesteryl ester inclusion bodies, which give the hallmark fatty streak its characteristic yellow appearance.

E. Macrophage-Colony Stimulating Factor

The differentiation of monocytes into macrophages occurs under the influence of macrophage-colony stimulating factor (M-CSF). M-CSF also increases expression of the acetyl-LDL receptor in cultured human monocytes (36) and thereby supports foam cell formation. Clinton and colleagues found increased mRNA for M-CSF in atheromatous human lesions and demonstrated increased expression of M-CSF mRNA in cytokine-activated endothelial and smooth muscle cells (36). Although M-CSF seems to be involved in the accumulation of foam cells in early atherosclerotic lesions, it may also function as part of a protective mechanism to limit accumulation of lipids in the subendothelial space: Yamada and colleagues (37) showed increased cholesterol efflux from tissues to plasma high-density lipoproteins in response to intravenous M-CSF treatment in Watanabe heritable hyperlipidemic (WHHL) rabbits; and Inoue and colleagues (38) showed that WHHL rabbits treated with M-CSF had dramatic reductions in atheromatous lesions of the aorta. Similarly, Schaub and colleagues (39) demonstrated cholesterol and cholesteryl ester clearance via the reticuloendothelial system and regression of carrageenan-induced granulomata in WHHL rabbits treated with M-CSF. This divergent effect is likely a consequence of the difference in M-CSF gradient: in atheromata the gradient is chemoattractant toward the lesion while in the treatment regimen M-CSF plasma concentrations are higher and facilitate efflux of macrophages (and cholesterol) from the lesions.

Gerrity (40) has demonstrated that foam cells can migrate from fatty streaks into the circulation, which represents a mechanism for clearing or recycling foam cells from these lesions in a swine model. Progression of lesions beyond the fatty streak stage to involve smooth muscle cells may entail a failure of this mechanism, ultimately leading to necrosis of foam cells with release of inflammatory mediators and proteolytic enzymes. This, coupled with the associated endothelial dysfunction that promotes thrombosis and recruitment of leukocytes, would lead to further injury and growth of the atherothrombotic plaque.

IV. ADHESION MOLECULES AND ATHEROSCLEROTIC DETERMINANTS

A. CAMs in Hypercholesterolemia

Endothelial dysfunction in hypercholesterolemia has been demonstrated in a number of animal models and in human coronary arteries and peripheral vessels (41–46), and impairment of signal transduction by the acetylcholine receptor may be one of the underlying mechanisms (43). Of interest, endothelial dysfunction as assessed by impaired vasodilator responses to muscarinic agonists is more a consequence of hypercholesterolemia per se than frank atherosclerosis (41).

A number of experimental animal models of atherosclerosis have documented that one of the earliest demonstrable changes in hypercholesterolemia is the adhesion of monocytes to the endothelial wall. This interaction is mediated by a number of adhesion molecules on endothelial cells and their counterreceptors on monocytes (10 (Table 1), and appears to be dependent on changes induced both in the endothelial cell (47–49) and the monocytes (50–52). Monocyte adhesion to endothelium has been shown in hypercholesterolemic rats, pigeons, rabbits, swine, and primates (49,53–56). Swine and primates develop atherosclerosis similar in morphology and distribution to that of humans and have been studied by Gerrity, Schwartz, and Faggiotto and colleagues (53–56). As early as 1979, Gerrity demonstrated adhesion of monocytes to lesion-prone areas in the Evans blue model in swine. Subsequent work (40) showed monocytes passing between endothelial cells and within the neointima, and suggested that these monocytes were transformed into macrophages and developed into foam cells. Work by Faggiotto et al. (54,55) in the primate *Macaca nemestrina* showed monocyte adherence to endothelial cells of aorta within 12 days of the initiation of an hypercholesterolemic diet, and monocyte attachment to endothelium overlying fatty streaks throughout their evolution. Schwartz and colleagues (56), working with the baboon *Papio cynocephalus*, found adherence of monocytes to endothelium overlying plaques. Of interest, the endothelium with which the monocytes were associated was morphologically intact.

Eight adhesion molecules present on endothelial cells are potentially involved in this phenomenon of monocyte adhesion in atherogenesis: vascular cell adhesion molecule-1 (VCAM-1), platelet-endothelial cell adhesion molecule-1 (PECAM-1), intercellular adhesion molecule-1 (ICAM-1), ICAM-2, ICAM-3, P-selectin, E-selectin, and IG9 protein. Of these, five have been demonstrated to be present in human atherosclerotic plaques: VCAM-1, ICAM-1, P-selectin, E-selectin, and PECAM-1 (57,58). P-selectin seems to be most specifically associated with atherosclerotic plaques, while the others have variable expression in both normal and atherosclerotic arterial segments; Johnson and colleagues (58) stained human atherosclerotic plaques for P-selectin and were able to identify immunoreactivity only in endothelium overlying active atherosclerotic plaques. P-selectin was not found on endothelium overlying inactive fibrous plaques nor in normal arterial segments. PECAM-1 is expressed equally in atherosclerotic and normal coronary arteries.

VCAM-1 is present on the endothelium overlying fatty streaks in rabbits (59). It is observed in similar distribution in plaque segments and in adjacent morphologically normal segments (21% vs. 27%, respectively) on endothelial cells of human coronary arteries taken from patients with atherosclerotic coronary artery disease. Interestingly, VCAM-1 is much more prevalent in the endothelial cells of adventitia of plaques compared to controls (88% vs. 27%, respectively). Furthermore, it is present to some extent in the intimal neovasculature of 71% of plaques. This finding does not provide a mechanism by which monocytes adhere to endothelial cells to initiate plaque formation, but suggests that recruitment of inflammatory cells in established plaques occurs via the intimal neovasculature (60).

Wood and colleagues (61) studied autopsy specimens of human coronary arteries and abdominal aorta and did not find VCAM-1 expression on intimal endothelial cells. They did, however, find strong staining for VCAM-1 in lymphoid aggregates present in the adventitia. E-selectin and ICAM-1 were expressed in intimal endothelium in both normal coronary arteries and atheromatous arteries. They also noted immunoreactivity for VCAM-1 in areas surrounding macrophage deposition.

Under resting conditions, leukocytes do not adhere to endothelial cells, but, upon stimulation by cytokines, as occurs in atherogenesis, endothelial cells can be induced to express a number of adhesion molecules that promote leukocyte adhesion. More recently, it has been shown in a number of models that hyperlipidemia per se is capable of activating endothelial cells. Li and colleagues (48) showed in a rabbit model that after 1 week of an atherogenic diet, the aortic endothelium showed focal expression of VCAM-1. Macrophage infiltration was noted at 3 weeks but not at 1 week after the initiation of the diet. Therefore, one can conclude that VCAM-1 expression in atherosclerotic lesions precedes infiltration of monocytes and is not a consequence of cytokine release by inflammatory cells in areas of lesions.

Oxidized LDL has a much higher content of lysophosphatidylcholine than unoxidized LDL, and is found in higher concentrations in atherosclerotic lesions of animals fed an atherogenic diet than normals. Recently, Kume and colleagues (47) showed that lysophosphatidylcholine selectively induces expression of VCAM-1 and ICAM-1 in cultured rabbit and human endothelial cells. Furthermore, this upregulation was associated with increased mononuclear leukocyte adhesion.

Similarly, Calderon and colleagues (62) exposed human umbilical vein endothelial cells (HUVEC) and cultured human aortic endothelial cells to minimally modified LDL, and induced the expression of a new adhesion molecule, IG9 protein, specific for monocytes. Thus, a direct link between hypercholesterolemia, lipoprotein oxidation, and monocyte adhesion, mediated by specific cellular adhesion molecules, is now apparent from these hypercholesterolemic models of atherosclerosis.

Monocytes in hypercholesterolemia are also altered and may contribute to the increase in their adhesive interactions with endothelial cells. Bath and colleagues (50) found monocytes from hypercholesterolemic patients to be more sensitive to stimulation by N-formyl-L-methionyl-L-leucyl-L-phenylalanine and to manifest increased chemokinesis and chemotaxis as well as increased adhesion to porcine aortic endothelial cells. Kelley and colleagues (52) demonstrated increased monocyte activation in response to exposure to lipoproteins, especially LDL.

B. CAMs in Diabetes

In animal models of diabetes, there is an impairment of endothelium-dependent vasodilation associated with increased levels of endothelin and thromboxane A2 and decreased levels of prostacyclin and EDRF/NO (63). This impairment is reversed by cyclooxygenase inhibitors, antioxidants, and free-radical scavengers, which suggests that free radicals modulate, in part, the endothelial dysfunction caused by hyperglycemia (64). Plasma from diabetic subjects has been shown to increase adhesion of neutrophils from diabetic and control subjects to bovine aortic endothelial cells (65). Increased levels of glucose are associated with an increase in diacylglycerol in cultured rat and bovine aortic endothelial cells and smooth muscle cells (66), and in the aorta of diabetic dogs (67). Increased levels of diacylglycerol activate protein kinase C, and protein kinase C activation has been shown in other models (68,69) to reduce nitric oxide synthase activity. This effect would, therefore, induce endothelial dysfunction by reducing NO synthesis, which would promote leukocyte adhesion.

Another potential mechanism for endothelial dysfunction in diabetes involves the nonenzymatic glycation of proteins such as collagen. These proteins can undergo chemical changes to form advanced glycation end products (AGE) (70), which have been shown to be chemotactic for monocytes and can induce migration of monocytes across an endothelial monolayer (71). In addition, Bucala and colleagues have demonstrated that AGEs can inactivate nitric oxide in vitro and impair endothelium-dependent vasodilation in rats (72).

Human umbilical vein endothelial cells exposed to high concentrations of glucose express the integrin $\alpha_5\beta_1$, which binds to fibronectin. In addition, retinal microvessels of diabetic patients manifest increased expression of β_1-integrins. Increased expression of these integrins would enhance cell-matrix interactions and may thereby impair the endothelial migration and replication necessary for reparative responses following vascular injury (73).

C. CAMs in Hypertension

Endothelium-dependent relaxation is also impaired in hypertension and may be related to decreased NO, increased prostaglandin H_2, increased endothelin expression, and impaired responsiveness to NO (74). Studies have documented impairment of endothelium-dependent relaxation in the forearm and coronary circulations of hypertensive patients (75–77) and in normotensive individuals with a family history of hypertension (78). Clozel and colleagues (79) showed that, in spontaneously hypertensive rats, the regional distribution of endothelial dysfunction and monocyte infiltration are associated, and that treatment with an ACE inhibitor decreases monocyte infiltration and improves endothelium-dependent relaxation. Studies of endotoxin- or cytokine-activated cerebrovascular endothelial cell cultures from hypertensive rats showed increased adhesion of monocytes compared to cells from normotensive rats. Similarly, monocytes from hypertensive rats were more adherent to stimulated endothelial cells than monocytes from normotensive rats (80). The adhesion molecules responsible for this behavior of monocytes have not been identified and, interestingly, in other studies, monocytes from hypertensive rats were shown to manifest decreased expression of the integrin $\alpha_M\beta_4$ (81). ICAM-1 may be involved at the level of the endothelial cells. One study in rat brain endothelial cells (82) found similar constitutive expression of ICAM-1 in hypertensive rats and controls; however, endothelial cells from hypertensive rats were more sensitive to IL-1-, TNFα- and interferon-γ-induced upregulation of ICAM-1, and manifested higher maximal ICAM-1 expression in response to endotoxin.

D. CAMs and Smoking

Smoking has been associated with decreased vascular response to endothelium-dependent vasodilators in most (83–85), but not all (86), studies of the forearm circulation. A vasoconstrictor response to acetylcholine has been documented in angiographically normal coronary arteries of heavy smokers (87). Celermajer and colleagues (88) studied the brachial artery response to reactive hyperemia (increased flow causes endothelium-dependent vasodilation) in smokers and former smokers compared to nonsmokers. Impaired vasodilation in the smokers was directly related to lifetime smoking dose and appeared to be potentially reversible with cessation of tobacco use.

In parallel with these abnormal vasodilator responses, increased expression of adhesion molecules has been shown to accompany cigarette smoke exposure. Kalra and colleagues (89) observed increased monocyte adhesion to cultured human and bovine endothelial cells exposed to cigarette smoke condensate. They also demonstrated increased expression of E-selectin and ICAM-1 on endothelial cells and increased expression of α_M (CD11b) on monocytes in response to cigarette smoke condensate. Klute and colleagues (90) exposed rabbits to cigarette smoke and demonstrated increased expression of CD11b/CD18 and decreased expression of L-selectin on intravascular neutrophils in the pulmonary microvasculature.

E. CAMs in Platelet Responses

The role of thrombosis in the pathogenesis of atherosclerosis was recognized as early as 1948 by Dugrid (91), who demonstrated that thrombi in atheromatous lesions become organized to form a thickened, fibrous intima. Hand and Chandler (92) showed that fragmented autologous blood clots injected into the pulmonary arteries of rabbits induced atheromatous changes, and suggested that platelets could serve as a source of cholesterol for foam cells. The incorporation of thrombi into subendothelial deposits and atherosclerotic lesions has also been demonstrated (93).

The role of platelets as one important thrombotic determinant in the initiation and propagation of atherosclerosis depends on specific cellular adhesion molecules that promote their interaction with endothelial cells, the subendothelial surface, and leukocytes. Mendelsohn and Loscalzo (94) have demonstrated that platelets serve as a source of cholesterol for macrophages and that the adhesion of platelets to macrophages in the process depends on the integrin Mac-1 ($\alpha_M\beta_4$). Mac-1 is internalized when occupied by fibrinogen, providing an alternative pathway for clearing fibrinogen from the injured vascular wall; however, internalization of fibrinogen complexed to platelets via $\alpha_{2b}\beta_3$ (glycoprotein IIb/IIIa) can serve as a mechanism for thrombophagocytosis and, therefore, contribute to the uptake of cholesterol-containing platelet particles by evolving macrophage foam cells.

As is the case with monocytes, platelet-endothelial interactions do not occur unless there is activation of the endothelium (even though both platelets and endothelial cells constitutively express PECAM-1, which can serve as its own counterreceptor). Platelets contain sialyl-Lewis x oligosaccharides that are able to bind P-selectin expressed on activated endothelial cells. Alternatively, platelets can bind to leukocytes adherent to endothelial cells via P-selectin expressed on activated platelets and sialyl Lewis x expressed on leukocytes. These heterotypic interactions are important in facilitating signaling between platelets and monocytes, such as those mediated by platelet-derived growth factor (PDGF) and thrombin. PDGF is important in monocyte recruitment, and thrombin, generated on the surface of platelets, is able to induce production of MCP-1 by monocytes and endothelial cells (95). Thrombin can also increase expression of ICAM-1 and P-selectin on endothelial cells (96).

Exposure of the subendothelial matrix to flowing blood permits binding of platelets to collagen or fibronectin via two integrins—$\alpha_2\beta_1$ (glycoprotein Ia/IIa), and $\alpha_5\beta_1$ (glycoprotein Ic/IIa), respectively. These receptors are constitutively expressed on platelets, and mediate the initial interactions of platelets with the subendothelium (97). Shear stress activates platelet glycoprotein Ib/IX, which facilitates the binding of Von Willebrand factor; glycoprotein Ib can also bind fibrin monomers, which then bind vWF (98). Von Willebrand factor, in turn, binds directly to exposed collagen in the subendothelial matrix. Within minutes of platelet activation, P-selectin, stored in the α granules, appears on the cell surface and binds monocytes, neutrophils, eosinophils, T cells, and natural killer cells (10,97). Glycoprotein IIb/IIIa ($\alpha_{2b}\beta_3$), constitutively expressed on platelets, changes conformation after platelet activation (97) and is able to bind fibrinogen, thus facilitating platelet-platelet interactions through the bivalent fibrinogen ligand, and platelet-fibrin interactions. In addition, glycoprotein IIb-IIIa binds fibronectin, vitronectin, and Von Willebrand factor, providing another means by which platelets can interact with the exposed vascular matrix (10,93).

V. THERAPEUTIC APPLICATIONS

Congenital, severe deficiencies in cellular adhesion molecules (CD11/CD18 in leukocyte adhesion deficiency (LAD) type 1 or in counterligands (sialyl Lewis x in LAD type II) (19) illustrate the potential dangers of inhibiting molecules involved in host defense. Significant blockade or inhibition of these interactions is unlikely to have beneficial effects unless employed for a short period of time; the effects of partial blockade for extended periods of time (as would be required theoretically in treating or preventing atherosclerosis) have not been investigated. Clearly, a balance between permitting the protective effects of the inflammatory response and reducing the inflammatory consequences of injury must be sought. As we learn more about the selective involvement of various adhesion molecules in different types of inflammatory responses, we will be able to design better-focused therapeutic interventions.

A. Monoclonal Antibodies

In animal models, antibodies against P-, E-, and L-selectins (99–103), β_2-integrins (103–105), and ICAM-1 (106–109) have been shown to decrease reperfusion injury in various organs including the heart (110). Other experimental models have proved the utility of blocking cellular adhesion molecules in achieving antiinflammatory effects, and some of these applications have led to successful implementation in human diseases. A mouse monoclonal antibody against the α subunit of LFA-1 was used to prevent graft failure in patients with immunodeficiency who received HLA-mismatched bone marrow transplants (111), and trials are under way to determine the efficacy of anti-ICAM-1 antibodies in kidney allograft rejection (112).

The concern of impairing host defenses has led investigators to study directly the effect of CAM blockade on host responses to infection. In a rabbit model, there was no attenuation of host defenses against specific pathogens such as *S. aureus* and *E. coli* (113); and Xu and colleagues (114) have shown that mice genetically deficient in ICAM-1 are actually more resistant to the lethal effects of endotoxin. In their model, wild-type and deficient mice had similar levels of TNFα and IL-1 in response to endotoxin; however, the latter manifested a significant decrease in neutrophil infiltration in the liver at 24 hours, suggesting that the decrease in tissue injury (and consequently, protection in septic shock) resulted from the absence of ICAM-1-dependent leukocyte-endothelial interactions.

B. Alternative Strategies

Some of the practical problems with monoclonal antibody therapy—namely, immunogenicity, immune complex diseases, and the need for intravenous administration (19)—will continue to limit its application to acute diseases. Other strategies will need to be developed for atherosclerosis specifically, such as targeting treatment to atheromata.

Soluble cellular adhesion molecules overcome the problem of immunogenicity somewhat and have been used to decrease neutrophil adhesion in vitro (P-selectin), and soluble selectin-IgG chimeras have been used to decrease lung injury in animal models (19,115). It is impractical to produce sufficient quantities of these polypeptides, so alternatives are being sought with the use of synthetic peptide derivatives or peptidomimetics. Fecondo and colleagues (116) showed that synthetic peptides based on the extracellular domains of ICAM-1 had similar inhibitory effects as anti-ICAM-1 monoclonal antibodies. Short peptides (24-amino acids) based on the extracellular domains of ICAM-1 (117) have been shown to block adhesion of leukocytes to human umbilical vein endothelial cells. Synthetic peptides that mimic the lectin domains of selectins were able to decrease recruitment of leukocytes into cerebrospinal fluid in a rabbit model of meningitis (118).

Intravenous administration of sialyl Lewis x has been very effective in reducing lung injury in a rat model, and heparin oligosaccharides that inhibit P- and L-selectin function in vitro have been shown to diminish neutrophil influx in vivo (119). Interest is focusing on creating biostable and bioactive carbohydrate analogues of sialyl Lewis x for oral administration to attempt to block selectin-mediated adhesion (120). Also, antisense oligonucleotides have been shown to inhibit selectively expression of VCAM-1, ICAM-1, and E-selectin by cultured human umbilical vein endothelial cells (121,122).

C. Antioxidants

Reactive oxygen species have emerged as an important risk factor in the development of atherosclerosis. Superoxide anion promotes adhesion of neutrophils to vascular endothelium (123). Recently, Gaboury and colleagues (124) demonstrated that CD18 mediates this increased adhesion, and P-selectin mediates the increased leukocyte flux induced by superoxide. Sellak and colleagues (125) also studied the molecules involved in adhesion induced by reactive oxygen species and did not find P-selectin to play a role, but did confirm the importance of CD18, and provided evidence for the involvement of ICAM-1 and carbohydrate ligands on endothelial cells.

Inhibition of nitric oxide synthesis increases neutrophil adhesion to cultured human endothelial cells (13,126) and in rat mesenteric venules (127,128), and NO donors prevent integrin-induced leukocyte adhesion in postischemic venules (14). This characteristic of nitric oxide is likely related to its ability to inactivate superoxide directly, to inhibit NADPH oxidase (129), and to inhibit activation of mast cells. Antioxidants may also have effects that are not directly related to inactivation of reactive oxygen species. Alpha-tocopherol decreased steady-state levels of E-selectin mRNA in IL-1-treated cultured human endothelial cells (130), and antioxidants have been shown to inhibit the induction of VCAM-1 on astrocytoma cells; it is possible that a similar mechanism may be active in vascular cells as well (131).

VI. CONCLUSIONS

The pathogenesis of atherosclerosis is currently understood as a complex set of interactions among blood cells and platelets, cells of the vessel wall, and the various determinants of thrombosis. These interactions are now being understood at the molecular level, as the factors that promote atherosclerosis are directly linked to expression of cell adhesion molecules involved in atherogenesis. In animal models of acute inflammatory injury, targeting adhesion molecules has met with some success. As advances in defining selectivity permit rational long-term therapy, therapeutic intervention in chronic disease states such as atherosclerosis may be possible.

REFERENCES

1. Ross R. The pathogenesis of atherosclerosis—an update. N Engl J Med 1986; 314:488–500.
2. Jonasson L, Holm J, Skalli O, Bondjers G, Hansson K. Regional accumulation of T cells, macrophages, and smooth muscle cells in the human atherosclerotic plaque. Arteriosclerosis 1986; 6:131–138.
3. Munro JM, van der Walt JD, Munro CS, Chalmers JAC, Cox EL. An immunohistochemical analysis of human fatty streaks. Hum Pathol 1987; 18: 375–380.
4. Hansson GK, Jonasson L, Lojsthed B, Stemme S, Kocher O, Gabbiani G. Localization of T lymphocytes in fibrous and complicated human atherosclerotic plaques. Arteriosclerosis 1988; 72:135–141.
5. Munro JM. Endothelial-leukocyte adhesive interactions in inflammatory diseases. Eur Heart J 1993; 14(suppl K):72–77.
6. Stemme S, Hansson GK. Immune mechanisms in atherosclerosis. Coronary Artery Dis 1994; 5:216–222.
7. Nieminen MS, Mattila K, Valtonen V. Infection and inflammation as risk factors for myocardial infarction. Eur Heart J 1993; 14(suppl K):12–16.
8. Fuster V, Badimon L, Badimon J, Chesebro JH. The pathogenesis of coronary artery disease and the acute coronary syndromes. N Engl J Med 1992; 326:242–250.
9. Levine GN, Keaney JF Jr, Vita JA. Cholesterol reduction in cardiovascular disease: clinical benefits and possible mechanisms. N Engl J Med 1995; 332(8):512–521.
10. Loscalzo J. The induction of cellular interactions in atherogenesis and their modulation. Biochem Soc Trans 1993; 21:656–659.
11. Valente AJ, Rozek MM, Sprague EA, Schwartz CJ. Mechanisms in intimal monocyte-macrophage recruitment: a special role for monocyte chemotactic protein-1. Circulation 1992; 86(suppl III):III20–III25.
12. Loscalzo J, Welch G. Nitric oxide and its role in the cardiovascular system. Prog Cardiovasc Dis 1995; 37(6):1–18.
13. Kubes P, Suzuki M, Granger DN. Nitric oxide: an endogenous modulator of leukocyte adhesion. Proc Natl Acad Sci USA 1991; 88:4651–4655.
14. Kubes P, Kurose I, Granger DN. NO donors prevent integrin-induced leukocyte adhesion but not P-selectin-dependent rolling in postischemic venules. Am J Phys 1994; 267:H931–H937.
15. Radomski MW, Palmer RM, Moncada S. Endogenous nitric oxide inhibits

human platelet adhesion to the vascular endothelium. Lancet 1987; 2:1057-1058.
16. Furchgott RF, Zawadzki JV. The obligatory role of endothelial cells in the relaxation of arterial smooth muscle by acetylcholine. Nature 1980; 228:373-376.
17. Meredith IT, Anderson TJ, Uehata A, Yeung AC, Selwyn AP, Ganz P. Role of endothelium in ischemic coronary syndromes. Am J Cardiol 1993; 72: 27C-32C.
18. Furchgott RF, Cherry PD, Zawadski JV, Jothianandan D. Endothelial cells as mediators of vasodilation of arteries. J Cardiovasc Pharm 1984; 6(suppl 2):S336-S343.
19. Albelda SM, Smith CW, Ward PA. Adhesion molecules and inflammatory injury. FASEB J 1994; 8:504-512.
20. Von Adrian UH, Chambers JD, Berg EL, et al. L-selectin mediates neutrophil rolling in inflamed venules through sialyl Lewis x-dependent and -independent recognition pathways. Blood 1993; 82:182-191.
21. Von Adrian UH, Arfors K. Neutrophil-endothelial interactions in vivo: a chain of events characterized by distinct molecular mechanisms. Agents Actions Suppl 1993; 41:153-164.
22. Von Adrian UH, Berger EM, Ramezani L, et al. In vivo behavior of neutrophils from two patients with distinct inherited leukocyte adhesion deficiency syndromes. J Clin Invest 1993; 91:2893-2897.
23. Lawrence MB, Springer TA. Neutrophils roll on E-selectin. J Immunol 1993; 151:6338-6346.
24. Lawrence MB, Springer TA. Leukocytes roll on a selectin at physiologic flow rates: distinction from and prerequisite for adhesion through integrins. Cell 1991; 65:859-873.
25. Lawrence MB, Smith CW, Eskin SJ, McIntire LV. Effect of venous shear stress on CD18-mediated neutrophil adhesion to cultured endothelium. Blood 1990; 75:227-237.
26. Bevilacqua M, Nelson RM. Selectins. J Clin Invest 1993; 91:379-387.
26a. Mayadas TN, Johnson RC, Rayburn H, Hynes RO, Wagner DD. Leukocyte rolling and extravasation are severely compromised in P selectin-deficient mice. Cell 1993; 74:541-554.
26b. Yamada S, Mayadas TH, Yuan F, Wagner DD, Hynes RO, Melder RJ, Jain RK. Rolling in P-selectin-deficient mice is reduced but not eliminated in the dorsal skin. Blood 1995; 86:3487-3492.
26c. Bullard DC, Kunkel EJ, Kubo H, Hicks MJ, Lorenzo I, Doyle NA, Doerschuk CM, Ley K, Beaudet AL. Infectious susceptibility and severe deficiency of leukocyte rolling and recruitment in E-selectin and P-selectin double mutant mice. J Exp Med 1996; 183:2329-2336.
26d. Kunkel EJ, Jung U, Bullard DC, Norman KE, Wolitzky BA, Vestweber D, Beaudet AL, Ley K. Absence of trauma-induced leukocyte rolling in mice deficient in both P-selectin and intercellular adhesion molecule-1. J Exp Med 1996; 183:57-65.
27. Jiang Y, Beller DI, Frendl G, Graves DT. Monocyte chemoattractant pro-

tein-1 regulates adhesion molecule expression and cytokine production in human monocytes. J Immunol 1992; 148:2423–2438.
28. Muller WA, Ratti CM, McDonnell SL, Cohn ZA. A human endothelial cell-restricted, externally disposed plasmalemmal protein enriched in intercellular junctions. J Exp Med 1989; 170(2):399–414.
29. Albelda SM, Oliver PD, Romer LH, Buck CA. EndoCAM: a novel endothelial cell-cell adhesion molecule. J Cell Biol 1990; 110(4):1227–1237.
30. Muller WA, Weigl SA, Deng X, Phillips DM. PECAM-1 is required for transendothelial migration of leukocytes. J Exp Med 1993; 178:449–460.
31. Vaporciyan AA, DeLisser HM, Yan H-C, et al. Involvement of platelet-endothelial cell adhesion molecule-1 in neutrophil recruitment in vivo. Science 1993; 262:1580–1582.
32. Yoshimura T, Leonard EJ. Identification of high affinity receptors for human monocyte chemoattractant protein-1 on human monocytes. J Immunol 1990; 145:292–297.
33. Valente AJ, Rozek MM, Schwartz CJ, Graves DT. Characterization of monocyte chemotactic protein-1 binding to human monocytes. Biochem Biophys Res Commun 1991; 176:309–314.
34. Cushing SD, Berliner JA, Valente AJ, et al. Minimally modified low density lipoprotein induces monocyte chemotactic protein-1 in human endothelial cells and smooth muscle cells. Proc Natl Acad Sci USA 1990; 87:5134–5138.
35. Vaddi K, Newton RC. Regulation of monocyte integrin expression by beta-family chemokines. J Immunol 1994; 153(10):4721–4732.
36. Clinton SK, Underwood R, Hayes L, Sherman ML, Kufe DW, Libby P. Macrophage colony-stimulating factor gene expression in vascular cells and in experimental and human atherosclerosis. Am J Pathol 1992; 140(2):301–316.
37. Yamada N, Ishibashi S, Shimano H, et al. Role of monocyte colony-stimulating factor in foam cell generation. Proc Soc Exp Biol Med 1992; 200(2):240–244.
38. Inoue I, Inaba T, Motoyoshi K, et al. Macrophage colony stimulating factor prevents the progression of atherosclerosis in Watanabe heritable hyperlipidemic rabbits. Atherosclerosis 1992; 93(3):245–254.
39. Schaub RG, Bree MP, Hayes LL, et al. Recombinant human macrophage colony-stimulating factor reduces plasma cholesterol and carrageenan granuloma foam cell formation in Watanabe heritable hyperlipidemic rabbits. Arteriosc Thromb 1994; 14(1):70–76.
40. Gerrity RG. The role of monocyte in atherogenesis. II. Migration of foam cells from atherosclerotic lesions. Am J Pathol 1981; 103:191–200.
41. Zeiher AM. Endothelium-mediated coronary blood flow modulation in humans. Effects of age, atherosclerosis, hypercholesterolemia, and hypertension. J Clin Invest 1993; 92(2):652–662.
42. Van Boven AJ, Jukema JW, Paoletti R. Endothelial dysfunction and dyslipidemia: possible effects of lipid lowering and lipid modifying therapy. Pharm Res 1994; 29(3):261–272.
43. Gilligan DM, Guetta V, Panza JA, Garcia CE, Quyyumi AA, Cannon RO

3d. Selective loss of microvascular endothelial function in human hypercholesterolemia. Circulation 1994; 90(1):35–41.
44. Raij L, Nagy J, Coffee K, DeMaser EG. Hypercholesterolemia promotes endothelial dysfunction in vitamin E- and selenium-deficient rats. Hypertension 1993; 22(1):56–61.
45. Drexler H, Zeiher AM, Meinzer K, Just H. Impaired endothelium-dependent vasolidation of forearm resistance vessels in hypercholesterolaemia. Lancet 1991; 340(8833):1430–1432.
46. Drexler H, Zeiher AM. Endothelial function in human coronary arteries in vivo. focus on hypercholesterolemia. Hypertension 1991; 18(4 suppl):II90–II99.
47. Kume N, Cybulsky MI, Gimbrone MA Jr. Lysophosphatidylcholine, a component of atherogenic lipoproteins, induces mononuclear leukocyte adhesion molecules in cultured human and rabbit arterial endothelial cells. J Clin Invest 1992; 90:1138–1144.
48. Li H, Cybulsky MI, Gimbrone MA Jr, Libby P. An atherogenic diet rapidly induces VCAM-1, a cytokine-regulatable mononuclear leukocyte adhesion molecule in rabbit aortic endothelium. Arterio Thromb 1993; 13:197–204.
49. Mahamad N, Hama SY, Nguyen TB, Fogelman AM. Monocyte adhesion and transmigration in atherosclerosis. Coronary Artery Dis 1994; 5:198–204.
50. Bath PM, Gladwin AM, Martin JF. Human monocyte characteristics are altered in hypercholesterolemia. Arteriosclerosis 1991; 90:175–181.
51. Stragliotto E, Camera M, Postiglione A, Sirtori M, Di Minno G, Tremoli E. Functionally abnormal monocytes in hypercholesterolemia. Arterio Thromb 1993; 13:944–950.
52. Kelley JL, Rozek MM, Suenram CA, Schwartz CJ. Activation of human peripheral blood monocytes by lipoproteins. Am J Pathol 1988; 130:223–331.
53. Gerrity RG. The role of the monocyte in atherogenesis. I. Transition of blood-borne monocytes into foam cells in fatty lesions. Am J Pathol 1981; 103:181–190.
54. Faggiotto A, Ross R, Harker L. Studies of hypercholesterolemia in the nonhuman primate. I. Changes that lead to fatty streak formation. Arteriosclerosis 1984; 4:323–340.
55. Faggiotto A, Ross R. Studies of hypercholesterolemia in the nonhuman primate. II. Fatty streak conversion to fibrous plaque. Arteriosclerosis 1984; 4: 341–356.
56. Schwartz CJ, Sprague EA, Kelley JL, Valente AJ, Suenram CA. Aortic intimal monocyte recruitment in the normo and hypercholesterolemic baboon (*Papio cynocephalus*). An ultrastructural study: implications in atherogenesis. Virchows Arch Pathol Anat Histopathol 1985; 405:175–191.
57. Davies MJ, Gordon JL, Gearing AJ, et al. The expression of the adhesion molecules ICAM-1, VCAM-1 PECAM, and E-selectin in human atherosclerotic atherosclerosis. J Pathol 1993; 171:223–229.
58. Johnson-Tidey RR, McGregor JL, Taylor PR, Poston RN. Increase in the adhesion molecule P-selectin in endothelium overlying atherosclerotic

plaques. Coexpression with intercellular adhesion molecule-1. Am J Pathol 1994; 144(5):952–961.
59. Cybulsky MI, Gimbrone MA Jr. Endothelial expression of a mononuclear leukocyte adhesion molecule during atherogenesis. Science 1991; 251:788–791.
60. O'Brien KD, Allen MD, McDonald TO, et al. Vascular cell adhesion molecule-1 is expressed in human coronary atherosclerotic plaques. J Clin Invest 1993; 92:945–951.
61. Wood KM, Cadogan MD, Ramshaw AL, Parums DV. The distribution of adhesion molecules in human atherosclerosis. Histopathology 1993; 22(5): 437–444.
62. Calderon TM, Factor SM, Hatcher VB, Berliner JA, Berman JW. An endothelial cell adhesion protein for monocytes recognized by monoclonal antibody IG9. Lab Invest 1994; 70:836–849.
63. Jensovsky J, Andel M, Stolba P. Endothelial dysfunction in diabetes mellitus. Casopsis Lekaru Ceskych 1994; 133:419–422.
64. Tesfamarian B, Cohen RA. Free radicals mediate endothelial cell dysfunction caused by elevated glucose. Am J Physiol 1992; 263:H321–H326.
65. Anderson A, Goldsmith GH, Spagnuolo PJ. Neutrophil adhesive dysfunction in diabetes mellitus; the role of cellular and plasma factors. J Lab Clin Med 1988; 111:275–285.
66. Inoguchi T, Xia P, Kunisaki M, Higashi S, Feener EP, King GL. Insulin's effect on protein kinase C and diacylglycerol induced by diabetes and glucose in vascular tissues. Am J Physiol 1994; 267(3 Pt 1):E369–E379.
67. Xia P, Inoguchi T, Kern TS, Engerman RL, Oates PJ, King GL. Characterization of the mechanism for the chronic activation of diacylglycerol-protein kinase C pathway in diabetes and hypergalactosemia. Diabetes 1994; 43: 1122–1129.
68. Davda RK, Chandler LJ, Guzman NJ. Protein kinase C modulates receptor-independent activation of endothelial nitric oxide synthase. Eur J Pharm 1994; 266(3):237–244.
69. Bredt DS, Ferris CD, Snyder SH. Nitric oxide synthase regulatory sites. J Biol Chem 1992; 267:10976–10981.
70. Brownlee M, Cerami A, Vlassara H. Advanced glycosylation end products in tissues and the biochemical basis of diabetic complications. N Engl J Med 1988; 318:1315–1321.
71. Kirstein M, Brett J, Radoff S, et al. Advanced protein glycosylation induced transendothelial human monocyte chemotaxis and secretion of platelet-derived growth factor: role in vascular disease of diabetes and aging. Proc Natl Acad Sci USA 1990; 87:9010–9014.
72. Bucala R, Tracey KJ, Cerami A. Advanced glycosylation products quench nitric oxide and mediate defective endothelium-dependent vasodilation in experimental diabetes. J Clin Invest 1991; 87(2):432–438.
73. Roth T, Podesta F, Stepp MA, Boeri D, Lorenzi M. Integrin overexpression induced by high glucose and by human diabetes: potential pathway to cell

dysfunction in diabetic microangiopathy. Proc Natl Acad Sci USA 1993; 90: 9640–9644.
74. Luscher TF. Heterogeneity of endothelial dysfunction in hypertension. Eur Heart J 1992; 13(suppl D):50–55.
75. Linder L, Kiowski W, Buhler FR, Luscher TF. Indirect evidence for release of endothelium-derived relaxing factor in human forearm circulation in vivo: blunted response in essential hypertension. Circulation 1990; 81:1762–1770.
76. Panza JA, Quyyumi AA, Brush JE, Epstein SE. Abnormal endothelium-dependent vascular relaxation in patients with essential hypertension. N Engl J Med 1990; 323:22–27.
77. Brush JE Jr, Faxon DP, Salmon S, Jacobs AK, Ryan TJ. Abnormal endothelium-dependent coronary vasomotion in hypertensive patients. J Am Coll Cardiol 1992; 19(4):809–815.
78. Taddei S, Virdis A, Mattei P, Arzilli F, Salvetti A. Endothelium-dependent forearm vasodilation is reduced in normotensive subjects with familial history of hypertension. J Cardiovasc Pharm 1992; 20(suppl 12):193–195.
79. Clozel M, Kuhn H, Hefti F, Baumgartner HR. Endothelial dysfunction and subendothelial monocyte macrophages in hypertension: effect of angiotensin converting enzyme inhibition. Hypertension 1991; 18:132–141.
80. McCarron RM, Wang L, Siren A-L, Spatz M, Hallenbeck JM. Monocyte adhesion to cerebromicrovascular endothelial cells derived from hypertensive and normotensive rats. Am J Physiol 1994; 267(6 Pt 2):H2491–H2497.
81. Arndt H, Smith CW, Granger DN. Leukocyte-endothelial cell adhesion in spontaneously hypertensive and normotensive rats. Hyperten Dallas 1993; 21:667–673.
82. McCarron RM, Wang L, Siren A-L, Spatz M, Hallenbeck JM. Adhesion molecules on normotensive and hypertensive rat brain endothelial cells. Proc Soc Exp Biol Med 1994; 205(3):257–262.
83. Kiowski W, Linder L, Stoschitzky K, et al. Diminished vascular response to inhibition of endothelium-derived nitric oxide and enhanced vasoconstriction to exogeneously administered endothelin-1 in clinically healthy smokers. Circulation 1994; 90:27–34.
84. Celermajer DS, Sorensen KE, Georgakopoulos D, et al. Cigarette smoking is associated with dose-related and potentially reversible impairment of endothelium-dependent dilation in healthy young adults. Circulation 1993; 88: 2149–2155.
85. Celermajer DS, Sorensen KE, Gooch VM, et al. Noninvasive detection of endothelial dysfunction in children and adults at risk of atherosclerosis. Lancet 1992; 340:1111–1115.
86. Jacobs MC, Lenders JW, Kapma JA, Smits P, Thien T. Effect of chronic smoking on endothelium-dependent vascular relaxation in humans. Clin Sci 1993; 85:51–55.
87. Nitenberg A, Antony I, Foult JM. Acetylcholine-induced coronary vasoconstriction in young, heavy smokers with normal coronary arteriographic findings. Am J Med 1993; 95:71–77.

88. Celermajer DS, Sorensen KE, Georgakopoulos D, et al. Circulation 1993; 88: 2149–2155.
89. Kalra VK, Ying Y, Deemer K, Natarajan R, Nadler JL, Coates TD. Mechanism of cigarette smoke condensate induced adhesion of human monocytes to cultured endothelial cells. J Cell Physiol 1994; 160(1):154–162.
90. Klute ME, Doerschuk CM, Van Eeden SF, Burns AR, Hogg JC. Activation of neutrophils within pulmonary microvessels of rabbits exposed to cigarette smoke. Am J Respir Cell Mol Biol 1993; 91:82–89.
91. Dugrid JB. Thrombosis as a factor in the pathogenesis of atherosclerosis. J Pathol Bacteriol 1948; 60:57–61.
92. Hand RA, Chandler AS. Atherosclerotic metamorphosis of autologous pulmonary thromboemboli in the rabbit. Am J Pathol 1962; 40:469.
93. White JG. Platelets and atherosclerosis. Eur J Clin Invest 1994; 24(suppl 1): 25–29.
94. Mendelsohn ME, Loscalzo J. Role of platelets in cholesteryl ester formation by U-937 cells. J Clin Invest 1988; 81:62–80.
95. Colotta F, Sciacca FL, Sironi M, Luini W, Rabiet MJ, Mantovani A. Expression of monocyte chemotactic protein-1 by monocytes and endothelial cells exposed to thrombin. Am J Pathol 1994; 144:975–985.
96. Sugama Y, Tiruppathi C, Offakidevi K, Anderson TT, Fenton JW 2d, Malik AB. Thrombin-induced expression on P-selectin and intercellular adhesion molecule-1: a mechanism for stabilizing neutrophil adhesion. J Cell Biol 1992; 119:935–944.
97. Rosen P, Schwippert B, Tschope D. Adhesive interactions in platelet-endothelial interactions. Eur J Clin Invest 1994; 24(suppl 1):21–24.
98. Loscalzo J, Inbal A, Handin RI. Von Willebrand protein facilitates platelet incorporation in polymerizing fibrin. J Clin Invest 1986; 78(4):1112–1119.
99. Davenpeck KL, Gauthier TW, Albertine KH, Lefer AM. Role of P-selectin in microvascular leukocyte-endothelial interaction in splanchnic ischemia-reperfusion. Am J Physiol 1994; 267:H622–H630.
100. Steinberg JB, Mao HZ, Niles SD, Jutila MA, Kapelanski DP. Survival in lung reperfusion injury is improved by an antibody that binds and inhibits L- and E-selectin. J Heart Lung Transpl 1994; 13:306–318.
101. Carden DL, Young JA, Granger DN. Pulmonary microvascular injury after intestinal ischemia-reperfusion: role of P-selectin. J Appl Physiol 1993; 75: 2529–2534.
102. Winn RK, Liggitt D, Vedder NB, Paulson JC, Harlan JM. Anti-P-selectin monoclonal antibody attenuates reperfusion injury to the rabbit ear. J Clin Invest 1993; 92:2042–2047.
103. Seekamp A, Till GO, Mulligan MS, et al. Role of selectins in local and remote tissue injury following ischemia and reperfusion. Am J Pathol 1994; 144: 592–598.
104. Kurose I, Anderson DC, Miyasaka M, et al. Molecular determinants of reperfusion-induced leukocyte adhesion and vascular protein leakage. Circ Res 1994; 74:336–343.
105. Seekamp A, Mulligan MS, Till GO, et al. Role of beta 2 integrins and

ICAM-1 in lung injury following ischemia-reperfusion of rat hind limbs. Am J Pathol 1993; 143:464–472.
106. Hess DC, Zhao W, Carroll J, McEachin M, Buchanan K. Increased expression of ICAM-1 during reoxygenation in brain endothelial cells. Stroke 1994; 25:1463–1467.
107. Suzuki S, Toledo-Pereyra LH. Monoclonal antibody to intercellular adhesion molecule 1 as an effective protection for liver ischemia and reperfusion injury. Transpl Proc 1993; 25:3325–3327.
108. Kukielka GL, Hawkins HK, Michael L, et al. Regulation of intercellular adhesion molecule-1 (ICAM-1) in ischemic and reperfused myocardium. J Clin Invest 1993;92:1504–1516.
109. Horgan MJ, Ge M, Gu J, Rothlein R, Malik A. Role of ICAM-1 in neutrophil-mediated lung vascular injury after occlusion and reperfusion. Am J Physiol 1991; 261:H1578–H1584.
110. Ma XL, Weyrich AS, Lefer DJ, et al. Monoclonal antibody to L-selectin attenuates neutrophil accumulation and protects ischemic reperfused cat myocardium. Circulation 1993; 88:649–658.
111. Fischer A, Griscelli C, Blanche S, et al. Prevention of graft failure by an anti-HLA-1 monoclonal antibody in HLA-mismatched bone marrow transplantation. Lancet 1986; 2:1058–1061.
112. Haug CE, Colvin RB, Delmonico FL, et al. A phase I trial of immunosuppression with anti-ICAM-1 (CD54) mAb in renal allograft recipients. Transplantation 1993; 55:766–773.
113. Sharar SR, Sasaki SS, Flaherty LC, Paulson JC, Harlan JM, Winn RK. P-selectin blockade does not impair leukocyte host defense against bacterial peritonitis and soft tissue infection in rabbits. J Immunol 1993; 151:4982–4988.
114. Xu H, Gonzalo JA, St. Pierre Y, et al. Leukocytosis and resistance to septic shock in ICAM 1-deficient mice. J Exp Med 1994; 180:95–100.
115. Mulligan MS, Watson SR, Fennie C, Ward PA. Protective effects of selectin chimeras in neutrophil-mediated lung injury. J Immunol 1993; 151:6410–6417.
116. Fecondo JV, Kent SB, Boyd AW. Inhibition of intercellular adhesion molecule 1-dependent biological activities by a synthetic peptide analog. Proc Natl Acad Sci USA 1991; 88(7):2879–2882.
117. Ross L, Hassman F, Molony L. Inhibition of Molt-4-endothelial adherence by synthetic peptides from the sequence of ICAM-1. J Biol Chem 1992; 267(12):8537–8543.
118. Rozdzinski E, Jones T, Burnette WN, Burroughs M, Tuomanen E. Anti-inflammatory effects of prokaryotic peptides that mimic selectins. J Infect Dis 1993; 168:1422–1428.
119. Nelson RM, Cecconi O, Roberts WG, Aruffo A, Linhardt RJ, Bevilacqua MP. Heparin oligosaccharides bind L- and P-selectin and inhibit acute inflammation. Blood 1993; 82:3253–3258.
120. Nelson RM, Dolich S, Aruffo A, Cecconi O, Bevilacqua MP. Higher affinity oligosaccharide ligands for E-selectin. J Clin Invest 1993; 91:1157–1166.

121. Bennett CF, Condon TP, Grimm S, Chan H, Chiang MY. Inhibition of endothelial cell adhesion molecule expression with antisense oligonucleotides. J Immunol 1994; 152(7):3530–3540.
122. Chiang MY, Chan H, Zounes MA, Freier SM, Lima WF, Bennett CF. Antisense oligonucleotides inhibit intercellular adhesion molecule-1 expression by two distinct mechanisms. J Biol Chem 1991; 266(27):18162–18171.
123. Del Maestro RF, Planker M, Arfork KE. Evidence for the participation of superoxide anion radical in altering the adhesive interaction between granulocytes and endothelium, in vivo. Int J Microcirc Clin Exp 1982; 1:105–120.
124. Gaboury JP, Anderson DC, Kubes P. Molecular mechanisms involved in superoxide-induced leukocyte-endothelial cell interactions in vivo. Am J Physiol 1994; 266:H637–H642.
125. Sellak H, Franzini E, Hakim J, Pasquier C. Reactive oxygen species rapidly increase endothelial ICAM-1 ability to bind neutrophils without detectable upregulation. Blood 1994; 83:2669–2677.
126. Niu XF, Smith CW, Kubes P. Intracellular oxidative stress induced by nitric oxide synthesis inhibition increases endothelial cell adhesion to neutrophils. Circ Res 1994; 74:1133–1140.
127. Kubes P, Kanwar S, Niu X-F, Gaboury J. Nitric oxide synthesis inhibition induces leukocyte adhesion via superoxide and mast cells. FASEB J 1993; 7: 1293–1299.
128. Arndt H, Russell JR, Curose I, et al. Mediators of leukocyte adhesion in rat mesenteric venules is elicited by inhibition of nitric oxide synthesis. Gastroenterology 1993; 105:675–680.
129. Clancy RM, Leszczynska-Piziak J, Abramson SB. Nitric oxide, an endothelial cell relaxation factor, inhibits neutrophil superoxide anion production via a direct action on the NADPH oxidase. J Clin Invest 1992; 90(3):1116–1121.
130. Faruqi R, de la Motte C, DiCorleto PE. Alpha-tocopherol inhibits agonist-induced monocytic cell adhesion to cultured human endothelial cells. J Clin Invest 1994; 94:592–600.
131. Moynagh PN, Williams DC, O'Neill LA. Activation of NF-kappa B and induction of vascular cell adhesion molecule-1 and intracellular adhesion molecule-1 expression in human glial cells by IL-1. Modulation by antioxidants. J Immunol 1994; 153:2681–2690.

16

Adhesion Molecules in Bone Remodeling

David E. Hughes and Donald M. Salter
Department of Pathology, University of Sheffield Medical School, Sheffield, England

I. INTRODUCTION

Bone is a complex tissue that, despite its superficially lifeless appearance, is populated by a variety of cell types that control its anatomy, metabolism, and function. Bone is capable of remodeling itself throughout life in response to the demands of skeletal growth, changes in mechanical loading, and mineral metabolism. The ability of bone to constantly remodel itself without loss of structural integrity is due to a cycle of microanatomical events, occurring simultaneously but asynchronously at many different sites in the skeleton. These cellular events are referred to as the bone remodeling cycle, the basis of which is resorption of bone matrix by osteoclasts followed by resynthesis of matrix at the same site by osteoblasts. The bone remodeling cycle has been extensively reviewed elsewhere (1,2), but, as an appreciation of the events of the bone remodeling cycle is essential for understanding the potential importance of cell-cell and cell-matrix interactions in bone physiology, a brief description will be included here (see also Table 1).

II. CELLULAR EVENTS IN THE BONE REMODELING CYCLE

At any point in time, most bone surfaces are metabolically quiescent, and are covered either by lining cells, which are thought to be inactive osteo-

Table 1 Sequence of Events During Bone Remodeling

1. Resting phase
2. Activation: recruitment of osteoclast precursors to bone surface
3. Resorption: destruction of bone matrix by osteoclasts
4. Reversal: osteoclasts are replaced by osteoblasts
5. Formation: osteoblasts synthesized unmineralized (protein) matrix
6. Mineralization: unmineralized matrix becomes calcified
7. Resting phase

blasts, or periosteal cells on the outer cortex, which resemble fibroblasts but which have osteogenic potential. The stimulus that initiates bone remodeling at any site is unknown, but the cellular events that follow initiation have been carefully studied. During the initiation phase, mononuclear osteoclast precursors arrive at the bone surface, become attached to the underlying matrix, and fuse with each other to form multinucleated mature osteoclasts. The lining cells may play an active role by retracting their cytoplasm to expose the underlying matrix, and possibly by secreting metalloproteinases to remove the most superficial unmineralized layer of the matrix, which is known as the lamina limitans (3). There is also growing evidence that cell-cell contact and/or secretion of paracrine factors by cells of the osteoblast lineage is necessary for the terminal stages of osteoclast differentiation to occur (4,5). The events that lead to osteoclast precursors arriving at the bone surface are mysterious. Osteoclasts are known to be derived from hematopoietic precursors (6–8) and are closely related to the mononuclear phagocyte lineage. There is evidence that osteoclast precursors circulate (9), although to date it has not been possible to distinguish osteoclast precursors from other mononuclear cells in peripheral blood. As there is also no firm evidence to suggest that osteoclast precursors are permanent residents at the bone surface, it is likely that they have a homing mechanism that allows them to cross bone marrow blood vessel walls and to gain access to sites where they are required.

Once at the bone surface, osteoclasts become polarized and develop their anatomical hallmarks—the sealing zone and the ruffled border. The sealing zone is a ringlike structure around the periphery of the osteoclast by which it attaches itself to the bone surface. The remainder of the surface in contact with the bone matrix is unattached and essentially functions as an extracellular lysosome into which the osteoclast secretes hydrogen ions and, presumably, proteolytic enzymes (2,10) to cause dissolution of the mineral and breakdown of the protein components of the matrix. The cytoplasm immediately adjacent to this area organizes itself into a complex series of folds known as the ruffled border. Osteoclasts may be found singly on the bone surface, but more typically exist in groups that resorb the underlying matrix to form a structure known as a Howship's lacuna. An example of an osteoclast in a resorption lacuna is shown in Figure 1. Just as osteoclasts appear suddenly, they disappear suddenly. Various fates have been proposed for the osteoclast, such as fission, change of phenotype, or phagocytosis by neighboring cells (2,11). Evidence now exists that at least under some circumstances osteoclasts undergo programmed cell death, or apoptosis (12,13).

The phase of the remodeling cycle during which osteoclasts disappear from the bone surface is referred to as reversal. When osteoclasts disappear from a Howship's lacuna, they are normally replaced by stromal cells that rapidly differentiate to form osteoblasts (see Fig. 2). These then synthesize new bone matrix. Histologically it is possible to identify where reversal has occurred as a thin layer of proteoglycan-rich matrix highlighted by toluidine blue staining. The precise function of this specialized matrix, and which cells produce it, are unclear.

As they synthesize new matrix, a proportion of osteoblasts become incorporated into the matrix to become osteocytes. It is not known what makes

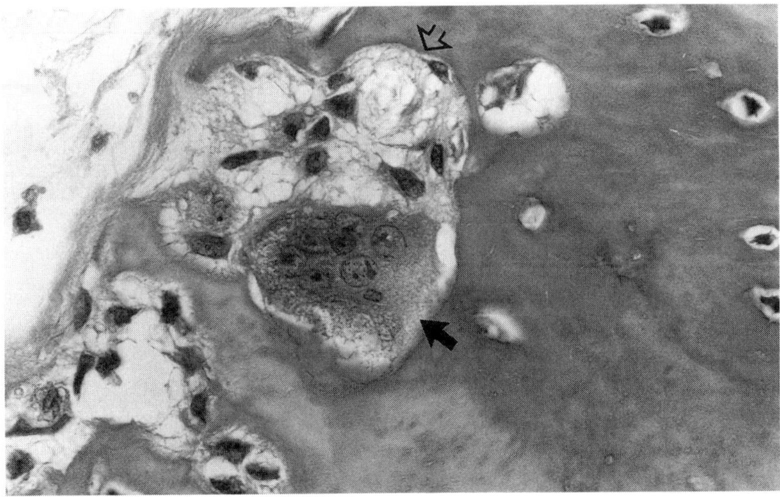

Figure 1 This photomicrograph demonstrates an osteoclast within a resorption (Howship's) lacuna (solid arrow). Adjacent to this cell is an area of bone surface which has been resorbed at which osteoclasts are no longer present (open arrow). This represents the reversal phase of the remodeling cycle. This surface will subsequently become lined by osteoblasts (see Fig. 2).

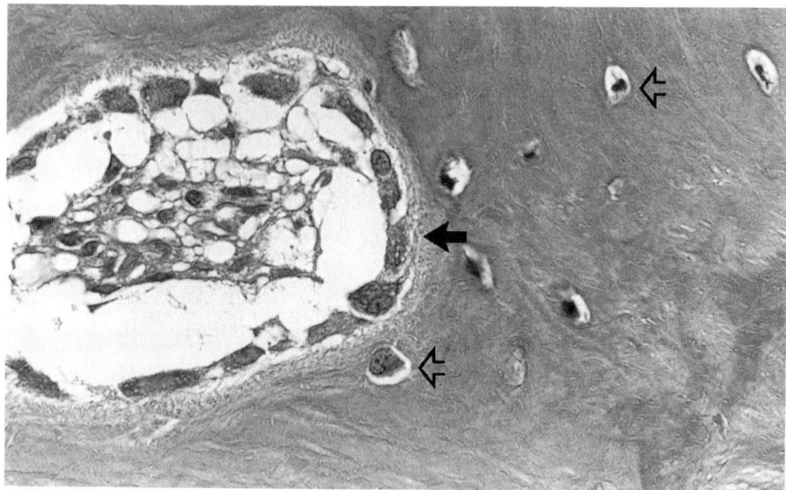

Figure 2 This photomicrograph demonstrates a Howship's lacuna from which the osteoclasts have disappeared and is now lined by active osteoblasts showing characteristic cuboidal morphology (solid arrow). These cells synthesize bone matrix in which a proportion of them become embedded and differentiate to form osteocytes (open arrows).

an osteoblast decide to differentiate into an osteocyte. Little is known of the function of osteocytes, although evidence is beginning to accumulate to support the theoretically attractive concept that the osteocyte functions as a stress transducer (14,15). The remainder of osteoblasts eventually lose their activity, and are thought to differentiate into lining cells. It is not known what determines how much matrix is synthesized by any group of osteoblasts.

Clearly, there are a number of cell-cell and cell-matrix interactions occurring during the bone remodeling cycle (see Table 2). It is likely that these are important control points in this process.

III. OSTEOCLAST ADHESION

Despite the theoretical importance of cell adhesion in bone outlined above, the initiation of research into this field came about by a slightly curious route. In the early 1980s, there was still much controversy regarding the ontogeny of the osteoclast. While a series of landmark studies had established that osteoclasts were derived from haemopoietic precursors (6–8), it was uncertain if osteoclasts were derived from the mononuclear phagocyte

Table 2 Cell-Cell and Cell-Matrix Interactions Known or Likely to Occur During the Bone Remodeling Cycle

Stage of cycle	Interactions
Initiation	1. changes in adhesion of lining cells to bone matrix
	2. osteoclast precursor-endothelial interactions
	3. osteoclast precursor-lining cell interactions
	4. osteoclast precursor-matrix adhesion
	5. osteoclast precursor fusion
Resorption	1. osteoclast-matrix adhesion (via sealing zone)
	2. continuing osteoclast fusion
	3. phagocytosis of osteocytes by osteoclasts
Reversal	1. loss of osteoclast-matrix adhesion
	2. phagocytosis of apoptotic osteoclasts
	3. osteoblast precursor-matrix adhesion
Synthesis	1. osteoblast-matrix adhesion
	2. osteocyte formation
	3. osteoblast-osteoblast interactions
	4. osteoblast-osteocyte interactions

lineage, or if they had a separate lineage of their own. One approach that was used in studying the relationship of osteoclasts to other cell types was the development of monoclonal antibodies against giant cell tumors of bone. These tumors are characterized by the formation of multinucleated cells phenotypically very similar to osteoclasts. A number of monoclonal antibodies that reacted with osteoclasts were raised by Horton and coworkers; two of them, 23C6 and 13C2, were used for further study (16). These two antibodies reacted with antigens expressed strongly on the plasma membranes of osteoclasts, but not by cells of the mononuclear phagocyte lineage or by osteoblasts. More significantly, these antibodies also inhibited the resportive activity of isolated osteoclasts in vitro (17).

Characterization of the proteins recognized by these antibodies revealed that 23C6 reacted with β_3-integrin and 13C2 reacted with α_V integrin (18). These two integrin chains in association with each other form the classical vitronectin receptor. Further study revealed that α_V and β_3 do associate with each other in osteoclasts (19). Thus these studies demonstrated that osteoclasts express certain members of the integrin family and that at least some of these proteins are functionally important. Further studies of integrin expression by osteoclasts have yielded inconsistent results, with expression of $\alpha_2\beta_1$, $\alpha_3\beta_1$, and $\alpha_5\beta_1$ (16,20,21) being reported in some studies but not others. However, strong expression of $\alpha_V\beta_3$ has consistently been found in all studies of osteoclast integrin expression.

The interaction of an $\alpha_V\beta_3$ integrin and vitronectin occurs via the arg-gly-asp (RGD) tripeptide sequence, in common with a number of other integrin-matrix protein interactions such as binding of $\alpha_5\beta_1$ integrin to fibronectin (22). The binding of osteoclasts to bone matrix can be inhibited by synthetic RGD peptides and the snake venom-derived disintegrins echistatin (23,24) and kistrin (25). Furthermore, the morphological change observed in osteoclasts following treatment with these peptides resembles that induced by the anti-β_3-integrin antibody 23C6. Vitronectin, the classic ligand for $\alpha_V\beta_3$, is not a prominent bone matrix constituent, thus raising the possibility that the major osteoclast $\alpha_V\beta_3$ ligand may be a bone matrix protein without a previously ascribed function. This, in combination with the evidence that RGD-mediated matrix binding may be important to osteoclast function, initiated a search for potential substrates among bone matrix proteins. A number of bone matrix proteins contain the RGD sequence, namely type I collagen, fibronectin, thrombospondin, bone sialoprotein II, and osteopontin (26–29). Of these proteins, it appears that osteopontin is the major ligand for osteoclast $\alpha_V\beta_3$ (30), although there is evidence that osteoclasts also bind to other matrix proteins such as bone sialoprotein II and fibronectin via this receptor (31).

Although there is now convincing evidence that RGD-mediated matrix binding is important in osteoclast function and that $\alpha_V\beta_3$ is the major integrin dimer expressed by osteoclasts, it is uncertain precisely how $\alpha_V\beta_3$-matrix binding influences osteoclast function, and whether other integrins may have accessory functions. It is becoming clear that binding of $\alpha_V\beta_3$ to suitable substrates influences intracellular signaling pathways in the osteoclast and may regulate other osteoclast-matrix interactions. For example, osteoclast binding to matrix-gla-protein is inhibited by RGD peptides, despite the fact that matrix-gla-protein does not contain an RGD sequence (32). A possible explanation for this is that RGD peptide-integrin interactions cause down-regulation of the expression or activity of other cell adhesion molecules.

Binding of RGD-containing peptides to isolated osteoclasts causes rapid but transient increases in nuclear calcium concentrations (33). While demonstrating that RGD-integrin binding leads to intracellular metabolic events, the precise significance of these changes in calcium is unclear. The snake venom-derived disintegrin, echistatin, does not cause changes in osteoclast nuclear calcium, but does cause osteoclast retraction and inhibits bone resorption in the same way as RGD peptides (34). The effects of $\alpha_V\beta_3$-ligand interactions on osteoclast shape indicate that there are interactions between this receptor and the actin cytoskeleton. $\alpha_V\beta_3$ is known to colocalize with vinculin and talin (35), which link other cell membrane proteins to actin. The interaction between $\alpha_V\beta_3$ and the cytoskeleton may also be mediated by other signaling pathways, as there is also evidence that binding of osteoclast $\alpha_V\beta_3$ to osteopontin leads to inositol trisphosphate-mediated actin reorganization, and activation of focal adhesion kinase (FAK) and the *c-src* product tyrosine kinase pp60$^{c\text{-}src}$ (36–38). The latter observation is of particular interest as *c-src* is essential for normal osteoclast function as demonstrated by the development of osteopetrosis in *c-src* knockout mice. In these mice osteoclasts resorb bone matrix very poorly due to an inability to form ruffled borders (39,40). The ruffled border is an area of complex convolution of the plasma membrane on the resorbing surface of the osteoclast, which facilitates the secretion of hydrogen ions and proteinases that cause the dissolution of bone matrix. It is therefore tempting to speculate that pp60$^{c\text{-}src}$ and one or more of its phosphorylation substrates may be involved in the signal transduction pathway used by $\alpha_V\beta_3$ integrin in osteoclasts. Cortactin (41), which appears to be a pp60$^{c\text{-}src}$ substrate in osteoclasts (42), is a potential candidate. It is possible that *c-src* knockout osteoclasts cannot form ruffled borders because the signal resulting from an $\alpha_V\beta_3$-ligand binding fails upstream of cytoskeletal reorganization. However, there is currently no direct evidence to support this, and it is likely that the consequences of integrin-ligand binding in the osteoclasts are complex, and involve a variety of pathways. Indeed evidence exists that different integrin-substrate interactions may have different functions. $\alpha_3\beta_1$/$\alpha_5\beta_1$-mediated fibronectin binding of giant cell tumor-derived osteoclasts is related to substrate adhesion, whereas $\alpha_V\beta_3$ mediates spreading and motility of these cells (21). To complicate matters further, local production of integrin ligands such as fibronectin and osteopontin, either by osteoblasts or by osteoclasts, adds an extra layer of complexity to the control of osteoclast adhesion and resorptive functions (21,43). Furthermore, osteoclasts can modify their adhesive substrates. Tartrate-resistant acid phosphatase, highly expressed in osteoclasts, can dephosphorylate osteopontin and bone sialoprotein. As the dephosphorylated forms of these proteins do not support osteoclast adhesion (44), this represents yet another potential level of control in osteoclast-matrix interactions.

Nevertheless it is a strong possibility that $\alpha_V\beta_3$-ligand interactions are essential for the development of a fully differentiated cytoarchitecture and function in osteoclasts. This view is supported by the growing weight of evidence that $\alpha_V\beta_3$-ligand interactions send signals to the cytoskeleton via the src family tryosine kinase pathways (45) and inositol phosphate cycle (36), and that these messages cause changes in cytoskeletal conformation (38). Another intriguing possibility is that $\alpha_V\beta_3$-ligand interactions influence osteoclast life span. Loss of integrin-mediated adhesion is associated with the induction of apoptosis (programmed cell death) in many cell types (46) including vascular endothelial cells where inhibition of $\alpha_V\beta_3$-ligand interactions induces apoptosis (47). As there is growing evidence that regulation of apoptosis may be an important determinant of osteoclast function (13,48), it is conceivable that osteoclast-matrix interactions control osteoclast function by regulating their life span.

Like many other cell types, osteoclasts appear to utilize nonintegrin cell adhesion molecules. Recent evidence suggests a role for E-cadherin in osteoclast fusion. Antibodies against E-cadherin and antagonistic peptides containing the histidine-alanine-valine (HAV) E-cadherin binding sequence inhibit osteoclast formation and bone resorption (49). Osteoclasts also express the immunoglobulin superfamily cell adhesion molecule ICAM-1 (50). This molecule is highly inducible by cytokines such as interleukin-1 (51) and mediates leukocyte-endothelial interactions in inflammation. Its function in osteoclasts is unknown, although there is evidence that inhibiting ICAM-1-LFA-1 interactions and VCAM-1-$\alpha_4\beta_1$ integrin interactions (VCAM-1 is another related immunoglobulin superfamily cell adhesion molecule) inhibit osteoclast formation in vitro (52,53). $\alpha_4\beta_1$ integrin is expressed by osteoblasts and a population of alkaline phosphatase-positive stromal cells adjacent to bone surfaces, which may be bone marrow stromal cells or preosteoblasts (20). It is becoming increasingly clear that osteoblasts/marrow stromal cells have an essential accessory function in the formation of osteoclasts (4,5). This accessory function appears to depend, at least partly, on cell-cell contact, and may therefore be mediated by the VCAM-1-$\alpha_4\beta_1$ interaction. Osteoclasts also express the multifunctional cell adhesion molecule CD44 (54,55). The possible functions of CD44 on osteoclasts include matrix adhesion, matrix macromolecule uptake, homing, or signal transduction leading to cytokine production, all of which have been described in other CD44 expressing cells (56,63).

IV. OSTEOBLAST/OSTEOCYTE ADHESION

Bone is formed by osteoblasts on the endosteal surfaces of bone, or by related periosteal cells on the outer surfaces of bone. Osteoblasts are derived from stromal stem cells, the precise phenotype of which is uncertain, but it is generally thought that these cells reside within the bone marrow rather than on the bone surfaces (64). Osteoblasts do not function singly, but are found in "seams," refilling Howship's lacunae previously resorbed by osteoclasts (see Fig. 2). It is thought that these precursors migrate to sites of reversal in bone remodeling units, attach to the bone surface, and differentiate to become mature osteoblasts. As mentioned above, during production of bone and refilling of Howship's lacunae, osteoblasts differentiate into osteocytes or inactive lining cells. At each stage of its life, the osteoblast may thus be involved in different cell-cell and cell-matrix interactions.

Just as interactions between osteoclasts and bone matrix appear to be vital for the function of these cells, interactions between cells of the osteoblast lineage and bone matrix may be important in influencing the function of these cells and, thus, bone formation. Much less is known about how cells of the osteoblast lineage interact with each other, the neighboring cells and the bone matrix than is known about osteoclast adhesion. This is largely due to the difficulty in identifying osteoblast precursors in vivo and the uncertainty of the precise phenotype of in vitro osteoblast models. Several studies have examined cell adhesion molecule expression in cells of the osteoblast lineage in vivo and in vitro. However, these studies have yielded conflicting results (20,65,66). This lack of agreement is probably due, at least in part, to the ability of cells of many lineages (presumably including the osteoblast lineage) to alter expression of cell adhesion molecules in vitro. To date there has been no definitive work published on whether osteoblast adhesive interactions modulate osteoblast function anyway. There is, however, evidence that integrin expression by cells of the osteoblast lineage can be regulated by cytokines (67,68). Further study will be required to establish whether such regulatory events have an influence on the function of osteoblasts.

The morphological appearance of active osteoblasts bears a slight resemblance to that of cuboidal epithelia, and indeed the cell-cell interactions of osteoblasts may in some respects resemble these cells. Epithelia and many other cell types adhere to neighboring cells via homotypic interactions, involving members of the calcium-dependent cell adhesion molecule family, the cadherins; osteoblasts may also utilize a novel N-cadherin-like protein (69). Microinjection studies using Lucifer yellow (70) and expression of the gap junction protein connexin 43 have shown that osteoblasts communicate with each other via gap junctions. Interestingly, gap junction communication in osteoblasts can be regulated by transforming growth factor β, which is a powerful regulator of osteoblast differentiation (71). How gap junctions influence osteoblast function is unclear, but the apparently coordinated activities of individual seams of osteoblast in vivo suggests that gap junction communication may be a central control mechanism. Indeed, the evidence that osteoblasts may also interact in this way with osteocytes (72), endothelium (73), and perhaps marrow stromal cells raises the possibility of a functional syncytium in bone, which coordinates remodeling events in response to the demands of growth or mechanical stress. This concept is in keeping with what is currently the most popular theory regarding osteocyte function.

The osteocyte is the most numerous and the least understood cell in bone. In a standard hematoxylin and eosin section, osteocytes appear as isolated single cells residing in lacunae within the bone matrix. Silver staining techniques or electron microscopy reveal that these cells are not isolated but communicate with neighboring osteocytes, osteoblasts, and lining cells on adjacent bone surfaces via a complex system of dendritic processes passing through tiny canaliculi in the bone matrix. Little is known of osteocyte function. Current thoughts are that they function as mechanoreceptors, act as coordinators of bone remodeling via their connections with cells on the bone surface, or mediate ion exchange (14,15).

Osteocytes and osteoblasts differ in expression of cell adhesion molecules in a number of respects which may reflect their different functions. Osteocytes express less α_V integrin than osteoblasts (20). In vivo studies have revealed that when an osteoblast becomes embedded in matrix to become an osteoid osteocyte, it strongly expresses CD44, a cell surface glycoprotein that is not expressed in detectable amounts by endosteal osteoblasts (54,55,74). CD44 consists of a family of >20 glycoproteins produced from a single gene by alternative splicing (56). The CD44 molecules appear to be multifunctional, mediating adherence to extracellular matrix substrates such as hyaluronan (62), fibronectin (61), and collagen (60), as well as mediating lymphocyte homing (75). The interactions between CD44 and its various ligands are dynamic, mediating uptake and degradation of hyaluronan (63), T cell activation (66,76), and cytokine production by macrophages (58,59). The function of osteocyte CD44 is unknown. A number of possibilities exist, including activity as a receptor for type VI collagen which is preferentially expressed in osteocytic lacunae but not in association with osteoblasts (personal observations). Human osteoblast-like cells, which acquire CD44 expression in vitro, express increased levels of this molecule when subjected to mechanical stimulation (78), although expression of integrins is not increased under these conditions. As β_1-integrins can act as mechanical receptors (79), it is tempting to speculate that signals transduced by β_1-integrins may regulate CD44 expression and therefore possibly osteocytic differentiation. Further study may therefore reveal CD44 to be a transducer of mechanical signals in osteocytes, signals that may then be transmitted to neighboring osteoblasts or lining cells. Such a mechanism would enable these cells to increase bone formation or perhaps initiate bone resorption by promoting osteoclast precursor recruitment.

V. POSSIBLE ROLE OF CELL ADHESION IN BONE DISEASE

The evidence that cell-matrix interactions may directly affect the function of osteoclasts has raised the possibility that interfering with these interactions may provide new strategies for treating bone disease characterized by increased osteoclastic activity such as Paget's disease of bone, skeletal metastases, renal osteodystrophy, and, most importantly, postmenopausal osteoporosis (see Table 3). Structure-function analyses suggest that RGD-containing oligopeptides with specificity for $\alpha_V\beta_3$ are the most potent inhibitors of osteoclast function (34). In addition to the ability of RGD-containing peptides to inhibit osteoclast function in vitro, it has been demonstrated that the RGD-containing snake venom disintegrin, echistatin, can reduce bone loss due to ovariectomy in vivo (80). The obvious danger of such peptides is that they are likely to interfere with the function of other cell types that utilize $\alpha_V\beta_3$ or β_3 integrins such as platelets. However, preliminary evidence suggests that echistatin inhibits bone resorption in vivo at doses that do not interfere with coagulation (80).

The ability of echistatin to inhibit postovariectomy osteoporosis together with the observations that estrogen therapy causes loss of adhesion of osteoclasts to the bone matrix (81), and the *c-src* product pp60$^{c\text{-}src}$ to dissociate from the actin cytoskeleton of the osteoclast (82), raises the possibility that the mechanism of action by which estrogen inhibits bone resorption may involve intracellular pathways that mediate $\alpha_V\beta_3$-ligand regulation of osteoclast function. While these pieces of evidence are indirect, the essential function of c-src in the osteoclast and evidence for involvement of pp60$^{c\text{-}src}$ in the transduction of $\alpha_V\beta_3$-ligand interactions suggests that this may be a fruitful area for future research. If estrogen controls osteoclast function through $\alpha_V\beta_3$-cytoskeleton pathways, possibilities exist for the development of estrogen alternatives, such as specific inhibitors of tyrosine kinases, for the treatment of postmenopausal osteoporosis. It is therefore interesting to note that herbimycin A, a tryosine kinase inhibitor that blocks the function of pp60$^{c\text{-}src}$, inhibits bone resorption in vitro and possibly also in vivo (83).

The above discussion suggests osteoclast-matrix interactions as the prime target for future development of drugs that interfere with bone resorption. However, there are other cellular events in bone resorption and remodeling that may be open to therapeutic manipulation. For example, if the molecu-

Table 3 Possible Targets for Therapeutic Intervention in Adhesion Events in Bone Resorption

1. Osteoclast precursor fusion	cadherin binding site antagonists—e.g., HAV peptides
2. Osteoclast-matrix adhesion	RGD peptides, disintegrins
3. Integrin signaling in osteoclasts	tyrosine kinase inhibitors (targeted against tyrosine kinases of the c-src pathway)

lar basis for osteoclast precursor homing was known, it might be possible to prevent osteoclasts forming by blocking the adhesion of their precursors to endothelial cells in the vessels of the bone marrow. Also, if cadherins are important mediators of osteoclast fusion, blocking cadherin-cadherin binding with specific antagonist peptides may also inhibit osteoclast formation, and thus bone resorption. However, development of such strategies awaits the results of research yet to be carried out.

Inhibition of bone resorption is currently seen as the major therapeutic weapon against osteoporosis. Indeed, the currently licensed treatments for diseases of increased bone resorption, such as estrogen replacement, bisphosphonates, calcitonin, and dietary calcium supplementation, are primarily inhibitors of bone resorption. While prevention is the ideal way of dealing with osteoporosis, this requires the use of effective screening programs which are not widely practiced and are unlikely to significantly benefit many of the 25 million or so U.S. citizens who have osteoporosis or significantly decreased bone mass. Unfortunately, rebuilding a porotic skeleton is likely to be a far greater therapeutic challenge than preventing it from becoming porotic in the first place. Our knowledge and understanding of the control of bone formation are not sufficiently well developed to judge to what extent osteoblast-matrix interactions regulate this process; filling in the gaps in our knowledge in this area may be of great value in further understand what controls how much bone matrix osteoblasts make, and therefore how this process can be therapeutically manipulated.

VI. CONCLUSIONS

In the past few years there has been a great increase in interest in cell-cell and cell-matrix interactions in bone. While much of our understanding of this subject is still rudimentary, there is now considerable evidence that cell-matrix interactions via the $\alpha_V\beta_3$ integrin play a central role in osteoclast function. There are also tantalizing signs that cell-cell and cell matrix interactions via integrins, cadherins, immunoglobulin superfamily adhesion molecules and CD44 may control other events in bone remodeling such as osteoclast precursor recruitment and differentiation, osteoblast and osteocyte recruitment and differentiation, and communication between these cells. It is likely that further work in this area in years to come will have the dual benefits of increasing our understanding of how bone functions as a tissue, and providing new avenues of therapeutic development in the treatment of osteoporosis and other diseases of bone.

ACKNOWLEDGMENTS

We would like to thank Susan Watson for preparing the manuscript.

REFERENCES

1. Frost HM. The laws of bone structure. Springfield, Ill: Charles C. Thomas, 1964.
2. Baron R, Vignery A, Horowitz M. Lymphocytes, macrophages and the regulation of bone remodelling. In: Peck WA, ed. Bone and Mineral Research, Annual 2. Amsterdam: Elsevier 1983:175–243.
3. Rodan GA, Martin TJ. Role of osteoblasts in hormonal control of bone resorption—a hypothesis. Calcif Tissue Int 1981; 33:349–351.
4. Hattersley G, Chambers TJ. Generation of osteoclasts from hemopoietic cells and a multipotential cell line in vitro. J Cell Physiol 1989; 140:478–482.
5. Fuller K, Gallagher AC, Chambers TJ. Osteoclast resorption-stimulating activity is associated with the osteoblast cell surface and/or the extracellular matrix. Biochem Biophys Res Commun 1991; 181:67–73.
6. Fischman DA, Hay ED. Origin of osteoclasts from mononuclear leucocytes in regenerating newt limbs. Anat Rec 1962; 143:329–338.
7. Walker DG. Control of bone resorption by hematopoietic tissue. The induction and reversal of congenital osteopetrosis in mice through use of bone marrow and splenic transplants. J Exp Med 1965; 142:651–663.
8. Scheven BAA, Visser JWM, Nijweide PJ. In vitro osteoclast generation from different bone marrow fractions, including a highly enriched haematopoietic stem cell population. Nature 1986; 321:79–81.
9. Kahn AJ, Simmons DJ, Krukowski M. Osteoclast precursor cells are present in the blood of preossification chick embryos. Dev Biol 1981; 84:230–234.
10. Silver IA, Murrills RJ, Etherington DJ. Microelectrode studies on the acid microenvironment beneath adherent macrophages and osteoclasts. Exp Cell Res 1988; 175:266–276.
11. Arey LB. The origin, growth and fate of osteoclasts and their relation to bone resorption. Am J Anat 1920; 26:315–345.
12. Fuller K, Owens JM, Jagger CJ, Wilson A, Moss R, Chambers TJ. Macrophage colony-stimulating factor stimulates survival and chemotactic behaviour in isolated osteoclasts. J Exp Med 1993; 178:1733–1744.
13. Wright KR, Hughes DE, Guise TA, et al. Osteoclasts undergo apoptosis at the interface between resorption and formation in bone remodelling units. J Bone Miner Res 1994; 9:S174.
14. Lanyon LE. Osteocytes, strain detection, bone modelling and remodelling. Calcif Tissue Int 1993; 53:S102–S107.
15. Aarden EM, Burger EH, Nijweide PJ. Function of osteocytes in bone. J Cell Biochem 1994; 55:287–299.
16. Horton MA, Davies J. Adhesion receptors in bone. J Bone Miner Res 1989; 4: 803–808.
17. Chambers TJ, Fuller K, Darby JA, Pringle JAS, Horton MA. Monoclonal antibodies against osteoclasts inhibit bone resorption in vitro. Bone Miner 1986; 1:127–135.
18. Davies J, Warwick J, Totty N, Philp R, Helfrich M, Horton M. The osteoclast functional antigen implicated in the regulation of bone resorption is biochemically related to the vitronectin receptor. J Cell Biol 1989; 109:1817–1826.

19. Nesbitt S, Nesbit A, Helfrich M, Horton M. Biochemical characterisation of human osteoclast integrins. Osteoclasts express alpha v beta 3, alpha 2 beta 1, and alpha v beta 1 integrins. J Biol Chem 1994; 268:16737–16745.
20. Hughes DE, Salter DM, Dedhar S, Simpson R. Integrin expression in human bone. J Bone Miner Res 1993; 8:527–533.
21. Grano M, Zigrino P, Colucci S, et al. Adhesion properties and integrin expression of cultured human osteoclast-like cells. Exp Cell Res 1994; 212:209–218.
22. Hynes RO. Integrins: versatility, modulation and signaling in cell adhesion. Cell 1992; 69:11–25.
23. Sato M, Sardana MK, Grasser WA, Garsky VM, Murray JM, Gould RJ. Echistatin is a potent inhibitor of bone resorption in culture. J Cell Biol 1990; 111:1713–1723.
24. Horton MA, Taylor ML, Arnett TR, Helfrich MH. Arg-Gly-Asp (RGD) peptides and the anti-vitronectin receptor antibody 23C6 inhibit dentine resorption and cell spreading by osteoclasts. Exp Cell Res 1991; 195:368–375.
25. King KL, D'Anza JJ, Bodary S, et al. Effects of kistrin on bone resorption in vitro and serum calcium in vivo. J Bone Miner Res 1994; 9:381–387.
26. Weiss RE, Reddi AH. Synthesis and localization of fibronectin during collagenous matrix-mesenchymal cell interaction and differentiation of cartilage and bone in vivo. Proc Natl Acad Sci USA 1980; 77:2074–2078.
27. Oldberg A, Franzen A, Heinegard D. Cloning and sequence analysis of rat bone sialoprotein (osteopontin) cDNA reveals an arg-gly-asp cell-binding sequence. Proc Natl Acad Sci USA 1986; 93:8819–8823.
28. Gehron Robey P, Young MF, Fisher LW, McClain TD. Thrombospondin is an osteoblast-derived component of mineralized extracellular matrix. J Cell Biol 1988; 108:719–727.
29. Heinegard D. Macromolecules in bone matrix. Connect Tissue Res 1989; 21: 333–344.
30. Reinholt FP, Hultenby K, Oldberg A, Heinegard D. Osteopontin a possible anchor of osteoclasts to bone. Proc Natl Acad Sci USA 1990; 87:4473–4475.
31. Helfrich MH, Nesbitt SA, Dorey EL, Horton MA. Rat osteoclasts adhere to a wide range of RGD (Arg-Gly-Asp) peptide-containing proteins, including the bone sialoproteins and fibronectin via a beta 3 integrin. J Bone Miner Res 1992; 7:335–343.
32. Loeser RF, Wallin R. Cell adhesion to matrix gla protein and its inhibition by an arg-gly-asp-containing peptide. J Biol Chem 1992; 267:9459–9462.
33. Shankar G, Davison I, Helfrich MH, Mason JW, Horton MA. Integrin receptor-mediated mobilization of intranuclear calcium in rat osteoclasts. J Cell Sci 1993; 105:61–68.
34. Horton MA, Dorey E, Nesbitt SA, et al. Modulation of vitronectin receptor-mediated osteoclast adhesion by arg-gly-asp peptide analogs: a structure-function analysis. J Bone Miner Res 1993; 8:239–247.
35. Zambonin-Zallone A, Teti A, Grano M, et al. Immunocytochemical distribution of extracellular matrix receptors in human osteoclasts: A β_3 integrin is colocalized with vinculin and talin in the podosomes of osteoclastoma giant cells. Exp Cell Res 1989; 182:645–652.

36. Hruska KA, Rolnick F, Huskey M. Occupancy of the osteoclast $\alpha v\beta 3$ integrin by osteopontin stimulates a novel src associated phosphatidylinositol 3 kinase (PI_3 kinase) resulting in phosphatidylinositol triphosphate (PIP_3) formation. J Bone Miner Res 1992; 9:S106.
37. Miyauchi A, Alvarez J, Greenfield EM, et al. Recognition of osteopontin and related peptides by an $\alpha v\beta 3$ integrin stimulates immediate cell signals in osteoclasts. J Biol Chem 1991; 266:20369–20374.
38. Chellaiah M, Hruska KA. Characterization of the osteoclast reorganization associated with integrin $\alpha v\beta 3$ stimulated bone resorption. J Bone Miner Res 1994; 9:S132.
39. Soriano P, Montgomery C, Geske R, Bradley A. Targeted disruption of the c-src proto-oncogene leads to osteopetrosis in mice. Cell 1991; 64:693–702.
40. Boyce BF, Yoneda T, Lowe C, Soriano P, Mundy GR. Requirement of $pp60^{c\text{-}src}$ expression for osteoclasts to form ruffled borders and resorb bone in mice. J Clin Invest 1992; 90:1622–1627.
41. Wu H, Parsons JT. Cortactin, and 80/85-kilodalton $pp60^{src}$ substrate, is a filamentous actin-binding protein enriched in the cell cortex. J Cell Biol 1993; 120:1417–1426.
42. Chen H, Reddy S, Feng J, et al. Antisense oligonucleotides to an 85kD src substrate (cortactin) inhibit osteoclastic resorption in vitro. J Bone Miner Res 1993; 8:S117.
43. Merry K, Dodds R, Littlewood A, Gowen M. Expression of osteopontin mRNA by osteoclasts and osteoblasts in modelling adult human bone. J Cell Sci 1993; 104:1013–1020.
44. Ek-Rylander B, Flores M, Wendel M, Heinegard D, Andersson G. Dephosphorylation of osteopontin and bone sialoprotein by osteoclastic tartrate-resistant acid phosphatase. J Biol Chem 1994; 269:14853–14856.
45. Baron R, Neff L, Yeh G, Stadel J, Soriano P, Levy J. RGD-induced tyrosine phosphorylation in osteoclasts requires c-src expression. J Bone Miner Res 1993; 9:S127.
46. Ruoslahti E, Reed JC. Anchorage dependence, integrins and apoptosis. Cell 1994; 77:477–478.
47. Brooks PC, Montgomery AMP, Rosenfield M, et al. Integrin $\alpha v\beta 3$ antagonists promote tumor regression by inducing apoptosis of angiogenic blood vessels. Cell 1994; 9:S138.
48. Hughes DE, Wright KR, Mundy GF, Boyce BF. TGFβ1 induces osteoclast apoptosis in vitro. J Bone Miner Res 1994; 9:S138.
49. Mbalaviele G, Niewolna M, Feng JQ, Bonewald LF, Mundy GR, Yoneda T. Expression of the cell adhesion molecule E-cadherin is required not only for formation of multinucleated osteoclasts but osteoclastic bone resorption. J Bone Miner Res 1994; 9:S131.
50. Athanasou NA, Quinn J. Immunophenotypic differences between osteoclasts and macrophage polykaryons; immunohistological distinction and implications for osteoclast ontogeny and function. J Clin Pathol 1990; 43:997–1003.
51. Bagby GC Jr. Interleukin-1 and hematopoiesis. Blood Rev 1989; 3:152–161.
52. Kurachi T, Morita I, Murota S. Involvement of adhesion molecules LFA-1

and ICAM-1 in osteoclast development. Biochim Biophys Acta 1993; 1178: 259–266.
53. Duong LT, Tanaka H, Rodan GA. VCAM-1 involvement in osteoblast-osteoclast interaction during oestoclast differentiation. J Bone Miner Res 1994; 9:S131.
54. Hughes DE, Salter DM, Simpson R. CD44 expression in human bone: a novel marker of osteocytic differentiation. J Bone Miner Res 1994; 9:39–44.
55. Nakamura H, Ozawa H. CD44 (hyaluronate receptor) localizes on the plasma membrane of osteoclasts and osteocytes. J Bone Miner Res 1994; 9:S171.
56. Haynes BF, Telen MJ, Hale LP, Denning SM. CD44; a molecule involved in leukocyte adherence and T cell activation. Immunol Today 1989; 10:423–428.
57. Miyake K, Underhill CB, Lesley J, Kincade PW. Hyaluronate can function as a cell adhesion molecule and CD44 participates in hyaluronate recognition. J Exp Med 1990; 172:69–75.
58. Webb DSA, Shimizu Y, Van Seventer GA, Shaw S, Gerrard TL. LFA-3, CD44 and CD45: physiologic triggers of human monocytes TNF and IL-1 release. Science 1990; 249:1295–1297.
59. Gruber MF, Webb DS, Gerrard TL. Stimulation of human monoctyes via CD45, CD44 and LFA-3 triggers macrophage colony-stimulating factor production. Synergism with lipopolysaccharide and IL-1β. J Immunol 1992; 148: 1113–1118.
60. Carter WG, Wayner EA. Characterisation of the class III collagen receptor, a phosphorylated, transmembrane glycoprotein expressed in nucleated human cells. J Biol Chem 1988; 263:4193–4201.
61. Jalkanen S, Jalkanen M. Lymphocyte CD44 binds the COOH-terminal heparin-binding domain of fibronectin. J Cell Biol 1992; 116:817–825.
62. Aruffo A, Stamenkovic I, Melnick M, Underhill CB, Seed B. CD44 is the principle cell surface receptor for hyaluronate. Cell 1990; 61:1303–1313.
63. Culty M, Nguyen HA, Underhill CB. The hyaluronate receptor (CD44) participates in the uptake and degradation of hyaluronan. J Cell Biol 1992; 116: 1055–1062.
64. Owen M. Cell population kinetics of an osteogenic tissue. J Cell Biol 1963; 19: 19–32.
65. Clover J, Dodds RA, Gowen M. Integrin subunit expression by human osteoblasts and osteoclasts in situ and in culture. J Cell Sci 1992; 103:267–271.
66. Moursi A, Globus R, Lull J, Zimmerman D, Damskey C. Regulated integrin expression and function during osteoblast differentiation. J Bone Miner Res 1994; 9:S300.
67. Dedhar S. Regulation of expression of the cell adhesion receptors, integrins, by recombinant interleukin 1β in human osteosarcoma cells: inhibition of cell proliferation and stimulation of alkaline phosphatase activity. J Cell Physiol 1989; 138:291–299.
68. Dedhar S. Signal transduction via the β1 integrins is a required intermediate in interleukin 1β induction of alkaline phosphatase activity in human osteosarcoma cells. Exp Cell Res 1989; 183:207–214.
69. Suva LJ, Towler DA, Harada S, La Fage M-H, Steuckle S, Rosenblatt M.

Hormonal regulation of osteoblast-derived cadherin expression in vitro and in vivo. J Bone Miner Res 1994; 9:S123.
70. Yamaguchi DT, Ma D, Lee A, Huang J, Gruber HE. Isolation and characterization of gap junctions in the osteoblastic MC3T3-E1 cell line. J Bone Miner Res 1994; 9:791–803.
71. Chiba H, Sawada N, Oyamada M, et al. Hormonal regulation of connexin 43 expression and gap junctional communication in human osteoblastic cells. Cell Struct Funct 1994; 19:173–177.
72. Jones SJ, Gray C, Sakamaki H, et al. The incidence and size of gap junctions between the bone cells in rat calvaria. Anat Embryol (Berl) 1993; 187:343–352.
73. Melchiore S, Huang J, Ma D, Brandi ML, Yamaguchi DT. Heterotypic gap junctional intercellular communication between bovine endothelial cells and MC3T3-E1 osteoblast-like cells. J Bone Miner Res 1994; 9:S237.
74. Jamal HH, Aubin JE. Expression of CD44 in bone: correlation with osteoblast maturation in fetal rat calvarial cells grown in vitro. J Bone Miner Res 1994; 9:S162.
75. Jalkanen S, Bargatze RF, Herron LR, Butcher EC. A lymphoid cell surface glycoprotein involved in endothelial cell recognition and lymphocyte homing in man. Eur J Immunol 1986; 16:1195–1202.
76. Huet S, Groux H, Caillou B, Valentin H, Pieur AM, Bernard A. CD44 contributes to T cell activation. J Immunol 1989; 143:789–801.
77. Shimizu Y, Van Seventer GA, Siraganian R, Wahl L, Shaw S. Dual role of the CD44 molecule in T cell adhesion and activation. J Immunol 1989; 143:2457–2463.
78. Keles AO, Schaffer JL, Gerstenfeld LC, Graves D, Stashenko P. Modulation of integrin and non-integrin adhesion molecules in normal human osteoblasts by spatially uniform biaxial strain in vitro. J Bone Miner Res 1994; 9:S305.
79. Wang N, Butler JP, Ingber DE. Mechanotransduction across the cell surface and through the cytoskeleton. Science 1993; 260:1124–1127.
80. Yamamoto M, Gentile M, Seedor JG, Balena R, Rodan S, Rodan GA. Echistatin prevents bone loss in ovariectomized mice. J Bone Miner Res 1994; 9: S201.
81. Liu C-C, Howard GA. Bone-cell changes in estrogen-induced bone-mass increase in mice: dissociation of osteoclasts from bone surfaces. Anat Rec 1991; 229:240–250.
82. Judd JW, Oursler MJ. Translocation of pp60src away from the osteoclast cytoskeleton after treatment with estrogen. J Bone Miner Res 1994; 9:S134.
83. Yoneda T, Lowe C, Lee C-H, et al. Herbimycin A, a pp60$^{c\text{-}src}$ tyrosine kinase inhibitor, inhibits osteoclastic bone resorption in vitro and hypercalcemia in vivo. J Clin Invest 1993; 91:2791–2795.

17

Adhesion Molecules in Tumor Growth and Metastasis

Gregory E. Hannigan
Department of Pathology, University of Toronto and Hospital for Sick Children, Toronto, Ontario, Canada

Shoukat Dedhar
Department of Medical Biophysics, University of Toronto, and Division of Cancer Biology Research, Sunnybrook Health Science Center, North York, Ontario, Canada

I. INTRODUCTION

Cancer has largely come to be seen as a disorder of cellular communication. A wealth of knowledge regarding the molecular responses of cells to a wide range of soluble growth stimulatory factors, growth-suppressing cytokines, and differentiation factors has provided key insights into the roles that dysregulated cell communications play in cancer etiology and progression. The fact that most of the known oncogene and tumor suppressor gene products function in some aspect of signal transduction, whether along extracellular-to-intracellular, cytoplasmic-to-nuclear, or intranuclear pathways, dramatically underscores this point. In addition to interactions with soluble factors, cell-cell and cell-extracellular matrix (ECM) interactions provide critical positional cues for cells, important in developmental processes such as tissue morphogenesis and cell migration. Pathologic alterations in cell-cell and cell-matrix adhesion properties underlie many of the phenotypic changes associated with tumor progression, including changes in cell morphology, migration, tissue invasiveness, and metastatic potential. The important concept has emerged, that cell adhesion molecules (CAMs) mediating these adhesion events function as informational as well as structural molecules, and that both aspects of CAM function contribute to tumor progression.

In this chapter, we will discuss molecular aspects of the structure and function of the major classes of adhesion receptors as they pertain to extracellular and intracellular processes relevant to tumor progression and metastasis, and the mechanisms whereby neoplastic cells uncouple these adhesion-mediated processes from normal regulatory influences. Some overlap with chapters on CAM structure, and CAMs in signal transduction are inevitable but necessary to provide context for the role of CAMs in tumor development. In particular, it will be appreciated that in vitro studies of CAM function and signaling are often relevant to an understanding of tumor behavior in situ and that, conversely, studies of human tumors can lead to important new insights into adhesion proteins that interact with, and influence, both the intracellular responses and the adhesive functions of CAMs. These interactions provide new targets for the screening of inhibitory compounds, and for the rational development of inhibitors (or activators) of CAM activity, as potential therapeutic agents.

II. INTEGRINS

A. Integrins: Regulation of Cell Growth by Extracellular Matrix Interactions

Integrins comprise a family of widely expressed transmembrane receptors for proteins of the ECM, such as fibronectin, laminin, vitronectin, and collagens. Integrins are obligate heterodimers, comprised of noncovalently associated α and β subunits, each of which spans the plasma membrane, and typically possesses a short (30 to 50 amino acids) cytoplasmic domain. Receptor diversity, and versatility in ligand binding is determined by the extracellular domains, through the specific pairing of nine β and 16 α subunits, to form a family of at least 24 recognized heterodimers. The extracellular domain of both α and β subunits contain a metal ion-dependent adhesion site, MIDAS. The MIDAS is a Mg^{2+} coordination site within the α subunit I domain, a stretch of approximately 200 amino acid residues that is required for integrin ligand binding. β subunits also contain a region of secondary structure with similarity to the I domain, and which is also required for ligand binding. Not all α subunits possess I domains, and residues located outside the I domain also contribute to ligand binding, notably those in domains resembling cation-binding, EF-hand motifs. The cytoplasmic domains of integrin subunits each contain conserved regions that may be important for integrin function. In the case of β subunits, an NPXY (single-letter amino acid code; X is any residue) motif regulates β_1 and β_3 affinity states, and an α subunit motif, KXGFFKR, also acts to regulate integrin function. Integrins are expressed in all cell types, mediating a wide range of cell-matrix and cell-cell interactions. However, the expression of certain integrins is tissue-restricted, most notably leukocyte expression of β_2 integrins such as $\alpha_L\beta_2$ (LFA-1) and $\alpha_M\beta_2$ (Mac-1), and platelet-specific expression of the $\alpha_{IIb}\beta_3$ integrin, gpIIb-IIIa. The leukocyte integrins mediate interactions with membrane-bound counterreceptors, such as ICAM-1, and in activated platelets $\alpha_{IIb}\beta_3$ binds fibrinogen and Von Willebrand factor, indicating the extreme versatility of integrin receptors in mediating a broad range of biological responses (1-7).

An extensive body of molecular and biochemical studies has demonstrated that integrins function in signal transduction from the ECM, and that intracellular signaling is mediated by integrin cytoplasmic domains. With respect to growth regulation and signaling, the most intensively studied integrin is the β_1 subunit. The β_1 cytoplasmic domain is evolutionarily conserved (>75% amino acid identity) among vertebrate species and in *Drosophila* (2,8), implying conservation of function. It is thought to interact directly with components of the actin cytoskeleton, localizing via these interactions in focal adhesion plaques (FAPs) which form cytoplasmically, at points of contact between integrin and the ECM. These FAPs represent the submembranous termini of actin stress fibres, indicating that integrins provide a structural bridge, linking the ECM and actin cytoskeleton. Integrins are also found in other cell-substratum contact points, where they could act to mediate cell migration (6). Synthetic peptides derived from the β_1 cytoplasmic tail have been shown to bind in vitro to actin cytoskeletal constituents of FAPs, such as the actin cross-linking protein α-actinin (9), and talin, which binds to a $\alpha_5\beta_1$ integrin in vitro (10). The β_2 integrin cytoplasmic domain has been reported to associate directly with filamin, which is not a component of FAPs, thus suggesting a distinct linkage of integrin subunits to cytoskeletal regulation (11).

In addition to structural proteins, FAPs contain a number of protein tyrosine kinases, such as the focal adhesion kinase $p125^{FAK}$, Src, and Csk, and ser/thr kinases such as protein kinase Cα. One of these, $p125^{FAK}$, has been reported to bind in vitro to peptides representing the cytoplasmic domain of β_1 integrin (12), although physiologic confirmation of this interaction is lacking. Nonetheless, the presence of regulatory protein kinases in FAPs indicates that these structures also function as integrin signal transduction complexes (reviewed in 13). An impression can be gained from the literature: that the structural and signaling aspects of integrin function are distinct. However, it is likely that, to some degree, these responses are overlapping, perhaps even interdependent, as evidenced by the ECM-dependent tyrosine phosphorylation of the actin cytoskeletal proteins paxillin (14) and tensin (15). Detailed molecular studies of the β_1 cytoplasmic domain have indicated that its requirement in integrin signaling also likely involves interactions with intracellular proteins. Studies with chimeric integrin subunits have shown that the β_1 cytoplasmic domain is required for integrin-induced phosphorylation of $p125^{FAK}$ (16,17), and that the β_1 cytoplasmic domain is both necessary and sufficient for localization to FAPs (18–20).

It has long been recognized that adherent cells require signals initiated by both soluble mitogens and adhesion for appropriate regulation of cell growth, and these requirements have been particularly well characterized in fibroblasts (21,22). More recently, it has been shown that fibroblasts will not proliferate in suspension, despite normal induction of c-fos and c-myc immediate early genes by serum growth factors (23). Integrins mediate the transduction of ECM signals regulating cell shape and motility, but also growth, survival, differentiation, and gene expression (3,13,24–26). Thus, in addition to providing a physical link between ECM and the cytoskeleton, integrin occupation elicits a variety of intracellular signaling events implicated in the regulation of cellular growth and survival (27), including stimulation of protein kinase C activity (28), Na^+/H^+ antiporter activity (29), and elevation of intracellular free Ca^{2+} (30). Importantly, integrin-initiated signals stimulate the tyrosine phosphorylation of a number of cellular proteins (31–36), and can result in the induction of nuclear events, such as metalloproteinase (37), and immediate to early (38) gene expression. Tyrosine kinase activity is required for at least some of these responses, in that inhibition of tyrosine phosphorylation blocks adhesion-induced gene expression (39) and localization of integrins to focal contact points (40). The inhibition of integrin mediated adhesion to ECM induces apoptotic signals in epithelial and endothelial cells (41–43). Adhesion-dependent signals are required for appropriate transit of fibroblasts through the G_1/S boundary (44), acting in concert with the soluble mitogenic signals required for entry of quiescent cells into G_1 (45). Cells may require ECM signals to exit G_1, as well as for ECM-dependent S phase transit, which appears to be related to adhesion-dependent cell spreading in hepatocytes (46). These results suggest multiple, cell type-specific differences in the adhesive inputs to cell cycle regulation. In adherent cells, the acquisition of anchorage-independent growth is a hallmark of neoplastic transformation (47), and, from the current perspective of integrins and signal transduction, implies dysregulation in growth signals emanating from the site(s) of cell-ECM interaction.

Recent work, using a variety of cell culture systems, has suggested that β_1 integrin-induced tyrosine kinase activity can activate the mitogenic Ras-MAPK cascade, thus indicating a significant overlap between integrin- and growth factor-mediated signaling pathways (36,48–50). However, studies of Rho-dependent ECM signaling in Swiss 3T3 fibroblasts raises significant questions regarding the extent of this overlap. In these cells, the mitogen lysophosphatidic acid or bombesin stimulates the assembly of FAPs in an ECM-dependent manner. In the absence of mitogenic stimulation, ECM and Rho cooperate to induce focal adhesion assembly, without activating the MAPK cascade (51,52). Thus, it is not clear to what extent growth factor and integrin signaling pathways overlap, and the resolution of this difficult question would be aided greatly by the development of a genetic model system. Another, more direct link between ECM and growth factor signaling pathways has been described recently. Signaling by $\alpha_6\beta_4$ integrin, a laminin receptor localized in hemidesmosomes, appears to occur via a direct interaction of the adapter protein Shc, with the cytoplasmic domain of the β_4 subunit, which is phosphorylated by an as yet unidentified protein tyrosine kinase. Shc in turn binds the Grb2 adaptor protein, which could then activate the Ras pathway, via Sos (53). In the case of β_1 and β_3 signaling, it is not known how overlapping growth regulatory pathways are integrated to coordinate appropriate responses, nor are the intracellular determinants of integrin signal specificity understood.

Integrin signal integration may hinge on the activity of the tyrosine kinase $p125^{FAK}$, which is itself tyrosine phosphorylated upon occupation of β_1 and β_3 integrins (32,33,54). As indicated above, ECM stimulates the tyrosine phosphorylation of paxillin and tensin, so FAK and other FAP tyrosine kinases appear to regulate actin cytoskeletal reorganization in response to integrin engagement (15,40). In addition, recently reported experiments suggest a mechanism by which FAK could act, in bridging adhesion- and tyrosine kinase-activated signaling pathways (50). In murine fibroblasts, integrin occupation by fibronectin stimulates the association of phosphotyrosyl FAK with the c-Src protein tyrosine kinase and the Grb2 adaptor protein, in complexes that also contain the Ras guanine exchange protein, Sos. Thus, FAK may act distal to Src, in the recruitment of these molecules to adhesion-dependent signaling complexes. Clearly, however, FAK does not function as an integrin-specific signal transducer, in that it is also rapidly tyrosine phosphorylated in response to activation of G-protein coupled receptors (55). Recently, a novel integrin-linked serine/threonine protein kinase, $p59^{ILK}$, has been cloned via its direct association with the cytoplasmic domain of β_1 integrin, in a yeast two-hybrid system (56). ILK is also present in anti-β_1 immunoprecipitates of epithelial cell lysates, demonstrating a physiologic association. Overexpression of $p59^{ILK}$ in epithelial cells suppresses cell-ECM interactions and induces anchorage-independent growth, suggesting that the supranormal level of ILK activity is supplying growth signals normally provided by adhesion. The fact that $p59^{ILK}$ complexes with β_1, and possibly β_3, integrin in vivo suggests that it functions as an integrin-specific signal transducer.

B. Integrins in Tumor Progression, Metastasis, and Angiogenesis

A number of studies (reviewed in 57) have documented changes in cellular integrin expression with the acquisition of tumorigenic or metastatic phenotypes. Although oncogenically transformed cells exhibit reduced substrate adhesion relative to nontransformed cells, no consistent pattern of altered integrin expression, vis-á-vis the tumorigenic/metastatic phenotype, has emerged. This presumably reflects the complexity of integrin interactions with ECM constituents, as well as cell-type-specific responses to these interactions. In some rodent fibroblasts, transformation with Rous sarcoma virus, or ras oncogenes, is associated with a reduction in expression of $\alpha_5\beta_1$, and not $\alpha_3\beta_1$ integrin, which act as high- and low-affinity fibronectin receptors, respectively (58). Conversely, in human PC3 prostatic carcinoma cells selected in vitro for a highly invasive phenotype, $\alpha_3\beta_1$ integrin expression was reduced dramatically, $\alpha_6\beta_4$ expression was induced, while $\alpha_5\beta_1$ (and other) integrin levels were the same as in noninvasive PC3 cells (59). Rous sarcoma virus-induced chicken sarcomas exhibited markedly elevated levels of fibronectin receptor (60), suggesting differences between in vivo and in vitro transformation, or species-specific responses of integrin biosynthesis to transformation. Of course it should be kept in mind that increases in integrin expression levels may not always parallel integrin function, and it is also likely that functional consequences would ensue from more subtle, transformation-associated alterations in the subcellular localization or distribution of integrins (61,62).

Functional studies involving gene transfer, anti-integrin antibodies, and adhesion-inhibiting peptides have been useful in demonstrating the potential for integrins to mediate tumorigenic or metastatic cell behavior. The cDNA-driven expression of $\alpha_5\beta_1$ integrin in CHO cells stimulates fibronectin secretion and matrix deposition in culture and suppresses CHO tumorigenicity in nude mice (63). The mechanism underlying this inhibition is not known, but it may involve integrin signaling. Transfection of an α_5 cDNA into α_5-minus HT29 colon carcinoma cells rendered the cells dependent on fibronectin for growth, leading to the suggestion that unoccupied $\alpha_5\beta_1$ receptor sends a growth inhibitory signal to the cell (64). Chemical transformation of human osteosarcoma cells, HOS, to a highly tumorigenic phenotype was associated with specific increases in the expression of functional $\alpha_6\beta_1$ laminin, and $\alpha_2\beta_1$ collagen receptors. Moreover, the transformed HOS showed increased in vitro invasiveness through matrigel, a reconstituted basement membrane containing laminin and collagen type IV (65). A rhabdomyosarcoma cell line, transfected with an α_2 integrin cDNA, exhibited increased adhesion to collagen and laminin substrates in vitro, and a marked increase in the number of lung metastases after intravenous or subcutaneous injection into nude mice (66). These results suggest that altered expression of, and perhaps signaling by, α integrin subunits may be important factors in the progression of some tumors.

The B16 murine melanoma model is relevant to some aspects of metastatic growth in human cancers (67), and has been used to examine the potential role of integrins in tumor metastasis. A synthetic peptide containing the Arg-Gly-Asp (RGD) core cell-binding peptide of fibronectin, GRDS, efficiently inhibits cell adhesion to fibronectin. Co-injection of this peptide with B16F10 cells, into the tail veins of C57BL/6 mice, inhibited experimental lung metastasis without affecting the intrinsic tumorigenicity of the cells (68), and resulted in prolonged survival of the injected mice (69). These experiments indicated the involvement of an integrin in mediating late stages of metastatic melanoma, most likely at the level of the interaction of tumor cells with pulmonary tissue. Data supportive of this notion has emerged from another study, wherein the α_4 integrin subunit was transfected into highly invasive, α_4-minus B16a melanoma cells (70). Expression of $\alpha_4\beta_1$ integrin on the B16a cell surface resulted in a significant reduction in matrigel invasiveness, and suppression of pulmonary metastases in C57BL/6 mice. Interestingly, specific suppression of spontaneous (subcutaneous injection), but not experimental metastases, was seen with B16a-α_4 transfectants, and it was shown that the α_4-expressing B16F10 cells do not form spontaneous metastases. This model suggests a minimum complexity for the involvement of integrin-ECM interactions in tumor progression, wherein one integrin (or lack of—e.g., $\alpha_4\beta_1$) influences earlier invasive behavior, with a distinct, RGD-sensitive integrin mediating subsequent tissue arrest and colonization.

As alluded to above, a critical step in the metastatic spread of tumor cells is the active penetration of basement membrane, and subsequent crossing of multiple tissue boundaries. Certain cell types express physiologic invasive behavior—e.g., trophoblasts and angiogenic endothelial cells—and it has been proposed that tumor cell invasiveness results from a loss of control of the normal invasive phenotype (71). With respect to integrins and physiologic invasiveness, it is known that the transendothelial migration of monocytes and lymphocytes is at least partly mediated by $\alpha_4\beta_1$ and $\alpha_L\beta_2$ integrins, respectively (reviewed in 4). As is the case with normal cells, tumor cell-ECM interactions exert a reciprocal, albeit aberrant, influence on malignant behavior. Transformed cells are often defective in secreting fibronectin and laying down an organized matrix (see 57), and the increased production of basement membrane-degrading metalloproteinases provides an obvious invasive advantage to the tumor cell (reviewed in 72). More surprisingly, it has been shown that nontumorigenic cells form locally invasive, highly vascularized tumors in nude mice, when coinjected with matrigel (73,74). This result demonstrates that tumorigenicity is not exclusively cell-autonomous, and suggests that a host or tumor cell integrin may mediate invasive behavior. In human melanoma cells, ligand occupation and antibody stimulation of the $\alpha_v\beta_3$ (vitronectin), but not $\alpha_5\beta_1$ (fibronectin) receptor, have each been shown to increase in vitro tumor cell invasiveness, and 72-kDa type IV collagenase expression (75). Interestingly, these treatments did not alter adhesion of the melanoma cells to vitronectin or matrix, suggesting that in these cells, reciprocal effects of the matrix and tumor cell might be linked via $\alpha_v\beta_3$ integrin-mediated signal transduction. Consistent with the notion of an active role for β_3 integrin in tumor invasiveness and/or metastasis, an immunohistochemical study of primary melanoma tumors demonstrated that expression of the β_3 subunit was exclusive to vertical growth phase and metastatic lesions, and was not detectable in benign nevi or radial growth phase tumors (76). Human melanoma cells isolated from lymph node metastases in nude mice exhibited increased cell surface expression of $\alpha_v\beta_3$ integrin and adhesion to human lymph node vitronectin (77). Consistent with these results, it was reported that melanoma cells selected for decreased $\alpha_v\beta_3$ expression were less tumorigenic than parental cells, and that tumorigenicity was restored by transfer of an α_v cDNA into the β_3-expressing cells (78). These data indicate that $\alpha_v\beta_3$ integrin may play a particularly important role in the development of metastatic melanoma (79).

Angiogenesis, the growth of new blood vessels from preexisting capillaries, is an important component of normal development, also playing a critical role in wound healing and inflammation. This process involves migration, local degradation of ECM, and proliferation of endothelial cells, all of which indicate an important role for cell adhesion molecules in neovascularization. Angiogenesis is often required for the growth of primary or metastatic tumors beyond a few millimeters in diameter. These new vessels also provide a connection between tumor cells and the general circulation, thus facilitating dissemination and metastasis (67). The net outcome of the angiogenic response appears to depend on the balance of activities between positive and negative regulators (80), and integrins have been shown in different systems to be positive mediators of angiogenesis. For example, migration of HUVEC endothelial cells on vitronectin and collagen was found to be mediated by $\alpha_v\beta_3$ and $\alpha_2\beta_1$ integrins, respectively (81). The expression of $\alpha_v\beta_3$ is increased in angiogenic blood vessels, both in human wound granulation tissue, and in chick chorioallantoic membranes treated with basic fibroblast growth factor, a positive angiogenic factor (82). The addition of tumor fragments to chorioallantoic membranes also induced angiogenesis which, along with tumor proliferation, was selectively inhibited by either of an anti-$\alpha_v\beta_3$ monoclonal antibody or specific peptide inhibitor of $\alpha_v\beta_3$ (83).

An interesting link has been made between leukocyte adhesion and angiogenesis in rheumatoid arthritis, indicating a novel mechanism for the involvement of integrins in neovascularization (84). As assayed in a rat cornea model, endothelial cell chemotaxis and angiogenesis are induced by rheumatoid synovial fluid and can largely be inhibited by antibodies to E-selectin and vascular cell adhesion molecule-1. Moreover, similar induction of angiogenesis by soluble E-selectin and VCAM-1 was inhibited by antibodies to their respective ligands, sialyl Lewis-x and $\alpha_4\beta_1$ integrin. An autocrine model was proposed, wherein leukocyte adhesion stimulates the release from endothelial cells of E-selectin and VCAM-1, which then act on adjacent endothelial cells via the above ligands, to exert a direct angiogenic effect. A similar mechanism contributing to tumor pathology has not been shown, but an interesting link between integrin function and angiogenesis in Kaposi sarcoma (KS) has been postulated recently. Induction of vascular cell (e.g., HUVEC) $\alpha_v\beta_3$ and $\alpha_5\beta_1$ integrin expression by inflammatory cytokines, found at elevated levels in KS lesions, results in these cells acquiring responsiveness to the Tat protein of HIV-1. Through interactions with these integrins, Tat stimulates KS cell growth and migration in vitro, similar to its effects on normal cytokine-treated endothelial cells, suggesting integrins could play a direct role in KS angiogenesis (85). Thus, in some tumors integrins may regulate neovascularization in response to soluble ligands, as well as mediating endothelial cell migration and survival on ECM.

III. CADHERINS

A. Cadherins: Intercellular Adhesion Receptors

Adherens junctions and desmosomes are the two major forms of intercellular junction, required for maintenance of tissue architecture and the epithelial differentiation state. Distinct members of the cadherin family of cell adhesion receptors are principal constituents of each type of junction, mediating Ca^{2+}-dependent homophilic intercellular adhesion via these structures. Many carcinomas display reduced intercellular adhesion and lose characteristics of differentiated epithelium, suggesting a critical role for cadherins in the malignant progression of carcinoma. The significance of altered cadherin function in epithelial to mesenchymal conversion and the development of invasive carcinoma has been reviewed recently (86).

The approximately 20 recognized cadherins and protocadherins are structurally related, integral membrane proteins. E-cadherin, the major epithelial cadherin, contains four conserved extracellular domains, a fifth extracellular domain possessing conserved cysteine residues, a transmembrane domain, and a cytoplasmic domain (87). The most N-terminal of the extracellular cadherin repeats contains a His-Ala-Val amino acid triplet, which comprises the cell adhesion recognition sequence required for homophilic association. The His and Val residues of this triplet are directly involved in the binding interface, with Ala buried in the hydrophobic core. Calcium binding sites lie between adjacent extracellular domains 1 to 4 (88). The cytoplasmic domain associates with cellular proteins that act to link cadherins to the actin cytoskeleton, as well as to signal transduction components. The desmosomal cadherins belong to two groups, the desmogleins and desmocollins, which share similarity with "classic (adherens junctions) cadherins in the extracellular domain but possess unique cytoplasmic structures (89). The desmosomal cadherins are apparently required for the formation of mature desmosomes (90), although it is not clear whether they provide for homophilic, or heterophilic interactions in these structures. These are not found in adherens junctions, and the classic adherens junction cadherins, notably E-cadherin and P-cadherin, are better studied with respect to their involvement in tumor progression. As is the case with integrins, cadherins are seen to be adhesion molecules that subsume informational, as well as structural functions.

B. Cadherins in Tumorigenesis and Metastasis

A body of experimental evidence has accumulated implicating cadherins in the suppression of tumor invasion and metastasis. This evidence derives from independent approaches involving analyses of cadherin expression in tumors of variable differentiation status, gene transfer experiments, mutational analysis of cadherins in tumors, and the identification of a cadherin-associated protein as the product of a tumor suppressor gene. With respect to the influences of intercellular adhesion on tumor development, the role of the epithelial E-cadherin (also known as uvomorulin, L-CAM, or cell-CAM 120/80), in carcinoma has been particularly well studied. In various carcinomas, expression of E-cadherin is low or absent in poorly differentiated, invasive tumors and tumor-derived cell lines, and epithelial cells become invasive for collagen gels and embryonic heart tissue, when intercellular adhesion is disrupted by anti-E-cadherin antibodies (91,92). Reversion of the invasive phenotype has been achieved by transfection of an E-cadherin cDNA into a dedifferentiated human breast carcinoma line (92), as well as highly invasive ras-transformed cell lines derived from canine kidney and murine mammary tumors (93). Moreover, in this latter study, transfectants reexpressed the invasive phenotype upon treatment with anti-E cadherin antibodies, or downregulation of endogenous E-cadherin mRNA expression by anti-sense RNA. Similar results have been obtained using a mouse model of epidermal carcinogenesis (94). In this system, levels of E-cadherin expression inversely correlated with the tumorigenicity of keratinocyte cell lines, representing different stages of epidermal carcinogenesis. This correlation was also seen in tumors, induced in nude mice by these cells. In particular, fibroblastoid lines, which induce spindle cell carcinomas in nude mice, had no detectable P-cadherin or E-cadherin expression. Expression of exogenous E-cadherin in a moderately differentiated, E-cadherin-negative epithelioid line partially suppressed its tumorigenicity. In human basal cell carcinomas, which generally express normal E-cadherin levels and low metastatic potential, reduced E-cadherin expression was restricted to a subset of invasive tumors (95).

As suggested above, clinical correlates of E-cadherin expression and tumor invasiveness are consistent with the in vitro data. Reduced or absent E-cadherin expression in poorly differentiated, invasive, or high grade tumors has been reported for sqamous cell carcinoma of the head and neck, basal cell carcinoma, gastric adenocarcinoma, and carcinomas of the breast (particularly lobular type), colon, and lung (reviewed in 86). In addition, reduced or absent E-cadherin expression is seen in bladder carcinoma (96) and female genital tract tumors (97). In breast carcinomas, a particularly high proportion of lymph node metastases was associated with reduced E-cadherin expression (98). Genetic support for a tumor, or invasion suppressor role for E-cadherin comes from studies documenting deletion of chromosome 16q22, encompassing the E-cadherin locus in prostate, ovarian, breast and hepatocellular carcinomas (99–104). Mutations resulting in exon skipping and loss of E-cadherin extracellular domain residues, were found in 50% of diffuse-type gastric carcinomas (105), and somatic mutations, with and without loss of E-cadherin gene heterozygosity, have been described in a small fraction of endometrial and ovarian carcinomas (106). These results provide strong support for the notion of E-cadherin functioning in tumor suppression.

Down-regulation of cadherin expression and function can arise by multiple mechanisms. In vivo footprinting experiments indicate that loss of E-cadherin expression in poorly differentiated carcinomas may result from alterations in chromatin structure, which render the cadherin gene promoter inaccessible to transcription factors (107). In addition, mesenchymal, and undifferentiated carcinoma cells showed hypermethylation of CpG doublets in the E-cadherin promoter, relative to epithelial cells expressing E-cadherin (107,108). These results suggest an attractive mechanism for the down-regulation of E-cadherin expression seen in advanced carcinomas, and could explain the low number of tumors harboring mutations in the cadherin structural gene. Nonetheless, in many of the studies mentioned above, the inverse correlation between E-cadherin expression and tumor progression is observed in less than 50% of cases. The high proportion of carcinomas which express normal levels of E-cadherin has, therefore, suggested that cadherin function can be dysregulated through mechanisms other than transcriptional down-regulation of, or mutations in, the cadherin gene. Transfection of mouse L-cells with E-cadherin cDNAs, encoding various cytoplasmic domain truncations, demonstrated that the cytoplasmic domain regulate the accumulation of E-cadherin at cell contact points, and is, moreover, required for homophilic cell binding (109).

Immunoprecipitation analyses of murine cells, expressing exogenous E-cadherin cytoplasmic domains identified three endogenous proteins, of apparent molecular weights of 102, 88, and 80 kDa, which did not coprecipitate in cells expressing cadherin cytoplasmic deletions. The 102, 88, and 80 kDa proteins were named catenins α, β, and γ, respectively, after the Latin catena, for chain, as it was thought they might function in linking E-cadherin with actin cytoskeletal elements (110). Some limited sequence similarity of α-catenin to vinculin has suggested that α-catenin links cadherin complexes to cytoskeletal components. β-catenin shares a high degree of similarity to the Drosophila segment polarity gene armadillo, a component of epithelial zonula adhaerens junctions, and plakoglobin, a major adherens junction protein that is thought to be identical to γ-catenin. Importantly, the associations between cadherin and catenins are functionally conserved in the Drosophila wingless, and Xenopus Wnt, developmental signaling pathways. These systems have provided genetic correlates to the biochemically and molecularly defined interactions, and have offered additional insights into cadherin-catenin signal transduction (111,112).

Mutant E-cadherin proteins lacking the extracellular domain acted in a dominant negative fashion to inhibit cell adhesion, presumably by sequestering catenins in inactive (at least with respect to adhesion) complexes (113). Molecular studies show that homophilic cell binding by E-cadherin requires β-catenin association, mediated by a 72-amino acid region in the cadherin cytoplasmic domain. Interaction with α-catenin seems to be critical for the linkage of cadherin and β-catenin to actin filaments (114). Functional support for the importance of these interactions was obtained when it was shown that the mutational loss of α-catenin, rather than down-regulation of E-cadherin expression, was correlated with reduced cell adhesion in a human lung carcinoma line (115). Moreover, transfection of an α-catenin cDNA into nonadhesive carcinoma cell induced tight cell-cell adhesion, and reversion to an epithelial morphology (116).

It has been shown that cellular transformation by v-src inhibits cadherin-mediated adhesion, in both rat and chick embryo fibroblasts. β-catenin is a major target for transformation-associated tyrosine phosphorylation in these cells, and inhibiting tyrosine kinase activity with herbimycin A restores cadherin-mediated aggregation to the transformed cell cultures (117,118). Similarly, transformation of canine kidney epithelial cells with a temperature-sensitive v-src mutant resulted in the conditional loss of cell-cell contacts and epithelial morphology, as well as the acquisition of an invasive phenotype. Concomitant with these changes, β-catenin was also seen to be a major target of tyrosine phosphorylation in cell junctions, where both v-Src and c-Src accumulate (119). Similar to E-cadherin expressing fibroblasts, src-transformed chick embryo fibroblasts (CEFs) lose N-cadherin-mediated cell adhesion, without loss of N-cadherin surface expression. In addition, N-cadherin-associated cellular proteins of 102 and 94 kDa, homologs of α-catenin and β-catenin respectively, are tyrosine phosphorylated in the transformed CEFs (118). Thus, catenins are relevant targets for dysregulation of cadherin function, either by mutational loss or alterations in post-translational regulation.

An additional, quantitative role for tyrosine kinase activity in the regulation of cadherin function is revealed by cell aggregation assays, wherein v-src transformation induces a shift in cadherin-mediated L-cell adhesion, from a strong to a weak state. Although β-catenin is heavily tyrosine phosphorylated in this system, this event is not required for the shift, suggesting there are β-catenin-dependent and -independent effects of tyrosine phosphorylation, influencing cadherin dysfunction and homophilic affinity, respectively (120). The potential contribution of the weak cadherin adhesive state to carcinoma progression is not presently understood, but could be significant, as extant data indicates that partial downregulation of E-cadherin expression is sufficient to confer an invasive phenotype on epithelial cells (93). As has been indicated in melanoma (121), it may hold that quantitative, rather than qualitative changes in cadherin function are germaine to carcinoma progression and metastasis.

The Src transformation-associated changes in cadherin function may reflect the physiological regulation of cadherin functional status by tyrosine kinases. Hepatocyte growth factor, acting through the met receptor tyrosine kinase, induces a switch from epithelial to mesenchymal phenotype, and HGF has recently been shown to induce the phosphorylation of β-catenin by Met (122). Treatment of epithelial cells with epidermal growth factor (EGF) reduces E-cadherin-mediated adhesion, and stimulates the formation of complexes containing tyrosine-phosphorylated EGF receptor and β-catenin. Recombinant EGF receptor phosphorylates β-catenin fusion proteins in vitro, suggesting that physiological tyrosine phosphorylation of β-catenin downregulates cadherin adhesion (123). Interestingly, $p120^{cas}$, a major cellular substrate of Src, has recently been identified as a member of the β/γ-catenin family (124). $P120^{cas}$ is detected in E-cadherin/catenin complexes, and this association is dependent on the catenin-binding region of E-cadherin. PTPμ, is a receptor tyrosine phosphatase that mediates homophilic adhesion through an immunoglobulin domain in the extracellular segment. The intracellular segment of PTPμ has a domain that is highly homologous to the cytoplasmic domain of cadherins (125). In lysates of rat heart, lung and brain tissues, PTPμ is found in a complex containing cadherins, α- and β-catenins, and in mink lung epithelial cells, PTPμ colocalizes with these proteins in regions of cell-cell contact. Moreover, the associated cadherins are tyrosine phosphorylated when PTPμ activity is inhibited by pervanadate, suggesting that cadherins may be PTPμ substrates (126). Taken together, these results indicate the dynamic regulation of cadherin function through intracellular phosphotyrosine signaling pathways, and suggest that perturbation of this regulatory mechanism could be important in the genesis of tumor invasiveness, and progression to metastasis.

A surprising connection, between cadherins, catenins and tumor progression, was made with the observation that the APC tumor suppressor protein and β-catenin form physiologic complexes (127,128). APC, β-catenin, and γ-catenin all possess armadillo repeat domains, which are comprised of up to 13 repeats of a 42 amino acid motif, originally described in the armadillo segment polarity gene of D. melanogaster. E-cadherin and APC compete, in vivo, for overlapping binding sites within the internal arm repeat region of β-catenin, and each of these complexes associates with α-catenin through the N-terminal domain of β-catenin. Ectopic expression of the arm repeat region of β-catenin is sufficient for induction of dorsoanterior axis duplication in Xenopus embryos, similar to the effect of full length β-catenin, indicating that the arm repeat mediates intracellular signaling independently of β-catenin regulation of cell adhesion (129). Although γ-catenin (plakoglobin) can substitute for β-catenin in either of the cadherin or APC complexes, it is not known whether the distinct arm repeats of these catenins regulate physiologic partitioning of the complexes (130,131). Interestingly, $p120^{cas}$ contains an arm repeat domain, which mediates its direct interaction with E-cadherin. Unlike β-catenin or plakoglobin, $p120^{cas}$ does not provide a direct link to the cytoskeleton, since it does not appear to interact with APC or α-catenin, and the binding of $p120^{cas}$ and β-catenin to E-cadherin is not mutually exclusive (132). The role of $p120^{cas}$, whether in cell adhesion, regulation of cadherin-catenin complexes, or as a divergent signal transduction pathway from cadherins is not yet known, however the presence of arm repeats in catenins, APC and $p120^{cas}$ underscores the importance of this motif in specifying structural and regulatory aspects of cadherin intracellular interactions.

The existence of mutually exclusive cadherin-catenin, and APC-catenin, complexes suggests that APC might not function in cadherin-mediated adhesion. Precancerous mutations in APC cluster close to the β-catenin binding region (133), yielding COOH-terminal truncated APC peptides that exhibit dysregulated interaction with β-catenin. Thus, chronology and molecular data suggest that APC mutations, and the alterations in cadherin expression or function that are associated with invasive carcinomas, are not directly linked. Nonetheless, the dynamic equilibrium between complexed and free, cytoplasmic β-catenin may play a role in regulatory interactions involving APC signaling and cadherin function. Of particular interest is a report on the effects of expressing a dominant negative N-cadherin variant in transgenic mice (134). Ectopic expression of an N-cadherin mutant protein lacking the extracellular domain, in the intestinal crypts of transgenic mice, leads to reduced E-cadherin expression in the crypts and the development of multiple, intestinal adenomas. Although a relationship between cadherin expression and APC signaling is not yet established experimentally, these results prompted the suggestion that a shift in the β-catenin equilibrium toward APC complexes may lead to the increased rates of proliferation and apoptosis, specifically observed in crypt epithelia expressing the mutant N-cadherin protein (134). Intriguingly, these lesions are similar to those resulting from a germline mutation of the murine APC homolog, in the Min mouse model of multiple intestinal neoplasia (135). It is thus possible that dysregulated signaling through mutant APC-β-catenin complexes contributes to the outgrowth of colonic polyps in familial adenomatous polyposis, and may also be a factor in a proportion of sporadic colon carcinomas. The identification of additional APC interactions may serve to clarify any relationship between cadherin function and signaling through APC-catenin complexes.

IV. IMMUNOGLOBULIN SUPERFAMILY CAMs

A. Deleted in Colorectal Cancer (DCC) Tumor Suppressor Gene

Genetic studies, indicating that a high proportion of colorectal carcinomas were associated with loss of heterozygosity (LOH) at a locus on human chromosome 18q, led to the cloning of the DCC candidate tumor suppressor gene at 18q21.1 (136). The DCC gene encodes a 1477-amino acid protein that shares overall structural similarity with the neural cell adhesion molecule (N-CAM), thus identifying DCC as a member of the immunoglobulin (Ig) superfamily of cell adhesion molecules. Similar to N-CAM, DCC has a large extracellular domain, comprised mainly of repeated (DCC = 4, N-CAM = 5) N-terminal Ig-like C2 domains, and multiple (DCC = 6, N-

CAM = 2) fibronectin (FN)-type III repeats. The deduced DCC protein contains a single transmembrane domain, and a 325-amino acid cytoplasmic domain, which does not share any significant sequence similarity with known protein domains, nor does it possess any known catalytic function (137). The presence of Ig-like domains, and FNIII repeats in the extracellular domain, strongly suggest that DCC is a receptor, mediating specific intercellular, or cell-matrix interactions.

Cancer-associated LOH involving a discrete chromosomal locus has been used to identify candidate tumor suppressor genes, which are functionally inactivated as a result of the deletional loss. The tumor suppressor role of DCC has thus been surmised, based on the LOH and allelic losses at 18q21, observed in sporadic colorectal cancers and advanced adenomas. Loss of DCC expression is associated with the progression of late-stage colon carcinomas (138). Other tumors have also been analyzed for allelic loss involving DCC, with 18p21 LOH observed in a significant proportion of gastric, pancreatic, breast, endometrial, esophageal and bladder carcinomas. In addition, reduced or absent DCC expression may be a factor in leukemias, of lymphoblastic and myelogenous origin (see 139, and references therein). Allelic loss on 18q has strong prognostic significance, in colorectal carcinoma patients with TNM stage II (Duke s stage B) disease. The five year survival rate for such patients carrying allelic 18q loss was 54%, relative to 93% for those with no 18q loss (140). The involvement of DCC allelic loss in a variety of carcinomas is consistent with DCC affecting late events in tumorigenesis, and indicates that its tumor suppressor role is not restricted to colonic epithelium. Recently, direct evidence for DCC as a tumor suppressor gene has been obtained, from transfection studies in human epithelial cells. Nitrosomethylurea (NMU) transformation of HPV-immortalized keratinocytes results in allelic loss of DCC, with concomitant loss of DCC expression and the acquisition of a tumorigenic phenotype, thus providing a model of late-stage tumorigenesis. Transfection of a full-length DCC cDNA into these NMU-transformed cells resulted in high levels of DCC protein expression, and suppression of the tumorigenic phenotype. Importantly, DCC expression was not detectable in revertant tumors. Cell-surface expression of DCC variants from truncated cDNAs, encoding proteins lacking most of the cytoplasmic domain, did not suppress the tumorigenic phenotype, suggesting that the DCC cytoplasmic domain mediates intracellular protein interactions important in epithelial differentiation (141).

Little is known about the physiologic function of DCC. Immunohistochemical localization in human tissues demonstrated tissue-restricted expression of DCC, including axonal processes of the spinal cord, CNS nerve fibres, and cerebellar Purkinje cells, with the highest levels of expression seen in mucin-secreting goblet cells of the colonic epithelium. Loss of DCC staining was seen in late adenomas, and was associated with poorly differentiated carcinomas. These results suggested that DCC is functionally associated with the differentiation of mucin-secreting cells. Furthermore, the low levels of DCC expression, relative to other adhesion molecules such as cadherins, suggests a primary role for DCC in differentiation-associated signal transduction (137). The identification of vertebrate DCC gene homologs indicates that its function is evolutionarily conserved. The Xenopus homolog, XDCCα, is predicted to possess greater than 80% overall amino acid identity with DCC and, in particular, the FNIII repeats and cytoplasmic domains share 86% to 87% identity. Developmentally regulated XDCCα expression is associated with neural induction, suggesting a role in brain morphogenesis (142). Interestingly, in adult chickens, cDCC (94% identity with DCC) expression is limited to the crypt region of intestinal villi, sqamous epithelium of skin, and mammary duct stem cells (143). Expression of cDCC in proliferating zones runs counter to the expected expression pattern of a tumor suppressor gene product, however the chicken model does provide important clues to the cell adhesion function of DCC. Antibodies against cDCC inhibited the aggregation of embryonic skin epithelial cells, in a manner that suggested that DCC mediates heterophilic binding.

Additional evidence for differentiation-associated DCC-mediated intercellular adhesion comes from studies in mammalian cells. Nerve growth factor (NGF)-induced neurite outgrowth from PC12 pheochromocytoma cells is a widely used model for sympathetic neuronal differentiation. NGF-stimulated outgrowth was inhibited by expression of DCC antisense constructs in PC12 cells, indicating a requirement for DCC in maintenance of the neuronal phenotype (144). NIH 3T3 cells expressing human DCC also stimulated neurite outgrowth in PC12 cells, in a response that seemed to require cell-cell contact, as transfectant cell supernatants did not stimulate outgrowth. Moreover in this study, NIH 3T3 transfectants expressing a truncated DCC variant, lacking most of the cytoplasmic domain, did not stimulate PC12 cell neurites. The significance of the cytoplasmic DCC interaction was not ascertained, but presumably reflects stabilization of cytoskeletal, or cytosolic interactions which regulate the DCC adhesive state (145). Similarly, co-culture of PC12 cells on NIH 3T3 cells transfected with an N-CAM cDNA, stimulated PC12 neurite outgrowth. The stimulation of neurite outgrowth by N-CAM was inhibited by pertussis toxin and an inhibitor of protein kinase activity, thus distinguishing N-CAM- from NGF-activated intracellular signaling pathways in this response (146). It is possible that DCC-mediated intercellular adhesion activates ligand-specific signaling in adjacent cells, representing one arm of a reciprocating signal transduction pathway. As yet, such a counter-receptor ligand for DCC has not been identified, and this remains one of the major questions that needs to be answered, in determining a physiologic role for DCC. It will also be important to ascertain whether, and to what extent, the cytoplasmic domain contributes to DCC-mediated intercellular adhesion.

B. Carcinoembryonic Antigen (CEA)

CEA is a cell surface glycoprotein member of the Ig superfamily (147,148), which can function in epithelial cells to mediate Ca^{2+}-independent, homotypic cell adhesion (149). The extracellular N-terminal domain of CEA contains an Ig V-like domain of 107-amino acid residues, which shares 89% identity with another family member, NCA, and 70% with a leukocyte member of the family, CGM6 (W272). Adjacent, and C-terminal to the V-like domain of CEA, are three repeated Ig C2-like domains—A1B1, A2B2, and A3B3. Only one C2-like domain is present in either of NCA or CGM6. The N-terminal domain of NCA is required for heterophilic binding to CGM6 (150), and transfection of truncated CEA cDNAs indicated that homophilic binding of CEA requires the N domain (151). Extensive mapping and inhibition studies with CEA mAbs further suggested that homophilic adhesion results from two bonds that are formed between antiparallel CEA molecules, involving residues in the N domain of one molecule and the A3B3 domain of the other (152). None of these CEA family members contain cytoplasmic or transmembrane domains, but are anchored in the external plasma membrane by a C-terminal, glycophosphatidyl-inositol moiety (153). CEA is overproduced in most colon carcinomas (154), as well as other carcinomas (155), and 10- to 100-fold elevated serum levels of CEA indicate a poor prognosis in patients with adenocarcinomas of colon, breast and lung. There are conflicting reports as to whether these increases in serum CEA represent increased release to the vasculature or cellular overproduction, but it is clear that overexpression of CEA in tumor epithelium, relative to surrounding mucosa, is a common feature of colon carcinoma (156), and both mechanisms could contribute to dysregulated function in carcinomas.

As indicated above, in contrast to the other CAMs, there appears to be no structural basis for a direct contribution of CEA signal transduction to cell growth regulation. However, ectopic expression of CEA in a model differentiation system, rat L6 myoblasts, completely inhibited myoblast fusion, along with other, early markers of myotube differentiation such as myogenin expression. The abrogation of myogenic differentiation was seen at relatively low levels of CEA expression, but required the N domain, as expression from an N-truncated cDNA did not block L6 differentiation. In addition, the treatment of CEA transfectants with N domain, and A3B3 adhesion-blocking peptides, reversed the block. Thus, in this system CEA-mediated adhesion seems to contribute indirectly to L6 signal transduction, possibly by facilitating interactions between unidentified growth-promoting surface receptors, or alternatively, blocking interactions between differentiation-promoting receptors (157).

In vitro, CEAwas shown to mediate homophilic binding of human colon adenocarcinoma cells, with cell aggregation kinetics correlating directly with levels of CEA production by the cells (149). Antibodies to the N domain blocked the adhesion of colon carcinoma cells (158) and normal coloncytes (159) to immobilized CEA. Freshly isolated, intestinal crypt coloncytes expressed CEA on their surface, and treatment of these cells with a phosphoinositide-specific phospholipase C markedly reduced surface CEA expression, and eliminated coloncyte adhesion to CEA (159). These studies clearly indicate that CEA can mediate adhesion events in intestinal epithelium, however some authors have argued that the low level of CEA protein, expressed on the luminal surface of columnar epithelium lining the upper crypts, is inconsistent with an in vivo adhesive role in adult epithelium. It was further suggested that the higher levels of basolateral CEA expression seen in fetal colonic epithelium, are consistent with a developmental role for CEA in intercellular adhesion, and that the expression of CEA in colon carcinomas may reflect a fetal pattern of expression that is disruptive of tissue architecture and differentiation (149).

Quantitative in situ hybridization analyses using CEA riboprobes, and immunoelectron microscopic localization of CEA, have suggested an in vivo role for CEA in adult colon, and in regulating adhesion in colorectal carcinoma (160). This latter study showed that CEA RNA and protein levels are detected in normal colon epithelium, and are increased at the invading front of tumors, relative to superficial regions. These authors postulate that CEA released from the tumor cells, and accumulating intercellularly at the invading front, may act as an antiadhesive factor to facilitate cell migration. Indeed, in an experimental metastasis model, systemic injection of mice with CEA resulted in the stimulation of liver cell metastases, by a weakly metastatic colorectal carcinoma cell line, suggesting that cell-free CEA does promote tumor migration and invasion. The step in metastasis that was enhanced by CEA was not identified (161). Other work suggests that CEA secreted by tumor cells may serve another anti-adhesive role, in effect protecting tumors from host defenses. Allogeneic lymphokine-activated killer (LAK) cells were shown to be inhibited from adhering in vitro to, and lysing monolayers of freshly isolated colorectal adenocarcinoma cells producing high levels of CEA. Three-dimensional spheroid cultures of these CEA-producing carcinoma cells were also resistant to in vitro tumor infiltration by LAK cells (162). This intriguing suggestion was the result of a correlative study wherein the involvement of CEA was not characterized, thus conclusions regarding this function of CEA must be interpreted with that caveat in mind. Nonetheless, although the mechanism(s) by which CEA contributes to carcinoma progression is still obscure, CEA overexpression clearly contrasts with the mutational inactivation or functional down-regulation, seen in other CAM-mediated intercellular adhesion systems. Thus, as a general consideration, progression to advanced carcinoma may require a nonrandom, temporally regulated series of events, involving down-regulation of the expression and function of one subset of CAMs, and up-regulation or increased expression levels of others (163).

V. CD44 AND RHAMM: HYALURONATE-BINDING PROTEINS

CD44 is a cell surface glycoprotein, identified as the major receptor-mediating cellular interactions with hyaluronate (HA), a glycosaminoglycan component of extracellular matrix (164,165). CD44 is widely expressed and exists in multiple forms, arising from alternatively spliced mRNA products of the single gene (166). Apparent differences in molecular weight forms are also due to highly variable glycosylation, in the CD44 extracellular region. The N-terminal extracellular region possesses significant amino acid identity with cartilage link, and proteoglycan core proteins (167,168). This extracellular domain mediates binding to HA, through a disulfide bond-stabilized loop structure. All CD44 isoforms contain a cytoplasmic domain, that may link CD44 to actin filaments through a direct interaction with moesin (169). The non-cartilage link portion of the extracellular domain is encoded, in part, by a group of 10 exons (v1 through v10), which are variably expressed in all but "standard CD44s, and define the different CD44 isoforms. Alternative splicing of these variable forms (CD44v) is regulated with tissue expression, or by antigen activation in lymphocytes. Presence or absence of alternative domains in the CD44 protein can influence its function. For example, CD44H (CD44s) mediates cell migration on HA, while a non-binding epithelial form, CD44E, does not support migration on the substrate. Cells expressing CD44H lacking the cytoplasmic domain are also unable to migrate on HA-coated substrates (170), and this requirement may reflect an interaction of the CD44 cytoplasmic domain with signaling molecules, rather than cytoskeletal proteins (171). Thus, the accumulation of HA in inflammatory sites (172), regions of tumor invasion (173), and in lymph nodes (174), could contribute to pathophysiologic migration or colonization by CD44v-expressing cells. However, HA binding does not provide a functional distinction between the metastasis-inducing CD44v and CD44s, and the role of HA binding in CD44-associated tumor progression is presently unclear (166). The tissue-restricted expression of CD44v forms presumably accounts for the implication of CD44 in a wide variety of functions, which include mediating lymphocyte adhesion to cells of high endothelial venules, cell migration, cytokine release, T cell activation and adhesion, and metastasis (170,175–178).

The metastasis-promoting ability of one CD44v isoform is of particular interest, as it may reflect the aberrant expression and regulation of a physiologic CD44v function. Transfection of this CD44v cDNA, encoding for a 162 amino acid insertion in the extracellular domain, resulted in the acquisition of a metastatic phenotype by rat pancreatic carcinoma cell (178). cDNA probes specific for the inserted sequences did not detect CD44 transcripts in selected non-metastatic variants of these cells, or the parental cells, and subsequent studies have confirmed the metastasis-associated expression of specific CD44v, in human carcinoma cell lines (179), as well as a positive correlation between CD44 expression levels and the metastatic potential of human melanoma cell (180). Western blot analyses of primary colon carcinomas and carcinoma cell lines demonstrated nearly undetectable CD44H expression, contrasting with high expression levels in normal colon mucosa. Significantly, transfection of a CD44H cDNA into the colon carcinoma cells restored CD44H expression and inhibited the tumorigenic phenotype of these cells, suggesting a growth advantage was conferred upon these cells by the loss of CD44H expression (181). Thus, it may be that loss of CD44H expression, with concomitant gain of CD44v expression through alternative splicing, are both required in the development of metastatic colorectal carcinoma. In the rat metastasis model, intravenous injection with a mAb specific for the metastasis-inducing CD44v domain, inhibited lung and lymph node metastases of the transfected pancreatic carcinoma line. Administration of the mAb after lymph node colonization did not inhibit further metastatic spread, supporting the notion that CD44 mediates the interaction of tumor cells with those of the draining lymph nodes (182). Interestingly, the metastasis-inducing CD44v was shown to be transiently induced in vivo, by antigen activation of rat T and B cells. Further analysis of this induced CD44v mRNA indicated the specific expression of exon v6, which is also included in the metastasis-inducing CD44v insertion, suggesting that a physiologic function of the exon v6 epitope is subverted in metastasis, perhaps as a result of its sustained expression (183,184).

The expression of CD44v has been shown to reflect the invasive and metastatic potential of some common human cancers. Increased expression of the exon v6 epitope was seen during the progression of colorectal tumors from hypertrophic, through late stages (Dukes C/D (185). Human homologs of the metastasis-inducing rat CD44v, were found to be overexpressed in invasive and metastatic colon carcinomas, as well as in regions of cellular dysplasia within adenomatous polyps, suggesting CD44v expression may be required, but is not sufficient for the induction of metastatic colon carcinoma (186). This conclusion is supported by a report of early CD44v expression in 41% (9/22) colorectal adenomas, prior to the acquisition of K-ras and p53 mutations, seen in later stage carcinomas (187).

The presence of CD44v containing exon v6 is also seen in aggressive, but not low-grade, non-Hodgkin's lymphoma (184). In a multivariate analysis of 245 non-Hodgkin's lymphomas, expression of CD44 protein in tumor tissue was found to have independent prognostic value, with tumors expressing low levels of CD44 indicating a favorable patient outcome. These tumors were also significantly less likely to be disseminated than CD44 expressing lymphomas (188,189). Breast carcinomas expressing >50% CD44-positive cells were associated with histologically aggressive tumors, and a greater mortality rate (190). Also, a multivariate analysis indicated that CD44v exon v6 expression has prognostic value for breast carcinoma patients, independent of progesterone receptor status, lymph node size or tumor grade. In this study, normal mammary ductal epithelium, and cells of hyperplastic regions did not express detectable exon v6 (191). These reports underscore the significance of the identification of CD44v exons in an animal metastasis model, and the direct application of this knowledge to the management of common human cancers. Conceivably, elucidation of the specific function of these metastasis-associated variant domains could allow a CD44-directed approach to treatment of some carcinomas.

One group has implicated another cellular HA binding protein, RHAMM, in the neoplastic transformation of mouse fibroblasts. RHAMM is a membrane-associated hyaluronate receptor, although it has no typical transmembrane domain and may associate in a complex with another membrane anchoring protein(s). RHAMM has been implicated in HA-mediated motility of Ras-transformed cells (192), and it has been suggested that RHAMM function is required for ras transformation of mouse fibroblasts (193). Fibroblasts, transformed by overexpression of RHAMM, formed fibrosarcomas when injected subcutaneously into mice, and 50% of these tumor-bearing mice developed spontaneous lung metastases. In this study the transforming capacity of RHAMM required the HA binding region. Expression of antisense RHAMM resulted in a loss of HA-regulated $p125^{FAK}$ phosphorylation, suggesting that HA-mediated signalling may target focal adhesion plaques indirectly, via RHAMM (193). To date, no involvement of mutated, or overexpressed RHAMM in primary human tumors has been reported.

VI. SUMMARY: THERAPEUTIC IMPLICATIONS

The dominant theme emerging from this discussion is the importance of protein-protein interactions in regulating cell adhesion and cellular responses to adhesion, and the consequences that dysregulation of protein interactions have for tumor progression. Binding interfaces provide attractive targets for therapeutic intervention, most likely involving the design, or screening of compounds, to inhibit specific protein interactions. From a drug delivery point of view, it is easiest to envision inhibitors of extracellular interactions. One example for such inhibitors is the development of RGD-based, peptide inhibitors of integrin-mediated cell-matrix interactions. Similarly, RGD-related synthetic peptides derived from the integrin binding regions of snake venom disintegrins, are potent inhibitors of platelet aggregation (194), however information about the potential antineoplastic effects of these disintegrin-derived peptides is lacking. This potential is illustrated by one report, showing that a synthetic, cyclic RGD peptide antagonist of $\alpha_v\beta_3$ integrin function was able to promote tumor regression, in an in situ model of angiogenesis (83), and by a study demonstrating in vivo inhibition of melanoma metastasis, by a synthetic RGD peptide (68). Interestingly, and perhaps important for therapeutic applications, disintegrins are about 1000-fold more potent inhibitors of β_1 and β_3 integrin function, than are the synthetic RGD peptides (195,196). A serious limitation to the therapeutic use of RGD peptides is the potential lack of specificity, as a number of integrin-ECM interactions are RGD-dependent and, ideally, it would be desirable to develop integrin- or subunit-specific inhibitors. In this regard, a highly selective peptide antagonist, exhibiting a non-RGD-based binding specificity for $\alpha_5\beta_1$ integrin, has been isolated from a phage display library (197). Importantly, disintegrins display a high degree of integrin specificity which can, moreover, be mimicked by small peptides (198). An optimized phage display strategy generated an antibody specific for the platelet integrin gpIIb-IIIa, starting from a synthetic RGD-containing antibody specific for $\alpha_v\beta_3$. Interestingly, the gpIIb-IIIa antibody contained an altered disintegrin-like recognition sequence (199). These results suggest the potential for developing integrin-specific therapeutic agents, using peptide and antibody engineering approaches based on naturally occurring integrin antagonists. Another group of anti-adhesion molecules, including SPARC, thrombospondin, tenascin (200), and the disintegrin-related, ADAM family of cellular proteins (201), might also provide models for the design of integrin-specific anti-adhesive drugs.

It may not be desirable to inhibit all function in a given CAM, or CAM subunit. The identification of intracellular effectors of CAM-mediated signalling is an area that holds great potential for drug development. For example, inhibitors designed to target one branch of a diverging CAM signal transduction pathway would leave intact other, essential functions of that CAM. Thus in principle, extracellular and intracellular interactions of CAMs are targets for intervention, however, attempts to address therapeutics to sites of intracellular interactions face considerable problems of drug delivery. This delivery problem applies for most conceivable forms of inhibitor, whether anti-sense oligonucleotides, mAbs, or peptide antagonists, as these do not traverse the plasma membrane. Effective gene replacement, for example of APC or catenins to restore cadherin function/ signaling, or of α_5 integrin to suppress carcinoma growth, will presumably rely on the development of safe viral vectors, or liposome-based technologies. Whatever delivery vehicle presents as the best solution to this problem, the further definition of specific CAM regulatory interfaces presents a rich potential for the development of relatively non-toxic and effective, CAM-specific therapeutic agents.

The considerations of CAM function in tumor progression have been presented in discrete sections here but, of course, the reality is more complex, as CAM systems do not function in isolation of one another and can, in fact, influence the function of other CAMs. This may occur intracellularly, as illustrated by a report of the effect of blocking cadherin function, on the regulation of integrin expression in differentiating keratinocytes (202). Thus, inhibition of one CAM system could have beneficial, or detrimental consequences with respect to tumor cell growth, by altering the expression or function of other tumor cell CAMs. Also, heterotypic cell interactions between members of different CAM families, such as integrins and ICAMs (4), or integrins and cadherins (203), suggest that therapeutic perturbation of one CAM system could have complex effects on tumor-host interactions. These effects could in theory antagonize, or cooperate with the targeted event, and an understanding of the physiologic and pathologic relationships between CAM systems is desirable, for the development of optimized intervention strategies.

In conclusion, it is accepted that dysregulation of CAM function, leading to altered cell-cell and/or cell-substratum interactions, is an important contributing factor to tumor progression. We have attempted to represent the molecular diversity underlying this dysregulation, for each of the major CAM families implicated in the later stages of tumor development. For example, in many carcinomas multiple mechanisms have been identified, which lead to loss or down-regulation of E-cadherin function. These include the mutational loss of E-cadherin, or loss of cadherin-associated intracellular proteins. Clearly, while in some carcinomas therapeutic upregulation of E-cadherin function might present an effective anti-cancer strategy, such an approach in a catenin-deficient tumor would likely prove fruitless. Thus, in the context of CAMs, optimal therapeutic approaches to these common cancers will require tumor-specific molecular profiles. It is hoped that our increasing ability to generate such profiles reflects a real potential for the improved management and treatment of what is yet a largely intractable clinical problem.

REFERENCES

1. Ruoslahti E, Pierschbacher MD. New perspectives in cell adhesion: RGD and integrins Science 1987; 238:491–497.
2. Hemler ME. Structures and functions of VLA proteins and related integrins. In: Mecham RP, McDonald JA, eds. Receptors for Extracellular Matrix. New York: Academic Press, 1991:255–299.
3. Hynes RO. Integrins: veratility, modulation and signaling in cell adhesion. Cell 1992; 69:11–25.
4. Stewart M, Thiel M, Hogg N. Leukocyte integrins. Curr Opin Cell Biol 1995; 7:690–696.
5. Hogg N, Landis RC, Bates PA, Standley P, Randi AM. The sticking point: how integrins bind to their ligands. Trends Cell Biol 1994; 4:379–382.
6. Huttenlocher A, Sandborg RR, Horwitz A. Adhesion in cell migration. Curr Opin Cell Biol 1995; 7:697–706.
7. Dedhar S. Novel functions for calreticulin: interaction with integrins and modulation of gene expression? Trends Biol Sci 1994; 19:269–271.
8. Grinblat Y, Zusman S, Yee G, Hynes RO, Kafatos FC. Functions of the cytoplasmic domain of the βPS integrin subunit during Drosophila development. Development 1994; 120:91–102.
9. Otey C, Pavalko FM, Burridge K. An interaction between α-actinin and the $\beta 1$ integrin subunit. J Cell Biol 1990; 268:21193–21197.
10. Horwitz A, Duggan, Buck C, Beckerle MC, Burridge K. Interaction of plasma membrane fibronectin receptor with talin—a transmembrane linkage. Nature 1985; 320:531–533.
11. Sharma CP, Ezzell RM, Arnaout MA. Direct interaction of filamin (ABP-280) with the $\beta 2$ integrin subunit CD18. J Immunol 1995; 154:3461–3470.
12. Schaller MD, Parsons JT. Focal adhesion kinase and associated proteins. Curr Opin Cell Biol 1994; 6:705–710.
13. Clark EA, Brugge JS. Integrins and signal transduction pathways: the road taken. Science 1995; 268:233–239.
14. Burridge K, Turner C, Romer L. Tyrosine phosphorylation of paxillin and $pp125^{FAK}$ accompanies cell adhesion to extracellular matrix: A role in cytoskeletal assembly. J Cell Biol 1992; 119:893–904.
15. Bockholt SM, Burridge K. Cell spreading on extracellular matrix proteins induces tyrosine phosphorylation of tensin. J Biol Chem 1993; 268:14565–14567.
16. Akiyama SK, Yamada SS, Yamada KM, LaFlamme SE. Transmembrane

signal transduction by integrin cytoplasmic domains expressed in single subunit chimeras. J Biol Chem 1994; 269:15961–15964.
17. Lukashev ME, Sheppard D, Pytela R. Disruption of integrin function and induction of tyrosine phosphorylation by the autonomously expressed $\beta 1$ integrin subunit cytoplasmic domain. J Biol Chem 1994; 269:18311–18314.
18. Solowska J, Guan JL, Marcantonio EE, Trevithick JE, Buck C, Hynes RO. Expression of normal and mutant avian integrin subunits in rodent cells. J Cell Biol 1989; 109:853–861.
19. Hayashi Y, Haimovich B, Reszka AA, Boettiger D, Horwitz A. Expression and function of chicken integrin $\beta 1$ subunit and its cytoplasmic domain mutants in mouse NIH 3T3 cells. J Cell Biol 1990; 110:175–184.
20. LaFlamme SE, Thomas LA, Yamada SS, Yamada KM. Single subunit chimeric integrins as mimics and inhibitors of endogenous integrin functions in receptor localization, cell spreading and migration, and matrix assembly. J Cell Biol 1994; 1126:1287–1298.
21. Otsuka H, Moskowitz MJ. Arrest of 3T3 cells in G1 phase by suspension culture. J Cell Physiol 1975; 87:213–220.
22. Benecke BJ, Ben-Ze ev A, Penman S. The control of mRNA production, translation and turnover in suspended and reattached anchorage-dependent fibroblasts. Cell 1978; 14:931–939.
23. Guadagno TM, Assoian RK. G1/S control of anchorage-independent growth in the fibroblast cell cycle. J Cell Biol 1991; 115:1419–1425.
24. Dedhar S. Integrins and tumor invasion. Bioessays 1991; 12:583–589.
25. Damsky CH, Werb Z. Signal transduction by integrin receptors for extracellular matrix: cooperative processing of extracellular information. Curr Opin Cell Biol 1992; 4:772–781.
26. Juliano RL, Haskill S. Signal transduction from the extracellular matrix. J Cell Biol 1993; 120:577–585.
27. Sastry SK, Horwitz AF. Integrin cytoplasmic domains; mediators of cytoskeletal linkages and extracellular initiated transmembrane signaling. Curr Opin Cell Biol 1993; 5:818–831.
28. Vuori K, Ruoslahti E. Activation of protein kinase C precedes $\alpha 5\beta 1$ integrin-mediated cell spreading on fibronectin. J Biol Chem 1993; 268:21459–21462.
29. Schwartz MA, Lechene C, Ingber DE. Insoluble fibronectin activates the Na/H antiporter by clustering and immobilizing integrin $\alpha 5\beta 1$, independent of cell shape. Proc Natl Acad Sci USA 1991; 88:7849–7853.
30. Schwartz MA. Spreading of human endothelial cells on fibronectin or vitronectin triggers elevation of intracellular free calcium. J Cell Biol 1993; 120: 1003–1010.
31. Golden A, Brugge JS, Shattil SJ. Role of platelet membrane glycoprotein IIb-IIIa in agonist induced tyrosine phosphorylation of platelet proteins. J Cell Biol 1990; 111:3117–3127.
32. Schaller MD, Borgman CA, Cobb BS, Vines RR, Reynolds AB, Parsons JT. $pp125^{FAK}$, a structurally distinctive protein-tyrosine kinase associated with focal adhesions. Proc Natl Acad Sci USA 1992; 89:5192–5196.

33. Hanks SK, Calalb MB, Harper MC, Patel SK. Focal adhesion protein-tyrosine kinase phosphorylated in response to cell spreading on fibronectin. Proc Natl Acad Sci USA 1992; 89:8487–8491.
34. Kornberg L, Earp HS, Parsons JT, Schaller MD, Juliano RL. Cell adhesion or integrin clustering increases phosphorylation of a focal adhesion associated protein kinase. J Biol Chem 1992; 267:23439–23442.
35. Guan JL, Trevithick J, Hynes RO. Fibronectin/integrin interaction induces tyrosine phosphorylation of a 120 kDa protein. Cell Regul 1991; 2:951–964.
36. Kapron-Bras C, Fitz-Gibbon L, Jeevaratnam P, Wilkins J, Dedhar S. Stimulation of tyrosine phosphorylation and accumulation of GTP-bound $p21^{ras}$ upon antibody-mediated $\alpha 2\beta 1$ integrin activation in T-lymphoblastic cells. J Biol Chem 1993; 268:20701–20704.
37. Werb Z, Tremble PM, Behrendtsen O, Crowley E, Damsky CH. Signal transduction through the fibronectin receptor induces collagenase and stromelysin gene expression. J Cell Biol 1989; 109:877–889.
38. Haskill S, Beg AA, Tompkins SM, et al. Characterization of an immediate-early gene induced in adherent monocytes that encodes I kappa B-like activity. Cell 65:1281–1289 (1991).
39. Lin TH, Yurochko A, Kornberg L, et al. The role of protein tyrosine phosphorylation in integrin-mediated gene induction in monocytes. J Cell Biol 1994; 126:1585–1593.
40. Parsons JT, Schaller MD, Hildebrand J, Leu TH, Richardson A, Otey C. Focal adhesion kinase: structure and signalling. J Cell Sci 1994; 18(suppl): 109–113.
41. Meredith JE, Fazeli B, Schwartz MA. The extracellular matrix survival factor. Mol Biol Cell 1993; 4:953–961.
42. Frisch SM, Francis H. Disruption of epithelial cell-matrix interactions induces apoptosis. J Cell Biol 1994; 124:619–626.
43. Boudreau N, Sympson CJ, Werb Z, Bissel MJ. Suppression of ICE and apoptosis in mammary epithelial cells by extracellular matrix. Science 1995; 267:891–893.
44. Guadagno TM. Ohtsubo M, Roberts JM, Assoian RK. A link between cyclin A expression and adhesion-dependent cell cycle progression. Science 1993; 262:1572–1575.
45. Han EKH, Guadagno TM, Dalton SL, Assoian RK. A cell cycle and mutational analysis of anchorage-independent growth: cell adhesion and TGF-$\beta 1$ control G1/S transit specifically. J Cell Biol 1993; 122:461–471.
46. Hansen LK, Mooney DJ, Vacanti JP, Ingber DE. Integrin binding and cell spreading on extracellular matrix act at different points in the cell cycle to promote hepatocyte growth. Mol Biol Cell 1994; 5:967–975.
47. Shin S, Freedman VH, Risser R, Pollack R. Tumorigenicity of virus-transformed cells in nude mice is correlated specifically with anchorage-independent growth in vitro. Proc Natl Acad Sci USA 1975; 72:4435–4439.
48. Guan JL, Shalloway D. Regulation of $pp125^{FAK}$ both by cellular adhesion and oncogenic transformation. Nature 1992; 358:690–692.

49. Chen Q, Kinch MS, Lin Th, Burridge K, Juliano RL. Integrin-mediated cell adhesion activates mitogen-activated protein kinases. J Biol Chem 1994; 269: 26602–26605.
50. Schlaepfer DD, Hanks SK, Hunter T, van der Geer P. Integrin-mediated signal transduction linked to Ras pathway by GRB2 binding to focal adhesion kinase. Nature 1994; 372:786–791.
51. Nobes CD, Hall A. Rho, rac and cdc42 regulate the assembly of multi-molecular focal complexes associated with actin stress fibres, lamellipodia and filopodia. Cell 1995; 81:53–62.
52. Hotchin NA, Hall A. The assembly of integrin adhesion complexes requires both extracellular matrix and intracellular rho/rac GTPases. J Cell Biol 1995; 131:1857–1865.
53. Mainiero F, Pepe A, Wary KK, et al. Signal transduction by the $\alpha6\beta4$ integrin: distinct $\beta4$ subunit sites mediate recruitment of Shc/Grb2 and association with the cytoskeleton of hemidesmosomes. EMBO J 1995; 14:4470–4481.
54. Richardson A, Parsons JT. Signal transduction through integrins: a central role for focal adhesion kinase? Bioessays 1995; 17:229–236.
55. Zachary I, Rozenqurt E. Focal adhesion kinase ($p125^{FAK}$: A point of convergence in the action of neuropeptides, integrins, and oncogenes. Cell 1992; 71: 891–894.
56. Hannigan GE, Leung-Hagesteijn C, Fitz-Gibbon L, et al. Regulation of cell adhesion and anchorage-dependent growth by a new $\beta1$-integrin-linked protein kinase. Nature 1996; 379:91–96.
57. Giancotti FG, Mainiero F. Integrin-mediated adhesion and signaling in tumorigenesis. Biochim Biophys Acta 1994; 1198:47–64.
58. Plantefaber LC, Hynes RO. Changes in integrin receptors on oncogenically transformed cells. Cell 1989; 56:281–290.
59. Dedhar S, Saulnier R, Nagle R, Overall CM. Specific alterations in the expression of $\alpha3\beta1$ and $\alpha6\beta4$ integrins in highly invasive and metastatic variants of human prostate carcinoma cells selected by in vitro invasion through reconstituted basement membrane. Clin Exp Metastasis 1993; 11:391–400.
60. Saga S, Chen WT, Yamada KM. Enhanced fibronectin receptor expression in Rous sarcoma virus-induced tumors. Cancer Res 1988; 48:5510–5513.
61. Giancotti FG, Comoglio PM, Tarone G. A 135,000 molecular weight plasma membrane glycoprotein involved in fibronectin-mediated cell adhesion. Immunofluorescence localization in normal and RSV-transformed fibroblasts. Exp Cell Res 1986; 163:47–62.
62. Akiyama SK, Nagata K, Yamada KM. Cell surface receptors for extracellular matrix components. Biochim Biophys Acta 1990; 1031:91–109.
63. Giancotti FG, Ruoslahti E. Elevated levels of the fibronectin receptor suppress the transformed phenotype of Chinese hamster ovary cells. Cell 1990; 60:849–859.
64. Juliano R. Signal transduction by integrins and its role in the regulation of tumor growth. Cancer Met Rev 1994; 13:25–30.
65. Dedhar S, Saulnier R. Alterations in integrin receptor expression on chemi-

cally transformed human cells: specific enhancement of laminin and collagen receptor complexes. J Cell Biol 1990; 110:481–489.
66. Chan BMC, Matsuura N, Takada Y, Zetter BR, Hemler ME. In vitro and in vivo consequences of VLA-2 expression in rhabdomyosarcoma cells. Science 1991; 251:1600–1602.
67. Folkman J. Angiogenesis in cancer, vascular, rheumatoid and other disease. Nature Med 1995; 1:27–31.
68. Humphries MJ, Olden K, Yamada KM. A synthetic peptide from fibronectin inhibits experimental metastasis of murine melanoma cells. Science 1986; 233:466–470.
69. Humphries MJ, Yamada KM, Olden K. Investigation of the biological effects of anti-cell adhesive synthetic peptides that inhibit experimental metastasis of B16-F10 murine melanoma cells. J Clin Invest 1988; 81:782–790.
70. Qian F, Vaux DL, Weissman I. Expression of the integrin on melanoma cells can inhibit the invasive stage of metastasis formation. Cell 1994; 77:335–347.
71. Aznavoorian S, Murphy AN, Stetler-Stevenson WG, Liotta LA. Molecular aspects of tumor cell invasion and metastasis. Cancer 1993; 71:1368–1383.
72. Stetler-Stevenson WG, Aznavoorian S, Liotta LA. Tumor cell interactions with the extracellular matrix during invasion and metastasis. Annu Rev Cell Biol 1993; 9:541–573.
73. Fridman R, Kibbey MC, Royce LS, et al. Enhanced tumor growth of both primary and established human and murine tumor cells in athymic nude mice after coinjection with matrigel. J Natl Cancer Inst 1991; 11:769–774.
74. Fridman R, Sweeney TM, Zain M, Martin GR, Kleinman H. Malignant transformation of NIH 3T3 cells with a reconstituted basement membrane (matrigel). Int J Cancer 1992; 51:740–744.
75. Seftor REB, Seftor EA, Gehlsen KR, et al. Role of $\alpha v\beta 3$ integrin in human melanoma cell invasion. Proc Natl Acad Sci USA 1992; 89:1557–1561.
76. Albelda SM, Mette SA, Elder DE, et al. Integrin distribution in malignant melanoma: association of the $\beta 3$ subunit with tumor progression. Cancer Res 1990; 50:6757–6764.
77. Nip J, Shibata H, Loskutoff DJ, Cheresh DA, Brodt P. Human melanoma cells derived for lymphatic metastases use integrin $\alpha v\beta 3$ to adhere to lymph node vitronectin. J Clin Invest 1992; 90:1406–1413.
78. Felding-Haberman B, Meuller BM, Romerdahl CA, Cheresh DA. Involvement of integrin αv gene expression in human melanoma tumorigenicity. J Clin Invest 1992; 89:2018–2022.
79. Nip J, Brodt P. The role of the integrin vitronectin receptor, $\alpha v\beta 3$ in melanoma metastasis. Cancer Met Rev 1995; 14:241–252.
80. Fidler IJ, Ellis LM. The implications of angiogenesis for the biology and therapy of cancer. Cell 1994; 79:185–188.
81. Leavesley DI, Schwartz MA, Rosenfeld M, Cheresh DA. Integrin $\beta 1$- and $\beta 3$-mediated endothelial cell migration is triggered through distinct signaling mechanisms. J Cell Biol 1993; 121:163–170.
82. Brooks PC, Clark RAF, Cheresh DA. Requirement of vascular $\alpha v\beta 3$ integrin for angiogenesis. Science 1994; 264:569–571.

83. Brooks PC, Montgomery AMP, Rosenfeld M, et al. Integrin $\alpha v\beta 3$ antagonists promote tumor regression by inducing apoptosis of angiogenic blood vessels. Cell 1994; 79:1157–1164.
84. Koch AE, Halloran MM, Haskell CJ, Shah M, Polverini PJ. Angiogenesis mediated by soluble forms of E-selectin and vascular cell adhesion molecule-1. Nature 1995; 376:517–519.
85. Albini A, Barillari G, Benelli R, Gallo RC, Ensoli B. Angiogenic properties of human immunodeficiency virus type I Tat protein. Proc Natl Acad Sci USA 1995; 92:4838–4842.
86. Birchmeier W, Behrens J. Cadherin expression in carcinomas: role in the formation of cell junctions and the prevention of invasiveness. Biochim Biophys Acta 1994; 1198:11–26.
87. Takeichi M. Cadherins in cancer: implications for invasion and metastasis. Curr Opin Cell Ciol 1993; 5:806–811.
88. Shapiro L, Fannon AM, Kwong PD, et al. Structural basis of cell-cell adhesion by cadherins. Nature 1995; 374:327–337.
89. Koch PJ, Franke WW. Desmosomal cadherins: another growing family of adhesion molecules. Curr Opin Cell Biol 1994; 6:682–687.
90. Troyanovsky SM, Eshkind LG, Troyanovsky RB, Leube RE, Franke WW. Contributions of cytoplasmic domains of desmosomal cadherins to desmosome assembly and intermediate filament anchorage. Cell 1993; 72:561–574.
91. Behrens J, Mareel MM, van Roy FM, Birchmeier W. Dissecting tumor cell invasion: Epithelial cells acquire invasive properties after the loss of uvomorulin-mediated cell-cell adhesion. J Cell Biol 1989; 108:2435–2447.
92. Frixen UH, Behrens J, Sachs M, et al. E-cadherin-mediated cell-cell adhesion prevents invasiveness of human carcinoma cells. J Cell Biol 1991; 113:173–185.
93. Velminckx K, Vakaet L, Mareel M, Fiers W, van Roy F. Genetic manipulation of E-cadherin expression by epithelial tumor cells reveals an invasion suppressor role. Cell 1991; 66:107–119.
94. Navarro P, Gomez M, Pizarro A, Gamallo C, Quintanilla M, Cano A. A role for the E-cadherin cell-cell adhesion molecule during tumor progression of mouse epidermal carcinogenesis. J Cell Biol 1991; 115:517–533.
95. Pizarro A, Benito N, Navarro P, et al. E-cadherin expression in basal cell carcinoma. Br J Cancer 1994; 69:157–162.
96. Otto T, Birchmeier W, Schmidt U, et al. Inverse relation of E-cadherin and autocrine motility factor receptor expression as a prognostic factor in patient with bladder carcinomas. Cancer Res 1994; 54:3120–3123.
97. Inoue M, Ogawa H, Miyata M, Shiozaki H, Tanizawa O. Expression of E-cadherin in normal, benign, and malignant tissues of female genital organs. Am J Clin Pathol 1992; 98:76–80.
98. Oka H, Shiozaki H, Kobayashi K, et al. Expression of E-cadherin cell adhesion molecules in human breast cancer tissues and its relationship to metastasis. Cancer Res 1993; 53:1696–1701.
99. Mansouri A, Spurr N, Goodfellow PN, Kemler R. Characterization and

chromosomal localization of the gene encoding the human cell adhesion molecule uvomorulin. Differentiation 1988; 38:67–71.
100. Natt E, Magenis RE, Zimmer J, Mansouri A, Scherer G. Reassignment of the human loci for uvomorulin (UVO) and chymotrypsinogen B (CTRB) with the help of two overlapping deletions in the long arm of chromosome 16. Cytogenet. Cell Genet 1989; 50:145–148.
101. Carter BS, Ewing CM, Ward WS, et al. Allelic loss of chromosomes 16q and 10q in human prostate cancer. Proc Natl Acad Sci USA 1990; 87:8751–8755.
102. Sato T, Saito H, Morita R, Koi S, Lee JH, Nakamura Y. Allelotype of human ovarian cancer. Cancer Res 1991; 51:5118–5122.
103. Devilee P, van Vliet M, van Sloun P, et al. Allelotype of human breast carcinomas: a second major locus for loss of heterozygosity is on chromosome 6q. Oncogene 1991; 6:1705–1711.
104. Tsuda H, Zhang WD, Shimosata Y, et al. Allele loss on chromosome 16 associated with progression of human hepatocellular carcinoma. Proc Natl Acad Sci USA 1990; 87:6791–6794.
105. Becker KF, Atkinson MJ, Reich U, et al. E-cadherin gene mutations provide clues to diffuse type gastric carcinomas. Cancer Res 1994; 54:3845–3852.
106. Risinger JI, Berchuk A, Kohler MF, Boyd J. Mutations in the E-cadherin gene in human gynecologic cancers. Nature Genet 1994; 7:98–102.
107. Hennig G, Behren J, Truss M, Frisch S, Reichmann, Birchmeier W. Progression of carcinoma cells is associated with alterations in chromatin structure and factor binding at the E-cadherin promotor in vivo. Oncogene 1995; 11: 475–484.
108. Yoshiura K, Kanai Y, Ochiai A, Shimoyama Y, Sugimura T, Hirohashi S. Silencing of the E-cadherin invasion-suppresor gene by CpG methylation in human carcinomas. Proc Natl Acad Sci USA 1995; 92:7416–7419.
109. Nagafuchi A, Takeichi M. Cell binding function of E-cadherin is regulated by the cytoplasmic domain. Cell 1988; 7:3679–3684.
110. Ozawa M, Baribault H, Kemler R. The cytoplasmic domain of the cell adhesion molecule uvomorulin associates with three independent proteins structurally related in different species. EMBO J 1098; 8:1711–1717.
111. Hulsken J, Behrens J, Birchmeier W. Tumor-suppresor gene products in cell contacts: the cadherin-APC-armadillo connection. Curr Opin Cell Biol 1994; 6:711–716.
112. Gumbiner B. Signal transduction by β-catenin. Curr Opin Cell Biol 1995; 7: 634–640.
113. Kintner C. Regulation of embryonic cell adhesion by the cadherin cytoplasmic domain. Cell 1992; 69:225–236.
114. Ozawa M, Ringwald M, Kemler R. Uvomorulin-catenin complex formation is regulated by a specific domain in the cytoplasmic region of the cell adhesion molecule. Proc Natl Acad Sci USA 1990; 87:4246–4250.
115. Shimoyama Y, Nagafuchi A, Fujita S, et al. Cadherin dysfunction in a human cancer cell line: possible involvement of loss of α-catenin expression in reduced cell-cell adhesiveness. Cancer Res 1992; 52:5770–5774.

116. Watabe M, Nagafuchi A, Tsukita S, Takeichi M. Induction of polarized cell-cell association and retardation of growth by activation of the E-cadherin-catenin adhesion system in a dispersed carcinoma line. J Cell Biol 1994; 127:247–256.
117. Matsuyoshi N, Hamaguchi M, Taniguchi S, Nagafuchi, Tsukita S, Takeichi M. Cadherin-mediated cell-cell adhesion is perturbed by v-src tyrosine phosphorylation in metastatic fibroblasts. J Cell Biol 1992; 118:703–714.
118. Hamaguchi M, Matsuyoshi N, Ohnishi Y, Gotoh B, Takeichi M, Nagai Y. P60$^{v\text{-}src}$ causes tyrosine phosphorylation and inactivation of the N-cadherin-catenin call adhesive system. EMBO J 1993; 12:307–314.
119. Behrens J, Vakaet L, Friis R, et al. Loss of epithelial differentiation and gain of invasiveness correlates with tyrosine phosphorylation of the E-cadherin/β-catenin complex in cells transformed with a temperature-sensitive v-SRC gene. J Cell Biol 120:757–766 (1993).
120. Takeda H, Nagafuchi A, Yonemura S, et al. V-src kinase shifts the cadherin-based cell adhesion from the strong to the weak state and β-catenin is not required for the shift. J Cell Biol 1995; 131:1839–1847.
121. Updyke TV, Nicholson GL. Malignant melanoma cell lines selected in vitro for increased homotypic adhesion properties have increased experimental metastatic potential. Clin Exp Met 1986; 4:273–284.
122. Shibamoto S, Hayakawa M, Takeuchi K, et al. Tyrosine phophorylation of β-catenin and plakoglobin enhanced by hepatocyte growth factor and epidermal growth factor in human carcinoma cells. Cell Adhesion Commun 1994; 1:295–305.
123. Hoscheutzky H, Aberle H, Kemler R. β-catenin mediates the interaction of the cadherin-catenin complex with epidermal growth factor receptor. J Cell Biol 1994; 127:1375–1380.
124. Shibamoto S, Hayakawa M, Takeuchi K, et al. Association of p120, a tyrosine kinase substrate, with E-cadherin/catenin complexes. J Cell Biol 1995; 128:949–957.
125. Brady-Kalnay SM, Tonks NK. Identification of the homophilic binding site of the receptor protein tyrosine phosphatase PTPμ. J Biol Chem 1994; 269: 28472–28477.
126. Brady-Kalnay SM, Tonks NK. Receptor protein tyrosine phosphatase PTPμ associates with cadherins and catenins in vivo. J Cell Biol 1995; 130:977–986.
127. Rubinfeld B, Souza B, Albert I, et al. Association of the APC gene product with β catenin. Science 1993; 262:1731–1733.
128. Su LK, Vogelstein B, Kinzler KW. Association of the APC tumor suppressor protein with catenins. Science 1993; 262:1734–1737.
129. Funayama N, Fagotto F, McCrea P, Gumbiner B. Embryonic axis formation by the armadillo repeat domain of β-catenin: evidence for intracellular signaling. J Cell Biol 1995; 128:959–968.
130. Hulsken J, Birchmeier W, Behrens J. E-cadherin and APC compete for the interaction with β-catenin and the cytoskeleton. J Cell Biol 1994; 127:2061–2069.

131. Rubinfeld B, Souza B, Albert I, Munemitsu S, Polakis P. The APC protein and E-cadherin form similar but independent complexes with α-catenin, β-catenin, and plakoglobin. J Biol Chem 1995; 270:5549–5555.
132. Daniel JM, Reynolds AB. The tyrosine kinase substrate p120^{cas} binds directly to E-cadherin but not to the adenomatous polyposis coli protein or α-catenin. Mol Cell Biol 1995; 15:4819–4824.
133. Polakis P. Mutations in the APC gene and their implications for protein structure and function. Curr Opin Genet Dev 1995; 5:66–71.
134. Hermiston ML, Gordon JI. Inflammatory bowel disease and adenomas in mice expressing a dominant negative N-cadherin. Science 1995; 270:1203–1207.
135. Su LK, Kinzler KW, Vogelstein B, et al. Multiple intestinal neoplasia caused by a mutation in the murine homolog of the APC gene. Science 1992; 256: 668–670.
136. Fearon ER, Cho KR, Nigro JM, et al. Identification of a chromosome 18q gene that is altered in colorectal cancers. Science 1990; 247:49–56.
137. Hedrick L, Cho KR, Fearon ER, Wu TC, Kinzler KW, Vogelstein B. The DCC gene in cellular differentiation and colorectal tumorigenesis. Genes Dev 1994; 8:1174–1183.
138. Kikuchi-Yanoshita R, Konishi M, Fukunari H, Tanaka K, Miyaki M. Loss of expression of the DCC gene during progression of colorectal carcinomas in familial adenomatous polypopsis and non-familial adenomatous polyposis. Cancer Res 1992; 52:3801–3803.
139. Cho KR, Fearon ER. DCC: linking tumor suppressor genes and altered cell surface interactions in cancer? Curr Opin Genet Dev 1995; 5:72–78.
140. Jen J, Kim H, Piantadosi S, et al. Allelic loss of chromosome 18q and prognosis in colorectal cancer. N Engl J Med 1994; 331:213–221.
141. Klingelhutz AJ, Hedrick L, Cho KR, McDougall JK. The DCC gene suppresses the malignant phenotype of transformed human epithelial cells. Oncogene 1995; 10:1581–1586.
142. Pierceall WE, Reale MA, Candia AF, Wright CVE, Cho KR, Fearon ER. Expression of a homologue of the deleted in colorectal cancer (DCC) gene in the nervous system of developing *Xenopus* embryos. Dev Biol 1994; 166:654–665.
143. Chuong CM, Jiang TX, Yin E, Widelitz RB. cDCC (chicken homologue to a gene deleted in colorectal carcinoma) is an epithelial adhesion molecule expressed in the basal cells and involved in epithelial-mesenchymal interaction. Dev Biol 1994; 164:383–397.
144. Lawlor KG, Narayanan R. Persistent expression of the tumor suppresor gene DCC is essential for neuronal differentiation. Cell Growth Diff 1992; 3:609–616.
145. Pierceall WE, Cho KR, Getzenberg RH, et al. NIH3T3 cells expressing the deleted in colorectal cancer tumor suppressor gene product stimulate neurite outgrowth in rat PC12 pheochromocytoma cells. J Cell Biol 1994; 124:1017–1027.

146. Doherty P, Ashton SV, Moore SE, Walsh FS. Morphoregulatory activities of N-CAM and N-Cadherin can be accounted for by G protein-dependent activation of L- and N-type neuronal Ca^{2+} channels. Cell 1991; 67:21–33.
147. Beauchemin N, Benchimol S, Cournoyer D, Fuks A, Stanners C. Isolation and characterization of full length functional cDNA clones for human carcinoembryonic antigen (CEA). Mol Cell Biol 1987; 7:3221–3230.
148. Paxton RJ, Mooser G, Pande H, Lee TD, Shively JE. Sequence analysis of carcinoembryonic antigen: identification of glycosylation sites and homology with the immunoglobulin super-gene family. Proc Natl Acad Sci USA 1987; 84:920–924.
149. Benchimol S, Fuks A, Jothy S, Beauchemin N, Shirota K, Stanners CP. Carcinoembryonic antigen, a human tumor marker, functions as an intercellular adhesion molecule. Cell 1989; 57:327–334.
150. Oikawa S, Inuzuka C, Kuroki M, et al. A specific heterotypic cell adhesion activity between members of carcinoembryonic antigen family, W272 and NCA, is mediated by N-domains. J Biol Chem 1991; 266:7995–8001.
151. Oikawa S, Inuzuka C, Kuroki M, Matsuoka Y, Kosaki G, Nakazato H. Homophilic association between Ig superfamily carcinoembryonic antigen molecules involves double reciprocal bonds. Biochem Biophys Res Commun 1989; 164:39–45.
152. Zhou H, Fuks A, Alcaraz G, Bolling TJ, Stanners CP.Homophilic association between Ig superfamily carcinoembryonic antigen molecules involves double reciprocal bonds. J Cell Biol 1993; 122:951–960.
153. Hefta SA, Hefta LJF, Lee TD, Paxton RJ, Shively JE. Carcinoembryonic antigen is anchored to a glycophosphatidylinositol moiety: identification of the ethanolamine linkage site. Proc Natl Acad Sci USA 1988; 85:4648–4652.
154. Shuster J, Thompson DMP, Fuks A, Gold P. Immunologic approaches to the diagnosis of disease. Prog Exp Tumor Res 1980; 25:89–139.
155. Zimmerman W, Weber B, Ortlieb B, et al. Chromosomal localization of the carcinoembryonic antigen gene family and differential expression in various tumors. Cancer Res 1988; 48:2443–2550.
156. Boucher D, Cournoyer D, Stanners CP, Fuks A. Studies on the control of gene expression of the carcinoembryonic antigen gene family in human tissue. Cancer Res 1989; 49:847–852.
157. Eidelman FJ, Fuks A, DeMarte L, Taheri M, Stanners CP. Human carcinoembryonic antigen, an intercellular adhesion molecule, blocks fusion and differentiation of rat myoblasts. J Cell Biol 1993; 123:467–475.
158. Jessup JM, Kim JC, Thomas P, et al. Adhesion to carcinoembryonic antigen by human colorectal carcinoma cells involves two epitopes. Int J Cancer 1993; 55:262–268.
159. Ishii S, Steele G, Ford R, et al. Normal colonic epithelium adheres to carcinoembryonic antigen and type IV collagen. Gastroenterology 1994; 106: 1242–1250.
160. Jothy S, Yuan SY, Shirota K. Transcription of carcinoembryonic antigen in normal colon and colon carcinoma. Am J Pathol 1993; 143:250–257.
161. Hostetter RB, Augustus LB, Mankarious R, et al. Carcinoembryonic antigen

as a selective enhancer of colorectal cancer metastasis. J Natl Cancer Inst 1990; 82:380–385.
162. Kammerer R, von Kleist S. CEA expression of colorectal adenocarcinoma is correlated with their resistance against LAK-cell lysis. Int J Cancer 1994; 57: 341–347.
163. Jothy S, Munro SB, LeDuy L, McClure D, Blaschuk OW. Adhesion or anti-adhesion in cancer: what matters more? Cancer Metastasis Rev 1995; 14: 363–376.
164. Aruffo A, Stamenkovic I, Melnick M, Underhill CB, Seed B. CD44 is the principal cell surface receptor for hyaluronate. Cell 1990; 61:1303–1313.
165. Miyake K, Underhill CB, Lesley J, Kincaide PW. Hyaluronate can function as a cell adhesion molecule and CD44 participates in hyaluronate recognition. J Exp Med 1990; 172:69–75.
166. Sherman L, Sleeman J, Herrlich P, Ponta H. Hyaluronate receptors: key players in growth, differentiation, migration and tumor progression. Curr Opin Cell Biol 1994; 6:726–733.
167. Stamenkovic I, Amiot M, Pesando JM, Seed B. A lymphocyte molecule implicated in lymph node homing is a member of the cartilage link protein family. Cell 1989; 56:1057–1062.
168. Goldstein LA, Zhou DFH, Picker LJ, et al. A human lymphocyte homing receptor, the Hermes antigen, is related to cartilage proteoglycan core and link proteins. Cell 1989; 56:1063–1072.
169. Tsukita S, Oishi K, Sato N, Sagara J, Kawai A, Tsukita S. ERM family members as molecular linkers between the cell surface glycoprotein CD44 and actin-based cytoskeletons. J Cell Biol 1994; 126:391–401.
170. Thomas L, Byers HR, Vink J, Stamenkovic I. CD44H regulates tumor cell migration on hyaluronate-coated substrate. J Cell Biol 1994; 118:971–977.
171. Murakami S, Shimabukuro Y, Miki Y, et al. Inducible binding of human lymphocytes to hyaluronate via CD44 does not require cytoskeleton association but does require new protein synthesis. J Immunol 1994; 152:467–477.
172. Weigel PH, Frost SJ, LeBoeuf RD, McGary CT. The specific interaction between fibrin(ogen) and hyaluronan: possible consequences in haemostasis, inflammation and wound healing. Ciba Found Symp 1989; 143:248–264.
173. Knudson CB, Knudson W. Similar epithelial-stromal interactions in the regulation of hyaluronate production during limb morphogenesis and tumor invasion. Cancer Lett 1990; 52:113–122.
174. Fraser JR, Kimpton WG, Laurent TC, Cahill RNP, Vakakis N. Uptake and degradation of hyaluronan in lymphatic tissue. Biochem J 1988; 256:153–158.
175. Jalkanen S, Bargatze RF, de los Toyos J, Butcher EC. Lymphocyte recognition of high endothelium: antibodies to distinct epitopes of an 85-95 kD glycoprotein antigen differentially inhibit lymphocyte binding to lymph node, mucosal, or synovial endothelial cells. J Cell Biol 1987; 105:983–990.
176. Webb DSA, Shimizu Y, van Seventer GA, Shaw S, Gerrard TL. LFA-3, CD44 and CD45: physiologic triggers of human monocyte TNF and IL-1 release. Science 1990; 249:1295–1298.
177. Shimizu Y, van Seventer GA, Siraganian R, Wahl L, Shaw S. Dual role of

the CD44 molecule in T cell adhesion and activation. J Immunol 1989; 143: 2457–2463.

178. Gunthert U, Hofmann M, Rudy W, et al. A new variant of glycoprotein CD44 confers metastatic potential to rat carcinoma cells. Cell 1991; 65:13–24.
179. Hofmann M, Rudy W, Zoller M, et al. CD44 splice variants confer metastatic behavior in rats: homologous sequences are expressed in human tumor cell lines. Cancer Res 1991; 51:5292–5297.
180. Birch M, Mitchell S, Hart IR. Isolation and characterization of human melanoma cell variants expressing high and low levels of CD44. Cancer Res 1991; 51:6660–6667.
181. Tanabe KK, Stamenkovic I, Cutler M, Takahashi K. Restoration of CD44H expression in colon carcinomas reduces tumorigenicity. Ann Surg 1995; 222: 493–503.
182. Seiter S, Arch R, Reber S, et al. Prevention of tumor metastasis formation by anti-variant CD44. J Exp Med 1993; 177:443–455.
183. Arch R, Wirth K, Hofmann M, et al. Participation in normal immune responses of a metastasis-inducing splice variant of CD44. Science 1992; 257: 682–685.
184. Koopman G, Heider KH, Horst E, et al. Activated human lymphocytes and aggressive non-Hodgkin s lymphoma express a homologue of the rat metastasis-associated variant of CD44. J Exp Med 1993; 177:897–904.
185. Wielenga VJM, Heider KH, Offerhaus GJA, et al. Expression of CD44 variant proteins in human colorectal cancer is related to tumor progression. Cancer Res 1993; 53:4754–4756.
186. Heider KH, Hofmann M, Hors E, et al. A human homologue of the rat metastasis-associated variant of CD44 is expressed in colorectal carcinoma and adenomatous polyps. J Cell Biol 1993; 120:227–233.
187. Kim H, Yang XL, Rosada C, Hamilton SR, August JT. CD44 expression in colorectal adenomas is an early event occurring prior to K-ras and p53 gene mutation. Arch Biochem Biophys 1994; 310:505–507.
188. Jalkanen S, Joensuu H, Soderstrom KO, Klemi P. Lymphocyte homing and clinical behavior of non-Hodgkin's lymphoma. J Clin Invest 1991; 87:1835–1840.
189. Pals ST, Horst E, Ossekoppele GJ, Fidgor CG, Scheper RJ, Meijer CJLM. Expression of lymphocyte homing receptor as a mechanism of dissemination in non-Hodgkin s lymphoma. Blood 1989; 73:885–888.
190. Joensuu H, Klemi PJ, Toikannen S, Jalkanen S. Glycoprotein CD44 expression and its association with survival in breast cancer. Am J Pathol 1993; 143:867–874.
191. Kaufmann M, Heider KH, Sinn HP, von Minckwitz G, Ponta H, Herrlich P. CD44 variant exon epitopes in primary breast cancer and length of survival. Lancet 1995; 345:615–619.
192. Hardwick C, Hoare K, Owens R, et al. Molecular cloning of a novel hyaluronan rceptor that mediates tumor cell motility. J Cell Biol 1992; 117:1343–1350.
193. Hall CL, Yang B, Yang X, et al. Overexpression of the hyaluronan receptor RHAMM is transforming and is also required for H-ras transformation. Cell 1995; 82:19–28.

194. Usami Y, Fujimari Y, Miura S, et al. A 28 kDa-protein with disintegrin-like structure (jararhagin-C) purified from Bothrops jararaca venom inhibits collagen- and ADP-induced platelet aggregation. Biochem Biophys Res Commun 1994; 201:331–339.
195. Gould RJ, Polokoff MA, Friedman PA, et al. Disintegrins: a family of integrin inhibitory proteins from viper venoms. Proc Soc Exp Biol Med 1990; 195:168–171.
196. Haas T, Plow EF. Integrin-ligand interactions: a year in review. Curr Opin Cell Biol 1994; 6:656–662.
197. Koivunen E, Wang B, Ruoslahti E. Isolation of a highly specific ligand for the integrin from a phage display library. J Cell Biol 1994; 124:373–380.
198. Scarborough RM, Naughton MA, Teng W, et al. Design of potent and specific integrin antagonists. Peptide antagonists with high specificity for glycoprotein IIb-IIIa. J Biol Chem 1993; 268:1066–1073.
199. Smith JW, Hu D, Satterthwait A, Pinz-Sweeney S, Barbas CF 3rd. Building synthetic antibodies as adhesive ligands for integrins. J Biol Chem 1994; 269: 32788–32795.
200. Sage EH, Bornstein P. Extracellular matrix proteins that modulate cell-matrix interactions. J Biol Chem 1991; 266:14831–14834.
201. Wolfsberg TG, Straight PD, Gerena RL, et al. ADAM, a widely distributed and developmentally regulated gene family encoding membrane proteins with a disintegrin and metalloprotease domain. Dev Biol 1995; 169:378–383.
202. Hodivala KJ, Watt FM. Evidence that cadherins play a role in the downregulation of integrin expression that occurs during keratinocyte terminal differentiation. J Cell Biol 1994; 124:589–600.
203. Cepek KL, Shaw SK, Parker CM, et al. Adhesion between epithelial cells and T lymphocytes mediated by E-cadherin and the $\alpha E\beta 7$ integrin. Nature 1994; 372:190–193.

18

Adhesion Molecules in Reperfusion Injury

Carol J. Cornejo
Department of Surgery, University of Washington, Seattle, Washington

John M. Harlan
Department of Medicine, Division of Hematology, University of Washington, Seattle, Washington

Robert K. Winn
Department of Surgery, Department of Physiology and Biophysics, University of Washington, Seattle, Washington

I. INTRODUCTION

Ischemia-reperfusion (I-R) injury is involved in a wide spectrum of disease processes such as stroke, myocardial infarction, mesenteric and peripheral vascular disease, organ transplantation, and circulatory shock. Therapy directed at reducing I-R injury could have a significant impact on the outcome of a large number of events in many patients. Traumatic injuries claim 160,000 lives per year in the United States. Half of the deaths occur at the scene, one-fourth occur within 24 hours of injury, and one-fourth occur later in the hospital course. These late deaths occur between a few days and a few weeks after injury as a result of multiple organ failure syndrome. These late deaths have been postulated to occur as a result of hemorrhagic shock that results in a systemic or whole-body I-R injury. Thus, this group of patients may respond to therapy directed to reduce I-R injury. This amount to approximately 25,000 to 40,000 lives per year that may benefit from this therapy in the trauma setting alone. Coronary artery disease can have an I-R component and is the leading cause of death in the United States, claiming almost 500,000 lives annually. The potential I-R portion of injury may be more evident with the onset of thrombolytic therapy. Treatment to reduce I-R injury following myocardial ischemia could potentially reduce the morbidity and mortality for a large number of patients.

Ischemia-reperfusion injury is thought to begin with an ischemic insult causing both the initial damage and starting a chain of events resulting in a reperfusion injury that is the result of an inflammatory response. The injury that occurs during reperfusion is mediated in part by neutrophils (PMN); PMN depletion reduces I-R injury in the heart (1,2), liver (3,4), and gut (5). In addition, generalized I-R injury that occurs following hemorrhagic shock is reduced by PMN depletion (6). For PMNs to induce injury, they must first adhere to the vascular endothelium. This leads to the formation of a microenvironment between the PMN and endothelial cell, where inflammatory mediators produced by the PMN can reach high concentrations and overcome local anti-inflammatory mechanisms allowing vascular injury to occur. Once adherent, PMNs may then emigrate to extravascular tissue, where they can mediate further organ injury.

The adherence of PMNs to vascular endothelial cells depends on the interaction of adhesion molecules expressed on the surfaces of both cell types. The adhesion molecules are divided into three families: the selectins, the integrins, and the immunoglobulin superfamily. PMNs express the β_2-integrin family and the leukocyte selectin, L-selectin. The β_2-integrins are heterodimers consisting of one common β subunit designated CD18 and one of three α subunits designated CD11a, CD11b, and CD11c. There are three adhesion molecules on the endothelial cell that bind to PMNs during I-R: ICAM-1, a member of the immunoglobulin superfamily; and two endothelial selectins, P- and E-selectin.

The PMN-endothelial cell adhesive interaction begins with the PMNs initially rolling along the surface of the endothelial cells followed by firm adherence. The initial interaction is mediated by the selectin family of adhesion molecules (7–9) and firm adherence is mediated by the β_2-integrins interacting with ICAM-1 (CD54) (7,10). Following firm adherence, the PMNs diapedese between the endothelial cells and gain access to the extravascular space.

Monoclonal antibodies (MAbs) have been developed that bind to functional epitopes on P- (CD62P), L- (CD62L), and E- (CD62E) selectin, CD18, CD11b, and ICAM-1. These MAbs have been shown to block PMN function in response to a variety of inflammatory stimuli in vitro. In addition, administration of these MAbs has been shown to reduce I-R in vivo in a variety of animal models. This chapter will review the current understanding of the effects of anti-adhesion therapy during I-R.

II. SELECTINS

All of the molecules in the selectin family have a lectin domain, a variable number of complement regulatory repeat sequences, an epidermal growth factor domain, a transmembrane domain, and a cytoplasmic domain (11). These molecules are thought to function by the binding of their lectin domain to specific carbohydrate moieties found on their counterstructure. The sialyl Lewis^x(SLe^x, CD15s) antigen and other sialylated, fucosylated glycoproteins and glycolipids are recognized by the selectin family (12,13).

Monoclonal antibodies to functional epitopes on the lectin domain prevent the initial rolling of PMNs along the endothelium in vitro (14). Likewise, administration of the carbohydrate Sle^x reduces rolling through competitive binding of the functional epitope (15,16). Prevention of the initial rolling of PMNs is postulated to reduce PMN-mediated injury by blocking the initial interaction of PMNs and endothelial cells, thus preventing PMN adherence.

A. P-selectin

P-selectin is located in the Weibel-Palade bodies of endothelial cells and the alpha granules of platelets (17). P-selectin is rapidly redistributed to the cell surface of endothelial cells in response to stimulation by histamine, thrombin, or phorbol esters (18). Once on the endothelial cell surface, it can interact with carbohydrate moieties expressed on glycoproteins or glycolipids on the surface of leukocytes causing rolling of the leukocytes. The leukocyte glycoprotein PSGL-1 (P-selectin glycoprotein 1) has recently been shown to be a high affinity ligand for P-selectin (19–21). The duration of expression on the cell surface following these stimuli is 15 min or less (18). However, Patel et al. (22) found that endothelial cell expression of P-selectin in response to H_2O_2, t-butylhydroperoxide, and menadione was prolonged up to 4 hours. Clearly, P-selectin expression could occur during the I-R injury due to the production of free radicals, thrombin, or histamine and blocking P-selectin might be protective in I-R injury.

Monoclonal antibodies that block functional epitopes on P-selectin have been found to reduce I-R injury in the heart (23), small intestine (24), skeletal muscle (25), and rabbit ear (26). I-R in the ears of rabbits was induced by partially amputating the ear at the base leaving only the central artery and vein intact. The central artery was occluded with a microvascular clamp producing complete ischemia. The clamp was removed after 6 hours of ischemia, and the ears allowed to reperfuse. Treatment was given immediately prior to reperfusion with either saline, a nonblocking P-selectin MAb, or a blocking P-selectin MAb. A quantitative determination of the I-R injury was determined by measuring the ear volume (edema) by water displacement daily for 7 days. There was a significant decrease in I-R injury (ear volume) in the animals treated with the blocking P-selectin MAb compared to animals treated with saline or the nonblocking P-selectin MAb (see

Fig. 1). Also, there was a significant decrease in estimated tissue necrosis at 7 days following ischemia in the blocking P-selectin MAb-treated rabbits as compared to the saline or nonblocking MAb-treated rabbits. As an indicator of PMN accumulation, myeloperoxidase (MPO) content in homogenates of the whole ear at 24 hours following ischemia was determined by colorimetric assay. The MPO content in the blocking P-selectin MAb-treated rabbits was significantly less than in the control rabbits. P-selectin expression was determined by immunohistochemistry and was found in ears subjected to ischemia but not in the nonischemic normal ear. Thus, P-selectin is expressed in I-R, and blocking P-selectin with MAb reduces PMN accumulation and I-R injury.

Winn et al. (27) also found that blockade of P-selectin with MAb reduced global I-R injury that occurs with hemorrhagic shock. Hypovolemic shock was induced in rabbits by withdrawing blood until the cardiac output was reduced to 33% of baseline for 90 min. Treatment was given at the time of resuscitation with either saline, a blocking P-selectin MAb, or a nonblocking P-selectin MAb. Resuscitation was initiated by infusion of

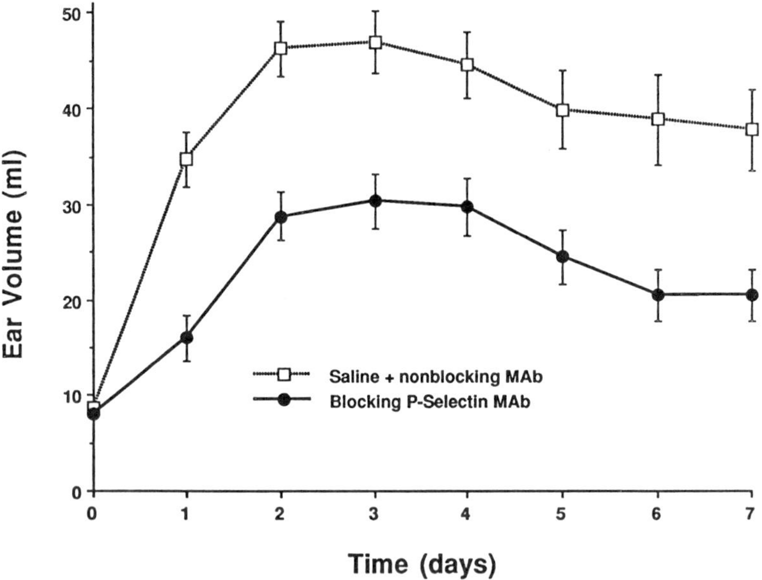

Figure 1 I-R injury of the rabbit ear. I-R injury was quantified by measuring ear volume (edema) by water displacement daily for 7 days. Rabbits were treated with either saline, nonblocking P-selectin MAb, or blocking P-selectin MAb at the time of reperfusion. Adapted from Winn et al. (26), with permission.

shed blood and then continued with Lactated Ringer's as necessary to maintain the cardiac output at >90% of baseline for 6 hours following shock. Fluid requirements in the blocking P-selectin MAb-treated groups were significantly less than that in the control group (12 ± 6 ml/kg vs. 54 ± 14 ml/kg).

B. L-selectin

L-selectin in constitutively expressed on the cell surface of most leukocytes, including PMNs, lymphocytes, and monocytes. L-selectin was first extensively studied as a homing receptor for lymphocytes binding to the high endothelial venules of lymph nodes (28). Subsequently, it was determined that L-selectin also mediated PMN recruitment to sites of inflammation and that blocking L-selectin with MAb inhibited PMN accumulation at sites of inflammation (29). It is postulated that blocking L-selectin adhesion inhibits rolling of PMNs on endothelial cells and might reduce the PMN-mediated injury that occurs during I-R. A number of studies have found this to be true in several tissues. MAb blockade of L-selectin was found to reduce local I-R injury in the heart (30), skeletal muscle (31), and rabbit ear (32). L-selectin MAb has also been shown to reduce remote injury in the lung following hind limb ischemia in the rat (31).

Milhecic et al. (32) induced I-R in rabbit ears and administered a MAb blocking L-selectin. They induced I-R in the ears of rabbits using the same protocol as described above for Winn et al. (26). After 6 hours of ischemia, rabbits were treated with saline, a nonblocking L-selectin MAb, or a blocking L-selectin MAb. The ears were then allowed to reperfuse. I-R injury determined by ear edema (volume) was found to be significantly less in the blocking L-selectin MAb-treated group (peak 25 ± 4 ml) compared to the saline-treated group (peak 40 ± 8 ml) and the nonblocking L-selectin group (peak 41 ± 6 ml). The estimated tissue necrosis of the rabbits treated with blocking L-selectin MAb was also significantly less than the rabbits treated with saline or nonblocking MAb. This decrease in I-R injury correlated with a reduction in MPO content. The MPO content in ears of the blocking L-selectin-treated group was significantly less than the MPO content in ears of the control groups. They concluded that blocking L-selectin adherence decreases I-R injury and PMN accumulation.

C. E-selectin

E-selection is expressed on the surface of endothelial cells following stimulation by several cytokines, bacterial endotoxin, and substance P (33–35). Stimulation by these substances induces transcription of the E-selectin gene which results in peak surface expression of E-selectin in 4–6 hours which decreases to basal levels by 24 hours (34,35). Blocking E-selectin MAbs have been found to reduce PMN accumulation in inflamed peritoneum and lung in the rat (36), presumably by inhibiting rolling of PMNs on the activated endothelial cells. There has been very little work done on the role of E-selectin during I-R due to the lack of blocking E-selectin MAb for most animal species.

Seekamp et al. (31) reported one of the few studies that investigated the role of E-selectin in I-R injury. In this study, I-R was induced in the hind limbs of rats that were treated with blocking MAbs to either P-, or L-, or E-selectin. The protocol that was used resulted in injury to muscle and lung and protection with MAb was examined in both. A tourniquet was placed on the hind limbs of rats at a pressure sufficient to block arterial blood flow. The tourniquet was left in place for 4 hours; it was then removed and the limb allowed to reperfuse for an additional 4 hours. The rats were given treatment with either a blocking MAb to E-selectin, a blocking MAb to P-selectin, or a blocking MAb to L-selectin at the onset of reperfusion. They measured vascular permeability (leakage of ^{125}I-labeled albumin), hemorrhage (extravasation of ^{51}Cr-labeled erythrocytes), and PMN accumulation (indicated by MPO content) in lungs and hindlimb muscle. They found that vascular permeability, hemorrhage, and MPO content in lungs were significantly less in the E-selectin MAb and L-selectin MAb-treated groups than in the nontreated groups. They did not find a decrease in these injury parameters in the P-selectin MAb-treated groups. However, when the hind limb ischemia was reduced to 1 hour and reperfusion to 1.5 hours, they did find a decrease in vascular permeability, hemorrhage, and MPO content in the lungs of the P-selectin MAb-treated group. These same parameters were used to examine injury to the hind limb muscle. A decrease in vascular permeability, hemorrhage, and MPO content was found in only the L-selectin MAb-treated group. The E-selectin MAb and P-selectin MAb-treated groups were not different from controls even if ischemia and reperfusion times were shortened. They concluded that both local (hindlimb muscle) and remote (lung) tissue injury following I-R is L-selectin-dependent, but only remote (lung) tissue injury following I-R is E-selectin and to a lesser extent P-selectin-dependent. The lack of protection in reducing local I-R injury by P-selectin MAb is in contrast to the previously sited studies where local tissue injury following I-R was found to be P-selectin-dependent in the heart (23), intestine (24), skeletal muscle (25), and rabbit ear (26).

Altavilla et al. (37) showed that E-selectin MAb reduced local I-R injury of the myocardium in rats. In this study, rats were subjected to temporary left main coronary artery ligation for 1 hour followed by 1 hour of reperfusion. The rats were treated with E-selectin MAb 3 hours before the onset of ischemia. Survival was higher in the E-selectin-treated group (80%) compared to the control group (50%) and infarct size (expressed as a percentage of area at risk) was smaller in the treated animals (26%) than in the control animals (70%). The reduced size of infarction was accompanied by a decrease in myocardial tissue MPO in the E-selectin MAb-treated rats.

D. Sle^x

The ligands for the selections are thought to be sialylated, fucosylated carbohydrate moieties contained on glycoproteins or glycolipids on the cell surface of target cells. In vitro studies have shown that the selectins recognize a common carbohydrate structure, the sialyl $Lewis^x$ oligosaccharide (Sle^x) (38) and that administration of Sle^x inhibits selectin-mediated PMN adherence to endothelial cells (15,16). Therefore, since P- and L-selectin MAbs are effective in preventing I-R injury, administration of Sle^x should be equally effective in reducing I-R injury. Thus, it is not surprising that Sle^x reduced local I-R injury in the heart (39), skeletal muscle (31), and rabbit ear (83). Seekamp et al. (31) (see above) induced 4 hours of ischemia in rat hindlimbs followed by reperfusion and measured vascular permeability, hemorrhage, and MPO content in both the lung and the hindlimb muscle. Remote and local tissue injury was examined following reperfusion and treatment with Sle^x or sialyl-N-acetyllactosamine pentasaccharide (SleN), a nonfucosylated nonblocking carbohydrate at 1, 2, and 3 hours after reperfusion. Treatment with Sle^x significantly reduced vascular permeability, hemorrhage, and MPO content in both lungs and in hindlimb muscle when compared to treatment with SleN or PBS. They concluded that treatment with Sle^x was protective in reducing both remote (lung) and local (hindlimb muscle) injury following I-R.

III. β_2-INTEGRINS

β_2-integrins are a family of molecules found on the cell surface of leukocytes that are responsible for their firm adherence to endothelial cells. There are three members of this family, and each member consists of an α-chain noncovalently linked to a β-chain that is identical for all β_2-integrins. These heterodimers are designated CD11a/CD18, CD11b/CD18, and CD11c/CD18. CD11a/CD18 is expressed on all leukocytes, whereas CD11b/CD18 and CD11c/CD18 are expressed on PMNs, monocytes, and natural killer cells but not lymphocytes (40). An increase in cell surface expression of CD11b/CD18 and CD11c/CD18 can be rapidly induced from intracellular stores by inflammatory mediators such as chemotactic compounds or cytokines (41). However, cell surface expression does not in itself initiate adhesion of leukocytes to endothelial cells. It is necessary for the β_2-integrin molecule to change confirmation or to otherwise become activated before adhesion occurs. Activation of β_2-integrins is induced by soluble agents (cytokines, chemotactic factors, coagulation factors) and by binding of cell surface receptors. The β_2-integrins bind to their counter-structures on endothelial cells. CD11a/CD18 and CD11b/CD18 both use ICAM-1 as their counter-structure to mediate firm adherence of leukocytes to endothelial cells; CD11b/CD18 also binds significantly to other, as yet unidentified, structures on endothelium. The most extensively studied of these molecules on PMNs is CD11b/CD18 and it was found to have a significant role in PMN mediated I-R injury. Monoclonal antibodies that recognize a functional epitope of either subunit of this molecule can reduce PMN accumulation and injury following I-R.

A. CD18

Monoclonal antibodies recognizing functional epitopes on CD18 have been found to reduce local I-R injury in the brain (42–44), heart (45–48), intestine (49), kidney (50), skeletal muscle (51–53), lung (54,55), and rabbit ear (56,57), as well as remote injury in the lung following I-R of the hind limb (58). In addition, MAb blockade of CD18 reduced global I-R injury that occurs with hemorrhagic shock (59–61).

Vedder et al. found that blockade of CD18 with MAb either prior to (60) or following (61) hypovolemic shock reduced organ injury and increased survival in rabbits. Mileski et al. (59) found similar results in primates when the CD18 MAb was administered following shock. In the latter experiments, hypovolemic shock was induced in rhesus monkeys by withdrawing blood until the cardiac output was reduced to 30% of baseline for 90 min. Treatment was given at the end of the 90-min shock period and consisted of either saline or a CD18 MAb. Resuscitation was begun by reinfusing the shed blood. The animals were resuscitated with a maintenance infusion of 4 ml/kg/hr lactated Ringer's solution and supplemented with additional lactated Ringer's solution as necessary to maintain their cardiac output at ≥90% of baseline for 24 hours following shock. Fluid requirement in the CD18 MAb-treated group was significantly less than that in the control group (see Fig. 2). In fact, fluid requirements in the MAb-treated group were essentially equal to the maintenance infusion whereas the control group required 260 ml/kg, which is equivalent to 17 L in a 70-kg man. In addition, all control animals had hemorrhagic gastritis on endoscopy, whereas none of the CD18 MAb-treated animals exhibited gastritis.

The breakdown of the mucosal barrier and bacterial translocation have been implicated in sustaining the inflammation associated with multiple

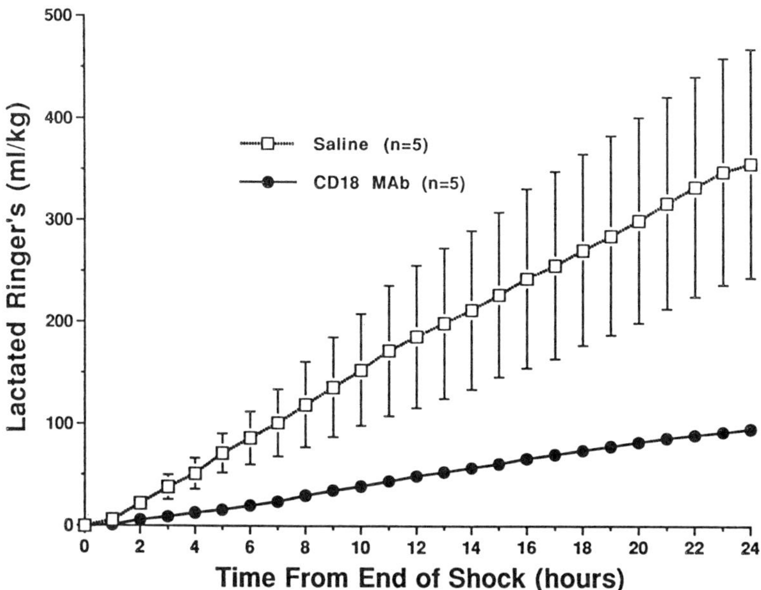

Figure 2 Fluid requirements following hypovolemic shock in monkeys. Treatment was with either saline or CD18 MAb at time of resuscitation. Animals were given maintenance fluid of 4 ml/kg/hr plus additional lactated Ringer's as necessary to keep cardiac output at ≥90% of baseline. Adapted from Mileski et al. (59), with permission.

organ failure syndrome. Preventing this cascade of events could greatly decrease the morbidity and mortality following hemorrhagic shock.

Blockage of CD18 with MAbs was equally protective in reducing I-R injury in the rabbit ear whether administered prior to ischemia or just prior to reperfusion. Thus, the injury that is prevented by CD18 blockade occurs during reperfusion (57). Sharar et al. (56) determined the effect of delayed administration of CD18 MAb following I-R of the rabbit ear. Ischemia was induced in the rabbit ears, and treatment was with either saline or CD18 MAb at reperfusion or at 1 hour, 4 hours, or 12 hours following reperfusion. The I-R injury (ear volume) was significantly less in the animals treated with CD18 MAb at reperfusion or at 1 hour or 4 hours following reperfusion as compared to the saline-treated animals (see Fig. 3). However, if treatment was delayed for 12 hours, the protective effect was completely lost. Tissue necrosis in those animals having reduced edema was also significantly less than the saline treated or 12 hours delayed treatment animals. In summary, CD18 blockade was protective when administered

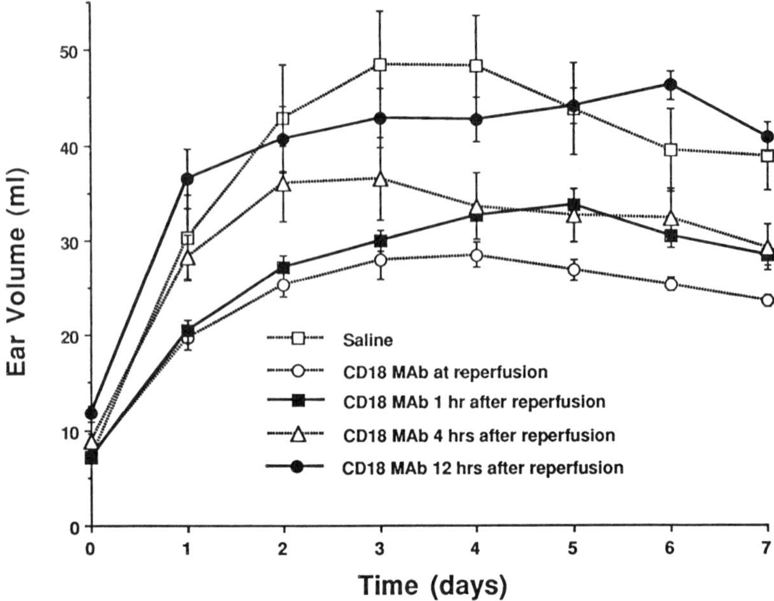

Figure 3 I-R injury in the rabbit ear with delayed CD18 MAb treatment. I-R injury was quantified by measuring ear volume (edema) daily for 7 days. Rabbits were treated with either saline or CD18 MAb at reperfusion or 1 hour, 4 hours, or 12 hours following reperfusion. Adapted from Sharar et al. (56), with permission.

within 4 hours of reperfusion but delaying treatment to 12 hours was not. These findings have important clinical implications since it may not be feasible to administer anti-adhesion therapy immediately upon reperfusion.

B. CD11b

Monoclonal antibody blockade of the CD11b subunit of β_2-integrins reduces local I-R injury in the heart (47,62,63) and liver (64), and remote injury in the lung and liver following gut I-R (58). In an early study by Simpson et al. (63), regional myocardial ischemia was induced in dogs by temporarily occluding the left circumflex artery for 90 min. Treatment with either a CD11b MAb, a nonspecific control MAb, or vehicle for the MAb was administered after 45 min of ischemia. They found a significant reduction in infarct size (expressed as a percentage of the area at risk for infarction) following 6 hours of reperfusion in the CD11b MAb-treated group (26%) as compared to the control group (48%). In a later study, Simpson et al. (62) found that the reduction in infarct size in the CD11b MAb-treated group was sustained for 72 hours when the MAb was administered at 45 min of ischemia and then again at 12, 24, 36, and 48 hours after reperfusion.

C. CD11a

Investigations of I-R injury following treatment with MAbs directed to CD11a are limited. Monoclonal antibody blockade of CD11a has been found to reduce I-R injury in the brain (42) and heart (47). Matsuo et al. (42) found that in rats treatment with CD11a MAb prior to transient middle cerebral artery occlusion and again immediately after reperfusion reduced cerebral edema formation and infarct size compared to control rats. They also found that PMN accumulation (as measured by histology and MPO content) was reduced in the CD11a MAb-treated rats. Yamazaki et al. (47) found protection of the myocardium following I-R of the heart in rats where treatment with CD11a MAb occurred prior to transient left coronary artery occlusion. Infarct size (as expressed as a percentage of left ventricle) was reduced in the treated rats as compared to control rats (8% versus 34%).

IV. ICAM-1

ICAM-1 is a member of the immunoglobulin superfamily and is constitutively expressed in low levels on endothelial cells. After stimulation of endothelial cells with cytokines (interleukin-1, tumor necrosis factor, interferon-γ), the expression of ICAM-1 is increased (66), reaching a peak expression at 12 to 24 hours with expression partially maintained for at least 72 hours (reviewed in Carlos et al., 67). ICAM-1 binds both CD11a/CD18 and CD11b/CD18 on leukocytes and this interaction mediates the firm adherence of leukocytes to endothelial cells. Therefore, blocking the functional epitope of ICAM-1 with MAb should reduce PMN adherence and thus I-R injury. In fact, MAb blockade of ICAM-1 has been found to reduce local I-R injury in the brain (42,68), heart (45,47,69), kidney (70), liver (71), skeletal muscle (25), and lung (72) and to reduce remote injury in the lung following hindlimb I-R (58).

Kelly et al. (70), examined the effects of MAb blockade of ICAM-1 on renal function following temporary occlusion of the renal artery in rats for 30 or 40 min. Treatment with an MAb to ICAM-1 was at the time of ischemia, or 0.5 hour, 2 hours, or 8 hours after ischemia. Treatment with ICAM-1 MAb at the time of ischemia was protective against renal dysfunction with a mean serum creatinine at 24 hours of 0.61 in the treated group versus 2.4 in the control group. Treatment with ICAM-1 MAb was also protective against renal failure when administered 30 or 120 min after reperfusion, but was not protective when administered 480 min after reperfusion. These findings again illustrate that antiadhesion therapy is protective even if administration is delayed for a short time following the onset of reperfusion but that longer delays in treatment are not protective.

V. SAFETY CONCERNS

As outlined above, there is strong evidence that antiadhesion therapy reduces I-R injury in a variety of animal models. However, antiadhesion therapy results in the blockade of at least some of the normal host defense systems, raising the concern that even temporary inhibition of these systems may result in a higher incidence of infectious complications. This concern arises in part from clinical observations of patients with two different genetic deficiencies in adhesion molecules, designated leukocyte adhesion deficiency type I (LAD I) or type II (LAD II) (reviewed by Etzioni, 73). The LAD I patients have a reduced or absent expression of β_2-integrins and therefore reduced firm adherence of leukocytes to endothelial cells. This syndrome is characterized by a profound defect in phagocyte emigration and these patients do not accumulate PMNs at sites of inflammation. Their clinical course is characterized by frequent bacterial infections that are life-threatening. The LAD II patients have a defect in their fucose metabolism resulting in a congenital absence of Sle^x, the receptor ligand for P- and E-selectin. Therefore, these patients have a defect in the initial rolling of leukocytes along the endothelium. Their clinical course is also characterized by recurrent bacterial infections, although their course is not as severe as the LAD I patients.

The adverse clinical courses in LAD I patients have led to several studies to try to determine the extent of increased risk for infectious complications following antiadhesion therapy. Rosen et al. (74) found that pretreating mice with a CD11b MAb significantly increased their mortality following inoculation with *Listeria monocytogenes* compared to a control group. In contrast, several other investigators have found no significant increase in infectious complications following antiadhesion therapy.

Mileski et al. (75) devascularized the appendix and treated with CD18 MAb. The appendix was removed 18 hours later and a second treatment with CD18 MAb given. There was no increase in mortality or infectious complications at 10 days in rabbits treated with CD18 MAb compared with control rabbits. However, there was a significant decrease in PMN emigration into the peritoneum at 18 hours in the CD18 MAb-treated rabbits. Using the same model of appendiceal devascularization, Thomas et al. (76) administered CD18 MAb at the time of appendectomy and again every

12 hours for 3 days. They found a significant increase in survival at 10 days in the CD18 MAb-treated rabbits (90%) as compared to the control rabbits (40%). Mileski et al. (77) studied the effect of CD18 MAb on PMN emigration into the peritoneum 4 hours following *Escherichia coli*-induced peritonitis in rabbits. They found a significant reduction in PMN emigration in the CD18 MAb-treated group. In contrast Sharar et al. (78) found no decrease in the PMN emigration into the peritoneum of rabbits treated with P-selection MAb using the same model of *E. coli*-induced peritonitis.

Sharar et al. (79) have also shown that treatment with CD18 MAb does not increase the incidence of abscess formation or size of abscesses in rabbits resulting from a clinically relevant skin inoculation of *Staphylococcus aureus* ($<10^8$ colony forming units; CFU). However, a bacterial inoculation of $>10^8$ CFU resulted in a higher incidence of abscess formation and a larger size of abscesses in the CD18 MAb-treated rabbits (see Fig. 4). There was no increase in the incidence of abscess formation or size of abscesses in rabbits treated with P-selectin MAb at the time of *Staph. aureus* skin inoculation even at $>10^8$ CFU (78). In another investigation of infectious risk following CD18 MAb treatment, Garcia et al. (80) found a significant

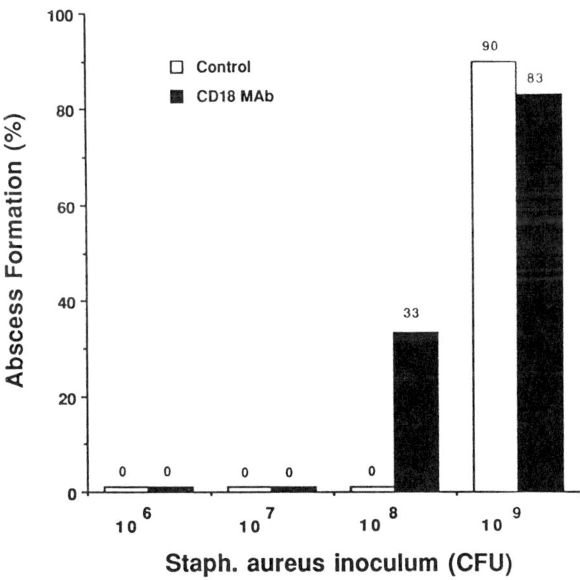

Figure 4 Percent of abscess formation following subcutaneous injection with *Staphylococcus aureus* bacteria. Rabbits were treated at the time of injection with either saline or CD18 MAb. Adapted from Sharar et al. (79), with permission.

increase in the incidence and severity of subcutaneous abscess formation in rabbits inoculated with *Pseudomonas aeruginosa*. However, there was no increase in incidence or severity of abscess in rabbits pretreated with L-selectin MAb. Mileski et al. (81) showed that treatment with ICAM-1 MAb did not increase the severity of abscess formation in rabbits given subcutaneous inoculations of either *Staph. aureus* or *P. aeruginosa*.

Tuomanen et al. (82) actually found that treatment with CD18 MAb reduced mortality in a rabbit model of bacterial meningitis by intracisternal injection of *Streptococcus pneumoniae*. They found a decrease in cerebral edema and CSF protein content due to protection of the blood-brain barrier in CD18 MAb-treated animals. The course of bacterial growth in the blood was delayed in the CD18 MAb-treated rabbits, but the course of bacterial growth in the cerebral spinal fluid was no different between the CD18 MAb-treated group and the control group. The inflammatory response to bacterial killing by ampicillin was decreased in the CD18 MAb-treated group.

As outlined above, it is still unclear to what extent anti-adhesion molecule therapy would increase the risk for infectious complications. It appears that treatment with CD18 MAb in intra-abdominal infections does not increase infectious complications or mortality. There have been no studies published to date on the effect of blockade of L- or E-selectin, ICAM 1, or CD11b on intra-abdominal infections. In *Staph. aureus* subcutaneous infections, it appears that neither CD18 nor P-selectin blockade increases the severity of infection at clinically relevant inoculations. However, CD18 MAb but not ICAM-1 or L-selectin MAb increased the severity of infection following subcutaneous injection of *P. aeruginosa*. Therefore, the potential benefits of anti-adhesion therapy are likely to outweigh the potential infectious complications.

VI. CONCLUSION

Anti-adhesion therapy has been found to significantly reduce I-R injury in a variety of animal models. Treatments that block either the initial rolling of PMNS on endothelial cells (selectin MAbs or Sle^x) or the firm adherence of PMNS to endothelial cells (β_2-integrin or ICAM-1 MAbs) have been found to decrease I-R injury and PMN accumulation. In addition, these MAbs have been found to be protective even if their administration is delayed until after perfusion is restored. This ability to delay administration has important clinical implications since it may not always be feasible to administer anti-adhesion therapy immediately upon reperfusion. There are numerous disease processes in humans that could be impacted by the use of anti-adhesion therapy to reduce I-R injury, and the number of patients who could potentially benefit from this therapy is quite large. The potential infectious complications will have to be considered before administrating anti-adhesion therapy to a large number of patients, but hopefully in future controlled clinical trials the potential benefits of antiadhesion therapy will be found to outweigh any possible complications.

REFERENCES

1. Engler RL, Schmid-Schonbein GW, Pavelec RS. Leukocyte capillary plugging in myocardial ischemia and reperfusion in the dog. Am J Pathol 1983; 111:98–111.
2. Romson JL, Hook BG, Kunkel SL, Abrams GD, Schork A, Lucchesi BR. Reduction of the extent of ischemic myocardial injury by neutrophil depletion in the dog. Circulation 1983; 67:1016–1023.
3. Langdale LA, Flaherty LC, Liggitt HD, Harlan JM, Rice CL, Winn RK. Neutrophils contribute to hepatic ischemia-reperfusion injury by a CD18-independent mechanism. J Leukoc Biol 1993; 53:511–517.
4. Jaeschke H, Farhood A, Smith CW. Neutrophils contribute to ischemia/reperfusion injury in rat liver in vivo. FASEB J 1990; 4:3355–3359.
5. Smith SM, Holm-Rutili L, Perry MA, et al. Role of neutrophils in hemorrhagic shock-induced gastric mucosal injury in the rat. Gastroenterology 1987; 93:466–471.
6. Barroso-Aranda J, Schmid-Schonbein GW, Zweifach BW, Engler RL. Granulocytes and no-reflow phenomenon in irreversible hemorrhagic shock. Circ Res 1988; 63:437–447.
7. Lawrence MB, Springer TA. Leukocytes roll on a selectin at physiologic flow rates: distinction from and prerequisite for adhesion through integrins. Cell 1991; 65:859–873.
8. Lawrence MB, Springer TA. Neutrophils roll on E-selectin. J Immunol 1993; 151:6338–6346.
9. Ley K, Gaehetgens P, Fennie C, Singer MS, Lasky LA, Rosen SD. Lectin-like cell adhesion molecule 1 mediates leukocyte rolling in mesenteric venules in vivo. Blood 1991; 77:2553–2555.
10. Von-Andrian UH, Chambers JD, McEvoy LM, Bargatze RF, Arfors KE, Butcher EC. Two-step model of leukocyte-endothelial cell interaction in inflammation: distinct roles for LECAM-1 and the leukocyte beta 2 integrins in vivo. Proc Natl Acad Sci USA 1991; 88:7538–7542.
11. Bevilacqua MP, Nelson RM. Selectins. J Clin Invest 1993; 91:379–387.
12. Phillips ML, Nudelman E, Gaeta FC, et al. ELAM-1 mediates cell adhesion by recognition of a carbohydrate ligand, sialyl-Le^x. Science 1990; 250:1130–1132.
13. Waltz G, Aruffo A, Kolanus W, Bevilacqua M, Seed B. Recognition by ELAM-1 of the sialyl-Le^x determinant on myeloid and tumor cells. Science 1990; 250:1132–1135.
14. Abbassi O, Kishimoto TK, McIntire LV, Anderson DC, Smith CW. E-selectin

supports neutrophil rolling in vitro under conditions of flow. J Clin Invest 1993; 92:2719–2730.
15. Asako H, Kurose I, Wolf R, et al. Role of H1 receptors and P-selectin in histamine-induced leukocyte rolling and adhesion in postcapillary venules. J Clin Invest 1994; 93:1508–1515.
16. Zimmerman BJ, Paulson JC, Arrhenius TS, Gaeta FC, Granger DN. Thrombin receptor peptide-mediated leukocyte rolling in rat mesenteric venules: roles of P-selectin and sialyl Lewis X. Am J Physiol 1994; 267:H1049–H1053.
17. McEver RP, Beckstead JH, Moore KL, Marshall CL, Bainton DF. GMP-140, a platelet alpha-granule membrane protein, is also synthesized by vascular endothelial cells and is localized in Weibel-Palade bodies. J Clin Invest 1989; 84:92–99.
18. Hattori R, Hamilton KK, Fugate RD, McEver RP, Sims PJ. Stimulated secretion of endothelial Von Willebrand factor is accompanied by rapid redistribution to the cell surface of the intracellular granule membrane protein GMP-140. J Biol Chem 1989; 264:7768–7771.
19. Moore KL, Stults NL, Diaz S, et al. Identification of a specific glycoprotein ligand for P-selectin (CD62) on myeloid cells. J Cell Biol 1992; 118:445–456.
20. Moore KL, Eaton SF, Lyons DE, Lichenstein HS, Cummings RD, McEver RP. The P-selectin glycoprotein ligand from human neutrophils displays sialylated, fucosylated, O-linked poly-N-acetyllactosamine. J Biol Chem 1994; 269: 23318–23327.
21. Norgard KE, Moore KL, Diaz S, et al. Characterization of a specific ligand for P-selectin on myeloid cells . A minor glycoprotein with sialylated O-linked oligosaccharides. J Biol Chem 1993; 268:12764–12774. (Erratum appears in J Biol Chem 1993; 268(30):22953.)
22. Patel KD, Zimmerman GA, Prescott SM, McEver RP, McIntyre TM. Oxygen radicals induce human endothelial cells to express GMP-140 and bind neutrophils. J Cell Biol 1991; 112:749–759.
23. Weyrich AS, Ma X-L, Albertine KH, Lefer AM. In vivo neutralization of P-selectin protects feline heart and endothelium in myocardial ischemia and reperfusion injury. J Clin Invest 1993; 91:2620–2629.
24. Davenpeck KL, Gauthier TW, Albertine KH, Lefer AM. Role of P-selectin in microvascular leukocyte-endothelial interaction in splanchnic ischemia-reperfusion. Am J Physiol 1994; 267:H622–H630.
25. Jerome SN, Dor'e M, Paulson JC, Smith CW, Korthuis RJ. P-selectin and ICAM-1-dependent adherence reactions: role in the genesis of postischemic no-reflow. Am J Physiol 1994; 266:H1316–H1321.
26. Winn RK, Liggitt D, Vedder NB, Paulson JC, Harlan JM. Anti-P-selectin monoclonal antibody attenuates reperfusion injury to the rabbit ear. J Clin Invest 1993; 92:2042–2047.
27. Winn RK, Paulson JC, Harlan JM. A monoclonal antibody to P-selectin ameliorates injury associated with hemorrhagic shock in rabbits. Am J Physiol 1994; 267:H2391–H2397.
28. Gallatin WM, Weissman IL, Butcher EC. A cell-surface molecule involved in organ-specific homing of lymphocytes. Nature 1983; 304:30–34.

29. Lewinsohn D, Bargatze R, Butcher E. Leukocyte-endothelial cell recognition: evidence of a common molecular mechanism shared by neutrophils, lymphocytes, and other leukocytes. J Immunol 1987; 138:4313–4321.
30. Ma XL, Weyrich AS, Lefer DJ, et al. Monoclonal antibody to L-selectin attenuates neutrophil accumulation and protects ischemic reperfused cat myocardium. Circulation 1993; 88:649–658.
31. Seekamp A, Till GO, Mulligan MS, et al. Role of selectins in local and remote tissue injury following ischemia and reperfusion. Am J Pathol 1994; 144:592–598.
32. Mihelcic D, Schleiffenbaum B, Tedder TF, Sharar SR, Harlan JM, Winn RK. Inhibition of leukocyte L-selectin function with a monoclonal antibody attenuates reperfusion injury to the rabbit ear. Blood 1994; 84:2322–2328.
33. Leuwenberg JFM, Von Asmuth EJU, Jeunhomme TMAA, Buurman WA. IFN-g regulates the expression of the adhesion molecule. ELAM-1 and IL-6 production by human endothelial cells in vitro. J Immunol 1990; 145:2110–2114.
34. Bevilacqua M, Pober J, Mendrick D, Cotran R, Gimbrone M. Identification of an inducible endothelial-leukocyte adhesion molecule. Proc Natl Acad Sci USA 1987; 84:9238–9242.
35. Bevilacqua M, Stenglin S, Gimbrone M Jr, Seed B. An inducible receptor for neutrophils related to complement regulatory proteins and lectins. Science 1989; 243:1160–1165.
36. Mulligan MS, Varani J, Dame MK, et al. Role of endothelial-leukocyte adhesion molecule 1 (ELAM-1) in neutrophil-mediated lung injury in rats. J Clin Invest 1991; 88:1396–1406.
37. Altavilla D, Squadrito F, Ioculano M, et al. E-selectin in the pathogenesis of experimental myocardial ischemia-reperfusion injury. Eur J Pharmacol 1994; 270:45–51.
38. Foxall C, Watson SR, Dowbenko D, et al. The three members of the selectin receptor family recognize a common carbohydrate epitope, the sialyl Lewis(x) oligosaccharide. J Cell Biol 1992; 117:895–902.
39. Buerke M, Weyrich AS, Zheng Z, Gaeta FCA, Forrest MJ, Lefer Am. Sialyl LewisX-containing oligosaccharide attenuates myocardial reperfusion injury in cats. J Clin Invest 1994; 93:1140–1148.
40. Arnaout MA. Structure and function of the leukocyte adhesion molecules CD11/CD18. Blood 1990; 75:1037–1050.
41. Carlos T, Harlan J. Membrane proteins involved in phagocyte adherence to endothelium. Immunol Rev 1990; 114:5–28.
42. Matsuo Y, Onodera H, Shiga Y, et al. Role of cell adhesion molecules in brain injury after transient middle cerebral artery occlusion in the rat. Brain Res 1994; 656:344–352.
43. Clark WM, Madden KP, Rothlein R, Zivin JA. Reduction of central nervous system ischemic injury in rabbits using leukocyte adhesion antibody treatment. Stroke 1991; 22:877–883.
44. Mori E, Del ZGJ, Chambers JD, Copeland BR, Arfors KE. Inhibition of polymorphonuclear leukocyte adherence suppresses no-reflow after focal cerebral ischemia in baboons. Stroke 1992; 23:712–718.

45. Byrne JG, Smith WJ, Murphy MP, Couper GS, Appleyard RF, Cohn LH. Complete prevention of myocardial stunning, contracture, low-reflow, and edema after heart transplantation by blocking neutrophil adhesion molecules during reperfusion. J Thorac Cardiovasc Surg 1992; 104:1589–1596.
46. Lefer DJ, Shandelya SM, Serrano CV Jr, Becker LC, Kuppusamy P, Zweier JL. Cardioprotective actions of a monoclonal antibody against CD-18 in myocardial ischemia-reperfusion injury. Circulation 1993; 88:1779–1787.
47. Yamazaki T, Seko Y, Tamatani T, et al. Expression of intercellular adhesion molecule-1 in rat heart with ischemia/reperfusion and limitation of infarct size by treatment with antibodies against cell adhesion molecules. Am J Pathol 1993; 143:410–418.
48. Ma X-L, Tsao PS, Lefer AM. Antibody to CD18 exerts endothelial and cardiac protective effects in myocardial ischemia and reperfusion. J Clin Invest 1991; 88:1237–1243.
49. Hernandez LA, Grisham MB, Twohig B, Arfors K-E, Harlan JM, Granger DN. Role of neutrophils in ischemia-reperfusion induced microvascular injury. Am J Physiol 1987; 253:H699–H703.
50. Thornton MA, Winn R, Alpers CE, Zager RA. An evaluation of the neutrophil as a mediator of in vivo renal ischemic-reperfusion injury. Am J Pathol 1989; 135:509–515.
51. Jerome SN, Smith CW, Korthuis RJ. CD18-dependent adherence reactions play an important role in the development of the no-reflow phenomenon. Am J Physiol 1993; 264:H479–H483.
52. Petrasek PF, Liauw S, Romaschin AD, Walker PM. Salvage of postischemic skeletal muscle by monoclonal antibody blockade of neutrophil adhesion molecule CD18. J Surg Res 1994; 56:5–12.
53. Weselcouch EO, Grove RI, Demusz CD, Baird AJ. Effect of in vivo inhibition of neutrophil adherence on skeletal muscle function during ischemia in ferrets. Am J Physiol 1991; 261:H1178–H1183.
54. Horgan MJ, Wright SD, Malik AB. Antibody against leukocyte integrin (CD18) prevents reperfusion-induced lung vascular injury. Am J Physiol 1990; 259:L315–L319.
55. Bishop MJ, Kowalski TF, Guidotti SM, Harlan JM. Antibody against neutrophil adhesion improves reperfusion and limits alveolar infiltrate following unilateral pulmonary artery occlusion. J Surg Res 1992; 52:199–204.
56. Sharar SR, Mihelcic DD, Han KT, Harlan JM, Winn RK. Ischemia-reperfusion injury in the rabbit ear is reduced by both immediate and delayed CD18 leukocyte adherence blockade. J Immunol 1994; 153:2234–2238.
57. Vedder NB, Winn RK, Rice CL, Chi E, Arfors K-E, Harlan JM. Inhibition of leukocyte adherence by anti-CD18 monoclonal antibody attenuates reperfusion injury in the rabbit ear. Proc Natl Acad Sci USA 1990; 81:939–944.
58. Seekamp A, Mulligan MS, Till GO, et al. Role of beta 2 integrins and ICAM-1 in lung injury following ischemia-reperfusion of rat hind limbs. Am J Pathol 1993; 143:464–472.
59. Mileski WJ, Winn RK, Vedder NV, Pohlman TH, Harlan JM, Rice CL.

Inhibition of CD18-dependent neutrophil adherence reduces organ injury after hemorrhagic shock in primates. Surgery 1990; 108:205–212.

60. Vedder NB, Winn RK, Rice CL, Chi E, Arfors K-E, Harlan JM. A monoclonal antibody to the adherence promoting leukocyte glycoprotein CD18 reduces organ injury and improves survival from hemorrhagic shock and resuscitation in rabbits. J Clin Invest 1988; 81:939–944.
61. Vedder NB, Fouty BW, Winn RK, Harlan JM, Rice CL. Role of neutrophils in generalized reperfusion injury associated with resuscitation from shock. Surgery 1989; 106:509–516.
62. Simpson PJ, Todd RF III, Mickelson JK, et al. Sustained limitation of myocardial reperfusion injury by a monoclonal antibody that alters leukocyte function. Circulation 1990; 81:226–237.
63. Simpson PJ, Todd RF III, Fantone JC, Mickelson JK, Griffin JD, Lucchesi BR. Reduction of experimental canine myocardial reperfusion injury by a monoclonal antibody (Anti-Mo-1, Anti-CD11b) that inhibits leukocyte adhesion. J Clin Invest 1988; 81:624–629.
64. Jaeschke H, Farhood A, Bautista AP, Spolarics Z, Spitzer JJ, Smith CW. Functional inactivation of neutrophils with a Mac-1 (CD11b/CD18) monoclonal antibody protects against ischemia-reperfusion injury in rat liver. Hepatology 1993; 17:915–923.
65. Hill J, Lindsay T, Rusche J, Valeri CR, Shepro D, Hechtman HB. A Mac-1 antibody reduces liver and lung injury but not neutrophil sequestration after intestinal ischemia-reperfusion. Surgery 1992; 112:166–172.
66. Pober JS, Gimbrone MA Jr, Lapierre LA, et al. Overlapping patterns of activation of human endothelial cells by interleukin 1, tumor necrosis factor, and immune interferon. J Immunol 1986; 137:1893–1896.
67. Carlos TM, Harlan FM. Leukocyte-endothelial adhesion molecules. Blood 1994; 84:2068–2101.
68. Bowes MP, Zivin JA, Rothlein R. Monoclonal antibody to the ICAM-1 adhesion site reduces neurological damage in a rabbit cerebral embolism stroke model. Exp Neurol 1993; 119:215–219.
69. Ma XL, Lefer DJ, Lefer AM, Rothlein R. Coronary endothelial and cardiac protective effects of a monoclonal antibody to intercellular adhesion molecule-1 in myocardial ischemia and reperfusion. Circulation 1992; 86:937–946.
70. Kelly KJ, Williams WW Jr, Colvin RB, Bonventre JV. Antibody to intercellular adhesion molecule 1 protects the kidney against ischemic injury. Proc Natl Acad Sci USA 1994; 91:812–816.
71. Suzuki S, Toledo-Pereyra LH. Monoclonal antibody to intercellular adhesion molecule 1 as an effective protection for liver ischemia and reperfusion injury. Transpl Proc 1993; 25:3325–3327.
72. Horgan MJ, Ge M, Gu J, Rothlein R, Malik AB. Role of ICAM-1 in neutrophil-mediated lung vascular injury after occlusion and reperfusion. Am J Physiol 1991; 261:H1578–H1584.
73. Etzioni A. Adhesion molecule deficiencies and their clinical significance. Cell Adhes Commun 1994; 2:257–260.

74. Rosen H, Gordon S, North RJ. Exacerbation of murine listeriosis by a monoclonal antibody specific for the type 3 complement receptor of myelomonocytic cells. Absence of monocytes at infective foci allows Listeria to multiply in nonphagocytic cells. J Exp Med 1989; 179:27–37.
75. Mileski WJ, Winn RK, Harlan JM, Rice CL. Transient inhibition of neutrophil adherence with anti-CD18 monoclonal antibody 60.3 does not increase mortality rates in abdominal sepsis. Surgery 1991; 109:497–501.
76. Thomas JR, Harlan JM, Rice CL, Winn RK. Role of leukocyte CD11/CD18 complex in endotoxic and septic shock in rabbits. J Appl Physiol 1992; 73: 1510–1516.
77. Mileski W, Harlan J, Rice C, Winn R. *Streptococcus pneumoniae*-stimulated macrophages induce neutrophils to emigrate by a CD18-independent mechanism of adherence. Circ Shock 1990; 31:259–267.
78. Sharar SR, Sasaki SS, Flaherty LC, Paulson JC, Harlan JM, Winn RK. P-selectin blockade does not impair leukocyte host defense against bacterial peritonitis and soft tissue infection in rabbits. J Immunol 1993; 151:4982–4988.
79. Sharar S, Winn R, Murry C, Harlan J, Rice C. A CD18 monoclonal antibody increases the incidence and severity of subcutaneous abscess formation after high-dose *Staphylococcus aureus* injection in rabbits. Surgery 1991; 110:213–220.
80. Garcia NM, Mileski WJ, Sikes P, et al. Effect of inhibiting leukocyte integrin (CD18) and selectin (L-selectin) on susceptibility to infection with *Pseudomonas aeruginosa*. J Trauma 1994; 36:714–718.
81. Mileski WJ, Sikes P, Atiles L, Lightfoot E, Lipsky P, Baxter C. Inhibition of leukocyte adherence and susceptibility to infection. J Surg Res 1993; 54:349–354.
82. Tuomanen EI, Saukkonen K, Sande S, Cioffe C, Wright SD. Reduction of inflammation, tissue damage, and mortality in bacterial meningitis in rabbits treated with monoclonal antibodies against adhesion-promoting receptors of leukocytes. J Exp Med 1989; 170:959–968.
83. Han KT, Sharar SR, Phillips ML, Harlan JM, Winn RK. Sialyl Lewis^x oligo0-saccharide reduces ischemia-reperfusion injury in the rabbit. J Immunol 1995; 155:4011–4015.

19

Cellular Adhesion Molecules in Neurology

Paul J. Marchetti and Paul O'Connor
Division of Neurology, University of Toronto at St. Michael's Hospital, Toronto, Ontario, Canada

I. INTRODUCTION

The research on cellular molecules (CAMs) neurodevelopment and neuropathology will have a major impact on theories of disease pathogenesis and perhaps treatment. Adhesion molecules are expressed on cell surfaces and result in a specific phenotype that is associated with a specific behavior pattern. Cell adhesion molecules interact with a ligand on another cell surface or within the extracellular matrix. They transduce signals to allow a neuron, for example, to respond with alteration of neurite outgrowth, cell migration, growth cone behavior, and synapse formation. In neurodevelopment, a vast neural network is generated and adhesion molecule-mediated interactions lead to distinct spatial differentiation. Unfortunately, the extensively differentiated elements of this complex network are restricted in their plasticity and are sensitive to insults. It follows that an understanding of cellular interactions will have important implications for the treatment of both developmental diseases and diseases that infiltrate the network.

CAM research is shedding light on mechanisms of abnormal neuronal migration and myelination. Some of the corresponding clinical expressions of deranged migration and myelination might include mental retardation, primary epilepsy, and peripheral and possibly central dysmyelination. There are also many acquired diseases which are specific to the nervous system; this specificity, in part, relates to unique CAMs and emphasizes the value of understanding pathogenetic cellular interactions and their therapeutic potential.

Two areas of neurology are seeking to benefit from this research. Multiple sclerosis (MS), an autoimmune-mediated central demyelination process, occurs in approximately 0.1% of the Canadian population. Through CAM manipulation, the pathological immune response might be targeted rather than exposing the organism to systemic immunosuppression. Similarly, specific CAMs are likely responsible for the behavior of invasive infiltrating tumors and neurotropic metastases.

This chapter will review the current data on the normal function of CAMs in the nervous system and their significance in MS and neurooncology.

Based on structural homology, several families of adhesion molecules are recognized: the immunoglobulin (Ig) superfamily, cadherins, selectins, and integrins (1) as discussed in several chapters in this volume.

The immunoglobulin supergene family includes the T-cell receptor (TRC), the T-cell markers CD4 and CD8, major histocompatibility (MHC) molecules, lymphocyte function antigens LFA-2 and its ligand LFA-3, intercellular and vascular cell adhesion molecules (ICAMs and VCAMs), neural CAMs (NCAM, L1, AMOG), and the myelin associated CAMs (P_0, PMP-22, and MAG). The cadherins are subclassified as types N, E, and P, and interact with cadherins bound to the extracellular matrix. The selectins are comprised of the lymphocyte homing receptor (L-selectin, formerly Mel-14/LAM-1), the endothelial leukocyte adhesion molecule (E-selectin, formerly ELAM-1, and the platelet-activation-dependent granule external membrane protein (P-selectin). They are mucinlike molecules, characterized by an N-terminal lectin-binding domain. Finally, the integrins are heterodimer transmembrane proteins comprised of genetically distinct alpha and beta subunits. The very late antigens (VLAs), which are expressed on T-cells after activation, and LFA-1 belong to the integrin subgroups identified respectively by beta-1 and beta-2 subunits.

II. NEURONAL CELL MIGRATION, DIFFERENTIATION, AND PLASTICITY

Located subventricularly, the germinal matrix is a region of undifferentiated neuroblasts, originating from the epithelial lining of the neural tube. There is a subpopulation of glial cells in the germinal matrix which have fine processes oriented radially, and connecting lumenal and pial surfaces. This provides a scaffolding for the movement of immature neurons. During normal development, neuroblasts migrate in an ameboid fashion to and fro along radial glial processes, while continuing in the cell cycle. They divide only upon returning to the germinal matrix. Eventually, the cell stops dividing; then, partly differentiated, it migrates a predetermined distance to aggregate with similar neurons. Its differentiation is subsequently influenced by axonal pathfinding, target recognition, and synaptogenesis (2).

The pathfinding and migration behavior is known as fasciculation. Axon-substrate interactions provide adhesion for the growth cone to proceed in response to a tropic stimulus. The adhesive interactions themselves also likely influence differentiation. Target-associated signals will further specify neuronal phenotype even before a synapse is made. Final differentiation requires synapse formation. Even then, apoptosis may occur if the created circuitry is inactive or inappropriate to the cell type. There is some evidence that synapses are stabilized and transmission is facilitated by polymerization of cellular adhesion molecules within the synapse (3).

A. Growth Cone

At the terminus of the growth cone, filopodia are protruded and are believed to function as an autonomous sensory unit to direct growth orientation (4). Filopodial movements include protrusion, lateral sweeping, and retraction. The mechanism of these movements is calcium-dependent, and a restricted local increase in cytosolic calcium concentration in the filopodia is thought to be mediated by L- and N-type calcium channels. An imposed electrical field will produce an increase in calcium in the growth cone at the cathode with subsequent orientation of the cone toward the cathode (5).

The CAMs, NCAM, N-cadherin, and L1 similarly mediate fluxes of calcium in the filopodia. Cell contact-dependent neurite outgrowth can be completely inhibited by N- and L-type calcium channel antagonists (6). Chelation of intracellular calcium and a calcium-poor bath also prevent CAM-induced neurite outgrowth. This is not a result of impaired CAM binding since the homophilic interactions of the Ig CAMs, NCAM and L1, are calcium-independent. Neurite outgrowth induced by nerve growth factor (NGF) or protein kinases A or C occurs by a different mechanism. It is unaffected by modulation of the voltage dependent calcium channels, L and N, or extracellular calcium concentration.

Arachidonic acid (AA) will mimic the neurite outgrowth response stimulated by CAMs (6). Doherty and Walsh proposed a model for signal transduction following the homophilic interaction of CAMs wherein the interaction of adhesion molecules would result in the synthesis of AA. They hypothesize a subsequent change in tertiary structure to allow an interaction between NCAM and the fibroblast growth factor receptor (FGFR). The activated tyrosine kinase domain of FGFR in turn activates phospholipase C-gamma via a second-messenger pathway. This enzyme produces diacylglycerol (DAG) as substrate for DAG lipase with the subsequent generation of arachadonic acid (AA). Phospholipase A2, another generator of AA, may also be activated through stimulated G proteins.

B. Ig Superfamily

Neural cell adhesion molecule (NCAM) was the first CNS adhesion molecule identified and belongs to the Ig superfamily. It was initially characterized as a molecule capable of neutralizing antibodies which prevented retinal cell aggregation in vitro (2). Several forms are known to exist: some with transmembrane proteins, another with membrane attachment, and one secreted form.

Ig CAMs expressed on opposing cell surfaces undergo homophilic binding. The mechanism of intercellular adhesion is calcium-independent occurring at the amino-terminal Ig-like domains. As a member of the Ig superfamily, NCAM undergoes posttranslational modification, particularly by sialylation. The extent of sialylation is threefold greater in embryos than in adults, contributing 30% of the molecule's mass.

Curiously, sialylation reduces the binding activity of Ig CAMs but enhances the neurite outgrowth response. The removal or blocking of polysialic acid residues by enzymes or antibodies respectively results in disrupted neural tube formation (7), defects in neural crest cell migration, and inhibition of motor axon outgrowth in chick embryos (8,9). The sialylation of NCAM has been measured in certain primary CNS tumors (10). Ependymomas, which are capable of seeding throughout the CSF pathways, preferentially express a low sialylated NCAM. However, this observation is of questionable significance since the neural cell tumor, medulloblastoma, expresses a high sialylated NCAM and is similarly notorious for seeding via CFS pathways.

In rat embryo, NCAM-dependent neurite outgrowth in retinal ganglion cells and hippocampus is lost contemporaneously with increased use of the so-called variable alternatively spliced exon (VASE) in NCAM synthesis (11). Again, this loss of neurite outgrowth induction is associated with increased NCAM binding activity in vitro. The reciprocal mechanisms of diminished postranslational sialylation and increased VASE incorporation are implicated in the changing role of CAMs to promote neurite outgrowth in early development and thereafter to stabilize synaptic connections and inhibit plasticity.

C. Mutations of Ig Superfamily CAMs

The importance of Ig superfamily CAMs in CNS development has been explored using null mutations. In mice, NCAM knockout produces only a mild phenotype with underdevelopment of the olfactory bulb and a lack of rostral migration of granule cells into the bulb (12). This was surprising given the presumed significance of NCAM in early neurodevelopment, and it suggests that compensatory molecules exist.

L1 is a member of the Ig superfamily that affects neurite outgrowth, fasciculation, migration, and synapse formation. Three mutations in the human L1 gene have been identified. The corresponding diseases are X-linked hydrocephalus due to stenosis of the Sylvian aqueduct (HSAS) (13,14), a form of hereditary spastic paraplegia (SPG1) (14), and congenital mental retardation with aphasia, shuffling gait, and adducted thumbs (MASA) (14,15). What remains unknown is the specific pathogenesis of each syndrome as it relates to the respective mutated domains.

The adhesion molecule on glia (AMOG) exhibits cell type-specific binding to neurons and increases neurite outgrowth. It is in fact the β_2 subunit of the enzyme Na/K-ATPase and, like its homologue the β_1 subunit, tightly associates with the α subunit. Expression begins in late embryonic stages, and is maximal in the mature adult brain. Knockout mutant mice behave normally until 2 weeks of age, when they experience a rapidly progressive neurodegenerative disorder. The bulk of disease is in the diencephalon and brainstem with the spongiform appearance of swollen axons and glia and prominent vacuoles. It is unknown whether the pathology represents failure of the molecule's function in cell-cell interaction or as part of an ion pump. One would expect that AMOG functions in a recognition-mediated triggering of ion pump activity in the maturing CNS (16).

Homophilic interactions of the myelin protein P zero (P_0) and peripheral myelin protein-22 (PMP-22) function in myelin compaction in the PNS (16–18). A P_0 null mutant mouse has hypomyelinated nerves with a phenotype known as the *shiverer* mouse. In humans, point mutations of the P_0 gene on chromosome 1q are responsible for the distal neuropathy of the CMT-1/HSMN-I and more severe congenital Dejerine-Sottas/HSMN-III phenotypes. A duplication mutation of the PMP-22 gene on chromosome 17q similarly causes a CMT-I phenotype, and a point mutation has been responsible for a Dejerine-Sottas phenotype. Each of these diseases is characterized by onion bulb pathology with multiple rings of myelin, revealing a history of repeated demyelination and remyelination. A deletion mutation of PMP-22 causes a hereditary hypertrophic neuropathy with susceptibility to pressure palsies, also known as tomaculous neuropathy. Tomacula refer to focal, sausagelike enlargements of myelinated axons which consist of redundant loops of myelin. That mutation of an inter-Schwann cell gap junction also causes a CMT-1 phenotype suggests a relationship between P_0 or PMP-22 and gap junction formation in the formation and maintenance of myelin (19).

Myelin-associated glycoprotein (MAG) similarly belongs to the Ig superfamily and affects neurite outgrowth and adhesion of oligodendrocytes and Schwann cells to neurons. In MAG-deficient mice, the mutant phenotype may be subtle (20). In the CNS, there is disorganization of the periaxonal cytoplasmic collar and hypermyelination. In the PNS, structure and function are unchanged perhaps because NCAM functions as a surrogate adhesion molecule. There is some speculation that MAG confers additional stability to the maintenance of myelin, but remains to be established in aging knockout mutants (17).

Alternative mRNA splicing introduces a specific limitation to the utility of null mutation experiments for the Ig superfamily. Different forms of CAMs from the same gene are expressed at different times and sites. Neuroglian, for example, has cytoplasmic domains of different lengths, depending on its expression in the central or peripheral nervous system (21).

D. Cadherins

The disaggregation of neural tissue may also be induced by the administration of antibodies to N-cadherin. The cadherins derive from multiple distinct genes. Again, the mechanism of intracellular adhesion is homophilic but calcium-dependent. The cytoplasmic domain binds to a family of proteins called catenins which interact with the cytoskeleton. Expression of N-cadherin begins at the time of neural induction and is maintained after neural differentiation (6).

Binding of the cytoplasmic domain to catenins is a requisite for extracellular homophilic binding. To demonstrate this, catenin-deficient mutant cell lines were created. Another approach was the production of a mutant which expressed only the cytoplasmic domain of N-cadherin. The molecule binds catenin and effectively blocks intercellular interactions from externally applied N-cadherin (22,23).

E. Integrins

The integrins offer heterophilic binding either to the extracellular matrix (ECM) or to members of the Ig superfamily. They are glycoproteins comprised of two noncovalently associated subunits, alpha and beta. At present, 10 different beta subunits subclassify the integrins (12). The β_1 subclass is designated VLA for very late antigen expression of VLA-2 and VLA-4 by T cells. VLA-4 binds to VCAM-1 as well as to fibronectin. The β_2 subclass includes LFA-1, which is expressed on all leukocytes and binds ICAM-1. In the extracellular matrix, fibronectin is secreted by fibroblasts and other mesenchymal cells and provides an adhesive surface for migrating neural crest cells and regenerating axons. Laminin is its counterpart in nonneural tissues and in the basal lamina of the endoneurium (24).

Targeted gene inactivation of the integrins has emphasized the important role they play in developmental processes (12). Beta_1 subunit inactivation has produced an embryonal lethal phenotype around the date of implantation. Alpha_4 or α_5 subunit inactivation similarly result in early embryonic death. At this very early stage, CNS development was undisturbed. Efforts are being made to restrict these experiments topographically at various ontogenic stages to explore specific effects on nervous system development. Beta_1 antisense retrovirus has been constructed for local injection. Antibodies and antisense oligonucleotides might be similarly used to interfere with ECM ligands (12).

F. Extracellular Matrix (ECM)

Altered expression of ECM components will affect function of their integrin ligand. They are not CAMs per se since they are not anchored to cell membrane. Congenital hypogonadism from deficiency of gonadotrophin-releasing hormone and anosmia associated with the absence of olfactory bulbs occur together in an X-linked disorder known as Kallman syndrome. The presumed migrational defect relates to neuroblasts within the olfactory placode. The responsible gene produces a protein known as KAL, which shares properties with other ECM glycoproteins (25).

Merosin is an isoform of the glycoprotein laminin. It belongs to a group of dystrophin-associated glycoproteins which establish a structural link between sarcolemma and the cytoskeleton. As exemplified in Duchenne's and Becker's muscular dystrophies, the complex has proven itself essential to the myocyte's structural integrity, and merosin, specifically, has been implicated in Fukuyama congenital muscular dystrophy (26).

S-laminin, another laminin isoform, has been inactivated by gene targeting to produce a perinatally lethal phenotype. The presynaptic membrane of the neuromuscular junction has a markedly diminished number of active zones (12). The concurrent finding of poorly developed postsynaptic junctional folds is likely secondary. A more subtle mutation of this gene may be implicated in various forms of congenital myasthenia.

Animals not expressing fibronectin die in an early embryonal stage. Therefore, its later function in neural development will require spatially restricted models.

There is reason to believe that ECM molecule accessibility as a CAM ligand is modified by its association with other ECM molecules, although the interplay among CAM systems remains largely unexplored.

G. Tyrosine Kinases (trks)

Tyrosine kinases are critical in signal transduction, particulary in regulating cell growth and proliferation. Neurotrophins bind and activate receptor trks. The extracellular domains of trks share sequence homology with CAMs (27). Trks influence cadherins and integrin adhesion through cytoskeletal phosphorylation. Remarkably, Src- and Fyn-deficient neurons have specific impairment of neurite outgrowth in response to L1 and NCAM-140, respectively. Fyn-deficient mice have impaired myelination which may be explained by a possible interaction with MAG (27).

III. MULTIPLE SCLEROSIS (MS)

MS is believed to be an autoimmune disease. Demyelination in multiple sclerosis occurs as in conjunction with perivascular lymphocytic infiltrates. The diagnosis requires clinical or paraclinical evidence of plaques of inflammatory demyelination disseminated in space and time. Axonal loss and incomplete remyelination are processes that limit the extent of recovery. The periventricular inflammation is characterized by an enlarging ring of CD4+ cells and myelinophagic macrophages/microglia with death of oligodendrocytes.

One of the most important early events in the pathogenesis of autoimmune demyelination is migration of the activated TH-cell across the blood-brain barrier. Its migration may be dived into three stages, known as the area code model (28,29). The first is a weak attachment of the T-cell to the endothelium, referred to as rolling and capture. The second is the upregulation and production of additional adhesion molecules, induced by cytokine release in the immediate vicinity. The third is diapedesis across the endothelium, as the T-cell follows chemotactic gradients.

Treatment with available immunosuppressive drugs has not been successful to date and tends to be limited by their systemic toxicity. Recent partial success with IFN-β has refined this treatment approach. Subcutaneous administration of interferon β_{1b} diminished recurrent attacks by one third and stabilized disease activity by MRI assessment (30). Among its several immune-modulating effects, IFN-β reduces the expression of proinflammatory cytokines such as IFN-γ and IL-1. The latter are known to induce cell surface concentrations of adhesion molecules. CAMs play a crucial role in intiation and activation of the immune response. Assaying serum levels of soluble CAMs (sCAMs) may provide an alternate means of quantifying disease activity in the assessment of investigational therapies.

A. Initiation of the Immune Response

Immunosurveillance of the CNS by migrating T-cells is a normal occurrence. In particular, memory T-cells show increased expression of L-selectin

(31). E-selectin is found exclusively on vascular endothelium (28). The interaction of these molecules may be the basis for T-cell rolling and capture on the endothelial surface. The expression of other adhesion molecules, particularly integrins, is induced by this initial interaction.

The adhesion molecule systems described generally include an Ig superfamily adhesion molecule expressed by endothelium coupled with an integrin expressed by leukocytes. While ICAM-1 is constitutively expressed at low levels on astrocytes, the expression of other adhesion molecules in the CNS depends on induction. Of the integrins, VLA-4 has shown increased affinity to the endothelially expressed adhesion molecules VCAM-1 and ICAM (32). Myelin basic protein-sensitized T-cells increase their expression of VLA-4, and antibodies to this integrin block the development of experimental allergic encephalomyelitis (EAE) (33).

The mechanisms of normal immunosurveillance yield a pathological response when antigen-specific clones are recruited. TH-cell activation is mediated in part through the formation of the trimolecular complex of processed antigen in conjunction with an MHC molecules on the antigen-presenting cell and the TCR. In the CNS, astrocytes, microglia, and perivascular cells are capable of antigen presentation. This interaction is specific to T-cells bearing the CD4 adhesion molecule; antibodies to MHC class II will inhibit the development of EAE (34).

CD8-positive T-cells interact with endogenous antigens presented in the context of class I MHC antigens. Increased endothelial expression of MHC class I occurs during inflammation, but the parenchymal elements of the CNS do not significantly express this MHC molecule (36).

Class II MHC is inducible in a population of macrophages and reactive microglia. However, even in the absence of T-cell infiltration, as occurs in chronic silent MS plaques, macrophages remain intensely immunoreactive for MHC class II. It is speculated that the prolonged expression of class II MHC makes these areas susceptible to recurrent inflammation (35).

Apart from their function in cellular migration across the blood-brain barrier, adhesion molecules are also crucial in lymphocyte proliferation (36). Lymphocytes taken from rats pretreated with anti-ICAM-1 monoclonal antibodies failed to proliferate in response to MBP. Furthermore, LFA-2 and LFA-3 interactions promote T-cell proliferation independent of TCR stimulation. So, binding not only facilitates contact between T-cells and antigen-presenting cells, but provides costimulatory signals for T-cell activation (37–39).

B. Adhesion Molecule Profile of the MS Plaque

The initial impetus for MS research into adhesion molecules was to identify CNS-specific molecules. To date, it seems that molecules expressed in inflammation in the CNS are no different from those in peripheral inflammation. Brosnan et al. have used immunohistochemistry of MS lesions to document cytokine and adhesion molecule localization and their temporal expression (40). Sixteen of 18 postmortem MS specimens were from patients with chronic progressive MS. Non-MS neuropathology was represented by inflammatory and noninflammatory diseases. TNF-α was centrally focused in the perivascular MS lesion, localizing primarily to macrophages, with minor expression on endothelial cells and astrocytes. IFN-γ expression on perivascular inflammatory cells was prominent but did not correlate with lesion age. IL-1 was expressed by macrophages both perivascularly and at the expanding margin of the lesion, suggesting a role for it as the forerunner of cytokine expression in immune activation.

In normal brain, the VCAM-1/VLA-4 and ICAM-1/LFA-1 systems were rarely detected. In all inflammatory CNS lesions, high expression of adhesion molecules was present, but a specific pattern in MS could not be identified. The ICAM-1/LFA-1 system was ubiquitous in plaques of all ages. VCAM-1/VLA-4 was expressed in established active and chronic silent plaques. These postmortem findings are consistent with in vitro experiments demonstrating induction of adhesion molecule expression in response to pro-inflammatory cytokines (IL-1, TNF-α, and IFN-γ).

Experimental allergic/autoimmune encephalitis is an animal model for MS wherein the animal is sensitized to myelin basic protein. Administration of monoclonal antibodies against adhesion molecules has produced a more severe course of disease with anti-LFA-1 (41,42), stabilization of active disease with anti-CAM-1 (36,41), and prevention of disease with anti-VLA-4 antibodies (33). As suggested by the temporal sequence of adhesion molecule expression, the response to the injection of monoclonal antibodies in EAE may depend highly on the timing of their administration (43).

Increased neutrophil and lymphocyte adhesion to endothelial cells had been demonstrated with interferon (IFN)-γ, tumor necrosis factor (TFN)-α, and interleukin (IL)-1 in vitro. Subsequently, upregulation of ICAM-1 in response to this preincubation was documented (29). TNF also induces expression of its receptor (TNFR). Shedding from cell surfaces produces a peak in plasma levels of the soluble receptor (sTNFR) by 4 weeks after a relapse (44). sTNFR may be capable of antagonizing TNF action in vivo. In the chronic relapsing model of EAE, disease was improved by administration of sTNFR.

The targeted depletion of lymphocytes can be accomplished by use of monoclonals against specific CAMs. Some early clinical trails have already yielded promising results in modulating the pathological immune response of MS. A depletion of CD4+ cells has been induced by administration of monoclonal anti-CD4 antibody. Counts remained depressed at 6 months, and no infectious complications occurred in the cohort of 29 patients with MS (45). Cdw52 antigen is present on all lymphocytes and some monocytes. An anti-CDw52 monoclonal antibody was administered to seven patients and induced pan-lymphocyte depletion (46). The number of active lesions on gadolinium-enhanced magnetic resonance images decreased in both study populations. More aggressive pan-lymphocyte depletion with OKT3 is limited by systemic toxicity and opportunistic infections (47). Plans are underway for clinical trials using monoclonals against VLA and TCR (48).

C. Soluble Adhesion Molecules in MS

CAMs are ultimately shed from the cell surface. The detection of soluble adhesion molecules may be valuable in quantifying the activity of inflammatory disease and marking its temporal progression. This has important ramifications in therapeutic trials, where the clinical estimate of disease activity can be difficult. In this regard, MRI of the brain has proven helpful but remains time-consuming and costly and correlates weakly with clinical scales of disability.

Preliminary data of CSF levels of soluble ICAM-1 has shown correlation with disease acitvity in multiple sclerosis (MS) (31). Although high levels of ICAM and VCAM have been observed in other inflammatory and degenerative neurologic diseases (40), the value of their measurement would be in quantifying disease activity rather than in establishing diagnosis. The finding that CSF T-cell expression of VLA-4 in MS is less than in normal controls is unexpected, since its ligand is VCAM-1 and monoclonal antibodies to VLA-4 abrogate EAE.

Soluble adhesion molecules may retain functional activity. Physiological levels of L-selectin have shown ability to impair lymphocyte adhesion to endothelial cells in vitro (49).

In a study involving 147 patients with MS, correlation was sought among circulating adhesion molecules, soluble TNF-receptor (sTNFR), and magnetic resonance imaging (MRI) (43). Neurologic disease controls included inflammatory (myasthenia gravis, viral encephalitis) and noninflammatory disease. Healthy controls were also included. 76 of the 147 MS patients satisfied the MRI criteria for active disease; 75 of this group were considered to be clinically active as well. Soluble CAM levels correlated modestly with MRI assessment of active disease: 0.47 for sL-selectin, 0.36 for sV-CAM-1, and 0.28 for sICAM-1. While the correlations are insufficient for clinical use, they do emphasize the relative specificity of L-selectin and VCAM-1 for active disease. The lack of better correlation may reflect the temporal sequence of adhesion molecule expression and the uncertain kinetics of their shedding from cell surfaces.

IV. CAMs IN NEURO-ONCOLOGY

Equally tantalizing is the research of cell surface molecules in primary CNS tumors. Of the gliomas, astrocytomas in particular extensively infiltrate the surrounding brain, making them refractory to surgical management. The purpose of adhesion molecule research in this field lies in clarifying the mechanisms of tumor infiltration and immune response.

A. Tumor Invasion and Metastases in the CNS

Invasion is thought to be mediated by adhesion molecules for specific cell membrane markers or distinct ECM constituents. In comparison with other tissues, brain ECM is a relatively amorphous matrix of glycosaminoglycans. In the process of invasion by tumor, the ECM is digested by secreted hydrolytic enzymes, and tumor cells migrate onto a proteolytically modified matrix. The rarity of extracranial metastases of the primary CNS tumors argues for their dependence on a specific migrational substrate.

In part, this may be attributable to an overexpression of the cell s basic phenotype. For example, CD44 is expressed on several cell surfaces including fibroblasts, granulocytes, macrophages, erythrocytes, and glia. In non-Hodgkin's lymphoma, Horst et al. demonstrated a worse prognosis among patients whose tumors had higher expression of CD44. CD44 binds hyaluronic acid, and high hyaluronate binding activity has been observed in several particularly invasive tumors (24). It also binds to a wide range of ECM components including chondroitin sulfate, fibronectin, laminin, collagen IV, and Matrigel (50). CD44 thus likely represents a nonspecific molecule for cellular adhesion to the ECM.

In normal brain, glial expression of CD44 occurs especially on perivascular astrocytes within the white matter. In the cortex, glia do not express CD44 (51), while in high-grade astrocytomas and glioblastoma multiforme, it is strongly expressed. However, the level of CD44 expression fails to correlate with tumor grade (50). More recently, splicing varients of CD44 (CD44v) have been recognized as metastasis-promoting factors. Comparing the immunohistochemistry of primary brain tumors to that of brain metastases, Li et al. (52) demonstrated a strong correlation. All of 56 sampled primary brain tumors expressed standard CD44, while 22 of 26 metastases expressed CD44v.

With regard to metastasis, E-selectin has also been shown to mediate adhesion of a colon carcinoma cell line (53). Similarly, an epitope of VCAM-1 supports adhesion of melanoma cells, and a correlation exists between ICAM-1 expression on melanoma cells and the risk of metastases (54). LFA-1 expression is characteristic of metastatic rests of lymphoma cells (55). This introduces a role for adhesion molecules in establishing cell rests of metastatic tumors which is further discussed in Chapter 17.

Other distinctive markers of glioblastoma include LFA-3 and ICAM-1 (55). NCAM expression may be associated with reduced tumor infiltration (57). With respect to tumor-associated angiogenesis, tenascin expression shows a moderate correlation with astrocytoma grade and vascular hyperplasia (58,59). Tenascin-c, with in vitro documentation of adhesive and neurotropic properties, has failed to show an essential function in developmental models with null mutations (12).

B. Immune Response

As in other areas of oncology, the aim of this research is to enhance immune surveillance and responsiveness to tumor cells. Tumor infiltrating lymphocytes (TILs) are described in most solid tumors (60,61), presumably responding to novel antigens. The TILs show phenotypic characteristics of both cytotoxic/suppressor and helper/inducer T-cells. In the case of the CNS, the blood-brain barrier (BBB) establishes an immune privileged environment and, other than microglia, which derive from monocyte lineage, leukocytes are excluded.

With tumor-associated angiogenesis there is rarely a complete BBB, the endothelial tight junctions being induced only by mature glia. This provides the opportunity for lymphocyte infiltration. The mechanisms of rolling and capture must be involved, and in fact, increased expression of E-selectin on intratumoral endothelial cells has been demonstrated (24).

The reason for its activation remains unclear. Human glioblastoma cells may retain their function as antigen-presenting cells, and 40% of malignant gliomas express MHC class II molecules (62). ICAM-1 expression may be an adequate surrogate for immune activation in the absence of MHC class II (63). ICAM-1 and LFA-1 monoclonal antibodies inhibited TIL and LAK cell binding to human glioblastoma cells in vitro (60). In a study of 12 glioma specimens, the lymphocyte infiltrate was more prominent in the seven that expressed ICAM-1 (64). Although there was no relationship between the extent of lymphocyte infiltration and clinical course of the gliomas, a larger sample size would be required to confirm the finding.

It was hoped that LAK cells would stimulate ICAM-1 expression, assuming impairment of expression in tumor cells (65). TILs generally do not express markers of activation such as ICAM-1 and LFA-3. Nevertheless, their constitutive expression of LFA-1 and increased ICAM-1 on intratumoral endothelium may be sufficient to promote tumor infiltration. Additional stimulus may be offered by microglia, which, being of monocyte derivation, are induced to secrete TNF-a and IL-1b upon engagement of LFA-3 and CD-44.

Since VCAM-1/VLA-4 interactions have been implicated in lymphocyte migration modulating VCAM-1 expression might have more immediate clinical implications. The upregulation of VCAM-1 expression on astrocytoma cells has been investigated in response to combinations of IFN-γ, TNF-α, and IL-2 (66). TNF-α in concert with IFN-γ provided the greatest stimulus for upregulation of VCAM-1.

Attempts at enhancing tumor infiltration by T-cells and subsequent activation are still in their infancy. In vivo, LAK cells will increase ICAM-1 expression on vascular endothelium. Unfortunately, even after administration of LAK cells and IL-2 into gliomas, the subsequent migration of effector immune cells was negligible. TILs are suppressed in their cytotoxic and proliferative capacities by antiinflammatory cytokines associated with tumors such as TGF β_2 and PGE2 (67,68). Furthermore, the direct effects of a given cytokine on astrocytoma cells can be either inhibitory or stimulatory depending on the specific array of cell surface antigens in the cell line (69).

V. CONCLUSIONS

CAMs offer insight into abnormal neural development. Growth cone behavior is strongly dependent on CAM interactions and exemplifies their dual roles of cell surface adhesion and signal transduction. The function of a given CAM will change based on the timing and localization of its expression. More ubiquitous adhesion molecules have proven essential in the development of multiple systems. In strong contrast, altered expression of nervous system-specific CAMs may produce only subtle neurodevelopmental abnormalities and varying degrees of dysmyelination from the myelin-associated CAMs. The presence of surrogate CAM systems is specualted.

In MS, memory T cells are implicated in the pathogenesis by virtue of high L-selectin expression. After capture on the endothelium, VLA-4 expession is increased and interacts with VCAM-1 and ICAM. Following migration into the CNS, antigen-presenting cells interact with the lymphocytes and induce cytokine release. In turn, cytokines induce antigen presentation and increase the expression of adhesion molecules. Lyphocytes proliferate and more lymphocytes are recruited. As LFA expression increases, proliferation may be antigen-independent. Microglia become myelinophagic and maintains a high expression of MHC class II, suggesting a perivascular region primed for recurrent or chronic inflammation. Immune modulation has been demonstrated with various monoclonal antibodies against CAMs. Soluble CAMs show promise in monitoring disease activity.

In the realm of neuro-oncology, NCAM may reduce the infiltrative behavior of glial tumors. Alternatively, it may only reflect a greater degress of glial differentiation. CD44 expression may promote infiltration into the parenchyma, while its splicing variants are associated with metastatic potential. E-selectin, ICAM-1, and LFA-1 have been similarly implicated in determining the metastatic potential of certain malignancies. Tumor growth is also dependent on its vascular supply, and to that end, the role of CAMs in tumor-associated angiogenesis is being pursued. Finally, immunosurveillance of tumor cells can be exploited. In spite of a population of TILs, the cells remain largely inactive. Their function in the tumor is unknown, but attempts are being made to enhance recruitment of peripheral immune cells and to activate existing TILs.

REFERENCES

1. Springer TA. Adhesion receptors of the immune system. Nature 1990; 346: 425–434.
2. Dodd J, Jessell TM. Axon guidance and the patterning of neuronal projections in vertebrates. Science 1988; 242:692–699.
3. Schmidt R. Cell-adhesion molecules in memory function. Behav Brain Res 1995; 66:65–72.
4. Davenport RW, Don P, Rehder V, et al. A sensory role for neuronal growth cone filopodia. Nature 1993; 361:721–723.
5. Bedlack RS, Wei MD, Loew LM. Localised membrane depolarisations and localised calcium influx during electric field-guided neurite growth. Neuron 1992; 9:393–403.
6. Doherty P, Walsh FS. Signal transduction events underlying neurite outgrowth stimulated by cellular adhesion molecules. Curr Opin Neurobiol 1994; 4:49–55.
7. Bronner-Fraser M, Wolf JJ, Murray BA. Effects of antibodies against N-cadherin and NCAM on the cranial neural crest and neural tube. Dev Biol 1992; 153:291–301.
8. Landmesser L, Dahm L, Tang JC, et al. Polysialic acid as a regulator of intramuscular nerve branching during embryonic development. Neuron 1990; 4:655–667.
9. Tang JC, Landmesser L, Rutishauser U. Polysialic acid influences specific pathfinding by avian motoneurons. Neuron 1992; 8:1031–1044.
10. Figarella-Branger DF, Durbec PL, Rougon GN. Differential spectrum of expression of neural cell adhesion molecule isoforms on human neuroectodermal tumours. Cancer Res 1990; 50:6364–6370.
11. Doherty P, Moolenaar CECK, Ashton SV, et al. The VASE exon downregulates the neurite growth promoting activity of NCAM 140. Nature 1992; 356: 791–793.
12. Muller U, Kypta R. Molecular genetics of neuronal adhesion. Curr Opin Neurobiol 1995; 5:36–41.
13. Jouet M, Rosenthal A, MacFarlane J, et al. Missense mutation confirms the L1 defect in HSAS. Nature Genet 1993; 4:331.

14. Jouet M, Rosenthal A, Armstrong G, et al. X-linked spastic paraplegia, MASA syndrome, and X-linked hydrocephalus result from mutations in the L1 gene. Nature Genet 1994; 7:402–407.
15. Vits L, VanCamp G, Coucke P, et al. MASA Syndrome is due to mutations in L1 CAM. Nature Genet 1994; 7:408–413.
16. Filbin MT, Tennekoon GI. Myelin Po-protein, more than just a structural protein? Bioessays 1992; 14:541–547.
17. Schachner M. Neural recognition molecules in disease and regeneration. Curr Opin Neurobiol 1994; 4:726–734.
18. Snipes GJ, Suter U, Shooter EM. The genetics of myelin. Curr Opin Neurobiol 1993; 3:694–702.
19. Bergoffen J, Scherer SS, Wang S, et al. Connexin mutations in X-linked CMT disease. Science 1993; 262:2039–2042.
20. Montag D, Giese KP, Bartsch U, et al. Mice deficient for the myelin-associated glycoprotein show subtle abnormalities in myelin. Neuron 1994; 13:1–20.
21. Grumet M. Cell adhesion molecules and their subgroups in the nervous system. Curr Opin Neurobiol 1991; 1:370–376.
22. Kinter C. Regulation of embryonic cell adhesion by the cadherin cytoplasmic domain. Cell 1992; 69:225–236.
23. Hirano S, Kimoto N, Shimoyama Y, et al. Identification of a neural alpha-catenin as a key regulator of cadherin function and multicellular organisation. Cell 1992; 70:293–301.
24. Couldwell WT, deTribolet N, Antel JP, et al. Adhesion molecules and malignant gliomas: implications for tumorigenesis. J Neurosurg 1992; 76:782–791.
25. Franco B, Guioli S, Pragliola A, et al. A gene deleted in Kallman s syndrome shares homology with neural cell adhesion and axonal path-finding molecules. Nature 1991; 353:529–536.
26. Hayashi YK, Engvall E, Arikawa-Hirasawa E, et al. Abnormal localization of laminin subunits in muscular dystrophies. J Neurol Sci 1993; 119:53–64.
27. Schneider R, Schweiger M. A novel mosaic of cell adhesion motifs in the extracellular domains of the neurogenic trk and trkB tyrosine kinase receptors. Oncogene 1991; 6:1807–1811.
28. Bevilaqua MP, Nelson RM. Selectins. J Clin Invest 1993; 91:379–387.
29. Springer TA. Traffic signals for lymphocyte recirculation and leukocyte emigration: the multiple step paradigm. Cell 1994; 76:301–314.
30. The interferon beta multiple sclerosis study group. Interferon beta-1b in the treatment of multiple sclerosis: final outcome of the randomised controlled trial. Neurol 1995; 45:1277–1285.
31. Svenningsson A, Hansson GK, Andersen O, et al. Adhesion molecule expression on cerebrospinal fluid T lymphocytes: evidence for common recruitment mechanisms in multiple sclerosis, aseptic meningitis, and normal controls. Ann Neurol 1993; 34:155–161.
32. Baron JL, Madri JA, Ruddle NH, et al. Surface expression of alpha-4 integrin by CD4 T cells is required for their entry into brain parenchyma. J Exp Med 1993; 177:57–68.
33. Yednok TA, Cannon LC, Fritz F, et al. Prevention of experimental autoim-

mune encephalomyelitis by antibodies against alpha4 beta1 integrin. Nature 1992; 356:63-66.
34. Steinman L. Autoimmunity and the nervous system. In: Goldstein RA, ed. Neuroimmune Disorders. Immunol Allergy Clin North Am 1988; 8:213-221.
35. Raine CS. Multiple sclerosis: immune system molecule expression in the central nervous system. J Neuropathol Exp Neurol 1994; 53:328-337.
36. Archelos JJ, Jung S, Maurer M, et al. Inhibition of experimental autoimmune encephalomyelitis by an antibody to the intercellular adhesion molecule ICAM-1. Ann Neurol 1993; 34:145-154.
37. Damle NK, Aruffo A. Vascular cell adhesion molecule 1 induces T-cell antigen receptor-dependent activation of CD4+ T lymphocytes. Proc Natl Acad Sci USA 1991; 88:6403-6407.
38. Dang LH, Michalek MT, Takei F, et al. Role of ICAM-1 in antigen presentation demonstrated by ICAM-1 defective mutants. J Immunol 1990; 144:4082-4091.
39. Dougherty GJ, Murdock S, Hogg N. The function of human intercellular adhesion molecule-1 in the generation of an immune response. Eur J Immunol 1988; 18:35-39.
40. Brosnan CF, Cannella B, Battistini L, et al. Cytokine localization in MS lesions: correlation with adhesion molecule expression and reactive nitrogen species. Neurol 1995; 45(S6):S16-S22.
41. Cannella B, Cross AH, Raine CS. Anti-adhesion molecule therapy in experimental autoimmune encephalomyelitis. J Neuroimmunol 1993; 46:43-56.
42. Welsh CT, Roose JW, Hill KE, et al. Augmentation of adoptively transferred experimental allergic encephalomyelitis by administration of a monoclonal antibody specific for LFA-1 alpha. J Neuroimmunol 1993; 43:161-168.
43. Hartung HP, Reiners K, Archelos JJ, et al. Circulating adhesion molecules and tumor necrosis factor receptor in MS: correlation with MRI. Ann Neurol 1995; 38:186-193.
44. Reickmann P, Weichselbraun I, Albrecht M, et al. Serial analysis of circulating adhesion molecules and TNF receptor in serum from patients with multiple sclerosis: cICAM-1 is an indicator for relapse. Neurology 1994; 44:1523-1526.
45. Lindsey JW, Hodgkinson S, Mehta R, et al. Phase 1 clinical trial of chimeric monoclonal anti-CD4 antibody in MS. Neurol 1994; 44:413-419.
46. Moreau T, Thorpe J, Miller D, et al. Preliminary evidence from MRI for reduction in disease activity after lymphocyte depletion in MS. Lancet 1993; 344:298-301.
47. Weinshenker BG, Bass B, Karlik S, et al. An open trial of OKT3 in patients with MS. Neurol 1991; 41:1047-1052.
48. Reingold SC. Clinical Trials of New Agents in MS, Planned, in Progress, Recently Completed. New York: National MS Society, 1995.
49. Schleiffenbaum B, Spertini O, Tedder TF. Soluble L-selectin is present in human plasma at high levels and retains functional activity. J Cell Biol 1992; 119:229-238.
50. Radotra B, McCormick D, Crockard A. CD44 plays a role in adhesive interac-

tions between glioma cells and extracellular matrix components. Neuropathol Appl Neurobiol 1994; 20:399–405.
51. Picker LJ, de los Toyos J, Telen MJ, et al. Monoclonal antibodies against the CD44 and Pgp-1 anitgens in man recognize the Hermes class of lymphocyte homing receptors. J Immunol 1989; 142:2046–2051.
52. Li H, Liu J, Hofmann M, et al. Differential CD44 expression patterns in primary brain tumours and brain metastases. Br J Cancer 1995; 72:160–163.
53. Rice GE, Bevilacqua MP. An inducible endothelial cell surface glycoprotein mediates melanoma adhesion. Science 1989; 246:1303–1306.
54. Johnson JP, Stade BG, Holzmann B, et al. De novo expression of ICAM-1 in melanoma correlates with increased risk of metastasis. Proc Natl Acad Sci USA 1989; 86:641–644.
55. Roossien FF, de Rijk D, Bikker A, et al. Involvement of LFA-1 in lymphoma invasion and metastasis demonstrated with LFA-1 deficient mutants. J Cell Biol 1989; 108:1979–1985.
56. Gingras MC, Roussel E, Bruner JM, et al. Comparison of cell adhesion molecule expression between glioblastoma multiforme and autologous normal brain tissue. J Neuroimmunol 1995; 57:143–153.
57. Edvardsen K, Pedersen PH, Bjerkvig R, et al. Transfection of glioma cells with the neural-cell adhesion molecule NCAM: effect on glioma-cell invasion and growth in vivo. Int J Cancer 1994; 58:116–122.
58. Zagzag D, Friedlander DR, Miller DC, et al. Tenascin expression in astrocytomas correlates with angiogenesis. Cancer Res 1995; 55:907–914.
59. Higuchi M, Ohnishi T, Arita N, et al. Expression of tenascin in human gliomas: its relation to histological malignancy, tumor dedifferentiation and angiogenesis. Acta Neuropathol 1993; 85:481–487.
60. Kuppner MC, van Meir E, Hamou MF, et al. Cytokine regulation of intercellular adhesion molecule-1 expression on human glioblastoma cells. Clin Exp Immunol 1990; 81:142–148.
61. Miescher S, Whiteside TL, de Tribolet N, et al. In situ characterization, clonogenic potential, and antitumor cytolytic activity of T lymphocytes infiltrating human brain cancers. J Neurosurg 1988; 68:438–448.
62. Dhib-jalbut S, Kufta CV, Flerlage M, et al. Adult human glial cells can present target antigens to HLA-restricted cytotoxic T-cells. J Neuroimmunol 1990; 29: 203–211.
63. Altmann DM, Hogg N, Trowsdale J, et al. Cotransfection of ICAM-1 and HLA-DR reconstitutes human antigen-presenting cell function in mouse L cells. Nature 1989; 338:512–514.
64. Yamanaka R, Tanaka R, Saito T. Immunohistochemical analysis of tumour-infiltrating lymphocytes and adhesion molecules (ICAM-1, NCAM) in human gliomas. Neurologia Medico-Chiurgica 1994; 34:583–587.
65. Barba D, Saris SC, Holder C, et al. Intra tumoral LAK cell and IL-2 therapy of human gliomas. J Neurosurg 1989; 70:175–182.
66. Rosenman SJ, Shrikant P, Dubb L, et al. Cytokine induced expression of VCAM-1 by astrocytes and astrocytoma cell lines. J Immunol 1995; 154:1888–1899.

67. Couldwell WT, Dore-Duffy P, Apuzzo MLJ, et al. Malignant glioma modulation of immune function: relative contribution of different soluble factors. J Neuroimmunol 1991; 33:89–96.
68. Kuppner MC, Hamou MF, Swamura Y, et al. Inhibition of lymphocyte function by glioblastoma-derived TGF beta-2. J Neurosurg 1989; 71:211–217.
69. Chen TC, Hinton DR, Apuzzo ML, et al. Differential effects of TNF-alpha on proliferation, cell surface antigen expression and cytokine interactions in

20

Adhesion Molecules in Inflammatory Lung Injury

Peter A. Ward and Michael S. Mulligan
Department of Pathology, University of Michigan Medical School, Ann Arbor, Michigan

I. INTRODUCTION

Over the past several years it has become apparent that both leukocytic and endothelial adhesion molecules play important roles in the recruitment of blood leukocytes into lung with resulting injury. This information has come from both the use of blocking antibodies and the use of genetically induced deletions ("knockouts"). It has become apparent that, depending on the stimulus and also the tissue or organ under study, participation of both cytokines and adhesion molecules may vary. It has also been demonstrated in lung that, depending on the bacterial species employed, recruitment of neutrophils into the alveolar compartment may be CD18-dependent or independent, whereas when the same microorganisms are used in subcutaneously implanted sponges in rabbits, there is a consistent dependency on CD18 for neutrophil influx (1). The information reviewed below will describe our results in which both cytokine and adhesion molecule requirements have been defined in four different models of lung injury.

A. Models of Inflammatory Lung Injury

Three different models of lung injury in rats have been employed. In the first, intraalveolar deposition of IgG immune complexes is induced by the intratracheal instillation of rabbit polyclonal IgG to bovine serum albumin (BSA), while the antigen is given intravenously. Over a 4-hour period an intense hemorrhage and neutrophil-enriched alveolitis develops in a manner that is complement-dependent. The injury is absolutely dependent upon the influx of neutrophils (2). The sequence of events appears to include formation in alveolar walls of the BSA anti-BSA complexes, complement activation, activation of pulmonary macrophages resulting in the release of TNFα and IL-1, upregulation of vascular adhesion molecules including E-selectin and ICAM-1, influx of neutrophils, and then injury induced by products of activated neutrophils as well as pulmonary macrophages (reviewed, 3). The ultimate tissue-damaging products of these cells appear to be oxidants derived either from molecular oxygen via the NADPH oxidase pathway or products of nitric oxide as well as the role of metalloproteinases. The second model of acute lung injury involves the intrapulmonary formation of IgA immune complexes, induced by intratracheal instillation of a MOPC-21 murine protein with antigen reactivity to dinitrophenol (DNP). This is accompanied by an intravenous injection of DNP coupled to bovine serum albumin, the result of which is also intraalveolar deposition of IgA immune complexes. In this situation, complement is also a requirement for the ultimate development of injury, but very little neutrophil recruitment occurs, with predominantly direct activation of pulmonary macrophages by the immune complexes that have interacted with complement (4,5). Both alveolar epithelial injury and vascular endothelial injury occurs, with oxidants of the type described above being released from stimulated pulmonary macrophages. As will be discussed below, the profile for involvement of cytokines in this reaction is very different from injury induced by intrapulmonary deposition of IgG immune complexes. The third type of experimental injury is induced by the intravenous injection of cobra venom factor, which induces massive activation via the alternative pathway of complement, resulting in intravascular stimulation of neutrophils and their aggregation with entrapment in the lung pulmonary capillaries (6,7). Simultaneously, lung vascular P-selectin is upregulated and this, in combination with the activated neutrophils, results in both vascular endothelial and adjacent alveolar epithelial injury, with interstitial and intra-alveolar edema and extensive intraalveolar hemorrhage (8). As implied, this model is complement and neutrophil dependent and devoid of any requirements of cytokines, especially in view of the fact that the peak of injury in this model occurs at 30 minutes. The model of ischemia reperfusion with secondary injury to rat lung will be discussed below.

II. ROLE OF β_2 INTEGRINS AND ICAM-1 IN INFLAMMATORY LUNG INJURY

In each of the three models of acute lung injury described above, the role of β_2 integrins and ICAM-1 have been determined by the use of blocking monoclonal antibodies. The targets of these antibodies have been CD11a, CD11b, CD18, and ICAM-1. The effects of these blocking antibodies are described in Table 1 in the three animal models. In the cobra venom factor model of lung injury, in which neutrophils play a dominant role, the infusion of approximately 200 μg of each of the blocking antibodies significantly reduced lung vascular injury (9). Anti-CD11a reduced injury (defined by increased permeability) by 30%, while anti-CD11b reduced it by 53% and anti-CD18 reduced injury by 74%. In parallel with this LFA-1 (CD11a/CD18) and Mac-1 (CD11b/CD15) dependency, not surprisingly, blocking of ICAM-1 reduced injury by 60%. In the IgG immune complex model of acute lung injury which is neutrophil-dependent, blocking of CD11a reduced injury by 61%, but blocking of CD11b did not produce a statistically significant reduction (16%) in lung injury while blocking of CD18 and ICAM-1 reduced injury by 55% and 61%, respectively (10). Thus, under the experimental conditions employed, lung injury induced by intrapulmonary deposition of IgG immune complexes is LFA-1 but not Mac-1-dependent. In the IgA immune complex model of acute lung injury which is neutrophil-independent, blocking of CD11a, CD11b, or CD18 reduced increased vascular permeability by 35%, 63%, and 89%, respectively. As expected, in view of the LFA-1 and Mac-1 requirements in this model of injury, ICAM blockade also reduced injury by 61%. Thus, there are differing patterns of β_2 integrin requirements as reflected by the differences in the CD11b requirements described in Table 1.

There is now recent evidence that β_2 integrins play a compartmentalized role in the development of lung injury and recruitment of neutrophils in the IgG immune complex model (11). The data underlying this conclusion are shown in Table 2. In each case, blocking antibodies to CD11a, CD11b, ICAM-1, or L-selectin have been given either intravenously or intratracheally. The effects on permeability and hemorrhage, parameters of injury, were determined as well as changes in neutrophil accumulation as defined

Table 1 Requirements for β_2 Integrins and ICAM-1 in Acute Inflammatory Tissue Injury

Model of lung injury	Neutrophil dependency	Protection (%)[a] due to blocking antibody:			
		CD11a	CD11b	CD18	ICAM-1
CVF	+	30	53	74	60
IgG immune complex	+	61	16	55	61
IgA immune complex	–	35	63	89	61

Sources: References 9, 10.
[a]Protection as defined by permeability changes.

Table 2 Compartmentalized Role for β_2 Integrins in IgG Immune Complex-Induced Lung Injury

Blocking antibody	Mode of administration	Reduction (%) in injury: Permeability	Reduction (%) in injury: Hemorrhage	Reduction (%) in MPO
CD11a	intravenous	63	48	31
	intratracheal	<5	<5	<5
Anti-CD11b	intravenous	<5	<5	<5
	intratracheal	59	52	75
Anti-ICAM-1	intravenous	62	64	42
	intratracheal	53	48	35
Anti-L-selectin	intravenous	68	70	33
	intratracheal	<5	<5	<5

Source: Reference 11.

by lung MPO content. The data show some very interesting trends, which strongly suggest a compartmentalized role for CD11a, CD11b, and L-selectin. The intravenous infusion of antibody to CD11a blocked permeability and hemorrhage by 63% and 48%, respectively, and reduced neutrophil accumulation (as reflected in MPO content) by 31%. In contrast, when the same blocking antibodies were given intratracheally with the anti-BSA, there was no reduction in injury and no change in MPO buildup. In the case of antibody to CD11b, the results were exactly the opposite. The intravenous infusion of antibody to CD11b does not significantly reduce the parameters of injury or MPO content, whereas the intratracheal instillation of the same antibody reduced permeability and hemorrhage by 59% and 52%, respectively, and MPO content by 75%. Thus, CD11a plays a central role in recruitment of neutrophils into the alveolar compartment in this model, whereas CD11b is operating in a similar manner but under conditions in which the antibody has protective effects only if it is administered intratracheally. It is likely that the anti-CD11b is blocking activation of pulmonary macrophages, as will be discussed below. Blocking antibodies to ICAM-1 achieve protective effects which are rather similar, whether the antibody is given intravenously or intratracheally (Table 2). It seems possible that both endothelial ICAM-1, which is upregulatable, and alveolar epithelial cell ICAM, which is constitutively expressed (12), play important roles—the former in the sequence of events leading to neutrophil influx into the lung, the latter perhaps having to do with the tethering of alveolar macrophages to alveolar epithelial type II cells, resulting in optimal effector function of the activated macrophages (production of oxidants, release of proteases). Finally, the intravenous infusion of anti-L-selectin reduced permeability and hemorrhage in this model of injury by 68% and 70%, respectively, and blocked MPO accumulation by 33%. The intratracheal administration of this antibody had no effect. This is consistent with the lack of L-selectin content on alveolar macrophages (personal observations) and suggests that L-selectin on neutrophils plays a vital role in their recruitment from the vascular compartment but not in events beyond this point.

III. CHANGES IN BRONCHOALVEOLAR LAVAGE TNFα IN THE PRESENCE OF BLOCKING ANTIBODIES

In view of the data in Table 2 suggesting that there is a compartmentalized effect and requirements for CD11a, CD11b and L-selectin in the IgG model of lung injury, it was next determined the extent to which in vivo these blocking antibodies affect TNFα production as measured in bronchoalveolar lavage fluids. The data for these experiments are shown in Table 3. Whether MOPC-21, a class matched (IgG-1) immunoglobulin, was given intravenously or intratracheally, the content of TNFα in the BAL fluids was similar with approximately 12 to 14 units/ml (with a total lavage volume of approximately 5 ml). Although anti-CD11a blocked neutrophil influx when given intravenously, neither its intravenous or intratracheal administration resulted in any significant reduction in BAL TNF content (11). In fact, for

Table 3 In Vivo Changes in BAL TNFα Levels Induced by Blocking Antibodies

Blocking antibody	Mode of administration	BAL TNFα content (units/ml)[a]
MOPC-21	intravenous	13.8 ± 1.86
	intratracheal	12.06 ± 2.57
Anti-CD11a	intravenous	27.7 ± 7.07 (NS)
	intratracheal	30.1 ± 1.98
Anti-CD11b	intravenous	12.9 ± 3.61
	intratracheal	1.45 ± 0.45
Anti-ICAM-1	intravenous	26.2 ± 7.5 (NS)
	intratracheal	15.8 ± 2.4 (NS)
Anti-L-selectin	intravenous	30.3 ± 2.9
	intratracheal	46.4 ± 3.8

Source: Data from Reference 11.
[a]In BAL fluids of normal controls, TNFα was <2 units/ml.

reasons that have not been defined, the intratracheal administration of this antibody resulted in a significant increase in BAL content of TNFα (Table 3). The intravenous infusion of anti-CD11b had no effect on the content of TNFα in the BAL fluids while the intratracheal administration of this antibody dramatically reduced the TNFα content in the BAL fluids, almost surely related to the ability of anti-CD11b (when given in this manner) to protect from lung injury. Neither the intravenous nor the intratracheal infusion of antibody to ICAM-1 resulted in any significant difference in the production of TNFα as measured in the BAL fluid contents. Finally, the intravenous or intratracheal infusion of antibody to L-selectin caused a significant increase in BAL content of TNFα for reasons that are not clear. These data would suggest that CD11b plays an important extravascular role in this model of acute lung injury and that one clearly defined role of CD11b is production of TNFα by stimulated pulmonary macrophages.

IV. LUNG VASCULAR ICAM AFTER INTRATRACHEAL DELIVERY OF BLOCKING ANTIBODIES

On the basis of the data in Tables 2 and 3, it was predicted that intratracheal instillation of anti-CD11b would result in a significant reduction in upregulation of lung vascular ICAM because of the nearly complete abolition of BAL TNFα (Table 3). This possibility was investigated by the use of a technology in which radioactive antibody to rat ICAM-1 was infused 15 minutes before sacrifice (at 3 hours and 45 minutes after induction of immune complex deposition), followed by extensive perfusion with saline of the pulmonary vasculature and assessment of the amount of antibody fixed to the lung (13). These studies require the companion use of a class-matched irrelevant antibody (MOPC-21) that is radiolabeled and used for leakage of the molecule due to vascular injury. The corrected counts are then determined, and the difference in lung vascular ICAM expression has been calculated. As shown by the data in Table 4, the intratracheal instillation of antibody to CD11a did not effect the level of upregulation of lung vascular ICAM as compared to intratracheal instillation of MOPC-21. In both cases, there was approximately a 180% increase in the level of lung vascular ICAM. In striking contrast, when blocking antibody to CD11b was given intratracheally, there was a 74% reduction in upregulation of lung vascular ICAM, such that under these conditions only a 47% increase in lung vascular ICAM was found (11). These data indicate that the blocking activity of antibody to CD11b when given intratracheally is linked to its suppression in the production and/or release of TNFα and the corresponding blockade in upregulation of lung vascular ICAM. On the basis of these

Table 4 Changes in Lung Vascular ICAM-1 After Intratracheal Delivery of Blocking Antibodies

Blocking antibody	Change (%) in lung vascular ICAM-1 relative to normal lung
MOPC-21	180
Anti-CD11a	178
Anti-CD11b	47

Source: Reference 11.

data, there appear to be clear indications that the β_2 integrin roles in this model of lung injury are compartmentalized (to the intravascular and extravascular areas).

V. REQUIREMENTS FOR CYTOKINES IN MODELS OF LUNG INJURY

We have investigated in the IgG anti-BSA immune complex model of lung injury as well as dermal vascular injury and the lung model of injury induced by IgA immune complexes with respect to the requirements for IL-1, TNFα, and MCP-1. The data are shown in Table 5. In the IgG immune complex model of lung injury, there were profound reductions in both permeability and hemorrhage when either blocking antibody to IL-1 or TNFα was given intravenously (14,15). These interventions caused similar reductions in lung buildup of MPO content. In striking contrast, blocking antibody to MCP-1 in this model had no measurable effects on any of these parameters (16). When companion studies were done in the same animals in skin to assess the profile of cytokine requirements, some very different results were found (17). Blocking antibody to IL-1 reduced permeability and hemorrhage parameters by 60% and 39%, respectively, and diminished by 41% the increase in MPO content (Table 5). In striking contrast, whereas blocking antibody to TNFα was highly protective in lung, it had no effect on the parameters of injury or the MPO levels in this model of dermal vascular injury. Thus, the cytokine profiles differ depending on the organ under study.

In the IgA immune complex model of lung injury, neither antibody to IL-1 nor TNFα had any protective effects, which is consistent with the fact that the BAL fluids have very low levels of these cytokines (reviewed in 19). In striking contrast to the case of IgG immune complex lung injury, blocking antibody to MCP-1 had highly protective effects in the IgA immune

Table 5 Cytokine Requirements in Inflammatory Injury

Model	Antibody-induced cytokine blockade	Reduction (%) in injury		Reduction (%) in MPO
		Permeability	Hemorrhage	
BSA anti-BSA IgG immune complex lung injury	IL-1	69	81	41
	TNFα	65	70	50
	MCP-1	<5	<5	<5
BSA anti-BSA IgG immune complex dermal vascular injury	IL-1	60	39	41
	TNFα	<5	<5	<5
DNP anti-DNP IgA immune complex lung injury	IL-1	<5	<5	
	TNFα	<5	<5	
	MCP-1	67	52	

Sources: References 14–17 and 19.

complex model of lung injury, reducing permeability and hemorrhage by 67% and 52%, respectively (16). MPO measurements were not determined in this model because relatively few neutrophils accumulate, and neutrophils play no role in the development of injury. On the basis of these studies, then, it can be concluded that depending on the stimulus and the organ under study, cytokine requirements have variable patterns reflective of the location and the stimulus.

VI. SELECTIN REQUIREMENTS IN INFLAMMATORY LUNG INJURY

Using the three models of acute lung injury described above, selectin requirements have been defined in each of the three cases using either blocking monoclonal antibodies or selectin-Ig chimeric proteins, which represent dimeric selectin molecules attached to the CH1 and CH2 regions of the Fc part of the human IgG (20–22). On the basis of these types of blocking strategies, it has been demonstrated that in the IgG BSA anti-BSA immune complex induced lung injury model, E-selectin plays an important role with its blockade resulting in approximately 70% reductions in permeability and hemorrhage and an 81% reduction in build up of MPO in lung (Table 6).

Table 6 Selectin Requirements in Inflammatory Injury

Model	Target of blockade	Reduction (%) in injury: Permeability	Reduction (%) in injury: Hemorrhage	Reduction (%) in MPO
BSA anti-BSA IgG	E-selectin	72	71	81
immune complex	L-selectin	33	33	43
lung injury	P-selectin	<5	<5	<5
DNP anti-DNP IgA	E-selectin	<5	<5	
immune complex	L-selectin	<5	<5	
lung injury	P-selectin	<5	<5	
CVF-induced lung	E-selectin	<5	<5	<5
injury	L-selectin	51	51	67
	P-selectin	51	70	50

Sources: References 8, 20–22.

Not surprisingly, L-selectin is also required in that its antibody-induced blockade reduced permeability and hemorrhage each by 33% and reduced MPO buildup by 43%. Finally, blocking antibody to P-selectin or P-selectin Ig chimeric protein had no measurable effect in this model of injury, nor did it have any effect on the accumulation of MPO in lung tissue. In striking contrast, in the IgA immune complex model of lung injury no selectin requirements have been defined (Table 6), which may be in keeping with the fact that it is assumed that in this model of lung injury the IgA immune complexes only activate residential phagocytic cells, with little or no recruitment of either neutrophils or monocytes from the circulation (20–22). In the CVF-induced model of acute lung injury, E-selectin plays no measurable role and its blockade by the use of E-selectin Ig chimeric protein failed to reduce the permeability and hemorrhage parameters and failed to reduce MPO accumulation (21). Even though E-selectin was not required in this model of lung injury, requirements for both L-selectin and P-selectin have been defined as demonstrated in Table 6. Blockade of L-selectin by antibody reduced by approximately 50% both permeability and hemorrhage changes, and by 67% the increase in MPO, while blockage of P-selectin had similar effects (8,20,21). It should be noted in this model of lung injury that immunostaining for P-selectin indicated lung vasculature presence of P-selectin as early as 5 minutes after the infusion of the cobra venom factor with a peaking at about 15 minutes, which would be consistent with rapid induction of P-selectin (8). It has recently been demonstrated using human umbilical vein endothelial cells that C5a has the ability to cause P-selectin upregulation, suggesting that the mechanism of P-selectin upregulation in this particular model may reflect a direct interaction of C5a with receptors on the endothelial cells (23).

VII. ROLE OF CYTOKINES AND ADHESION MOLECULES IN SECONDARY INJURY OF LUNG AFTER ISCHEMIA/REPERFUSION

In this model of lung injury, the blood flow to each of the lower extremities in rats was interrupted by arterial ligation. Depending on the time of ischemia, reperfusion occurs; and there was a gradual increase in evidence of lung injury over a 4-hour period of reperfusion (24). During the early reperfusion (from 1 to 2.5 hours), there was evidence of the appearance of both TNF and IL-1 as well as IL-6 in the plasma. Lung injury developed over the period of 2 to 4 hours of reperfusion and was associated with a requirement for both complement and neutrophils (25). Lung injury was associated with the accumulation of neutrophils in the pulmonary vasculature and with hemorrhage into the alveolar compartment.

In defining the role of cytokines, blocking antibodies to TNFα or IL-1 as well as the soluble TNFα receptor-1 or the IL-1 receptor antagonist were employed, and the effects on lung injury after 4 hours of ischemia and 4 hours of reperfusion were identified (24). The results of these studies are shown in Table 7. Whether TNFα was blocked with antibody or with the infusion of the soluble TNFα receptor 1, there were dramatic reductions in permeability and hemorrhage and similar reductions in MPO accumulations in lung. With respect to IL-1, blocking antibody as well as the IL-1 receptor antagonist also had substantial blocking effects by reducing permeability and hemorrhage parameters and reducing the buildup of MPO in lung tissue (Table 7). Accordingly, it is clear that in this model of secondary lung injury related to ischemia reperfusion events involving the lower ex-

Table 7 Role of Cytokines in Ischemia-Reperfusion Secondary Injury of Lung

	Reduction (%)		
Intervention	Permeability	Hemorrhage	MPO
Anti-TNFα	75	83	82
TNFα R-I	67	77	56
Anti-IL-1	52	66	26
IL-1 receptor antagonist	67	85	39

Source: Reference 24.

tremities of rats, changes in lung resulting in injury are both TNFα- and IL-1-dependent.

VIII. ROLE OF ADHESION MOLECULES IN ISCHEMIA REPERFUSION SECONDARY INJURY OF LUNG

Using blocking antibodies to the β_2 integrin components as well as to ICAM, the role of these molecules was determined as described by the data in Table 8 (26). After 4 hours of ischemia and reperfusion, in the presence of antibody to CD11a, permeability and hemorrhage were reduced by 44% and 70%, respectively, while MPO accumulation was reduced by 61%. Similar effects were found with blocking antibodies to CD11b and to CD18 (Table 8). Not surprisingly, in view of these requirements for LFA-1 (CD11a/CD18) and Mac-1 (CD11b/CD18), ICAM was found to be an important adhesion molecule requirement as demonstrated by the protective effects of its blockade, and by the 72% reduction in MPO as a result of blocking of this adhesion molecule. With respect to the selectin requirements in this model, anti-E-selectin reduced permeability and hemorrhage by 55% and 79%, respectively and reduced MPO accumulation by 73% (27). Antibody to P-selectin after 4 hours of ischemia and 4 hours of reperfusion had no protective effects in terms of reducing permeability and hemorrhage, and did not affect MPO accumulation in lung. However, if the ischemia period was limited to 1 hour and the reperfusion period limited to 1.5 hours, there was a 93% reduction in permeability and 94% reduction in hemorrhage and an 87% reduction in MPO (Table 8). It should be

Table 8 Role of Adhesion Molecules in Ischemia Reperfusion Secondary Injury of Lung

	Reduction (%)		
Blocking intervention	Permeability	Hemorrhage	MPO
Anti-CD11a	44	70	61
Anti-CD11b	64	81	74
Anti-CD18	60	88	90
Anti-ICAM-1	64	72	72
Anti-E-selectin	55	79	73
Anti-P-selectin (at 4 hr)	12 (NS)	4 (NS)	3 (NS)
Anti-P-selectin (at 1.5 hr)	93	94	87
Anti-L-selectin	51	62	58

Sources: References 26 and 27.

stressed, however, that under these abbreviated conditions of ischemia and reperfusion, the extent of tissue injury was much below that occurring when ischemia and reperfusion were each held at 4 hours. Finally, L-selectin appeared to be a requirement in that its blockade with antibody reduced permeability and hemorrhage by 51% and 62%, respectively, and MPO by 58% (Table 8).

IX. CHANGES IN β_2 INTEGRINS ON NEUTROPHILS DURING ISCHEMIA REPERFUSION INJURY

Since it has been demonstrated that in the course of ischemia and reperfusion (4 hours each) of rat lower extremities there was evidence of progressive consumptive depletion of complement (25), a series of studies was undertaken to evaluate blood neutrophil content of CD18, CD11a, and CD11b under various conditions. If rat blood neutrophils were isolated from normal donors, as shown by the data in Table 9, addition of human recombinant C5a or phorbol myristate acetate each substantially increased the CD18 content of neutrophils, failed to alter the CD11a content, and also significantly increased the CD11b content, consistent with other studies that have demonstrated that these agonists are potent upregulators of Mac-1 (CD11b/CD18). After ischemia of the lower extremities for 4 hours and no reperfusion, there were no changes in CD18, CD11a, or CD11b on blood neutrophils (Table 9). However, during the first 60 minutes of reperfusion, there was no statistically significant change in the neutrophil expression of these integrins. At 4 hours there was a clear increase in CD18, no change in CD11a and a clear increase in CD11b. These data are consistent with the fact that after ischemia and during reperfusion systemic acti-

Table 9 Upregulation of β_2 Integrins During Ischemia Reperfusion Injury

		Mean channel fluorescence		
Neutrophil donor	Additions	CD18	CD11a	CD11b
Normal	none	49.6 ± 2.9	24.1 ± 1.1	9.5 ± 0.7
Normal	C5a (100 nM)	90.4 ± 2.5	23.8 ± 2.1	17.4 ± 0.9
Normal	PMA (32 nM)	114 ± 4.1	25.3 ± 1.6	22.8 ± 2.2
Ischemia (4 hr)	none	50.4 ± 2.0	25.1 ± 1.9	9.6 ± 0.3
Ischemia (4 hr), reperfusion (1 hr)	none	51.3 ± 2.9	24.6 ± 1.2	10.1 ± 0.5
Ischemia (4 hr), reperfusion (4 hr)	none	66.2 ± 3.4	25.8 ± 2.2	16.3 ± 0.5

Source: Reference 26.

vation of complement is occurring, causing up regulation of Mac-1 on circulating blood neutrophils. It also appears likely that the presence in the circulation of TNFα and IL-1 (24) leads to upregulation of ICAM-1 and E-selectin on the pulmonary vascular endothelium. Under these circumstances, blood neutrophils would have been activated within the vascular compartment and ligands for adhesive interactions between neutrophils upregulated on the endothelium (E-selectin and ICAM-1), setting the stage for damage of the pulmonary vascular and alveolar epithelial compartments, with neutrophils becoming adherent to the altered pulmonary vascular endothelium and injuring endothelial and epithelial cells on the basis of their oxidants that would be expected to be produced.

X. CONCLUSIONS

Depending on the nature of the stimulus, these studies conclusively demonstrate the requirements for cytokines and adhesion molecules in inflammatory lung injury which involves the recruitment of neutrophils. Cytokine requirements correlate with ICAM-1 and E-selectin requirements for neutrophil recruitment into lung. A direct association between lung TNFα expression and upregulation of lung vascular ICAM-1 has been demonstrated. In turn, there are the expected β_2 integrin requirements. Inflammatory conditions that are associated with recruitment of neutrophils consistently show requirement for L-selectin. In the IgG immune complex model of lung injury, there is also evidence of a compartmentalized role for CD11a and CD11b—the former occurring in the vascular compartment, the latter occurring in the extravascular compartment. The role of CD11b appears to be related to the requirements of lung macrophages for CD11b-dependent production of TNFα after stimulation by IgG immune complexes. In the CVF model of lung injury, the very rapid accumulation of neutrophils and induction of injury are consistent with roles for P- and L-selectin and of β_2 integrins (LFA-1 and Mac-1) and constitutive ICAM-1, and the lack for requirement for TNFα and IL-1.

In the IgA immune complex model of lung injury, in view of the apparent absence of involvement of blood monocytes and neutrophils, the demonstrated lack of requirements for any selectins is to be expected. In this model, the requirement for MCP-1 but not for TNFα or IL-1 suggests that lung macrophages become activated and predominantly produce MCP-1 which may function as an autocrine stimulator. The β_2 integrin and ICAM-1 requirement in this model may relate to adhesive interactions between alveolar macrophages and alveolar epithelial cell ICAM-1. Finally, injury of lung after cessation of blood flow (followed by reperfusion) in the lower extremities follows the pattern of IgG immune complex-induced injury of lung, namely, cytokine expression, and upregulation of endothelial and leukocytic adhesion molecules, leading ultimately to lung injury. Understanding the diverse pathways leading to inflammatory lung injury should provide important clues to therapeutic approaches for blocking of adhesion molecules.

REFERENCES

1. Doerschuk CM, Winn RK, Coxson HO, Harlan JM. CD18-dependent and -independent mechanisms of neutrophil emigration in the pulmonary and systemic microcirculation of rabbits. J Immunol 1990; 144:2327–2320.
2. Johnson KJ, Ward PA. Acute immunologic pulmonary alveolitis. J Clin Invest 1974; 54:349–357.
3. Ward PA, Mulligan MS. Molecular mechanisms in acute lung injury. Adv Pharmacol 1993; 24:275–292.
4. Johnson KJ, Wilson BS, Till GO, Ward PA. Acute lung injury in rat caused by immunoglobulin A immune complexes. J Clin Invest 1984; 74:358–369.
5. Johnson KJ, Ward PA, Kunkel RG, Wilson BS. Mediation of IgA induced lung injury in the rat. Role of macrophages and reactive oxygen products. Lab Invest 1986; 54:499–506.
6. Till GO, Johnson KJ, Kunkel R, Ward PA. Intravascular activation of complement and acute lung injury Dependency on neutrophils and toxic oxygen metabolites. J Clin Invest 1982; 69:1126–1135.
7. Ward PA, Till GO, Kunkel R, Beauchamp C. Evidence for role of hydroxyl radical in complement and neutrophil-dependent tissue injury. J Clin Invest 1983; 72:789–801.
8. Mulligan MS, Paulson JC, De Frees S, Zheng Z-L, Lowe JB, Ward PA. Protective effects of oligosaccharides in P-selectin-dependent lung injury. Nature 1993; 364:149–151.
9. Mulligan MS, Smith CW, Anderson DC, et al. Role of leukocyte adhesion molecules in complement-induced lung injury. J Immunol 1993; 150:2401–2406.
10. Mulligan MS, Wilson GP, Todd RF, et al. Role of β_1, β_2 integrins and ICAM-1 in lung injury following deposition of IgG and IgA immune complexes. J Immunol 1993; 150:2407–2417.
11. Mulligan MS, Vaporciyan AA, Warner RL, et al. Compartmentalized roles for leukocytic adhesion molecules in lung inflammatory injury. J Immunol 1995. In press.
12. Christensen PJ, Kim S, Simon RH, Toews GB, Paine R III. Differentiation-related expression of ICAM-1 by rat alveolar epithelial cells. Am J Respir Cell Mol Biol 1993; 8:9–15.
13. Mulligan MS, Vaporciyan AA, Miyasaka M, Tamatani T, Ward PA. Tumor necrosis factor alpha regulates in vivo intrapulmonary expression of ICAM-1. Am J Pathol 1993; 142:1739–1749.
14. Warren JS, Yabroff KR, Remick DG, et al. Tumor necrosis factor participates

in the pathogenesis of acute immune complex alveolitis in the rat. J Clin Invest 1989; 84:1873–1882.
15. Warren JS. Intrapulmonary interleukin-1 mediates acute immune complex alveolitis in the rat. Biochem Biophys Res Commun 1991; 175:604–610.
16. Jones ML, Mulligan MS, Flory CM, Ward PA, Warren JS. Potential role of monocyte chemoattractant protein 1/JE in monocyte/macrophage-dependent IgA immune complex alveolitis in the rat. J Immunol 1992; 149:2147–2154.
17. Mulligan MS, Ward PA. Immune complex-induced lung and dermal vascular injury. Differing requirements for tumor necrosis factor-alpha and IL-1. J Immunol 1992; 149:331–339.
18. Mulligan MS, Jones ML, Vaporciyan AA, Howard MC, Ward PA. Protective effects of IL-4 and IL-10 against immune complex-induced lung injury. J Immunol 1993; 151:5666–5674.
19. Ward PA, Warren JS, Varani J, Johnson KJ. PAF, cytokines, toxic oxygen products and cell injury. Mol Aspects Med 1991; 12:169–174.
20. Mulligan MS, Miyasaka M, Tamatani T, Jones ML, Ward PA. Requirements for L-selectin in neutrophil-mediated lung injury in rats. J Immunol 1994; 152: 832–840.
21. Mulligan MS, Watson SR, Fennie C, Ward PA. Protective effects of selectin chimeras in neutrophil-mediated lung injury. J Immunol 1993; 151:6410–6417.
22. Mulligan MS, Varani J, Dame MK, et al. role of endothelial-leukocyte adhesion molecule 1 (ELAM-1) in neutrophil-mediated lung injury in rats. J Clin Invest 1991; 88:1396–1406.
23. Foreman KE, Vaporciyan AA, Bonish BK, et al. C5a-induced expression of P-selectin in endothelial cells. J Clin Invest 1994; 94:1147–1155.
24. Seekamp A, Warren JS, Remick DG, Till GO, Ward PA. Requirements for tumor necrosis factor-α and interleukin-1 in limb ischemia/reperfusion injury and associated lung injury. Am J Pathol 1993; 143:453–463.
25. Seekamp A, Mulligan MS, Till GO, Ward PA. Requirements for neutrophil products and L-arginine in ischemia-reperfusion injury. Am J Pathol 1993; 142:1217–1226.
26. Seekamp A, Mulligan MS, Till GO, et al. Role of β_2-integrins and ICAM-1 in lung injury following ischemia-reperfusion of rat hind limbs. Am J Pathol 1993; 143:464–472.
27. Seekamp A, Till GO, Paulson JC, et al. Role of selectins in local and remote tissue injury following ischemia and reperfusion. Am J Pathol 1994; 144:592–598.

21

Role of Adhesion Molecules in Intestinal Inflammation

Brent Johnston
Department of Medical Physiology, University of Calgary, Calgary, Alberta, Canada
D. Neil Granger
Department of Physiology, Louisiana State University Medical Center, Shreveport, Louisiana
Paul Kubes
Department of Medical Physiology, University of Calgary, Calgary, Alberta, Canada

I. INTRODUCTION

There is a growing body of evidence suggesting an important role for polymorphonuclear leukocytes (PMNs) in mediating the tissue injury and dysfunction associated with various pathophysiological or inflammatory conditions in the gastrointestinal tract. The evidence for a PMN component in various models of intestinal inflammation is threefold: 1. there is an increase in PMN infiltration into afflicted intestinal tissues; 2. PMN depletion attenuates the tissue injury; and 3. preventing PMN recruitment into the target tissue also abrogates the injurious process. Table 1 summarizes several intestinal inflammatory conditions wherein PMNs infiltrate the tissue and appear to contribute to subsequent tissue injury. We will use ischemia/reperfusion (I/R) of the small bowel as a prototype to address the mechanisms underlying the leukocyte recruitment and subsequent tissue injury. Additionally, we will elaborate on other models of gastrointestinal inflammation, including nonsteroidal anti-inflammatory drug (NSAID)-induced gastropathy and bacterial toxin-induced intestinal dysfunction. Although we cannot cover all of the intestinal models of injury wherein adhesion molecules have been evoked, Table 1 (1–15) refers the reader to important literature describing the role of PMN adhesion in alcohol-induced intestinal injury, stress-induced gastric ulceration, and, more recently, the importance of adhesion molecules in *Helicobacter pylori*-induced inflammation.

Table 1 Intestinal Inflammatory States Associated With PMN-Mediated Injury

Inflammatory states	Reference
Ischemia/reperfusion	1–3
Hemorrhagic shock	4–6
Platelet activating factor	7,8
Ethanol	9
Experimental colitis	10
NSAID-induced gastric ulceration	11,12
Helicobacter pylori infection	13
Clostridium difficile toxin A	14
Cold stress-induced gastritis	15

II. MECHANISMS OF LEUKOCYTE RECRUITMENT

The evidence implicating the selectins and integrins as the underlying mechanisms for PMN-endothelial cell interactions is reviewed extensively elsewhere in this book. Nevertheless, before examining the role of PMN adhesion in intestinal inflammation, a very brief overview of the mechanisms of leukocyte infiltration into tissue is warranted. It has been well established that leukocyte infiltration is a multistep mechanism which requires that a leukocyte moving at very high speeds in the mainstream of blood makes initial contact with the endothelial cells lining the vessel wall and roll along the vessel at a greatly reduced velocity relative to red blood cells. This initial leukocyte-endothelial cell interaction is termed leukocyte rolling and is largely dependent upon the selectin family of adhesion molecules. L-selectin is constitutively expressed on the surface of PMNs and appears to be essential for the ability of a leukocyte to initiate rolling in the microcirculation. Two other selectins, P-selectin (induced in minutes) and E-selectin (4 to 6 hours for maximal induction) expressed on activated endothelium also contribute significantly to the rolling event. Whether L-selectin is a ligand for the two endothelial cell selectins is an area of controversy (16,17).

When activated, the rolling PMN firmly adheres to the endothelium and ultimately emigrates out of the vasculature. This event is mediated by the integrins found on leukocytes, and in the case of the PMN the β_2-integrin (CD11/CD18). There are two important points to keep in mind when considering this scheme for PMN recruitment. First, this is an interrelated cascade of events in which the rolling event is a necessary prerequisite to PMN adhesion and subsequent emigration. Secondly, rolling, adhesion, and emigration transpire primarily within the postcapillary venules; very rarely do investigators observe PMN-endothelial cell interactions in other vessels.

III. ISCHEMIA/REPERFUSION

A. Models of Intestinal Ischemia/Reperfusion

One common model of intestinal ischemia/reperfusion employs a period of low-flow localized ischemia. The preparation of choice for this model is the cat as an arterial circuit can be established between the femoral and superior mesenteric arteries with a flow probe interposed within the circuit to monitor blood flow (1,18–20). In studies utilizing this method, blood flow to the intestinal circulation is reduced to approximately 20% of control during the ischemic phase by partially occluding the flow circuit with an adjustable clamp. Reperfusion is later achieved by removing the clamp. A more severe model of local ischemia/reperfusion involves complete ischemia. In these studies, intestinal blood flow is abolished during the ischemic phase by complete occlusion of the arteries supplying blood to the splanchnic circulation. If ischemia is maintained for >60 min prior to reperfusion, a state of splanchnic artery occlusion (SAO) shock ensues, characterized by a significant decrease in systemic arterial blood pressure, release of cardiotoxic myocardial depressant factor, and high mortality (21). In contrast to models of localized ischemia, hemorrhagic shock models induce systemic ischemia through the removal of up to 80% of the total blood volume from the circulation. Reperfusion is achieved by reinfusing the experimental animal with the previously shed blood. Although ischemia in this model is systemic, we will focus on the effects observed in the intestine.

B. Neutrophils Infiltrate the Postischemic Intestine

Initial evidence demonstrating that neutrophils do indeed infiltrate the postischemic intestine was based on the measurement of tissue myeloperoxidase (MPO) levels (1,2). This enzyme is found primarily in PMNs and therefore gives a reasonable estimate of tissue PMN levels. Using the MPO assay to monitor PMN kinetics, we observed that PMNs infiltrate into the small bowel during ischemia/reperfusion. This PMN infiltration was increased in all layers of the intestine (2). The MPO activity in the mucosa doubled during ischemia and increased fourfold during reperfusion, whereas the MPO activity in the submucosa was found to increase significantly only during reperfusion (3.5-fold). The intestinal muscle tissue seemed most sensitive to I/R-induced granulocyte infiltration as MPO activity increased sixfold and 10-fold during ischemia and reperfusion, respectively. It is possible that tissue-specific variations in granulocyte recruitment result from differences in the production of chemotactic agents that recruit neutrophils. However, some caution must be used when directly comparing different layers of the intestine inasmuch as baseline MPO levels are far greater in the mucosa than in other layers of the intestine. Therefore, the fold increase in PMN recruitment during reperfusion may be greater in the muscle layer, but the net PMN recruitment is in fact larger in the mucosa. This assessment also revealed that the pattern of PMN recruitment into the postischemic mesentery is similar to events in the mucosa. This latter point is of significance inasmuch as the mesentery is used to directly visualize PMN recruitment and the data are often used to reflect events in the mucosal tissue.

Although the MPO assay has been extremely important in establishing that PMNs infiltrate postischemic tissue, it tells us little about the behavior and mechanism of PMN recruitment into postischemic vessels. Much of the evidence for the multistep recruitment of PMNs into postischemic vessels is based on intravital videomicroscopy, which allows investigators to visualize PMN-endothelial cell interactions within the microcirculation. Briefly, the intestine is exteriorized and the optically clear mesentery is placed over a viewing pedestal to directly visualize 20 to 40 μm postcapillary venules. Following the induction of ischemia/reperfusion in the cat intestine (or other animal models), a very dramatic increase in PMN rolling, adhesion, and ultimately PMN emigration into the surrounding tissue is observed (Fig. 1). This has permitted the study of each of the individual events that make up the PMN recruitment cascade. This technique can provide a real-time quantitative measure of white cell infiltration. Furthermore, histological assessment has established the phenotype of adhering and emigrating cells as PMNs (19); however, to date it has been impossible to determine the phenotype of the rolling leukocytes.

C. Targeting the β_2-Integrin (CD11/CD18) to Prevent PMN Adhesion and Emigration

One of the more notable advances that made possible the study of PMNs in ischemia/reperfusion was the development of monoclonal antibodies directed against various adhesion molecules. One of the first monoclonal antibodies (MAb) raised against a leukocyte adhesion molecule was IB_4. This MAb immunoneutralized the β-subunit of the CD11/CD18 adhesion glycoprotein complex (22). Crossreactivity of this antibody with CD11/CD18 on cat PMNs was established by its ability to prevent stimulated feline neutrophils from adhering to serum-coated plastic (23). Reperfusion of the ischemic intestine was associated with a very significant increase in leukocyte adhesion, which could be prevented when animals were pretreated with MAb IB_4 (19,23). These results suggest that the adhesion was entirely mediated by the CD18 glycoprotein complex. Although PMN emi-

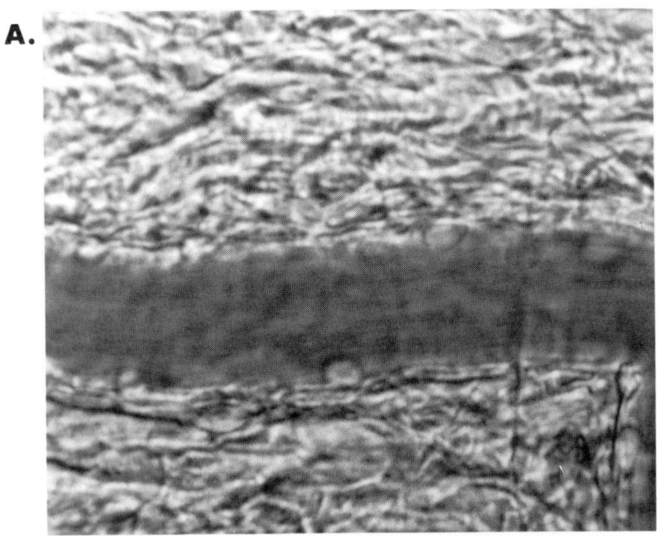

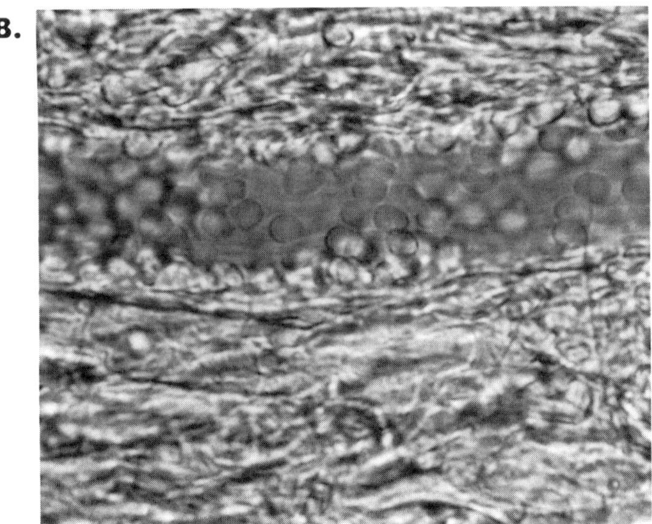

Figure 1 Leukocyte influx into a 38-μm cat mesenteric venule (A) under control conditions, and (B) after 5 min of reperfusion following 60 min of low-flow ischemia. Under control conditions few leukocytes can be seen within the venule, whereas reperfusion induced the recruitment of over 90 cells within the field of view.

gration was also completely prevented, it remained unclear whether the emigration was dependent on CD18 or simply on the ability of the PMN to adequately adhere. The observation that administration of MAb IB_4 at 60 min of reperfusion reversed the adhesion to postcapillary venules within 10 min (23) may be of clinical relevance. Clearly, the PMN-endothelial cell interaction can be disengaged therapeutically.

Similar results were obtained by measuring MPO as an index of PMN influx (2). Acute treatment with an anti-CD18 MAb (IB_4) prior to the experiment prevented the reperfusion-induced increase in mucosal MPO levels in all layers of the intestine, implicating a β_2-integrin-dependent (CD18) mechanism of neutrophil recruitment during I/R. Posttreatment with IB_4 at 1 hour of reperfusion immediately reduced MPO activity by 40%, 35%, and 20% in the mucosa, submucosa, and mesentery, respectively. This posttreatment reduction represents the contribution of adherent cells within the vasculature to the total MPO activity of the tissue. Interestingly, posttreatment with IB_4 did not significantly decrease MPO activity in the muscle layer. The authors attributed this observation to the phenomenon of no-reflow which occurs when activated PMNs plug the microvasculature and prevent blood flow through vessels (24–26). This phenomenon seems to be prevalent in postischemic muscle tissue.

With the advent of a monoclonal antibody (CL26) that immunoneutralized rat CD18, Kurose et al. (3) used a rat model of ischemia/reperfusion to confirm the initial observations in the cat; i.e., reperfusion-induced PMN adhesion was indeed dependent upon CD11/CD18. These experiments were further extended to establish a role for the CD11b (MAb 1B6c) subunit of CD11/CD18 adhesion complex as a modulator of reperfusion-induced PMN adhesion. Moreover, a MAb (1A29) against endothelial ICAM-1 (one of the ligands for CD18) also significantly reduced reperfusion-induced PMN adhesion and emigration. The ICAM-1 data were recently confirmed in the cat model of low-flow ischemia/reperfusion using the monoclonal antibody RR1/1 (27).

D. Targeting Leukocyte Rolling

Based on the premise that inhibiting PMN rolling will reduce PMN adhesion, a number of investigators have begun to target the selectins as a mechanism of reducing PMN infiltration into postischemic vessels. Using intravital microscopy, it was observed that anti-P-selectin MAb (PB1.3) administered at the time of reperfusion reduced PMN rolling by approximately 60% (28). An identical result was observed in animals given anti-L-selectin MAb (DREG 200) prior to reperfusion; the number of rolling leukocytes was again decreased by approximately 60% at both 10 and 60 min of reperfusion. In animals given both anti-L-selectin and anti-P-selectin MAbs the response was not cumulative; a 60% reduction in rolling was still observed. The lack of additive effect of tandem antibody therapy was interpreted as either indicating that P-selectin and L-selectin pathways worked in concert, i.e., as counterligands, or that they mediate different components of leukocyte rolling in a sequential manner so that one depends on the other (17).

In another series of experiments, the fucosylated carbohydrate fucoidin (25 mg/kg) essentially abolished (>90%) leukocyte rolling in postischemic vessels, suggesting a significant (30% to 40%) population of leukocytes that roll in postischemic vessels, independent of L-selectin or P-selectin. These data suggest that there exists an L- and P-selectin-independent fucoidin-inhibitable rolling pathway during reperfusion. An alternative explanation to the L- and P-selectin-independent leukocyte rolling is invoked by E-selectin, the third member of the selectin family and a glycoprotein also postulated to induce leukocyte rolling (29,30). It is unlikely, however, that E-selectin participates in this rolling event during 1 hour of reperfusion as E-selectin is expressed on the surface of feline postischemic endothelium only at 4 hours of reperfusion and in very small amounts (15% to 20% of all vessels) (31). Moreover, the leukocyte rolling in the feline intravital microscopy study was abolished by fucoidin, yet fucoidin has been shown not to bind to E-selectin (32,33). Although the identity of the fucoidin-sensitive pathway remains unknown, it is conceivable that fucoidin, a heavily sulfated polysaccharide, may interfere with the ability of leukocytes to interact with sulfate-containing proteoglycans on the surface of vascular endothelium. Although presently speculative, various other sulfated molecules have been shown to strongly interfere with leukocyte rolling in vivo (34–36) and leukocyte adhesion in vitro (37).

A somewhat surprising observation was the fact that the 60% reduction in leukocyte rolling flux with the antiselectin antibodies failed to reduce leukocyte adhesion. These data suggest that there exists a surplus of rolling cells in postischemic vessels and the reduction in PMN rolling has to reach a critical level to impact on PMN adhesion. Indeed, this hypothesis is supported by the observation that adhesion was significantly reduced only when leukocyte rolling was reduced by >90% with fucoidin. These data suggest that antiselectin therapy requires a high level of efficiency in this particular model of I/R before adhesion is affected.

It is important to note that this observation may not extend to all models of ischemia/reperfusion. For example, in the rat mesentery exposed to complete ischemia and reperfusion, Kurose et al. (3) did not report an increase in PMN rolling but did observe increased adhesion and emigration from baseline. Under these conditions, immunoneutralizing P-selectin did not diminish PMN adhesion. In contrast, Davenpeck et al. (38), using a similar model and identical reagents, reported an increase in PMN rolling at 30 min of reperfusion that returned to baseline values by 60 min. In these animals, adhesion was elevated throughout the experimental protocol. In animals treated with PB1.3, the anti-P-selectin antibody prevented the rise in PMN rolling and abolished the elevated PMN adhesion. These data are in direct contrast with those of Kurose et al. (3) and raise some important issues regarding technical aspects of the intravital microscopy model. We have observed that in preparations where baseline PMN rolling is higher than 20 cells/min, inducers of P-selectin such as histamine do not increase PMN rolling, whereas preparations wherein PMN rolling is below 20 cells/min, P-selectin inducers cause a very profound increase in PMN rolling (39,40). One potential explanation for this observation is that high baseline rolling is a reflection of increased P-selectin expression, making it difficult to further mobilize P-selectin to the surface of endothelium and elicit additional PMN rolling. It should also be noted that although Kurose et al. (3) failed to see a reduction in PMN adhesion with P-selectin antibody, they did observe a reduction in platelet-leukocyte aggregate formation. These aggregates move through the vasculature at reduced speeds and many reflect another type of PMN-endothelial cell interaction that may be critical to the overall pathogenesis of reperfusion injury. Clearly, the role of P-selectin in I/R-induced leukocyte rolling and subsequent adhesion merits further investigation.

A final point that needs to be taken into consideration is the role of shear forces in ischemia/reperfusion-induced PMN rolling. For example, even when PMN rolling was reduced by 90% with fucoidin in cat mesenteric venules, a significant proportion of leukocytes adhered in venules if reperfusion shear rates were below 70% of control (28). Clearly, the efficiency of leukocyte adhesion during low rolling states was significantly improved when shear rates were compromised. This is an important factor to consider since inflammation is characterized by heterogeneous shear responses within venules; i.e., some vessels have high shear and others have reduced shear. In the aforementioned study, the leukocyte adhesion during lower hydrodynamic dispersal forces was not simply a “stopping phenomenon” as a result of insufficient shear but rather a CD18-dependent adhesion inasmuch as an antibody directed against the β_2-integrin prevented the leukocyte adhesion (28). Although the reason for the reduced blood flow response to reperfusion in some but not all vessels remains unclear, the low shear forces were not due to elevated levels of rolling and adhering PMNs. This contention is based on the observation that fucoidin plus anti-CD18 antibody treatment abolished PMN rolling and adhesion but did not affect the low hydrodynamic dispersal forces. This observation questions the importance of PMN rolling and adhesion as an important contributor to reduced shear rates in postischemic venules.

E. Targeting I/R-Induced Injury

Although we will focus on the postischemic mechanisms that mediate injury directly to the intestine, there is significant work that suggests that ischemia/reperfusion of the intestine can lead to distal organ injury. For example, the adhesive mechanisms responsible for lung injury following ischemia/reperfusion of the intestine have been extensively studied (41,42).

Initial studies to establish that PMNs contributed in a significant manner to postischemic intestinal injury made use of polyclonal antisera to remove PMNs from the circulation. Depletion of neutrophils by administration of an antineutrophil serum (ANS) significantly attenuated the increased microvascular permeability associated with ischemia/reperfusion (18). ANS treatment also attenuated reperfusion-induced gastric bleeding in models of hemorrhagic shock (4). These data suggested that PMNs were responsible for a significant portion of the injury in postischemic tissues of the gastrointestinal tract.

A model of ischemia/reperfusion in the feline small bowel has been used to characterize the importance of CD18 in I/R-induced intestinal dysfunction (1). Prevention of PMN adhesion to postcapillary venules in the cat intestinal circulation with anti-CD18 antibodies (MAb 60.3 or IB_4) attenuated the increased microvascular permeability associated with I/R. These data suggested that neutrophil adherence or a neutrophil adherence-dependent event (e.g., emigration) was a rate-limiting step in neutrophil-mediated microvascular dysfunction. The role of circulating PMNs in microvascular dysfunction has recently been confirmed by examining FITC-albumin leakage from single venules in the rat mesenteric vasculature. Kurose et al. (3) found that vascular protein leakage of FITC-albumin was highly correlated with both the number of adherent cells in postcapillary venules and the number of cells emigrated from the vasculature. Treatment of animals with MAbs against CD18 (CL26), CD11b (1B6c), or ICAM-1 (1A29) significantly attenuated reperfusion-induced increases in PMN adhesion, emigration, and FITC-albumin leakage. These data provide strong evidence that vascular protein leakage was a secondary response to leukocyte adherence and/or emigration.

These data were extended further using phalloidin, a stabilizer of F-actin. This agent significantly attenuated both protein leakage and PMN emigration without reducing PMN adhesion, suggesting that vascular dysfunction is associated primarily with PMN emigration rather than PMN adhesion. These data need to be interpreted with some caution inasmuch as phalloidin could conceivably affect PMN function independent of PMN adhesion—i.e., superoxide production and/or release of proteases. Finally, it should be noted that PMN emigration is not the only cause of I/R-induced endothelial permeability. This is best exemplified by the fact that MAbs to CD18 completely prevent I/R-induced adhesion and emigration but do not completely eliminate vascular leakage (3,19). Additional dysfunction and injury may result from anoxia and/or the action of proinflammatory mediators released from PMN-independent sources such as the mast cell.

In addition to the microvascular barrier however, the intestine also has a mucosal or epithelial barrier that prevents translocation of noxious stimuli into the host environment. Associated with the increased PMN accumulation within the mucosa and the increased edema formation following reperfusion of the ischemic intestine, there was a reduction in villus height and crypt depth as well as a reduction in mucosal thickness with epithelial lifting down the sides of the villi and a disruption of the lamina propria (43). These morphological alterations translated into a dramatic increase in mucosal permeability (barrier dysfunction) to ^{51}Cr-EDTA, a marker of epithelial permeability (1). However, acute treatment with an anti-CD18 MAb (IB_4) prior to ischemia/reperfusion did not reduce the epithelial permeability to ^{51}Cr-EDTA during ischemia/reperfusion despite complete prevention of PMN infiltration into postischemic mucosa (assessed as mucosal MPO levels). This suggested a CD18-independent mechanism of epithelial barrier dysfunction during I/R. It should be noted that the intestinal mucosa has one of the highest levels of baseline MPO activity (2), which indicates that there is a significant population of MPO-positive PMNs always present within the mucosa. Chronic administration of the anti-CD18 antibody (IB_4) over 3 days essentially eliminated both baseline and reperfusion-induced MPO values within the mucosal layer of the bowel (1). This regimen entirely abolished the rise in epithelial permeability following ischemia/reperfusion and suggests that there may be two distinct populations of PMNs responsible for reperfusion-induced intestinal dysfunction (Fig. 2); a circulating PMN population mediates the microvascular dysfunction and a resident mucosal population underlies the mucosal injury associated with ischemia/reperfusion. This scenario seems to make sense from an anatomical perspective; the proximity of resident granulocytes to the epithelium makes them a likely candidate to induce mucosal injury in the initial stage of reperfusion.

These data also suggest that there is a continuous CD18-dependent influx of PMNs into the mucosal interstitium under normal conditions. This study also reveals that the turnover rate for neutrophils in the intestinal intersti-

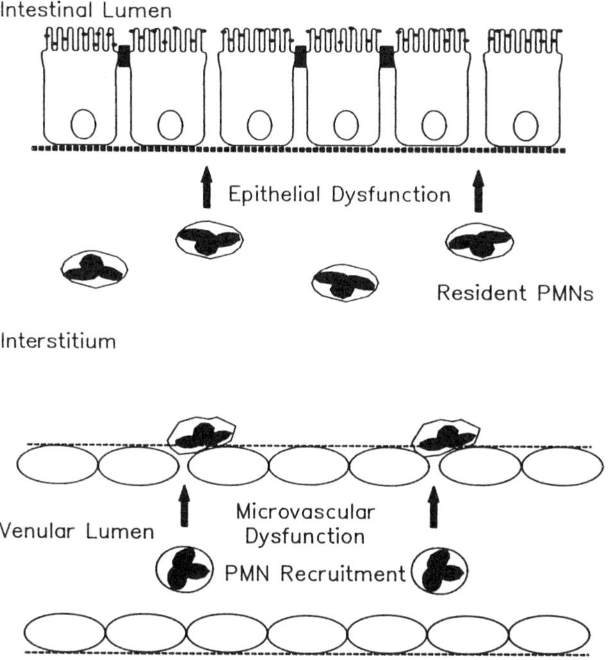

Figure 2 Resident PMN populations are situated in proximity to the intestinal epithelium and likely mediate ischemia/reperfusion-induced mucosal dysfunction, whereas newly recruited PMNs are likely to mediate microvascular dysfunction by adhering and emigrating across the vasculature. During the first hour of reperfusion, the intravascular leukocytes do not contribute to reperfusion-induced mucosal barrier dysfunction.

tium is less than 3 days; however, the physiologic importance of this ongoing recruitment of PMNs to the mucosa remains unknown. It is tempting to predict that these PMNs contribute to the continuous homeostatic defence against invading foreign particles and organisms.

Targeting leukocyte rolling as a method of reducing vascular dysfunction has revealed some very peculiar results. Davenpeck et al. (38) have reported an increase in plasma aminonitrogen concentrations (an indicator of total plasma proteolysis) following splanchnic occlusion and reperfusion which was not evident in animals pretreated with an anti-P-selectin antibody. Kurose et al. (3) reported subtle protection in single postischemic venules of antibodies directed against L-selectin or P-selectin. It is very interesting that the P-selectin antibody did not reduce PMN influx into the same postischemic vessels. Although these data seem inconsistent with the premise that PMN infiltration is responsible for subsequent vascular dysfunction, they follow a similar trend reported by other investigators. Indeed, PMN influx and tissue protection did not coincide in a recent study wherein P-selectin antibody protected the lung from distant reperfusion injury without reducing PMN infiltration (42). Moreover, soluble sialyl Lewis^x (SLe^x) inhibits FITC-albumin leakage from mesenteric venules exposed to *Clostridium difficile* toxin A (see next section) without affecting PMN emigration.

Work in progress from our laboratory has revealed that antiselectin therapy reduces ischemia/reperfusion-induced microvascular dysfunction independent of leukocyte infiltration (44). Animals were pretreated with either an anti-E- and L-selectin antibody (EL-246), fucoidin, or an SLe^x analog, respectively, and PMN adhesion/emigration and FITC-albumin leakage were examined in single 20- to 40-μm venules. The increased rolling was reduced by 85%, 65%, and 0% with fucoidin, EL-246, and the SLe^x analog, respectively. Fucoidin significantly reduced PMN adhesion, but all of the interventions significantly decreased the microvascular dysfunction associated with reperfusion. Moreover, fucoidin, at a concentration that was ineffective as an inhibitor of PMN function, still reduced vascular permeability significantly. Although the underlying mechanism remains unclear, these interventions may have direct inhibitory effects on microvascular permeability alterations associated with ischemia/reperfusion, or they may suppress PMN activation independent of the rolling/adhesion/emigration process. Another possibility is that the selectins may affect other components of the inflammatory response including platelet function. Clearly, the complexity of the inflammatory cascade underscores the importance of visualizing PMN behavior and vascular dysfunction in the same post-ischemic vessels in vivo.

F. Low-Flow vs. Hemorrhagic Ischemia

Vedder et al. (5) demonstrated that treatment of rabbits with an anti-CD18 MAb (60.3) prior to induction of hemorrhagic shock significantly increased their survival rate over that seen in control animals which received a saline treatment (100% vs. 29% survival, respectively, at 5 days). Blocking of CD18-mediated interactions also markedly attenuated histologically assessed damage to the gastric mucosa, including accumulation of neutrophils, endothelial injury, edema, hemorrhage, and necrosis, suggesting that anti-CD18 antibody therapy can provide some protection against tissue injury in this global ischemia/reperfusion model. Unfortunately, the effect of MAb 60.3 on intestinal dysfunction in this hemorrhagic shock model remains unexplored.

In comparing local low-flow and hemorrhage-induced ischemia/reperfusion, Perry and Granger (6) found that leukocyte adhesion and emigration responses in both models were nearly identical. However, administration of an anti-CD18 MAb (IB_4) was more effective in preventing reperfusion-induced leukocyte adhesion and emigration than in the model of local ischemia compared with hemorrhagic ischemia. This difference could not be explained by differences in hemodynamic parameters (shear rates) or other obvious changes. Whether the hemorrhage model activates other leukocytes (monocytes, lymphocytes, eosinophils) that are known to adhere via CD18-independent mechanisms or whether hemorrhage induces the expression of novel adhesion molecules on endothelium and/or PMNs remains to be investigated. It was also observed that after reperfusion, the leukocyte rolling velocity increased in the hemorrhage model while velocity remained depressed in the ischemia model. The authors proposed that higher levels of circulating pro-inflammatory agents in the hemorrhage model may increase the degree of L-selectin shedding from leukocytes, lowering the number of adhesive interactions between rolling PMNs and the endothelium, resulting in a higher rolling velocity. However, this seems unlikely inasmuch as the reduced rolling velocity in local ischemia/reperfusion appears to be unrelated to L-selectin (26). Alternatively, there may be an increased expression of endothelial and or PMN ligands in local vs. global ischemia/reperfusion. Whether vascular permeability is altered to the same degree in these different models was not determined in this study and remains an important unanswered question.

G. Intestinal Transplantation

One obvious clinical extension of the intestinal ischemia/reperfusion work presented herein is the potential for improvement of intestinal transplantation. While intestinal transplants are a potential treatment for severe intestinal disease, the clinical success of such procedures is low perhaps due to the contribution of ischemia/reperfusion to tissue injury during preparation, transport, and storage of the donor tissue. A recent study by Slocum et al. (45) demonstrated that even preparation of the small bowel for immediate transplantation caused increased mucosal permeability, increased vascular protein and fluid leakage, and reduced transmucosal water absorption. Hypothermic ischemia for as little as 2 hours prior to transplant and reperfusion exacerbated this injury severely. Treatment of transplant recipients with anti-CD18 MAb (IB_4) but not antibodies against ICAM-1 or P-selectin significantly attenuated vascular protein clearance but did not reduce mucosal permeability, suggesting a role for circulating PMNs in the microvasculature but not mucosal dysfunction. As already mentioned (Fig. 2), it may be likely that mucosal clearance in this model is due to actions of tissue-resident PMN populations rather than circulating PMNs. Clearly, depleting resident PMNs from donor tissue may be a viable approach for successful intestinal transplantation. Methods or drugs that selectively and rapidly remove PMNs from the mucosa (rather than 48-hour pretreatment with IB_4) or significantly impair PMN activation could be very attractive agents as therapeutic interventions in intestinal transplantation.

IV. NONSTEROIDAL ANTI-INFLAMMATORY DRUGS

The prolonged use of NSAIDs for the treatment of chronic inflammatory conditions such as rheumatoid arthritis is often limited by drug-induced ulceration of the gastrointestinal tract, most notably in the stomach. Although a number of mechanisms (including ischemia, impaired mucus production, and an altered acid-neutralizing capacity) have been implicated in the pathogenesis of NSAID-induced gastric mucosal injury, recent attention has focused on the potential contribution of PMNs (46,47). The concept that PMNs may mediate NSAID injury evolved from the observation that exposure of rat gastric mucosa to aspirin results in the appearance of "white thrombi," presumably consisting of neutrophils, in mucosal capillaries (48). The view that NSAIDs may elicit the activation and margination of PMNs is supported by several recent reports that describe an enhanced accumulation of PMNs within postcapillary venules that are directly exposed to NSAIDs (12,49,50). In addition, it has been shown that human neutrophils adhere more avidly to monolayers of cultured human umbilical vein endothelial cells (HUVEC) after exposure of both cell types to aspirin (51). The dose-dependent adhesion response elicited by aspirin is observed when the neutrophils, but not HUVEC monolayers alone, are exposed to the NSAID, suggesting that direct activation of neutrophils is the dominant action of aspirin in this in vitro model.

Exposure of rat mesenteric venules to either aspirin or indomethacin results in an accumulation of firmly adherent PMNs but does not increase PMN emigration into the adjacent interstitial compartment (12,49,50). The NSAID-induced adhesion response is dose-dependent and it is often accompanied by a reduction in venular shear rate. However, the decline in shear rate is not sufficient to explain the significant level of PMN adhesion induced by either NSAID. A more likely mechanism for the NSAID-induced leukocyte adhesion relates to the ability of these agents to inhibit cyclooxygenase activity and thereby shunt arachidonic acid through the 5-lipoxygenase pathway to produce leukotrienes. A leukotriene B_4-dependent mechanism for the recruitment and activation of PMNs following aspirin or indomethacin treatment is supported by several lines of evidence: 1. LTB_4

production in rat mesentery is increased after exposure to either aspirin or indomethacin (49,50); 2. the PMN-endothelial cell adhesion elicited by aspirin or indomethacin in postcapillary venules and the aspirin-induced neutrophil adhesion to HUVEC are largely prevented by pretreatment with either an LTB_4 receptor antagonist or a 5-lipoxygenase inhibitor (49,50); 3. replenishment of tissue cyclooxygenase products with exogenous prostacyclin (PGI_2) or misoprostol (PGE_1) blunts the adhesion responses normally induced by aspirin or indomethacin (49,50); 4. salicylate, which is structurally similar to aspirin but lacks the cyclooxygenase inhibitory activity, does not elicit PMN-endothelial cell adhesion (49); and 5. tenidap, which effectively inhibits both cyclooxygenase and 5-lipoxygenase activities, does not cause PMNs to adhere within postcapillary venules (52). The latter findings indicate that the deleterious inflammatory actions of NSAIDs on the gastric mucosa may be alleviated by the use of agents that block both arms of arachidonic acid metabolism, rather than cyclooxygenase alone.

Studies of aspirin-mediated adhesion of human neutrophils to HUVEC have revealed the involvement of the β_2-integrins (CD11a/CD18, CD11b/CD18) on neutrophils and ICAM-1 on endothelial cells (51). Incubation of neutrophils with aspirin leads to an increased surface expression of CD11b/CD18. Furthermore, MAbs against either ICAM-1 or P-selectin, but not E-selectin, significantly attenuate indomethacin-induced PMN adherence in mesenteric venules (12). Although ICAM-1 expression on cultured endothelial cells is not increased following brief exposure to aspirin (51), there is some in vivo evidence that supports the view that inhibition of prostaglandin synthesis with NSAIDs elicits rapid upregulation of ICAM-1 on endothelial cells in postcapillary venules in rat gastric mucosa (53). The number of ICAM-1-stained vessels/mm^2 increased about fourfold within 30 min after exposure of the gastric mucosa to either aspirin or indomethacin. The PGE_1 analog misoprostol, which has been shown to largely abolish the leukocyte adhesion elicited by either aspirin or indomethacin in mesenteric venules, appears to be very effective in blunting the increased NSAID-induced ICAM-1 expression in rat gastric mucosa. Overall, these findings indicate that NSAIDs promote PMN-endothelial cell adhesion by increasing the surface expression of adhesion glycoproteins on both neutrophils (CD11b/CD18) and endothelial cells (ICAM-1). The mechanism by which misoprostol inhibits ICAM-1 expression on activated endothelial cells remains unclear, but this finding suggests that prostaglandin analogs may have significant therapeutic potential in the management of chronic inflammatory diseases.

The pathophysiologic relevance of the PMN-endothelial cell adhesion induced by NSAIDs is exemplified by observations that agents that prevent or attenuate the accumulation of inflammatory cells in the microvasculature significantly blunt NSAID gastropathy (46,47). Rats made neutropenic either by administration of antineutrophil serum (ANS) or methotrexate are resistant to the damaging effects of indomethacin or naproxen, both of which inhibit gastric cyclooxygenase activity by >95%, in the gastric fundus (11). This protective action appears to be limited to the fundic region of the gastric mucosa inasmuch as neutropenia does not prevent indomethacin-induced ulceration of the gastric antrum (54). The lack of a role for neutrophils in indomethacin-induced ulceration of the rat gastric antrum is further supported by histological evidence that the vascular damage and mucosal ulceration caused by indomethacin precedes the infiltration of PMNs (55). An early PMN-independent phase of indomethacin-induced mucosal injury has also been demonstrated in rat jejunum, wherein there is rapid cyclooxygenase inhibition and corresponding microvascular alterations (56).

Protection of the gastric fundus against the mucosal damaging actions of indomethacin can also be demonstrated in rabbits that receive a monoclonal antibody (MAb IB_4) directed against the β-subunit of the leukocyte adhesion glycoprotein, CD11/CD18 (57). The vascular congestion, PMN margination, and hemorrhagic lesions normally observed in the gastric mucosa of indomethacin-treated rabbits were not observed in animals receiving MAb IB_4. Monoclonal antibodies directed against endothelial cell adhesion molecules (ICAM-1, P-selectin, and E-selectin) also reduce the severity of mucosal damage induced by indomethacin (12). While the MAb directed against ICAM-1 reduced indomethacin-induced gastric mucosal injury and PMN adherence by a similar extent (~75%), the E-selectin MAb significantly reduced gastric mucosal injury without exerting a comparable inhibitory effect on PMN adherence (12). It is noted that leukocyte adherence was examined for a period of only 30 min, whereas the severity of gastric injury was examined 3 hours after administration of indomethacin. It is possible that leukocyte adhesion mediated by E-selectin occurred later than that mediated via other adhesion molecules resulting in protective effects of anti-E-selectin antibodies at this later time point. A P-selectin MAb attenuated indomethacin-induced PMN adhesion by ~50% and reduced mucosal injury by 35%. Nevertheless, these findings are consistent with the view that the major molecular determinants of the PMN infiltration elicited by NSAIDs are CD11/CD18 on neutrophils and ICAM-1 on venular endothelial cells. The effectiveness of these MAbs in blunting NSAID-induced gastric injury lends credence to the therapeutic strategy of coadministering NSAIDs with low-molecular-weight agents that interfere with PMN-endothelial cell adhesion. It remains unclear whether misoprostol represents such an agent.

V. BACTERIAL TOXINS

The gastrointestinal lumen is populated by a large number and wide variety of bacteria, many of which are capable of eliciting a local inflammatory response through the release of cell membrane-associated exotoxins. In recent years, much attention has been devoted to defining the mechanisms that underlie the inflammatory responses elicited by bacterial toxins within the gastrointestinal tract. Both *Helicobacter pylori* and *Clostridium difficile* have been implicated in the pathogenesis of gastric and duodenal ulceration. Eradication of the *H. pylori* microorganism in patients with ulcers results in resolution of the disease while reinfection is associated with recurrence of ulcers. It has recently been shown that a water extract of *H. pylori* promotes the adherence of isolated human PMNs to HUVEC monolayers and increases PMN adherence and emigration in rat mesenteric venules (13). The in vitro studies indicate that the *H. pylori* extract promotes PMN adhesion in a dose-dependent fashion, and that preincubation of endothelial cells alone does not elicit a response. Furthermore, the extract-induced PMN adhesion to HUVEC is significantly diminished by MAbs against either CD11a, CD11b, or CD18 on neutrophils or ICAM-1 on endothelial cells (13). In vivo studies also implicate a role for CD11/CD18 and ICAM-1 in mediating the PMN adherence and emigration elicited in postcapillary venules following topical application of the *H. pylori* extract (58). Intravital microscopic analysis of the responses of the mesenteric microvasculature to the extract revealed an increased albumin leakage from venules which is initially associated with mast cell degranulation and then accompanied by PMN adherence and emigration. Mast cell stabilizers effectively blunt the early rise in albumin leakage, while MAbs directed against either CD11/CD18 or ICAM-1 reduce the later component of albumin extravasation. The *H. pylori* extract also induced the formation of platelet-leukocyte aggregates within postcapillary venules, a process that is inhibited by a MAb directed against P-selectin (presumably expressed by platelets). Overall, these findings indicate that the inflammatory responses induced by exotoxins released from *H. pylori* involve a complex interaction of mediators and adhesion molecules associated with mast cells, platelets, endothelial cells, and PMNs.

C. difficile toxin A (Tx-A) has been implicated in the pathogenesis of antibiotic-associated pseudomembranous colitis in humans and experimental animals. The mechanisms responsible for the inflammatory reactions elicited by Tx-A were recently studied using intravital videomicroscopy (14). Topical application of Tx-A to rat mesenteric venules results in the recruitment of adherent and emigrating PMNs, with a corresponding increase in the rate of extravasation of albumin. These responses are accompanied by mast cell degranulation and the formation of platelet-leukocyte aggregates. The Tx-A induced PMN adhesion/emigration and albumin leakage were significantly attenuated by prior administration of MAbs against CD11/CD18, ICAM-1, or P-selectin. Sialyl Lewis^x, a putative counterreceptor for P-selectin, mast cell stabilizers, histamine H_1 (but not H_2) receptor antagonists, and diamine oxidase (histaminase) were all comparably effective in blunting the Tx-A-induced leukocyte adhesion and albumin leakage responses. These observations indicate that Tx-A causes inflammation and a PMN-dependent leakage of albumin via a mechanism that involves mast cells, endothelial cells, and platelets. Mast cell-derived histamine appears to mediate at least part of the PMN adhesion and platelet-leukocyte aggregation by engaging H_1 receptors on endothelial cells and platelets to increase the expression of P-selectin.

VI. SUMMARY

In this brief review we have attempted to demonstrate the importance of adhesion molecules in the pathogenesis of intestinal disease. Although anti-CD18 antibodies appear to consistently and efficiently inhibit PMN infiltration and intestinal injury, significant complications exist that make it difficult to administer anti-CD18 antibodies to patients. Clearly, substances are needed that temporarily impair CD18 function or selectively impair inflammatory processes without affecting the ongoing fight against infection.

ACKNOWLEDGMENTS

Supported by grants from the Alberta Heritage Foundation for Medical Research and the Medical Research Council of Canada.

REFERENCES

1. Kubes P, Hunter J, Granger DN. Ischemia/reperfusion-induced feline intestinal dysfunction: importance of granulocyte recruitment. Gastroenterology 1992; 103:807–812.
2. Kurtel H, Tso P, Granger DN. Granulocyte accumulation in postischemic intestine: role of leukocyte adhesion glycoprotein CD11/CD18. Am J Physiol 1992; 262:G878–G882.
3. Kurose I, Anderson DC, Miyasaka M, et al. Molecular determinants of reperfusion-induced leukocyte adhesion and vascular protein leakage. Circ Res 1994; 74:336–343.
4. Smith SM, Holm-Rutili L, Perry MA, et al. Role of neutrophils in hemor-

rhagic shock-induced gastric mucosal injury in the rat. Gastroenterology 1987; 93:466–471.
5. Vedder NB, Winn RK, Rice CL, Chi EY, Arfors K-E, Harlan JM. A monoclonal antibody to the adherence-promoting leukocyte glycoprotein, CD18, reduces organ injury and improves survival from hemorrhagic shock and resuscitation in rabbits. J Clin Invest 1988; 81:939–944.
6. Perry MA, Granger DN. Leukocyte adhesion in local versus hemorrhage-induced ischemia. Am J Physiol 1992; 263:H810–H815.
7. Kubes P, Ibbotson G, Russell J, Wallace JL, Granger DN. Role of platelet-activating factor in ischemia/reperfusion-induced leukocyte adherence. Am J Physiol 1990; 259:G300–G305.
8. Kubes P, Arfors KE, Granger DN. Platelet-activating factor-induced mucosal dysfunction: role of oxidants and granulocytes. Am J Physiol 1991; 260:G965–G971.
9. Kvietys PR, Perry MA, Gaginella TS, Granger DN. Ethanol enhances leukocyte-endothelial interactions in mesenteric venules. Am J Physiol 1990; 259: G578–G583.
10. Wallace JL, Higa A, McKnight GW, MacIntyre DE. Prevention and reversal of experimental colitis by a monoclonal antibody which inhibits leukocyte adherence. Inflammation 1992; 16:343–354.
11. Wallace JL, Keenan CM, Granger DN. Gastric ulceration induced by nonsteroidal anti-inflammatory drugs is a neutrophil-dependent process. Am J Physiol 1990; 259:G462–G467.
12. Wallace JL, McKnight W, Miyasaki M, et al. Role of endothelial adhesion molecules in NSAID-induced gastric mucosal injury. Am J Physiol 1993; 265: G993–G998.
13. Yoshida N, Granger DN, Evans DJ Jr., et al. Mechanisms involved in *Heliobacter pylori*-induced inflammation. Gastroenterology 1993; 105:1431–1440.
14. Kurose I, Pothoulakis C, LaMont JT, et al. *Clostridium difficile* toxin A-induced microvascular dysfunction: role of histamine. J Clin Invest 1994; 94: 1919–1926.
15. Coskun T, Alican I , Yegen BC, Sand T, Cetinel S, Kurtel H. Cyclosporin A reduces the severity of cold-restraint-induced gastric lesions: role of leukocytes. Digestion. 1995; 56:214–219.
16. Picker LJ, Warnock RA, Burns AR, Doerschuk CM, Berg EL, Butcher EC. The neutrophil selectin LECAM-1 presents carbohydrate ligands to the vascular selectins ELAM-1 and GMP-140. Cell 1991; 66:921–933.
17. Lawrence MB, Bainton DF, Springer TA. Neutrophil tethering to and rolling on E-selectin are separable by requirement for L-selectin. Immunity 1994; 1: 137–145.
18. Hernandez LA, Grisham MB, Twohig B, Arfors KE, Harlan JL, Granger DN. Role of neutrophils in ischemia-reperfusion-induced microvascular injury. Am J Physiol 1987; 253:H699–H703.
19. Oliver MG, Specian RD, Perry MA, Granger DN. Morphologic assessment of leukocyte-endothelial cell interactions in mesenteric venules subjected to ischemia and reperfusion. Inflammation 1991; 15:331–346.

20. Kubes P, Kurose I, Granger DN. NO donors prevent integrin-induced leukocyte adhesion but not P-selectin-dependent rolling in postischemic venules. Am J Physiol 1994; 267:H931–H937.
21. Karasawa A, Guo J-P, Ma X-L, Tsao PS, Lefer AM. Protective roles of a leukotriene B_4 antagonist in splanchnic ischemia and reperfusion in rats. Am J Physiol 1991; 261:G191–G198.
22. Wright SD, Rao PE, Van Voorhis, et al. Identification of the CRbi receptor of human monocytes and macrophages by using monoclonal antibodies. Proc Natl Acad Sci USA 1983; 80:5699–5703.
23. Suzuki M, Inauen W, Kvietys PR, et al. Superoxide mediates reperfusion-induced leukocyte-endothelial cell interactions. Am J Physiol 1989; 257: H1740–H1745.
24. Barroso-Aranda J, Schmid-Schonbein G, Zweifach BW, Engler RL. Granulocytesand no-reflow phenomenon in irreversible hemorrhagic shock. Circ Res 1988; 63:437–447.
25. Carden DL, Smith JK, Korthuis RJ. Neutrophil-mediated microvascular dysfunction in postischemic canine skeletal muscle: role of granulocyte adherence. Circ Res 1990; 66:1436–1444.
26. Jerome SN, Smith CW, Korthuis RJ. CD18-dependent adherence reactions play an important role in the development of the no-reflow phenomenon. Am J Physiol 1993; 264:H479–H483.
27. Bienvenu K, Granger DN. Molecular determinants of shear rate-dependent leukocyte adhesion in postcapillary venules. Am J Physiol 1993; 264:H1504–H1508.
28. Kubes P, Jutila M, Payne D. Therapeutic potential of inhibiting leukocyte rolling in ischemia/reperfusion. J Clin Invest. 1995; 95:2510–2519.
29. Kishimoto TK, Warnock RA, Jutila MA, et al. Antibodies against human neutrophil LECAM-1 (Lam-1/Leu-8/DREG-56 antigen) and endothelial cell ELAM-1 inhibit a common CD18-independent adhesion pathway in vitro. Blood 1991; 78:805–811.
30. Lawrence MB, Springer TA. Neutrophils roll on E-selectin. J Immunol 1993; 151:6338–6346.
31. Weyrich AS, Buerke M, Albertine KH, Lefer AM. Time course of coronary endothelial adhesion molecule expression during reperfusion of the ischemic feline myocardium. J Leuk Biol 1995; 57:45–55.
32. Bevilacqua MP, Nelson RM. Selectins. J Clin Invest 1993; 91:379–287.
33. Nelson RM, Dolich S, Aruffo A, Cecconi O, Bevilacqua MP. Higher-affinity ligands for E-selectin. J Clin Invest 1993; 91:1157–1166.
34. Tangelder GJ, Arfors K-E. Inhibition of leukocyte rolling in venules by protamine and sulfated polysaccharides. Blood 1991; 77:1565–1571.
35. Ley K, Cerrito M, Arfors K-E. Sulfated polysaccharides inhibit leukocyte rolling in rabbit mesentery venules. Am J Physiol 1991; 260:H1667–H1673.
36. Arfors K-E, Ley K. Sulfated polysaccharides in inflammation. J Lab Clin Med 1993; 121:201–202.
37. Cecconi O, Nelson RM, Roberts WG, et al. Inostitol polyanions: noncarbohy-

drate inhibitors of L- and P-selectin that block inflammation. J Biol Chem 1994; 269:15060–15066.
38. Davenpeck KL, Gauthier TW, Albertine KH, Lefer AM. Role of P-selectin in microvascular leukocyte-endothelial interaction in splanchnic ischemia-reperfusion. Am J Physiol 1994; 267:H622–H630.
39. Kubes P, Kanwar S. Histamine induces leukocyte rolling in post-capillary venules: a P-selectin-mediated event. J Immunol 1994; 152:3570–3577.
40. Asako H, Kurose I, Wolf R, et al. Role of H_1 receptors and P-selectin in histamine-induced leukocyte rolling and adhesion in postcapillary venules. J Clin Invest 1994; 93:1508–1515.
41. Klausner JM, Anner H, Paterson IS, et al. Lower torso ischemia-induced lung injury is leukocyte dependent. Ann Surg 1988; 208:761–767.
42. Carden DL, Young JA, Granger DN. Pulmonary microvascular injury after ischemia-reperfusion: role of P-selectin. Am J Physiol 1993; 75:2529–2534.
43. Parks DA, Granger DN. Contributions of ischemia and reperfusion to mucosal lesion formation. Am J Physiol 1986; 250:G749–G753.
44. Payne D, Jutila M, Kubes P. Anti-selectin therapy reduces ischemia/reperfusion-induced microvascular dysfunction independent of leukocyte infiltration. FASEB J. 1995; 9:A420.
45. Slocum MM, Granger DN. Early mucosal and microvascular changes in feline intestinal transplants. Gastroenterology 1993; 105:1761–1768.
46. Wallace JL. Gastric ulceration: critical events at the neutrophil-endothelium interface. Can J Physiol Pharmacol 1993; 71:98–102.
47. Wallace JL, Granger DN. Pathogenesis of NSAID gastropathy: are neutrophils the culprits? TiPS 1992; 13:129–131.
48. Kitahora T, Guth PH. Effect of aspirin plus hydrocholric acid on the gastric mucosal microcirculation. Gastroenterology 1987; 93:810–817.
49. Asako H, Kubes P, Wallace J, Wolf RE, Granger DN. Modulation of leukocyte adhesion in rat mesenteric venules by aspirin and salicylate. Gastroenterology 1992; 103:146–152.
50. Asako H, Kubes P, Wallace J, Gaginella T, Wolf RE, Granger DN. Indomethacin-induced leukocyte adhesion in mesenteric venules: role of lipoxygenase products. Am J Physiol 1992; 262:G903–G908.
51. Yoshida N, Takemura T, Granger DN, et al. Molecular determinants of aspirin-induced neutrophil adherence to endothelial cells. Gastroenterology 1993; 105:75–724.
52. Panes J, Russell JM, Wolf RE, Wallace JL, Granger DN. Effects of tenidap on leukocyte-endothelial cell adhesion in mesenteric venules. J Rheumatol. 1995; 22:444–449.
53. Andrews FJ, Malcontenti-Wilson C, O'Brien PE. Effect of nonsteroidal anti-inflammatory drugs on LFA-1 and ICAM-1 expression in gastric mucosa. Am J Physiol 1994; 266:G657–G664.
54. Trevethick MA, Clayton NM, Bahl AK, Sanjar S, Strong P. Neutrophil depletion does not prevent indomethacin-induced ulceration of the rat gastric antrum. Agents Actions 1994; 41:C226–C227.

55. Doble ATA, Bahl AK. Histological evidence that vascular damage and mucosal ulceration caused by indomethacin precede neutrophil infiltration in the rat gastric antrum. Agents Actions 1994; 41:C228–C230.
56. Nygård G, Anthony A, Piasecki C, et al. Acute indomethacin-induced jejunal injury in the rat: early morphological and biochemical changes. Gastroenterology 1994; 106:567–575.
57. Wallace JL, Arfors K-E, McKnight GW. A monoclonal antibody against the CD18 leukocyte adhesion molecule prevents indomethacin-induced gastric damage in the rabbit. Gastroenterology 1991; 100:878–883.
58. Kurose I, Granger DN, Evans DJ Jr., et al. *Helicobacter pylori*-induced microvascular protein leakage in rats: role of neutrophils, mast cells, and platelets. Gastroenterology 1994; 107:70–79.

22

Adhesion Molecules in Rheumatoid Arthritis

Elisabeth Bloemena, Anna C.H.M. van Dinther-Janssen, and Chris J.L.M. Meijer
Department of Pathology, Free University Hospital, Amsterdam, Netherlands

I. INTRODUCTION

Rheumatoid arthritis (RA) is a systemic, chronic disabling disease in which the joint manifestations dominate the clinical picture. Histopathologically, the rheumatoid synovial membrane shows synovial lining cell hyperplasia and inflammatory changes: edema, hyperemia, and infiltration with mononuclear cells that are diffusely distributed or nodularly arranged. In the latter configuration, compartmentalization occurs with the formation of a T-cell area, a transformation zone, and a plasma cell-rich area. Sometimes secondary lymphoid follicles with germinal centers are formed (1,2). Synovial fluid collects in the joint space. The hyperplastic synovial lining overgrows and destroys the articular cartilage, thereby forming the pannus. This ultimately results in complete loss of cartilage and sometimes even bone, leading to joint malformation (3,4).

The etiology and pathogenesis of RA are largely unknown. The disease is associated with certain HLA haplotypes, suggesting that immune-mediated mechanisms play an important role (5–8). It has been hypothesized that an immune response against certain antigens in susceptible individuals is of pivotal importance in initiating the joint inflammation. The interest in the role of extrinsic antigens originates from the observation that infections with certain microorganisms, e.g., mycobacteria and other slow-growing bacteria, can mimic the symptoms observed in RA (9). A bacterial etiology has clearly been established in reactive arthritis. In this case, an infection of the gastrointestinal or genitourinary tract is followed by arthritis, and antigens of several microorganisms, such as *Chlamydia*, *Yersinia*, *Salmonella*, and *Shigella*, have been detected in the joints of affected individuals (10–13). The nature of the antigen(s) involved in RA has thus far remained elusive. Much attention has focused on T-cell responses to the 65-kD heat shock protein of mycobacteria (14–16), since this protein was found to contain the epitope recognized by arthritogenic T-cell clones in animal models of RA (17). Cross-reactivity of T-cells isolated from the synovial fluid of RA patients with cell wall antigens of mycobacteria from the gut and proteoglycans of joint cartilage has been observed (18,19).

It seems probable that arthritis in RA is triggered by the entrance of an arthrotropic antigen into the joint. In this process, circulating antigen-presenting cells, e.g., dendritic cells, loaded with antigen(s) are thought to play a crucial role (20–24). After extravasation of dendritic cells into the joints, a local immune response is initiated. Next, lymphocytes extravasate from blood vessels into the synovial membrane, possibly attracted by locally produced cytokines (25–28). This in turn augments the local immune response, leading to B-cell stimulation and the production of certain immunoglobulins, known as rheumatoid factors (3,29,30). T-cell reactivity against the initial antigen can become autoreactive by molecular mimicry between this antigen and autoantigens present in the joint (8,31). Although the described sequence of events is largely hypothetical, sequential histopathological observations in joint lesions early in the course of RA support this concept.

Therefore, extravasation of immune-competent cells from the blood is a major event in the pathogenesis of RA. Understanding the mechanisms by which mononuclear cells adhere to and cross the endothelium and enter the synovial membrane is important in providing tools to therapeutically interfere with the joint disease in RA.

II. LYMPHOCYTE-ENDOTHELIAL CELL INTERACTION

Recirculation of lymphocytes through the body is essential for an adequate immune response. Lymphocytes migrate into lymphoid organs via specialized postcapillary venules. These venules are lined with high cuboidal endothelial cells (ECs) and have therefore been designated high endothelial venules (HEV) (32,33). The homing of lymphocytes is determined by interaction of adhesion molecules on lymphocytes with specific receptors for adhesion molecules on endothelial cells.

Different families of adhesion molecules involved in leukocyte extravasation have been discerned: the selectins, the integrins, the Ig superfamily,

CD44, and the sialomucins. Table 1 summarizes these adhesion molecules and individual family members, relevant to the current review.

Adhesion of lymphocytes to HEV is a multistep process in which several adhesion mechanisms are sequentially involved. This process has been elucidated for neutrophils, but it has recently been established that similar mechanisms play a role in lymphocyte extravasation. First, rolling of leukocytes is mediated by selectins that interact with their carbohydrate ligands on the EC. Next, the adhesion is affirmed when β_1-integrin molecules on the leukocytes become activated and bind with high affinity to their Ig superfamily receptors on endothelium. In this process of activation, locally produced cytokines, chemokines, and chemoattractants are involved. These

Table 1 Families of Adhesion Molecules

		Distribution	Ligand
Integrins			
$\alpha_4\beta_1$	CD49d/CD29	broad	fibronectin (CS1)
			VCAM-1
$\alpha_5\beta_1$	CD49e/CD29	broad	fibronectin (RGD)
			laminin
$\alpha_L\beta_2$	CD11a/CD18	leukocytes	ICAM-1,2,3
$\alpha_4\beta_7$		lymphocytes	MAdCAM-1
			fibronectin (CS1)
			VCAM-1
Immunoglobin superfamily			
MHC class I		broad	CD8
MHC classII		broad	CD4
ICAM-1	CD54	broad	LFA-1
			CR3
			CD43
VCAM-1	CD106	endothelium	$\alpha_4\beta_1$
		dendritic cells	$\alpha_4\beta_7$
Selectins			
L-selectin	CD62L	leukocytes	GlyCAM-2
E-selectin	CD62E	endothelium	silayl Lewis X
			CLA
Unclassified			
CD44	CD44	broad	hyaluronic acid
Sialomucins			
GlyCAM-2	CD34	broad	L-selectin
MAdCAM-1		endothelium	$\alpha_4\beta_7$
			(L-selectin)

factors are produced by tissue macrophages and activated EC, as the latter cells are able to produce IL-8 and MIP-1β, which can activate integrin molecules on leukocytes. Besides these soluble factors, adhesion molecules on EC can directly stimulate leukocytes. Finally, after firm adhesion, the leukocytes transmigrate across the endothelium. In the latter process, β_2-integrins play an important role (reviewed in 34–36).

Lymphocytes show preferential homing patterns into lymphoid organs—e.g., the peripheral lymph nodes and the mucosa-associated lymphoid tissues. The organ-specific trafficking of lymphocytes depends on the interaction of homing receptors on lymphocytes with organ-specific receptors (vascular addressins) on HEV endothelium. In mice, both the vascular addressins for peripheral lymph nodes and the mucosal lymphoid tissues have been characterized, as well as their counterreceptors on lymphocytes.

In man, the mucosal vascular addressin is unknown. Homing to peripheral lymph nodes is mediated by the interaction of L-selectin on lymphocytes with the peripheral node addressin (PNAd) CD34 (GlyCAM-2) on EC (37–41). In the gut, lymphocyte homing is mediated by binding of the integrin $\alpha_4\beta_7$ on lymphocytes to MAdCAM-1 on EC (42–46). It has recently been shown that homing characteristics of lymphocytes can be fine-tuned by differences in glycosylation of the vascular addressin. Modification of the mucinlike domain in MAdCAM-1 by oligosaccharides adds an L-selectin binding site to the molecule. MAdCAM-1 in mesenteric lymph nodes and Peyer's patches possesses this binding site for L-selectin, while MAdCAM-1 on EC in the lamina propria of the gut lacks it and can only interact with $\alpha_4\beta_7$ (47).

Likewise, migration of lymphocytes into sites of inflammation depends on the interaction with endothelial cells. In many inflammatory processes, among which rheumatoid arthritis, small vessels with morphological and histochemical resemblance to HEV are present (24,33,48–50). From in vitro experiments with cultured human umbilical vein EC (HUVEC), it has become clear that cytokines, such as IL-1, IFN-γ, and TNF, and combinations thereof, can induce morphological changes (51) and the expression of several receptors for adhesion molecules on the EC (52–56) (Fig. 1). However, the susceptibility of endothelium for the cytokines differs, depending on the source of origin of the endothelium. IFN-γ alone increases the expression of ICAM-1 in synovial endothelium in vitro, whereas IFN-γ alone has no effect on HUVEC (57,58).

Organ specificity of lymphocyte homing has been observed in inflammatory processes—e.g., in recirculation through lymphoid tissues. In allergic and delayed-type hypersensitivity skin reactions, as well as in cutaneous discoid lupus erythematodes, expression of E-selectin on the EC of cutaneous vessels (56,59) coincides with infiltration by a subpopulation of memory T

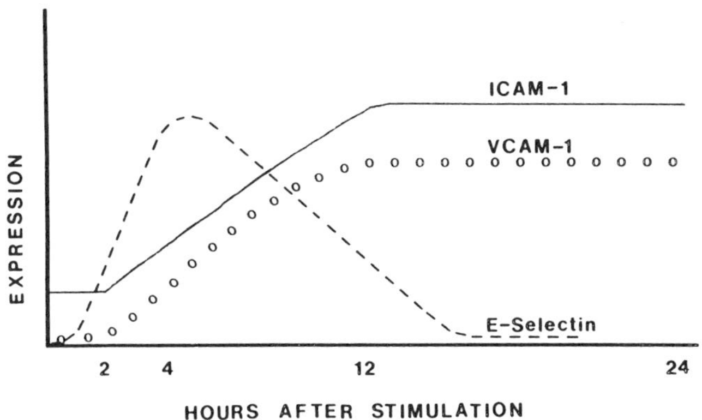

Figure 1 Induction of endothelial cell adhesion molecules by IL-1 and TNF-α on cultured human umbilical vein endothelial cells.

lymphocytes that express the cutaneous lymphocyte-associated antigen (CLA) recognized by the monoclonal antibody (mAb) HECA 452 (60). From these data, it has been hypothesized that E-selectin represents the vascular addressin of the skin (61) and is the ligand for CLA, the cutaneous homing receptor (62). Based on the in vitro observation that a lymph node and a mucosa HEV-specific B lymphoblastoid cell line did not adhere to synovial HEV (63), a synovial-specific adhesion mechanism has been postulated.

The organ specificity of lymphocyte recirculation is also regulated by the expression of adhesion molecules on the lymphocytes, the affinity of which is determined by the activation state of the cells. Naive T-cells preferentially bind to HEV in peripheral lymph nodes, but in memory T-cells the expression of the peripheral lymph node homing receptor L-selectin is decreased, resulting in a diminished binding to the EC of the HEV in peripheral lymph nodes in vitro (39,64). Thus, it is thought that activation of lymphocytes by antigen at specific sites leads to an alteration in the expression and activation state of various adhesion molecules, creating a homing pattern that is specific for the site where the antigen was encountered (65).

The molecular mechanisms involved in the lymphocyte-EC adhesion pathways known to be operative in rheumatoid arthritis will be discussed in more detail.

A. E-selectin/Sialyl-Lewis X

E-selectin is not constitutively expressed on EC, but is rapidly induced on cultured HUVEC after incubation with IL-1, TNF, or LPS. Expression is maximal at 4 to 6 hours after activation but declines thereafter, despite the continued presence of cytokines (52,54) (Fig. 1.). Although E-selectin is only transiently expressed in vitro, its presence on EC has been observed in several chronic inflammatory processes, among them rheumatoid arthritis (66,67), and soluble E-selectin is released in the synovial fluid during the active phase of the inflammatory process (68).

The binding site of E-selectin consists of a lectinlike domain, while its counterstructure on leukocytes is a carbohydrate moiety. The fucosylated tetrasaccharide sialyl-Lewis X (sLe^x is the best-known ligand for E-selectin (69–71). SLe^x is expressed on neutrophils, monocytes, and natural killer cells (72). In polymorphonuclear cells, L-selectin can be modified by sLe^x and thus serves as a ligand for E-selectin. On peripheral blood lymphocytes, sLe^x is expressed only after activation (73), but binding of L-selectin to E-selectin via sLe^x does not seem to occur (74). Alternatively, the CLA-positive skin-homing subset of T lymphocytes binds to E-selectin via HECA-452. The HECA-452 antigen is sialylated and fucosylated and is thus closely related to sLe^x (62).

In vitro, T-cells isolated from the synovial membrane and the synovial fluid of the joints of RA patients strongly adhere to purified E-selectin (75). However, it remains to be elucidated whether the binding of T-cells to vascular E-selectin in the synovial membrane contributes to the extravasation of lymphocytes in RA. The counterreceptor on T lymphocytes involved is presently not defined. SLe^x, the prototypical receptor for E-selectin, is only expressed on a small minority of T-lymphocytes in the synovial membrane (67,72), although a high expression would be expected as the majority of the T-cells display an activated phenotype (67,75–79). Secondly, in the synovial membrane of RA patients, no HECA-452 positive lymphocytes are found (24). Thirdly, since the T lymphocytes in the synovial membrane are activated, L-selectin expression is downregulated (39,64,67), irrespective of the fact that in T-cells L-selectin probably cannot serve as a counterreceptor for E-selectin. Thus, none of the known ligands of E-selectin are abundantly expressed on T-cells in the rheumatoid synovial membrane, casting doubt on the importance of this adhesion pathway in T-lymphocyte–EC interaction in RA, although the presence of an as yet unidentified receptor for E-selectin on synovial-homing T-cells cannot be excluded.

B. VCAM-1/$\alpha_4\beta_1$

VCAM-1 is an adhesion molecule that belongs to the Ig superfamily, and is induced on HUVEC in vitro after activation with IL-1, LPS, and TNF (53,55). In contrast to E-selectin, expression is maximal 12 hours after stimulation and remains high in the presence of cytokines (Fig. 1). The ligand for VCAM-1 is the α_4 chain of the integrin VLA-4 ($\alpha_4\beta_1$) on lymphocytes (80–83). VLA-4 is expressed on peripheral blood T and B lymphocytes, and its expression is increased on T memory cells and after T-cell activation, as encountered in the synovial membrane and synovial fluid in RA (64,67,84,85).

In the inflamed synovial membrane of RA patients, a high percentage of VCAM-1-positive HEV is present (64,66,67,86). In in vitro adhesion experiments with frozen sections of inflamed RA synovial membrane, we have demonstrated that peripheral blood T cells adhere to HEV. This adhesion is strongly inhibited by mAbs against VCAM-1 and α_4 (64) (Fig. 2). Relative adherence of in vitro activated T-cells and T-cells isolated from the RA synovial membrane is two- to threefold increased compared to resting peripheral T-cells (Table 2). Also, the binding of activated and memory T-cells to synovial EC of HEV is strongly inhibited by mAbs against VCAM-1 and α_4 (Fig. 3). In contrast, in parallel experiments using peripheral lymph node sections, no inhibitory effect of mAbs against VCAM-1 or α_4 are observed. These data indicate that the binding of lymphocytes to synovial HEV largely depends on the interaction of VLA-4 with VCAM-1, while in the adhesion of lymphocytes to peripheral lymph nodes other mechanisms are apparently operative (64).

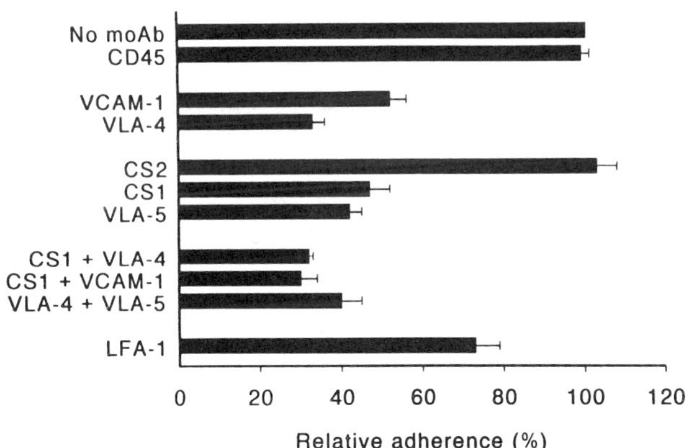

Figure 2 Inhibition of adhesion of normal peripheral blood T-lymphocytes to synovial HEV by mAbs against various adhesion molecules and the synthetic peptide CS1. Adhesion is expressed as % of the control (no mAb). The values represent the mean ± SEM of at least three experiments. MAbs against CD45 and the fibronectin-derived peptide CS2 are used as controls.

Table 2 Binding of Resting and Activated T-cells to Synovial Membrane HEV

Cell type	Synovial membrane HEV
Resting peripheral T-cells	5.75 ± 0.52
Synovial membrane T-cells	15.28 ± 0.91
αCD3-activated T-cells	15.39 ± 3.50

The results are expressed as the number of T-cells/mm HEV length. Each value represents the mean ± SEM of at least three experiments.

C. Fibronectin/$\alpha_4\beta_1$ and Fibronectin/$\alpha_5\beta_1$

VLA-4 is a unique member of the β_1-integrin family in its binding properties since it mediates both cell-cell and cell-matrix interaction (80). Besides to VCAM-1, VLA-4 also binds to the extracellular matrix protein fibronectin, a heterodimer with two cell-binding domains (87). The central domain, containing the sequence RGD, is involved in binding of VLA-5, the classical fibronectin receptor (88–90). VLA-4 adheres to fibronectin via a non-RGD sequence in an alternatively spliced domain of fibronectin, the connecting

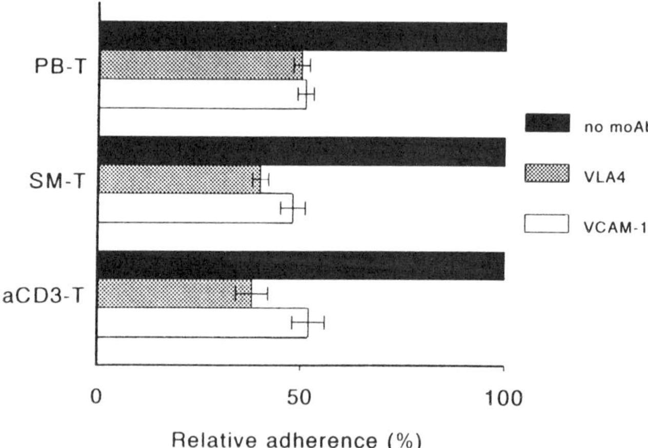

Figure 3 Inhibition of adhesion of activated T-lymphocytes (SM-T = RA synovial membrane T-cells; aCD3-T = T cells activated in vitro by anti-CD3 mAb) by mAbs against the integrin α_4 chain or VCAM-1. Adhesion is expressed as % of the control (no mAb). The values represent the mean ± SEM of at least three experiments.

segment 1 (CS1) (81–83,91,92). Fibronectin may play an important role in lymphocyte recirculation. In RA synovial specimens, fibronectin can immunohistochemically be demonstrated in vessel walls, basement membranes, and on the luminal surface of the vascular endothelium (93) (Fig. 4). Moreover, at this site fibronectin contains the CS1 segment (94).

In vitro lymphocyte adhesion to HEV in frozen sections of the RA synovial membrane can be blocked by the CS1 peptide (93,94). Also, a mAb against the α-chain of VLA-5 significantly reduces the binding of T cells, albeit to a lesser extent than the anti-α_4 mAb (93) (Fig. 2). Combination of the anti-α_4 mAb against α_4 with CS1 does not augment the inhibition of lymphocyte adhesion to synovial HEV, which might be explained by the fact that the epitope recognized by the anti-α_4 mAb affects both binding to VCAM-1 and the CS1-domain of fibronectin (81). As can be expected, the effects of CS1 and anti-VCAM-1 mAb on lymphocyte binding are additive. Moreover, combining anti-α_4 and anti-α_5 mAbs does not increase the inhibitory effect of either (93) (Fig. 2). This implies that in the homing to the inflamed joint in RA, binding of the lymphocytes to the endothelium also

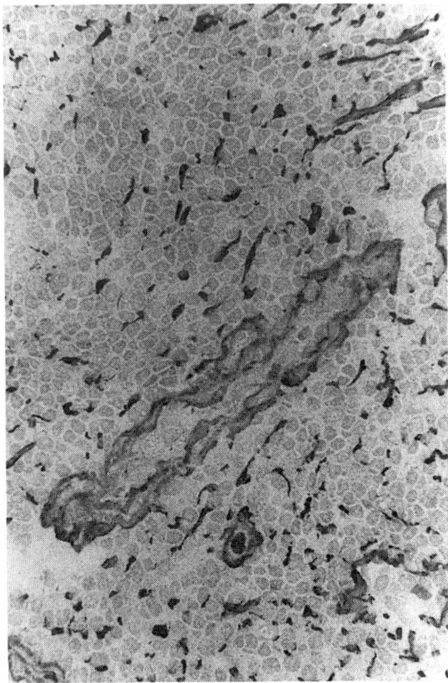

Figure 4 Expression of fibronectin on rheumatoid synovial HEV. Magnification ×400.

depends on fibronectin and that both the CS1/VLA-4 and the RGD/VLA-5 pathways are used. The fact that the combination of anti-α_4 and anti-α_5 mAbs has no additive effect on the inhibition of binding indicate that both pathways are used sequentially.

D. VCAM-1/$\alpha_4\beta_7$ and Fibronectin/$\alpha_4\beta_7$

It has recently been described that a large percentage of synovial membrane T-lymphocytes bear the $\alpha_4\beta_7$ integrin chain (95). In mice, this integrin has been described as the organ-specific lymphocyte receptor for gut homing (43–46). The counterstructure for $\alpha_4\beta_7$ in the gut, the mucosal vascular addressin MAdCAM-1 has until now only been identified in mice (42). In humans, it has been demonstrated that $\alpha_4\beta_7$ can bind to VCAM-1 and to the CS1 domain of fibronectin (82,96,97). Thus, in this respect, the $\alpha_4\beta_7$ integrin resembles $\alpha_4\beta_1$. It is unknown whether the binding only depends on the α-chain or to which extend the β-chain contributes to this interaction. In contrast to $\alpha_4\beta_1$, human $\alpha_4\beta_7$ adheres to purified mouse MAdCAM-1 in vitro (98). Therefore, currently unidentified counterreceptors, notably related to MAdCAM-1, might be operative in the adhesion of lymphocytes to HEV in the synovial membrane in humans.

E. ICAM-1/LFA-1

LFA-1 belongs to the β_2-integrin family, and its expression is restricted to leukocytes. ICAM-1, one of its counterreceptors on endothelium, is constitutively expressed by endothelial cells, albeit its expression is upregulated after cytokine stimulation (99). The kinetics of ICAM-1 expression are similar to that of VCAM-1. The interaction of LFA-1 with its receptor is critically important in acute inflammation, as demonstrated in patients deficient for the β_2 integrin (LAD type I syndrome). These patients show a defect in the accumulation of neutrophils into sites of inflammation, but the migration of lymphocytes into inflamed tissue is not impaired, indicating that the LFA-1/ICAM-1 interaction is not the only adhesion mechanism involved in lymphocyte homing into sites of inflammation (100,101). Besides its function in lymphocyte-endothelial cell adhesion, in vitro experiments have provided evidence for the involvement of the LFA-1/ICAM-1 pathway in the transendothelial migration of T-lymphocytes (102–104).

In the synovial membrane of RA patients, ICAM-1 is detected on endothelium, macrophages, infiltrating lymphocytes, fibroblasts, and synovial lining cells (66). LFA-1 is present on lymphocytes, on monocytes, and in some areas of synovial lining cells (67,105). In vitro adhesion of lymphocytes to HEV in frozen sections of the RA synovial membrane is only partially inhibited by anti-LFA-1 mAb (64) (Fig. 2). However, in vivo, mAbs against LFA-1 can suppress arthritis in animal models of RA (106). Also, encouraging results have been described for the treatment of RA patients with anti-ICAM-1 mAbs (107–110). These data indicate that, besides the β_1-integrin VLA-4, the β_2-integrin LFA-1 plays a role in the inflammatory process in RA, but that the interaction of the latter with ICAM-1 is probably not of primary importance in the adhesion of lymphocytes to EC.

F. Others

Recently, novel endothelial cell adhesion molecules have been described, defined by mAbs derived from immunization of mice with stromal fragments and synovial cells from synovial membranes of RA patients. The human endothelial cell adhesion molecule, defined by mAb 1B2 has been designated VAP-1. Detailed analysis shows that the molecule recognized by this mAb differs from all known endothelial adhesion molecules. VAP-1 is present on HEV in synovial membrane and on the endothelium in tonsil, appendix, and lamina propia of the gut. It is absent from HUVEC, also after in vitro stimulation with various cytokines. In vitro binding of lymphocytes to HEV in RA synovial membrane can be inhibited up to ±50% by 1B2 (111).

Another novel human endothelial cell adhesion molecule, also present on a subpopulation of venules in the synovial membrane, is recognized by the mAb 4G4. The molecule, L-VAP-2, is also expressed by venules in lymphoid tissue, among which are some HEV and venules in a wide variety of nonlymphoid tissues. It can be demonstrated on cultured HUVEC, but its expression cannot be upregulated by activation of the HUVEC with IL-1β, TNFα, or LPS. Moreover, the molecule is present on a subpopulation of peripheral blood T and B lymphocytes. In vitro, L-VAP-2 is functionally involved in the adhesion of lymphocytes to cultured HUVEC (112).

MAb 4A11 recognizes arteriolar and venular EC in synovial tissue, from both RA patients and controls. Also, 4A11-positive vessels are present in lymphoid tissue, some tumors, and inflammatory processes. It is expressed by cultured HUVEC, but the expression is unaffected by stimulation with cytokines. The molecules identified by 4A11 are the blood group antigens Le^y and H-5-2. The functional role of these molecules in lymphocyte homing has not yet been investigated (113).

The counterstructures of these novel endothelial molecules on lymphocytes are unknown. Moreover, their relative importance in lymphocyte migration to inflammatory sites, in particular inflamed synovial membranes, still has to be established. However, it can be envisaged that in the near future novel molecular mechanisms of lymphocyte-EC adhesion in the synovial membrane in RA will be unravelled.

III. DENDRITIC CELLS

As already discussed in the Introduction, dendritic cells, loaded with antigen, are thought to enter the synovial membrane from the blood, thereby initiating the immune response in RA locally (20–24). Although in this concept the extravasation of dendritic cells is the first step in developing arthritis, we started our discussion with the mechanisms of homing of lymphocytes, since the mechanisms that operate in extravasation of dendritic cells are largely unknown. Moreover, no in vitro studies have been performed with dendritic cells, and most of the data have been derived from in situ immunohistochemical studies.

In lymphoid tissue, dendritic cells are found in the T-cell areas. In several chronic inflammatory processes, they are frequently encountered in connection with T-cells (22,114–116). Morphologically, dendritic cells have long cytoplasmic protrusions via which they interact with T cells to form cell clusters. Both in vitro and in vivo, dendritic cells are potent antigen-processing cells and stimulators of T lymphocytes (117–120). Considering the colocalization of dendritic cells and T lymphocytes in inflammation, it can be assumed that dendritic cells contribute to the local T-cell activation.

The notion that dendritic cells might play an important role in the induction of arthritis stems from observations in antigen-induced arthritis in the rat in which accumulation of large numbers of Ia+ dendritic cells occurs at the onset of the inflammatory response (23). Therefore, we have investigated the distribution of dendritic cells in the RA synovial membrane (1,24). Although the presence of cells with a dendritic morphology has been described in the synovial lining, the identification of dendritic cells has been hampered by the lack of specific markers. We have defined dendritic cells on the basis of positive staining for certain antigens (i.e., RFDI, L25, and MHC class II), absence of acid phosphatase activity, and morphological criteria. Interestingly, a small subpopulation of these cells also expresses the HECA-452 antigen.

In the synovial lining, HECA-452+ dendritic cells are also present, comprising about 10% of the synovial lining cells. In the stroma, the distribution pattern depends on the stage of the disease and the extent of the lymphoid infiltrate. Lymphocytes are found either diffusely scattered or in nodularly arranged infiltrates. The latter configuration shows compartmentalization with a T-cell area, a transformation zone, and a plasma cell-rich area. Sometimes secondary lymphoid follicles with a germinal center are formed. A gradual transition from the diffuse infiltrate to the organoid nodular infiltrate is apparent in most rheumatoid synovial membranes. Stromal HECA-452+ dendritic cells display two patterns of distribution, corresponding to the architecture of the lymphoid infiltrate. In diffusely scattered infiltrates, HECA-452+ dendritic cells are present near small vessels without characteristics of HEVs. Often the dendritic cells are in close contact with CD4+/MHC class II+ T cells. The number of HECA-452+ dendritic cells is relatively high compared to the number of lymphocytes present. In nodular lymphoid infiltrates, HECA-452+ dendritic cells lie around blood vessels in the periphery of the nodules. In this configuration, the endothelium of the vessels is cuboid and the vessels resemble HEV. In both diffusely scattered and nodularly arranged infiltrate, HECA-452-negative cells are also present.

It is noteworthy that dendritic cells are never encountered in noninflamed synovial membrane, although some MHC class II+/acid phosphatase+ cells with "dendritic" morphology are present. However, these cells do not express the dendritic cell markers RFDI and L25.

These data suggest that HECA-452+ dendritic cells represent a population of activated antigen-presenting cells. These cells are already present in the synovial membrane when the lymphocytic infiltrate is relatively scarce and diffusely scattered. Also, it seems likely that in the course of the disease the diffuse infiltrates precede the nodularly arranged infiltrates. In the latter HEV are formed (48), probably under the influence of continuous local cytokine production, thereby facilitating the entrance of lymphocytes into the synovial membrane.

IV. LYMPHOCYTE-SYNOVIAL LINING CELL INTERACTION

Proliferation of synovial lining cells (SLCs) with formation of a pannus leads to the destruction of cartilage and is thus very important in the pathogenesis of the joint disease in RA (3). The synovial lining consists mainly of two cell types that differ in morphology and acid phosphatase reactivity (121–123). They are distinguished in a macrophagelike type A cell, and a fibroblastlike type B cell. In vitro, isolated SLCs can be induced to proliferate by cytokines (124). In vivo, this process is thought to be mediated by local cytokine production by lymphocytes and macrophages. It has been hypothesized that interaction of T lymphocytes with SLCs can lead to mutual cellular activation, resulting in proliferation of SLCs and ongoing stimulation of T cells (125). Moreover, activated SLCs produce prostaglandins and proteases, involved in the breakdown of proteoglycans from the cartilage (126,127). Also, in activated SLCs enzymatically active metalloproteases are produced from their proenzymes (4,128). Secretion of the metalloproteases further contributes to the destruction of matrix proteins in the joint. Therefore, it is important to understand the mechanisms by which immune-competent cells interact with and stimulate SLCs.

In RA synovial membrane, the SLCs express several adhesion molecules. Both type A and type B SLCs are strongly MHC class II-positive (129,130). In vitro, these cells have been shown to be efficient antigen-presenting cells. Moreover, both SLC cell types are ICAM-1 positive, whereas ICAM-2 is not expressed. VCAM-1 is present on the fibroblastlike type B cell, while the acid phosphatase type A cell expresses VCAM-1 to a considerably lesser extent (66,86,131). In sharp contrast, type A SLCs from noninflamed synovial membrane are only weakly positive for MHC class II and ICAM-1, whereas type B cells are completely negative (131) (Table 3).

From in vitro experiments with cultured synovial fibroblasts, it has become clear that VCAM-1 and ICAM-1 can be upregulated by several (combinations of) cytokines. TNFα, IL-1β, and IFN-γ increase the expression of both ICAM-1 and VCAM-1 (86,132–134), whereas IL-4 upregulates the expression of VCAM-1 only (86,135). When SLCs are cultured in the presence of monocytes, cell-cell contact is not needed to induce adhesion molecules, indicating that the expression of adhesion molecules depends on the production of soluble factors by the monocytes (133). In sharp contrast, in cultures of SLCs with T lymphocytes cell-cell contact is critically important in the induction of VCAM-1 and ICAM-1 on the synovial fibroblasts (125).

In vitro adhesion studies with cultured synovial fibroblasts have shown that activated T lymphocytes and T lymphocytes isolated from the synovial fluid of RA patients bind to synovial cells (86,105,132,134,135). Using frozen sections of synovial membranes, we demonstrated that both resting and activated peripheral blood T lymphocytes bind to the synovial layer in RA synovial membrane, but not in control synovial membrane (131). The adhesion is mainly restricted to the fibroblastlike type B cells. In this in vitro adhesion assay, we have investigated the mechanisms involved in the

Table 3 Expression of Adhesion Molecules by Synovial Lining Cells (SLCs) in Normal and RA Synovial Membrane

		MHC II	ICAM-1	ICAM-2	E-selectin	VCAM-1
non-RA SLC	type A	+	±	–	–	–
	type B	–	–	–	–	–
RA SLC	type A	+ +	+ + +	–	–	±
	type B	+ + +	+ +	–	–	+ + +

Expression of molecules was detected immunohistochemically on acetone-fixed cryostat sections of non-RA and rheumatoid synovial membrane, using a two-step immunoperoxidase staining method. The staining was graded from – (no reactivity) to + + + (strong reactivity).

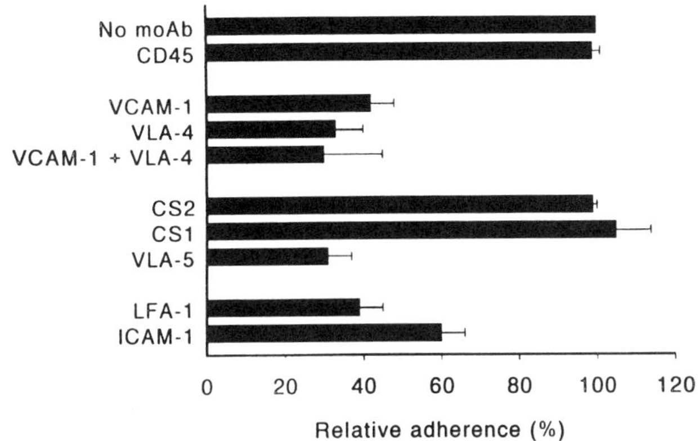

Figure 5 Inhibition of adhesion of anti-CD3 activated peripheral blood T-lymphocytes to type B synovial lining cells by mAbs against various adhesion molecules and the synthetic peptide CS1. Adhesion is expressed as % of the control (no mAb). The values represent the mean ± SEM of at least three experiments. MAbs against CD45 and the fibronectin-derived peptide CS2 are used as controls.

adhesion of T cells to SLCs. Inhibition experiments with anti-α_4 and anti-VCAM-1 mAbs showed that the VLA-4/VCAM-1 pathway is important in the binding of T-lymphocytes to SLCs (Fig. 5). Although fibronectin staining is positive on SLCs, the VLA-4/CS1 pathway is not involved in the lymphocyte binding to SLCs since the purified CS1 peptide has no effect on lymphocyte adhesion to SLCs in frozen sections. A mAb against the classical fibronectin receptor $\alpha_5\beta_1$ (VLA-5) strongly reduces T-cell binding to SLCs (131) (Fig. 5). This indicates that fibronectin at this site can support adhesion of T lymphocytes but that the alternatively spliced domain containing CS1 is not present.

Anti-LFA-1 mAb inhibits the binding of resting and activated T lymphocytes to SLCs. However, although ICAM-1 is strongly expressed by RA SLCs, mAbs against ICAM-1 only partially block T-cell binding to SLCs in frozen sections (131) (Fig. 5) and to cultured synovial fibroblasts (132). It is unknown whether another LFA-1 ligand is also implicated in the adhesion of T cells to SLCs. Involvement of ICAM-2 in the binding of lymphocytes to SLCs is very unlikely, since ICAM-2 is not expressed in synovial lining cells (Table 3). Although ICAM-3 is present on SLCs, its function in lymphocyte adhesion to SLCs in the RA synovial membrane still has to be established (136).

V. CONCLUSIONS

The pathogenesis of joint disease in RA is very complicated and at present poorly understood. The entrance of immune-competent cells into the joint is of pivotal importance in the initiation and maintenance of the inflammatory response, but the mechanisms involved are still not completely clarified. Studies have mainly centered around the adhesion of polymorphonuclear granulocytes and lymphocytes to endothelial cells. Moreover, in many instances, data obtained in various inflammatory processes have been translated into general mechanisms, ignoring the fact that in various organs differences in fine-tuning of adhesion can occur.

Knowledge of the molecular mechanisms of extravasation of immune-competent cells from the blood can provide novel tools to therapeutically interfere in inflammatory processes. In several animal models of autoimmune diseases and inflammation, amelioration of the process has been observed using blocking mAbs against adhesion molecules in vivo (137–140). In animal models of arthritis, beneficial effects of anti-LFA-1 mAbs have been reported (106). Furthermore, preliminary studies in man indicate that the administration of anti-ICAM-1 can reduce the severity of the arthritis in RA patients (107–110).

We and others have hypothesized that extravasation of dendritic cells loaded with antigen in the joint initiates the inflammation (20–24). It can be presumed that dendritic cells enter the synovial membrane from the blood (partly) via the same mechanisms as have been described for polymorphonuclear granulocytes and lymphocytes. However, as mentioned, no in vitro experiments with dendritic cells have been performed, and the molecular basis of dendritic cell homing is completely unknown. We have postulated that HECA-452+ dendritic cells, found in the joint in the early phases of inflammation, are activated antigen-presenting cells (24). With regard to the putative role of these cells in initiating the inflammatory response in the joint, the recently proposed hypothesis that activation of professional antigen-presenting cells, like dendritic cells, leads to spreading of epitopes is of great interest. According to this postulate, activation of dendritic cells induces alterations in their endosomal enzymes which results in the expression of former cryptic epitopes of the processed antigen. Alternatively, other cell types, like synovial lining cells in the joint, can become antigen-presenting cells by upregulation of MHC class II by inflammatory cytokines (129–131). Also, these activated nonprofessional antigen-presenting cells can express epitopes which lead to stimulation of autoreactive T-lymphocytes (141). Both the crucial role of activated dendritic cells and the crossreactivity of T cells to exogenous and endogenous antigens, described in patient with RA, fit very well into this concept.

Recirculation through the joint by a small number of lymphocytes presumably also occurs in the absence of inflammation. However, in the presence of antigen-loaded dendritic cells, antigen-specific lymphocytes are trapped in the synovial membrane. This leads to local immune activation with the production of cytokines (25–28,142). Cytokines induce morphological changes in EC and the expression of adhesion molecules, further facilitating the homing of lymphocytes into the joint. In the synovial membrane in RA, extravasation of lymphocytes is mediated by binding of α_4 on the lymphocyte to VCAM-1 on the EC. Adhesion of lymphocytes to fibronectin present on the EC is also implicated in homing to the inflamed joint. Both the RGD sequence and the domain containing CS1 are involved at this site, interacting with $\alpha_5\beta_1$ and $\alpha_4\beta_1$ on the T-lymphocyte, respectively. However, since mAbs against α_4 and α_5 have no additive inhibitory effect on the T-cell binding in vitro, these pathways are probably used sequentially (93). The presence of locally produced cytokines provides the prerequisite for the activation of the integrin molecules on the lymphocyte, necessary for strengthening of the adhesion. In the extravasation process, sLe^x/E-selectin and LFA-1/ICAM-1 are apparently of minor significance (64).

However, these mechanisms do not explain the homing specificity that has been postulated for the synovial membrane (63). In this respect it is interesting that crossreactivity between adhesion mechanisms in the synovial membrane and the mucosa-associated lymphoid tissue has been observed (143). Moreover, in RA a large population of synovial T-cells express the β_7-integrin chain (95,144), in mice known to be the homing receptor for the gut (43–46). It is not known how the specificity of homing to the synovial membrane is regulated. Probably, new vascular addressins will be discovered to clarify this point. It seems possible that the human homolog of the murine mucosal vascular addressin (presently unknown in man) is also expressed in the synovial membrane. In this respect, the recent observation that the mucosal addressin in mice (MAdCAM-1) shows differences in glycosylation at different sites in the body is important, since it provides possibilities for subtle differences in homing characteristics for various organs (47). Moreover, in mice, de novo expression of MAdCAM-1 on endothelial cells after activation in vitro (145) and at sites of inflammation (146–148) has been described.

Although ICAM-1 is expressed on the EC in the synovial membrane, it can be concluded from our data that the interaction of the β_2-integrin LFA-1 on the lymphocyte with ICAM-1 is not of primary importance in the adhesion of lymphocytes to the endothelial cells in the synovial membrane (64). However, since anti-ICAM-1 mAbs show promising results in the treatment of RA patients (108–110), the LFA-1/ICAM-1 pathway must be important in other steps of the inflammatory process. It has been demonstrated that β_2-integrins are involved in the transmigration of lymphocytes through the vessel wall (102–104). Additionally, the binding of T-lymphocytes to dendritic cells largely depends on the interaction of LFA-1 and ICAM-1 (149). These observations indicate that the effect of treatment of RA patients with anti-ICAM-1 mAb interferes with the inflammatory process at a stage after firm adhesion of the lymphocyte to the HEV in the synovial membrane.

REFERENCES

1. Meijer CJLM, van de Putte LBA, Eulderink F, Kleinjan R, Lafeber GJM, Bots GTAM. Characteristics of mononuclear cell populations in chronically inflamed synovial membranes. J Pathol 1977; 121:1–8.
2. Meijer CJLM, De Graaff-Reitsma CB, Lafeber GJM, Cats A. In situ localization of lymphocyte subsets in synovial membranes of patients with rheumatoid arthritis with monoclonal antibodies. J Rheumatol 1982; 9:359–365.
3. Harris ED Jr, Gallin JI, Goldstein IM, Snyderman R. Pathogenesis of rheumatoid arthritis: a disorder associated with dysfunctional immunoregulation. In: Inflammation: Basic Principles and Clinical Correlates. New York: Raven Press, 1988:751–773.
4. Firestein GS. Mechanisms of tissue destruction and cellular activation in rheumatoid arthritis. Curr Opin Rheumatol 1992; 4:348–354.
5. Albani S, Roudier J. Molecular basis for the association between HLA DR4 and rheumatoid arthritis. From the shared epitope hypothesis to a peptidic model of rheumatoid arthritis. Clin Biochem 1992; 25:209–212.
6. Lanchbury JS. The HLA association with rheumatoid arthritis. Clin Exp Rheumatol 1992; 10:301–304.
7. Wordsworth P. Rheumatoid arthritis. Curr Opin Immunol 1992; 4:766–769.
8. Wicks I, McColl G, Harrison L. New perspectives on rheumatoid arthritis. Immunol Today 1994; 15:553–556.
9. Rook GAW, Stanford JL. Slow bacterial infection or autoimmunity? Immunol Today 1992; 13:160–164.
10. Sieper J, Braun J, Brandt J, et al. Pathogenetic role of *Chlamydia*, *Yersinia* and *Borrelia* in undifferentiated oligoarthritis. J Rheumatol 1992; 19:1236–1242.
11. Mäki-Ikola O, Granfors K. *Salmonella*-triggered reactive arthritis. Lancet 1992; 339:1096–1098.
12. Nikkari S, Merilahti-Palo R, Saario R, et al. *Yersinia*-triggered reactive arthritis. Use of polymerase chain reaction and immunocytochemical staining in the detection of bacterial components from synovial specimens. Arthritis Rheum 1992; 35:682–687.
13. Kingsley G, Sieper J. Current perspectives in reactive arthritis. Immumol Today 1993; 14:387–391.
14. Strober S, Holoshitz J. Mechanisms of immune injury in rheumatoid arthri-

tis: evidence for the involvement of T cells and heat-shock protein. Immunol Rev 1990; 118:233–255.
15. Res P, Thole J, de Vries R. Heat-shock proteins and autoimmunity in humans. Springer Semin Immunopathol 1991; 13:81–98.
16. van Eden W. Heat shock proteins as immunogenic bacterial antigens with the potential to induce and regulate autoimmune arthritis. Immunol Rev 1991; 21:5–28.
17. van den Broek MF, Hogervorst EM, van Bruggen MCJ, van Eden W, van der Zee R, van den Berg WB. Protection against streptococcal cell wall-induced arthritis by pretreatment with the 65-kD myobacterial heat shock protein. J Exp Med 1989; 170:449–466.
18. Holoshitz J, Drucker I, Yaretzky T, et al. T lymphocytes of rheumatoid arthritis patients show augmented reactivity to a fraction of mycobacteria cross-reactive with cartilage. Lancet 1986; ii:305–309.
19. Res PCM, Schaar CG, Breedveld FC, et al. Synovial fluid T cell reactivity against 65 kD heat shock protein of mycobacteria in early chronic arthritis. Lancet 1988; ii:478–480.
20. Schumacher HR, Kitridou RC. Synovitis of recent onset: a clinicopathologic study during the first month of disease. Arthritis Rheum 1972; 15:465–484.
21. Janossy G, Panayi G, Duke O, Bofill M, Poulter LW, Goldstein G. Rheumatoid arthritis: a disease of T-lymphocyte/macrophage immunoregulation. Lancet 1981; ii:839–842.
22. Poulter LW, Duke O, Hobbs S, Janossy G, Panayi G, Seymour G. The involvement of interdigitating (antigen presenting) cells in the pathogenesis of rheumatoid arthritis. Clin Exp Immunol 1983; 51:247–254.
23. Verschure PJ, Van Noorden CJF, Dijkstra CD. Macrophages and dendritic cells during the early stages of antigen-induced arthritis in rats: immunohistochemical analysis of cryostat sections of the whole knee joint. Scand J Immunol 1989; 29:371–381.
24. van Dinther-Janssen ACHM, Pals ST, Scheper RJ, Breedveld FC, Meijer CJLM. Dendritic cells and high endothelial venules in the rheumatoid synovial membrane. J Rheumatol 1990; 17:11–17.
25. Brennan FM, Zachariae CO, Chantry D, et al. Detection of interleukin 8 biological activity in synovial fluids from patients with rheumatoid arthritis and production of interleukin 8 mRNA by isolated synovial cells. Eur J Immunol 1990; 20:2141–2144.
26. Peichl P, Ceska M, Effenberger F, Haberhauer G, Broell H, Lindley IJ. Presence of NAP-1/IL-8 in synovial fluid indicates a possible pathogenic role in rheumatoid arthritis. Scand J Immunol 1991; 34:333–339.
27. Koch AE, Kunkel SL, Burrows JC, et al. Synovial tissue macrophage as a source of the chemotactic cytokine IL-8. J Immunol 1991; 147:2187–2195.
28. Brennan FM, Feldmann M. Cytokines in autoimmunity. Curr Opin Immunol 1992; 4:754–759.
29. Steinitz M. B cell clones in rheumatoid arthritis. Springer Semin Immunopathol 1988; 10:181–188.
30. Moynier M, Abderrazik M, Didry C, Sany J, Brochier J. The B cell repertoire

in rheumatoid arthritis. III. Preferential homing of rheumatoid factor-producing B cell precursors in the synovial fluid. Arthritis Rheum 1992; 35: 49–54.

31. Tsuchiya N, Williams RC Jr. Molecular mimicry—hypothesis or reality? West J Med 1992; 157:133–138.
32. Ford WL, Gowans JL. The traffic of lymphocytes. Semin Hematol 1969; 6: 67–83.
33. Kraal G, Duijvestijn AM, Hendriks HH. The endothelium of the high endothelial venule: a specialized endothelium with unique properties. Exp Cell Biol 1987; 55:1–10.
34. Springer TA. Traffic signals for lymphocyte recirculation and leukocyte emigration: the multistep paradigm. Cell 1994; 76:301–314.
35. Picker LJ. Control of lymphocyte homing. Curr Opin Immunol 1994; 6:394–406.
36. Carlos TM, Harlan JM. Leukocyte-endothelial adhesion molecules. Blood 1994; 84:2068–2101.
37. Gallatin WM, Weissman I, Butcher EC. A cell-surface molecule involved in organ-specific homing of lymphocytes. Nature 1983; 403:30–34.
38. Jalkanen S, Reichert R, Gallatin WM, Bargatze RF, Weissman I, Butcher EC. Homing receptors and the control of lymphocyte migration. Immunol Rev 1986; 91:39–60.
39. Kishimoto TK, Jutila MA, Butcher EC. Identification of a human peripheral lymph node homing receptor. Proc Natl Acad Sci USA 1990; 87:2244–2248.
40. Berg EL, Robinson MK, Warnock RA, Butcher EC. The human peripheral lymph node vascular addressin is a ligand for LECAM-1, the peripheral lymph node homing receptor. J Cell Biol 1991; 114:343–349.
41. Baumhueter S, Singer MS, Henzel W, et al. Binding of L-selectin to the vascular sialomucin CD34. Science 1993; 262:436–438.
42. Streeter PR, Berg EL, Rouse B, Bargatze RF, Butcher EC. A tissue-specific endothelial cell module involved in lymphocyte homing. Nature 1988; 331: 41–46.
43. Holzmann B, Weissman IL. Peyer's patch-specific lymphocyte homing receptor consist of a VLA-4-like α chain associated with either of two integrin β chains, one of which is novel. EMBO J 1989; 8:1735–1741.
44. Hu MC, Crowe DT, Weissman IL, Holzmann B. Cloning and expression of mouse integrin βp (β7): a functional role in Peyer's patch-specific lymphocyte homing. Proc Natl Acad Sci USA 1992; 89:8254–8258.
45. Berlin C, Berg EL, Briskin MJ, et al. $\alpha 4\beta 7$ integrin mediates lymphocyte binding to the mucosal vascular addressin MAdCAM-1. Cell 1993; 74:185–195.
46. Hamann A, Andrew DP, Jablonski-Westrich D, Holzmann B, Butcher EC. Role of α4-integrins in lymphocyte homing to mucosal tissues in vivo. J Immunol 1994; 152:3282–3293.
47. Berg EL, McEvoy LM, Berlin C, Bargatze RF, Butcher EC. L-selectin-mediated lymphocyte rolling on MAdCAM-1. Nature 1993; 366:695–698.

48. Freemont AJ, Jones CJP, Bromley M, Andrews P. Changes in vascular endothelium related to lymphocyte collections in diseased synovia. Arthritis Rheum 1983; 26:1427-1433.
49. Freemont AJ. Functional and biosynthetic changes in endothelial cells of vessels in chronically inflamed tissues: evidence for endothelial control of lymphocyte entry into diseased tissues. J Pathol 1988; 155:225-230.
50. Ziff M. Role of the endothelium in chronic inflammatory synovitis. Arthritis Rheum 1991; 11:1345-1352.
51. Pober JS, Lapierre LA, Stolpen AH, et al. Activation of cultured human endothelial cells by recombinant lymphotoxin: comparison with tumor necrosis factor and interleukin 1 species. J Immunol 1987; 133:3319-3324.
52. Bevilacqua MP, Pober JS, Mendrick DL, Cotran RS, Gimbrone MA, Jr. Identification of an inducible endothelial-leukocyte adhesion molecule. Proc Natl Acad Sci USA 1987; 84:9238-9242.
53. Osborn L, Hession C, Tizard R, et al. Direct expression cloning of vascular adhesion molecule 1, a cytokine-induced endothelial protein that binds to lymphocytes. Cell 1989; 59:1203-1211.
54. Leeuwenberg JFM, Von Ashmuth EJU, Jeunhomme TMAA, Buurman WA. IFN-γ regulates the expression of the adhesion molecule ELAM-1 and IL-6 production by human endothelial cells in vitro. J Immunol 1990; 145:2110-2114.
55. Rice GE, Munro JM, Bevilacqua MP. Inducible cell adhesion molecule 110 (INCAM-110) is an endothelial receptor for lymphocytes. A CD11/CD18 independent adhesion mechanism. J Exp Med 1990; 171:1369-1374.
56. Barker JNWN, McDonald DM. Cutaneous lymphocyte trafficking in the inflammatory dermatoses. Br J Dermatol 1992; 126:211-215.
57. Gerritsen ME, Niebala MJ, Szczepanski A, Carley WW. Cytokine activation of human macro- and microvessel-derived endothelial cells. Blood Cells 1993; 19:325-339.
58. Gerritsen ME, Kelley KA, Ligon G, et al. Regulation of the expression of intercellular adhesion molecule-1 in cultured human endothelial cells derived from rheumatoid synovium. Arthritis Rheum 1993; 36:593-601.
59. Cotran RS, Gimbrone MA Jr, Bevilacqua MP, Mendrick DL, Pober JS. Induction and detection of a human endothelial activation antigen in vivo. J Exp Med 1986; 164:661-666.
60. Picker LJ, Michie SA, Rott LS, Butcher EC. A unique phenotype of skin-associated lymphocytes in humans. Am J Pathol 1990; 136:1053-1068.
61. Picker LJ, Kishimoto TK, Smith CW, Warnock RA, Butcher EC. ELAM-1 is an adhesion molecule for skin-homing T cells. Nature 1991; 349:796-799.
62. Berg EL, Yoshino T, Rott LS, et al. The cutaneous lymphocyte antigen is a skin lymphocyte homing receptor for the vascular lectin endothelial cell-leukocyte adhesion molecule 1. J Exp Med 1991; 174:1461-1466.
63. Jalkanen S, Steere AC, Fox RI, Butcher EC. A distinct endothelial cell recognition system that controls lymphocyte traffic into inflamed synovium. Science 1986; 233:556-558.

64. van Dinther-Janssen ACHM, Horst E, Koopman G, et al. The VLA-4/VCAM-1 pathway is involved in lymphocyte adhesion to endothelium in rheumatoid synovium. J Immunol 1991; 147:4207–4210.
65. Mackay CR. Homing of naive, memory and effector lymphocytes. Curr Opin Immunol 1993; 5:423–427.
66. Koch, AE, Burrows JC, Haines GK, Carlos TM, Harlan JM, Leibovich SJ. Immunolocalization of endothelial and leukocyte adhesion molecules in human rheumatoid and osteoarthritic synovial tissues. Lab Invest 1991; 64: 313–320.
67. Takahashi H, Söderström K, Nilsson E, Kiessling R, Patarroyo M. Integrins and other adhesion molecules on lymphocytes from synovial fluid and peripheral blood of rheumatoid arthritis patients. Eur J Immunol 1992; 22:2879–2885.
68. Carson CW, Beall LD, Hunder GG, Johnson CM, Newman W. Soluble E-selectin is increased in inflammatory synovial fluid. J Rheumatol 1994; 21: 605–611.
69. Phillips ML, Nudelman E, Gaeta FCA, et al. ELAM-1 mediates cell adhesion by recognition of a carbohydrate ligand, sialyl-Le X. Science 1990; 250:1130–1132.
70. Lowe JB, Stoolman LM, Nair RP, Larsen RD, Berhend TL, Marks RM. ELAM-1-dependent cell adhesion to vascular endothelium determined by a transfected human fucosyltransferase cDNA. Cell 1990; 63:475–484.
71. Tiemeyer M, Swiedler SJ, Ishihara M, et al. Carbohydrate ligands for endothelial-leukocyte adhesion molecule-1. Proc Natl Acad Sci USA 1991; 88: 1138–1142.
72. Munro JM, Lo SK, Corless C, et al. Expression of sialyl-Lewis X, an E-selectin ligand, in inflammation, immune processes and lymphoid tissues. Am J Pathol 1992; 141:1397–1408.
73. Ohmori K, Takada A, Yoneda T, et al. Differentiation-dependent expression of sialyl stage-specific embryonic antigen-1 and I-antigens on human lymphoid cells and its implications for carbohydrate-mediated adhesion to vascular endothelium. Blood 1993; 81:101–111.
74. Picker LJ, Warnock RA, Burns AR, Doerschuk CM, Berg EL, Butcher EC. The neutrophil selectin LECAM-1 presents carbohydrate ligands to the vascular selectins ELAM-1 and GMP-140. Cell 1991; 66:921–933.
75. Postigo AA, Garcia-Vicuna R, Diaz-Gonzalez F, et al. Increased binding of synovial T lymphocytes from rheumatoid arthritis to endothelial-leukocyte adhesion molecule-1 (ELAM-1) and vascular cell adhesion molecule-1 (VCAM-1). J Clin Invest 1992; 89:1445–1452.
76. Burmester GR, Yu DT, Irani A-M, Kunkel HG, Winchester RJ. Ia+ T cells in synovial fluid and tissues of patients with rheumatoid arthritis. Arthritis Rheum 1981; 24:1370–1376.
77. Poulter LW, Duke O, Panayi G, Hobbs S, Raftery MJ, Janossy G. Activated T lymphocytes of the synovial membrane in rheumatoid arthritis and other arthropathies. Scand J Immunol 1985; 22:683–690.
78. Laffon A, Sanchez-Madrid F, de Landazuri MO, et al. Very late activation

antigen on synovial fluid T cells from patients with rheumatoid arthritis and other rheumatic diseases. Arthritis Rheum 1989; 32:386–392.
79. Potonick AJ, Kinne R, Menninger H, Zacher J, Emmrich F, Kroczek RA. Expression of activation antigens on T cells in rheumatoid arthritis patients. Scand J Immunol 1990; 31:213–224.
80. Hemler ME, Elices MJ, Parker CM, Takada Y. Structure of the integrin VLA-4 and its cell-cell and cell-matrix adhesion functions. Immunol Rev 1990; 114:47–65.
81. Elices MJ, Osborn L, Takada Y, et al. VCAM-1 on activated endothelium interacts with the leukocyte integrin VLA-4 at a site distinct from the VLA-4/fibronectin binding site. Cell 1990; 60:577–584.
82. Chan BMC, Elices MJ, Murphy E, Hemler ME. Adhesion to vascular cell adhesion molecule 1 and fibronectin. Comparison of $\alpha 4\beta 1$ (VLA-4) and $\alpha 4\beta 7$ on the human B cell line JY. J Biol Chem 1992; 267:8366–8370.
83. Masumoto A, Hemler, ME. Multiple activation states of VLA-4. Mechanistic differences between adhesion to CS1/fibronectin and to vascular cell adhesion molecule-1. J Cell Biol 1993; 268:228–234.
84. Laffon A, Garcia-Vicuna R, Humbria A, et al. Unregulated expression and function of VLA-4 fibronectin receptors of human activated T cells in rheumatoid arthritis. J Clin Invest 1991; 88:546–552.
85. Ueki Y, Eguchi K, Shimada H, et al. Increase in adhesion molecules on CD4+ cells and CD4+ cell subsets in synovial fluid from patients with rheumatoid arthritis. J Rheumatol 1994; 21:1003–1010.
86. Morales-Ducret J, Wayner E, Elices MJ, Alvaro-Garcia JM, Zvaifler NJ, Firestein GS. $\alpha 4\beta 1$ integrin (VLA-4) ligands in arthritis. Vascular cell adhesion molecule-1 expression in synovium and on fibroblast-like synoviocytes. J Immunol 1992; 149:1424–1431.
87. Potts JR, Campbell ID. Fibronectin structure and assembly. Curr Opin Cell Biol 1994; 6:648–655.
88. Liao N, St. John J, Tak Cheung H. Adhesion of lymphoid cell lines to fibronectin-coated substratum; biochemical and physiological characterization and identification of a 140-kDa fibronectin-receptor. Exp Cell Res 1987; 171:306–320.
89. Hynes RO. Integrins: A family of cell surface receptors. Cell 1987; 48:549–554.
90. Ruoslahti E, Pierschbacher MD. New perspectives in cell adhesion: RGD and integrins. Science 1987; 238:491–497.
91. Wayner EA, Garcia-Pardo A, Humphries MJ, McDonald JA, Carter WG. Identification and characterization of the lymphocyte receptor for an alternative cell attachment domain (CS1) in plasma fibronectin. J Cell Biol 1989; 109:1321–1330.
92. Ager A, Humphries MJ. Use of synthetic peptides to probe lymphocyte-high endothelial cell interactions. Lymphocytes recognize a ligand on the endothelial surface which contains the CS1 adhesion motif. Int Immunol 1990; 2: 921–928.
93. van Dinther-Janssen ACHM, Pals ST, Scheper RJ, Meijer CJLM. Role of

the CS1 adhesion motif of fibronectin in T cell adhesion to synovial membrane and peripheral lymph node endothelium. Ann Rheum Dis 1993; 52: 672–676.
94. Elices MJ, Tsai V, Strahl D, et al. Expression and functional significance of alternatively spliced CS1 fibronectin in rheumatoid arthritis microvasculature. J Clin Invest 1994; 93:405–416.
95. Lazarovits AI, Karsh J. Differential expression in rheumatoid synovium and synovial fluid of $\alpha4\beta7$ integrin. A novel receptor for fibronectin and vascular cell adhesion molecule-1. J Immunol 1993; 151:6482–6489.
96. Ruegg C, Postigo AA, Sikorski EE, Butcher EC, Pytela R, Erle DJ. Role of integrin $\alpha4\beta7/\alpha4\beta P$ in lymphocyte adherence to fibronectin and VCAM-1 and in homotypic clustering. J Cell Biol 1992; 117:179–189.
97. Postigo AA, Sanchez-Mateos P, Lazarovits AI, Sanchez-Madrid F, de Landazuri MO. $\alpha4\beta7$ integrin mediates B cell binding to fibronectin and vascular cell adhesion molecule-1. Expression and function of $\alpha4$ integrins on human B lymphocytes. J Immunol 1993; 151:2471–2483.
98. Erle DJ, Briskin MJ, Butcher EC, Garcia-Pardo A, Lazarovits AI, Tidswell M. Expression and function of the MadCAM-1 receptor, integrin $\alpha4\beta7$, on human leukocytes. J Immunol 1994; 153:517–528.
99. Dustin ML, Rothlein R, Bhan AK, Dinarello CA, Springer TA. Induction by IL-1 and IFN-γ: tissue distribution, biochemistry and function of a natural adhesion molecule (ICAM-1). J Immunol 1986; 137:245–254.
100. Anderson DC, Schmalsteig FC, Finegold MJ, et al. The severe and moderate phenotypes of inheritable Mac-1, LFA-1 deficiency: their quantitative definition and relation to leukocyte dysfunction and clinical features. J Infect Dis 1985; 152:668–689.
101. Anderson DC, Springer TA. Leukocyte adhesion deficiency: an inherited defect in the Mac-1, LFA-1 and p150,95 glycoproteins. Annu Rev Med 1987; 38:175–194.
102. Oppenheimer-Marks N, Davis LS, Lipsky PE. Human T lymphocyte adhesion to endothelial cells and transendothelial migration. Alteration of receptor use relates to the activation status of both the T cell and the endothelial cell. J Immunol 1990; 145:140–148.
103. Luscinskas FW, Cybulsky MI, Kiely J-M, Peckins CS, Davis V., Gimbrone MA Jr. Cytokine-activated human endothelial monolayers support enhanced neutrophil transmigration via a mechanism involving both endothelial-leukocyte adhesion molecule-1 and intercellular adhesion molecule-1. J Immunol 1991; 146:1617–1625.
104. Oppenheimer-Marks N, Davis LS, Tompkins Bogue D, Ramberg J, Lipsky PE. Differential utilization of ICAM-1 and VCAM-1 during the adhesion and transendothelial migration of human T lymphocytes. J Immunol 1991; 147:2913–2921.
105. Haynes BF, Hale LP, Denning SM, Le PT, Singer KH. The role of leukocyte adhesion molecules in cellular interactions: implications for the pathogenesis of inflammatory synovitis. Springer Semin Immunopathol 1989; 11:163–185.
106. Jasin HE, Lightfoot E, Davis LS, Rothlein R, Faanes RB, Lipsky PE. Ame-

lioration of antigen-induced arthritis in rabbits treated with monoclonal antibodies to leukocyte adhesion molecules. Arthritis Rheum 1992; 35:541–549.
107. Moreland LW, Heck LW Jr, Sullivan W, Pratt PW, Koopman WJ. New approaches to the therapy of autoimmune diseases: rheumatoid arthritis as a paradigm. Am J Med Sci 1993; 305:40–51.
108. Elliot MJ, Maini RM. New directions for biological therapy in rheumatoid arthritis. Int Arch Allergy Immunol 1994; 104:112–125.
109. Kavanaugh AF, Davis LS, Nichols LA, et al. Treatment of refractory rheumatoid arthritis with a monoclonal antibody to intercellular adhesion molecule 1. Arthritis Rheum 1994; 37:992–999.
110. Gorski A. The role of cell adhesion molecules in immunopathology. Immunol Today 1994; 15:251–255.
111. Salmi M, Jalkanen S. A 90-kilodalton endothelial cell molecule mediating lymphocyte binding in humans. Science 1992; 257:1407–1409.
112. Airas L, Salmi M, Jalkanen S. Lymphocyte-vascular adhesion protein-2 is a novel 70-kDa molecule involved in lymphocyte adhesion to vascular endothelium. J Immunol 1993; 151:4228–4238.
113. Koch AE, Nickoloff BJ, Holgersson J, et al. 4A11, a monoclonal antibody recognizing a novel antigen expressed on abberant vascular endothelium. Upregulation in an in vivo model of contact dermatitis. Am J Pathol 1994; 144:244–259.
114. Kabel PJ, Voorbij HAM, De Haan M, Van der Gaag RD, Drexhage HA. Intrathyroidal dendritic cells. J Clin Endocrinol Metab 1988; 65:199–207.
115. Seldenrijk CA, Drexhage HA, Meuwissen SGM, Pals ST, Meijer CJLM. Dendritic cells and scavenger macrophages in chronic inflammatory bowel disease. Gut 1989; 30:484–491.
116. Edwards JC, Wilkinson LS, Speight P, Isenberg DA. Vascular cell adhesion molecule-1 and $\alpha4\beta1$ integrins in lymphocyte aggregates in Sjögren's syndrome and rheumatoid arthritis. Ann Rheum Dis 1993; 52:806–811.
117. Knight SC, Mertin J, Stackpoole A, Clark J. Induction of immune responses in vivo with small numbers of veiled (dendritic) cells. Proc Natl Acad Sci USA 1983; 80:6032–6035.
118. Boog CJP, Kast WM, Timmers HTM, Boes J, De Waal LP, Melief CJM. Abolition of specific immune response defect by immunization with dendritic cells. Nature 1985; 318:59–62.
119. Breel M, Mebius R, Kraal G. The interaction of interdigitating cells and lymphocytes in vitro and in vivo. Immunology 1988; 63:331–337.
120. Austyn JM. Morris PJ. T-cell activation by dendritic cells: CD18-dependent clustering is not sufficient for mitogenesis. Immunology 1988; 63:537–543.
121. Barland P, Novikoff AB, Hamerman D. Electron microscopy of the human synovial membrane. J Cell Biol 1962; 14:207–216.
122. Burmester GR, Dimitru-Bona A, Waters SJ, Winchester RJ. Identification of three major synovial lining cell populations by monoclonal antibodies directed to Ia antigens and antigens associated with monocytes/macrophages and fibroblasts. Scand J Immunol 1983; 17:69–82.
123. Wilkinson LS, Pitsillides AA, Worrall JG, Edwards JCW. Light microscopic

characterization of the fibroblast-like synovial intimal cell (synoviocyte). Arthritis Rheum 1992; 35:1179–1184.
124. Hamilton JA, Butler DM, Stanton H. Cytokine interactions promoting DNA synthesis in human synovial fibroblasts. J Rheumatol 1994; 21:797–803.
125. Bombara MP, Webb DL, Conrad P, et al. Cell contact between T cells and synovial fibroblasts causes induction of adhesion molecules and cytokines. J Leukoc Biol 1993; 54:399–406.
126. Werb Z, Mainardi CL, Vater CA, Harris ED Jr. Endogenous activation of collagenase secreted by rheumatoid synovial cells: evidence for the role of plasminogen activator. N Engl J Med 1977; 296:1017–1023.
127. Mizel SB, Dayer JM, Krane SM, Mergenhagen SE. Stimulation of rheumatoid synovial cell collagenase and prostaglandin production by partially purified lymphocyte-activating factor (IL-1). Proc Natl Acad Sci USA 1981; 78: 2474–2477.
128. Okada Y, Nagase H, Harris ED Jr. A metalloproteinase from human rheumatoid synovial fibroblasts that digests connective tissue matrix components. Purification and characterization. J Biol Chem 1986; 261:14245–14255.
129. Shiozawa F, Shiozawa K, Fujita T. Presence of HLA-DR antigens on synovial type A and B cells: an immunoelectron microscopic study in rheumatoid arthritis, osteoarthritis and normal traumatic joints. Immunology 1983; 50: 587–594.
130. Burmester GR, Jahn B, Rohwer P, Zacher J, Winchester RJ, Kalden JR. Differential expression of Ia antigens by rheumatoid synovial lining cells. J Clin Invest 1987; 80:595–604.
131. van Dinther-Janssen ACHM, Kraal G, van Soesbergen RM, Scheper RJ, Meijer CJLM. Immunohistochemical and functional analysis of adhesion molecule expression in the rheumatoid synovial lining layer. Implications for synovial lining cell destruction. J Rheumatol 1994; 21:1998–2004.
132. Krzesicki RF, Fleming WE, Winterrowd GE, Hatfield CA, Sanders ME, Chin JE. T lymphocyte adhesion to human synovial fibroblasts. Role of cytokines and the interaction between intercellular adhesion molecule-1 and CD11a/CD18. Arthritis Rheum 1991; 34:1245–1253.
133. Blue ML, Conrad P, Webb DL, Sarr T, Macaro M. Interacting monocytes and synoviocytes induce adhesion molecules by a cytokine-regulated process. Lymphokine Cytokine Res 1993; 12:213–218.
134. Cicuttini FM, Martin M, Boyd AW. Cytokine induction of adhesion molecules on synovial type B cells. J Rheumatol 1994; 21:406–412.
135. Shimada H, Eguchi K, Ueki Y, et al. Interleukin 4 increases human synovial cell expression of VCAM-1 and T cell binding. Ann Rheum Dis 1994; 53:601–607.
136. Szekanecz Z, Haines GK, Lin TR, et al. Differential distribution of intercellular adhesion molecules (ICAM-1, ICAM-2 and ICAM-3) and the MS-1 antigen in normal and diseased human synovia. Their possible pathogenetic and clinical significance in rheumatoid arthritis. Arthritis Rheum 1994; 37:221–231.
137. Issekutz TB. Inhibition of in vivo lymphocyte migration to inflammation and

homing to lymphoid tissues by the TA-2 monoclonal antibody. A likely role for VLA-4 in vivo. J Immunol 1991; 147:4178–4184.
138. Issekutz TB. Dual inhibition of VLA-4 and LFA-1 maximally inhibits cutaneous delayed-type hypersensitivity-induced inflammation. Am J Pathol 1993; 143:1286–1293.
139. Scheynius A, Camp RL, Pure E. Reduced contact sensitivity reactions in mice treated with monoclonal antibodies to leukocyte function-associated molecule-1 and intercellular adhesion molecule-1. J Immunol 1993; 150:655–663.
140. Pizcueta P, Luscinskas FW. Monoclonal antibody blockade of L-selectin inhibits mononuclear leukocyte recruitment to inflammatory sites in vivo. Am J Pathol 1994; 145:461–469.
141. Elson CJ, Barker RN, Thompson SJ, Williams NA. Immunologically ignorant autoreactive T cells, epitope spreading and repertoire limitation. Immunol Today 1995; 16:71–75.
142. Romagnani S. Lymphokine production by human T cells in disease states. Annu Rev Immunol 1994; 12:227–257.
143. Kadioglu A, Sheldon P. Adhesion of rheumatoid peripheral blood and synovial fluid mononuclear cells to high endothelial venules of gut mucosa. Ann Rheum Dis 1992; 51:126–127.
144. Jörgensen C, Travaglio-Encinoza A, Bologna C, D'Angeac AD, Reme T, Sany J. Human mucosyl lymphocyte marker expression in synovial fluid lymphocytes of patients with rheumatoid arthritis. J Rheumatol 1994; 21: 1602–1607.
145. Sikorski EE, Hallman R, Berg EL, Butcher EC. The Peyer's patch high endothelial receptor for lymphocytes, the mucosal vascular addressin, is induced on a murine endothelial cell line by tumor necrosis factor-α and IL-1. J Immunol 1993; 151:5239–5250.
146. Hänninen A, Taylor C, Streeter PR, et al. Vascular addressins are induced on islet vessels during insulitis in nonobese diabetic mice and are involved in lymphoid cell binding to islet endothelium. J Clin Invest 1993; 92:2509–2515.
147. Lee M-S, Sarvetnick N. Induction of vascular addressins and adhesion molecules in the pancreas of IFN-γ transgenic mice. J Immunol 1994; 152;4597–4603.
148. Faveeuw C, Gagnerault M-C, Lepault F. Expression of homing and adhesion molecules in infiltrated islets of Langerhans and salivary glands of nonobese diabetic mice. J Immunol 1994; 152:5969–5978.
149. Scheeren RA, Koopman G, van der Baan S, Meijer CJLM, Pals ST. Adhesion receptors involved in the early clustering of blood dendritic cells and T lymphocytes. Eur J Immunol 1991; 21:1101–1105.

23

Adhesion Molecules in the Immunopathogenesis of Skin Diseases

Onno J. de Boer
Department of Cardiovascular Pathology, Academic Medical Centre, University of Amsterdam, Amsterdam, The Netherlands

Pranab K. Das
Departments of Cardiovascular Pathology and Dermatology, Academic Medical Centre, University of Amsterdam, Amsterdam, The Netherlands

I. INTRODUCTION

A. Skin Immune System: A Preface

The skin is the largest organ of the human body and forms a physical barrier to pathogens. It consists of several cell types that play important roles in cutaneous immune reactions. Streilein referred to them collectively as the skin-associated lymphoid tissue (SALT), which includes keratinocytes (KC), lymphocytes (Ly), Langerhans cells (LC), dermal dendritic cells, dermal microvascular endothelial cells (DMVEC), and the skin draining regional lymph nodes (1). Each cell type in the SALT, resident or recruited, contributes to the local immune defense system in a distinctive manner. KC, which are the major resident cell type in the epidermis, form a microenvironment that is favorable for antigen trapping and uptake, and they are able to produce various cytokines, such as IL-1, IL-3, IL-6, GM-CSF, and TGF-β_1, which regulate local immune responses (2). All resident epidermal cells (KC, LC, and melanocytes: MC) are able to process and present antigens to lymphocytes (3–6), to respond to cytokines, and to express various adhesion or activation molecules. Adhesion molecules are important for local immune reactions due to their ability to direct the migration of immunocompetent cells from the blood through the dermal vessels and matrix into the skin and to provide primary and costimulatory signals. Moreover, adhesion molecules are important in cell-cell and cell-matrix interactions for the maintenance of tissue morphology and function (7,8). Conversely, adhesion molecule dysregulation is often encountered in skin cancers (9,10) and nonmalignant proliferative skin disorders such as psoriasis (11).

The current review deals with selected adhesion molecules that are considered particularly important for the skin immune system in health and disease. Their role is discussed with a focus on cell migration into the skin, immune responses, cell motility, and tissue degeneration.

B. The Skin Immune System

The normal human skin harbors lymphocytes, almost exclusively of the T-cell type. The majority of them are localized around the postcapillary venules, and almost all are of the memory (CD45RO+) type with an even distribution among the CD4 and CD8 subsets (12). The infiltrates of most inflammatory skin diseases, except granulomatous and granulocytic diseases, consist predominantly of CD4 positive lymphocytes, as observed in psoriasis, cutaneous lupus erythematosus, allergic contact dermatitis, atopic dermatitis, lichen rubber planus, and pityriasis rosea (12). In lepromatous leprosy, the majority of T-cells are of the CD8 phenotype, whereas in tuberculoid leprosy most of them are CD4-positive (13). In diseases such as vaculitis, where there may be local deposition of autoantibodies and/or immune complexes, the infiltrating lymphocytes are also predominantly T-cells (Fig. 1a, b), whereas in a disease such as immunocytoma the skin infiltrates sometimes contain B-cells as well as T-cells (14). The cellular infiltrate in skin malignancies usually consists of approximately equal numbers of CD4+ and CD8+ lymphocytes and varying numbers of B lymphocytes and monocytes. In addition, large numbers of macrophages (CD68+, CD14+) or other antigen-presenting cells may be found at sites of tissue infiltration (15). Neutrophilic granuloctyes or eosinophils are found in leukocytoclastic vasculitis, pyoderma gangrenosum, erythema elevatum diutinum, and Sweet's syndrome, as well as in psoriasis. Furthermore, there are some skin diseases in which cellular infiltrates are not always seen, although they have been implicated in the loss of tissue integrity and tissue destruction. The best example is vitiligo, where loss of melanocytes has been attributed to their interactions with immunocompetent cells (16). It has been postulated that cutaneous T-lymphocytes, macrophages, and Langerhans cells, when appropriately activated, are able to influence the differentiation of epidermal cells, which may result in hyper-or parakeratosis of keratinocytes as observed in psoriasis (17,18).

The influx of immunocompetent cells occur via the microvascular system. The formation and regeneration of the cutaneous microvasculature

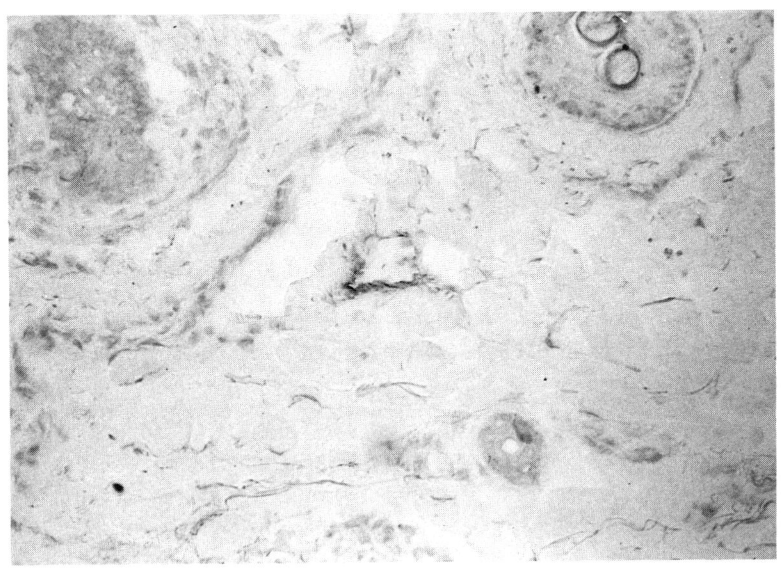

(a)

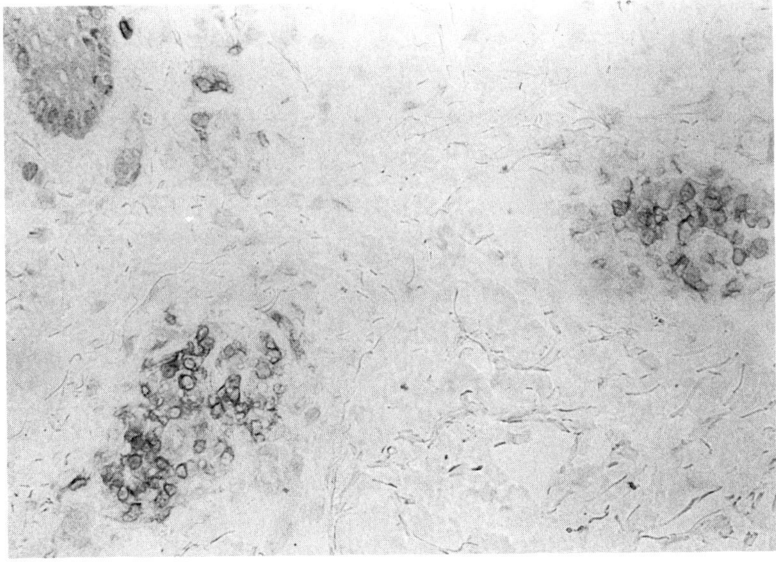

(b)

Figure 1 Indirect immunoperoxidase staining of a 6-μm cryostat section of a skin biopsy from a patient with SLE (a) deposition of immunoglobulin IgG/IgM isotype, and (b) infiltrates of CD3+ TCR-$\alpha\beta$ cells on a section from the same biopsy.

often parallel the cell proliferation of endothelial cells from preexisting vessels. Matrix molecules such as fibronectin, vitronectin, and fibrinogen potentially serve as a scaffold for endothelial cell migration into sites infiltrated by immunocompetent cells involved in neovascularization and local immune responses. Although the target antigens are known in many allergic, infectious, and autoimmune diseases, in certain idiopathic dermatoses such as psoriasis, the antigens recognized by the infiltrating lymphocytes remain unidentified. Discordant expression of certain types of adhesion receptors may play a role in the pathogenesis of some of these diseases (19).

II. ADHESION MOLECULES: GENERAL ASPECTS

Adhesion molecules are divided into different families on the basis of structural homologies: the immunoglobulin superfamily, the integrin family, the selectins, the cadherins, and the CD44 family (20–24). The immunoglobulin superfamily members share an immunoglobulin domain containing 90 to 100 amino acids, arranged in a sandwich of two antiparallel β-strands, usually stabilized by a disulphide bond in its center. Two members of this family, the B-cell receptor (immunoglobulins) and the T-cell receptor (TCR), are known to contain variable regions that undergo somatic diversification. Other members of this family are ICAM-1/2/3, VCAM-1, CD2, LFA-3, and MHC class I and II molecules.

Integrins are heterodimers consisting of α and β subunits that are noncovalently associated (21). They are divided into "subfamilies" on the basis of common β-chains. For example, β_1 (CD29) integrins (also known as the VLA family) are involved in cell-matrix and cell-cell (e.g., immune cell-endothelial cell, and immune cell-epidermal cell) interactions. β_2 (CD18) integrins are involved in cell-cell interactions during costimulation as well as migration processes; the best-known member of this family is LFA-1. Several other β-subunits have been identified. Some of the integrin α-subunits associate with one of these different β-subunits; α_4 (CD49d), for example, combines with either β_1 (VLA-4) or with β_7 to form an integrin receptor (25). An additional level of integrin diversity is generated by alternative splicing of certain α and β units to produce isoforms—e.g., α_{3a}, α_{3b}—that differ in their cytoplasmic or extracellular domain (21). In addition to functioning as cell and matrix receptors, integrins can also direct the assembly of the ECM (26).

The selectins consist of a tandem array of three kinds of domains: an amino-terminal, calcium-dependent lectinlike domain; a domain that resembles epidermal growth factor; and a domain that contains variable numbers of repeated units, conforming to the short consensus repeat of complement regulatory proteins. The family contains three members: E (CD62E),

L (CD62L), and P (CD62P) selectin, or, per their former names, ELAM-1, LECAM-1, and PADGEM (22). The selectins are involved in leukocyte endohelial interactions during cell migration to the skin. Adhesion molecules not only are important for the influx of T cells, but are also involved the differential migration of macrophages and Langerhans cells from the skin to the skin draining lymph nodes and vice versa. In the lymph node, the emigrated antigen-presenting cells present the antigen that they acquired in the skin to lymphocytes to elicit antigen-specific T-and B-cell responses (27). Members of the CD44 family of adhesion molecules have also been implicated in the lymphocyte migration into the skin and the subsequent antigen-specific T-cell activation (28).

A. Leukocyte Recirculation

The discussion in the preceding sections suggests that among various cells emigrating to the skin, lymphocytes and various professional APCs form the vast majority. Lymphocytes continuously migrate between the blood and the lymph, surveying the various lymphoid and extralymphoid microenvironments for the presence of an antigen that can bind to their receptors. The first step in this process is the adherence of lymphocytes to specialized endothelial venules, the high endothelial venules (HEV) in the lymph nodes, and the postcapillary venules. This traffic is not a random migration process, as lymphocytes display preferential homing regions, depending on their activation state (29). In general, naive, virgin T-cells (the CD45RA+ subset) leave the blood at the HEV of the lymph nodes, and extravasate poorly in extra lymphoid (tertiary) sites, even after inflammatory provocation. In contrast, memory/effector lymphocytes or lymphoblasts (the CD45RO+ subset) efficiently localize in tertiary sites, such as the skin (30).

The principal mechanism of extravasation of bloodborne cells to skin requires at least four successive steps. Primary adhesion, a relatively unstable event under shear stress conditions, is transient unless it is followed by the second step, adhesion triggering. During this, signals originating from the endothelial cells or surrounding tissue may lead to the participation of activation-dependent lymphocyte-endothelial adhesion pathways. This condition is called secondary adhesion, which is stable under shear stress conditions and leads to leukocyte transmigration (31,32). Once inside the tissue, the cells may subsequently migrate to the sites within immune or inflammatory activity, either along the vascular bed of the skin or along the matrix components, e.g., collagen.

The first step, primary adhesion (rolling), is a process usually mediated by selectins. It is in this phase that adhesion molecules for tissue-selective homing discriminate between rolling (possibly followed by transmigration) and continued circulation. L-selectin (CD62L) is involved in lymphocyte migration into peripheral lymph nodes, where it recognizes glycoprotein ligands (termed peripheral lymph nodes addressins (PNAd)). E-selectin (CD62E) is expressed by cytokine-activated endothelial cells (33) and interacts with sialyl Lewis^x (sLe^x), high levels of nonsialyated fucooligosaccharides (34), and sulfated oligosaccharides (35). In addition, cutaneous lymphocyte associated antigen (CLA) is a counterreceptor for E-selectin (36). There is one integrin that is also involved in primary adhesion, and that is the $\alpha_4\beta_7$ integrin. This integrin and its ligand MAdCAM-1 mediate tissue-specific homing to the gut (37,38). However, whether $\alpha_4\beta_7$ is really a gut-homing receptor remains controversial. We observed that under certain conditions, e.g., in Leishmaniasis and histiocytosis, both LCs and T-cells of skin infiltrates also express $\alpha_4\beta_7$ (unpublished observation). In the mouse, a subpopulation of skin-associated T-cells express this integrin molecule (G. Kraal, personal communication). Secondary adhesion requires activation of the adhesion receptors and involves adhesion molecules from the integrin and immunoglobulin family: LFA-1 ($\alpha_1\beta_2$ integrin) and its ligands ICAM-1 (CD54) and ICAM-2 (CD102), VLA-4 ($\alpha_4\beta_1$) and VCAM-1. After secondary adhesion, leukocytes migrate along the lining of the endothelial cells under the influence of chemotactic factors (chemokines) toward the site of immune provoked injury.

1. Cutaneous Leukocyte-Associated Antigen (CLA)

As discussed in the preceding section, tissue-specific leukocyte traffic is guided by specific interactions between tissue homing receptors and corresponding ligands on endothelial cells. Most T-lymphocytes in various dermatoses express the antigen recognized by MAb HECA-452, suggesting that the antigen identified by this monoclonal antibody (MAb) is a homing receptor for lymphocyte migration into the skin, which was designated the cutaneous lymphocyte associated antigen (CLA) (39). Approximately 80% to 90% of lymphocytes in inflammatory skin diseases are HECA-452+, while only approximately 15% of the peripheral blood lymphocytes are HECA-452+. In lymphoid tissue (lymph nodes, spleen, gut-associated lymphoid tissue, and the thymus) and other organs such as the gut, lung, heart, synovium, and liver, up to 10% to 15% of lymphocytes are HECA-452+ (39). HECA-452 is a rat MAb originally found to identify high endothelial venules (40). Western blot analysis of the HECA-452 antigen showed that the molecular weight of the antigen varied among different cell types. Lymphocytes showed predominant bands at 200 kDa whereas a 125-kDa species is displayed by monocytes. In contrast, PMNs showed multiple intense bands ranging from 75 kDa to 160 kDa, and a faint band in the region of 55 kDa to 60 kDa (39). The exact biochemical structure of CLA is not known, but HECA-452 is reported to recognize carbohydrate structures sLe^x, the sialylated form of Lewis-x (Le^x; CD15), sLe^a, an isomeric form of sLe^x, and closely related carbohydrate moieties (36,41).

Antigen isolated from tonsillar lysates with anti-CLA antibody binds E-selectin-transformed cell lines. In addition, anti-CLA antibody blocks T-lymphocyte binding to E-selectin transfectants (36). These observations indicate that E-selectin is a ligand for CLA. The aforementioned studies were performed with either E-selectin transfectants or recombinant E-selectin, but not with endothelial cells or CLA-positive leukocytes. When studies are performed under adequate conditions, it is possible to inhibit the adhesion of CLA lymphocytes and neutrophils to TNF-α stimulated endothelial cells with HECA-452. Antibodies against sLe^x (CSLEX1) showed similar kinetics in blocking the adhesion of these cells to endothelium (42). Further recent studies suggest that skin T-cells are highly enriched for the CLA+/L-selectin+/$\alpha_e\beta_7$ integrin/memory/effector subset (43). In an elegant study using in vitro generated T-cell clones from diseased skin it was demonstrated that the accumulation of T-cells in skin infiltrates indeed involves the interaction of sLe^x determinants on the CLA molecule with E-selectin on the dermal microvasculature (44).

An interesting issue is the mechanism by which lymphocytes acquire the CLA phenotype. The amount of CLA-positive CD45RA + /RO+ T-lymphocytes (lymphocytes that are in transition from virgin cells to memory-type cells) is increased in skin-associated peripheral lymph nodes as compared to the mucosal microenvironment of the appendix. It appears that the local microenvironment, where T-cells are activated for the first time and where they undergo their virgin-to-memory transition, is critical for acquiring the CLA phenotype (45). From in vitro studies using mitogen-activated T-cells it became clear that accessory signals are needed for the acquisition of the CLA phenotype. Transforming growth factor-β_1 (TGF-β_1) and interleukin-6, both cytokines produced by keratinocytes, may be such signals (45). In addition, TGF-β_1 and IL-6 are also able to increase the expression of L-selectin on T-cells (45). This is in line with the observation that T-cells in the skin are CLA+ L-selectin+.

The data presented indicate that the interaction between emigrating cells to skin and endothelial cells involves paired molecules like CLA/E-selectin, LFA-1/ICAM-1, and/or VLA-4/VCAM-1 (46). These molecular pairs function in an orchestrated manner under the influence of cytokine networks for the emigration of T-cells as well as their local activation in eliciting immune responses (47,48). Similar molecular events might also apply to other skin emigrating cells, such as LCs, monocytes, macrophages, and neutrophils. The principle adhesion molecular pairs responsible for mediating leukocyte migration are summarized in Table 1.

Table 1 Cellular Adhesion Molecules and Ligands Associated in Skin Extravasation of Leukocytes

Leukocyte AR	Endothelial cell ligands	Mechanistic step in extravasation
CLA (carbohydrate structure)	E-selectin (selectin family)	initial contact of leukocyte (rolling)
Sialyl Lewis-X (carbohydrate structure)		
VLA-4 (CD49d/CD29) (β_1 integrin)	VCAM-1 (Ig supergene family)	Leukocyte activation → strong binding
LFA-1 (CD11a/CD18) (β_2 integrin)	ICAM-1 (Ig supergene family)	diapedesis
Mac-1 (CD11b/CD18) (β_2 integrin)	ICAM-1 (Ig supergene family)	"skin homing"

B. Costimulatory Adhesion Molecules Related to Skin Immune Reactivity

One of the most powerful antigen-presenting cells in the skin are Langerhans cells (3). Besides LC, resident KC and MC have also been shown to function as APC in vitro (5,6,49). In the in vivo situation the professional LC functions as APC, as demonstrated by immunohistochemical techniques in the elicitation phase of allergic contact dermatitis and other inflammatory dermatoses where HLA-DR+ LC are frequently seen in juxtaposition to activated T-cells (50–52). Furthermore, in vivo studies on the mechanism of contact allergenicity have shown that LCs are the key APCs of the skin (53). Other CD1a− dendritic cells (for example, factor XIIIa+ cells) are important professional APC populations in the skin (54); skin cells like MCs may be functional as APCs in vivo, as is evident from our in situ studies on the immunopathology of vitiligo (55).

The role of antigen-presenting cells in T cell activation primarily depends on the interaction of the trimolecular complex of TCR and MHC in conjunction with antigen as well as their ability to deliver costimulatory signals. These costimulatory signals are provided by the interaction between adhesion receptors and their counterstructures on T-cells and APC, respectively. The primary costimulatory molecules involved in skin are the paired molecules: ICAM-1/LFA-1, CD2/LFA-3, and to a lesser extent VCAM-1/VLA-4 (5,56). Most recently new families of costimulatory molecules known as the B7 and CD28 family have been shown to play a prominent role. T-cell stimulation in the absence of crosslinking of B7 and its ligand CD28 leads to T-cell anergy (57,58). In skin, B7 family molecules show low constitutive expression on LC, macrophages, KC, and MC; however, upon stimulation their level of expression can be upregulated (59–64). Lymphocytes are also able to express B7 molecules (65). The in situ expression of these costimulatory molecules in different dermatoses appears to be related to disease specificity and will be discussed later.

C. Adhesion Molecules in Cell-Matrix Interactions in Skin

Major advances have been made in the understanding of the molecular mechanism of cell adhesion in morphogenesis and skin integrity, in the elucidation of intracellular/cell substratum adhesion molecules (66–69), and their corresponding ligands in the extracellular matrix (ECM) (70,71). The contact of the epidermal unit, particularly KC and MC underlying the basal lamina and their intracellular adhesion points, is essential in maintaining both morphological and functioning integrity of the skin. The molecular events involved in both inter-and intracell-cell and cell-matrix interactions are important in the in vivo regulation of KC and MC proliferation, maturation, motility, differentiation, physiological function, and the migration of LCs to and from the epidermis and the lymph nodes (72–75). The stratified epidermis is derived from a proliferating basal cell layer that is attached to a basement membrane (BM) by hemidesmosomes (76,77). The phenotypic, proliferative polarity and in situ structural morphogenesis of keratinocytes are maintained through an ingenious interaction between the cell surface adhesion molecules (integrins), the cytoskeletal structural complex (actin and vinculin), and the ECM components (e.g., collagen and the noncollagenous proteins (laminin, tenascin, heparan sulfate, epiligrin/kalinin) (8,21,78–81). Additionally, the cadherin family of adhesion receptors is important for selective KC-KC and KC-LC interactions (23,82,83). The integrin family members are important in regard to the physiologic morphology of the skin as well as cell-ECM interactions. We have recently observed the expression of CD44 variants in the epidermis (unpublished observation), a finding substantiated by the recent observation that CD44 and hyaluronan are expressed in interfacing keratinocytes and leukocytes (84). The accumulation of cellular integrins guides the formation of focal contacts at points of cell adhesion to ECM, where these molecules bind through ECM domains to the matrix and through their cytoplasmic domains to the cytoskeleton, albeit not all integrins form focal contacts. In this regard, integrins are important for 1. the adhesion of normal KC and MC to the underlying BM; 2. the physical interaction between KC and MC as well as LC with KC, resulting into the formation of the epidermal unit; 3. stem cells in the epidermis committed for differentiation, which are

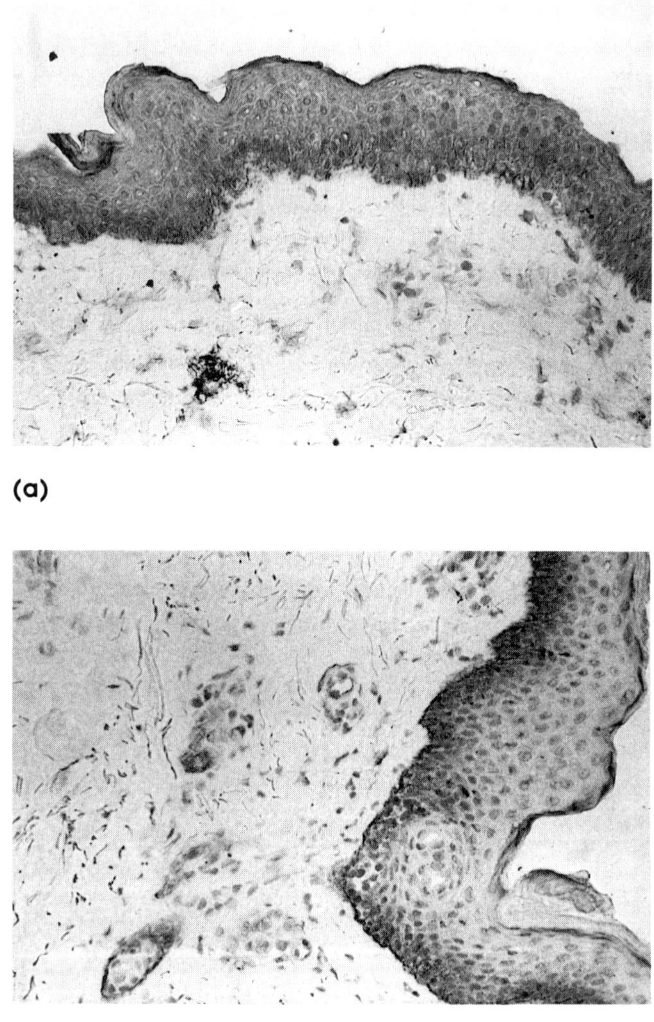

Figure 2 Representative immunoperoxidase staining pattern of 6-μm cryostat sections of a skin biopsy from a healthy donor showing the distribution of integrin (a) VLA-2; (b) VLA-3; (c) VLA-6.

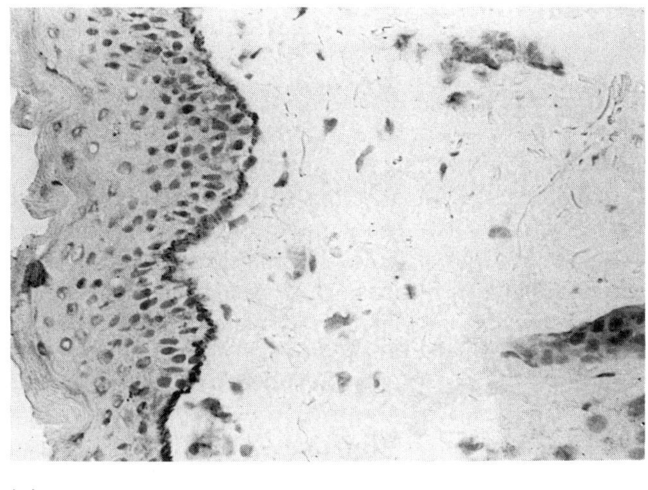

(c)

Figure 2 Continued.

entirely dependent on the activation or inactivation of integrin ECM receptors on their surface (85); and 4. the migration of the epidermal resident cells during the host sensitation to allergens (86), during wound healing (87), repigmentation, and during the loss of melanocytes (88). Furthermore $\alpha v\beta 3$ (vitronectin receptor) is important for microvascular endothelial interaction with ECM for the maintenance of skin function (89). The distribution of the integrins in the skin is illustrated in Figure 2a–c). The polarity of the integrin receptors is particularly prominent in relation to normal epidermal integrity. The dynamic cell-cell and cell-matrix interactions are regulated and dysregulated by microenvironmental cytokines (e.g., TGF-β_1, TNF-α etc.) (90).

III. ADHESION MOLECULES IN SKIN DISEASES

In the preceding sections, the molecular basis underlying leukocyte trafficking in the skin and maintenance of skin physiological architecture has been discussed. This section will briefly discuss inflammatory dermatoses and cutaneous neoplasms. Attention will focus on diseases relating to adhesion molecules in cellular recruitment and costimulation, adhesion molecules and ECM cell interaction, and adhesion molecules as targets for therapeutics and diagnostics.

A. Cutaneous Inflammation (Inflammatory Dermatoses)

As stated earlier, the critical determinants in the histopathology of most inflammatory dermatoses is the composite of cellular infiltrates of T-lymphocytes, antigen-presenting cells, neutrophils, granulocytes, hyperproliferation of epidermis cells, acanthosis, parakeratosis, and occasionally epidermolysis. Naturally, the adhesion receptors responsible for leukocyte trafficking alter the histomorphology of the skin in varying degrees in different diseases. Cellular infiltrates are accompanied by the in situ generation of cytokine cascades which in turn influence the expression of different classes of adhesion molecules on endothelial cells and migrating leukocytes differently in a disease-specific manner (91,92). Consequently, parallelling both the extent and the nature of the local immune response and depending on the composite of the skin cellular infiltrates or the immune deposits (93), the adhesion molecules involved in adherence of leukocytes to DMVEC, ECM, and epidermal cells (KC, LC, and MC) appear to be differentially expressed. The differential expression pattern of different adhesion molecules and ECM are reflected in the histopathology of different inflammatory diseases (94). Interestingly, in some idiopathic dermatoses (e.g., Sweet's syndrome, erythema multiforme) and autoimmune diseases like epidermolysis bullosa, some epitopes of ECM and/or cell adhesion receptors are the target molecules (95,96). In dermatoses with inflammation and antibody deposits, differences in histopathology are correlated with different composition of infiltrates. In some diseases, differences are seen in adhesion receptor expression for tissue cells such as KC, LC, and DMVEC. Varying degrees of local activation of microvascular endothelium by cytokines released by APCs and leukocytes stimulate the expression of EC– adhesion molecules differently, which in turn influences the extent of extravasation and localization of leukocytes at the inflammatory site. To get an insight into the role of adhesion molecules in the pathophysiology of dermatoses, immunohistochemical analysis of adhesion molecules/ligands on DMVEC, KC, LC, migrating leukocytes, and ECM receptors or in vitro adhesion assays have been used. In the following paragraph different patterns of expression of adhesion receptors on different skin cells and the corresponding ECM components in several inflammatory dermatoses are briefly described.

1. Endothelial Cell Adhesion Molecules

Endothelial cells in normal skin are ICAM-1 positive, express low levels of E-selectin (19,97), and are VCAM-1 negative (19). The expression of E-selectin in normal skin is remarkable, because in vitro studies have shown that cultured endothelial cells express E-selectin only after stimulation with the pro-inflammatory cytokines IL-1 and TNF-α (33). This indicates that recruitment of inflammatory cells can occur only during the concerted activation of normal DMVEC by cytokines. Consequently, in various skin diseases, the expression of various endothelial adhesion receptors is upregulated in parallel with infiltration by inflammatory cells, but it remains unclear whether the altered CAM expression causes disease or is just an epiphenomenon of inflammation. In psoriasis, endothelial cells show increased expression of HLA-DR, ICAM-1, E-selectin, and VCAM-1 (19,97, 98). Particularly, double immunoenzyme staining revealed a close relationship between E-selectin expression and neutrophil margination, suggesting a functional link (97,99,100). Besides psoriasis, increased expression of E-selectin by endothelial cells has been reported in atopic dermatitis, allergic contact dermatitis, photodermatoses, and systemic sclerosis (97,101–105).

Although endothelial cells in lesional psoriasis skin show the characteristics of high endothelial venules (106,107), they are HECA-452 negative (28). Interestingly, in psoriasis it has been found that vessels in lesional and nonlesional skin show abnormalities, like increased luminal and endothelial cell volume (107) and increased expression of adhesion receptors (18,19). These observations suggest that the noninvolved skin of psoriasis patients is also in an activated state and that these activated endothelial cells can recruit increased numbers of immunocompetent cells, a process that may easily lead to psoriatic lesions, especially after trauma (Koebner phenomenon).

Although E-selectin positive endothelial are regarded to be the preferred sites of leukocyte trafficking into the inflammatory skin, the differential expression of other adhesion molecules like ICAM-1 and VCAM-1 also play a critical role in immunologic responses and initial trafficking of memory T-cells and other leukocytes in cutaneous inflammation. Varying levels and kinetics of expression of these adhesion molecules in different disease states are regarded to reflect local cytokine cascades and have increased our understanding of diseases such as psoriasis, contact dermatitis, Lichen planus, and infectious diseases like leprosy (103,108). Furthermore, studies of atopic dermatitis have shown selective accumulation of eosinophils and neutrophils at sites of induced upregulation of E-selectin and ICAM-1 on DMVEC (109). On the other hand, lack of expression of ICAM-1 by endothelial cells in the skin of patients with lepromatous leprosy attests to the inability of these patients to develop an immune granuloma in response to *Mycobacterium leprae* (108).

We found that the level of VCAM-1 expression on endothelial cells in skin lesions with contact dermatitis is higher than in psoriasis lesions, whereas levels of E-selectin and ICAM-1 were comparable (102). The difference in adhesion receptor expression in these diseases may be the result of differences in cytokines produced by T-cells (Th1-Th2-Th0) in the lesions. Furthermore, keratinocytes in allergic contact dermatitis specimens focally express high levels of HLA-DR, whereas HLA-DR expression by keratinocytes in psoriasis is relatively rare, suggesting differences in IFN-γ production by the lymphocytes infiltrating the epidermis (51). The type of infiltrate in an inflammatory lesion may determine the level and type of adhesion receptors expressed by endothelial cells and their corresponding ligands on infiltrating leukocytes. For example, the peripheral lymph node addressin (PNAd), a counterreceptor for L-selectin-positive lymphocytes (110), is often expressed by venules in inflammatory dermatoses like psoriasis, allergic contact dermatitis, and lichen planus, while ICAM-1 is upregulated in other inflammatory dermatoses, like allergic contact dermatitis, and in purpura pigmentosa chronica (111). Taken together, adhesion receptors expressed by endothelial cells can be differentially regulated in different skin disorders, either by cytokines or by other exogenous factors. These differences may result in the recruitment of different subsets of immunocompetent cells.

2. Leukocyte Adhesion Molecules

Since vascular adhesion molecules act as adhesion ligands for recirculating and recruited cells like neutrophils, dendritic cells, and eosinophils, the extent of expression of counterreceptors on these cells may indicate inflammatory activation. Various in vivo and in vitro studies have suggested that expression of the counterreceptors on leukocytes is important in the pathogenesis of inflammatory dermatoses. Immunohistochemical studies of human skin have shown that approximately 40% of lymphocytes in normal skin express CLA (112), compared with 16% of peripheral blood lymphocytes (39). Increased numbers of CLA-positive T-cells in normal skin may be attributed to the expression of E-selectin by endothelial cells in the normal skin (97). In normal skin, the HECA-452 antigen can also be expressed by a small population of macrophages and (CD1a+) Langerhans cells (112). sLe^x is also expressed by lymphocytes and Langerhans cells in normal skin (27,113). In many inflammatory skin diseases like psoriasis, allergic contact dermatitis, and lichen planus, the frequency of CLA+ T-lymphocytes is significantly increased compared to peripheral blood and normal skin (39). Increased numbers of these cells may also be attributed to the expression of E-selectin by the capillaries in the inflamed skin. Increased frequency of CLA+ lymphocytes has been reported in psoriasis lesions as well as in nonlesional psoriatic skin (19). Similarly, increased numbers of HECA-452+ LC and macrophages have been found in those skin specimens. The expression pattern of CLA on different immunocompetent cells differs in different diseases (18).

Besides CLA molecules, the integrins VLA-4 and LFA-1 are also upregulated on recirculating and recruited immunocompetent cells, particularly on APC in inflamed tissues. Langerhans cells normally do not express LFA-1, but express it in psoriatic skin (19, 114).

As alluded to, adhesion receptors involved in costimulation have recently gained attention. CD28, a costimulatory molecule, is the ligand for the B7 family of molecules and is expressed by a minority of T-cells in normal human skin. In contrast, in allergic contact dermatitis and Lichen planus, most skin-invading T-cells are CD28-positive (61). Furthermore, besides CD28, most T-cells in psoriasis lesions express B7-1, raising the possibility of T-T interactions (115).

3. Epidermal Cell Adhesion Molecules

The major cell type of the epidermis is the keratinocyte. Only the basal keratinocytes proliferate, whereas the suprabasal keratinocytes undergo terminal differentiation. During this differentiation, the keratinocytes change in several aspects, including the expression of adhesion molecules. Keratinocytes are characterized by discrete membrane domains (116); the $\alpha_2\beta_1$ and $\alpha_3\beta_1$ intergins in the epidermis are able to bind laminin and collagen that line the apicolateral domain of the basal cells. In addition to KC and MC, the pigment cell is also located within the mesh of basal keratinocytes along the lining of the dermal-epidermal junction.

When activated, both KC and MC express several activation markers, and may participate in immune reactions by interacting directly with recruited immunocompetent cells. IFN-γ induces expression of MHC class II molecules on KC and MC in vitro (117–120) as well as the costimulatory molecule B7-3, recognized by MAb BB1 (121). B7-1 (CD80) and B7-2 (CD86) are expressed on MC after IFN-γ treatment (unpublished observation). Although KC provide costimulatory signals to T-cells, blocking of CD28, a ligand of the B7 family, does not inhibit keratinocyte-mediated T-cell activation in vitro, suggesting that other activation pathways are operative (121). Both KC and MC express ICAM-1 in vitro after stimulation with TNF-α or IL-1. Although the major molecules involved in AR-ligand interaction between KC and APC include ICAM-1/LFA$_1$ and CD2/LFA$_3$, it was recently reported that CD44 can also act as a costimulatory signal in T-cell activation (122). Interestingly, intense staining for hyaluronan (ligand for CD44) on dermal capillary loops in combination with reduced expression of CD44/hyaluronan in KC-leukocyte interfaces is seen in psoriatic skin (84). However, the CD44 class of adhesion receptors is principally involved in binding skin cells to ECM compounds, and immunohistochemical studies have shown a special distribution pattern of CD44 on KC in certain rare skin diseases (123). Other ARs involved in ECM interactions include tenascin and cadherins. Schalkwijk et al. (124) have shown that the in situ expression patterns of tenascin in skin diseases differs from other known ECM compounds. Furthermore, desmosomal cadherin molecules, linked to the intermediate keratin filaments network via plakoglobin and desmoplakin, are involved in KC-KC and KC-MC interactions (82,125). A disease like pemphigus has been shown to be an anticadherin autoimmune disease.

The epidermis expresses constitutively low ICAM-1 levels, which may suggest that resting epidermis is resistant to interactions with leukocytes. The regulated expression of adhesion molecules in the epidermis is vital for the initiation and evolution of local inflammatory processes (126). Induction of ICAM-1 in the epidermis may be an important factor in the induction of leukocyte-dependent damage to the epidermis (127). Induced expression of ICAM-1 on epidermal KC has been shown in various inflammatory dermatoses such as contact dermatitis, Lichen planus, pemphigus, fixed drug eruption, erythema multiforma, subacute cutaneous lupus erythermatosus, and pigmentary disorders (23,128–131). Their differential expressions in different skin diseases may reflect the pathogenic importance of these molecules. It is now recognized that at least three critical adhesion pathways—B7/CD28, CD2/LFA-3, and LFA-1/ICAM-1—are important in leukocyte and epidermal cell interactions in different dermatoses (121, 132).

Most dermatoses of immunologic or inflammatory origin show epidermal hyperplasia, parakeratosis, acanthosis, sponzioses, neovascularization, dermal aberration, fibroblast proliferation, and increased turnover of ECM. The changes in histomorphology of the diseased skin can be attributed to the changes in individual CAMs as well as changes in the expression of adhesion molecules involving epidermis-ECM interactions. Diseases with cutaneous inflammation that affect epidermal-ECM-related adhesion molecules include psoriasis, scleroderma, systemic lupus erythematosus, systemic sclerosis, bullous pemphigus, and epidermolysis bullosa (8,11,96,105, 133).

B. Adhesion Molecules in Cutaneous Neoplasms

The role of adhesion molecules in skin cancers has been studied in malignant melanoma, basal cell carcinoma, cutaneous lymphoma, and histiocytosis. The ability of melanoma cells to invade through basement membranes and to penetrate the ECM, including collagen, fibronectin, and laminin, is dependent on the interaction with and migration through these matrices (134). The development of malignant melanomas is also considered to be the result of an aberration in cell-cell interaction and stromal reactions (135).

In skin malignancies, tumor cells originating from resident skin cells undergo changes in their expression pattern of adhesion molecules. Increased expression of several adhesion molecules has been reported in malignant melanoma. Most melanomas, as well as benign nevi, show expression of ICAM-1 (9,136,137), which is not correlated with clinical parameters such as metastasis and prognosis. Integrin expression by squamous cell and basal cell carcinomas has been correlated to metastasis (10).

The interaction of tumor cells with neighboring cells and with components of the extracellular matrix is important in tumor invasion and metastasis (135). Since metastas involves adhesion of malignant cells to DMVEC, several studies have focused on the AR-ligand pathways for dissemination of melanoma cells; Lee et al. suggested that besides ICAM-1, VCAM-1, and E-selectin, other, yet unknown proteins may be involved (138). Aberrant expression of CAM in Langerhans cell histiocytosis indicates that such molecules may play a pathogenetic role (139,140). CAM may also play a role in leukocyte infiltration into malignant tumors. CLA is expressed by T-lymphocytes infiltrating primary cutaneous neoplasms, such as carcinomas and primary cutaneous melanomas, while cellular infiltrates in noncutaneous primary neoplasms or metastases are HECA-452-negative (141), suggesting that primary tumors evoke responses that are similar to the inflammatory responses. A study of primary cutaneous and noncutaneous T-non-Hodgkin's lymphoma (T-NHL) showed that most of the tumor cells of cutaneous T-NHL were HECA-452 positive, while almost all noncutaneous T-NHL were HECA-452 negative (142). Besides CLA, other molecules, such as ICAM-1, VLA-4, and B7, are also expressed on the leukocytes that invade dermal tumors.

C. Targeting Adhesion Molecules (Therapeutic and Diagnostic)

The foregoing description suggests that in situ immunohistochemical detection of aberrantly expressed CAM in skin diseases may be important in categorizing disease activity. In addition, some data suggest that if one blocks EC-leukocyte binding sites or inhibits CAM molecule expression (42), progressive inflammation can be inhibited. On the other hand, CAM may serve as the target autoantigen, as shown in pemphigus vulgaris epidermolysis bullosa (82,96,143). Therapeutic strategies to be considered include administration of monoclonal antiadhesion antibodies, soluble forms of adhesion molecules, peptide antagonists, or other low-molecular-weight drugs (82,144–149). This strategy is primarily based on down-regulating those CAM considered to be responsible for the pathogenesis of the disease.

Since activation of cells leads to upregulation of cell surface and intracellular AR (150), several studies have measured serum levels of circulating adhesion molecules together with other parameters like autoantibodies in an attempt to correlate these findings with disease activity or prognosis (151–154). Soluble ICAM-1 and E-selectin levels together with ECM components (e.g., procollagen III peptide) are reliable markers of disease severity in psoriasis (155,156). Soluble T-cell surface molecules such as CD27 and the interleukin-2 receptor are also useful for the follow-up of psoriasis patients (157,158). The measurement of soluble adhesion molecules is also useful for the follow-up of patients with skin cancers such as melanoma (137,159). Hansen et al. (136) observed significantly increased levels of soluble ICAM-1 in stage 4 melanoma patients as compared to those with stages 1 to 3.

IV. EPILOGUE

Recently, research has been aimed at understanding the relationship between systemic immunity and local immune responses at the tissue level. The role of cell-cell and cell-matrix interactions, the migration of bloodborne immunocompetent cells to a specific organ, and their local interactions in triggering disease-specific immune responses have been foci of research. Altered expression of cell adhesion molecules is probably critical to the changes in cutaneous vascular and lymphatic endothelial cell lining and cell-matrix interactions, resulting in loss of normal skin structure and function. Understanding this expanding field of skin biology may lead to the development of preventive, therapeutic, and diagnostic tools. For example, the differential expression of different adhesion molecules in a disease-specific manner may assist in making the appropriate diagnosis and may be useful for monitoring disease activity. In addition, either anti-CAM antibodies or synthetic analogs of adhesion molecules may modify leukocyte trafficking into the skin during inflammation. Modifiers of cell–extracellular matrix adhesion molecules or ligands may also be beneficial. Much more understanding of these complex interactions is needed before such therapies can applied in standard practice.

REFERENCES

1. Streilein JW. Lymphocyte traffic, T cell malignancies and the skin. J Invest Dermatol 1978; 71:167–171.

2. Barker JNW, Mitra RS, Griffiths CEM, Dixit VM, Nickoloff BJ. Keratinocytes as initiators of inflammation. Lancet 1991; 337:211-214.
3. Bjercke S, Elgo J, Braathen L, Thorsby E. Enriched epidermal Langerhans cells are potent antigen presenting cells for T cells. J Invest Dermatol 1984; 83:286-289.
4. Poole IC, Van den Wijngaard RMJGJ, Westerhof W, et al. Phagocytosis by normal human melancoytes in vitro. Exp Cell Res 1993; 205:388-395.
5. Poole IC, Mutis T, Van den Wijngaard RMJGJ, et al. A novel, antigen-presenting function of melanocytes and its possible relation to hypopigmentary disorders. J Immunol 1993; 151:7284-7292.
6. Nickoloff BJ, Turka LA. Immunological functions of non-professional antigen presenting cells: new insights from studies of T cell interactions with keratinocytes. Immunol Today 1994; 15:464-469.
7. Edelman GM. Cell adhesion molecules in the regulation of animal form and tissue. Annu Rev Cell Biol 1986; 2:81-116.
8. Chuong CM, Chen HM. Enhanced expression of neural cell adhesion molecules and tenascin (cytotactin) during wound healing. Am J Pathol 1991: 138: 427-440.
9. Van Duinen CM, Van den Broek LJ, Vermeer BJ, Fleuren GJ, Bruijn JA. The distribution of cellular adhesion molecules in pigmented skin lesions. Cancer 1994; 73:2131-2139.
10. Rosen K, Dahlstrom KK, Mercurio AM, Wewer UM. Expression of the alpha 6 beta 4 integrin by squamous cell carcinomas and basal cell carcinomas: possible relation to invasive potential?. Acta Dermatol Venereol 1994; 74: 101-105.
11. Pellegrini G, de Luca M, Orecchia G, et al. Expression, topography and function of integrin receptors are severely altered in keratinocytes from involved and non involved psoriatic skin. J Clin Invest 1992; 89:1783-1795.
12. Bos JD, Kapsenberg ML. Lymphocyte subpopulations of the skin immune system. In. Jan D Bos, ed. Skin Immune System. Boca Raton: CRC Press, 1990.
13. Das PK, Grange JGJ. Mycobacteria, tissue immune response, and pathogenesis. J Med Microbiol 1993; 4:15-23.
14. Ralfkiaer E, Lange Wantzin G, Mason DY, Stein H, Thomsen K. Characterization of benign cutaneous lymphocytic infiltrates by monoclonal antibodies. Br J Dermatol 1984; 111:635-642.
15. Rowden G. Monocytes, macrophages, histiocytes and dendritic cells. In Bos JD, ed. Skin Immune System (SIS). Boca Raton. CRC Press 1990:125-157.
16. Abdel Nasr MB, Krüger-Krasagakes S, Krasagakes K, Gollnick A, Orfanos CE. Further evidence for both cell mediated and humoral immunity in generalized vitiligo. Pigment Cell Res 1994; 7:1-8.
17. Streilein JW. Speculations on the immunopathogenesis of psoriasis: T cell violation of a keratinocyte sphere of influence. J Invest Dermatol 1990; 95: 201-215.
18. de Boer OJ, Verhagen CE, Visser A, Bos JD, Das PK. Cellular interactions in psoriasis skin. Acta Dermatol Venereol 1994; 186(suppl):15-18.

19. de Boer OJ, Wakelkamp IMMJ, Pals ST, Claessen N, Bos JD, Das PK. Increased expression of adhesion receptors in both involved and non involved psoriatic skin. Arch Dermatol Res 1994; 286:304–311.
20. Springer, TA. Adhesion receptors of the immune system. Nature 1990; 346: 425–434.
21. Hynes RO. Integrins: versality, modulation and signalling in cell adhesion. Cell 1992; 69:11–25.
22. Lasky AL. Selectins: interpreters of cell specific carbohydrate information during inflammation. Science 1992; 258:964–969.
23. Furukawa F, Takigawa M, Matsuyoshi N, et al. Cadherins in cutaneous biology. J Dermatol 1994; 21:802–813.
24. Herrlich P, Zoller M, Pals ST, Ponta H. CD44 splice variants: metastases meet lymphocytes. Immunol Today 1993; 14:395–399.
25. Holzman B, Weisman IL. Peyers patch specific lmphocyte homing receptors consist of a VLA-4 like α-chain associated with either two integrin β-chains, one of which is novel. EMBO J 1989; 8:1735–1741.
26. Damsky CH, Werb Z. Signal transduction by integrin receptors for extracellular matrix. Curr Opin Cell Biol 1992; 4:772–781.
27. Ross EL, Barker JN, Allen MH, CHU AC, Groves RW, MacDonald DM. Langerhans' cell expression of the selectin ligad, sialyl Lewis x. Immunology 1994; 81:303–308.
28. Jalkanen S, Saari S, Kalimo H, et al. Lymphocyte migration into the skin: the role of lymphocyte homing receptor (CD44) and endothelial cell antigen (HECA-452). J Invest Dermatol 1990; 94:786–792.
29. Picker LJ. Control of lymphocyte homing. Curr Opin Immunol 1994; 6:394–406.
30. Mackay CR, Martson WL, Dudler L. Naive and memory T cells show distinct pathways of lymphocyte recirculation. J Exp Med 1990; 171:801–817.
31. Lawrence MB, Springer TA. Leucocytes roll on a selectin at physiologic flow rates: distinction from and prerequisite for adhesion through integrins. Cell 1991; 65:859–873.
32. Butcher EC. Leucocyte endothelial cell recognition: Three (or more) steps to specificity and diversity. Cell 1991; 67:1033–1036.
33. Bevilacqua MP, Pober JS, Mendrick DL, Cotran RS, Gimbrone MA Jr. Identification of inducible endothelial-leucocyte adhesion molecule. Proc Natl Acad Sci USA 1987: 84:9238–9242.
34. Larkin M, Ahern TJ, Stoll MS, et al. Spectrum of sialyated and non sialyated fuco-oligosaccharides bound by the endothelial-leucocyte adhesion molecule E-selectin. Dependence of the carbohydrate binding activity on E-selectin density. J Biol Chem 1992; 67:13661.
35. Yuen CT, Lawson AM, Chai W, et al. Novel sulfated ligands for the cell adhesion molecule E-selectin revealed by the neoglycolipid technology among O-linked oligosaccharides on an ovarian cystadenoma glycoprotein. Biochemistry 1992; 31:9126.
36. Berg EL, Yoshino T, Rott LS, et al. The cutaneous lymphocyte antigen is a

skin lymphocyte homing receptor for the vascular lectin endothelial cell-leukocyte adhesion molecule 1. J Exp Med 1991; 174:1461–1466.

37. Berlin C, Berg EL, Briskin MJ, et al. $\alpha 4\beta 7$ integrin mediates lymphocyte binding to the mucosal vascular addressin MadCAM-1. Cell 1993; 74:185–195.
38. Hamann A, Andrews DP, Jablonski-Westrich D, Holzman B, Butcher EC. The role of $\alpha 4$ integrins in lymphocyte homing to mucosal tissues in vivo. J Immunol 1994; 152:3282–3293.
39. Picker LJ, Michie SA, Rott LS, Butcher EC. A unique phenotype of skin-associated lymphocytes in humans. Am J Pathol 1990; 136:1053–1068.
40. Duijvestijn AM, Horst E, Pals ST, et al. High endothelial differentiation in human lymphoid and inflammatory tissues defined by monoclonal antibody HECA-452. Am J Pathol 1988; 130:147–155.
41. Berg EL, Robinson MK, Mansson O, Butcher EC, Magnani JL. A carbohydrate domain common to both sialyl Le(a) and sialyl Le(X) is recognized by the endothelial cell leukocyte adhesion molecule ELAM-1. J Biol Chem 1991; 266:14869–14872.
42. de Boer OJ, Horst E, Pals ST, Bos JD, Das PK. Functional evidence that the HECA-452 antigen is involved in the adhesion of human neutrophils and lymphocytes to tumor necrosis factor-α stimulated endothelial cells. Immunology 1994; 81:359–365.
43. Picker LJ, Martin RJ, Trumble A, et al. Differential expression of lymphocyte homing receptors by human memory/effector T cells in pulmonary versus cutaneous immune effector sites. Eur J Immunol 1994; 24:1269–1277.
44. Rossiter H, van Reijsen F, Mudde GC, et al. Skin disease-related T cells bind to endothelial selectins: expression of cutaneous lymphocyte antigen (CLA) predicts E-selectin but not P-selectin binding. Eur J Immunol 1994; 24:205–210.
45. Picker LJ, Treer JR, Ferguson-Darnell B, Collins PA, Bergstresser PR, Terstappen LW. Control of lymphocyte recirculation in man. II. Differential regulation of the cutaneous lymphocyte-associated antigen, a tissue-selective homing receptor for skin-homing T cells. J Immunol 1993; 150:1122–1136.
46. Santamaria Babi LF, Moser R, Perez Soler MT, Picker LJ, Blaser K, Hauser C. Migration of skin homing T cells across cytokine activated human endothelial cell layers involves the interaction of the cutaneous leucocyte associated antigen (CLA), the very late antigen-4 (VLA-4) and the lymphocyte function associated antigen-1 (LFA-1). J Immunol 1995; 154:1543–1550.
47. Buchsbaum ME, Kupper TS, Murphy GF. Differential induction of intercellular adhesion molecule-1 in human skin by recombinant cytokines. J Cutan Pathol 1993; 20:21–27.
48. Uccini S, Ruco LP, Monardo F, La Parola IL, Cerimele D, Baroni CD. Molecular mechanisms involved in intraepithelial lymphocyte migration: a comparative study in skin and tonsil. J Pathol 1993; 169:413–419.
49. Mutis T, Debueger M, Bakker A, Ottenhof THM. HLA class II human keratinocytes presents *Mycobacterium leprae* antigens to CD4+ and cytotoxic and proliferative T-cells. Scand J Immunol 1993; 37:43–51.

50. Avnstorp C, Ralkier E, Jorgensen J, Wantzin GL. Sequential immunophenotypic study of lymphoid infiltrate in allergic and irritant reactions. Contact Derm 1987; 16:239–245.
51. de Boer OJ, CM Van der Loos, Hamerlinck F, Bos JD, Das PK. Reappraisal of in situ immunophenotypic analysis of psoriasis skin: interacting HLA-DR+ immunocompetent cells are a major feature of psoriatic lesions. Arch Dermatol Res 1994; 286:87–96.
52. Rambukkana A, Das PK, Krieg S, Faber WR. Association of mycobacterial 30 kDa region proteins with the cutaneous infiltrates of leprosy lesions: evidence for the role of secreted proteins in the local tissue immune response of leprosy. Scand J Immunol 1992; 36:36–48.
53. Cumberbatch M, Illingworth I, Kimber I. Antigen bearing dendritic cells in draining lymph nodes of contact sensitized mice: cluster formation with lymphocytes. Immunology 1991; 74:139–143.
54. Nestle FO, Zheng XG, Thompason CB, Turka LA, Nickoloff BJ. Characterization of dermal dendritic cells obtained from normal skin reveals phenotypic and functionally distinctive subsets. J Immunol 1993; 151:6365–6545.
55. Van den Wijngaard RMJGJ, Poole IC, Menko W, Westerhof W, Das PK. Involvement of T cells in vitiligo: immunohistochemical evidence (abstract). Melanoma Res 1994; 4:23.
56. Teunissen MBM, Rongen HAH, Bos JD, Function of adhesion molecules LFA-3 and ICAM-1 on human epidermal Langerhans cells in antigen specific cell activation. J Immunol 1994; 152:3400–3409.
57. Gimmi CD, Freeman GJ, Gribben JG, Gray G, Nadler LM. Human T-cell clonal anergy is induced by antigen presentation in the absence of B7 costimulation. Proc Natl Acad Sci USA 1993; 90:6586–6590.
58. Linsley PS, Ledbetter JA, Thomson CB. Role of CD28 receptor during T cell responses to antigen. Annu Rev Immunol 1993; 11:191–212.
59. Razi-Wolf Z, Freeman GJ, Galvin F, Benacerraf B, Nadler L, Reiser H. Expression and function of the murine B7 antigen, the major costimulatory molecule expressed by peritoneal exudate cells. Proc Natl Acad Sci USA 1992; 89:4210–4214.
60. Freedman AS, Freeman GJ, Rhynart K, Nadler LM. Selective induction of B7/BB-1 on interferon-γ stimulated monocytes: a potential mechanism for amplification of T cell activation through the CD28 pathway. Cell Immunol 1991; 137:429–437.
61. Simon JC, Dietrich A, Mielke V, et al. Expression of the B7/BB1 activation antigen and its ligand CD28 in T-cell-mediated skin diseases. J Invest Dermatol 1994; 103:539–543.
62. Caux C, Vanbervliet B, Massacrier C, et al. B70/B7-2 is identical to CD86 and is the major functional ligand for CD28 expressed on human dendritic cells J Exp Med 1994; 180:1841–1847.
63. Symington FW, Bradly W, Linsley PS. Expression and function of B7 on human epidermal Langerhans cells. J Immunol 1993; 150:1286–1295.
64. Fleming TE, Mirando WS, Trefzer U, Tubesing KA, Elmets CA. In situ

expression of a B7-like adhesion molecule on keratinocytes from human epidermis. J Invest Dermatol 1993; 101:754–758.

65. Azuma M, Yssel H, Phillips JH, Spits H, Lanier LL. Functional expression of B7/BB1 on activated T lymphocytes J Exp Med 1993; 177:845–850.
66. Buck CA, Horwitz AF. Cell surface receptors for extracellular matrix molecules. Annu Rev Cell Biol 1987; 3:179–205.
67. Albeda SM, Buck CA. Integrins and other cell adhesion molecules. FASEB J 1990; 4:2869–2880.
68. Rusolati E. Integrins. J Clin Invest 1991; 87:1–5.
69. Marchiso PC, Bondanza S, Cremona O, Cancedda R, de Luca M. Polarized expression of integrin receptors ($\alpha_6\beta_4$, $\alpha_2\beta_1$ and $\alpha_5\beta_5$) and their relationship with the cytoskeleton and basement membrane matrix in cultured keratinocytes. J Cell Biol 1991; 112:761–773.
70. Yurchenko PD, Schittny JC. Molecular architecture of basement membranes. FASEB J 1990; 4:1577–1590.
71. Beck K, Hunter I, Engel J. Structure and function of laminin: anatomy of a multidomain glycoprotein. FASEB J 1990; 4:148–160.
72. Adams JC, Watt FM. Changes in keratinocyte adhesion during terminal differentiation: reduction of fibronectin binding precedes $\alpha_3\beta_1$ integrin loss from the cell surface. Cell 1990; 63:425–435.
73. De Luca M, Tamura RN, Kaiji S, et al. Polarized integrins mediates keratinocyte adhesion to basal lamina. Proc Natl Acad Sci USA 1990; 87:6888–6892.
74. Zambruno G, Marchiso PC, Melchiori A, Bondanza S, Cancedda R, De Luca M. Expression of integrin receptors and their role in adhesion, spreading, and migration of normal human melanocytes. J Cell Sci 1993; 105:179–190.
75. Steinman R, Hofman L, Pope M. Maturation and migration of cutaneous dendritic cells. J Invest Dermatol 1995; 105:2s–7s.
76. Jones JCR, Kurpakas MA, Cooper HM, Quaranta V. A function for the integrin $\alpha_6\beta_4$ in the hemidesmosomes. Cell Regul 1991; 2:427–438.
77. Kurpakas MA, Jones JCR. A novel hemisdesmosomal plaque component: tissue distribution and incorporation into assembling hemidesmosomes in an in vitro model. Exp Cell Res 1991; 194:139–146.
78. Hotchin NA, Kovach NL, Watt FM. Functional downregulation of integrins is reversible but commitment to terminal differentiation. J Cell Sci 1993; 106: 1131–1138.
79. Jones PH, Waat FM. Separation of human epidermal stem cells from the transit amplifying cells on the basis of differences in integrin function and expression. Cell 1993; 73:713–724.
80. Aumailly M, Krieg T. Structure and function of the cutaneous extracellular matrix. Eur J Dermatol 1994; 4:271–286.
81. Carter WG, Ryan MC, Gahr PJ. Epiligrin, a new cell adhesion ligand for integrin $\alpha 3\beta 1$ in epithelial basement membranes. Cell 1991; 65:599–610.
82. Amagai MC. Adhesion molecules. I. Keratinocyte-keratinocyte interactions; cadherins and pemphigus. J Invest Dermatol 1995; 104:146–142.
83. Tang AM, Amagai M, Granger LG, Stanley JR, Udey MC. Adhesion of

epidermal Langerhans cells to keratinocytes mediated by E-cadherin. Nature 1993; 361:82–85.
84. Tammi R, Paukkonen K, Wang C, Horsmanheimo M, Tammi M. Hyaluronan and CD44 in psoriatic skin. Intense staining for hyaluronan on dermal capillary loops and reduced expression of CD44 and hyaluronan in keratinocyte-leucocyte interfaces. Arch Dermatol Res 1994; 286:21–29.
85. Jones PH, Watt FM. Separation of human epidermal stem cells from transit amplifying cells on the basis of differences in integrin function and expression. Cell 1993; 73:713–724.
86. Mathias G, Meinardus-Hager G, Roth J, Goerd S, Sorg C. Nickel chloride and cobalt chloride, two common sensitizers directly induce expression of ICAM-1, VCAM-1, and ELAM-1 by endothelial cells. J Invest Dermatol 1993; 100:754–767.
87. Covani A, Zambrun G, Manca V, et al. Integrin expression during wound healing. J Invest Dermatol 1992; 98:538–544.
88. Morelli JG, Yohn JJ, Zekman T, Norris DA. Melanocyte movement in vitro. Role of matrix proteins and integrin receptors. J Invest Dermatol 1993; 101: 605–608.
89. Sepp NT, Li L-J, Lee KH, et al. Basic fibroblast growth factor increases the expression of $\alpha v\beta 3$ integrin complex on human vascular endothelial cells. J Invest Dermatol 1994; 103:295–299.
90. Kagami S, Border WA, Ruoslahti E, Noble NA. Coordinated expression of $\beta 1$ integrins and transforming growth factor β induced matrix proteins in glomerulonephritis. Lab Invest 1993; 69:68–76.
91. Schröder JM. Inflammatory mediators and chemoattractants. Clin Dermatol 1995; 13:137–150.
92. Smith CH, Barker JNWN. Cell trafficking and role of adhesion molecules in psoriasis. Clin Dermatol 1995; 13:151–160.
93. Black C, Briggs D, Welsh K. The immunogenic background of scleroderma—an overview. Clin Exp Dermatol 1992; 17:73–78.
94. Viac J, Guerniche A. Adhesion molecules and cutaneous inflammation. Eur J Dermatol 1993; 3:381–382.
95. Von den Driesch P, Gruschwitz M, Hornstein OP, Sterry W. Adhesion molecules modulatiion in Sweet's syndrome compared to erythema multiforma. Eur J Dermatol 1993; 3:393–397.
96. Jonkman MF, de Jong MCJM, Heeres K, Sonnenberg A. Expression of integrin $\alpha 6\beta 4$ in junctional epidermolysis bullosa. J Invest Dermatol 1992; 99:489–496.
97. Groves RW, Allen MH, Barker JN, Haskard DO, MacDonald DM. Endothelial leucocyte adhesion molecule-1 (ELAM-1) expression in cutaneous inflammation. Br J Dermatol 1991; 124:117–23.
98. Lisby S, Ralfkiaer E, Rothlein R, Vejlsgaard GL. Intercellular adhesion molecule-1 (ICAM-1) expression correlated to inflammation. Br J Dermatol 1989; 12:479–484.
99. Wakita H, Takigawa M. E-selectin and vascular cell adhesion molecule-1 are

critical for initial trafficking of helper-inducer/memory T cells in psoriatic plaques. Arch Dermatol 1994; 130:457–463.

100. Rhode D, Schluter-Wigger W, Mielke V, Von den Driesch P, Von Gaudecker B, Sterry W. Infiltration of both T cells and neutrophils in the skin is accompanied by the expression of endothelial leukocyte adhesion molecule-1 (ELAM-1): an immunohistochemical and ultrastructural study. J Invest Dermatol 1992; 98:794–799.
101. Kyan-Aung U, Haskard DO, Poston RN, Thornhill MH, Lee TH. Endothelial leukocyte adhesion molecule-1 and intercellular adhesion molecule-1 mediate the adhesion of eosinophils to endothelial cells in vitro and are expressed by endothelium in allergic cutaneous inflammation in vivo. J Immunol 1991; 146:521–528.
102. Das PK, de Boer OJ, Visser A, Verhagen CE, Bos JD, Pals ST. Differential expression of ICAM-1, E-selectin and VCAM-1 by endothelial cells in psoriasis and contact dermatitis. Acta Dermatol Venereol Suppl (Stockh) 1994; 186: 21–22.
103. Groves RW, Ross EL, Barker JN, MacDonald DM. Vascular cell adhesion molecule-1: expression in normal and diseased skin and regulation in vivo by interferon gamma. J Am Acad Dermatol 1993; 29:67–72.
104. Norris PG, Barker JN, Allen MH, et al. Adhesion molecule expression in polymorphic light eruption. J Invest Dermatol 1992; 99:504–508.
105. Gruschwitz M, Von den Driesch P, Kellner I, Hornstein OP, Sterry W. Expression of adhesion proteins involved in cell-cell and cell-matrix interactions in the skin of patients with progressive systemic sclerosis. J Am Acad Dermatol 1992; 27:169–177.
106. Heng MCY, Allen SG, Chase DG. High endothelial venules in involved and uninvolved psoriatic skin: recognition by homing receptors on cytotoxic T-lymphocytes. Br J Dermatol 1988; 118:315–326.
107. Barton SP, Abdullah MS, Marks R. Quantification of microvascular changes in the skin in patients with psoriasis. Br J Dermatol 1992: 126:569–574.
108. Moncada B, Torres-Alvarez MB, Gonzalez-Amaro R, et al. Lack of expression of intercellular adhesion molecule ICAM-1 in lepromatous leprosy patients. Int J Lepr Other Mycobact Dis 1993; 61:581–585.
109. Walsh LJ, Trinchieri G, Waldorf HA, Whittaker D, Murphy GF. Human dermal mast cells contain and release tumour necrosis factor alpha, which induces endothelial leucocyte adhesion molecule 1. Proc Natl Acad Sci USA 1991: 88:4220–4224.
110. Michie SA, Streeter PR, Bolt PA, Butcher EC, Picker LJ. The human periphral lymph node vascular addressin. An inducible endothelial antigen involved in lymphocyte homing. Am J Pathol 1993; 143:688–698.
111. Von den Driesch P, Simon M Jr. Cellular adhesion antigen modulation in purpura pigmentosa chronica. J Am Acad Dermatol 1994; 30:193–200.
112. Bos JD, de Boer OJ, Tibosch E, Das PK, Pals ST. Skin-homing T lymphocytes: detection of cutaneous lymphocyte-associated antigen (CLA) by HECA-452 in normal human skin. Arch Dermatol Res 1993; 285:179–183.

113. Koszik F, Strunk D, Simonitsch I, Picker LJ, Stingl G, Payer E. Expression of monoclonal antibody HECA-452 defined E-selectin ligands on Langerhans cells in normal and diseased skin. J Invest Dermatol 1994; 102:773–780.
114. McGregor JM, Barker JNWN, Ross EL, MacDonald DM. Epidermal dendritic cells in psoriasis possess a phenotype associated with antigen presentation: in situ expression of beta2 integrins. J Am Acad Dermatol 1992; 27: 383–388.
115. Nickoloff BJ, Nestle FO, Zheng XG, Turka LA. T lymphocytes in skin lesions of psoriasis and mycosis fungoides express B7-1: a ligand for CD28. Blood 1994; 83:2580–2586.
116. De Luca M, Pellegrini G, Zambruno G, Marchisio PC. Role of integrins in cell adhesion and polarity in normal keratinocytes and human skin pathologies. J Dermatol 1994; 21:821–828.
117. Messadi DV, Pober JS, Murphy GF. Effects of recombinant gamma-interferon on HLA-DR and DQ expression by skin cells in short-term organ culture. Lab Invest 1988; 58:61–67.
118. Nilsson H, Johansson C, Sandberg K, Funa K, Alm GV, Scheynius A. Induction of mRNA for HLA-DR beta in human keratinocytes cocultured with interferon gamma. Arch Dermatol Res 1989; 281:260–266.
119. Al Badri AMT, Foulis AK, Todd PM, et al. Abnormal expression of MCH class II molecules and ICAM-1 by melanocytes in vitiligo. J Pathol 1993; 169: 203–206.
120. Smit NPM, Le Poole IC, Wijngaard RMJGJ, Tigges AJ, Westerhof W, Das PK. Expression of different immunological markers by cultured human melanocytes. Arch Dermatol Res 1993; 285:365–376.
121. Nickoloff BJ, Mitra RS, Lee K, et al. Discordant expression of CD28 ligands, BB-1, and B7 on keratinocytes in vitro and psoriatic cells in vivo. Am J Pathol 1993; 142:1029–1040.
122. Bruynzeel I, Koopman G, Van der Raaij LM, Pals ST, Willemze R. CD44 antibody stimulates adhesion of peripheral blood T cells to keratinocytes through the leukocyte function-associated antigen-1/intercellular adhesion molecule-1 pathway. J Invest Dermatol 1993; 100:424–428.
123. Harada M, Hashimoto K, Fujiwara K Immunohistochemical distribution of CD44 and desmoplakin I & II in Hailey-Hailey's disease and Darier's disease. J Dermatol 1994; 21:389–393.
124. Schalkwijk J, Vlijmen I van, Oosterling B, et al. Tenascin expression in hyperproliferative skin diseases. Br J Dermatol 1991; 124:13–20.
125. Nakazawa K, Bonnard M, Damour O, Collombel C. Functional role of E-cadherin in melanocyte-keratinocyte adhesion in vitro (abstract). Melanoma Res 1995; 5:40.
126. Gaughman SW, Li JJ, Degitz K. Human intracellular adhesion molecule-1 gene and its expression in the skin. J Invest Dermatol 1992; 98:61s–65s.
127. Bebbion SD, Middleton MH, David-Bajar KM, Brice S, Norris DA. In three types of interface dermatitis different patterns of expression of intercellular adhesion molecule-1 (ICAM-1) indicate different triggers of disease. J Invest Dermatol 1995; 105:71s–79s.

128. Konter U, Kellner I, Hoffmeister B, Sterry W. Induction and upregulation of adhesion receptors in oral and dermal lichen planus. J Oral Pathol Med 1990; 19:459–63.
129. Karashima T, Hachisuka H, Okubo K, Sasai Y. Epidermal keratinocytes of bullous pemphigoid express intracellular adhesion molecule-1 (ICAM-1). J Dermatol 1992; 19:82–86.
130. Teraki Y, Siohara T, Nagashima M, Nishikawa T. Prurigo pigmentosa: role of ICAM-1 in the localization of the eruption. Br J Dermatol 1991; 125:360–363.
131. Poole IC, Wijngaard Van dern RMJG, Westerhof W, Das PK. Presence of T cells and macrophages in inflammatory vitiligo skin parallels melanocyte disappearance. Am J Pathol. 1996; 148:1219–1228.
132. Singer KH. Interactions between epithelial cells and T lymphocytes: role of adhesion molecules. J Leucoc Biol 1990; 48:367–374.
133. Needleman BW. Increased expression of intercellular adhesion molecule 1 on the fibroblasts of scleroderma patients. Arthritis Rheum 1990; 33:1847–1851.
134. Liotta LA, Wewer K, Rao NC, et al. Biochemical mechanisms of tumour invasion and metastases. Adv Exp Med Biol 1988; 233:161–169.
135. Van Duinen CM, Fleuren GJ, Bruijn JA. The extracellular matrix in pigmented skin lesions: an immunohistochemical study. Histopathology 1994; 24:33–40.
136. Hansen NL, Ralfkiaer E, Hou Jensen K, Drzewiecki KT, Rothlein R, Vejlsgaard GL. Expression of intercellular adhesion molecule-1 in benign naevi and malignant melanomas. Acta Derm Venereol 1991; 71:48–51.
137. Kageshita T, Yoshii A, Kimura T, et al. Clinical relevance of ICAM-1 expression in primary lesions and serum of patients with malignant melanoma. Cancer Res 1993; 53:4927–4932.
138. Lee KH, Lawley TJ, Xu YL, Swerlick RA. VCAM-1-, ELAM-1-, and ICAM-1-independent adhesion of melanoma cells to cultured human dermal microvascular endothelial cells. J Invest Dermatol 1992; 98:79–85.
139. Graaf JH, Tamminga RYJ, Kamps WA, Timens W. Langerhans' cells histiocytosis: expression of leucocyte cellular adhesion molecules suggest abnormal homing and differentiation. Am J Pathol 1994; 144:466–472.
140. de Graaf JH, Tamminga RYJ, Kamps WA, Timens W. The role of cellular adhesion molecules in the pathogenesis of Langerhans' cell histiocytosis. Am J Pathol 1995; 147:1161–1171.
141. Gelb AB, Smoller BR, Warnke RA, Picker LJ. Lymphocytes infiltrating primary cutaneous neoplasms selectively express the cutaneous lymphocyte associated antigen (CLA). Am J Pathol 1993; 142:1556–1564.
142. Noorduyn LA, Beljaards RC, Pals ST, et al. Differential expression of the HECA-452 antigen (cutaneous lymphocyte associated antigen) in cutaneous and non cutaneous T cell lymphomas. Histopathology 1992; 21:59–64.
143. Baudoin C, Miquel C, Blanchet-Bardon C, Gambini C, Meneguzzi G, Ortonne JP, Herlitz junctional epidermolysis bullosa keratinocytes display heterogeneous defects of nicein/kalinin gene expression. J Clin Invest 1994; 93: 862–869.

144. Harlan JM, Winn RK, Vedder NB, Doerschuk CM, Rice CL. In vivo models of leucocyte adherence to endothelium. In Harlan JM, Liu DY, eds. Adhesion: Its Role in Inflammatory Disease. New York: W. H. Freeman, 1992; 117–150.
145. Haug CE, Colvin RB, Delmonico FL, et al. A phase I trial of immunosuppression with anti ICAM-1 (CD54) mAb in renal allograft recipients. Transplantation 1993; 55:766–772.
146. Cagnoni ML, Ghersetich I, Lotti T, Pierleoni M, Landi G. Treatment of psoriasis vulgaris with topical calcipotriol: is the clinical improvement of lesional skin related to a down-regulation of some cell adhesion molecules? Acta Dermatol Venereol Suppl (Stockh) 1994; 186:55–57.
147. Foster CA, Dreyfuss M, Mandak B, et al. Pharmacological modulation of endothelial cell associated adhesion molecule expression: implications for future treatment of dermatological diseases. J Dermatol 1994; 21:19–26.
148. Cornelius LA, Sepp N, Li LJ, et al. Selective upregulation of intercellular adhesion molecule (ICAM-1) by ultraviolet B in human dermal microvascular endothelial cells. J Invest Dermatol 1994; 103:23–28.
149. Edwards BD, Andrew SM, O'Driscoll JB, Chalmers RJ, Ballardie FW, Freemont AJ. Changes in numbers of epidermal cell adhesion molecules caused by oral cyclosporin in psoriasis. J Clin Pathol 1993; 46:713–717.
150. Leeuwenberg JFM, Smeets EF, Neefjes JJ, Shaffer MA, Cinek T, Jeunhomme TMAA. E-selectin and intercellular adhesion molecule-1 are released by activated endothelial cells. Immunology 1992; 77:543–549.
151. Wang CR, Liu MF, Tsai RT, Chuang CY, Chen CY. Circulating intercellular adhesion molecules-1 and autoantibodies including anti-endothelial cell, anticardiolipin, and anti-neutrophil cytoplasma antibodies in patients with vasculitis. Clin Rheumatol 1993; 12:375–380.
152. Gearing AJH, Newman W. Circulating adhesion molecules in disease. Immunol Today 1993; 14:506–512.
153. Lobb RR, Chi-Rosso G, Leone DR, Rosa MD, Bixler S, Newman BM. Expression and functional characterization of a soluble form of endothelial-leukocyte adhesion molecule 1. J Immunol 1991; 147:124–129.
154. Schopf RE, Naumann S, Rehder M, Morshes B. Soluble intercellular adhesion molecule-1 levels in patients with psoriasis. Br J Dermatol 1993; 128:34–37.
155. Ameglio F, Bonifati C, Carducci M, Alemanno L, Sacerdoti G, Fazio M. Soluble intercellular adhesion molecule-1 and procollagen III peptide are reliable markers of disease severity in psoriasis. Acta Dermatol Vehereol Suppl (Stockh) 1994; 186:19–20.
156. Bonifati C, Trento E, Carducci M, et al. Soluble E-selectin and soluble tumour necrosis factor receptor (60 kD) serum levels in patients with psoriasis. Dermatology 1995; 190:128–131.
157. Hintzen RQ, De Jong R, Lens SM, Van Lier RA. CD27: marker and mediator of T-cell activation? Immunol Today 1994; 15:307–311.
158. De Rie MA, Hamerlinck F, Hintzen RQ, Bos JD, Van Lier RA. Quantif-

ication of soluble CD27, a T-cell activation antigen, and soluble interleukin 2 receptor in patients with psoriasis. Arch Dermatol Res 1991; 283:533–534.

159. Kageshita T, Yoshii A, Kimura T, Ono T. Analysis of expression and soluble form of intercellular adhesion molecule-1 in malignant melanoma. J Dermatol 1992; 19:836–840.

24

Adhesion Molecules in Renal Diseases

Andrey V. Cybulsky
Department of Medicine, Royal Victoria Hospital, McGill University, Montreal, Québec, Canada

I. INTRODUCTION

Until recently, there has been little information on the role of adhesion molecules in regulating functions of renal cells or in facilitating infiltration of inflammatory leukocytes in pathological disorders of the kidney. Adhesion of cells to extracellular matrices (ECMs) or to neighboring cells is facilitated by specific cell surface receptors, which include integrins, selectins, cadherins, proteoglycans, molecules of the immunoglobulin superfamily, and others. A number of comprehensive reviews on cell adhesion molecules have appeared in recent years, reflecting the intensive investigation in this field (a partial list includes refs. 1–17). In addition to molecules that facilitate adhesion, it has also been recognized that cells may produce "anti-adhesive" proteins, such as tenascin, SPARC (secreted protein acidic and rich in cysteine; also termed osteonectin or BM-40), and others (reviewed in refs. 18,19). Several reviews have focused on the expression of adhesion molecules in the kidney and the potential roles of these molecules in renal development, renal inflammation, and other disorders (20–27). The reader is referred to these reviews for details on the structural characteristics, binding of various ligands, and tissue distribution of these proteins. This chapter will highlight some recent studies on the role of adhesion molecules in renal diseases and in signal transduction in renal cells.

II. ADHESION MOLECULES AND LEUKOCYTE-ENDOTHELIAL INTERACTIONS IN GLOMERULONEPHRITIS

Acute and chronic inflammation of tissues, including the kidney, is often characterized by leukocyte infiltration. This process is frequently initiated by antibody-antigen interactions and local generation of chemoattractants and cytokines. Leukocytes emigrate toward the inflammatory focus, where they produce additional chemoattractants and cytokines and release lysosomal enzymes and oxygen free radicals, which lead to tissue damage. Infiltration of leukocytes involves their adhesion to vascular endothelial cells, followed by migration through endothelium and basement membranes, processes that appear to be mediated by adhesion molecules on the surfaces of both endothelial cells and leukocytes. The major endothelial cell adhesion molecules include intercellular adhesion molecule-1 (ICAM-1; CD54), vascular cell adhesion molecule-1 (VCAM-1), P-selection (CD62), and E-selectin (ELAM-1). ICAM-1, a member of the immunoglobulin superfamily, is a ligand for all leukocytes. ICAM-1 binds to β_2 integrins, which include lymphocyte function-associated antigen-1 (LFA-1; $\alpha_L\beta_2$; CD11a/CD18) and Mac-1 ($\alpha_M\beta_2$; CD11b/CD18). VCAM-1 binds lymphocytes, monocytes, and eosinophils, but not neutrophils, through interaction with $\alpha_4\beta_1$ integrin (VLA-4; CD49d/CD29). P-selectin and E-selectin bind neutrophils, monocytes, and some T cells, via sialyl oligosaccharides. ICAM-1 and VCAM-1 are expressed on a variety of cell types, whereas P-selectin is restricted to endothelial cells and platelets, and E-selectin to endothelial cells. Cytokines (e.g., interleukin-1 and tumor necrosis factor) can upregulate expression of E-selectin, ICAM-1, and VCAM-1. Also, P-selectin can be mobilized from intracellular compartments to the cell surface (thus enhancing surface expression).

Recent studies support an in vivo model in which selectins mediate the initial attachment and rolling of neutrophils along the vascular endothelium, while integrins are responsible for subsequent tight association and emigration of the cells. This model may also be relevant to other leukocytes. Chemoattractants (including C5a, platelet activating factor, leukotriene B4), and chemokines (e.g., interleukin-8) can rapidly upregulate the affinity of leukocyte integrins, allowing them to interact with endothelial counterreceptors, or can enhance surface expression of leukocyte integrins by inducing mobilization from intracellular compartments to the cell surface.

There are a number of reports on the expression of adhesion molecules in normal kidney, mainly in humans and rodents (reviewed in 23–27). ICAM-1 is expressed constitutively by glomerular and peritubular capillary endothelial cells, by cells in renal arterioles, some cells in Bowman's capsule, and in the mesangium. VCAM-1 does not appear to be expressed within normal glomeruli but may be found in some cells in Bowman's capsule. E-selectin is not expressed in normal kidneys, and so far there appear to be no data on P-selectin expression.

A typical example of an inflammatory disease in the kidney is acute glomerulonephritis. In humans, this encompasses postinfectious glomerulonephritis, antiglomerular basement membrane (GBM) nephritis (Goodpasture nephritis), membranoproliferative glomerulonephritis, and IgA nephropathy/Henoch-Schönlein purpura. In these disorders, glomerular injury is manifested as hematuria, proteinuria, and/loss of renal function. Mediators of injury in various forms of glomerulonephritis have been defined by use of experimental animal models (28). Among the inflammatory models, experimental anti-GBM nephritis (the counterpart of human anti-GBM or Goodpasture nephritis) has been studied most extensively. Autoimmune anti-GBM nephritis is produced by active immunization of host species with isolated glomeruli or purified GBM (29). Animals develop circulating antibodies reactive with the α_3 chain of type IV collagen (Goodpasture antigen), which deposit along the glomerular capillary wall. Alternatively, the more widely employed passive model (also known as nephrotoxic serum nephritis) is induced with an intravenous injection of heterologous anti-GBM antiserum (nephrotoxic serum) (29). This antiserum is classically raised against crude preparations of glomeruli or GBM, and appears to be reactive with multiple GBM, endothelial and epithelial cell wall constituents. In classic anti-GBM nephritis (nephrotoxic serum nephritis), heterologous phase proteinuria develops within the first 24 hours and resolves after 48 to 72 hours. By specifically depleting animals of individual humoral mediators or circulating inflammatory cell types, it has been demonstrated that heterologous phase injury, assessed by proteinuria, can be induced by antibody alone; antibody and complement; or antibody, complement, and neutrophils (28,29). Autologous phase proteinuria develops subsequently, in association with a host immune response to the heterologous antiserum. The autologous phase is generally mediated by macrophges, independently of complement (28,29). There are, however, several variations to this classic model, which may depend on the host species and origin of the anti-GBM antiserum. Moreover, the host species and source of antiserum may influence the glomerular histopathology, which may appear normal or, alternatively, demonstrate infiltration of leukocytes, or even fibrocellular crescents (28,29).

The presence of crescents in glomerulonephritis is associated with a poor prognosis. Crescent formation appears to depend on infiltration of the glomerular urinary space (Bowman's space) with mononuclear leukocytes. Thus, interaction of leukocytes with glomerular endothelium may be an important early event in the generation of crescents. Nishikawa et al. (30) tested this hypothesis in a model of active anti-GBM crescentic glomerulonephritis in Wistar-Kyoto rats. Two weeks after immunization with bovine GBM, rats developed circulating anti-GBM antibodies reactive with the α_3 chain of type IV collagen, deposited the anti-GBM antibodies in the GBM, increased the expression of ICAM-1 in glomerular endothelial cells, and the glomeruli became infiltrated with LFA-1-positive T cells and monocytes/macrophages. After 5 weeks, diffuse fibrocellular crescents, glomerular sclerosis, tubulointerstitial lesions, and renal insufficiency were present (Fig. 1). Administration of antibodies to ICAM-1 and the α subunit of LFA-1 reduced the severity of renal disease (Fig. 1). When administered three times weekly, starting 2 days prior to immunization with GBM and continuing for 14 days, glomerular lesions and proteinuria were virtually absent at 2 weeks and only mild changes were present at 5 weeks. When anti-ICAM-1 and anti-LFA-1 were administered three times weekly, starting 2 weeks after immunization with GBM, the progression of the disease was retarded. Anti-ICAM-1 and anti-LFA-1 antibodies did not affect circulating leukocyte counts. The two antibodies had relatively small effects on the titers of anti-GBM antibodies in the circulation, suggesting that the anti-ICAM-1 and anti-LFA-1 antibodies prevented development of glomerulonephritis, most likely by interfering with the interactions between leukocytes and activated glomerular endothelial cells. However, in models based on active immunization, it is difficult to exclude effects of antiadhesion molecule antibodies on the immune response—i.e., on the production of nephritogenic autoantibodies.

Glomerular injury in passive anti-GBM nephritis (nephrotoxic serum nephritis) is also dependent on cell-adhesion molecule interactions. Mulligan et al. (31) administered heterologous anti-GBM antiserum to Long-Evans rats; in this model, nephritis develops within 24 hours and is characterized by linear deposition of antibody along the GBM in conjunction with complement fixation, neutrophil infiltration, and the development of proteinuria. The authors studied glomerular changes after 6 hours, and urine protein excretion after 24 hours. Following administration of anti-GBM antibody, control rats developed marked proteinuria, neutrophil infiltration, and upregulation of ICAM-1, VCAM-1, and E-selectin, while animals treated with antibodies to ICAM-1, β_2 integrin subunit, α subunit of Mac-1, or α subunit of VLA-4 had significantly reduced proteinuria and neutrophil infiltration. These antibodies were administered together with anti-GBM antibody and/or at later time points. Administration of antibodies to the α subunit of LFA-1 or to E-selectin were ineffective, and none of the antibodies affected circulating leukocyte counts. The authors also demonstrated that neutrophil infiltration and proteinuria were abolished

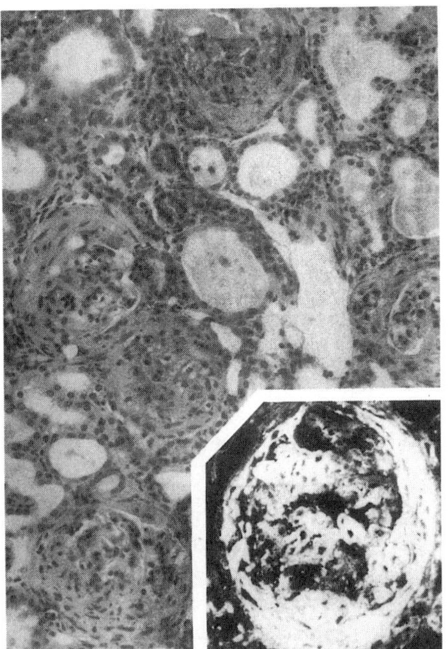

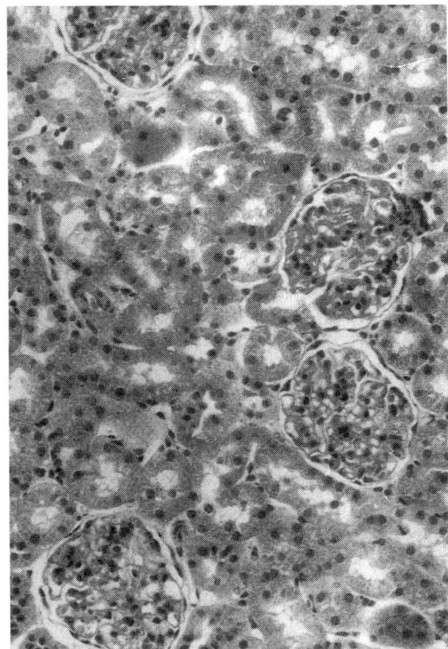

Figure 1 Effect of antibodies of ICAM-1 and LFA-1 in rat autoimmune anti-GBM glomerulonephritis. Left panel: Photomicrograph showing a kidney section from a rat 5 weeks after immunization with GBM (light microscopy). Crescentic sclerosing glomerulonephritis and tubular atrophy is present (×220). The inset (×315) shows deposition of fibrinogen/fibrin in glomerular tufts and in a crescent (immunofluorescence microscopy). Right panel: Section of kidney tissue obtained 14 days after immunization with GBM from a rat injected with anti-ICAM-1 and anti-LFA-1 antibodies three times per week starting 2 days prior to immunization with GBM. Glomeruli, tubules, and interstitium appear normal (×210). Reprinted from ref. 30 by copyright permission of Rockefeller University Press.

after administration of an antibody to tumor necrosis factor-α, suggesting that this cytokine is involved in upregulating the expression of the relevant adhesion molecules in the kidney, thus facilitating adhesion of neutrophils to endothelium. The authors did not, however, assess potential effects of the various anti-adhesion molecule antibodies on the glomerular binding of anti-GBM antibodies. In addition, the role of VLA-4 in this model of anti-GBM nephritis is unexpected, since VLA-4 is not found on neutrophils, and it is therefore possible that monocytes have also been involved in the mediation of glomerular injury.

Further support for the role of adhesion molecules in anti-GBM nephritis was provided by Kawasaki et al. (32). The authors' model in WKY rats is characterized by early infiltration of CD8-positive NK cells, followed by an influx of monocytes/macrophages into the glomeruli. One day after administration of anti-GBM antiserum, there is upregulation of ICAM-1 in glomerular endothelium, and expression of LFA-1 by infiltrating cells. At day 3, glomeruli manifest hypercellularity, and after day 6, there is crescent formation as well as development of proteinuria. In rats that received anti-GBM antiserum together with antibodies to ICAM-1 or to the α subunit of LFA-1, there were dose-dependent reductions in urine protein excretion. In addition, both antibodies reduced glomerular hypercellularity and crescent formation. The antibodies to ICAM-1 and LFA-1 had no effect on glomerular binding of anti-GBM antibody or on circulating leukocyte counts.

There are several other reports on adhesion molecules in anti-GBM glomerulonephritis. Wada et al. (33) administered anti-ICAM-1 antibody to Wistar rats just prior to anti-GBM antibody, and subsequently for 12 days. Proteinuria and glomerular leukocyte infiltration were inhibited in rats receiving anti-ICAM-1, as compared to control animals. In addition, there was upregulation of ICAM-1 expression in glomeruli of nephritic control rats. Lefkowith and Wu (34) administered antibody to the α subunit of Mac-1 (CD11b) 30 min or 16 hours prior to induction of rat anti-GBM nephritis. Anti-Mac-1 decreased proteinuria by 50% (30 min) or 80% (16 hours), and glomerular neutrophil infiltration by 50% (16 hours), as compared to control rats. Glomerular macrophage infiltration and circulating leukocyte numbers were not affected. Tipping et al. (35) studied complement-independent, neutrophil-dependent glomerular injury in a mouse model of anti-GBM nephritis. The authors demonstrated rapid induction of P-selectin on glomerular endothelium, within 30 min of antibody administration. This was associated with glomerular neutrophil infiltration, which peaked at 1 hour, as well as significant proteinuria. Treatment with anti-P-selectin antibody inhibited both neutrophil infiltration and proteinuria. Glomerular binding of anti-GBM antibody was not affected.

Changes in the expression of adhesion molecules have also been demonstrated in an accelerated model of anti-GBM nephritis, produced by preimmunizing rats with rabbit IgG, followed by administration of rabbit anti-GBM antiserum (36,37). Such rats developed proteinuria, renal dysfunction, glomerular hypercellularity, and crescents (36). The kidneys of these rats showed a marked upregulation of glomerular and tubular ICAM-1 expression, significant infiltration of glomeruli with macrophages, and infiltration of the interstitium with macrophages and T cells (37). Treatment with interleukin-1 receptor antagonist from the day prior to induction of glomerulonephritis resulted in a partial reduction in glomerular ICAM-1 expression and leukocyte infiltration, a marked reduction in interstitial expression of ICAM-1 and leukocyte infiltration, abolition of crescent formation, reduction in proteinuria, and maintenance of renal function. The authors concluded that interleukin-1-mediated ICAM-1 upregulation is associated with leukocyte infiltration and damage.

The above studies provide support for ICAM-1 and its ligands in mediating leukocyte infiltration and glomerular injury in acute inflammatory models of glomerulonephritis, although some of the studies did not rigorously exclude other potential effects of blocking antibodies on the immune response or on the binding of glomerular nephritogenic antibodies. P-selectin, VCAM-1 and α_4 integrin may also be key mediators in at least some experimental models. Although antibodies to adhesion molecules have not been employed in human glomerulonephritis, there is indirect evidence to support a role for adhesion molecules in these conditions. As in animal models, ICAM-1 expression is increased in several human glomerulonephritides, including crescentic glomerulonephritis, membranoproliferative glomerulonephritis, IgA nephropathy, Henoch-Schönlein purpura, and lupus nephritis. Changes in expression occur mainly on glomerular endothelial cells and in the mesangium (reviewed in 24–26). There may also be de novo expression of ICAM-1 in proliferating glomerular epithelial cells (GEC) in crescents, as well as in tubular epithelium, particularly when glomerulonephritis is accompanied by tubulointerstitial inflammation. Increased expression of glomerular and tubular VCAM-1 has been reported in crescentic glomerulonephritis, IgA nephropathy and lupus nephritis. De novo E-selectin expression has also been observed in these nephropathies.

Studies of adhesion molecules in humans have generally relied on immunofluorescence or immunohistochemistry techniques, and the results of such studies must be interpreted with caution. Although enhanced adhesion molecule expression suggests a pathogenetic role, further studies are required to confirm that their induction is functionally important for leukocyte infiltration and inflammation. It should be noted that since many adhesion molecules can also exist in soluble form, the appearance of such molecules in biopsy specimens could reflect passive trapping from the circulation. Moreover, increased expression of ICAM-1 has also been reported in glomerular diseases that are believed to be noninflammatory, including minimal change disease and membranous nephropathy.

III. INTEGRINS AS TARGETS OF NEPHRITOGENIC ANTIBODIES

In addition to the mediating leuckocyte-endothelial interactions, adhesion molecules may themselves be targets of nephritogenic antibodies in glomerulonephritis. The visceral GEC is the primary site of injury in several immunological and toxin-induced experimental models of glomerular disease, and damage to GEC (e.g., effacement of foot processes, loss of filtration slit diaphragms, or detachment from GBM) appears sufficient to cause proteinuria (38,39). The best-characterized model of GEC injury is experimental membranous nephropathy (passive Heymann nephritis in the rat), in which antibodies directed at GEC membrane antigens lead to the assembly of the complement C5b-9 membrane attack complex in GEC plasma membranes (40,41). C5b-9 injures GEC without recruitment of leukocytes. There are also other models of GEC injury, in which antibodies may bind to GEC membrane antigens and induce proteinuria in the absence of complement activation or glomerular leukocyte infiltration (28). Recent studies have demonstrated that two well-known nephritogenic antisera—anti-rat GBM antiserum (nephrotoxic antiserum; discussed above), and anti-rat Fx1A (the nephritogenic antibody of passive Heymann nephritis)—react with β_1 integrin adhesion molecules (42,43). It should be noted that in normal human adult kidney, $\alpha_3\beta_1$ is the major integrin present on the basal and lateral surface of GEC foot processes, suggesting that $\alpha_3\beta_1$ mediates attachment of GEC to the GBM (20,44). Other glomerular β_1 integrins include $\alpha_1\beta_1$, found on mesangial cells, and $\alpha_2\beta_1$, on endothelial cells.

Although the classical animal model of anti-GBM glomerulonephritis (discussed above) is mediated by antibody, complement, and leukocytes, certain anti-GBM antisera can induce glomerular injury independently of these other mediators. The capability of anti-GBM antibody to mediate glomerular injury independently was demonstrated convincingly in a study where rat kidneys, perfused ex vivo with a physiological solution containing only anti-GBM antibody, developed proteinuria (45). These findings suggest that the mechanism for proteinuria may involve antibody-dependent injury of visceral GEC. Anti-GBM antibody (nephrotoxic antiserum) is actually a polyspecific antiserum, which reacts with many cell surface antigens, in addition to GBM components (29). Using immunofluorescence microscopy, O'Meara et al. (42) demonstrated that nephritogenic anti-GBM antibody binds to GEC in culture. Moreover, anti-GBM antibody–coated tissue culture plates provide an adhesive substrate for these cells, and anti-GBM antibody can block GEC adhesion to collagen I, collagen IV, laminin, and fibronectin substrata. To further characterize anti-GBM antibody, the authors subjected solubilized GEC membrane proteins to affinity chromatography on a column containing immobilized anti-GBM antibody. The eluate from this column contained a protein doublet of 118 and 140 kDa, and furthermore, anti-β_1 integrin antibody immunoprecipitated the same protein doublet from this eluate (Fig. 2). The same protein doublet was also immunoprecipitated with anti-GBM antibody directly from cultured GEC

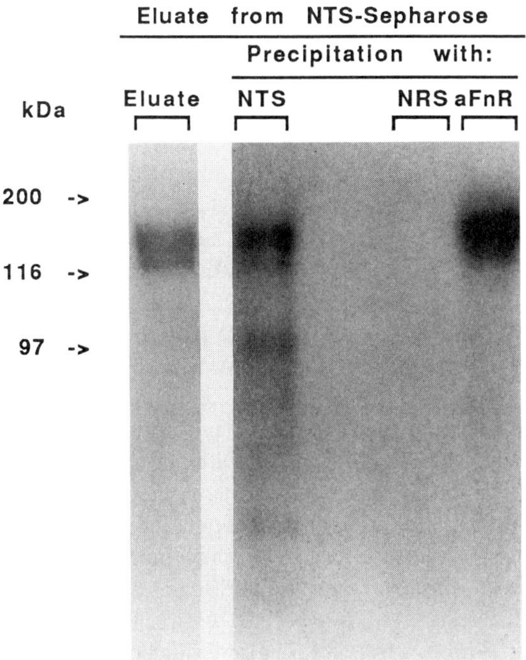

Figure 2 Anti-GBM antibody (nephrotoxic serum; NTS) reacts with β_1 integrin. Sodium dodecyl sulfate polyacrylamide gel electrophoresis and autoradiography of ^{125}I-labeled rat GEC membrane proteins eluted from a sheep NTS-affinity column (left lane), and then immunoprecipitated with NTS, or an antibody that identifies the β_1 integrin subunit of the fibronectin receptor (aFnR). Note that the eluate is highly enriched in proteins of 118 and 140 kDa, and that the same two bands are identified by both antibodies. Control immunoprecipitation with normal rabbit serum (NRS) is negative. Reprinted from ref. 42 by copyright permission of the American Physiological Society.

and isolated glomeruli. Together, these results indicate that anti-GBM antibody reacts with a β_1 integrin on the surface of glomerular epithelial cells, and suggest that anti-integrin activity in vivo may lead to alterations in GEC morphology (e.g., cell detachment from GBM) and consequent proteinuria.

Adler and Chen (43) studied the effects of the nephritogenic antibody of passive Heymann nephritis, anti-Fx1A, on GEC interactions with ECM. Anti-Fx1A is prepared by immunization of animals with a proximal tubular brush border fraction of rat kidney (29). The antiserum is typically polyspecific, and crossreacts with antigens on visceral GEC in vivo and in culture. Affinity chromatography of surface labeled proteins from rat GEC in culture demonstrated that antibodies to β_1 or α_3 integrin subunits specifically immunoprecipitated proteins from the eluate of a column containing immobilized anti-Fx1A, indicating that anti-Fx1A reacts with $\alpha_3\beta_1$. Anti-Fx1A inhibited adhesion of GEC to collagen I, collagen IV, laminin, and fibronectin substrata, and induced rounding and detachment of GEC adherent to substrata. Competition studies showed no additive effects of anti-Fx1A and anti-β_1 in inhibiting adhesion suggesting that the effect of anti-Fx1A on adhesion is attributable to its anti-β_1 activity. It should be noted that injection of rats with anti-Fx1A IgG does not result in proteinuria, unless complement is activated and there is formation of the C5b-9 membrane attack complex. However, proteinuria can be induced independently of complement when anti-Fx1A F(ab′)$_2$ or Fab′ are administered in vivo (46). The reason for this discrepancy is unknown at present, but may be related to the kinetics or relative accessibility of the various antibody fragments to the GEC. These studies support the view that alterations of GEC adhesion to the GBM (e.g., foot process effacement, or detachment) are important factors in the induction of proteinuria (38,39). Furthermore, they suggest that β_1 integrins may be major mediators of GEC attachment to the GBM. Interestingly, in a recent report, it was shown that anti-β_1 antibody IgG or F(ab′)$_2$ can increase glomerular permeability to albumin in an in vitro model employing isolated glomeruli (47).

It should be recognized that while antisera that contain reactivities to β_1 integrins can induce GEC injury and proteinuria in animal models, the extent to which similar antibodies are relevant to the pathogenesis of human nephropathies is not known. A recent preliminary report has demonstrated that a subset of patients with systemic lupus erythematosus (~20% of patients) have circulating autoantibodies directed to β_1 integrin (48), although the nephritogenic potential of such antibodies was not studied. The importance of the results in experimental animal models is primarily to provide insights into the structure and function of the GEC and of the glomerular capillary wall under normal conditions, and in proteinuric diseases.

IV. ANTI-ADHESIVE PROTEINS IN GLOMERULONEPHRITIS

Recent evidence suggests that inhibition of cell adhesion and cell spreading on substrata may be as important as the facilitation of these processes in the control of cellular functions. Based on studies in cell culture, at least two proteins—SPARC and tenascin—can block cell spreading (18,19). Although the inhibitory mechanisms of these proteins have not been established precisely, the proteins are able to induce rounding of adherent cells and partial detachment from substrata. An analogous process may occur in experimental membranous nephropathy (Heymann nephritis in the rat) in vivo. As discussed above, the visceral GEC is the target of complement-mediated injury in the passive and active forms of this model (40,41). Specifically, in experimental membranous nephropathy, GEC injury is characterized by morphological changes, including vacuolization, effacement of foot processes, loss of filtration slit diaphragms, and occasional detachment from the GBM. Increased transglomerular passage of proteins has been localized to the sites of GEC detachment.

Floege et al. (49) demonstrated that SPARC mRNA and protein are expressed constitutively in normal rat glomeruli and immunolocalized SPARC primarily to the cytoplasm of visceral GEC. Immunostaining for SPARC increased in GEC of rats with passive Heymann nephritis in parallel with the development of proteinuria. The increase in immunostaining was associated with an increase in glomerular SPARC mRNA. Complement depletion prior to the induction of passive Heymann nephritis prevented proteinuria, as well as the increases in GEC immunostaining for SPARC and glomerular SPARC mRNA. Enhanced SPARC expression was also demonstrated in active Heymann nephritis, but not in several other models of glomerular injury, such as anti-GBM nephritis. In normal rats, positive immunostaining for tenascin occurred in GEC, similar to the staining observed for SPARC. Small increases in tenascin immunostaining were demonstrated in rats with passive or active Heymann nephritis. The authors concluded that in experimental membranous nephropathy, increased glomerular expression of SPARC (and tenascin) is likely to reflect a cellular response to immunological injury. Furthermore, they speculated that the significant increase in SPARC expression could facilitate focal detachment of GEC from the underlying GBM or possibly augment other mechanisms that may be involved in cell detachment, such as altered expression or function of integrins. Further studies will be required to define the functional consequences of increased SPARC expression in nephropathies.

V. ADHESION MOLECULES IN ACUTE RENAL FAILURE

Acute renal failure in humans (so-called acute tubular necrosis) is most commonly caused by transient ischemia or nephrotoxic drugs. Several animal models have been developed to facilitate studies of this disorder, including clamping of the renal artery and administration of nephrotoxins (reviewed in 23). Following the transient ischemic or toxic insult to the kidney, there is a reperfusion/maintenance phase, which may be characterized morphologically by tubular epithelial injury, including cell necrosis, denudation of the tubular basement membrane due to loss of epithelial cells, tubular dilation, and formation of intratubular casts (23). Regenerative changes including mitoses in tubular cells may also be evident. Functionally, acute renal failure is characterized by a rise in intratubular pressure (which may lead to a decrease in glomerular filtratin), and tubular backleak of glomerular filtrate (23). Interestingly, some of the detached renal tubular epithelial cells can be recovered in the urine as viable cells (22).

Multiple factors may contribute to the pathophysiology of acute renal failure and to its recovery process. In regard to cell adhesion mechanisms, experimental models of acute renal failure are characterized by alterations in adhesion molecules on renal tubular epithelial cells. Second, there is infiltration of the tissue by leukocytes, a process that may be mediated by adhesion molecules similar to those in glomerulonephritis. The maintenance of normal tubular architecture is most likely due to integrin-dependent interactions of epithelial cells with basement membranes and with each other. In kidney sections, immunohistochemistry has demonstrated that cells in the proximal tubule contain mainly the α_6 integrin subunit, as well as α_v and possibly α_3 (reviewed in 23). Alpha$_3$ integrin appears to be present in the distal tubule, along with α_2; α_1 integrin is found in the interstitium, while α_4 and α_5 appear to be absent. The presence of these integrin subunits on tubular cells has been verified in primary renal tubular cell cultures and in cultured renal tubular cell lines (23). In vitro studies, using the cultured tubular cell lines BS-C-1, have assessed the effects of oxidant stress (a common pathogenetic denominator for ischemic and nephrotoxic causes of acute renal failure) on integrin distribution and function (50). Under normal culture conditions, BS-C-1 cells contain predominantly α_3, α_v, and β_1 integrin subunits. The cells adhere to ECM proteins and form focal adhesions, which contain integrins and cytoskeletal proteins (vinculin and talin), on the basal cell surface. When subjected to oxidant stress, BS-C-1 cells respond with disorganization of focal adhesion complexes. Furthermore, although there was no loss of the major integrin subunits from cells, there was a reversal in the polarity of the α_3 integrin subunit from a predominantly basal location to a random distribution on both apical and basal surfaces. In an assay that measured adhesion of BS-C-1 cells in suspension to BS-C-1 monolayers, it was shown that adhesion of the suspended cells was enhanced when the monolayers were first subjected to oxidant stress, inducing apical expression of integrins (51). These cell-cell interactions were partially inhibited with the RDG-containing peptide GRGDNP (Gly-Arg-Gly-Asp-Asn-Pro). The apically expressed integrins in oxidant-stressed cells were also able to bind laminin-coated latex beads, which "activated" the cells—i.e., triggered increases in cytosolic free Ca^{2+} concentration (51).

As a result of these studies, it has been proposed that following renal ischemia, there may be lethal and sublethal injury to tubular epithelial cells. Sublethally injured cells may become detached from basement membranes, in part, due to loss of integrins from the basal surfaces to the apical membranes (23). Detached cells are then washed through the tubule with the flow of tubular fluid, and some may be recovered in the urine. Morphologically, this process is represented by occasional gaps in the epithelial integrity of the tubules, and functionally, it may account for the tubular backleak of glomerular filtrate observed in acute renal failure. Moreover, the expression of integrins on the apical surface of tubular epithelial cells may facilitate a pathological interaction between these cells and the desquamated epithelial cells within tubular lumena. These cell-cell interactions may explain the morphological findings of cellular debris impacting the lumen. Functionally, this process would represent the increased intratubular pressure observed in acute renal failure. Thus, the hypothesis assumes that intratubular casts are at least initially made up of viable tubular epithelial cells, and their formation is regulated by adhesion molecules.

The functional importance of inappropriately expressed α_3 integrin on the apical surface of tubular epithelium in vivo was tested in rats subjected to unilateral renal artery clamping to induce ischemia (51). Following relief of the clamp, rats were infused with the RGD-containing peptide GRGDNP, a control peptide (GRGESP), or vehicle. In control peptide-or vehicle-treated rats, a marked increase in intratubular pressure was evident 1 hour after relief of renal artery occlusion; however, this increase was prevented with the RGD peptide. The authors thus concluded that in acute renal failure, integrins may mediate cell-cell adhesion, which leads to tubular obstruction. While these results are intriguing, further confirmation for a role for integrins in the pathogenesis of tubular cell detachment and cell-cell adhesion will require blocking studies with integrin subunit-specific antibodies or antibody fragments, in order to define precisely the specific subunits and their relevant counter receptors.

Another prominent feature of ischemic acute renal failure is renal medullary vascular congestion, which may result from leukocyte-endothelial interactions, and consequent obstruction of peritubular capillaries by leukocytes, or leukocyte-mediated increases in endothelial permeability. This process could account for sustained ischemia in the outer medulla during the reperfusion phase. Some (52,53) but not all studies (54,55) have shown that depletion of circulating neutrophils with antineutrophil antisera decreases renal ischemia/reperfusion injury, suggesting that neutrophils participate in the pathogenesis of acute renal failure. Based on this observation, it was reasonable to test if inhibition of neutrophil adhesion to vascular endothelium would accelerate recovery from renal injury. Although an earlier study showed no effect of anti-β_2 integrin subunit-specific antibody on the course of acute ischemic renal failure in a rabbit model (55), a more recent study in rats has demonstrated that administration of antibody to ICAM-1 markedly attenuated the course of renal failure (56). In the rat model, renal ischemia was produced by clamping of the renal artery. Following release of the clamp (i.e., during the reperfusion phase), renal function declined, and returned to baseline levels after 72 hours. Administration of anti-ICAM-1 antibody at the time of ischemia, or at 30 or 120 min after ischemia, largely prevented the decline in renal function observed in control rats. Furthermore, mortality of rats was reduced from 16% in controls to 0% with anti-ICAM-1. Antibody to the α subunit of LFA-1 provided modest protection against ischemic injury; however, the combination of anti-LFA-1 together with a subprotective dose of anti-ICAM-1 provided complete protection. This result suggests that in this model of acute renal failure, ICAM-1 interacts with LFA-1 as well as with other ligand molecules on neutrophils. Morphologically, 48 hours after ischemic injury, kidneys manifested necrosis in the tubules of the outer medulla, with some regenerative changes. In addition, tubules were filled with obstructing debris (Fig. 3). In rats treated with anti-ICAM-1, necrosis was significantly reduced, and tubules remained largely patent (Fig. 3). The protective effect of anti-ICAM-1 was also associated with a marked reduction in renal neutrophil infiltration, but the antibody had no effect on circulating neutrophil counts. This study supports a role for leukocyte-endothelial adhesion in the pathophysiology of acute renal failure. As in glomerulonephritis, it is likely that activated neutrophils release cytokines, reactive oxygen species, proteases, and other enzymes, which damage renal tissue directly, increase vascular permeability to other inflammatory cells, or alter renal hemodynamics unfavorably.

Results of the above study were confirmed in a uninephrectomized model of ischemic acute renal failure. In a preliminary report, Rabb et al. (57) treated rats with anti-ICAM-1 antibody prior to induction of transient renal ischemia (60 min). The authors showed that after 24 hours, impairment in renal function and morphological injury to the kidney were significantly reduced in antibody-treated rats, as compared to controls. Furthermore, in this study (57), it was shown that pretreatment with antibody to P-selectin also inhibited renal injury, similar to anti-ICAM-1. Thus, the two endothelial adhesion molecules contribute to the pathogenesis of acute renal failure in this model. Rabb et al. (58) also administered antibodies to the α subunit of LFA-1 (CD11a) and the α subunit of Mac-1 (CD11b) in the uninephrec-

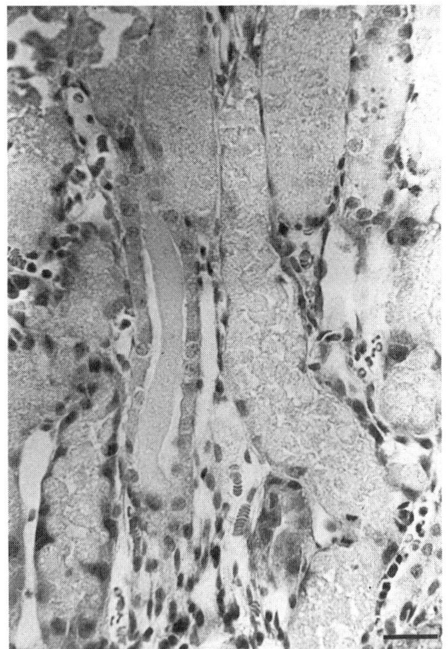
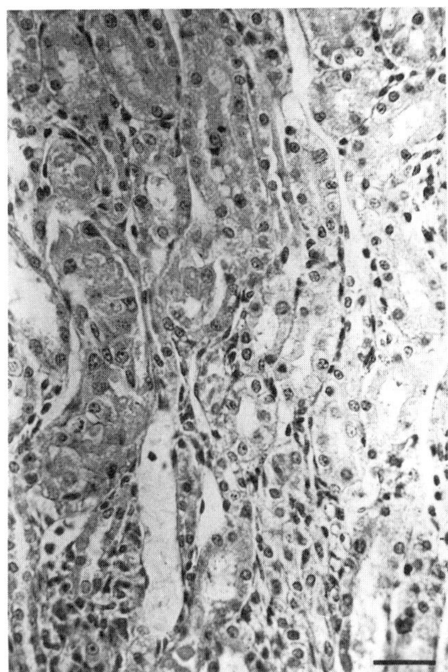

Figure 3 Effect of antibody to ICAM-1 on histopathology of renal ischemia shown by light microscopy of rat kidney outer medulla 48 hour after 30 min of renal ischemia. Left panel: In control rats, extensive necrosis of the tubules of the outer zone of the medulla is present, with regenerative changes of mitosis. Tubules are filled with cellular debris. Right panel: In animals treated with anti-ICAM-1 immediately after relief of ischemia, minimal individual cell necrosis of the outer medulla is present. Tubules are generally patent. (Bar = 50 μm.) Reprinted from ref. 56 by copyright permission of the National Academy of Sciences USA.

tomized model of ischemic acute renal failure. The authors demonstrated that the two antibodies, when given together 30 min prior to induction of renal ischemia, partially reduced the decline in renal function and morphological injury at 24 hours as compared to control rats. The two antibodies did not deplete the number of circulating leukocytes. It was concluded that LFA-1 and Mac-1 mediate renal injury but that other leukocyte adhesion molecules are also involved since anti-LFA-1 and anti-Mac-1 provided only partial protection.

Linas et al. employed an isolated perfused rat kidney model to study the role of neutrophils and ICAM-1 in renal ischemia. Control kidneys or kidneys made ischemic in situ were perfused ex vivo by the isolated kidney technique with untreated neutrophils or neutrophils primed with lipopolysaccharide or phorbol myristate acetate (59). While primed neutrophils had no effect on control kidneys, and untreated neutrophils did not affect injury to ischemic kidneys, perfusion of ischemic kidneys with primed neutrophils markedly accentuated postischemic functional renal impairment (59). In a second study (60), rats were treated with anti-ICAM-1 antibody, and kidneys were made ischemic and then perfused ex vivo with a solution containing neutrophils with or without anti-ICAM-1 antibody. Perfusion with neutrophils in the absence of anti-ICAM-1 worsened the glomerular filtration rate, but this decline did not occur when anti-ICAM-1 was included in the neutrophil perfusate. The authors also assessed the role of oxygen metabolites in ICAM-1-mediated injury. Perfusion of ischemic kidneys with neutrophils and catalase prevented a decline in renal function, as compared to neutrophils alone. Coversely, perfusion of nonischemic kidneys with neutrophils and H_2O_2 reduced the glomerular filtration rate, as compared to neutrophils alone, and this reduction was prevented with anti-ICAM-1. It was concluded that oxygen metabolites produced during reperfusion of ischemic kidneys lead to increased expression of ICAM-1, which mediates neutrophil retention in kidneys and worsening of ischemic injury.

Recently, Kelly et al. (61) adopted an alternate strategy to provide further evidence for the involvement of ICAM-1 in acute ischemic renal failure. The authors induced acute renal ischemia in control mice and mice rendered deficient in ICAM-1 expression by homologous recombination ("knockout mice"). Compared to control, the ICAM-1-deficient animals had a smaller rise in plasma creatinine, reduced histological evidence of cell injury and a lower mortality. These results are consistent with the earlier studies showing that anti-ICAM-1 antibody is protective in acute renal failure.

In addition to elucidating the pathophysiology of acute renal failure, the above studies provide insights into potential modes of therapy for this disorder. Thus, potential therapeutic strategies to consider in acute renal failure may include administration of RGD peptides or blocking antibodies to specific cell adhesion molecules. However, the pathophysiology of acute renal failure may vary among species; consequently, confirmation of these mechanisms in species other than rats or mice will be required.

VI. INTEGRIN SIGNALING IN GLOMERULAR EPITHELIAL CELLS

Studies of cell adhesion molecules, including those in renal models discussed above, have mainly focused on defining the structure of adhesion molecules and characterizing their interactions with ligands on cells or ECM proteins under normal or pathological conditions. In recent years, "outside-in" signal transduction by adhesion of cells to ECM is emerging as a major area of research. An example of this is provided by studies in cultured rat GEC, derived as primary cultures from explants of rat glomeruli. The pathways activated as a result of GEC adhesion to ECM (discussed below) include breakdown of inositol phospholipids, mediated at least in part via β_1 integrins (62,63). Other examples of ECM/integrin-mediated signals have been presented in recent years (reviewed in refs. 3-6,9,12, and in this volume). In platelets, collagens can interact with $\alpha_2\beta_1$ integrin to induce hydrolysis of inositol phospholipids, and platelet integrins regulate Na^+/H^+ exchange and protein tyrosine phosphorylation. Induction of collagenase and stromelysin gene expression and activation of the Na^+/H^+ antiporter is mediated through fibronectin receptors in fibroblasts, and fibronectin also augments anti-CD3-induced proliferation of human T cells via $\alpha_3\beta_1$ and $\alpha_4\beta_1$ integrins. Ligation of the β_2 integrin leads to the activation/tyrosine phosphorylation of phospholipase C-γ1 (64). Adhesion of cells to fibronectin, or antibody-crosslinking of β_1 integrins, induces a rise in cytosolic Ca^{2+} concentration (65), tyrosine phosphorylation of proteins, and activation of protein tyrosine kinases, including focal adhesion kinase and mitogen-activated protein kinase (66). Several other examples of signal transduction associated with adhesion molecules may be found in the literature.

Studies in rat GEC have demonstrated that adhesion to ECM modulates proliferation and phospholipid turnover (62,63). In the mature kidney there are two types of GEC, visceral and parietal. Both types of GEC are of common embryological origin and are in contact with ECM. In normal kidneys, GEC adhere to basement membranes, which contain mainly type IV collagen, laminin, and heparan sulfate proteoglycans; in pathological states, type I collagen may also be present in glomeruli. While the functions of parietal GEC have not been defined, visceral GEC are normally involved in the synthesis of GBM components and maintenance of glomerular permselectivity and architecture. In normal kidneys, there appears to be little turnover of GEC, but proliferation of parietal and possibly visceral GEC may occur as a result of glomerular injury—e.g., in focal segmental sclerosis or crescentic glomerulonephritis. Studies were thus undertaken to gain a better understanding of factors that might regulate proliferation and differentiation functions of GEC. In cultured GEC, epithelial cell growth factors (e.g., epidermal growth factor) stimulated proliferation of cells adherent to type I or IV collagen matrices. Independently of growth factors, adhesion to collagen matrices resulted in the breakdown of inositol phospholipids, consistent with activation of phospholipase C. This enzyme hydrolyzes inositol phospholipids to yield 1,2-diacylgycerol (an endogenous activator of protein kinase C) and inositol phosphates (67,68). When compared to GEC adherent to plastic, GEC on collagen had increased levels of diacylglycerol (Fig. 4), and increased inositol phosphates. Adhesion to collagen also stimulated release of free arachidonic acid, due to metabolism of diacylglycerol, and activation of phospholipase A_2. The changes in diacylglycerol and arachidonic acid occurred soon after plating of cells onto substrata, and were maintained in monolayers of cells. Furthermore, protein kinase C appeared to be activated in parallel with elevated diacylglycerol, and the activation of protein kinase C downregulated proliferation of GEC on collagen. Thus, adhesion to ECM provided two opposing signals for GEC proliferation. First, adhesion was necessary for growth factors to induce a mitogenic response. Second, activation of the diacylglycerol/protein kinase C pathway by ECM downregulated the mitogenic effect of growth factors.

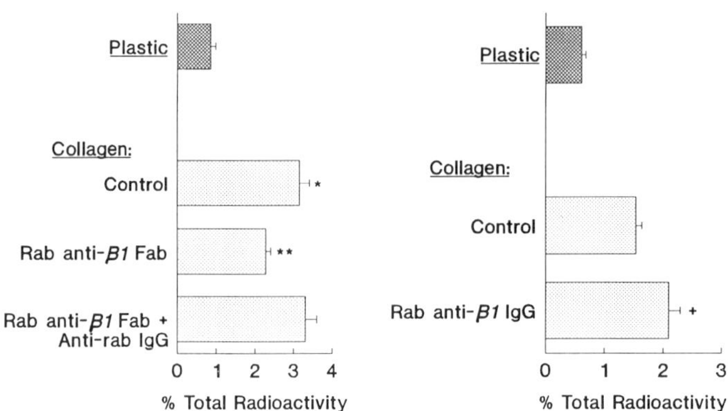

Figure 4 Effect of adhesion to collagen and anti-β_1 integrin antibody on 1,2-diacylglycerol production in rat GEC. Phospholipids were labeled with [^{3}H]arachidonic acid to isotopic equilibrium, and cells were then plated onto plastic or collagen gels. Left panel: Medium contained rabbit (rab) anti-β_1 Fab and goat anti-rabbit IgG. Nonimmune rabbit Fab with goat anti-rabbit IgG or nonimmune rabbit IgG was added to control incubations. Right panel: Medium contained rabbit anti-β_1 IgG, or nonimmune IgG in controls. Lipids were extracted after 2 hour incubation and [^{3}H]arachidonoyl-diacylglycerol was separated by thin-layer chromatography. Levels of [^{3}H]diacylglycerol are expressed as a % of total radioactivity in cellular lipids. Values are means ± SEM three to six experiments. Significant differences were present among groups. *$P < .005$ collagen (control) vs. plastic; **$P < .01$ anti-β_1 Fab vs. control (collagen); $^+P < .02$ anti-β_1 IgG vs. control (collagen). Reprinted from ref. 63 by copyright permission of the American Physiological Society.

The increases in diacylglycerol and free arachidonic acid and the decrease in inositol lipids were partially inhibited when anti-β_1 integrin antibody Fab was added to culture medium (Fig. 4), indicating that β_1 integrins were involved in collagen-induced breakdown of inositol phospholipids. Conversely, intact antibody IgG promoted diacylglycerol production; furthermore, crosslinking of anti-β_1 integrin Fab with a secondary antibody reversed the inhibition of diacylglycerol production (Fig. 4). Reduction of collagen-mediated diacylglycerol production by anti-β_1 Fab tended to increase GEC proliferation, while enhancement of diacylglycerol production by anti-β_1 IgG was associated with decreased proliferation, and this effect was abolished in GEC that had been functionally depleted of protein kinase C. GEC integrin collagen receptors may include $\alpha_2\beta_1$ and $\alpha_3\beta_1$ (21,63,69); however, further characterization of the α subunits involved in phospholipase activation by ECM will require development of rat subunit-specific antibodies that are functionally inhibitory.

Detachment of collagen gels with adherent GEC monolayers from tissue culture plates resulted in a reduction in stimulated levels of diacylglycerol and arachidonic acid (63). ECM proteins, immobilized on a solid surface, can increase tension within adherent cells, and they resist tensile forces generated by cytoskeletal elements, such as microfilaments, possibly leading to alterations in polymerization within the cytoskeleton (3). Release of collagen gels from culture plates would have reduced matrix-induced tensile forces. Thus, the reduction in diacylglycerol and arachidonic acid levels following detachment of ECM from a solid surface supports the involvement of the cytoskeleton in inositol lipid breakdown. Another way of reducing the amount of tension transmitted across cells is to reduce the density of collagen molecules immobilized onto substrata (3). Stimulation of diacylglycerol was shown to be less effective with lower matrix collagen concentrations, as compared to the higher concentrations in collagen gels. Thus in GEC, β_1 integrin interactions with ECM proteins appear to mediate ECM-induced mechanical forces acting on the cytoskeleton and on breakdown of inositol lipids. Evidence for the role of tension in ECM-dependent signaling has been demonstrated in other systems, including capillary tube formation and expression of differentiation-specific genes in vitro (3), suggesting that integrin-cytoskeletal interactions are a common "upstream" event in multiple transmembrane modes of signaling.

The biochemical mechanism of inositol phospholipid breakdown, which occurs after the interaction of β_1 integrin with ECM, has not been fully characterized. Initially, adhesion of GEC to collagen may induce aggregation of $\beta 1$ integrins on plasma membranes, and clustering of receptors may then facilitate inositol lipid breakdown. Recent studies (70) suggest that the phospholipase C-$\gamma 1$ isoform (68) is relevant to this process. Adhesion of

GEC to ECM enhanced the catalytic activity of phospholipase C-γ1, and induced translocation of phospholipase C-γ1 from the cytosol to the membrane fraction, which may facilitate inositol lipid hydrolysis by increasing the proximity of phospholipase C-γ1 and substrate. Furthermore, in GEC adherent to collagen, there appeared to be increased synthesis of substrate for phospholipase C (i.e., phosphatidylinositol-4,5-bisphosphate).

One can speculate on the role of ECM/integrin-mediated breakdown of inositol lipids in GEC. Cells that are performing differentiated functions have relatively lower rates of proliferation than less differentiated cells. Thus, a possible role of β_1 integrins in ECM-induced breakdown of inositol lipids might be to suppress proliferation and promote differentiated functions of GEC in the kidney. Such functions may include synthesis of basement membrane components and maintenance of glomerular permselectivity. In addition, production of prostaglandins, as a result of ECM-induced arachidonate release, may regulate glomerular hemodynamics. Alternatively, in glomerulonephritis, glomeruli may become infiltrated with inflammatory cells (which are a source of epithelial growth factors), and proliferation of GEC may increase. Subsequent accumulation of interstitial collagen in glomeruli may stimulate breakdown of inositol lipids and a reduction in GEC proliferation during the fibrotic phase of glomerular injury. Further studies on the interactions of GEC and ECM will be required to define the regulation of GEC functions in the normal kidney and in glomerular diseases.

ACKNOWLEDGMENTS

The author would like to thank Dr. D. Baran for helpful discussion, and Drs. G. Andres, J. V. Bonventre and D. J. Salant for providing prints. Work from the author's laboratory and the preparation of this manuscript were supported by the Medical Research Council of Canada, the Kidney Foundation of Canada, the Baxter Extramural Grants Program, and the Fonds de la Recherche en Santé du Québec.

REFERENCES

1. Albelda SM, Buck CA. Integrins and other cell adhesion molecules. FASEB J 1990; 4:2868–2880.
2. Springer TA. Adhesion receptors of the immune system. Nature 1990; 346: 425–434.
3. Ingber D. Integrins as mechanochemical transducers. Curr Opin Cell Biol 1991; 3:841–848.

4. Hynes RO. Integrins: versatility, modulation, and signaling in cell adhesion. Cell 1992; 69:11-25.
5. Zachary I, Rozengurt E. Focal adhesion kinase ($p125^{FAK}$): a point of convergence in the action of neuropeptides, integrins, and oncogenes. Cell 1992; 71: 891-894.
6. Schwartz MA. Transmembrane signalling by integrins. Trends Cell Biol 1992; 2:304-308.
7. Varki A. Selectins and other mammalian sialic acid-binding lectins. Curr Opin Cell Biol 1992; 4:257-266.
8. Ginsberg MH, Du X, Plow EF. Inside-out integrin signalling. Curr Opin Cell Biol 1992; 4:766-771.
9. Damsky CH, Werb Z. Signal transduction by integrin receptors for extracellular matrix: cooperative processing of extracellular information. Curr Opin Cell Biol 1992; 4:772-781.
10. Wight TN, Kinsella MG, Qwarnström EE. The role of proteoglycans in cell adhesion, migration and proliferation. Curr Opin Cell Biol 1992; 4:793-801.
11. McEver RP. Leukocyte-endothelial interactions. Curr Opin Cell Biol 1992; 4: 840-849.
12. Juliano RL, Haskill S. Signal transduction from the extracellular matrix. J Cell Biol 1993; 120:577-585.
13. Grunwald GB. The structural and functional analysis of cadherin calcium-dependent cell adhesion molecules. Curr Opin Cell Biol 1993; 5:797-805.
14. Williams MJ, Hughes PE, O'Toole TE, Ginsberg MH. The inner world of cell adhesion: integrin cytoplasmic domains. Trends Cell Biol 1994; 4:109-112.
15. Loftus JC, Smith JW, Ginsberg MH. Integrin-mediated cell adhesion: the extracellular surface. J Biol Chem 1994; 269:25235-25238.
16. Haas TA, Plow EF. Integrin-ligand interactions: a year in review. Curr Opin Cell Biol 1994; 6:656-662.
17. Rosen SD, Bertozzi CR. The selectins and their ligands. Curr Opin Cell Biol 1994; 6:663-673.
18. Chiquet-Ehrismann R. Anti-adhesive molecules of the extracellular matrix. Curr Opin Cell Biol 1991; 3:800-804.
19. Lane TF, Sage EH. The biology of SPARC, a protein that modulates cell-matrix interactions. FASEB J 1994; 8:163-173.
20. Korhonen M, Laitinen L, Ylänne J, Gould VE, Virtanen I. Integrins in developing, normal and malignant kidney. Kidney Int 1992; 41:641-644.
21. Adler S. Integrin receptors in the glomerulus: potential role in glomerular injury. Am J Physiol 1992; 262:F697-F704.
22. Racusen L. Alterations in tubular epithelial cell adhesion and mechanisms of acute renal failure. Lab Invest 1992; 67:158-165.
23. Goligorsky MS, Lieberthal W, Racusen L, Simon E. Integrin receptors in renal tubular epithelium: new insights into pathophysiology of acute renal failure. Am J Physiol 1993; 264:F1-F8.
24. Wuthrich RP. Intercellular adhesion molecules and vascular cell adhesion molecule-1 and the kidney. J Am Soc Nephrol 1992; 3:1201-1211.
25. Briscoe DM, Cotran RS. Role of leukocyte-endothelial cell adhesion molecules

in renal inflammation: in vitro and in vivo studies. Kidney Int 1993; 44:S27-S34.
26. Brady HR. Leukocyte adhesion molecules: potential targets for therapeutic intervention in kidney diseases. Curr Opin Nephrol Hypertens 1993; 2:171-182.
27. Rabb HAA. Cell adhesion molecules and the kidney. Am J Kidney Dis 1994; 23:155-166.
28. Couser WG. Mediation of immune glomerular injury. J Am Soc Nephrol 1990; 1:13-29.
29. Salant DJ, Cybulsky AV. Experimental glomerulonephritis. In: DiSabato G, ed. Methods in Enzymology, Vol. 162. New York: Academic Press, 1988:421-461.
30. Nishikawa K, Guo YJ, Miyasaka M, et al. Antibodies to intercellular adhesion molecule 1/lymphocyte function-associated antigen 1 prevent crescent formation in rat autoimmune glomerulonephritis. J Exp Med 1993; 177:667-677.
31. Mulligan MS, Johnson KJ, Todd RF, et al. Requirements for leukocyte adhesion molecules in nephrotoxic nephritis. J Clin Invest 1993; 91:577-587.
32. Kawasaki K, Yaoita E, Yamamoto T, Tamatani T, Miyasaka M, Kihara I. Antibodies against intercellular adhesion molecule-1 and lymphocyte function-associated antigen-1 prevent glomerular injury in rat experimental crescentic glomerulonephritis. J Immunol 1993; 150:1074-1083.
33. Wada J, Makino H, Shikata K, et al. Role of intercellular adhesion molecule-1 (ICAM-1) in nephrotoxic serum nephritis. J Am Soc Nephrol 1992; 3:647. Abstract.
34. Wu X, Pippin J, Lefkowith JB. Attenuation of immune-mediated glomerulonephritis with an anti-CD11b monoclonal antibody. Am J Physiol 1993; 264: F715-F721.
35. Tipping PG, Huang XR, Berndt MC, Holdsworth SR. A role for P selectin in complement-independent neutrophil-mediated glomerular injury. Kidney Int 1994; 46:79-88.
36. Lan HY, Nikolic-Paterson DJ, Zarama M, Vannice JL, Atkins RC. Suppression of experimental crescentic glomerulonephritis by the interleukin-1 receptor antagonist. Kidney Int 1993; 43:479-485.
37. Nikolic-Paterson DJ, Lan HY, Hill PA, Vannice JL, Atkins RC. Suppression of experimental glomerulonephritis by the interleukin-1 receptor antagonist: inhibition of intercellular adhesion molecule-1 expression. J Am Soc Nephrol 1994; 4:1695-1700.
38. Daniels BS. The role of the glomerular epithelial cell in the maintenance of the glomerular filtration barrier. Am J Nephrol 1993; 13:318-323.
39. Salant DJ. The structural biology of glomerular epithelial cells in proteinuric diseases. Curr Opin Nephrol Hypertens 1994; 3:569-574.
40. Salant DJ, Quigg RJ, Cybulsky AV. Heymann nephritis: mechanisms of renal injury. Kidney Int 1989; 35:976-984.
41. Cavallo T. Membranous nephropathy. Insights from Heymann nephritis. Am J Pathol 1994; 144:651-658.
42. O'Meara YM, Natori Y, Minto AWM, Goldstein DJ, Manning EC, Salant

DJ. Nephrotoxic antiserum identifies a β1 integrin on rat glomerular epithelial cells. Am J Physiol 1992; 262:F1083–F1091.

43. Adler S, Chen X. Anti-Fx1A antibody recognizes a β1-integrin on glomerular epithelial cells and inhibits adhesion and growth. Am J Physiol 1992; 262: F770–F776.
44. Kerjaschki D, Ojha PP, Susani M, et al. A β1-integrin receptor for fibronectin in human kidney glomeruli. Am J Pathol 1989; 134:481–489.
45. Couser WG, Darby C, Salant DJ, Adler S, Stilmant M, Lowenstein LM. Anti-GBM antibody-induced proteinuria in isolated perfused rat kidney. Am J Physiol 1985; 249:F241–F250.
46. Salant DJ, Madaio MP, Adler S, Stilmant MM, Couser WG. Altered glomerular permeability induced by $F(ab')_2$ or Fab′ antibodies to rat renal tubular epithelial antigens. Kidney Int 1981; 21:36–43.
47. Adler S, Sharma R, Savin VJ, Abbi R, Eng B. Alteration of glomerular permeability to macromolecules induced by cross-linking of β_1 integrin receptors. Am J Pathol 1996; 149:987–996.
48. Puccetti A, Sabbatini A, Carbone G, et al. Lupus autonantibodies bind β1 integrin subunit. J Am Soc Nephrol 1994; 5:764. Abstract.
49. Floege J, Alpers CE, Sage EH, Pritzl P, Gordon K, Johnson RJ, Couser WG. Markers of complement-dependent and complement-independent glomerular visceral epithelial cell injury in vivo. Expression of antiadhesive proteins and cytoskeletal changes. Lab Invest 1992; 67:486–497.
50. Gailit J, Colflesh D, Rabiner I, Simone J, Goligorsky MS. Redistribution and dysfunction of integrins in cultured renal epithelial cells exposed to oxidative stress. Am J Physiol 1993; 264:F149–F157.
51. Goligorsky MS, DiBona GF. Pathogenetic role of arg-gly-asp-recognizing integrins in acute renal failure. Proc Natl Acad Sci USA 1993; 90:5700–5704.
52. Klausner JM, Paterson IS, Goldman G, et al. Postischemic renal injury is mediated by neutrophils and leukotrienes. Am J Physiol 1989; 256:F794–F802.
53. Hellberg POA, Källskog ÖT, Öjteg G, Wolgast M. Peritubular capillary permeability and intravascular RBC aggregation after ischemia: effects of neutrophils. Am J Physiol 1990; 258:F1018–F1025.
54. Paller MS. Effect of neutrophil depletion on ischemic renal injury in the rat. J Lab Clin Med 1989; 113:379–386.
55. Thornton MA, Winn R, Alpers CE, Zager RE. An evaluation of the neutrophil as a mediator of in vivo renal ischemia-reperfusion injury. Am J Pathol 1989; 135:509–515.
56. Kelly KJ, Williams WW, Colvin RB, Bonventre JV. Antibody to intercellular adhesion molecule 1 protects the kidney against ischemic injury. Proc Natl Acad Sci USA 1994; 91:812–816.
57. Rabb H, Mendiola C, Saba SR, et al. Antibodies to P-selectin and ICAM-1 protect kidneys from ischemic-reperfusion injury. J Am Soc Nephrol 1994; 5: 907. Abstract.
58. Rabb HA, Mendiola CC, Dietz J, et al. Role of CD11a and CD11b in ischemic acute renal failure in rats. Am J Physiol 1994; 267:F1052–F1058.
59. Linas SL, Whittenburg D, Parsons PE, Repine JE. Mild renal ischemia acti-

vates primed neutrophils to cause acute renal failure. Kidney Int 1992; 42:610–616.
60. Linas SL, Whittenburg D, Parsons PE, Repine JE. Ischemia increases neutrophil retention and worsens acute renal failure: role of oxygen metabolites and ICAM 1. Kidney Int 1995; 48:1584–91.
61. Kelly KJ, William WW, Colvin RB, Meehan SM, Springer TA, Gutierrez-Ramos JC, Bonventre JV. Intercellular adhesion molecule-1-deficient mice are protected against ischemic renal injury. J Clin Invest 1996; 97:1056–63.
62. Cybulsky AV, Bonventre JV, Quigg RJ, Wolfe LS, Salant DJ. Extracellular matrix regulates proliferation and phospholipid turnover in glomerular epithelial cells. Am J Physiol 1990; 259:F326–F337.
63. Cybulsky AV, Carbonetto S, Cyr MD, McTavish AJ, Huang Q. Extracellular matrix-stimulated phospholipase activation is mediated by $\beta 1$ integrin. Am J Physiol 1993; 264:C323–C332.
64. Kanner SB, Grosmaire LS, Ledbetter JA, Damle NK. $\beta 2$-integrin LFA-1 signaling through phospholipase C-$\gamma 1$ activation. Proc Natl Acad Sci USA 1993; 90:7099–7103.
65. Schwartz MA, Denninghoff K. αv integrins mediate the rise in intracellular calcium in endothelial cells on fibronectin even though they play a minor role in adhesion. J Biol Chem 1994; 269:11133–11137.
66. Chen Q, Kinch MS, Lin TH, Burridge K, Juliano RL. Integrin-mediated cell adhesion activates mitogen-activated protein kinases. J Biol Chem 1994; 269: 26602–26605.
67. Majerus PW, Connolly TM, Deckmyn H, et al. The metabolism of phosphoinositide-derived messenger molecules. Science 1986; 234:1519–1526.
68. Cockcroft S, Thomas GMH. Inositol-lipid-specific phospholipase C isoenzymes and their differential regulation by receptors. Biochem J 1992; 288:1–14.
69. Mendrick DL, Kelly DM. Temporal expression of VLA-2 and modulation of its ligand specificity by rat glomerular epithelial cells in vitro. Lab Invest 1993; 69:690–702.
70. Cybulsky AV, McTavish AJ, Papillon J. Extracellular matrix stimulates production and breakdown of inositol phospholipids. Am J Physiol 1996; 271: F579–F587.

25

Adhesion Molecules and Artificial Membranes

Alfred K. Cheung
Departments of Medicine and Medical Service, Veterans Affairs Medical Center and University of Utah, Salt Lake City, Utah

Syed Fazal Mohammad
Department of Pathology and Artificial Heart Laboratory, University of Utah, Salt Lake City, Utah

John Kenneth Leypoldt
Departments of Medicine, Bioengineering, and Research Service, Veterans Affairs Medical Center and University of Utah, Salt Lake City, Utah

I. INTRODUCTION

Alteration and activation of blood cells and plasma proteins commonly occur when blood contacts the artificial surfaces contained in biomedical devices. Clinical treatment modalities involving extracorporeal circuits, such as hemodialysis (HD) and cardiopulmonary bypass (CPB), have been most extensively studied in this context. Analytical techniques for quantifying cell surface adhesion molecules have been applied to identify the interactions of leukocytes and platelets with artificial membranes. The studies serve at least two purposes. First, they have relevance to the problems encountered by patients treated with these devices and will likely lead to the development of biomaterials that minimize blood cell alterations. Second, they serve as a convenient acute model to investigate responses by activated blood cells. This chapter will describe the role of adhesion molecules in the interactions of leukocytes and platelets with artificial membranes during HD and CPB, two commonly used clinical modalities involving extracorporeal circulation of blood.

II. MATERIALS FOR ARTIFICIAL MEMBRANES AND COMPLEMENT ACTIVATION

In vitro contact of isolated blood cells with artificial surfaces in the absence of plasma can activate the cells, but the in vivo and clinical significance of this phenomenon is questionable because artificial surfaces are invariably and rapidly coated with plasma proteins upon exposure to whole blood. Cell activation on the surface is therefore largely determined by the nature of the protein layer present at the interface. Notable examples of the effects of plasma proteins are the deposition of coagulation proteins for platelet activation and the deposition of complement proteins for neutrophil activation.

Membranes employed during clinical HD can be classified into one of two categories—cellulose-derived (cellulosic), or synthetic polymers (1). Presumably because of the carbohydrate moieties on their surface, membranes regenerated from cellulose tend to activate the plasma complement system more vigorously, predominantly via the alternative pathway of activation (2). The complement system is important in discussions of adhesion molecules since complement activation during HD is often intense (3–5) and activation products, such as C5a and iC3b, have well-known leukocyte-directed properties (6).

Cellulosic HD membranes made from cuprophan have been used clinically for almost half a century and are the prototype of complement activators among HD membranes (2–5). These membranes are also potent activators of leukocytes (7,8). For these reasons, cuprophan is often used as a model surface in studies of biocompatibility of HD membranes. While synthetic membranes tend to be relatively weaker activators of complement than cellulosic membranes, in general, all clinically employed HD membranes have been shown to activate complement, and there are overlaps in activation potential between the two categories (9–11). Most oxygenators currently used in CPB are membrane oxygenators; bubble oxygenators are rarely employed. All membranes used in clinical CPB circuits are made from synthetic polymers and tend to activate complement modestly (12–14). The factors initiating and controlling complement activation on synthetic membranes are poorly understood.

III. NEUTROPHILS AND ARTIFICIAL MEMBRANES

Neutrophils, the most abundant leukocyte in the circulation, play a major role in cellular defense mechanisms that do not require prior exposure to antigens. Alterations of neutrophils during HD are especially pronounced, and we shall devote attention largely to data obtained from clinical HD. These alterations include the adherence of cells to the artificial membrane surface (15), transient aggregation and adherence to vascular endothelial cells in the microcirculation (16–18), enhanced production of reactive oxygen species (ROS) (19), and degranulation with the extracellular release of granular proteins (7,20–22). In addition, there are changes in a number of cell surface molecules, such as C5a receptors (23), fMLP receptors (24), and Mac-1 (25), that are involved in mediating the functional activities of these cells.

A. Neutrophil Adherence to Hemodialysis Membranes

During the activation of C3 via the alternative pathway of complement on cuprophan membranes, C3b is deposited onto the membrane surface and is subsequently degraded by plasma regulatory proteins into iC3b (2) (Fig. 1). The fragments C3b and iC3b have been identified by SDS-polyacrylamide

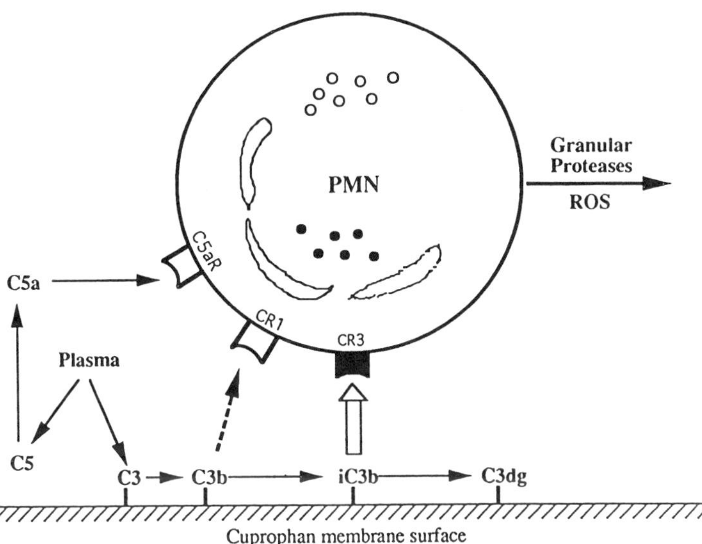

Figure 1 Schematic diagram of proposed mechanisms by which neutrophils adhere to cellulosic hemodialysis membranes. Contact of plasma with cuprophan membrane leads to activation of complement C3 and deposition of C3b onto the membrane surface. Dialysis membrane-associated C3b is degraded into iC3b and C3dg by plasma complement regulatory proteins. Membrane-bound iC3b interacts with CR3 (Mac-1) on the neutrophil surface, thereby promoting adherence of the cells to the dialysis membrane. The interaction between cuprophan membrane-bound C3b and neutrophil surface CR1 appears to be relatively weak but may potentiate the effect of CR3 in promoting cell adhesion under certain conditions. Ligation of CR3 appears to be important in stimulating neutrophils to release intragranular proteases. The C5a that is generated as a result of C5 activation stimulates the upregulation of CR3 on neutrophils and enhances adherence. C5a also promotes degranulation and ROS production. (Adapted from reference 26, with permission.)

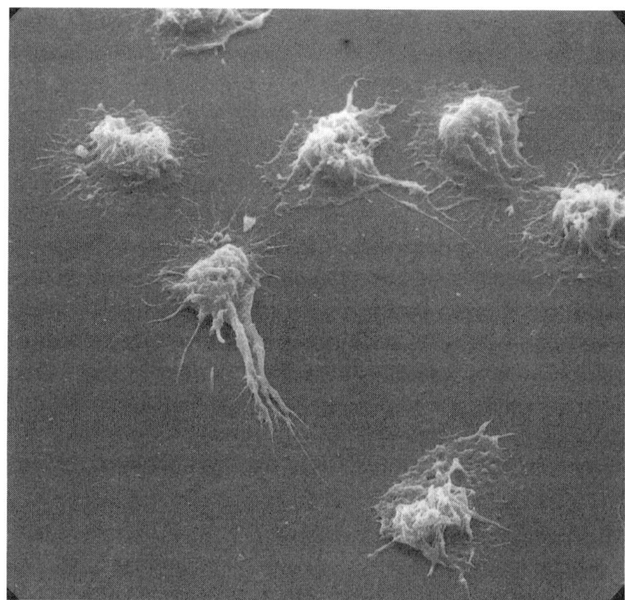

Figure 2 Scanning electronmicrograph of neutrophils on cuprophan hemodialysis membranes. Adherent cells displayed extensive lamellipod and retraction fiber formation toward the cuprophan membrane. (Adapted from reference 26, with permission.)

gel electrophoresis and immunoblotting of the proteins eluted from cuprophan membranes after incubation with human plasma (26). Major receptors for C3b and iC3b present on the surface of neutrophils, monocytes, and other cells are complement receptor type 1 (CR1 or CD35) and Mac-1 (complement receptor type 3, CR3, or CD11b/CD18), respectively (6). When cuprophan membranes are incubated with human blood in vitro, leukocytes adhere and spread on the membrane surface (26) (Fig. 2). The adherent cells are predominantly neutrophils. Experiments using isolated neutrophils show that the presence of plasma and divalent cations is essential for the adherence and that this process is primarily mediated by the binding between Mac-1 on the cell surface and iC3b on the cuprophan membrane (Fig. 1).

Adherence of neutrophils to cuprophan coil dialyzers (27) and cellulose acetate (cellulosic membranes where a fraction of the hydroxyl groups are acetylated) hollow fiber membrane dialyzers (28,29) has also been demonstrated during clinical HD. The observation that leukocyte adherence to membranes is less during HD with cellulose acetate or hemophan (another type of cellulosic membrane where certain hydroxyl groups are substituted by DEAE moieties) membranes (30), which are weaker activators of complement than the unsubstituted cuprophan membrane (2,5,10), further supports the importance of complement proteins in governing neutrophil adherence to cellulosic membranes. Cell adherence to HD membrane surfaces may be important for the release of ROS and degranulation by neutrophils, as discussed below.

B. Hemodialysis-Induced Neutropenia

Transient peripheral leukopenia is often observed during HD (Fig. 3). Maximum leukopenia usually occurs between 5 and 20 min after the onset of the procedure, the intensity of which is dependent on the type of HD membrane employed (3–5,9–11,16–19,25). When cuprophan membranes are used, neutrophils practically disappear from the circulation. Other granulocytes and monocytes are also affected, but lymphocytopenia is rarely observed (31). Although neutrophils adhere to the HD membrane surface (18,26–30), predominant evidence indicates that margination of the cells inside the patients' body is largely responsible for their disappearance from

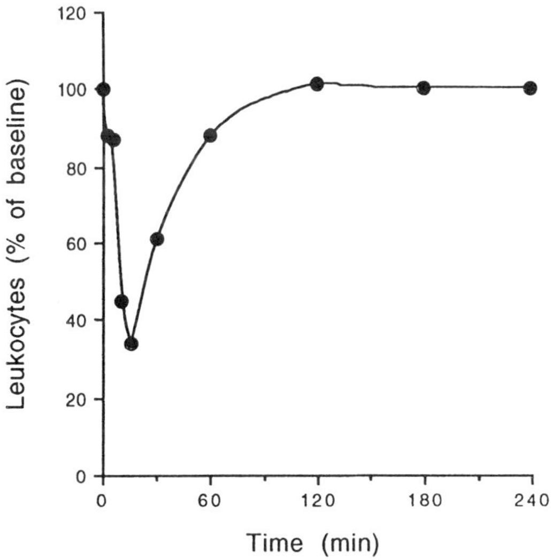

Figure 3 Total peripheral leukocyte counts during hemodialysis using cuprophan membranes. Each data point represents the mean of 12 observations on different chronic hemodialysis patients. (Adapted from reference 3, with permission.)

the circulation (3,18). Animal studies have shown that the pulmonary microcirculation is the site of neutrophil aggregation and sequestration (17), and studies conducted during clinical HD using radiolabeled cells also suggest that the pulmonary microcirculation is the primary sequestration site (18). The leukocyte count returns to baseline or sometimes overshoots by 60 min after initiation of the HD session. The mechanisms of HD-induced neutropenia and its recovery have been a subject of intense investigation. The interest stems from the drastic nature of the event and its frequent use as a standard indicator of HD membrane biocompatibility.

Activation products of complement and upregulation of Mac-1 have been most frequently incriminated in the pathogenesis of HD-induced neutropenia. Upregulation of Mac-1 on neutrophils during HD (25,31–34) and CPB (35–37) has been consistently demonstrated by numerous investigators. Mac-1 is upregulated within minutes during clinical HD with cuprophan membranes (Fig. 4) because Mac-1 is stored in the secondary granules of neutrophils and can be readily translocated to the cytoplasmic membrane surface (38). Since the rapid upregulation of Mac-1 on neutrophils is coincident with HD-induced neutropenia and since upregulation of Mac-1 expression induces neutrophil aggregation, Arnaout and co-workers postulated

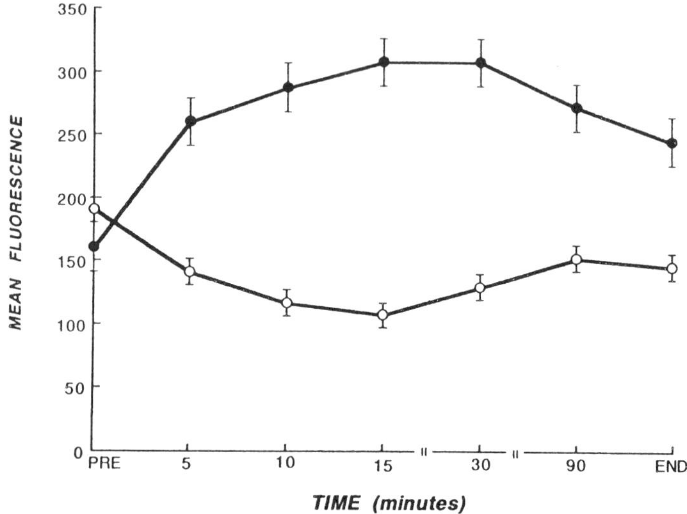

Figure 4 Alterations in Mac-1 and LAM-1 expression on neutrophil surface during hemodialysis using cuprophan membranes. Each data point represents the mean ± SE of 10 observations on different chronic hemodialysis patients. Closed circles = Mac-1; open circles = LAM-1. (Adapted from reference 32, with permission.)

that this adhesion molecule is an important mediator of HD-induced neutropenia (25).

The mechanisms governing the upregulation of Mac-1 expression on neutrophils have also been investigated. Effluent plasma from a cuprophan membrane dialyzer induced more Mac-1 expression on normal human neutrophils than afferent plasma, suggesting that soluble factors generated during passage of blood through the dialyzer are mediating this event (39). A prime candidate is the anaphylatoxin C5a, a fragment of C5 that is known to possess both spasmogenic and leukocyte-directed activities (6). Incubation of whole blood with C5a purified from human serum enhances the expression of Mac-1 on neutrophils and monocytes (40), although laboratories using recombinant C5a have reported inconsistent results (32,33). Furthermore, in vitro HD experiments using whole blood have demonstrated that inhibition of complement activation by recombinant soluble complement receptor type 1 (sCR1) also inhibited Mac-1 upregulation on neutrophils (41). Therefore, the importance of complement activation in Mac-1 upregulation on neutrophils during HD appears to be established.

Further support for the role of complement proteins in HD-induced neutropenia can be inferred from clinical observations that intradialytic plasma C3a levels correlate positively with Mac-1 expression on neutrophils (42) and inversely with neutrophil counts (3,25,32,42). For example, plasma C3a levels and Mac-1 expression on neutrophils are higher during HD with new cuprophan membranes than with either cuprophan membranes that have been reused (4), hemophan membranes (42), or synthetic membranes made of polysulfone (42), polyamide (43), or polyacrylonitrile (AN69) (33).

Some studies have suggested that platelet-activating factor (PAF) may also be involved in triggering HD-induced neutropenia (44,45). Generation of PAF has been demonstrated during HD (21). Sakaguchi et al. recently reported that the increase in the circulating plasma PAF level occurred after HD-induced neutropenia, suggesting that PAF release is the consequence rather than the cause of neutrophil activation (46). Data from an earlier study, however, refute this temporal relationship (21). Inasmuch as PAF has been shown to upregulate Mac-1 (47), PAF released by activated neutrophils may further facilitate the development of HD-induced neutropenia.

The mechanisms governing the recovery from neutropenia are more controversial. Radiolabeled leukocyte studies during clinical HD have demonstrated that recovery from neutropenia is primarily due to the release of cells that had been sequestered, although recruitment of cells from the bone marrow also contributed to some extent (16). The reasons for the release of sequestered cells are, however, unclear. Studies using functional assays (23) or fluorescein-labeled C5a binding techniques (40) suggested that C5a receptors on neutrophils were downregulated (or occupied) during HD. It was proposed that this downregulation decreased cell adhesiveness, rendering them resistant to further stimulation by C5a that was believed to be generated throughout the HD session (23). Other studies, however, did not support this hypothesis. It was observed that C5a generation during HD was transient, occurring primarily in the first half hour (5,25,48–50), and that C5a receptors on neutrophils were not downregulated (51). Whether or not C5a receptors are downregulated, this event alone does not appear to provide an adequate explanation for the release of cells sequestered in the microcirculation.

Since upregulation of Mac-1 expression has been proposed as the cause of HD induced neutropenia, the fate of this adhesion molecule during the recovery phase may be important. Several studies reported that increased Mac-1 expression on neutrophils persists throughout the entire dialysis treatment (32,42,43) (Fig. 4). This persistent rather than transient upregulation of Mac-1 expression seems to be incompatible with its role in regulating the attachment to and subsequent detachment from the endothelial cells. The selectin LAM-1 (CD62L) has been observed to become downregulated during HD (32,52) (Fig. 4) and CPB (36), yielding the hypothesis that the low LAM-1 state contributes to the cell's hypoadhesiveness (32,53).

Recently, transient upregulation of MoF11Ag on neutrophil surface has been noted to correlate inversely with neutrophil counts during HD with cuprophan membrane (43). MoF11Ag is overexpressed on neutrophils upon stimulation with either fMLP or phorbol esters and has been proposed to be a useful marker for activated neutrophils. Presumably, the simultaneous upregulation of both Mac-1 and MoF11Ag was necessary to induce neutropenia and that the subsequent loss of MoF11Ag expression resulted in the release of sequestered cells. Dialysis using the synthetic polyamide membrane was not associated with changes in neutrophil MoF11Ag expression and, correspondingly, induced only relatively mild neutropenia.

Another neutrophil adhesion molecule, sialyl-Lewis x (CD15), has been recently proposed to be involved in HD-induced neutropenia (33). CD15 is a ligand for selectins and mediates the binding of neutrophils to endothelial cells and platelets. Expression of CD15 remained unaffected during HD with synthetic polyacrylonitrile (AN69) membranes when no neutropenia occurred. During HD with cuprophan membranes, however, CD15 was upregulated (at 15 min), and its expression throughout the entire HD treatment correlated better with changes in intradialytic neutrophil counts than the expression of Mac-1.

It should be emphasized that the role of Mac-1, or any other cell surface adhesion molecule studied to date, in the genesis of HD-induced neutrophenia is only inferential. Most of the hypotheses proposed are based on the known biological activities of the adhesion molecules in other settings and the temporal relationship between their quantitative changes and neutrophil counts during HD. Conformational or functional changes in these molecules have not been addressed. Indeed, certain studies appear to refute the importance of Mac-1 upregulation in governing neutropenia. For example, transient neutropenia induced by the systemic injection of zymosan-activated plasma, which contains various activated complement fragments, into rabbits was not influenced by treatment with an anti-CD18 antibody (54). Furthermore, neutropenia observed during CPB in experimental piglets could not be prevented by leumedin, a Mac-1 inhibitor (35). Conclusive evidence for the importance of all these surface adhesive molecules will require more rigous mechanistic investigations. Conceivably, modulation in C5a receptors, Mac-1, LAM-1, CD15, and MoF11Ag plays some (perhaps interrelated) role in the development and recovery of HD-induced neutropenia. Regardless of the adhesion molecules involved, it is clear that neutrophils obtained from patients after the nadir in leukocyte count during HD with cuprophan membranes aggregate poorly (55) and adhere subnormally to human umbilical vein endothelial cells (32), indicating that loss of cell adhesiveness is a key factor in recovery from leukopenia.

C. Production of ROS and Degranulation by Artificial Surface-Activated Neutrophils

Among the important functions of neutrophils in host defense are their abilities to generate ROS, release intragranular proteases, and ingest foreign particles. While phagocytosis of the artificial membrane is impossible, degranulation (7,20,36,37) and ROS release (19,31,35) by activated neutrophils have been observed during clinical HD and CPB. Because of the known biological activities of C5a (57,58), this complement fragment has also been postulated to be a major mediator of these cellular events during HD. Observations during clinical HD, however, have shown that complement activation using plasma C3a as marker does not necessarily correlate well with plasma levels of released neutrophil granular proteins (20,21), suggesting that HD-induced neutrophil degranulation is partly mediated by noncomplement mechanisms.

A role for both complement and Mac-1 in governing neutrophil degranulation has been investigated in vitro. Incubation of cuprophan membranes and normal plasma with human neutrophils that were congenitally deficient in CD18 (the β chain of Mac-1) resulted in a 90% decrease in degranulation compared to similar experiments using normal cells (59). The importance of Mac-1 is further illustrated by a positive correlation between Mac-1 expression and elastase release during circulation of blood through an in vitro CPB circuit (37). Inhibition of complement activation by sCR1 during circulation of blood through cuprophan membranes in vitro reduced neutrophil ROS production by only 50%, although complement activation, upregulation of Mac-1, and shedding of LAM-1 were completely abolished (41). These latter data suggest that Mac-1 upregulation was totally dependent on complement activation but that ROS generation under these conditions depends only partially on complement. In separate studies in which complement activation was prevented by employing either complement-depleted plasma (59) or sCR1 (60), neutrophil degranulation induced by cuprophan membranes and plasma decreased by 45% to 70%. Although there appear to be some quantitative inconsistencies among the results of these studies, together they suggest that HD membrane-induced neutrophil degranulation partially depends on complement and Mac-1, and that ROS production is partially dependent on complement activation. Presumably, some other plasma constituents are activated or altered upon exposure to artificial membranes, which could also in turn stimulate neutrophils by interactions with Mac-1 or other cell surface receptors. Although there are a number of candidates for this role (e.g., kallikrein and immunoglobulins), these noncomplement mediators have not been definitively identified.

Although cell adherence to the endothelium via Mac-1 is a common initial step for neutrophil-mediated oxidative tissue damage in vivo (61), there is another potentially important mechanism by which neutrophils induce damage during extracorporeal circulation. The extracorporeal circulation is unique in that, besides inducing the generation of peptides (such as C5a and PAF) that enhance Mac-1 expression, the artificial membrane also provides a large surface on which the bound iC3b can serve as a ligand for Mac-1, thereby initiating Mac-1-dependent cellular events (Fig. 1).

HD using synthetic polysulfone membranes did not induce either Mac-1 expression or ROS production by neutrophils (31). The lack of neutrophil activation when using these particular membranes may be largely attributed to their low complement activation potential.

D. Other Integrins on Neutrophils During HD

Although not strictly classified as an adhesion molecule, it is worth mentioning that complement receptor type 1 (CR1) on neutrophils is substantially upregulated during HD and that this upregulation appears to depend on the intensity of complement activation (62). Limited data on integrins other than Mac-1 show that there are no changes in the expression of CD11a (LFA-1) and CD11c on neutrophils during clinical HD with cuprophan membranes (32,33,43), while one report claimed that CD11c was upregulated (52). Studies quantitating the expression of surface molecules on neutrophils during HD are summarized in Table 1.

Table 1 Expression of Molecules on Neutrophil (PMN) Surface During Hemodialysis

1st author, year (ref)	Membrane	PMN count	CD 11b (Mac-1)	CD 11a	CD 11c	CD 62L (LAM-1)	CD 43	CD 15	CD 54 (ICAM-1)	CD 45	MoF 11	CD 49d
Arnaout, 1985	Cuprophan	↓↓[a]	↑↑									
(25)	Cupro, reused	0	0									
Jacobs, 1989 (34)	Cuprophan	↓↓	↑↑									
Alvarez, 1991	Cuprophan	↓↓	↑↑	0	↑	↓↓	↓↓					
(52)	AN69[b]	0	0	0		0	0					
Himmelfarb,	Cuprophan		↑↑			↓↓		0				
1992 (32)	Cupro, reused		0			0						
Thylen, 1992	Cuprophan	↓↓	↑↑									
(42)	Hemophan	↓	↑									
	Polysulphone	↓	↑									
Lundahl, 1992 (39)	Cuprophan	↓↓	↑↑									
Tielemans,	Cuprophan	↓↓	↑↑	0					0	↑↑		
1993 (68)	CA[b]	↓	↑↑	0					0	↑↑		
	AN69	0	↑	0					0	0		
Cristol, 1994	Cuprophan	↓↓	↑↑									
(31)	Polysulfone	0	0									
Combe, 1994	Cuprophan	↓↓	↑↑	0							↑↑	
(43)	Polyamide	0	0	0							0	
Stuard, 1995	Cuprophan/CA	↓↓	↑↑	0	0			↑↑	0			0
(33)	AN69	0	0	0	0			0	0			0

[a] ↓↓ = decrease, ↑↑ = increase, ↓ = marginal decrease, ↑ = marginal increase, 0 = no change.
[b] AN69 = polyacrylonitrile; CA = cellulose acetate.

IV. MONOCYTES AND LYMPHOCYTES IN EXTRACORPOREAL CIRCUITS

The expression of adhesion molecules on blood monocytes during extracorporeal circulation has not been studied as extensively as that on neutrophils, although many studies have convincingly demonstrated that monocytes are activated in this setting (33,42,63–69). Activation of peripheral blood monocytes during HD is manifested by monocytopenia (33,42,68), change in cell surface molecules (33,42,68), and the production of monokines (63–67). The degree of activation varies depending on the type of HD membrane (66,67), as well as on other factors such as contamination of the dialysate by bacterial endotoxins (69).

The degree of monocytopenia during HD does not always correlate with the complement activation potential of the membrane (9,70). Preliminary data furthermore suggest that the use of cellulosic membranes reprocessed without bleach ameliorates both complement activation and neutropenia but not monocytopenia (11), suggesting that the mechanisms governing HD-induced monocytopenia and neutropenia are different. Consistent with this hypothesis are several recent studies which have shown that adhesion molecule expression on monocytes during HD is different from that on neutrophils. Thylen et al. studied the expression of Mac-1 during HD using cuprophan, hemophan, and polysulfone membranes and found that the degree of monocytopenia and neutropenia during the initial 15 min of HD was greater for the stronger complement activator, cuprophan, than the others (42). While increased expression of Mac-1 on neutrophils was observed during the initial 15 min of HD, no changes in its expression on monocytes were observed during this time interval. While there was increased Mac-1 expression on monocytes later in the HD session (after 180 min), the upregulation was small and coincided temporally with the recovery from monocytopenia, rather than its development. Tielemans et al. determined cell counts and the expression of adhesion molecules on both monocytes and neutrophils during HD using cuprophan, cellulose acetate, and polyacrylonitrile (AN69) membranes (68). While transient neutropenia occurred only with cuprophan and cellulose acetate membranes, monocytopenia occurred with all three types of membranes. Although Mac-1 expression on monocytes increased during the initial 15 min when monocytopenia occurred, the increase was mild. Similar to the observation by Thylen et al., Mac-1 expression on monocytes increased toward the end of the HD session when the monocyte counts returned to baseline. Incidentally, the time course of Mac-1 expression on neutrophils and monocytes during CPB parallels that during HD (71). These results therefore cast doubt on the importance of Mac-1 in HD-induced monocytopenia.

Instead, Stuard et al. proposed that CD15 may play a role in mediating HD-induced monocytopenia (33). These workers demonstrated that monocytopenia occurred 15 min after the start of HD when using cuprophan membranes but not when using polyacrylonitrile (AN69) membranes. During this time interval, an upregulation of CD15, but not Mac-1, was observed on monocytes and a strong negative correlation was noted between CD15 expression and monocyte counts.

Another surface molecule on monocytes that may be important in extracorporeal circulation is CD14. CD14 serves as a receptor for bacterial lipopolysaccharides and initiates monokine production (72). It also contributes to the adherence of monocytes to cytokine-stimulated endothelial cells (73). Downregulation of CD14 on monocytes has been described during HD with cuprophan membranes (74). Although complement fragments C3a and C5a promote monokine production (by either inducing mRNA transcription (75) or protein generation (76)), it is unclear whether the alteration in CD14 expression on monocytes during HD is a result of complement activation. It is also uncertain whether changes in CD14 expression are related to the development of transient monocytopenia. The relative importance of CD14 in mediating cytokine release in the extracorporeal circuit is likely to depend on both the contamination of the circuit by bacterial products and the complement activation potential of the artificial membrane employed. A summary of changes in monocyte surface molecules during HD is presented in Table 2.

Information on lymphocyte adhesion molecules during extracorporeal circulation is very limited, and the available literature indicates that the changes are either absent or very mild. A summary of these changes during HD is presented in Table 3.

V. PLATELETS AND ARTIFICIAL MEMBRANES

Platelet abnormalities appear to have a more profound acute clinical impact during CPB than during HD; thus, there is a larger volume of literature on platelets in CPB, compared to HD, where greater efforts in recent years have been devoted to the study of leukocytes. This section will deal with some general principles of plasma protein adsorption and platelet adhesion on artificial surfaces, and then concentrate on clinical data obtained primarily during CPB.

A. Platelet Adhesion to Artificial Surfaces

Platelets are activated during extracorporeal circulation as evidenced by alteration in surface receptors, release of granular and other cellular contents, formation of aggregates, adherence to the artificial surface, and re-

Table 2 Expression of Molecules on Monocyte (Mono) Surface During Hemodialysis

1st author, year (ref)	Membrane	Mono count	CD 11b (Mac-1)	CD 11a	CD 45	CD 54 (ICAM-1)	CD 25	CD 11c	CD 62L (LAM-1)	CD 15	CD 49d
Thylen, 1992 (42)	Cuprophan	↓[a]	0								
	Hemophan	↓	0								
	Polysulphone	↓	0								
Tielemans, 1993 (68)	Cuprophan	↓	↑	0	↑	0					
	CA[b]	↓	↑	0	↑	0					
	AN69	0	0	0	↑	0					
Cristol, 1994 (31)	Cuprophan	↓	↑				↑				
	Polysulfone	↓	↑				0				
Stuard, 1995 (33)	Cuprophan/CA	↓	↑	0		0		0	0	↑↑	0
	AN69	0	0	0		0		0	0	0	0

[a]↓ = marginal decrease, ↑ = marginal increase, ↑↑ = increase, 0 = no change.
[b]AN69 = polyacrylonitrile; CA = cellulose acetate.

Table 3 Expression of Molecules on Lymphocyte Surface During Hemodialysis

1st author, year (ref)	Membrane	Lymphocyte count	CD 62L (LAM-1)	CD 11b (Mac-1)	CD 25
Himmelfarb, 1992 (32)	Cuprophan		0[a]		
	Cupro, reused		0		
Cristol, 1994 (31)	Cuprophan	$\downarrow$		0	$\uparrow$
	Polysulfone	0		0	0

[a]$\downarrow$ = marginal decrease, $\uparrow$ = marginal increase, 0 = no change.

duction in responsiveness to soluble activators after the procedure (77). Loss of platelets during CPB is well recognized and can be attributed, at least in part, to their adhesion to the artificial surfaces. Although the changes in platelet count during CPB vary significantly depending on a number of factors, the area of artificial surface with which blood comes into contact and the duration of the procedure appear to be important determinants. The larger the surface area and the longer the procedure, the greater the decrease in platelet count.

The mechanisms by which artificial surfaces attract platelets are not fully understood. It is believed, however, that a number of factors, including the interfacial properties of the surface, flow conditions, and various blood components influence the disappearance of platelets from the circulation. The nature (smooth or textured) and composition (e.g., hydrophobicity and availability of specific chemical groups) of the surface are important because they affect the reactions with plasma proteins that the surface would initiate and sustain. Flow rate plays a critical role by determining the cell-artificial surface contact time while flow geometry may facilitate contact in areas of disturbed flow.

Platelets in circulating blood assume a discoid shape maintained by the microtubules and other contractile elements in their cytoplasm. In the presence of minute amounts of appropriate stimuli, they lose their discoid shape and rapidly form pseudopods (78). These events are followed by the formation of aggregates, or, upon contact with a surface under appropriate conditions, the platelets may adhere to the surface. Following their activation, platelets release a number of potent agonists, such as adenosine diphosphate (ADP) and thromboxane A_2, which help to sustain these reactions and, at the same time, activate other platelets in the vicinity to enable them to participate in the thrombotic process. Platelets are equipped with a unique surface canalicular system, which is presumably formed by infolding of the plasma membrane extending deep into the cytoplasm (79). The presence of this system permits the platelets to transport the agonists from their granules to the outside rapidly. Immediately after their contact with an artificial surface, adherent platelets further extend their pseudopodia to fill the intercellular gaps, thus completing the spreading process. Platelets, which are normally approximately 3 μm in diameter in circulating blood, may cover 10 μm or more when they are fully spread (Fig. 5). Only the first layer of platelets adheres to the surface; all subsequent layers attach to the surface-bound platelets which may be at various stages of spreading.

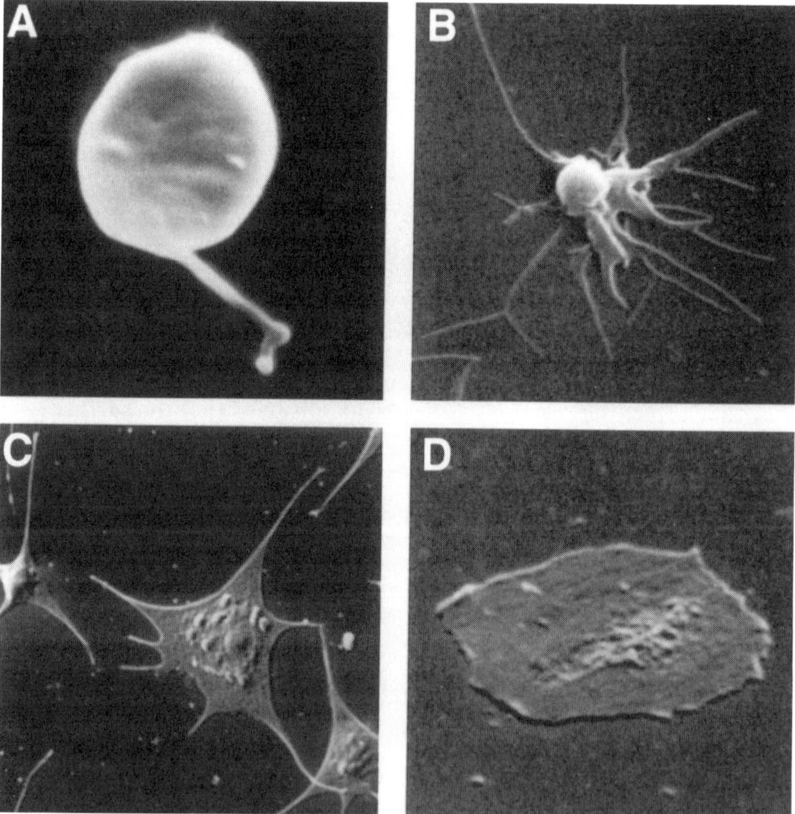

Figure 5 Scanning electronmicrographs of platelets on glass. Panels A–D show platelets with increasing degree of spreading on glass.

B. Adsorption of Plasma Proteins on Artificial Surfaces

There is general consensus that adsorption of plasma proteins to the artificial surface precedes platelet adhesion, and the resulting composition of the interface plays a critical role both in initiating platelet adhesion and in activation of the coagulation pathway (80). Upon contact of blood with many types of artificial surfaces, plasma fibrinogen is immediately adsorbed (81–84). Based on published adsorption isotherms for fibrinogen, it appears that this adsorption is a complex process, partially depending on the concentrations of fibrinogen and other proteins in the medium (84). In diluted plasma, fibrinogen preferentially adsorbs to the surface. In normal plasma, however, other proteins not only compete with fibrinogen for the initial binding, but there is an attempt to reshape the adsorbed layer of proteins in a dynamic manner—a phenomenon known as the Vroman effect (85). For example, Vroman noted that the adsorbed fibrinogen was later replaced by plasma high-molecular-weight kininogen (HK) (86). Confirming this phenomenon, Brash reported that the amount of fibrinogen adsorbed at the interface was significantly less when normal plasma rather than HK-deficient plasma was used in the experiment. Moreover, the adsorbed fibrinogen may be degraded and thus removed from the surface by plasmin (87). These observations suggest that the rapid adsorption of fibrinogen may be of greater significance for platelet adhesion during the early stages of blood-material contact. At the later stages, when surface-bound fibrinogen has been replaced, other plasma proteins may assume more importance in platelet adhesion. These other proteins probably include von Willebrand factor (vWF), fibronectin, vitronectin, thrombin, and other coagulation factors. Many of these proteins have been shown to influence platelet adhesion to artificial surfaces in vitro (80,88); however, their role in platelet adhesion during extracorporeal circulation has not been critically examined.

C. Glycoprotein Receptors Involved in Platelet Adhesion to Artificial Surfaces

Although the precise mechanisms by which platelets adhere to an artificial surface remain to be elucidated, cell adhesion molecules associated with platelets are attracting attention because these glycoproteins may provide important clues not only for understanding the molecular basis of platelet adhesion but also in setting the stage for regulating these reactions to prevent undesirable events during extracorporeal circulation. On the platelet surface are a number of glycoprotein receptors, which mediate their adherence to other platelets, leukocytes, blood vessel wall, and artificial membranes (89,90). Through these receptors, platelets may also bind specific coagulation factors and, in this manner, participate in the activation of the coagulation cascade resulting in fibrin formation. Available evidence indicates that vWF, a large multimeric glycoprotein in circulating blood, readily adsorbs to artificial surfaces (91). Adsorbed vWF contains functional binding domains for coagulation factor VIII, heparin, collagen, and two platelet surface glycoproteins—GPIb, and GPIIb-IIIa (92). The GPIb binding domain on vWF appears to be cryptic in normal circulation and becomes expressed only when the vWF is associated with extracellular matrix, fibrin, or an artificial surface, although the latter has not been conclusively demonstrated (93). Using a parallel plate perfusion chamber and a specific monoclonal antibody, Baruch et al. observed that, by inhibiting the interaction between vWF and GPIb, platelet adhesion to components of subendothelium in vitro was inhibited by approximately 45% (94). This suggests that the binding of platelet GPIb to vWF is a major contributing factor in platelet adhesion to subendothelium. Results from other laboratories support this conclusion (95). Given the significance of vWF in platelet adhesion, it is not surprising that vWF is also stored in the α granules of platelets. The concentration of vWF in platelets has been reported to be 60 times greater than the vWF concentration in circulating plasma (96). vWF released from activated platelet functions in an autocrine fashion to facilitate their adhesion and spreading on surfaces (97).

Another plasma protein that may have a significant effect on platelet adhesion is fibrinogen. Fibrinogen readily binds to GPIIb-IIIa, one of the most widely studied receptors that is expressed in large numbers on activated platelets (98–100). The relative importance of fibrinogen, vWF, and their respective primary receptors (GPIIb-IIIa and GPIb) in platelet adhesion on artificial surfaces is controversial. Several arguments support the hypothesis that the interaction between vWF and GPIb is a contributor to platelet adhesion in this setting. First, vWF readily adsorbs onto artificial surfaces and platelets adhere to surfaces coated with vWF (91). Second, vWF promotes platelet adhesion at higher shear stress (101,102), and high shear stresses are certainly encountered when prosthetic devices (e.g., hollow fiber hemodialyzers) are used. Third, while platelets also bind to fibrinogen-coated surfaces, fibrinogen that is initially adsorbed onto artificial surfaces upon blood contact is rapidly replaced by other plasma proteins or is degraded (85–87).

Recent evidence, however, supports the notion that GPIIb-IIIa is a more important adhesion molecule mediating platelet adherence on artificial surfaces. For example, monoclonal antibodies directed against GPIIb-IIIa inhibit platelet adhesion to plasma-coated polystyrene plates (98,103). Further, blockade of GPIIb-IIIa is effective in inhibiting platelet adhesion to polyethylene, even under flow conditions (104). One could argue that GPIIb-IIIa could be binding to surface-bound vWF instead of surface-bound fibrinogen (98). However, at least one study showed that platelets bound poorly to polystyrene plates that were coated with afibrinogenemic plasma (103), suggesting that platelets bind via interactions between platelet GPIIb-IIIa and fibrinogen on artificial surfaces. In addition, platelet adherence to plasma-coated plates could be inhibited by antibodies to other plasma proteins (e.g., fibronectin (98)). Taken together, these observations suggest that fibrinogen is an important surface-bound ligand for platelets, although platelet adherence to artificial surfaces may be mediated by multiple interactions between various plasma proteins and cell surface receptors. It should be emphasized that most of the studies published so far have employed microtiter plates or other polystyrene and polyethylene surfaces in the absence of flow. Since the kinetics of protein binding to these surfaces may be substantially different from those occurring on CPB and HD membranes, caution must be exercised to extrapolate results obtained from one setting to another. Nonetheless, these recent observations strongly suggest that platelet adherence to artificial surfaces is different from that on the subendothelium.

In addition to mediating platelet binding to the artificial surface, GPIIb-IIIa may also help to anchor nonadherent platelets to the surface by allowing fibrinogen to form a bridge between two platelets. In this context, the role of GPIIb-IIIa is important in that this may be the primary mechanism by which platelets adhere to each other, a process essential for the progression of thrombogenesis. Therefore, suppression of GPIIb-IIIa expression or inhibition of this glycoprotein by antibodies or antagonists may be effective in preventing thrombosis in extracorporeal circulation. A schematic diagram of some proposed mechanisms by which platelets adhere to and aggregate on artificial surfaces is presented in Figure 6.

Granular membrane protein-140 (GMP-140, PADGEM, CD62P) is a constitutent of the α granules in platelets. When platelets are activated, this protein is translocated onto the cytoplasmic membrane surface (105). GMP-140 mediates binding between platelets and leukocytes, presumably via CD15 on the neutrophil and monocyte surface (106). Expression of GMP-140 has been used as a marker of platelet degranulation during extracorporeal circulation, as described below.

D. Alterations in Platelet Adhesion Molecules During Extracorporeal Circulation

Using antibodies and flow cytometry, several studies have shown the expression of GPIb and GPIIb-IIIa on platelets to be decreased during CPB (107–111). These results are corroborated by the demonstration that bind-

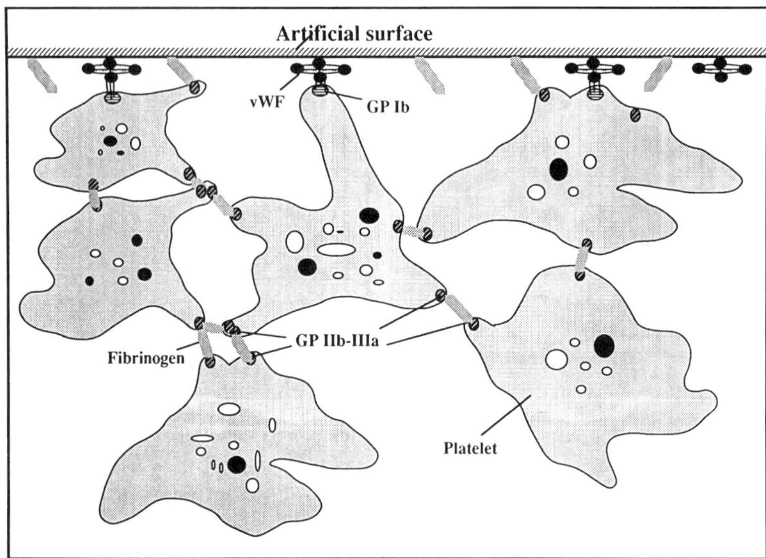

Figure 6 Schematic diagram of proposed mechanisms of platelet adhesion to artificial surface. For simplicity, platelets are shown to adhere to the surface via vWF-GPIb interaction and fibrinogen-GPIIb-IIIa interaction only (see text for details). Other interactions involving the same molecules, other plasma proteins, and other adhesion molecules may be contributory but are not depicted here. Fibrinogen molecules form bridges between GPIIb-IIIa across two platelets in the multiple cell layers.

ing of radiolabeled fibrinogen (the primary ligand for GPIIb-IIIa) to platelets is also decreased (110,112,113). Some investigators have shown that this event was accompanied by an increased plasma concentration of platelet membrane microparticulates, but not accompanied by evidence of degranulation (108,109). Such results suggest that the decreased GPIb and GPIIb-IIIa expression is a result of platelet fragmentation by mechanical trauma rather than cell activation. In contrast, other investigators demonstrated a substantial increase in the expression of GMP-140 on the platelet surface during CPB (62,107,114,115), which correlated with the decrease in GPIb expression (107). Further, an increase in plasma soluble GMP-140 and β-thromboglobulin (another platelet granular protein) has been reported (116). These observations strongly suggest that platelets are activated and that the changes in their surface receptors cannot be attributed to cell framentation alone. Rinder et al. showed that GPIb was downregulated on both activated and nonactivated platelets; however, the activated platelets had a greater loss of GPIb than the unactivated cells (107). Altogether, these data suggest that both mechanisms, cell fragmentation and cell activation, are operative to induce the loss of intrinsic membrane adhesion receptors on the platelet surface during CPB.

It should be noted that some investigators have failed to find changes in adhesion molecule expression on platelets during CPB. In the study of Kestin et al. in which sufficient heparin was employed to maintain a markedly prolonged activating clotting time, there was no downregulation of platelet GPIb and GPIIb-IIIa (117). Expression of GMP-140 on platelet surfaces was also negligible. This study suggests that when thrombin generation and platelet activation are prevented by heparin, platelet surface markers can be preserved. A summary of changes in molecules on platelet surface during CPB is presented in Table 4.

Activation and degranulation of platelets, as indicated by the release of granular proteins β-thromboglobulin and platelet factor 4, have been well documented during HD (118). More recently, a gradual increase in plasma thrombospondin (another platelet α granular glycoprotein) during HD has been reported (119).

VI. LEUKOCYTE-PLATELET AGGREGATION DURING EXTRACORPOREAL CIRCULATION

Since GMP140 is expressed when platelets are activated and since GMP-140 functions to bind platelets to leukocytes, it would be reasonable to postulate that the upregulation of GMP-140 would lead to the formation of platelet-leukocyte aggregates during extracorporeal circulation. Using flow cytometry techniques to detect GMP-140 on leukocytes, the binding of platelets to neutrophils and monocytes has been demonstrated during CPB (115) (Fig. 7). Platelet-monocyte aggregates increased by more than twofold, while platelet-neutrophil aggregates increased only slightly. An increase in platelet-lymphocyte aggregation was not observed, consistent with the in vitro observation that GMP-140 does not bind lymphocytes (120).

Using GMP-140 as marker, Stuard et al. also observed the formation of platelet-leukocyte aggregates during HD using cuprophan membranes but not with synthetic polyacrylonitrile (AN69) membranes (33). In contrast, using GPIIb-IIIa as marker, Gawa et al. found an increase in platelet-leukocyte aggregation during HD using three different types of synthetic membranes; however, the magnitude of increases was small (121).

Presumably, platelet-leukocyte binding facilitates communications and cross-signaling between the two cells, resulting in an enhancement in inflammatory response. An example of this phenomenon is the utilization of platelet-derived arachidonate by neutrophils to generate leukotrienes (122).

Table 4 Expression of Molecules on Platelet Surface During Cardiopulmonary Bypass

1st author, year (ref)	Platelet count	CD 42 (GPIb)	CD 41 (GPIIb-IIIa)	CD 62P (GMP-140)	Thrombospondin	GPIV
George, 1986 (109)	↓↓[a]	↓↓	↓↓	0	0	
Dechavanne, 1987 (108)	↓↓		↓↓	0	0	
Wenger, 1989 (110)	↓↓		↓↓			
Van Oeveren, 1990 (151)		↓↓	0			
Rinder, 1991 (114)	↓↓			↑↑		
Rinder, 1991 (107)	↓↓	↓↓	↓↓	↑↑		0
Kestin, 1993 (117)		0	0	0		
Kondo, 1993 (111)	↓↓	↓↓	0	↑↑		

[a]↓↓ = decrease, ↑↑ = increase, 0 = no change.

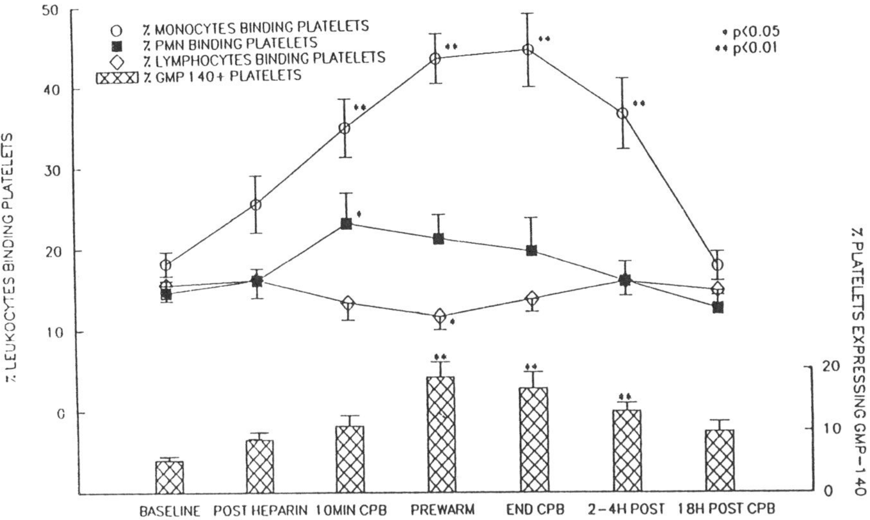

Figure 7 Aggregates between platelets and leukocytes and the expression of GMP140 on platelet surfaces during CPB. Platelet-leukocyte aggregates were measured in whole blood before surgery (BASELINE), 5 minutes after heparinization (POST HEPARIN), 10 min after start of CPB (10MIN CPB), before warming (PREWARM), before separation from CPB (END CPB), 2 to 4 hours after termination of CPB (2-4H POST), and 18 hours after CPB (18H POST CPB). The percentage of circulating platelets expressing GMP140 is displayed in the bar graphs for the same time points. All values represent means ± SE for 17 patients. (Reproduced from reference 115, with permission.)

VII. CLINICAL IMPLICATIONS OF CELL ADHESION MOLECULES AND ARTIFICIAL MEMBRANES

A. Clinical Implications of Neutrophil Adhesion Molecules and Artificial Membranes

Most of the published studies regarding adhesion molecules in extracorporeal circulation are largely phenomenological in nature, simply describing the alterations in expression of various molecules on cell surfaces. As described above, rather limited efforts have been made to examine their potential functional consequences on a cellular level. The clinical implications of these changes, as well as those resulting from other bioincompatible events, are even more difficult to define. Nonetheless, based on the known biological activities of cell adhesion molecules and the observed cellular events, some speculations can be advanced.

Activation of the peripheral blood cells during HD or CPB is a generally undesirable side effect of the treatment modality, although it probably represents a well-orchestrated event by the blood elements as a defense mechanism in response to invasion by the artificial surface. There are two general consequences from this unintentional activation of leukocytes. The first is pro-inflammatory events as a result of leukocyte activation. The second is an altered response of the cells to subsequent stimuli.

1. Pro-Inflammatory Events

A variety of organ damage collectively termed the "postperfusion syndrome" is often observed immediately after CPB and usually lasts for several days. A prominent feature of this syndrome is capillary leak in the lungs, leading to pulmonary dysfunction and hypoxemia. Other features include shock, intravascular hemolysis, leukopenia, and thrombocytopenia (12,35,123,124). The cause of this syndrome is not completely clear, but it has been attributed to activation of plasma humoral systems (complement, coagulation, and fibrinolytic proteins) and blood cells (neutrophils and platelets). The role of neutrophil adhesion molecules is underscored by recent studies in which upregulation of Mac-1 and Mac-1-dependent neutrophil-endothelium interactions during CPB using a bubble oxygenator in piglets were partially inhibited by a leumedin (35). Treatment with this agent resulted in less radical-mediated peroxidation of plasma lipids; lower pulmonary vascular resistance; lower lung tissue myeloperoxidase content; less intravascular leukosequestration, interstitial edema, and intraalveolar hemorrhage in the lungs; and higher arterial oxygen tension, compared to control animals. Interestingly, there was no difference in the peripheral neutrophil counts between the two groups, suggesting that the lower Mac-1 expression and activity resulting from the incomplete blockade by leumedin are sufficient to induce cell adhesion or that Mac-1 is unimportant for cell adhesion under these conditions.

Acute anaphylatoid reactions can occur during HD, although the incidence is low (<1% of all HD sessions). The etiologies are controversial and most likely involve several different mechanisms (125,126). Although anaphylatoxins C3a and C5a may play a role in some of these reactions, adhesion molecules have been neither incriminated nor studied.

Plasma protein catabolism has been demonstrated during clinical HD in uremic patients (127) and sham HD in normal human subjects (128) employing cuprophan membranes. The mechanisms by which cuprophan membranes might induce protein catabolism are unclear but some investigators have suggested that the membranes' ability to induce the release of granular proteases from neutrophils (129) and cytokines from monocytes (128,130) play a role. Although Mac-1 appears to be important in governing neutrophil degranulation induced by cuprophan membrane in the presence of plasma in-vitro, as described above, its role in plasma protein catabolism during clinical HD has not been established.

ROS released by activated neutrophils can enhance the peroxidation of plasma lipids and lipids in cell membranes (131). Macrophages laden with oxidized low-density lipoproteins (LDL) are prominent components of early atherosclerotic plaques (132). Peroxidation of phospholipids on erythrocyte membranes predisposes the cells to lysis. ROS provides a permissive role for protein degradation by neutrophil proteases even in the presence of protease inhibitors (133). This combined action of ROS and proteases may therefore induce organ injury in patients during extracorporeal circulation. Increased susceptibility of uremic lipoproteins to oxidation (134), enhanced lipid peroxidation during HD (135), the presence of autoantibodies to oxidized LDL (136), decreased half-life of circulating erythrocytes (137), delayed recovery of acute renal failure in patients (138), and premature atherosclerosis (139) have all been demonstrated in patients undergoing HD, consistent with ROS effects. The cause-and-effect relationships of these events with neutrophil activation and enhanced expression of adhesion molecules during HD, however, have not been established.

2. Altered Response of Cells

The second general consequence of cell activation during extracorporeal circulation is modulation of the cellular response to subsequent stimuli. In many instances, the cells become hyporesponsive or refractory. For example, leukocytes obtained from patients during HD with cuprophan membranes exhibit subnormal aggregation (23,140), adhesiveness to endothelial cells (32) or nylon fibers (55,141), and chemotaxis (23,140); diminished surface receptor expression upon stimulation (142); impaired oxidative response to soluble stimuli (24) or phagocytic challenge (143); and decreased production of ROS upon exposure to bacteria (19). The hyporesponsiveness may be due to downregulation of surface receptors or depletion of intracellular signaling mechanisms or metabolic machineries that are necessary to perform these cellular functions. On the other hand, priming of the neutrophils during ex vivo (144) or in vitro (145) exposure to HD membranes such that they become hyperresponsive to subsequent stimuli has also been reported. Although frequently discussed in this light, it is at present difficult to ascertain the extent to which activation of cells during extracorporeal circulation contributes to the immunodeficiency state of uremic and postoperative patients.

B. Clinical Implications of Platelet Adhesion Molecules and Artificial Membranes

Similar to the leukocytes, activation of platelets during extracorporeal circulation has direct and indirect consequences. Aggregated platelets deposited on HD or CPB oxygenator membranes may activate the coagulation (146) and complement (147) pathways. Activation of the coagulation pathway, in conjunction with adherent platelets, sets the stage for thrombus formation. Adherent thrombi impede blood flow and reduce the effectiveness of HD or oxygenation of blood. Although coagulation is usually confined to the extracorporeal circuit, significant thrombosis requires cessation of the treatment and prohibits the multiple use of hemodialyzers. Dislodgment of the thrombus from the extracorporeal device can lead to thromboembolism and blood lost in the circuit necessitates transfusion of blood products following CPB. Activated platelets may also release a number of potent vasoactive agents that have hemodynamic effects (148), and complement activation leads to the proinflammatory responses described above.

The indirect consequences of platelet activation in the extracorporeal circuit are decreases in platelet number and function. Thrombocytopenia is common after CPB, and published studies have documented that the number of circulating platelets may decrease by as much as 85% during bypass (149). Qualitative defects are frequently found in the platelets that remain in the circulation. These defects are manifested clinically as prolonged bleeding time and increased blood loss. Although other events such as hypothermia during CPB are probably contributory, a major etiologic factor appears to be abnormalities induced by blood-artificial surface interactions (150). Adhesion molecules are perhaps involved in the pathogenesis of the platelet dysfunction in two ways.

First, by initiating and sustaining cell adhesion and aggregation, adhesion molecules promote platelet activation. As a result of the activation, platelets degranulate and the contents of the platelet granules become reduced. These events turn the platelets into a refractory state as evidenced by a decrease in aggregation in response to adenosine diphosphate (ADP) and other agonists (114,150).

Second, the downregulation of cell surface adhesion molecules may directly contribute to the refractory state of the platelets, although this defect does not appear to be as pronounced as the depletion of granular contents. Both the thrombocytopenia and the qualitative defect in the remaining platelets enhance the risk of hemorrhage in patients undergoing CPB. In this context, administration of the protease inhibitor, aprotinin, has been reported to be associated with preservation of GPIb expression on platelets and decreased blood loss during and after CPB (151,152).

The decrease in platelet count during HD is usually absent or mild, although severe thrombocytopenia has been reported (153). Decreased aggregation of platelets in response to ADP has also been demonstrated after HD (154). This deterioration is superimposed on the preexisting platelet dysfunction that results from uremia.

ACKNOWLEDGMENTS

This work was supported by U.S. PHS grants RO1-DK-45575 to Dr Cheung, RO1-HL-42555 to Dr Mohammad, and the U.S. Department of Veterans Affairs Medical Research Funds.

REFERENCES

1. Lyman DJ. Membranes. In: Drukker W, Parsons FM, Maher JF, eds. Replacement of Renal Function by Dialysis. Boston: Martinus Nijhoff, 1983: 97–105.
2. Cheung AK, Parker CJ, Wilcox L, Janatova J. Activation of the alternative pathway of complement by hemodialysis membranes. Kidney Int 1989; 36: 257–265.
3. Chenoweth DE, Cheung AK, Henderson LW. Anaphylatoxin formation during hemodialysis: effects of different dialyzer membranes. Kidney Int 1983; 24:764–769.
4. Chenoweth DE, Cheung AK, Ward DM, Henderson LW. Anaphylatoxin formation during hemodialysis: comparison of new and reused dialyzers. Kidney Int 1983; 24:770–774.
5. Hakim RM, Fearon DT, Lazarus JM. Biocompatibility of dialysis membranes: effects of chronic complement activation. Kidney Int 1984; 26:194–200.
6. Goldstein IM. Complement: biological active product. In: Gallin JI, Goldstein IM, Snyderman R, eds. Inflammation: Basic Principles and Clinical Correlates. New York: Raven Press, 1988:55–74.
7. Hörl WH, Steinhauer HB, Riegel W, Schollmeyer P, Schafer RM, Heidland A. Effect of different dialyzer membranes on plasma levels of granulocyte elastase. Kidney Int 1988; 33(suppl 24):S90–S91.
8. Zaoui P, Green W, Hakim RM. Hemodialysis with cuprophane membrane modulates interleukin-2 receptor expression. Kidney Int 1991; 39:1020–1026.
9. Smeby LC, Widerøe W-E, Balstad T, Jørstad S. Biocompatibility aspects of cellophane, cellulose acetate, polyacrylonitrile, polysulfone and polycarbonate hemodialyzers. Blood Purif 1986; 4:93–101.
10. Falkenhagen D, Bosch T, Brown GS, et al. A clinical study on different cellulosic dialysis membranes. Nephrol Dial Transplant 1987; 2:537–545.
11. Cheung AK, Leypoldt JK, Deeter RB. Discordance between neutropenia and monocytopenia during hemodialysis (HD). (Abstract) J Am Soc Nephrol 1995; 6:525.

12. Chenoweth DE, Cooper SW, Hugli TE, Stewart RW, Blackstone EH, Kirklin JW. Complement activation during cardiopulmonary bypass. N Engl J Med 1981; 304:497–503.
13. Deleuze PH, Intrator L, Liou A, Contremoulins A, Cachera JP, Loisance DY. Complement activation and use of a cell saver in cardiopulmonary bypass. ASAIO Trans 1990; 36:M179–M181.
14. Gu YJ, Van Oeveren W, Boonstra PW, De Haan J, Wildevuur CRH. Leukocyte activation with increased expression of CR3 receptors during cardiopulmonary bypass. Ann Thorac Surg 1992; 53:839–843.
15. Mason RG, Zucker WH, Bilinsky RT, Shinoda MA, Sharp DE, Mohammad SF. Blood components deposited on used and reused dialysis membranes. Biomat Med Dev Art Org 1976; 4:333–358.
16. Brubaker LH, Nolph KD. Mechanisms of recovery from neutropenia induced by hemodialysis. Blood 1971; 38:623–631.
17. Craddock PR, Fehr J, Dalmasso AP, Brigham KL, Jacob HS. Hemodialysis leukopenia: pulmonary vascular leukostasis resulting from complement activation by dialyzer cellophane membranes. J Clin Invest 1977; 59:879–888.
18. Dodd NJ, Gordge MP, Tarrant J, Parsons V, Weston MJ. A demonstration of neutrophil accumulation in the pulmonary vasculature during haemodialysis. Proc Eur Dial Transplant Assoc 1983; 20:186–189.
19. Himmelfarb J, Ault KA, Holbrook D, Leeber DA, Hakim RM. Intradialytic granulocyte reactive oxygen species production: a prospective, crossover trial. J AM Soc Nephrol 1993; 4:178–186.
20. Hörl WH, Schaefer RM, Heidland A. Effect of different dialyzers on proteinases and proteinase inhibitors during hemodialysis. Am J Nephrol 1985; 5:320–326.
21. Tetta C, Segoloni G, Pacitti A, et al. The production of platelet-activating factor during hemodialysis. Int J Artif Organs 1989; 12:766–772.
22. Hallgren R, Venge P, Danielson BG. Neutrophil and eosinophil degranulation during hemodialysis are mediated by the dialysis membrane. Nephron 1982; 32:329–334.
23. Skubitz KM, Craddock PR. Reversal of hemodialysis granulocytopenia and pulmonary leukostasis: a clinical manifestation of selective down-regulation of granulocyte responses to $C5a_{desarg}$. J Clin Invest 1981; 67:1383–1391.
24. Cohen MS, Elliot DM, Chaplinski T, Pike MM, Niedel JE. A defect in oxidative metabolism of human polymorphonuclear leukocytes that remain in circulation early in hemodialysis. Blood 1982; 60:1283–1289.
25. Arnaout MA, Hakim RM, Todd RF III, Dana N, Colten HR. Increased expression of an adhesion-promoting surface glycoprotein in the granulocytopenia of hemodialysis. N Engl J Med 1985; 312:457–462.
26. Cheung AK, Hohnholt M, Gilson J. Adherence of neutrophils to hemodialysis membranes: role of complement receptors. Kidney Int 1991; 40:1123–1133.
27. Marshall JW, Ahearn DJ, Nothum RJ, Esterly J, Nolph KD, Maher JF. Adherence of blood components to dialyzer membranes: morphological studies. Nephron 1974; 12:157–170.

28. Cheung AK. Interactions between plasma proteins and hemodialysis membranes. Adv Nephrol 1993; 22:417–437.
29. Aljama P, Martin-Malo A, Garin JM, et al. Granulocyte adherence changes: an index of biocompatibility. Kidney Int 1988; 33(suppl 24):S68–S72.
30. Debrand-Passard A, Lajous-Petter A, Schmidt R, et al. Thrombogenicity on dialyzer membranes as assessed by residual blood volume and surface morphology at different heparin dosages. Contrib Nephrol 1989; 74:2–9.
31. Cristol JP, Canaud B, Rabesandratana H, Gaillard I, Serre A, Mion C. Enhancement of reactive oxygen species production and cell surface markers expression due to haemodialysis. Nephrol Dial Transplant 1994; 9:389–394.
32. Himmelfarb J, Zaoui P, Hakim R, Holbrook D. Modulation of granulocyte LAM-1 and MAC-1 during dialysis: a prospective, randomized controlled trial. Kidney Int 1992; 41:388–395.
33. Stuard S, Carreno M-P, Poignet J-L, Albertzaii A, Haeffner-Cavaillon N. A major role for CD62P/CD15s interactions in leukocyte margination during hemodialysis. Kidney Int 1995; 48:93–102.
34. Jacobs AA Jr, Ward RA, Wellhausen SR, McLeish KR. Polymorphonuclear leukocytes function during hemodialysis: relationship to complement activation. Nephron 1989; 52:119–124.
35. Gillinov AM, Redmond JM, Zehr KJ, et al. Inhibition of neutrophil adhesion during cardiopulmonary bypass. Ann Thorac Surg 1994; 57:126–133.
36. El Habbal MH, Carter H, Smith LJ, Elliott MJ, Strobel S. Neutrophil activation in paediatric extracorporeal circuits: effect of circulation and temperature variation. Cardiovasc Res 1995; 29:102–107.
37. Kappelmayer J, Bernabei A, Gikakis N, Edmunds LH Jr. Colman RW. Upregulation of Mac-1 surface expression on neutrophils during extracorporeal circulation. J Lab Clin Med 1993; 121:118–126.
38. Silber R, Moldow CF. Biochemistry and function of neutrophils, composition of neutrophils. In: Williams WJ, Brutler E, Erslev A, Lichtman MA, eds. Hematology. 3d ed. New York: McGraw-Hill, 1983:726–734.
39. Lundahl J, Hed J, Jacobson SH. Dialysis granulocytopenia is preceded by an increased surface expression of the adhesion-promoting glycoprotein Mac-1. Nephron 1992; 61:163–169.
40. Werfel T, Oppermann M, Schulze M, Krieger G, Weber M, Götze O. Binding of fluorescein-labeled anaphylatoxin C5a to human peripheral blood, spleen, and bone marrow leukocytes. Blood 1992; 79:152–160.
41. Himmelfarb J, McMonagle E, Holbrook D, Toth C. Soluble complement receptor 1 inhibits both complement and granulocyte activation during ex vivo hemodialysis. J Lab Clin Med. 1996; 126:392–400.
42. Thylen P, Lundahl J, Fernvik E, Hed J, Svenson SB, Jacobson SH. Mobilization of an intracellular glycoprotein (Mac-1) on monocytes and granulocytes during hemodialysis. Am J Nephrol 1992; 12:393–400.
43. Combe C, Pourtein M, de Précigout V, et al. Granulocyte activation and adhesion molecules during hemodialysis with cuprophane and a high-flux biocompatible membrane. Am J Kidney Dis 1994; 24:437–442.

44. Camussi G, Pacitti A, Tetta C, et al. Mechanisms of neutropenia in hemodialysis (HD). Trans Am Soc Artif Intern Organs 1984; 30:364–368.
45. Camussi G, Segoloni G, Rotunno M, Vercellone A. Mechanism involved in acute granulocytopenia in hemodialysis: cell-membrane direct interactions. Int J Artif Organs 1978; 3:123–127.
46. Sakaguchi K, Morimoto S, Chen Y-H, Nakamoto Y, Ogihara T. Increases in circulating level of platelet-activating factor lag behind transient neutropenia during hemodialysis with cuprophane membranes. Nephron 1991; 59:455–460.
47. Shalit M, Allmen CV, Atkins PC, Zweiman B. Platelet activating factor increases expression of complement receptors on human neutrophils. J Leuk Biol 1988; 44:212–217.
48. Enia G, Catalano C, Misefari V, et al. Complement activated leucopenia during hemodialysis: effect of pulse methyl-prednisolone. Int J Artif Organs 1990; 13:98–102.
49. Aljama P, Martin-Malo A, Castillo D, et al. Anaphylatoxin C5a generation and dialysis-induced leukopenia with different hemodialyzer membranes. Blood Purif 1986; 4:88–92.
50. Bingel M, Arndt W, Schulze M, et al. Comparative study of C5a plasma levels with different hemodialysis membranes using an enzyme-linked immunosorbent assay. Nephron 1989; 51:320–324.
51. Lewis SL, Van Epps DE, Chenoweth DE. Leukocyte C5a receptor modulation during hemodialysis. Kidney Int 1987; 31:112–120.
52. Alvarez V, Pulido R, Campanero MR, Paraiso V, de Landazuri MO, Sanchez-Madrid F. Differentially regulated cell surface expression of leukocyte adhesion receptors on neutrophils. Kidney Int 1991; 40:899–905.
53. Kishimoto TK, Jutila MA, Berg EL, Butcher EC. Neutrophil Mac-1 and MEL-14 adhesion proteins inversely related by chemotactic factors. Science 1989; 245:1238–1241.
54. Lundberg C, Wright S. Relation of the CD11/CD18 family of leukocyte antigens to the transient neutropenia caused by chemoattractants. Blood 1990; 76:1240–1245.
55. Spagnuolo PJ, Bass JH, Smith MC, Danviriyasup K, Dunn MJ. Neutrophil adhesiveness during prostacyclin and heparin hemodialysis. Blood 1982; 60: 924–929.
56. Addonizio VP, Strauss JF III, Chang L-F, Fisher CA, Colman RW, Edmunds LH Jr. Release of lysosomal hydrolases during simulated extracorporeal circulation. J Thorac Cardiovasc Surg 1982; 84:28–34.
57. Webster RO, Hong SR, Johnston RB Jr, Henson PM. Biological effects of the human complement fragments C5a and $C5a_{des\ Arg}$ on neutrophil function. Immunopharmacology 1980; 2:201–219.
58. Sacks T, Moldow CF, Craddock PR, Bowers TK. Oxygen radicals mediate endothelial cell damage by complement-stimulated granulocytes. J Clin Invest 1978; 61:1161–1167.
59. Cheung AK, Parker CJ, Hohnholt M. β_2 integrins are required for neutrophil degranulation induced by hemodialysis membranes. Kidney Int 1993; 43:649–660.

60. Cheung AK, Parker CJ, Hohnholt M. Soluble complement receptor type 1 inhibits complement activation induced by hemodialysis membranes in-vitro. Kidney Int 1994; 46:1680–1687.
61. Shappell S, Toman C, Anderson D, Taylor A, Entman M, Smith C. Mac-1 (CD11b/CD18) mediates adherence-dependent hydrogen peroxide production by human and canine neutrophils. J Immunol 1990; 144:2702–2711.
62. Lee J, Hakim RM, Fearon DT. increased expression of the C3b receptor by neutrophils and complement activation during haemodialysis. Clin Exp Immunol 1984; 56:205–214.
63. Mathew JP, Rinder CS, Tracey JB, et al. Acadesine inhibits neutrophil CD11b up-regulation in-vitro and during in-vivo cardiopulmonary bypass. J Thorac Cardiovasc Surg 1995; 109:448–456.
64. Luger A, Kovarik J, Stummvoll H-K, Urbanska A, Luger TA. Blood-membrane interaction in hemodialysis leads to increased cytokine production. Kidney Int 1987; 32:84–88.
65. Pereira BJG, Shapiro L, King AJ, Falagas ME, Strom JA, Dinarello CA. Plasma levels of IL-1β, TNFα and their specific inhibitors in undialyzed chronic renal failure, CAPD and hemodialysis patients. Kidney Int 1994; 45: 890–896.
66. Haeffner-Cavaillon N, Cavaillon J, Ciancioni C, Bacle F, Delons S, Kazatchkine MD. In-vivo induction of interleukin-1 during hemodialysis. Kidney Int 1989; 35:1212–1218.
67. Schiller B, Ziegler-Heitbrock HWL, Meyer N, Schmidt B, Blumenstein M. Monocyte phenotype and interleukin-1 production in patients undergoing haemodialysis. Nephron 1991; 59:573–579.
68. Tielemans CL, Delville J-PC, Husson CP, et al. Adhesion molecules and leukocyte common antigen on monocytes and granulocytes during hemodialysis. Clin Nephrol 1993; 39:158–165.
69. Laude-Sharp M, Caroff M, Simard L, Pusineri C, Kazatchkine MD, Haeffner-Cavaillon N. Induction of IL-1 during hemodialysis: transmembrane passage of intact endotoxins (LPS). Kidney Int 1990; 38:1089–1094.
70. Heierli C, Markert M, Lambert PH, Kuwahara T, Wauters JP. On the mechanisms of haemodialysis-induced neutropenia: a study with five new and re-used membranes. Nephrol Dial Transplant 1988; 3:773–783.
71. Rinder CS, Bonan JL, Rinder HM, Mathew J, Hines R, Smith BR. Cardiopulmonary bypass induces leukocyte-platelet adhesion. Blood 1992; 79:1201–1205.
72. Wright SD, Ramos RA, Tobias PS, Ulevitch RJ, Mathison JC. CD14, a receptor for complexes of lipopolysaccharide (LPS) and LPS binding protein. Science 1990; 249:1431–1433.
73. Beekhuizen H, Blokland I, Corsel–Van Tilburg AJ, Koning F, Van Furth R. CD14 contributes to the adherence of human monocytes to cytokine-stimulated endothelial cells. J Immunol 1991; 147:3761–3767.
74. Mege JL, Sanguedolce MV, Purgus R, et al. Chronic and intradialytic effects of high-flux hemodialysis on tumor necrosis factor-α production: relationship to endotoxins. Am J Kidney Dis 1992; 20:482–488.

75. Schindler R, Gelfand JA, Dinarello CA. Recombinant C5a stimulates transcription rather than translation of IL-1 and TNF: priming of mononuclear cells with recombinant C5a enhances cytokine synthesis induced by LPS, IL-1 or PMA. Blood 1990; 76:1631–1635.
76. Haeffner-Cavaillon N, Cavaillon J-M, Laude M, Kazatchkine M. C3a(C3a$_{desArg}$) induces production and release of interleukin 1 by cultured human monocytes. J Immunol 1987; 139:794–799.
77. Addonizio VP, Coleman RW. Platelets and extracorporeal circulation. Biomaterials 1982; 3:9–15.
78. White, JG. The ultrastructure and regulatory mechanisms of blood platelets. In: Lasslo A, ed. Blood Platelet Function and Medicinal Chemistry. New York: Elsevier Biomedical, 1984:15–60.
79. White JG. Is the canalicular system the equivalent of the muscle sarcoplasmic reticulum?. Hemostasis 1975; 4:185–191.
80. Salzman EW, Merrill EW. Interaction of blood with artificial surfaces. In: Colman RW, Marder VJ, Salzman EW, Hirsh J, eds. Hemostasis and Thrombosis: Basic Principles and Clinical Practice. 2d ed. Philadelphia: J.B. Lippincott, 1987:1335–1347.
81. Brash JL, Ten Hove P. Effect of plasma dilution on adsorption of fibrinogen to solid surfaces. Thromb Haemostas 1984; 51:326–330.
82. Horbett TA. Mass action effects on the adsorption of fibrinogen from hemoglobin solutions and from plasma. Thrombos Haemostas 1984; 51:174–181.
83. Lindon JN, McNanama G, Kushner L, Merrill EW, Salzman EW. Does the conformation of adsorbed fibrinogen dictate platelet interactions with artificial surfaces? Blood 1986; 68:355–362.
84. Brash JL. The fate of fibrinogen following adsorption at the blood-biomaterial interface. Ann NY Acad Sci 1987; 516:206–222.
85. Cuypers PA, Willems GM, Hemker HC, Hermens WT. Adsorption kinetics of protein mixtures: a tentative explanation of the Vroman effect. Annu NY Acad Sci 1987; 516:244–252.
86. Vroman L, Adams AL, Fischer GC, Munoz PC. Interaction of high molecular weight kininogen, factor XII and fibrinogen in plasma at interfaces. Blood 1980; 55:156–159.
87. Brash JL, Chan BM, Szota P, Thibodeau JA. Degradation of adsorbed fibrinogen by surface-generated plasmin. J Biomed Mater Res 1985; 19:1017–1029.
88. Young BR, Lambrecht LK, Albrecht RM, Mosher DF, Cooper SL. Platelet-protein interactions at blood-polymer interfaces in the canine test model. Trans Am Soc Artif Intern Organs 1983; 29:442–447.
89. Steiner M. Platelet adhesion. In: Richardson PD, Steiner M, eds. Principles of Cell Adhesion. Boca Raton: CRC Press, 1995:307–316.
90. Kieffer N, Phillips DR. Platelet membrane glycoproteins: functions in cellular interactions. Annu Rev Cell Biol 1990; 6:329–357.
91. Danton MC, Zaleski A, Nichols WL, Olson JD. Monoclonal antibodies to platelet glycoproteins Ib and IIb/IIIa inhibit adhesion of platelets to purified solid-phase von Willebrand factor. J Lab Clin Med 1994; 124:274–282.

92. Coller BS, Peerschke EI, Scudder LE, Sullivan CA. Studies with a murine monoclonal antibody that abolishes ristocetin-induced binding of vWF to platelets: additional evidence in support of GP Ib as a platelet receptor for vWF. Blood 1983; 61:99–110.
93. Wagner DD, Fay PJ, Sporn LA, Sinha S, Lawrence SO, Marder VJ. Divergent fates of vWF and its polypeptide (vWF antigen II) after secretion from endothelial cells. Proc Natl Acad Sci USA 1987; 84:1955–1959.
94. Baruch D, Denis C, Marteaux C, Schoevaert D, Coulombel L, Meyer D. Role of von Willebrand factor associated to extracellular matrices in platelet adhesion. Blood 1991; 77:519–527.
95. Savage B, Shattil SJ, Ruggeri ZM. Modulation of platelet function through adhesion receptors: a dual role for glycoprotein IIb-IIIa (integrin aIIbβ3) mediated by fibrinogen and glycoprotein Ib-von Willebrand factor. J Biol Chem 1992; 267:11300–11306.
96. Bouma BN, Hordik-Hos JM, De Graaf S, Sixma JJ. Presence of VIII related antigen in blood platelets of patients with von Willebrands disease. Nature 1975; 257:510–512.
97. George JN, Shattil SJ. The clinical importance of acquired abnormalities of platelet function. N Engl J Med 1991; 324:27–39.
98. DiFazio LT, Stratoulias C, Greco RS, Haimovich B. Multiple platelet surface receptors mediate platelet adhesion to surfaces coated with plasma proteins. J Surg Res 1994; 57:133–137.
99. Fitzgerald LA, Phillips DR. Platelet membrane glycoprotein. In: Colman RW, Marder VJ, Salzman EW, Hirsch J, eds. Hemostasis and Thrombosis: Basic Principles and Clinical Practice. 2d ed. Philadelphia: J.B. Lippincott, 1987:572–593.
100. Phillips DR, Charo IF, Praise LV, Fitzgerald LA. The platelet membrane glycoprotein IIb-IIIa complex. Blood 1988; 71:831–843.
101. Lawrence JB, Kramer WS, McKeown LP, Williams SB, Gralnick HR. Arginine-glycine-aspartic acid and fibrinogen gamma chain carboxyterminal peptides inhibit platelet adherence to arterial subendothelium at high wall shear rate: an effect dissociable from interference with adhesive protein binding. J Clin Invest 1990; 86:1715–1722.
102. Aleviadou BR, Moake JL, Turner NA, et al. Real-time analysis of shear-dependent thrombus formation and its blockade by inhibitors of von Willebrand factor binding to platelets. Blood 1993; 81:1263–76.
103. Nagai H, Handa M, Kawai Y, Watanabe K, Ikeda Y. Evidence that plasma fibrinogen and platelet membrane GPIIb-IIIa are involved in the adhesion of platelets to an artificial surface exposed to plasma. Thromb Res 1993; 71:467–477.
104. Sheppeck RA, Bentz M, Dickson C, et al. Examination of the roles of glycoprotein Ib and glycoprotein IIb/IIIa in platelet deposition on an artificial surface using clinical antiplatelet agents and monoclonal antibody blockade. Blood 1991; 78:673–680.
105. Steinberg PE, McEver RP, Shuman MA, Jacques YV, Bainton DF. A platelet alpha-granule membrane protein (GMP-140) is expressed on the plasma membrane after activation. J Cell Biol 1985; 101:880–886.

106. Larsen E, Palabrica T, Sajer S, et al. PADGEM-dependent adhesion of platelets to monocytes and neutrophils is mediated by lineage specific carbohydrate, LNFIII (CD15). Cell 1990; 63:467–484.
107. Rinder CS, Mathew JP, Rinder HM, Bonan J, Ault KA, Smith BR. Modulation of platelet surface adhesion receptors during cardiopulmonary bypass. Anesthesiology 1991; 75:563–570.
108. Dechavanne M, Ffrench M, Pages J, et al. Significant reduction in the binding of a monoclonal antibody (Lyp18) directed against the IIb/IIIa glycoprotein complex to platelets of patients having undergone extracorporeal circulation. Thromb Haemostas 1987; 57:106–109.
109. George JN, Pickett EB, Saucerman S, et al. Platelet surface glycoproteins: studies on resting and activated platelet membrane microparticles in normal subjects, and observations in patients during adult respiratory distress syndrome and cardiac surgery. J Clin Invest 1986; 78:340–348.
110. Wenger RK, Lukasiewicz H, Mikuta BS, Niewiarowski S, Edmunds LH. Loss of platelet fibrinogen receptors during clinical cardiopulmonary bypass. J Thorac Cardiovasc Surg 1989; 97:235–239.
111. Kondo C, Tanaka K, Takagi K, et al. Platelet dysfunction during cardiopulmonary bypass surgery: with specific reference to platelet membrane glycoproteins. ASAIO J 1993; 39:M550–M553.
112. Gluszko P, Rucinski B, Musial J, et al. Fibrinogen receptors in platelet adhesion to surfaces of extracorporeal circuits. Am J Physiol 1987; 252: H615–H621.
113. Musial J, Niewiarowski S, Hershock D, Morinelli TA, Colman RW, Edmunds LHJ. Loss of fibrinogen receptors from the platelet surface during simulated extracorporeal circulation. J Lab Clin Med 1985; 105:514–522.
114. Rinder CS, Bohnert J, Rinder HM, Mitchell J, Ault K, Hillman R. Platelet activation and aggregation during cardiopulmonary bypass. Anesthesiology 1991; 75:388–393.
115. Rinder CS, Bonan JL, Rinder HM, Mathew J, Hines R, Smith BR. Cardiopulmonary bypass induces leukocyte-platelet adhesion. Blood 1992; 79:1201–1205.
116. Komai H, Haworth SG. Effect of Cardiopulmonary bypass on the circulating level of soluble GMP-140. Ann Thorac Surg 1994; 58:478–82.
117. Kestin AS. Valeri CR, Khuri SF, et al. The platelet function defect of cardiopulmonary bypass. Blood 1993; 82:107–117.
118. Ireland H, Lane DA, Curtis JR. Objective assessment of heparin requirements for hemodialysis in humans. J Lab Clin Med 1984; 103:643–652.
119. Gawaz MP, Ward RA. Effects of hemodialysis on platelet-derived thrombospondin. Kidney Int 1991; 40:257–265.
120. Moore KL, Varki A, McEver RP. GMP-140 binds to a glycoprotein receptor on human neutrophils: evidence for a lectin-like interaction. J Cell Biol 1991; 112:491–499.
121. Gawaz MP, Mujais SK, Schmidt B, Gurland HJ. Platelet-leukocyte aggregation during hemodialysis. Kidney Int 1994; 46:489–495.
122. Marcus AJ, Broekman MJ, Safier LB, et al. Formation of leukotrienes and

other hydroxy acids during platelet-neutrophil interactions in-vitro. Biochem Biophys Res Commun 1982; 109:130–137.

123. Kirklin JW. Open-heart surgery at the Mayo Clinic: the 25th anniversary. Mayo Clin Proc 1980; 55:339–341.
124. Ratliff NB, Young WG Jr, Hackel DB, Mikat E, Wilson JW. Pulmonary injury secondary to extracorporeal circulation. J Thorac Cardiovasc Surg 1973; 65:425–432.
125. Cheung AK. Dialyzer biocompatibility: practical considerations. In: Nissenson AR, Fine RN, ed. Dialysis Therapy. 2d ed. Philadelphia: Hanley & Belfus, 1993:75–77.
126. Walton DF, Cheung AK. Membrane biocompatibility. In: Nissenson AR, Fine RN, Gentile DE, eds. Clinical Dialysis. 3d ed. Norwalk: Appleton & Lange, 1995:93–120.
127. Borah MF, Schoenfeld PY, Gotch FA, Sargent JA, Wolfson M, Humphreys MH. Nitrogen balance during intermittent dialysis therapy of uremia. Kidney Int 1978; 14:491–500.
128. Gutierrez A, Alvestrand A, Wahren J, Bergström J. Effect of in-vivo contact between blood and dialysis membranes on protein catabolism in humans. Kidney Int 1990; 38:487–494.
129. Hörl WH, Heidland A. Evidence for the participation of granulocyte proteinases on intradialytic catabolism. Clin Nephrol 1984; 21:314–322.
130. Dinarello CA. Cytokines: agents provacateurs in hemodialysis? Kidney Int 1992; 41:683–694.
131. Slater TF. Free-radical mechanisms in tissue injury. Biochem J 1984; 222:1–15.
132. Steinberg D, Parthasarathy S, Carew TE, Khoo JC, Witztum JL. Beyond cholesterol: modifications of low-density lipoprotein that increase its atherogenicity. N Engl J Med 1984; 320:915–924.
133. Weiss SJ, Regiani S. Neutrophils degrade subendothelial matrices in the presence of alpha-1-proteinase inhibitor: cooperative use of lysosomal proteinases and oxygen metabolites. J Clin Invest 1984; 73:1297–1303.
134. Maggi E, Bellazzi R, Falaschi F, et al. Enhanced LDL oxidation in uremic patients: an additional mechanism for accelerated atherosclerosis? Kidney Int 1994; 45:876–883.
135. Maher ER, Wickens DG, Griffin JFA, Kyle P, Curtis JR, Dormandy TL. Increased free-radical activity during haemodialysis? Nephrol Dial Transplant 1987; 2:169–171.
136. Maggi E, Bellazzi R, Gazo A, Seccia M, Bellomo G. Autoantibodies against oxidatively-modified LDL in uremic patients undergoing dialysis. Kidney Int 1994; 46:869–876.
137. Eklund SG, Johansson SV, Strandberg O. Anemia in uremia: Hemolysis evaluated by common methods and application of the bilirubin turnover method. Acta Med Scand 1971; 190:435–443.
138. Hakim RM, Wingard RL, Parker RA. Effect of the dialysis membrane in the treatment of patients with acute renal failure. N Engl J Med 1994; 331:1338–1342.

139. Charney DI, Walton DF, Cheung AK. Atherosclerosis in chronic renal failure. Curr Opin Nephrol Hypertens 1993; 2:876–882.
140. Klempner MS, Gallin JI, Balow JE, Van Kammen DP. The effect of hemodialysis and C5ades arg on neutrophil subpopulations. Blood 1980; 55:777–783.
141. Lespier-Dexter LE, Guerra C, Ojeda W, Martinez-Maldonado M. Granulocyte adherence in uremia and hemodialysis. Nephron 1979; 24:64–68.
142. Pulido R, Alvarez V, Mollinedo F, Sanchez-Madrid F. Biochemical and functional characterization of the leucocyte tysoine phosphatase CD45 (CD45RO, 180 kD) from human neutrophils: in-vivo upregulation of CD45RO plasma membrane expression on patients undergoing haemodialysis. Clin Exp Immunol 1992; 87:329–335.
143. Vanholder R, Ringoir S, Dhondt A, Hakim RM. Phagocytosis in uremic and hemodialysis patients: a prospective and cross sectional study. Kidney Int 1991; 39:320-327.
144. Ward RA, McLeish KR. Hemodialysis with cellulose membrane primes the neutrophil oxidative burst. Artif Organ 1995; 19:801–807.
145. Cheung AK, Wei S, Leypoldt J, Masaki T. Variable response by neutrophils (PMN) to stimuli following exposure to cuprophan membrane (CuM). (Abstract). J Am Soc Nephrol 1995; 6:525.
146. Claggett GP. Artificial devices in clinical practice. In: Colman RW, Marder VJ, Salzman EW, Hirsh J, eds. Hemostasis and Thrombosis: Basic Principles and Clinical Practice. 2d ed. Philadelphia: J.B. Lippincott, 1987:1348-1365.
147. Blajchman MA, Özge-Anwar AH. The role of the complement system in hemostasis. Prog Hematol 1986; 14:149–182.
148. Addonizio VP, Smith JB, Strauss JF, Colman RW, Edmunds LH Jr. Thromboxane synthesis and platelet secretion during cardiopulmonary bypass with bubble oxygenator. J Thorac Cardiovasc Surg 1980; 79:91–96.
149. Hope AF, Heynes A, Lötter MG, et al. Kinetics and sites of sequestration of ^{111}indium labelled platelets during cardiopulmonary bypass. J Thorac Cardiovasc Surg 1981; 81:880–886.
150. Woodman RC, Harker LA. Bleeding complications associated with cardiopulmonary bypass. Blood 1990; 76:1680–1697.
151. Van Oeveren W, Harder MP, Roozendaal KJ, Eijsman L, Wildevuur CRH. Aprotinin protects platelets against the initial effect of cardiopulmonary bypass. J Thorac Cardiovasc Surg 1990; 99:788–797.
152. Carrel T, Bauer E, Laske A, von Segesser L, Turnia M. Low-dose aprotinin for reduction of blood loss after cardiopulmonary bypass. Lancet 1991; 337:673.
153. Vicks SL, Gross ML, Schmitt GW. Massive hemorrhage due to hemodialysis-associated thrombocytopenia. Am J Nephrol 1983; 3:30–33.
154. Levin RD, Kwaan HC, Ivanovich P. Changes in platelet function during hemodialysis. J Lab Clin Med 1978; 92:779–786.

26

Cellular and Extracellular Matrix Adhesion Molecules in Organ Transplantation

Jerzy W. Kupiec-Weglinski
Department of Surgery, Brigham and Women's Hospital, and Harvard Medical School, Boston, Massachusetts

Leendert C. Paul
Department of Medicine, University of Toronto at St. Michael's Hospital, Toronto, Ontario, Canada

I. BACKGROUND

The adhesion of cells to other cells or to particular components of the tissue microenvironment is a basic function of cell recognition and migration and underlies many biological processes, including transplant rejection. Cell adhesion molecules mediate many cell-cell interactions and direct leukocyte trafficking between anatomical compartments, and cell recruitment to sites of tissue inflammation. They are involved in antigen recognition, cell activation and transformation, and the execution of cytotoxic effector functions. Advances in the understanding of T cell activation have placed particular emphasis on the interactions between the T cell receptor and the antigen presenting cell in the context of MHC antigens, but little consideration has been given to the activation in the context of extracellular matrix (ECM) glycoproteins. Indeed, the possibility that ECM proteins may have a determining role in lymphocyte adhesion, tissue localization, and function is now more plausible in view of the growing evidence indicating that integrins and other T cell antigens bind ECM components and exert synergistic effects on T cell activation (1–5). In addition to the role of adhesion molecules in antigen-dependent events, they also mediate antigen-independent events that lead to graft damage and remodelling.

II. ADHESION MOLECULES IN CELL-CELL INTERACTIONS IN GRAFT REJECTION

In vitro interactions between circulating blood cells and endothelial cells are largely controlled by four families of adhesion molecules: the selectins, the integrins, the immunoglobulin supergene family and CD44 (6). The selectins are involved in the initiation of adhesion of circulating leukocytes to both high endothelial venules in the lymphoid organs and to activated endothelial cells in inflamed tissues. L-selectin functions as a lymphocyte recirculation receptor and contributes to granulocyte and lymphocyte emigration at sites of tissue inflammation. E-selectin is transiently expressed on endothelial cells several hours after stimulation with inflammatory agents such as IL-1 and mediates a granulocyte adhesion pathway distinct from that mediated by ICAMs and integrins. P-selectin is stored in platelet α granules and endothelial Weibel-Palade bodies and is rapidly mobilized to the cell surface after stimulation by products of the clotting cascade. Interactions of the selectins with their receptors result in slowing down and rolling of leukocytes along the endothelium, allowing the cells to engage in other adhesion events and further activation.

The integrins are a family of transmembrane molecules comprised of two noncovalently linked polypeptide chains, the α and β chains (2), and are expressed on a variety of cell types. The extracellular domains of the molecules bind to matrix glycoproteins, complement components, and proteins on the surface of other cells, and they may serve as signal transduction molecules and mechanoreceptors. The cytoplasmic domains of the molecules interact with cytoskeletal components, such as vinculin, talin, actin, α actine, and tropomyosin; coordinate the binding of extracellular proteins to the cytoskeleton; and are responsible for signal transduction (5). The β_2 integrins, also known as the CD11/CD18 or leucam family, were identified by monoclonal antibodies that blocked adhesion-dependent lymphocyte functions, such as killing of target cells by cytotoxic lymphocytes. Both CD11b and CD11c containing integrins mediate leukocyte attachment to endothelial cells, extravasation, and function as complement receptors on phagocytic cells.

Most immunoglobulin gene superfamily members are integral plasma membrane proteins with widely divergent cytoplasmic tails with no homology to one another. Members of this family include the T cell receptor, the MHC molecules, CD2, CD3 γ (δ,ϵ), CD4, CD8, Thy-1, FcRII, NCAM, ICAM, VCAM-1, and PDGF receptors. While the selectins are mostly involved in the slowing down of circulating blood cells and the integrins strengthen the cellular interactions, the Ig supergene family members confer fine specificity to the cellular interactions.

The CD44 molecule is a widely expressed cell surface protein of 90 kDa with structural homology to cartilage link proteins. Anti-CD44 monoclonal antibodies inhibit lymphocyte binding to high endothelial venules and activated endothelium.

Studies on T lymphocyte-endothelial cell interactions have shown that the first step in the interaction is very likely mediated by the L- and E-selectins, followed by adhesions that are controlled by multiple integrin adhesion pathways. The importance of each of these pathways is dependent on the differentiation and activation state of the T cells, as well as the activation state of the endothelium (7). Naive and memory T cells both express LFA-1 and VLA-4, but memory T cells have severalfold higher levels of expression of these molecules (8). Acute activation of T cells results in increased binding affinities of both LFA-1 and VLA-4 without quantitative changes in their level of expression. At the endothelial cell level, activation by pro-inflammatory lymphokines results in rapid increases in expression of ICAM-1 and de novo expression of both VCAM-1 and ELAM-1. Based on in-vitro inhibition studies with monoclonal antibodies it has been shown that both resting and activated T cells adhere to activated endothelial cells through LFA-1, VCAM-1, and the adhesion pathway mediated by ELAM-1.

III. ADHESION MOLECULES AND EXTRACELLULAR MATRIX LIGANDS

The most extensively characterized ECM glycoproteins are fibronectin (FN), laminin (LN), and the collagens (COLL), molecules that are crucial in adhesion and activation (1–3). LN, a 900-kDa trimeric glycoprotein with a crosslike structure, is involved in cell attachment/locomotion and regulation of growth. The positive charges of the six continuous amino acids in the carboxyl part of the molecule may explain earlier findings suggesting the importance of charge in cell migration through the high endothelial venules (HEV) of postcapillary veins in lymph nodes (9,10). Similarly, FN and COLL, dimeric glycoproteins of the vascular basement membrane and interstitium, express domains that mediate cell-ECM interactions (1,11). As LN is part of a complex interstitial matrix that includes FN and COLL, the ECM may constitute a unifying principle for lymphocyte migration through the endothelium and distinct tissue microenvironment (12,13).

The cellular recognition of and adhesion to distinct ECM components is coordinated primarily by β_1 integrin molecules (VLAs) and cell surface proteoglycans (Table 1). Members of the VLA β_1 (CD29) subfamily contain a common β_1 chain associated with at least 10 distinct α chains (α_1 to α_9 and α_v). These molecules mediate the binding of T and B lymphocytes, among

Table 1 Leukocyte Adhesion Receptors and Their ECM Ligands

Receptor/subunit	Ligands
β_1 INTEGRIN (VLA) FAMILY	
VLA-1 (CD49a; $\alpha_1\beta_1$)	LN (COLL)
VLA-2 (CD49b; $\alpha_2\beta_1$)	COLL (LN)
VLA-3 (CD49c; $\alpha_3\beta_1$)	FN, LN, COLL, epiligrin, entactin
VLA-4 (CD49d; $\alpha_4\beta_1$)	FN, (IIICS, V25), VCAM-1
VLA-5 (CD49e; $\alpha_5\beta_1$)	FN
VLA-6 (CD49f; $\alpha_6\beta_1$)	LN, merosin, kalinin
CD51 ($\alpha_7/\alpha_8/\alpha_9/\alpha_v\beta_1$)	FN, VN, LN
β_2 INTEGRIN (LEUCAM) FAMILY	
LFA-1 (CD11a; $\alpha_L\beta_2$)	ICAM-1, 2, 3
Mac-1 (CD11b; $\alpha_M\beta_2$)	FB, iC3b, ICAM-1, Factor X
p150/95 (CD11c; $\alpha_X\beta_2$)	FB, iC3b
β_3 INTEGRIN (CYTOADHESIN) FAMILY	
IIB/IIIA (CD41; $\alpha_{IIB}\beta_3$)	FN, FB, VN, VWF
VNR (CD51; $\alpha_V\beta_3$)	VN, FB, VWF, LN, TSP, FN, denaturated COLL, TN, OPN
CD49f ($\alpha_6\beta_4$)	LN, kalinin
CD51 ($\alpha_v\beta_5 = \beta_x = \beta_s$)	VN
CD51 ($\alpha_v\beta_6$)	FN
CD49d ($\alpha_4\beta_7 = \beta_p$; LPAM-1)	FN, VCAM-1, Peyer's patch HEV (MAdCAM-1)
	OTHERS
CD44 (cartilage link protein)	HA, HEV
CD26 (serine protease)	COLL
CD73 (5′ nucleotidase)	LN

Abbreviations: COLL, collagen; FB, fibrinogen; FN, fibronectin; HA, hyaluronic acid; HEV, high endothelium venules; LN, laminin; MAdCAM-1, mucosal addressin cell adhesion molecule 1; VN, vitronectin; VWF, Von Willebrand's factor; TSP, thrombospodin.

other cells, to major ECM components. VLAs may thus play a major role in directing cell migration through tissues. Their expression on cells that cross endothelial barriers is important in controlling their localization at sites of inflammation. Indeed, cells that traverse lymph node postcapillary veins in vitro are more activated than noncirculating cells as judged by the expression of CD26 and ^{3}H-uridine incorporation (11). Moreover, memory cells express multifold higher amounts of VLAs than naive cells (14) and therefore may bind more efficiently to ECM. Unlike most VLA proteins, VLA-4 has two distinguishable binding domains: as a matrix receptor it binds to the alternatively spliced connecting segment domain of FN (CS-1) and as a "homing receptor" it binds to endothelial VCAM-1 (Table 1).

Thus, VLA-4 may mediate by distinct molecular mechanisms cell-ECM and cell-cell adhesion and recirculatory events, respectively. Although both VLA-4 and VLA-5 represent T cell receptors for FN (2,14), adhesion between B cells and FN depends exclusively on VLA-4. VLA-3 and VLA-5 interact with the arginyl-glycyl-aspartic acid (RGD) FN binding region. Of the three β_2 (CD18) subfamily members (LFA-1, Mac-1, and p150/95) involved in cell-cell aggregation and function, only Mac-1 binds to fibrinogen (FB). Some murine T cell lines express a homolog of β_3 integrin, which mediate the binding to FN, FB and vitronectin (VN).

Other leukocyte adhesion receptors for ECM include CD44, CD26, and CD73 (2,14). CD44 (Pgp-1, Hermes) mediates binding of lymphocytes to high endothelial venules and interacts with its ligand hyaluronic acid (HA), one of the carbohydrates in which ECM proteins are immersed. HA, in conjunction with CD3/TCR-mediated signals, is costimulatory for human T cell proliferation, for IL-2 production by T helper cell clones, and for activation of the lytic machinery of cytotoxic T cell clones (15). T cell activation is associated with a transient ability to bind HA, a process that depends on the interaction with CD44. CD26 interacts with FN, and may also mediate adhesion to COLL, whereas CD73 interacts with FN and LN (2,14). Both antigens are expressed on activated T and B cells, and monoclonal antibodies against CD26 and/or CD73 exert costimulatory effects on T cell proliferation in vitro.

Interest in the regulation of lymphocyte entry into the lymph has focused primarily on the interactions of lymphocytes with postcapillary venules found in lymph nodes of higher vertebrates (16), but in species without lymph nodes, lymphocytes also enter the lymph; furthermore lymphocyte recirculation is well established in fetal life (17). Thus, lymphocyte trafficking and positioning are regulated by specialized lymphocyte-endothelial interactions as well as other interactions. There is a selective distribution of some ECM components in lymphoid organs, as recently shown for tenascin in the T cell-dependent zones of the human thymus (18), which may serve as a basis for a preferential T cell migration through these zones. It has been suggested that abnormal expression of tenascin in nonlymphoid organs may serve as a basis for abnormal T lymphocyte migration in autoimmunity (M. Chilosi, personal communication). Additional evidence suggesting abnormal cell-ECM interactions in autoimmunelike lesions comes from the recent observations in TGF-β_1-deficient mice (19) and bacterial cell wall-induced erosive polyarthritis in rats (20); increased leukocyte adhesion in these models has been corrected by the administration of synthetic FN peptides.

Changes in the expression of ECM components have been also described during the course of rejection of cardiac allografts in rats (21,22). The earlier demonstration that LN is not confined to basement membranes but is also present in reticular fibers (23), led us to test the effects of an anti-LN antibody on the migration of ^{111}In-oxine-labeled specifically sensitized lymphocytes in T cell-deficient rats (24); in normal animals, putative LN-binding sites in lymph node stroma may be fully occupied. The T cell-deficient recipients of cardiac allografts received an injection of rabbit anti-rat LN antibody 30 min before transfer of radiolabeled lymph node cells and were killed 6 hours later. Lymphocyte recovery from the peripheral lymph nodes and from the grafts was significantly diminished compared to controls pretreated with normal rabbit serum, while no changes were detected in the amount of radioactivity recovered from host spleens, native hearts, or other organs (Fig. 1). Thus, adoptively transferred cells failed to enter the lymph nodes and the allograft, possibly "blinded" by prior administration of the antibody. This may be linked to the fact that several integrins bind LN (2,14), which in turn suggests that LN may play a role in lymphocyte homing. It is possible that during extravasation of lymphocytes, LN and/or other basement membrane components not only serve as a substratum for adhesion and migration but also regulate cell activation. Thus, T and B lymphocytes are equipped with an array of surface molecules that can determine their binding to vascular endothelium, to each other, and to distinct components of the ECM. In vivo lymphocyte migration studies have shown that LN-coated surfaces or "matrigel" (25), an ECM substrate that contains 60% LN and that resembles biochemically, structurally, and biologically normal basement membrane, enhances lymphocyte migration. These haptokinetic effects were blocked in a concentration-dependent manner by anti-LN but not by anti-Con A Fab fragments (Fig. 2).

IV. ROLE OF CELL ADHESION MOLECULES IN ISCHEMIA-REPERFUSION INJURY IN VASCULARIZED GRAFTS

Although the focus in this chapter is on the role of cell adhesion molecules in organ graft rejection, all vascularized organ grafts experience variable degrees of ischemia. As reviewed elsewhere in this book, ischemia in non-transplanted organs may cause alterations in adhesion molecules on parenchymal and vascular cells, leading, for example, in the kidney, to cell detachments, tissue infiltration with leukocytes, and vascular congestion (26,27). Polymorphonuclear neutrophils (PMN) are crucial in the injury associated with reperfusion of ischemic tissues (28). An early and rate-limiting event in PMN accumulation in postischemic tissues is their ability to firmly adhere to endothelial cells (29); indeed, PMN adherence may

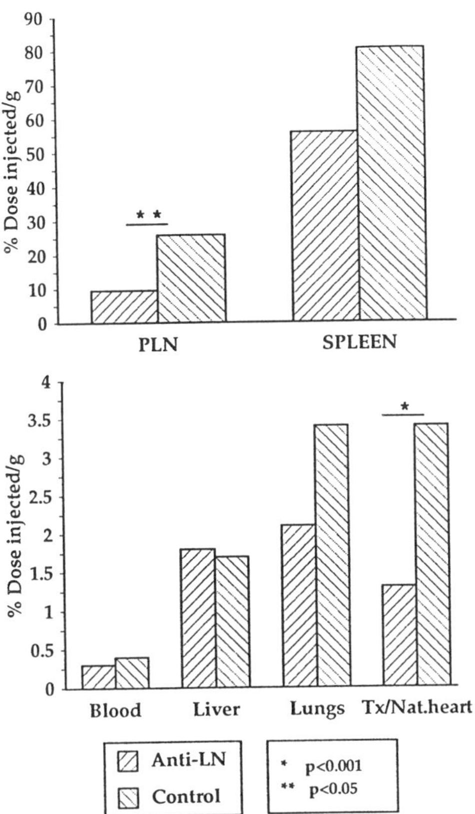

Figure 1 Migration of ^{111}In-labeled specifically sensitized peripheral lymph node cells in syngeneic T cell-deficient rat recipients of cardiac allografts pretreated with anti-LN antibody or control serum. Note: significantly diminished recovery of adoptively transferred cells from peripheral lymph nodes (PLN) and cardiac allografts (Tx) in recipients conditioned with anti-LN antibody as compared to controls. (Adapted from ref. 24.)

increase as much as 35-fold, depending on the degree and duration of ischemia.

The relative contribution of the adhesion molecules on PMN and on the endothelial cells that leads to injury following reperfusion of ischemic tissues has become a focus of attention. Pretreatment of animals with anti-CD18 antibody prevented by 95% PMN adhesion and emigration in the ischemic mesenteric venules, whereas pretreatment with anti-CD11a or anti-CD11b antibodies reduced PMN adhesion by 65% and emigration by 50%;

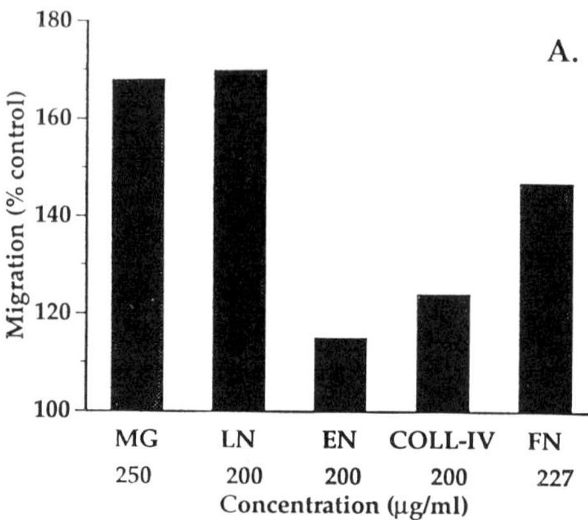

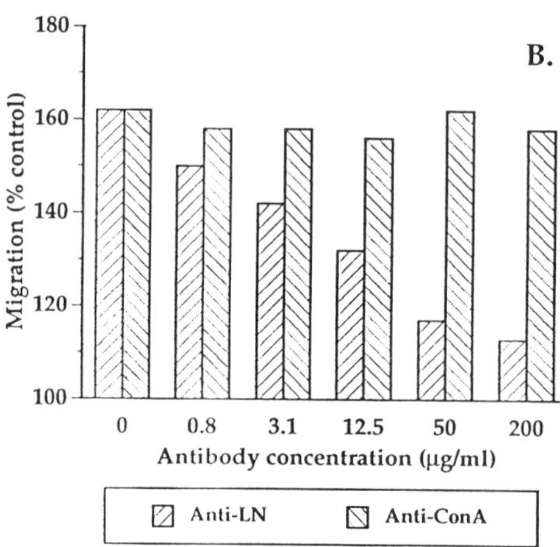

Figure 2 (A) Migration of lymphocytes on surface coated with matrigel and its components. Wells of migration plates were coated with matrigel (MG), laminin (LN), entactin (EN), type IV collagen (COLL IV), and fibronectin (FN) at concentrations indicated. Note: MG promotes lymphocyte migration in vitro, and LN is its component responsible for haptokinesis. (B) Effects of anti-LN treatment on lymphocyte migration in vitro. Anti-LN or anti-Con A Fab fragments, at concentrations indicated, were added to the wells coated with MG, and incubated before conducting the migration assay. Note: The haptokinetic effect of MG may be partially blocked in a dose-dependent manner by anti-LN but not by anti-Con A Fab fragments, supporting the role of LN in promoting cell migration in vitro. (Adapted from ref. 25.)

the anti-ICAM-1 (CD54) antibody reduced adhesion and emigration by 40% and 50%, respectively (30). Although this approach has not been successful in preventing ischemic renal failure in a rabbit model (31), a more recent study in rats demonstrated that an antibody against ICAM-1 markedly attenuated the course of renal failure (32). An antibody to the α subunit of LFA-1 provided modest protection but the combination of anti-LFA-1 together with subprotective doses of anti-ICAM-1 provided complete protection. The role of adhesion molecules in ischemia/reperfusion-induced leukocyte adhesion has furthermore been studied in in vitro models that utilized monolayers of cultured human umbilical vein endothelial cells (33). Endothelial cells exposed to anoxia/reoxygenation release soluble factor(s) that augment the expression and/or activation of CD11b/CD18 and increased PMN adherence as a result from interactions between CD11a/CD18 and CD11b/CD18 integrins on PMN and CD54 on endothelial cells. In liver transplants, preoperative ischemia may induce diffuse hepatocellular induction of ICAM-1 (34).

Preventing in vivo PMN adhesion by immunoblocking either the CD18 complex or its CD54 ligand reduced myocardial infiltration and edema, and markedly improved reflow and ventricular function after heart preservation and transplantation (35).

V. ROLE OF CELL ADHESION MOLECULES IN ORGAN ALLOGRAFT REJECTION

A. Hyperacute Rejection

Hyperacute rejection is the failure of a vascularized organ transplant within minutes or hours after transplantation. Such rejections are encountered following transplantation of a discordant xenograft or following transplantation of an allograft in a recipient with high titers of donor directed antibodies against graft antigens as a result of previous allosensitizations. In the xenogeneic situation, hyperacute rejection occurs in the presence or absence of donor-specific antibodies; the hyperacute rejection in the absence of high titers of donor-specific antibodies is believed to result from lack of inactivation of inordinate complement activation on the xenogeneic surface. The histopathology of hyperacute rejection is characterized by intravascular platelet aggregation, thrombosis, and influx of PMN, leading to ischemic necrosis. These lesions are initiated by the binding of recipient antibodies to graft endothelial cells, followed by complement activation, vasoconstriction, endothelial cell activation, and damage. In some xenogeneic combinations such as guinea pig hearts transplanted into rat recipients, this sequence of events occurs within 15 min and results from direct complement activation. Immunohistological studies of such grafts have shown focal induction of expression of P-selectin on the surface of arterioles and venules and occasional capillary endothelial cells as well as on the surface of platelets within the microthrombi of the graft (36), consistent with translocation of the proteins from the Weibel-Palade bodies to the endothelial cell surface.

Incubation of porcine endothelial cells with human TNF-α for 4 hours results in high levels of porcine E-selectin mRNA and a marked increase in binding to human neutrophils. Expression of recombinant porcine E-selectin in COS cells results in surface expression of the protein and increased binding to human neutrophils, showing that porcine E-selectin mediates adhesive interactions between porcine endothelial cells and human leukocytes (37). In vitro studies of porcine renal epithelial cells and human lymphocytes have furthermore shown that mitogen-activated lymphocytes use CD11a, CD18, and CD49d for cell adhesion with porcine cells (38). Thus, adhesion molecules may operate in a cross-species manner and may contribute in the pathophysiology of xenograft rejection.

B. Acute Rejection

Acute rejection is a form of tissue inflammation that depends on T cell recognition of alloantigens, trafficking of immune and inflammatory cells into the graft, and execution of cell-mediated effector functions. Cell kinetic studies have shown more than one log order increases in the influx and outflow of leukocytes to and from the graft during acute rejection (39) in conjunction with increased binding affinities of both the lymphocytes and the graft vascular endothelium (38). Immunohistologic studies of renal transplants with acute rejection have shown upregulation of various cell adhesion molecules. Expressed in small amounts on endothelium and in glomeruli of normal kidneys, ICAM-1 was upregulated during acute rejection on graft endothelial and tubular cells as well as on infiltrating mononuclear cells (41), and a similar induction pattern has been found for VCAM-1 (42–47). Induced expression is usually found in association with infiltration with $CD45^+$ and $CD3^+$ leukocytes, suggesting that they may contribute to the recruitment of immune and inflammatory cells and render the graft more susceptible to cell-mediated injury (47). In vitro studies of pooled sera from highly sensitized transplant patients have shown that such sera contain IgG antibodies against class I MHC-like molecules that induce up to 14-fold increases in endothelial ICAM-1 expression (48), an observation that may explain the increased graft rejection rate in this patient group. Immunohistology of renal transplants with acute rejection or CsA nephrotoxicity have shown that the interstitial infiltrate in both conditions consists of LFA-1^+

cells in combination with increased ICAM-1 expression on endothelial cells while VLA-4/VCAM-1 staining was mostly present in acute rejection but absent in CsA toxicity. VCAM-1 induced on renal epithelial cells is functionally capable of binding VLA-4, thereby enhancing the adhesion of potentially graft-damaging lymphoid cells (49,50).

A circulating form of the membrane bound ICAM-1 molecule has recently been identified and characterized in normal human serum. It is structurally similar to the membrane-associated form except that it lacks the cytoplasmic tail and the transmembrane region. Its concentration may increase in serum during inflammation, especially when the inflammatory process affects the vascular bed but circulating sICAM-1 levels have no diagnostic value in renal transplant patients as the baseline levels are elevated directly after transplantation, very likely as a result of ischemic tissue damage (51,52). Similarly, no significant changes were found in circulating VCAM-1 levels in acute rejection although the mean E-selectin serum concentration increased significantly in patients with acute rejection (52). On the other hand, urine levels of ICAM-1 and sVCAM-1 are significantly higher in acute, steroid-resistant rejection compared with steroid sensitive rejection or patients with stable graft function (53). Immunohistological comparisons showed correlations between soluble adhesion molecules in the urine and their tubular expression.

Several studies have investigated the expression of cell adhesion molecules over time in endomyocardial heart transplant biopsies (54–58). Intense ICAM-1 and VCAM-1 staining was found in all biopsies during the first 3 weeks after transplantation, in conjunction with elevated serum levels of troponin T, a sensitive marker of ischemic myocyte injury. Baseline ICAM-1 and VCAM-1 expression returned to normal within 3 to 4 weeks, as did serum troponin levels in all patients who did not develop rejection (57). All rejection episodes were associated with intense ICAM-1 staining while most acute rejections were associated with VCAM-1 and ELAM-1 staining (57,59). Weak endothelial LFA-3 induction has been observed during acute rejection (54,55) while E-selectin and P-selectin expression was rarely observed (57–61). Persistence of increased ICAM-1 and VCAM-1 staining after treated rejection episodes predicted a recurrent rejection episode within 2 months. Induced VCAM-1 and ELAM-1 expression has also been reported in association with cytomegalovirus infection (59).

Data from animal experiments have suggested that radiolabeled anti-ICAM-1 antibody can be used for imaging purposes to detect cardiac allograft rejection (62). Serum levels of sICAM-1 do not correlate with endomyocardial biopsy results (56).

Immunohistological studies of rejected human corneal transplants have shown focal ICAM-1 expression on epithelial cells, keratinocytes, and corneal and vascular endothelial cells, particularly at sites of dense leukocytic infiltration. E-selectin was found on vascular endothelial cells in the stroma while VCAM-1 expression was found predominantly on the inflammatory macrophages (63).

ICAM-1 and VCAM-1 expression on hepatocytes and endothelium respectively increases at the early stages of acute liver graft rejection (64–66). ICAM-1 expression also increases in association with CMV or bacterial infections and is therefore not a very specific marker for acute rejection (65). Elevated bile levels of sICAM-1, rather than serum sICAM-1 levels, correlate closely with rejection, and the levels decrease with steroid treatment (64). Finally, immunohistologic and Northern blot surveys of small bowel transplants have shown increased expression of ICAM-1 and CD44 in acute rejection (67–69).

Exposing the recipient to graft cellular and ECM constituents represents an unique experimental system to examine the role of ECM in an immune response whose effector arm is dependent on the migration of alloreactive cells into the foreign tissue. Immunohistochemical stainings of rat cardiac allografts have shown markedly increased intermyocyte and endothelial deposition of both LN and FN, preceding cellular infiltration (21,22), while only marginal and transient increases of ECM deposition were found in syngeneic grafts (22). Well-functioning allografts in recipients treated with an anti-TNF-α antiserum (70) or rapamycin (Kupiec-Weglinski, unpublished) had almost normal LN and FN levels. Increased FN expression has also been reported in a pig cardiac transplant model (71). The earliest phase of inflammation is characterized by extravasation and deposition of plasma FN (72,73), and the earliest cells appearing at the graft site are monocytes/macrophages with FN mRNA and protein expression (Fig. 3) (70). Macrophages are one of many cell types capable of synthesizing FN (74–77).

Rejecting cardiac allografts in rats show progressive cellular infiltration with ED-1$^+$ macrophages, OX-8$^+$ T cytotoxic lymphocytes, and VLA-4$^+$ (CDw49d) cells (70). As FN expression by macrophages occurs within the same interval as cellular infiltration, FN may act as an ECM component "signal" for selective homing of recirculating lymphocytes into the graft. Strong expression of FN, and to a lesser extent of tenascin, and COLL type VI, may be found in connective tissue around cellular infiltrates in human liver and kidney grafts with acute or chronic rejection (78,79). Cells infiltrating rejecting cardiac allografts in rats are consistently detected in close proximity to FN- rather than LN-containing interstitial and perivascular sites, as shown by immunohistochemistry and confirmed by three-color staining and confocal microscopy (22). These data support the hypothesis that FN plays a key role as an in vivo adhesive or chemoattractant factor

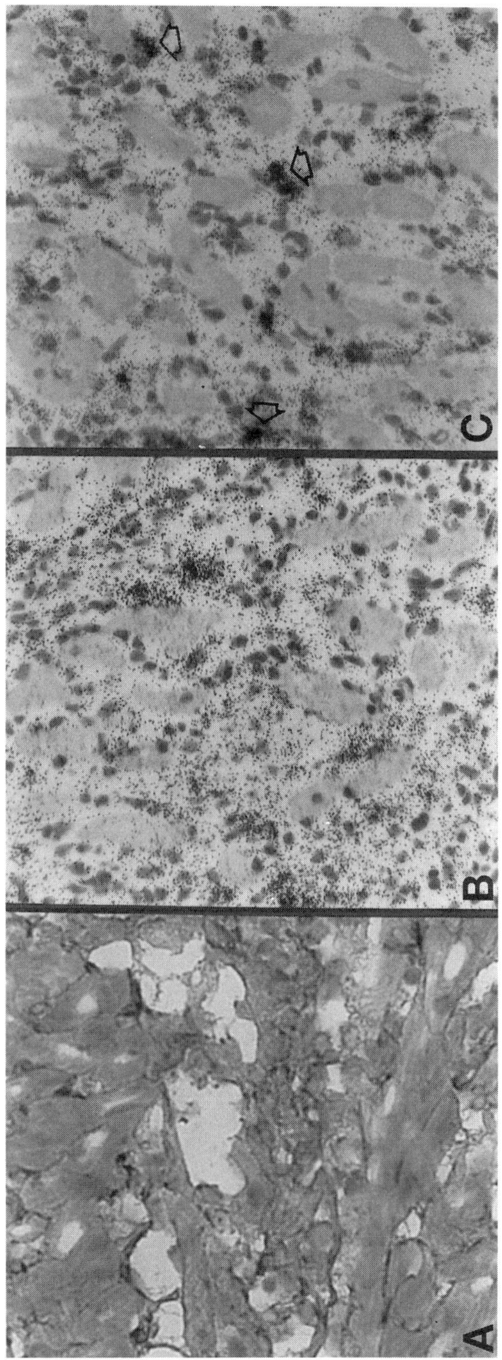

Figure 3 Immunohistochemical and in situ hybridization analysis of rat cardiac allografts undergoing acute rejection at day 4 after transplantation. (A) Dense interstitial and perivascular deposition of FN detected by immunoperoxidase and staining with a rabbit antirat FN antibody. (B) In situ hybridization for FN mRNA expression by cells in the myocardium using a probe that reacts with all forms of FN. (C) In situ hybridization for lysozyme mRNA demonstrates that the large proportion of interstitial cells are macrophages. (Adapted from ref. 70.)

for sensitized lymphocytes. As distinct FN isoforms exist, further research is needed to evaluate whether the different FN splicing variants mediate lymphocyte recirculation and facilitate local cell recruitment.

The preferential interaction of mononuclear cells with FN may have consequences that go beyond a simple function in cell positioning. Indeed, FN type III repeats found in human MHC class II region molecules not only have antigen presenting functions, but may also costimulate activation and signal transduction in T cells (80). There is also evidence for a FN-mediated role in the generation of specific cytotoxic T cell responses (81).

CsA treatment in rats may lead to graft "endothelialitis"—i.e., inflammation of the vascular intima with T cell infiltration in the subendothelial space, influx of smooth muscle cells, and accumulation of COLL type I and FN (82,83). Altered T cell-ECM interactions could play a role in the pathogenesis of CsA-induced transplant arteriopathy. CsA therapy in renal transplant patients with stable graft function inhibits the costimulatory effects of COLL type I and FN, but not of COLL type IV on CD-3-mediated T cell proliferation, while acute rejection is associated with augmented co-stimulation by all three ECM proteins (Fig. 4) (84). CsA-induced inhibition of costimulation by COLL type I and FN, while maintaining strong COLL type IV-dependent activation, may facilitate the passage of T cells through basement membranes to the subendothelial space and their local proliferation. Thus, CsA may reduce interstitial T cell activation and infiltration by abolishing local T cell-ECM interactions while the strong costimulatory signals derived from basement membrane-associated COLL type IV may trigger T cells migrating through the endothelium and their local proliferation. Indeed, the degree of mononuclear cell infiltration in acutely rejecting human renal allografts correlates with enhanced reactivity of graft infiltrating T cells to costimulatory signals mediated by COLL type IV (85). The majority of the infiltrating T cells were, in fact, the VLA-2^+ memory ($CD45RO^+$) cells, and only a negligible fraction of peripheral blood T cells expressed VLA-2, a classic COLL receptor. Moreover, physically damaged ECM may induce cytokine secretion by T cells, recruiting additional immune cells, and perpetuating local tissue injury (86).

Data from in vitro systems indicate that cell-ECM interactions may trigger production of various cytokines (87,88). Thus, adherence of macrophages to FN induces transcription of GM-CSF, while adhesion of monocytes to FN and COLL triggers the induction of CSF-1 and IL-1 receptor antagonist, or IL-1 and TNF-α genes, respectively. Moreover, certain cytokines modify cell-adhesive ECM constituents, with TGF-β stimulating fibrosis in vivo, IFN-γ enhancing macrophage binding to LN, and TNF-α promoting cell adhesion to FN in vitro. Alternatively, ECM proteins regu-

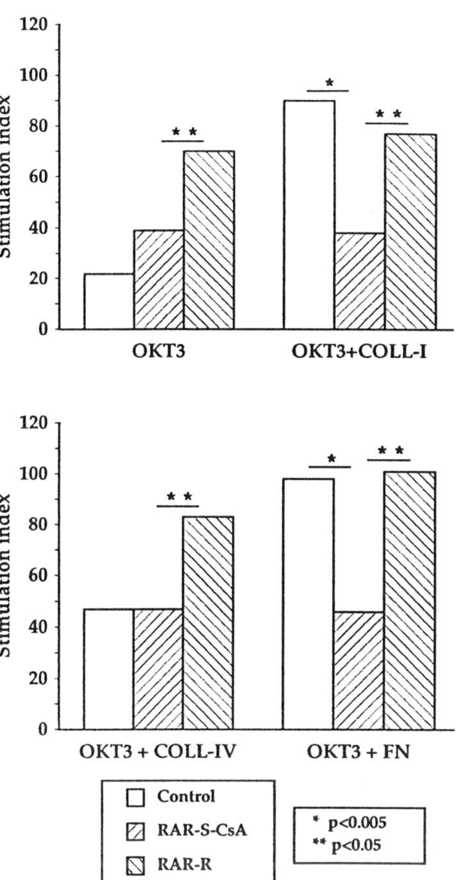

Figure 4 Costimulation of CD3-mediated T cell proliferation in vitro by COLL type I, IV, and FN in distinct groups of renal graft recipients (RAR), vs. proliferation of T cells from normal healthy donors. Note: In patients with a stable graft function (RAR-S), CsA therapy markedly decreases co-stimulatory effects of Type I COLL and FN, but not of Type IV COLL as compared to controls. Rejection (RAR-R) is associated with augmented T-cell responses to OKT3, and increased costimulation with all three ECM components. (Adapted from ref. 84.)

late secretion of TNF-α in culture, and may bind and present cytokines to other cells, leading ultimately to their sequestration from the circulation. Indeed, increased intragraft transcription of MIP-2 and KC genes—i.e., analogues of human IL-8 with neutrophil chemotactic/activation properties coincides with enhanced expression of FN in rejecting rat cardiac allografts (89). It is conceivable that, analogous to the in vitro situation, certain ECM components may present cytokines to cells, which ultimately leads to their sequestration from the circulation. Moreover, the effects of cytokines can be affected by the ECM milieu in which most cells are normally embedded (88). This modulation, called "control by context," emphasizes the role of the tissue environment in the control of lymphocyte traffic. It has not yet been proven whether such regulatory pathways involving ECM and cytokine networks operate in vivo. However, the observation that some ECM proteins show partial homology with cytokines and their receptors (87,88) supports such a possibility.

If T cell activation in vivo requires costimulation by ECM proteins, factors that influence ECM could influence the activation sequence. Some support for this possibility comes from cases of persistent cellular infiltration in well-functioning allografts in immunosuppressed rats (90,91). Thus, while C3 and fibrin deposition in cardiac grafts transplanted into presensitized recipients treated with CsA was diminished, the extent and degree of hypercellularity was similar to that in rejecting allografts, as was the expression of MHC antigens (90). Similarly, unaltered expression of MHC antigens coinciding with augmented cellular infiltration has been reported in rats immunosuppressed by a preoperative blood transfusion (91). Anti-IL-2R monoclonal antibodies eliminate or inhibit IL-2R$^+$ cells, but such antibodies did not inhibit graft infiltration by class II-positive cells and activated mononuclear phagocytes, the cells intimately associated with the deposition of α-chain of crosslinked fibrin (90). Only combined therapy of anti-IL-2R antibody and CsA, synergistically depressed these cellular components and often led to graft acceptance (92).

Another ambiguity of transplantation immunology may begin to unravel by relating ECM effects in the context of T cell activation and different rejection times for various organs transplanted across identical MHC barriers. For example, skin and liver allografts elicit very different immune responses in rat recipients, the former being most difficult to maintain despite continuous immunosuppression (93), and the latter surviving for long periods with minimal or no therapeutic intervention (94). As the skin has a very high proportion of COLL content by weight (ca. 70%) compared with the liver (<5%) (95), it is conceivable that the ability of these organs to provide costimulatory signals is markedly different.

C. Chronic Rejection

Chronic rejection is the most important cause of transplant loss after the first year. Its histopathologic features consist of atherosclerotic vessel wall changes, interstitial fibrosis and a variety of organ-specific lesions. Although the designation "rejection" implies that donor-specific immune mechanisms are pivotal in the progressive demise of structure and function, the protracted time course of the rejection reaction(s) involved allows the emergence of nonimmune adaptations in response to the progressive loss of renal mass as well as tissue remodeling as part of ongoing tissue repair mechanisms (96,97). The importance of each of these mechanisms is presently only partly understood.

Immunohistologic studies of organs undergoing chronic rejection have shown increased expression of cell adhesion molecules that may be involved in the regulation of the allogeneic immune responses and tissue inflammation, as in acute rejection (98,99). Levels of sICAM in urine in renal transplant patients with chronic rejection are higher than in patients with a stable graft function (53).

VI. CELL ADHESION MOLECULES—TARGETS FOR IMMUNOSUPPRESSIVE THERAPY

Consistent with the observations of increased expression of intragraft cell adhesion molecules in the early phase of rejection, treatment with antibodies against these structures should be of therapeutic advantage. Ideally, such therapies should block in vivo interactions between recipient immune or inflammatory cells and graft parenchymal or vascular cells or the graft ECM, as well as the interactions between immune cells themselves. Anti-LFA-1 antibodies exert immunosuppressive effects in human bone marrow transplant recipients (100). However, a monoclonal antibody against the α chain of LFA-1 molecule (CD11a) was not effective in reversing ongoing kidney graft rejection in humans (101) although the same antibody was very effective in preventing early rejection when used prophylactically (102). The treatment was well tolerated and the low level of host sensitization did not result in any deleterious side effects. A recent Phase I trial has established the dosing schedule and the clinical safety of an anti-ICAM-1 (CD54) antibody in patients who received renal transplants at high risk for delayed function (103). Graft survival was 78% at 16 to 30 months in the antibody treated group versus 56% in recipients who received conventional therapy. Both studies did not find depletion of circulating cells and postulated that the efficacy of treatment resulted very likely from inhibition of sensitization and/or leukocyte adhesion.

A short prophylactic CD54-directed therapy has been also tested as the sole immunosuppressive drug in cynomolgus renal allograft recipients (104). In the treated group, ischemic damage was limited and graft survival was prolonged from 9 days to about 24 days. Biopsies during antibody therapy showed decreased T cell infiltration and endothelial inflammation but no major changes were found in the frequency of circulating T cells compared to rejecting controls. In another primate study, anti-LFA-1 (CD11a) antibody alone or in concert with anti-LFA-2 (CD2) antibody, but not anti-LFA-2 antibody alone, significantly prolonged the survival of skin grafts in cynomolgus monkeys, with one animal bearing a viable allogeneic skin for >185 days (105). Antibody therapy did not affect circulating T cell levels and skin biopsies revealed comparable T cell infiltration in both treated and control animals.

The effects of targeting various components of the cell adhesion molecule network have furthermore been tested in a number of murine models. The efficacy of anti-P-selectin antibodies was tested in a rat model of accelerated cardiac allograft rejection in which the recipient was preimmunized with a donor-type skin graft. Treatment with such an antibody immediately prior to vascularization did not prolong graft survival while the same antibody gave a modest prolongation of graft survival in a discordant xenograft model (106). Anti-ICAM-1 antibody given to mice at the time of corneal transplantation delayed the influx of cells to the graft site, which was associated with slightly improved graft viability in one study (107) but not in another (108). The antibody did not reduce the incidence of rejection, but it prolonged graft survival time and inhibited both the cytotoxic T cell and delayed-type hypersensitivity responses to donor antigens. On the other hand, anti-LFA-1 antibody reduced the incidence of graft rejection from 90% in untreated recipients to 47% in treated mice. Neither anti-ICAM-1 nor anti-LFA-1 antibody prevented the rejection of corneal grafts in previously immunized mice (108). An anti-ICAM-1 antibody administered for 5 days to rats with a small bowel transplant delayed graft infiltration by inflammatory cells (109). Of interest is a study that showed that infusion of mouse anti-rat ICAM-1 into the portal vein of the liver graft before transplantation resulted in *accelerated* vascular rejection (110).

A single injection of anti-LFA-1 (CD11a) antibody uniformly prevented thyroid allograft rejection in mice, with 50% of recipients retaining intact grafts for >400 days (111). Similarly, a perioperative short course of anti-LFA-1 (CD11a) but not anti-LFA-2 (CD2) antibody treatment induced unresponsiveness to islet allografts in mice without causing any adverse effects (112). In contrast, treatment with anti-LFA-1 only resulted in a very modest prolongation of pancreas graft or pancreas islet graft survival in rat models (113,114). Similarly, a short treatment with either an anti-LFA-1 or an anti-ICAM-1 antibody prolonged the survival of primarily vascularized heart allografts in mice only marginally but combined therapy with both antibodies resulted in long-term graft acceptance in conjunction with donor-specific unresponsiveness (115). A similar approach in rats either failed to prolong the survival of vascularized cardiac grafts or resulted in a modestly prolonged graft survival although the combination of these antibodies with subtherapeutic doses of CsA resulted in permanent graft acceptance (116,117).

Intravenous administration of an antisense oligonucleotide specific for ICAM mRNA inhibits its expression and is associated with a modest prolongation of cardiac graft survival in mice. On the other hand, the antisense oligonucleotide in combination with a monoclonal antibody against LFA-1 induced prolonged graft survival and donor-specific tolerance (118).

Treatment with anti-LFA-1 antibodies alone potently and effectively prolonged the survival of nonprimarily vascularized (ear-pinna) murine heart grafts (119). The LFA-1/ICAM-1 interactions have furthermore been explored in a concordant xenograft model of rat pancreatic islets transplanted underneath the kidney capsule of diabetic mice. While anti-mouse or anti-rat ICAM-1 alone or in combination with anti-mouse or anti-rat LFA-1 respectively gave no or at best a slight prolongation of graft survival, the combination of anti-mouse LFA-1 (recipient) and anti-rat ICAM-1 (donor) gave a significant prolongation of graft survival from 9.0 ± 2.2 days in untreated recipients to 32.1 ± 18.4 days in treated animals (120).

Finally, moderately potent immunosuppressive effects have been reported in mice following targeting of endothelial VCAM-1 (121,122). Anti-VCAM-1 antibodies abrogated acute rejection and prolonged cardiac allograft survival in a dose-dependent manner to about 20 days in one study and to 60 days in another study. Grafts from antibody-treated animals showed a mild cellular infiltrate comprised of T cells which did not express IL-2R or transcribe IL-2, IL-4 or IFN-γ mRNAs, although they did express IL-6 and TNF mRNAs. The most distinguishing histologic feature of long surviving grafts after anti-VCAM-1 therapy was the development of interstitial and perivascular fibrosis, distinct from the classic chronic rejection fibrosis patterns. However, the demonstration of persistent T cell infiltration in long-term cardiac allografts despite anti-VCAM-1 antibody treatment and the transient T cell infiltrate devoid of VCAM-1 in isografts (121–123) strongly suggest a role for VCAM-1-independent collateral adhesion system(s) to mediate lymphocyte sequestration at the graft site.

Data on the use of anti-adhesion antibodies in rat and other transplantation models are limited. A marginal prolongation of vascularized cardiac allograft survival was recorded in rat recipients conditioned with anti-ICAM-1 (CD54) and/or anti-LFA-1β (CD18) antibodies, but only in a weakly histoincompatible donor-recipient strain combination while neither antibody regimen was able to abrogate acute rejection across strong MHC barriers (124,125). In contrast, treatment of rats with anti-LFA-1 (CD11a) antibody only, extended cardiac allograft survival from 7 days to about 24 days (126). Interestingly, administration of anti-VLA-4 antibody was less effective (graft survival about 14 days), whereas the combination of both modalities was not more effective than anti-LFA-1 antibody alone. The latter may surprise, considering earlier described additive effects of blocking β_1 (VLA-4) and β_2 (LFA-1) integrins in the in vitro lymphocyte adhesion assay (127). Although each antibody alone prevented or delayed expression of MHC class II antigens on graft endothelium, only the combined antibody therapy prevented completely early intragraft vasculitis (126). Finally, the only study of rabbit cardiac grafts, assessed after 7 days of treatment with anti-CD18 antibody, recorded significant reduction in the degree of interstitial cellular infiltration, while anti-VLA-4 antibody treatment actually enhanced the interstitial cellular infiltration (128).

Blocking of β_1 (VLA-4) or β_2 (LFA-1) integrins each had a marginal effect on the survival of islet allografts transplanted underneath the renal capsule in the rat, but the combination of the two antibodies resulted in long-term graft survival in five of seven animals (113). Since the same combination of antibodies failed to induce long-term survival of cardiac grafts (124), it is conceivable that acute rejection of different organ grafts proceeds using different combinations of cell adhesion molecules in different organs, much alike the inflammation in nontransplanted organs.

Although the efficacy of cell adhesion-directed antibodies has not yet been fully elucidated, studies published to date support the role of these surface proteins as attractive targets for novel therapeutic regimens in organ transplantation. The somewhat conflicting in vivo effects reported may relate to differences in experimental models and/or differences in antibodies used. The latter relate to divergent binding characteristics, different epitope specificities, and differences in isotypes of the antibodies studied.

The use of antibodies has the disadvantage that they usually are immunogenic, leading to antibody formation in the recipient that may subsequently interfere with their biological activity. Alternative strategies such as the use of specific enzyme inhibitors or the synthesis of adhesion site-binding peptides, soluble receptor molecules, or genetically engineered antibody fragments to portions of adhesion molecules may become available to permit long-term treatment regimens.

Castanospermine (CAST) is a glucosidase I inhibitor that inhibits the processing of the N-linked oligosaccharide moieties of glycoproteins in the rough endoplasmic reticulum. CAST administered by continuous subcutaneous infusion results in a decrease in CD11a expression on peripheral blood cells as well as reduced expression of ICAM-1 on capillary and arterial endothelium and on graft-infiltrating cells (129). CAST reduces intragraft cell migration and prolongs kidney, heart, and pancreas graft survival in rats in a dose-dependent manner (129–132). Mycophenolate mofetil is the 2-morpholinethyl ester of mycophenolic acid (MPA) and has recently undergone extensive testing as a new immunosuppressive drug. MPA is a potent noncompetitive inhibitor of inosine monophosphate dehydrogenase and inhibits the de novo pathway of guanosine nucleotide synthesis. Because lymphocytes are critically dependent for their proliferation on de novo synthesis of purines while other cell types can utilize salvage pathways, MPA has a more selective effect on lymphocytes compared with other cells. Since GTP is also required for the activation of mannose and fucose for glycoprotein synthesis, MPA may also affect expression of cell adhesion molecules. In vitro adhesion of T cells with activated endothelial cells was indeed inhibited by pretreatment of T cells and endothelial cells with MPA. Immunoprecipitation studies using monoclonal antibodies against VLA-4 and LFA-1 showed that MPA inhibited the incorporation of sugars of PHA-activated human peripheral blood cells (133).

As disruption of interactions between mononuclear cells and the ECM protein network may be vital for the arrest and extravasation of alloreactive cells at the graft site, additional strategies may be required to effectively prevent graft rejection. Galaptin 14-1 (GL-14-1) a galactose-binding lectin (14 kDa) associated with basement membrane in the vessel wall, interacts with galactose-terminating glycoproteins expressed by certain T cells. Both prophylactic and therapeutic treatment with this compound are effective in murine models of rheumatoid arthritis, multiple sclerosis, and myasthenia gravis, and early results of ongoing experiments in rat transplant models are encouraging (J. W. Kupiec-Weglinski, unpublished). The migration of sensitized lymph node cells transferred into either untreated or r.GL-14-1 treated rat recipients of skin allografts revealed that the recovery of cells from peripheral, mesenteric, and the draining axillary lymph nodes in rats given r.GL-14-1 is selectively depressed compared to controls. Secondly, treatment of normal rats with r.GL-14-1 abrogates acute rejection and significantly prolongs the survival of MHC-incompatible skin allografts. Thus, the modulation of basement membrane components may affect host immune events leading to graft rejection. Although the mechanism of immunosuppressive activity needs to be elucidated, GL-14-1 treatment may induce an abnormal adhesion in vivo, in which alloimmune cells become bound to endothelial cells lining the blood vessels, removing them from the circulation and preventing their eventual migration into the specific lymphoid organs. Alternatively, GL-14-1 may block putative extracellular ligands that mediate normal lymphocyte migration patterns.

The further definition of cell adhesion molecules and strategies to interfere with them during the appropriate time window in the posttransplant immune response may open new strategies to achieve donor specific nonresponsiveness, the "holy grail" in transplantation, and result in long-term graft survival without chronic use of immunosuppressive drugs.

REFERENCES

1. Shimizu Y, Shaw S. Lymphocyte interactions with extracellular matrix. FASEB J 1991; 5:2292–2299.
2. Hynes RO. Integrins: versatility, modulation and signalling in cell adhesion. Cell 1992; 69:11–25.
3. Lin CQ, Bissell MJ. Multi-faceted regulation of cell differentiation by extracellular matrix. FASEB J 1993; 7:737–743.
4. Levasque JP, Hatzfeld A, Hatzfeld J. Mitogenic properties of major extracellular proteins. Immunol Today 1991; 12:258–262.
5. Miyamoto S, Akiyama SK, Yamada KM. Synergistic roles of receptor occupancy and aggregation in integrin transmembrane function. Science 1995; 267:883–886.
6. Springer TA. Adhesion molecules in the immune system. Nature 1990; 346: 425–433.
7. Shimizu Y, Newman W, Tanaka Y, Shaw S. Lymphocyte interactions with endothelial cells. Immunol Today 1992; 13:106–109.
8. Pitzalis C, Kingsley G, Haskard D, Panayi G. The preferential accumulation of helper-inducer T lymphocytes in inflammatory lesions: evidence for regulation by selective endothelial and homotypic adhesion. Eur J Immunol 1988; 18:1397–1404.
9. Martin GR, Timpl R. Laminin and other basement membranes components. Annu Rev Cell Biol 1987; 3:57–85.
10. Freitas AA, de Sousa M. Control mechanisms of lymphocyte traffic. A study of the action of two sulphated polysaccharides on the distribution of ^{51}Cr and $[H^3]$adenosine labelled mouse lymph node cells. Cell Immunol 1976; 31:62–76.
11. Masuyama J-I, Berman JS, Cruikshank WW, Morimoto C, Center DM. Evidence for recent as well as long term activation of T cells migrating through endothelial cells monolayers in vitro. J Immunol 1992; 148:1367–1374.
12. De Sousa M, Tilney NL, Kupiec-Weglinski JW. Recognition of self within self: specific lymphocyte positioning and the extracellular matrix. Immunol Today 1991; 12:262–266.
13. De Sousa M, Da Silva MT, Kupiec-Weglinski JW. Collagen, the circulation and positioning of lymphocytes: a unifying clue? Scand J Immunol 1990; 31: 249–256.
14. Hemler ME. VLA proteins in the integrin family: structures, functions and their role on leukocytes. Annu Rev Immunol 1990; 8:365–400.
15. Galandrini R, Galluzzo E, Albi N, Grossi CE, Velardi A. Hyaluronate is costimulatory for human T cell effector functions and binds to CD44 on activated T cells. J Immunol 1994; 153:21–31.
16. Jalkanen S, Reichert RA, Gallatin WM, Borgatze RF, Weissman IL, Butcher EC. Homing receptors and the control of lymphocyte migration. Immunol Rev 1986; 91:39–60.

17. De Sousa M. Lymphocyte circulation: Experimental and Clinical Aspects. Chicester, UK: John Wiley & Sons, 1981.
18. Chilosi M, Lestani M, Benedetti A, et al. Constitutive expression of tenascin in T-dependent zones of human lymphoid tissues. Am J Pathol 1993; 143: 1348–1355.
19. Hines KL, Kulkarni AB, McCarthy JB, et al. Synthetic fibronectin peptides interrupt inflammatory cell infiltration in transforming growth factor $\beta 1$ knockout mice. Proc Natl Acad Sci USA 1994; 91:5187–5191.
20. Wahl SM, Allen JB, Hines KL, et al. Synthetic fibronectin peptides suppress arthritis in rats by interrupting leukocyte adhesion and recruitment. J Clin Invest 1994; 94:655–662.
21. Kupiec-Weglinski JW, Coito AJ, Binder J, de Sousa M. The expression of extracellular matrix proteins represents an integral part of the host immune response in organ transplantation. Surg Forum 1993; 44:415–418.
22. Coito AJ, Binder J, de Sousa M, Kupiec-Weglinski JW. The expression of extracellular matrix proteins during accelerated rejections of cardiac allografts in sensitized rats. Transplantation 1994; 57:599–605.
23. Kramer RH, Rosen SD, McDonald KA. Basement membrane components associated with the extracellular matrix of the lymph node. Cell Tissue Res 1988; 252:367–375.
24. Kupiec-Weglinski JW, De Sousa M. Lymphocyte traffic is modified in vivo by anti-laminin antibody. Immunology 1991; 72:312–313.
25. Li YY, Cheung HT. Basement membrane and its components on lymphocyte adhesion, migration, and proliferation. J Immunol 1992; 149:3174–3181.
26. Goligorsky MS, Lieberthal W, Racusen L, Simon E. Integrin receptors in renal tubular epithelium: new insights into pathophysiology of acute renal failure. Am J Physiol 1993: 264:F1–F8.
27. Cybulsky AV. Adhesion molecules in renal diseases. In: Paul LC, Issekutz T, eds. Targeting Cell Adhesion Molecules for Therapeutic Application. New York: Marcel Dekker (this volume).
28. Romson JL, Hook BG, Kunkel SL, Abrams GD, Schork MA, Lucchesi BR. Reduction in the extent of ischemic myocardial injury by neutrophil depletion in the dog. Circulation 1983; 67:1016–1023.
29. Lehr HA, Guhlmann A, Nolte D, Keppler D, Messmer K. Leukotrienes as mediators in ischemia-reperfusion injury in a microcirculation model in the hamster. J Clin Invest 1991; 87:2036–2041.
30. Granger DN, Russel M, Arfors KE, Rothlein R, Anderson DC. Role of CD11/CD18 and ICAM-1 in ischemia/reperfusion-induced leukocyte adherence and emigration in mesenteric venules. FASEB J 1991; 5:A1753.
31. Thornton MA, Winn R, Alpers CE, Zager RE. An evaluation of the neutrophil as a mediator of in vivo renal ischemia-reperfusion injury. Am J Pathol 1989; 135:509–515.
32. Kelly KJ, Williams WW, Colvin RB, Bonventre JV. Antibody to intercellular adhesion molecule 1 protects the kidney against ischemic injury. Proc Natl Acad Sci USA 1994; 91:812–816.

33. Yoshida N, Granger DN, Anderson DC, Rothlein R, Lane C, Kvietys PR. Anoxia/reoxygenation-induced neutrophil adherence to cultured endothelial cells. Am J Physiol 1992; 262:H1891–H1898.
34. Scoazec J-Y, Durand F, Degott C, et al. Expression of cytokine-dependent adhesion molecules in postreperfusion biopsy specimens of liver allografts. Gastroenterology 1994; 107:1094–1102.
35. Byrne JG, Smith WJ, Murphy MP, Couper GS, Appleyard RF, Cohn LH. Complete prevention of myocardial stunning, contracture, low-reflow, and edema after heart transplantation by blocking neutrophil adhesion molecules during reperfusion. J Thorac Cardiovasc Surg 1992; 104:1589–1596.
36. Blakeley ML, Van der Werf WJ, Berndt MC, Dalmasso AP, Bach FH, Hancock WW. Activation of intragraft endothelial and mononuclear cells during discordant xenograft rejection. Transplantation 1994; 58:1059–1066.
37. Rollins S, Evans MJ, Johnson KK, et al. Molecular and functional analysis of porcine E-selectin reveals a potential role in xenograft rejection. Biochem Biophys Res Commun 1994; 204:763–771.
38. Pleass HCC, Kirby JA, Forsythe JLR, Proud G, Taylor RMR. Adhesion molecule blockade in a porcine xenograft model. Transplant Proc 1994; 26: 1162–1163.
39. Nemlander A, Soots A, von Willebrand E, Husberg B, Hayry P. Distribution of renal allograft-responding leukocytes during rejection. J Exp Med 1982; 156:1087–1100.
40. Rekonen R, Turunen JP, Rapola J, Hayry P. Characterization of high endothelial-like properties of peritubular capillary endothelium during acute renal allograft rejection. Am J Pathol 1990; 137:643–651.
41. Moolenaar W, Bruijn JA, Schrama E, et al. T-cell receptors and ICAM-1 expression in renal allografts during rejection. Transplant Int 1991; 4:140–145.
42. Fuggle SV, Sanderson JB, Gray DWR, Richardson A, Morris PJ. Variation in expression of endothelial adhesion molecules in pretransplant and transplanted kidneys—correlation with intragraft events. Transplantation 1993; 55:117–123.
43. Brockmeyer C, Ulbrecht M, Schendel DJ, et al. Distribution of cell adhesion molecules (ICAM-1, VCAM-1, ELAM-1) in renal tissue during allograft rejection. Transplantation 1993; 55:610–615.
44. Briscoe DM, Pober JS, Harmon WE, Cotran RS. Expression of vascular cell adhesion molecule-1 in human renal allografts. J Am Soc Nephrol 1992; 3: 1180–1185.
45. Andersen CB, Ladefoged SD, Larsen S. Acute kidney graft rejection. A morphological and immunohistological study on "zero-hour" and follow-up biopsies with special emphasis on cellular infiltrates and adhesion molecules. APMIS 1994; 102:23–37.
46. Alpers CE, Hudkins KL, Davis CL, et al. Expression of vascular cell adhesion molecule-1 in kidney allograft rejection. Kidney Int 1993; 44:805–816.
47. Gibbs P, Berkley LM, Bolton EM, Briggs JD, Bradley JA. Adhesion mole-

cule expression (ICAM-1, VCAM-1, E-selectin and PECAM) in human kidney allografts. Transplant Immunol 1993; 1:109–113.
48. Hosenpud JD, Shipley GD, Morris TE, Hefeneider SH, Wagner CR. The modulation of human aortic edothelial cell ICAM-1 (CD-54) expression by serum containing high titers of anti-HLA antibodies. Transplantation 1993; 55:405–411.
49. Mampaso F, Sanchez-Madrid F, Marcen R, et al. Expression of adhesion molecules in allograft renal dysfunction. A distinct diagnostic pattern in rejection and cyclosporine nephrotoxicity. Transplantation 1993; 56:687–691.
50. Lin Y, Kirby JA, Browell DA, et al. Renal allograft rejection: expression and function of VCAM-1 on tubular epithelial cells. Clin Exp Immunol 1993; 92: 145–151.
51. John S, Neumayer H-H, Weber M. Serum circulating ICAM-1 levels are not useful to indicate active vasculitis or early renal allograft rejection. Clinical Nephrol 1994; 42:369–375.
52. Lebranchu Y, Kapahi P, Al Najjar A, et al. Soluble E-selectin, ICAM-1, and VCAM-1 in levels in renal allograft recipients. Transplant Proc 1994; 26: 1873–1874.
53. Bechtel U, Scheuer R, Landgraft R, König A, Feucht HE. Assessment of soluble adhesion molecules (sICAM-1, s VCAM-1, s-ELAM-1) and complement cleavage products (sC4d, s-C5b-9) in urine. Clinical monitoring of renal allograft recipients. Transplantation 1994; 58:905–911.
54. Rose ML, Page C, Hengstenberg C, Yacoub MH. Identification of antigen presenting cells in normal and transplanted human heart: importance of endothelial cells. Hum Immunol 1990; 28:179–185.
55. Steinhoff G, Behrend M, Haverich A. Signs of endothelial inflammation in human heart allografts. Eur Heart J 1991; 12(suppl D):141–143.
56. Tanio JW, Basu CB, Albelda Sm, Eisen HJ. Differential expression of the cell adhesion molecules ICAM-1, VCAM-1, and E-selectin in normal and posttransplantation myocardium. Cell adhesion molecule expression in human cardiac allografts. Circulation 1994; 89:1760–1768.
57. Herskowitz A, Mayne AE, Willoughby SB, Kanter K, Ansari AA. Patterns of myocardial cell adhesion molecule expression in human endomyocardial biopsies after cardiac transplantation. Induced ICAM-1 and VCAM-1 related to implantation and rejection. Am J Pathol 1994; 145:1082–1094.
58. Briscoe DM, Yeung AC, Schoen EL, et al. Predictive value of inducible endothelial cell adhesion molecule expression for acute rejection of human cardiac allografts. Transplantation 1995; 59:204–211.
59. Koskinen PK. The association of the induction of vascular adhesion molecule 1 with cytomegalovirus antigenemia in human heart allografts. Transplantation 1993; 56:1103–1108.
60. Briscoe DM, Schoen FJ, Rice GE, Bevilacqua MP, Ganz P, Pober JS. Induced expression of endothelial-leukocyte adhesion molecules in human cardiac allografts. Transplantation 1991; 51:537–539.

61. Ferran C, Peuchmaur M, Desruennes M, et al. Implications of de novo ELAM-1 and VCAM-1 expression in human cardiac allograft rejection. Transplantation 1993; 55:605–609.
62. Isobe M, Ohtani H, Yagita H, Okumura K, Strauss HW, Yazaki Y. Detection of cardiac rejection in mice by radioimmune scintigraphy using ^{123}iodine-labeled anti-ICAM-1 monoclonal antibody. Acta Cardiol 1993; 48:235–243.
63. Phillipp W. Leukocyte adhesion molecules in rejected corneal allografts. Graefe's Arch Clin Exp Ophthalmol 1994; 232:87–95.
64. Adams DH, Hubscher SG, Shaw J, Rothlein R, Neuberger JM. Intercellular adhesion molecule 1 on liver allografts during rejection. Lancet 1989; II: 1122–1125.
65. Lautenschlager IT, Hockerstedt KA. ICAM-1 induction on hepatocytes as a marker for immune activation of acute liver allograft rejection. Transplantation 1993; 56:1495–1499.
66. Bacchi CE, Marsh CL, Perkins JD, et al. Expression of vascular cell adhesion molecule (VCAM-1) in liver and pancreas allograft rejection. Am J Pathol 1993; 142:579–591.
67. Reid SD, Uff CR, Wood RFM, Pockley AG. Increased ICAM-1 and LFA-1 expression with developing rejection following rat small bowel transplantation. Transplant Proc 1994; 26:1519.
68. Uff CR, Reid SD, Wood RFM, Pockley AG. CD44 expression on enterocytes: an indicator of rejection following rat small bowel transplantation. Transplant Proc 1994; 26:1553.
69. Quan D, Grant DR, Zhong RZ, et al. Altered gene expression of cytokine, ICAM-1, and class II molecules precedes mouse intestinal allograft rejection. Transplantation 1994; 58:808–816.
70. Coito AJ, Binder J, Van De Water L, Brown L, De Sousa M, Kupiec-Weglinski JW. Anti-TNF-α antibody treatment downregulates the expression of fibronectin and decreases cellular infiltration of cardiac allografts in rats. J Immunol 1995; 154:2949–2958.
71. Clausell N, Molossi S, Rabinovitch M. Increased interleukin-1β and fibronectin expression are early features of the development of the postcardiac transplant coronary arteriopathy in piglets. Am J Pathol 1993; 142:1772–1786.
72. Brown LF, Yeo KT, Berse B, et al. Expression of vascular permeability factor (vascular endothelial growth factor) by epidermal keratinocytes during wound healing. J Exp Med 1992; 176:1375–1379.
73. Clark RAF, Dvorak HF, Colvin RB. Fibronectin in delayed-type hypersensitivity skin reactions: associations with vessel permeability and endothelial cell activation. J Immunol 1981; 126:787–793.
74. Hershkoviz R, Alon R, Gilat D, Lider O. Activated T lymphocytes and macrophages secrete fibronectin which strongly supports cell adhesion. Cell Immunol 1992; 141:352–361.
75. Brown LF, Dubin D, Lavigne L, Logan BL, Dvorak HF, Van De Water L. Macrophages and fibroblasts express "embryonic" fibronectins during cutaneous wound healing. Am J Pathol 1993; 142:793–801.

76. Barnes JL, Hastings RR, De la Garza MA. Sequential expression of cellular fibronectin by platelets, macrophages, and mesangial cells in proliferative glomerulonephritis. Am J Pathol. 1994; 145:585–597.
77. Yamauchi K, Martinet Y, Crystal RG. Modulation of fibronectin gene expression in human mononuclear phagocytes. J Clin Invest 1987; 80:1720–1727.
78. Hoshino K, Nashan B, Steinhoff, Pichlmayr R. Expression of cell-matrix molecules and integrin receptors in human liver grafts during chronic rejection. Transpl Int 1994; suppl 1:S637–S6340.
79. Gould VE, Martinez-Lacabe V, Virtanen I, Sahlin KM, Schwartz MM. Differential distribution of tenascin and cellular fibronectins in acute and chronic renal allograft rejection. Lab Invest 1992; 67:71–79.
80. Yamada A, Nikaido T, Nojima Y, Schlossman SF, Morimoto C. Activation of human CD4 T lymphocytes. Interaction of fibronectin with VLA-5 receptor on CD4 cells induces the AP-1 transcription factor. J Immunol 1991; 146: 53–56.
81. Kubota K, Tamauchi H. Three-dimensional collagen matrices as cell culture substrata affect the generation of alloreactive cytotoxic T lymphocytes. Immunol Lett 1988; 18:119–124.
82. Mennander A, Paavonen T, Häyry P. Cyclosporine-induced endothelialitis and accelerated arteriosclerosis in chronic allograft rejection. Transplant Proc 1992; 24:341.
83. Paavonen T, Mennander A, Lautenschlager I, Häyry P. Endothelialitis in accelerated allograft arteriosclerosis in human cardiac transplant recipients. Transplant Proc 1992; 24:342–343.
84. Gorski A, Mrowiec T, Paczek L, Stepien-Sopniewska B. Abnormal T-cell costimulation by proteins of the extracellular matrix in human renal allografting. Res Immunol 1993; 144:305–309.
85. Korczak-Kowalska G, Stepien-Sopniewska B, Mrowiec T, Gorski A. Rejection-associated abnormalities of T cell interactions with extracellular matrix. Transplant Proc 1995; 27:903–906.
86. Herskovitz R, Cahalon L, Gilat D, Miron S, Miller A, Lider O. Physically damaged extracellular matrix induces TNF-α secretion by interacting resting $CD4^+$ T cells and marcophages. Scand J Immunol 1993; 37:111–115.
87. Sterzel BR, Schulze-Lohoff E, Weber M, Goodman SL. Interactions between glomerular mesangial cells, cytokines, and extracellular matrix. J Am Soc Nephrol 1992; 2:S126–127.
88. Nathan C, Sporn M. Cytokines in context. J Cell Biol 1991; 113:981–986.
89. Wieder KJ, Hancock WW, Schmidbauer G, et al. Rapamycin treatment depresses intragraft expression of KC/MIP-2, granzyme B, and IFN-γ in rat recipients of cardiac allografts. J Immunol 1993; 151:1158–1566.
90. Hancock WW, DiStefano R, Braun P, Schweizer RT, Tilney NL, Kupiec-Weglinski JW. Cyclosporine and anti-interleukin 2 receptor monoclonal antibody therapy suppress accelerated rejection of rat cardiac allografts through different effector mechanisms. Transplantation 1990; 49:416–421.
91. Dallman MJ, Wood KJ, Morris PJ. Specific cytotoxic T cells are found in

the nonrejected kidneys of blood-transfused rats. J Exp Med 1987; 165:566–571.

92. Ueda H, Hancock WW, Cheung YC, Diamantstein T, Tilney NL, Kupiec-Weglinski JW. The mechanism of synergistic interaction between anti-interleukin 2 receptor monoclonal antibody and cyclosporine therapy in rat recipients of organ allografts. Transplantation 1990; 50:545–550.
93. Towpik E, Kupiec-Weglinski JW. Use of cyclosporine in transplantation of nonprimarily vascularized tissues. Transplant Rev 1987; 1:85–100.
94. Kamada N. Liver transplantation in the rat. In: Experimental Liver Transplantation. Boca Raton, FL: CRC Press, 1988:39–97.
95. Schultz RM, Liebman MN. Proteins. In: Devlin TM, ed. Biochemistry with Clinical Correlation. New York: John Wiley and Sons, 1992:25–134.
96. Paul LC. Chronic renal allograft dysfunction. Kidney Int 1995; 47:1491–1499.
97. Paul LC, Benediktsson H. Chronic transplant rejection: Magnitude of the problem and pathogenesis mechanisms. Transplant Rev 1993; 7:96–113.
98. Hancock WW, Whitley WD, Tullius SG, et al. Cytokines, adhesion molecules and the pathogenesis of chronic rejection of rat renal allografts. Transplantation 1993; 56:643–650.
99. Duijvenstijn A, Kok M, Miyasaka M, Van Breda Vriesman P. ICAM-1 and LFA-1/CD18 expression in chronic renal allograft rejection. Transplant Proc 1993; 25:2867–2868.
100. Fisher A, Griscelli C, Blanche S. Prevention of graft failure by anti-LFA1 monoclonal antibody in HLA-mismatched bone marrow transplantation. Lancet 1986; II:1058–1061.
101. Le Mauff B, Hourmant M, Rougier JP, et al. Effect of anti-LFA1 (CD11a) monoclonal antibodies in acute rejection in human kidney transplantation. Transplantation 1991; 52:291–296.
102. Hourmant M, Le Mauff B, Le Meur Y, et al. Administration of an anti-CD11a monoclonal antibody in recipients of kidney transplantation. A pilot study. Transplantation 1994; 58:377–379.
103. Haug CE, Colvin RB, Delmonico FL, et al. A Phase I trial of immunosuppression with anti-ICAM-1 (CD54) mAb in renal allograft recipients. Transplantation 1993; 55:766–773.
104. Cosimi AB, Conti D, Delmonico FL, et al. In vivo effects of monoclonal antibody to ICAM-1 (CD54) in nonhuman primates with renal allografts. J Immunol 1990; 144:4604–4612.
105. Berlin PJ, Bacher JD, Sharrow SO, Gonzales C, Gress RE. Monoclonal antibodies against human T cell adhesion molecules—modulation of immune function in nonhuman primates. Transplantation 1992; 53:840–849.
106. Coughlan AF, Berndt MC, Dunlop LC, Hancock WW. In vivo studies of P-selectin and platelet activating factor during endotoxemia, accelerated allograft rejection, and discordant xenograft rejection. Transplant Proc 1993; 25:2930–2931.
107. Guymer RH, Mandel TE. Monoclonal antibody to ICAM-1 prolongs murine heterotopic corneal allograft survival. Aus NZ J Ophthalmol 1991; 19:141–144.

108. He YG, Mellon J, Apte R, Niederkorn JY. Effect of LFA-1 and ICAM-1 antibody treatment on murine corneal allograft survival. Invest Ophthalmol Vis Sci 1994; 35:3218–3225.
109. Yamataka T, Kobayashi H, Yagita H, Okumura K, Tamatani T, Miyasaka M. The effect of anti-ICAM-1 monoclonal antibody treatment on the transplantation of the small bowel in rats. J Pediatr Surg 1993; 28:1451–1457.
110. Omura T, Ishikura H, Nakajima Y, et al. Accelerated rejection of allografted rat liver perfused with anti-ICAM-1 monoclonal antibody. Immunobiol 1992; 186:241–245.
111. Talento A, Nguyen M, Blake T, et al. A single administration of LFA-1 antibody confers prolonged allograft survival. Transplantation 1993; 55:418–422.
112. Gotoh M, Fukuzaki T, Monden M, et al. A potential immunosuppressive effect of anti-lymphocyte function-associated antigen-1 monoclonal antibody on islet transplantation. Transplantation 1994; 57:123–126.
113. Yang H, Issekutz TB, Wright JR. Prolongation of rat islet allograft survival by treatment with monoclonal antibodies against VLA-4 and LFA-1. Transplantation 1995; 60:71–76.
114. Kawabe A, Suzuki H, Kimura T, Harada Y, Tamatani T, Miyasaka M. Immunosuppressive effects of monoclonal antibodies against adhesion molecules in rat pancreas allografts. Transplant Proc 1994; 26:1937–1938.
115. Isobe M, Yagita H, Okumura K, Ihara A. Specific acceptance of cardiac allograft after treatment with antibodies to ICAM-1 and LFA-1. Science 1992; 255:1125–1127.
116. Komori A, Nagata M, Ochiai T, et al. Role of ICAM-1 and LFA-1 in cardiac allograft rejection of the rat. Transplant Proc 1993; 25:831–832.
117. Kobayashi H, Miyano T, Yamataka A, et al. Prolongation or rat cardiac allograft survival by a monoclonal antibody: anti-rat intercellular adhesion molecule 1. Cardiovasc Surg 1993; 1:577–582.
118. Stepkowski SM, Tu Y, Condon TP, Bennett CF. Blocking of heart allograft rejection by intercellular adhesion molecule-1 antisense oligonucleotides alone or in combination with other immunosuppressive modalities. J Immunol 1994; 153:5336–5346.
119. Nakakura EK, McCabe SM, Zheng B, et al. Potent and effective prolongation of anti-LFA-1 monoclonal antibody monotherapy of non-primarily vascularized heart allograft survival in mice without T cell depletion. Transplantation 1993; 55:412–417.
120. Fukuzaki T, Gotoh M, Monden M, et al. Role of adhesion molecules in islet xenograft rejection. Transplant Proc 1994; 26:1113.
121. Orosz CG, Ohye RG, Pelletier RP, et al. Treatment with anti-vascular cell adhesion molecule 1 monoclonal antibody induces long-term murine cardiac allograft acceptance. Transplantation 1993; 56:453–460.
122. Isobe M, Suzuki J, Yagita H, et al. Immunosuppression to cardiac allografts and soluble antigens by anti-vascular cellular adhesion molecule-1 and anti-very late antigen-4 monoclonal antibodies. J Immunol 1994; 153:5810–5818.

123. Pelletier RP, Morgan CJ, Sedmak DD, et al. Analysis of inflammatory endothelial changes, including VCAM-1 expression, in murine cardiac grafts. Transplantation 1993; 55:315–320.
124. Kameoka H, Ishibashi M, Tamatani T, et al. The immunosuppressive action of anti-CD18 monoclonal antibody in rat heterotopic heart allotransplantation. Transplantation 1993; 55:665–667.
125. Kobayashi H, Miyano T, Yamataka A, et al. Prolongation of rat cardiac allograft survival by a monoclonal antibody: anti-rat intercellular adhesion molecule-1. Cardiovasc Surg 1993; 1:577–582.
126. Paul LC, Davidoff A, Benediktsson H, Issekutz TB. The efficacy of LFA-1 and VLA-4 antibody treatment in rat vascularized cardiac allograft rejection. Transplantation 1993; 55:1196–1199.
127. Issekutz TB. Effect of anti-LFA-1 and anti-VLA-4 on T lymphocyte migration in inflammation and lymphoid tissues. FASEB J 1992; 6:A1144.
128. Sadahiro M, McDonald TO, Allen MD. Reduction in cellular and vascular rejection by blocking leukocyte adhesion molecule receptors. Am J Pathol 1993; 142:675–683.
129. Grochowicz PM, Hibberd AD, Clark DA, Cowden WB, Willenborg DO. Castanospermine modifies expression of adhesion molecules in allograft recipients. Transplant Proc 1993; 25:2900–2901.
130. Grochowicz PM, Hibberd AD, Bowen DA, et al. Castanospermine, an alpha glucosidase inhibitor, prolongs renal allograft survival in the rat. Transplant Proc 1990; 22:2117–2118.
131. Grochowicz PM, Bowen DA, Hibberd AD, Clark DA, Cowden WB, Willenborg DO. Castanospermine, an inhibitor of glycoprotein processing, prolongs pancreaticoduodenal allograft survival. Transplant Proc 1992; 24: 2295–2296.
132. Grochowicz PM, Bowen DA, Hibberd AD, et al. Interference with intracellular carbohydrate processing by castanospermine prolongs heart allograft survival. Transplant Proc 1993; 25:743–744.
133. Product monograph CellCept (mycophenolate mofetil), Hoffmann La Roche, Ltd., Missisauga, Ontario, Canada.

27

Adhesion Molecules in Defense of the Peritoneum

Clifford J. Holmes
Renal Division, Baxter Healthcare, McGraw Park, Illinois

Nicholas Topley
Institute of Nephrology, University of Wales College of Medicine, Cardiff Royal Infirmary, Cardiff, Wales, United Kingdom

I. INTRODUCTION

Classically, peritonitis is divided into two distinct types. Acute primary or spontaneous peritonitis is usually caused by an infection with a single organism, such as *Escherichia coli* or *Streptococcus pneumoniae*, where no identifiable source of contamination can usually be identified. Secondary, or surgical, peritonitis arises from an injury or lesions of the gastrointestinal tract, the biliary system, the pancreas, or the genitourinary tract and is predominantly polymicrobic in nature (1). To this classification peritoneal dialysis-associated peritonitis must be added; peritoneal dialysis-associated peritonitis, which is more closely related to spontaneous peritonitis, has many distinctive features of its own (2). Continuous ambulatory peritoneal dialysis (CAPD) and various forms of automated peritoneal dialysis (APD) are therapies for the treatment of chronic renal failure. Peritoneal dialysis treatment takes advantage of the semipermeable characteristics of the peritoneal membrane to permit clearance of uremic toxins and to achieve adequate fluid removal via ultrafiltration. One major complication of the therapy, however, is peritoneal infection. In the 1989 final report of the U.S. National CAPD Registry the overall incidence of peritonitis for peritoneal dialysis patients was 1.4 episodes per patient year (3). The recent introduction of improved dialysis solution delivery systems has minimized the potential for touch-related contamination and the current overall incidence of peritonitis is 0.8 episodes per year or better (4,5). However, the realization that there are over 60,000 patients on some form of chronic peritoneal dialysis therapy, a figure that is growing at approximately 10% per year, explains why there has recently been a tremendous resurgence of research into the cellular and molecular mechanisms of the peritoneal inflammatory response and host defense mechanisms.

The objective of this chapter is to provide an overview of peritoneal host defense mechanisms in the setting of peritoneal dialysis with particular emphasis on the involvement of adhesion molecules, cytokines, and chemokines in the host's response to peritoneal infection.

II. HOST DEFENSE MECHANISMS OF PERITONEUM IN PERITONEAL DIALYSIS

The first manifestation of peritonitis is a massive recruitment of neutrophils into the peritoneum from the vascular system, which in peritoneal dialysis patients appears as a turbid dialysate. Microbial pathogens invading the peritoneal cavity have traditionally been believed to initially encounter a "first line" of host defense provided by peritoneal macrophages. The interaction between the pathways and the macrophages are facilitated by opsonins present within the peritoneal fluid (6). The fundamental stages of this first-line defense putatively consist of opsonization of the bacteria by IgG, C3b, and C3bi, followed by generation of chemotactic factors such LTB_4, C5a, and f-met-leu-phe peptides following the initial interaction with the host, resulting in direct macrophage migration to the site of bacterial insult. Receptor-mediated attachment and phagocytosis involving potentially a wide array of receptor-ligand interactions ensues, followed by intracellular killing via both oxidative and non-oxidative mechanisms (7). The role of peritoneal lymphocytes, which represent a minor but significant proportion of the peritoneal leukocyte population, is poorly understood, but some evidence exists that they may augment macrophage killing via lymphokine activation (8).

This scenario assumes that the failure of first-line mechanisms to contain microbial colonization and dissemination results in a second line of host defense—i.e., neutrophil recruitment from the circulation. However, with the development of techniques to isolate and propagate human peritoneal mesothelial cells (HPMC) in vitro, there have been several recent observations that place the peritoneal membrane at the center of the host defense and the control of the intraperitoneal inflammatory responses (9,10). Consequently, a more contemporary view of peritoneal host defense mechanisms is emerging which proposes that, rather than a discrete line of phagocyte host defense, it is more likely that a continuum of signaling, activation, and recruitment events exists that involves many if not all cells of the peritoneal cavity. In this scenario, cytokine-driven expression of adhesion molecules, chemoattractants, and chemokines plays a pivotal role in peritoneal host defense (11).

This review will expand on each of the above-mentioned aspects of peritoneal host defense.

III. PERITONEAL FLUID OPSONINS AND LEUKOCYTES–THE BASIS OF THE FIRST LINE OF HOST DEFENSE MODEL

A. Opsonic Activity

IgG is the primary immunoglobulin found in normal peritoneal fluid and CAPD dialysate and is considered to be an important opsonin-mediating phagocyte-bacterium attachment and phagocytosis via Fc_γ receptors. Normal peritoneal fluid concentrations of IgG are similar to those reported for plasma (mean ~ 1250 mg/dl). CAPD dialysate, however, contains ~2 to 50 mg/dl (0.2% to 5% of normal) presumably due to its continuous removal and dilution with 2 to 3 L of dialysis fluid. In addition to IgG-mediated phagocytosis, complement C3b- and C3bi-mediated phagocyte-bacterium attachment is considered the second important form of opsonin-mediated interactions. Bacterium bound C3b binds to CR1 while the receptor for C3bi is the integrin CR3 (Mac 1, CD11b/CD18). CR1 and CR3 interactions with bacteria permit adhesion but a second signal, delivered by surface bound fibronectin and the fibronectin receptor VLA-5 (CD49e/CD29), is required for internalization of the particle (12). Complement levels are also dramatically reduced in CAPD dialysate relative to that of normal peritoneal fluid and plasma. Typical dialysate concentrations of C3 range from <1 to 3 mg/dl as compared to ~80 mg/dl for normal peritoneal fluid and plasma. The fibronectin levels in dialysate range from <1 to ~5 μg/ml, which is substantially lower than the mean plasma value of 245 μg/ml in CAPD patients (8).

B. Peritoneal Leukocyte Populations

In healthy women undergoing laparoscopy, total cell yields obtained from 3 to 15 ml of peritoneal fluid range from 7 million to 12 million cells which consist of 90% macrophages, 5% to 10% lymphocytes, and less than 5% polymorphonuclear neutrophils (PMN). Leukocyte recovery from 1 to 3 L of dialysate fluid from uninfected CAPD patients varies from less than 1 million to as many as 45 million cells (8,13–16). Differential counts from uninfected dialysis effluent range from 20% to 95% for macrophages, 2% to 84% for lymphocytes, and 0% to 27% for PMN. In general, however, it is reported that the majority of peritoneal leukocytes in CAPD patients are macrophages (≈60% to 80%), with lymphocytes representing a smaller population (≈15% to 30%) while PMN usually only constitute a small percentage (≈0% to 5%).

IV. PERITONEAL MACROPHAGE PHAGOCYTOSIS–OPSONINS AND ADHESION MOLECULES

Despite the significant dilution of peritoneal opsonins during peritoneal dialysis, a large number of studies have attempted to correlate dialysate effluent opsonic activity with the incidence of CAPD-associated peritonitis (11). Many of these studies have assayed dialysate IgG alone as a measure of opsonic activity, while others have included C3 or have assessed the total dialysate opsonic activity using a functional assay such as the phagocytic uptake of radiolabeled bacteria. There is no consensus on the existence of an inverse relationship between opsonin content of the dialysate and peritonitis incidence. Reasons for this are unclear but may involve the degree to which nonopsonic phagocytosis occurs, including phagocyte/integrin/bacterial cell wall interactions. Phagocyte attachment via nonopsonic mechanisms has been well described and involves interactions between structures such as protein A on *Staphylococcus aureus* and macrophage cytophilic IgG and between type I fimbriae on *Escherichia coli* and mannosyl/fucosyl receptors on phagocytes (6,17). Phagocyte attachment to microorganisms can also occur via the β_2 integrin molecules, Mac-1 (CR3/CD11b/CD18), LFA-1 (CD11a/CD18) and p150,95 (CD11c/CD18) by interaction with the diglucosamine phosphate of lipopolysaccharide and the β-glucan component of zymosan (18,19). Mac-1 also mediates phagocytosis involving several ligands other than its classical opsonin C3bi, including particles bound via CR1 and the IgG $Fc_{\gamma}R$ (20,21). Graham et al. (22) have demonstrated that the role of Mac-1 in PMN adhesion is distinct from its role in phagocytosis and probably utilizes a different epitopes on the CD11/CD18 complex for each function. The role of adhesion molecules in defense against infectious pathogens has been reviewed in detail in a previous chapter, and thus will not be further discussed here. The point to be made is that a correlation between opsonic activity and peritonitis rates may be weakened as the incidence of non-opsonic phagocytosis increases.

V. PERITONEAL MACROPHAGE INTRACELLULAR KILLING AND LYMPHOKINE-MEDIATED ACTIVATION

A. Macrophage Intracellular Killing

Mac-1 is particularly important in the initiation of the respiratory burst, the release of secretory lysosomal granules, and the activation of arachidonic acid metabolism in phagocytes (23). Although little information is available on Mac-1 expression on human peritoneal macrophages, Lewis et al. (24) found no difference in CR3 expression between peritoneal macrophages of CAPD patients and their own peripheral blood monocytes or cells isolated from normal volunteers.

Macrophages (PMØ) isolated from the peritoneal cavity of CAPD patients display many features characteristic of immature and/or activated cells, including altered chemotactic responses and cell surface receptor and antigen expression patterns (11,25). As many as 5×10^7 PMØ are removed each day during the process of dialysate exchange (26). This continual removal of macrophages has led to the suggestion that the bone marrow is required to replace them at an excessive rate, resulting in the release of immature cells into the circulation and the peritoneal cavity. The immaturity of these cells is paradoxically associated with some increased activation characteristics (27). The majority of studies to date have nevertheless concluded that the phagocytic capacity, intracellular killing, and oxidative metabolism (H_2O_2 and stimulated chemiluminescence response) of CAPD peritoneal macrophages are largely intact in most patients (11). In a recent study, however, peritoneal macrophage function was studied 24 to 48 hours prior to the onset peritonitis (12). Both the number and the phagocytic capacity of PMØ were reduced relative to control during this period. This suggests that this temporal macrophage dysfunction may be responsible for the progression to clinical peritonitis.

B. Peritoneal Lymphocyte Trafficking and Lymphokine-Mediated Macrophage Activation

As described above, lymphocytes constitute a minor but not insignificant proportion of the total leukocyte population in the peritoneal dialysate. The precise mechanisms by which lymphocyte trafficking from the submesothelial interstitium to the peritoneum occurs is poorly defined. A critical mechanism regulating lymphocyte influx, however, will be the coordinated expression of adhesion molecules interacting with corresponding lymphocyte integrin receptors. Lymphocyte-endothelial interactions have been well described in the literature (28). Several recent reports have demonstated that mesothelial cells are one of a restricted number of cell types that express vascular cell adhesion molecule-1 (VCAM-1) as well as the more widely distributed ICAM-1 (CD54) (29–33). The β_1 integrin $\alpha_4\beta_1$ or VLA-4 (CD49d/CD29) expressed by lymphocytes mediates adhesion partly through recognition of VCAM-1 and partly by binding to at least two regions of fibronectin, the CS-1, and the CS-5 regions (34). Peripheral blood lymphocytes express also β_2 integrins but predominantly LFA-1 (CD11a/CD18) and therefore can also interact with its counter receptor ICAM-1. In 1993

Jonjic et al. (30) reported that mesothelial cells derived from ovarian cancer ascites and grown in vitro constitutively expressed VCAM-1 and ICAM-1, the levels of which could be increased by exposure to TNFα and IFNγ. More recently Cannistra et al. (32) have confirmed that mesothelial cells obtained from malignant ascites of patients with ovarian cancer express cytokine-inducible VCAM-1. In contrast to the observations of Jonjic et al. (30), Cannistra et al. (32) were unable to detect VCAM-1 in unstimulated cells. The discrepancy between these findings may be related to the fact that in the studies of Jonjic et al. mesothelial cells were grown in the presence of conditioned medium of PHA-stimulated mononuclear cells prior to resting for 2 to 3 days; thus the cells may not have been in a truly resting state. Mesothelial VCAM-1 was found to have a molecular mass of 110 kDa and an mRNA transcript of $\approx$3.2 kb, consistent with the predominant VCAM-1 species of activated endothelium. Recent data using cultured human peritoneal cells derived from normal omentum suggest that these cells express both the normal and alternatively spliced form of VCAM-1 (35) (Liberek and Topley, unpublished observations).

Blocking experiments using antibodies directed against CD29, CD18, and α_4 demonstrated that lymphocytes prestimulated with PMA bind to human mesothelial cells in vitro via both CD18- and CD29-dependent interactions. Significant inhibition of binding only occurred when the two integrins were blocked simultaneously (Fig. 1A). Neutralizing antibodies to VCAM-1 and α_4 confirmed that lymphocyte VLA-4 mediated adhesion to human mesothelial cells occurs predominantly through VCAM-1 binding and not via binding to fibronectin (Fig. 1B). Sironi et al. (31) have also recently reported that human peritoneal mesothelial cell VCAM-1 expression in vitro can be upregulated with the anti-inflammatory cytokine IL-13. Confirmation that both ICAM-1 and VCAM-1 expression occurs in the peritoneal membrane comes from the work of Suassuna et al. (33). Using histochemical analysis of peritoneal biopsies from normal, uremic and CAPD patients, both with and without peritonitis all expressed ICAM-1 and VCAM-1 with no significant differences in staining intensity between the study populations.

VI. PERITONEAL MACROPHAGE CHEMOTAXIS, TRANSMIGRATION, AND SURFACE PHAGOCYTOSIS–THE ROLE OF MESOTHELIAL CELL ADHESION MOLECULES AND CHEMOKINES

The instillation of large volumes of dialysis solution into the peritoneum reduces peritoneal leukocyte concentrations to $\approx 10^3$ to 10^4/ml from their normal range of approximately 5×10^5 to 10^6/ml. The low macrophage

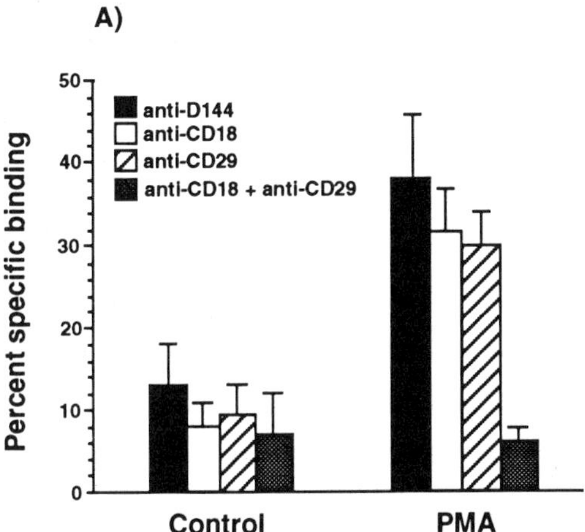

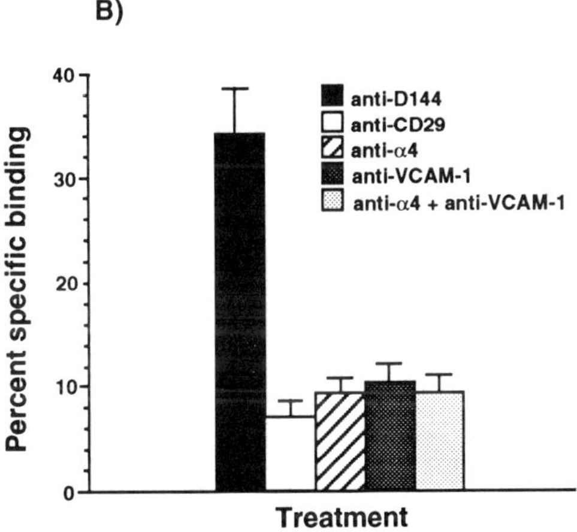

Figure 1 Inhibition of T lymphocyte adhesion to cultured human peritoneal mesothelial cells. Data are expressed as mean ± SEM percent specific binding of four separate experiments. Modified from Cannistra et al. (32). (A) The enhancement of T cell binding in PMA stimulated cells is inhibited in the combined presence of anti-CD29 and anti-CD18 antibodies. (B) The role of $\alpha_4\beta_1$ and VCAM-1 in T lymphocyte binding to peritoneal mesothelium. To neutralize the effects of β_2 integrins, these experiments were performed in the presence of anti-CD18 antibody. No additive effect is seen in the combined presence of anti-$\alpha_4\beta_1$ and VCAM-1 antibody.

concentrations in the fluid make it unlikely that they interact in the fluid phase. In vitro experiments have demonstrated that PMØ are incapable of controlling bacterial growth under such conditions (36). Despite this, peritoneal macrophages from clinically uninfected CAPD patients have been isolated with ingested bacteria (38), suggesting that the site of interaction between invading microorganisms and PMØ is most likely on the surface of the mesothelium (37). PMØ stimulated with bacteria also have the potential to recruit other macrophages via the production of LTB_4 and monocyte chemotactic protein-1 (MCP-1) (unpublished observations, Li and Topley). MacKenzie et al. (38) have shown that PMØ are potent sources of LTB_4 upon stimulation with staphylococci.

Direct examination of peritoneal biopsy tissue shows very few macrophages in the submesothelial membrane tissue and almost complete absence of cells adherent to the apical surfaces of mesothelial cells. The loss of apically adherent cells during tissue processing cannot, however, be ruled out. Recently, Suassuna et al. (33), using immunohistochemical staining techniques, observed a population of submesothelial macrophages with dendritic morphology in uremic, nondialyzed individuals. These PMØ residing close to the peritoneal basement membrane are thus potentially available to respond to chemotactic signals and transmigrate into the peritoneal cavity (33).

The involvement of mesothelial cells in macrophage recruitment is suggested by recent findings of chemotactic factor and adhesion molecule expression by these cells. In 1992, Jonjic et al. (30) reported the transcription of mRNA for MCP-1 and the secretion of MCP-1 protein by resting mesothelial cells from ovarian cancer ascites which was increased upon stimulation with either IL-1β or TNFα/IFNγ. Recently, Visser et al. (39) have shown that normal human peritoneal mesothelial cells transcribe mRNA for both MCP-1 and IP-10 after stimulation with either TNFα or IFNγ. MCP-1 is a chemoattractant for monocytes and basophils, and IP-10 attracts monocytes and a subset of T lymphocytes. Our own data suggest that human peritoneal mesothelial cells express mRNA and secrete the mononuclear cell-specific C-C chemokines MCP-1 and RANTES. The directed secretion of these chemokines appears to create a chemotactic gradient, facilitating the transmigration of mononuclear leukocytes across mesothelial cell monolayers (40).

The mechanism by which mononuclear cells interact with mesothelial cells facilitating their transmigration into the peritoneal cavity remains poorly defined. Preliminary studies by Jonjic et al. (30) using neutralizing antibodies to VLA-4 and CD18 revealed that monocyte adhesion to activated and resting human peritoneal mesothelial cells could be inhibited only by the anti-CD18 antibody (Fig. 2). This suggests that ICAM-1/β_2 integrin

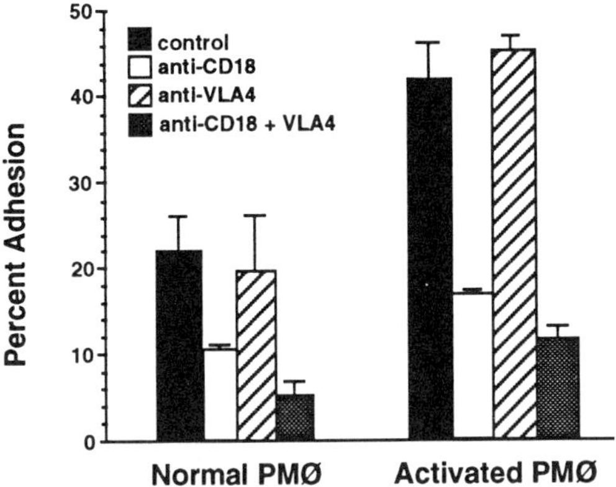

Figure 2 Binding of resting and activated human monocytes to human peritoneal cells. Monocytes were activated by culture for 20 hours with IFNα (500 U/ml) and LPS (100 ng/ml). Binding was determined using cytokine (TNFα/IFNα)-activated mesothelium as substratum. Reproduced from J Exp Med 1992; 176:1165–1174.

interactions predominate in the adhesion and possibly transmigration events of macrophages and mesothelial cells in the peritoneum.

Thus, the first line of peritoneal host defense appears to involve surface adherent macrophages which, upon stimulation with bacteria or their secreted products, secrete pro-inflammatory cytokines capable of activating adjacent mesothelial cells to produce macrophage specific chemotactic factors. This may result in the recruitment of additional macrophages from the submesothelial tissue. The transmigration of these cells into the peritoneal cavity is facilitated by the expression of both ICAM-1 and VCAM-1 on mesothelial cells (Fig. 3).

VII. POLYMORPHONUCLEAR LEUKOCYTE IMMIGRATION INTO THE PERITONEUM

A. Leukocyte Kinetics During Peritonitis

During CAPD-associated peritonitis, the dialysate differential count changes from one where neutrophils represent $<5\%$ of the total leukocyte population to one where they constitute $>50\%$ after several days and often reach over 80% (Fig. 4) (13–15). The magnitude of this response is such that this influx of leukocytes can sometimes approach the total circulating blood pool number (7×10^{10}) (16). It must also be noted, however, that the influx of PMN is accompanied by a concomitant increase in the number

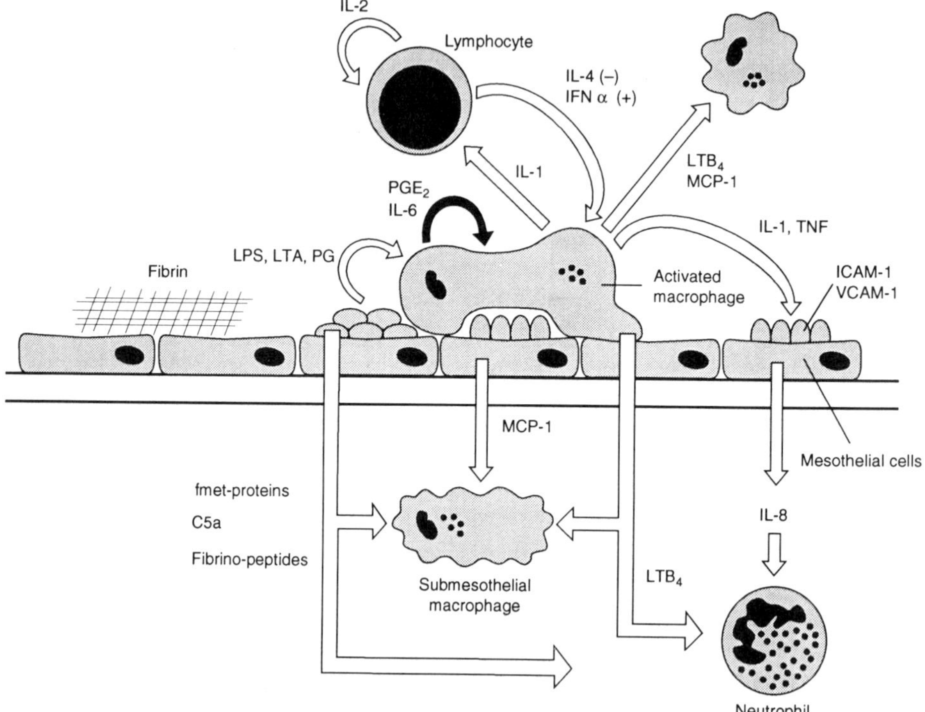

Figure 3 Schematic representation of a contemporary view of peritoneal first-line host defense mechanisms of the peritoneum. Reproduced from Holmes (11).

of macrophages, which eventually becomes the predominant cell population during the resolution phase of the infection. This phenomenon will be discussed later in this chapter.

Animal models of peritoneal inflammation have provided insight into the time course of PMN recruitment. Using intraperitoneal injection of zymosan suspended in saline in rabbits, Forrest et al. (41) demonstrated that peritoneal accumulation of PMN could be detected by as early as 2 hours after the injection which then increased slowly up to 4 hours (Fig. 4), whereupon it reached a maximum.

B. Neutrophil Recruitment as a First-line Phagocyte in the Peritoneum

The site of interaction between invading microorganisms and the peritoneal phagocytes would appear to be on the peritoneal membrane, as discussed

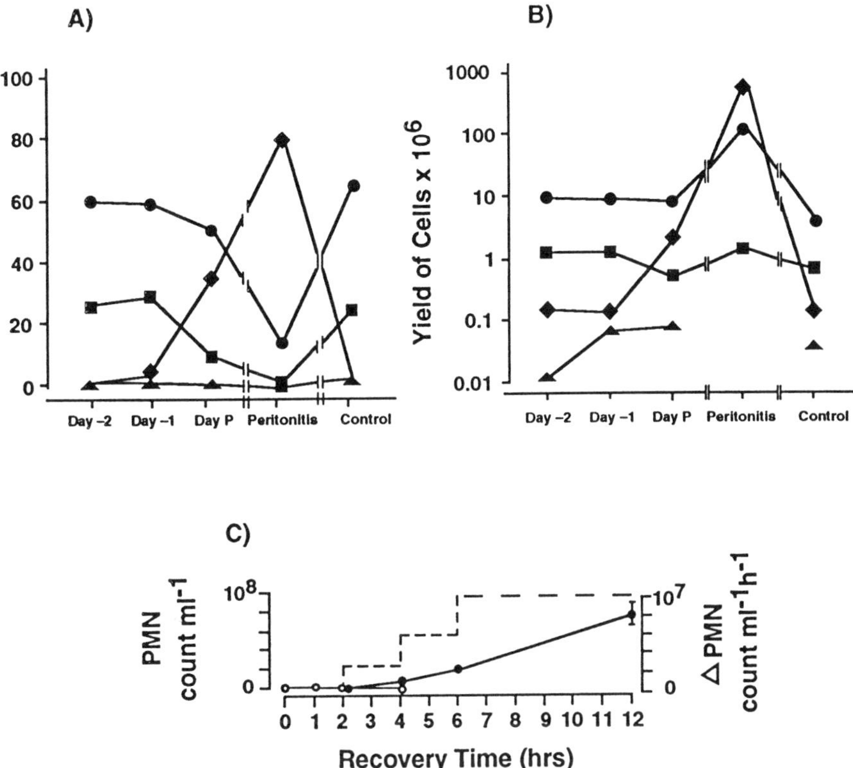

Figure 4 Peritoneal fluid leukocyte kinetics during peritonitis. (A) Changes in overnight dialysate effluent leukocyte differential count in CAPD patients. Results are shown from 48 hours prior to peritonitis, during overt clinical peritonitis and compared to uninfected periods. Symbols: ● macrophages; ■ lymphocytes; ♦ PMN; ▲ mesothelial cells. (B) Absolute cell yield from (A) above. (C) Kinetics of leukocyte influx into the peritoneum of rabbits injected IP with either zymosan in saline (10 mg/ml) or saline alone. The PMN content of exudates was determined by light microscopy. The rate of PMN accumulation (stippled line) induced by zymosan is plotted on a linear scale. Each point is the mean of six rabbits. Figures 4A and 4B are modified from Betjes et al. (13). Figure 4C is modified from Forrest et al. (41).

previously. Muijsken et al. (42) employed human peritoneal mesothelial cells to study surface phagocytosis of *S. aureus* by peripheral blood monocytes, neutrophils, and peritoneal macrophages; unlike peripheral blood monocytes or neutrophils, peritoneal macrophages were unable to phagocytose mesothelial adherent bacteria. This observation suggests that the role of peritoneal macrophages may lie more in their ability to recruit peripheral neutrophils via local cytokine and chemokine production than as first-line phagocytic cells. Consequently, the possibility exists that neutrophil immigration into the peritoneal cavity is the only effective supply of phagocytic cells operating in this environment (13).

C. Mechanisms of PMN Extravasation

While the mechanism controlling the recruitment, adherence, and movement of PMNs across endothelium are relatively well defined, the mechanisms by which PMNs migrate into the peritoneal cavity are less well understood (43). It is assumed that signals generated during intraperitoneal inflammation result in the recruitment of PMNs from the circulation. These cells then extravasate across the microvascular endothelium of the peritoneal membrane, cross the submesothelial stroma, and enter into the peritoneal cavity through mesothelial cell tight junctions. The mechanisms by which leukocytes are recruited and transmigrate across the mesothelium will be discussed later in this review.

D. Chemoattractants Involved in Peritoneal PMN Recruitment

Chemoattractants that can provide a chemotactic gradient from the peritoneum for neutrophils include the classical chemoattractants, LTB_4, complement C5a, PAF, N-formyl methionyl-capped bacterial peptides, membranes of the C-X-C chemokine family, IL-8 and gro/MGSA, and MIP-1α,β of the C-C chemokine family (Table 1). Interleukin 8 is a potent neutrophil chemoattractant that is synthesized by several cell types includ-

Table 1 Summary of Human Peritoneal Mesothelial Cell Chemoattractant, Chemokine, Cytokine, and Adhesion Molecule Expression

	Name	Reference
C-X-C chemokines	IL-8/NAP-1	39,48,49
	gro/MGSA	40
C-C chemokines	MCP-1	30,39,40
	IP-10	39
	RANTES	39,40
Adhesion molecules	ICAM-1	30,33,51,70,71,74
	VCAM-1	30,32,33
Cytokines	IL-6	75
	IL-1	76

ing macrophages, T lymphocytes, and fibroblasts upon exposure to LPS, lipoteichoic acid, or cytokines such as IL-1β and TNFα. Zemel et al. (44) have reported that TNFα is significantly elevated during the early phase of peritonitis. This cytokine is biologically active and is locally produced. As mesothelial cells do not express mRNA for TNFα, this protein is most likely secreted by PMØ, suggesting that this macrophage-derived cytokine plays a critical role in the initiation of the inflammatory response of the peritoneum.

Elevated levels of IL-8 have also been detected in dialysate during peritonitis and correlate closely with neutrophil numbers. These levels fall steadily as the infection subsides with treatment, reaching baseline levels by approximately 1 week (45–48). Recently, Lin et al. (46) have shown that both peritoneal macrophages and PMN isolated from peritonitis dialysate can be a potent source of IL-8, which may facilitate neutrophil recruitment during the inflammatory response.

While the source of intraperitoneal IL-8 may be of leukocyte origin, Topley et al. (48) and Betjes et al. (49) have reported that human mesothelial cells also synthesize IL-8 in response to both IL-1β and TNFα. The ability of these human peritoneal cells to express both mRNA and IL-8 protein, together with the observation that this chemokine is secreted in a predominantly apical direction, suggests that the mesothelial cell plays a central role in the recruitment of neutrophils (50,51). This is especially so as the mesothelial cell is the major cell type within the uninfected peritoneal cavity.

Extracellular matrix glycosaminoglycans can bind cytokines and growth factors and modify their biological activity. Webb et al. (52) have recently described the ability of heparan sulfate to specifically enhance the chemotactic ability of IL-8 for PMN using the Boyden chamber method. This result suggests that heparan sulfate, present on both endothelial and mesothelial apical and basolateral cell surfaces, may participate in the IL-8-dependent transmigration of PMN. As mesothelial cells secrete a series of proteoglycans, including heparan sulfate (53), we speculate that mesothelial secreted heparan sulfate may play an important role in the promotion of mesothelial derived IL-8-dependent transmigration. In a similar manner, Tanaka et al. (54) have demonstrated that proteoglycan bound MIP-1α facilitates lymphocyte adherence to endothelium.

E. Adhesion Molecules in Animal Models of Peritonitis

Much of the information leading to the understanding of adhesion molecules in endothelial rolling, firm adhesion, and transmigration steps has been derived from the use of antibodies that block the specific action of the target molecules. The understanding of peritoneal leukocyte recruitment has been advanced using animal models of peritonitis. Antiadhesion molecule studies have until recently concentrated on the role of the leukocyte integrins CD18, CD11a, and CD11b (55). These investigations clearly established the importance of the CD11/CD18 complex in peritoneal immigration of PMN into the peritoneal cavity.

1. CD18-Independent Versus CD18-Dependent PMN Migration

Using monoclonal antibodies to CD18 (β_1 integrin) and CD49d (β_2 integrin), Winn and Harlan (56) have reported in rabbits that there is an early CD18-dependent PMN immigration in response to either *E. coli* or protease peptone. In contrast, later PMN immigration (24 hours) appeared to be CD18-independent. Previous studies by this group had established a CD18-independent PMN immigration into the peritoneum of rabbits after the initial immigration of macrophages (57). As we have previously discussed in the human peritoneum, the maximal rate of macrophage influx occurs after the initial recruitment of PMN. Inhibition of PMN immigration was most effectively inhibited at 24 hours with the combination of anti-CD18 and anti-CD49d antibodies, supporting the importance of macrophage-dependent late PMN influx into the peritoneal cavity.

One explanation for these observations might lie in the sequential generation of different neutrophil chemoattractants in rabbits as described eloquently by Williams and colleagues (58,59). Using intraperitoneal zymosan as the inflammatory stimulus, they demonstrated that the acute (2 to 4 hours) immigration of PMN is dependent on local fluid phase activation of C5a; the subsequent influx of PMN was dependent upon locally secreted IL-8. As intraperitoneal IL-8 generation is likely to be driven by macrophage/monocyte-derived cytokines, prevention of mononuclear immigration into the peritoneum may explain why the late PMN influx is macrophage-dependent.

Other recent studies have concluded that PMN recruitment using CD18-dependent mechanisms into the peritoneum is related to the type of stimulus employed. Conlan et al. (60) used a murine model and found that *Salmonella typhimurium* elicited a CD18-dependent recruitment at 12 hours, in contrast to *Listeria monocytogenes*, which was CD18-independent. The same study reported that anti-CD11b treatment could almost completely prevent PMN accumulation in response to sodium casein, but only partially reduced PMN influx in response to thioglycollate and protease peptone. Clearly, more research is needed to elucidate the influence of the stimulus on adhesion molecule expression and leukocyte emigration in general.

2. Selectin Dependency of Leukocyte Recruitment Into the Peritoneum

Pretreatment of rats with anti-E selectin $F(ab')_2$ caused a 70% reduction in oyster glycogen-induced PMN accumulation (61). P-selectin blockade using an anti-P-selectin mAb in rabbits challenged with *E. coli* did not impair PMN emigration into the peritoneum (62). However, simultaneous blocking of L- and P-selectin completely inhibits PMN recruitment when using thioglycollate as the inflammatory stimulus (63,64). It is clear, however, that the nature of the response is dependent on the type of stimulus employed, which indicates that several different mechanisms exist whereby leukocytes transmigrate into the peritoneal cavity. Using L-selectin-deficient mice, Arbones et al. (65) demonstrated that thioglycollate-induced PMN recruitment into the peritoneal cavity was reduced by 72% at 2 hours and 36% at 4 hours. This observation confirmed a similar observation made in 1991 by Watson et al. using the anti-L-selectin mAb MEL 14 (67).

In an eloquently designed study, Bosse et al. (64) evaluated the ability of E-, P-, and L-selectin antibodies, both singly and in combinations, to inhibit PMN accumulation in the peritoneum of mice using thioglycollate as the stimulus. Although significant inhibition was observed with each antibody type when used alone, only the combination of L- and P-selectin blockade completely inhibited immigration. This study indicates that while E-selectin is involved in PMN immigration into the peritoneum, it is P- and L-selectins that play the dominate role.

In a 1995 publication, Bullard and colleagues (67) reported their findings on PMN emigration using a series of single and double mutant mice deficient for either P-selectin, ICAM-1, or both. In these experiments *Streptococcus pneumoniae* was used to induce peritonitis. In agreement with previous studies of single mutant mice, both the ICAM-1 and the P-selectin mutant groups significantly inhibited PMN accumulation after 4 hours (64% and 61%, respectively) (68). However, only the double mutant was able to completely block recruitment. Of particular interest was the finding that PMN immigration into the pulmonary alveoli was completely normal, revealing the existence of organ-specific differences. These latter two studies suggest that, while there is redundancy at each step of rolling, firm adherence, and transmigration, complete inhibition can be achieved by blockade of two adhesive interactions requiring specific combinations of adhesion molecules.

F. PMN Adhesive Interactions With Human Peritoneal Mesothelial Cells

1. Role of ICAM-1 in PMN Adhesion

As mentioned previously, having traversed the endothelium and submesothelial stroma, PMN must transmigrate across the mesothelium via tight junctions and be in contact with the mesothelial cell surface for at least a short period of time [69].

ICAM-1 expression levels, although the adherence could be partially inhibited with anti-ICAM-1 $F(ab')_2$ fragments. As with endothelium, there was no evidence in this study that VCAM-1/VLA-4 interactions played a role in PMN adherence to mesothelial cells.

In contrast to human vascular endothelial cells, this and previous studies observed no expression of E-selectin (protein or mRNA). This difference in adhesion molecule expression pattern may be in part related to the function of mesothelial cells. From a teleological perceptive, it is presumed that selectins exist to be able to slow the flow of passing leukocytes and to allow firm adhesion to occur. In contrast, leukocyte transmigration across mesothelium occurs in a basolateral to apical direction, under static conditions. This pattern of adhesion molecule expression observed with peritoneal mesothelial cells is similar to that described for intestinal epithelium—another organ system in which transmigration occurs in the basolateral to apical direction (72,73).

2. PMN Transmigration Across Peritoneal Mesothelial Cells

Using human peritoneal mesothelial cells grown on porous membranes of tissue culture inserts, Davenport et al. (74) examined PMN transmigration in the basolateral to apical direction using a combination of electrical resistance measurements and light and electron microscopy to confirm integrity of the membrane (Fig. 7). Apical prestimulation of mesothelial cells with either IL-1 or TNF resulted in a time- and dose-dependent PMN transmigration, which could also be mimicked by the apical addition of IL-8. Dependence on ICAM-1 was demonstrated by the significant inhibition of transmigration with anti-ICAM-1 $F(ab')_2$ antibody fragments. The importance of other adhesion molecules, such as CD11/CD18, has not yet been reported.

VIII. SUMMARY

Since the early 1980s, the classical model describing first line of host defense mechanisms of the peritoneal cavity during peritoneal dialysis has progressed from a model focusing on the phagocytosis of invading bacteria by peritoneal macrophages to one involving dialysate macrophages, submesothelial macrophages, lymphocytes, neutrophils, and mesothelial cells in an orchestrated response to a microbial insult (Fig. 3). Adhesion molecules are critically involved in each phase of the peritoneal host defense response to pathogen invasion. The CD18 integrin family are now recognized as playing an important role in bacterial attachment, while Mac-1 not only can mediate phagocytosis initiated by Fc and CR1 interaction, but also is involved in the initiation of the respiratory burst, secretion of lysosomal enzymes, and

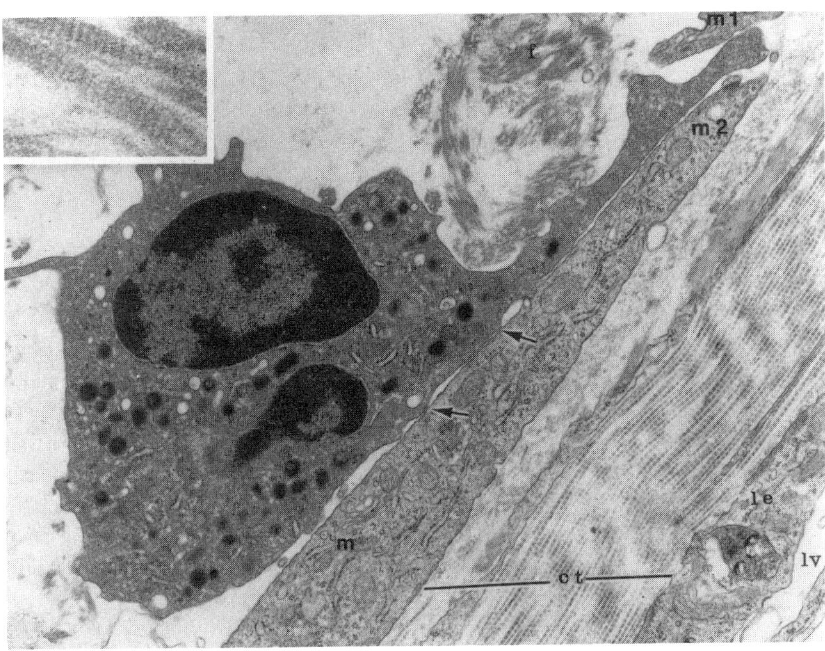

Figure 7 Electron micrograph illustrating neutrophil passage between two mesothelial cells (m 1 and m 2). Cells are closely apposed at several points of adherence (dark arrows). (×20,500.) Reproduced from Leak LV. Lab Invest 1983; 48:479–491.

arachidonic acid metabolism. Although much information has been available for several years, depicting the role of leukocyte and endothelial cell adhesion molecules and their respective counterligands in PMN rolling, firm adhesion, and transmigration across the endothelium, recent studies in animal models of peritonitis using either antiadhesion molecule monoclonal antibodies or gene knockout mice have provided new insights into the relative importance of specific adhesion molecules. These studies suggest that despite the recognized redundancy in adhesion molecule function, complete blockade of PMN emigration to the peritoneum can be achieved only when the function of P-selectin in addition to either L-selectin or ICAM-1 is removed.

Recent interest in the biology of peritoneal mesothelial cells has provided much new information on their potentially pivotal role as mediators of the inflammatory response in the peritoneal cavity. PMN adherence to and transmigration across normal human peritoneal mesothelial cells is mediated by ICAM-1, whose expression is upregulated by inflammatory cytokines. The selectin family do not appear to be expressed by peritoneal mesothelial cells, as with other cell types, such as intestinal epithelium, where conditions of shear stress do not need to be overcome and transmigration occurs from the tissue interstitium across the cells in a basolateral to apical direction. The transmigration of leukocytes is also facilitated by the creation of a chemotactic gradient across the mesothelial cell. In this respect, mesothelial cells are a potent source of PMN and mononuclear cell-specific chemokines, suggesting that these molecules (and other chemotactic agents) play a pivotal role in the recruitment of leukocytes into the peritoneum (Fig. 8). Peritoneal mesothelial cells also express VCAM-1, which in concert with ICAM-1 mediates the adherence of lymphocytes and mononuclear phagocytes. Taken together these data add to the growing support for a direct role of mesothelial cells in mononuclear cell recruitment into the peritoneum.

It is anticipated that over the next few years, continued research into the role of adhesion molecules and chemoattractant molecules will provide a more complete understanding of their participation in the host defense mechanisms of the peritoneum.

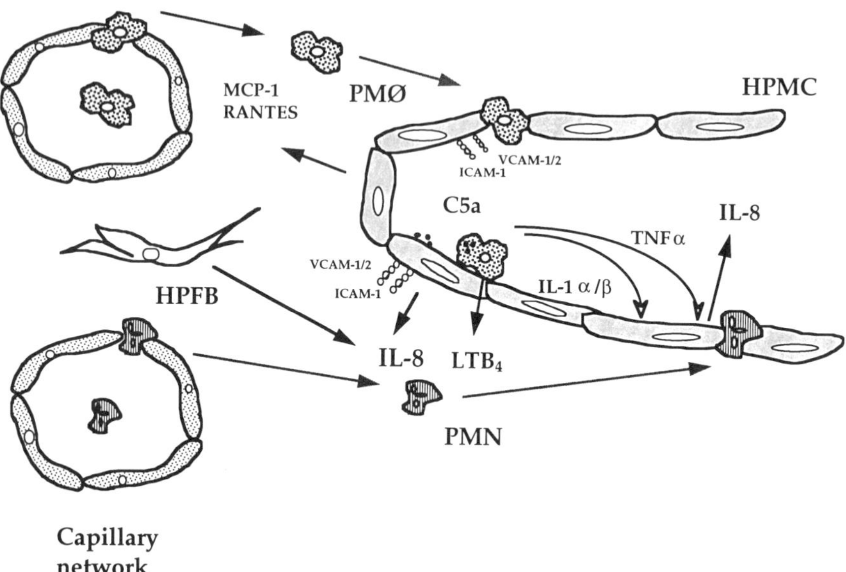

Figure 8 A schematic representation of leukocyte recruitment into the peritoneal cavity under the control of chemokines and cytokines.

REFERENCES

1. Hau T, Ahrenholz DH, Simmons RL. Secondary bacterial peritonitis: the biologic basis of treatment. In: Current Problems in Surgery. Chicago: Year Book Medical Publishers, 1979; 16:1–65.
2. Dobbie JW. Surgical peritonitis: its relevance to the pathogenesis of peritonitis in CAPD. Perit Dial Int 1988; 8(4):241–248.
3. Lindblad AS, Novak JW, Nolph KD. Continuous Ambulatory Peritoneal Dialysis in the USA: Final Report of the National CAPD Registry 1981–1988. Dordrecht: Kluwer Academic Publishers, 1989.
4. Port FK, Held PJ, Nolph KD, Turenne MN, Wolfe RA. Risk of peritonitis and technique failure by CAPD connection technique: a national study. Kidney Int 1992; 42:967–974.
5. Dasgupta MK, Fox S, Gagnon D, Bettcher K, Ulan RA. Significant reduction of peritonitis rate by the use of twin-bag system in a Canadian regional CAPD program. In: Khanna R, Nolph KD, Prowant BF, Twardowski ZJ, Oreopoulos DG, eds. Advances in Peritoneal Dialysis. Toronto: Peritoneal Dialysis Bulletin, 1992:223–229.
6. Verbrugh HA, Keane WF, Hoidel JF, Freiberg MR, Elliot GR, Peterson PK. Peritoneal macrophages and opsonins: antibacterial defense in patients undergoing chronic peritoneal dialysis. J Infect Dis 1983; 1476:1018–1029.
7. Holmes CJ, Lewis S. Host defense mechanisms in the peritoneal cavity of continuous ambulatory dialysis patients. 2. Humoral defenses. Perit Dial Int 1991; 11:112–117.
8. Lewis S, Holmes CJ. Host defense mechanisms in the peritoneal cavity of continuous ambulatory peritoneal dialysis patients. Part one. Perit Dial Int 1991; 11:14–21.
9. Stylianou E, Jenner LA, Davies M, Coles GA, Williams JD. Isolation, culture and characterization of human peritoneal mesothelial cells. Kidney Int 1990; 37:1563–1570.
10. Topley N, Mackenzie R, Coles GA, Williams JD. Cytokine networks in CAPD: interactions of resident cells during inflammation in the peritoneal cavity. Perit Dial Int 1993; 13:S282–S285.
11. Holmes CJ. Peritoneal host defense mechanisms in peritoneal dialysis. Kidney Int 1994; 46(suppl 48):S58–S70.
12. Wright SD, Craigmyle LS, Silverstein SC. Fibronectin and serum amyloid P component stimulate C3b- and C3bi-mediated phagocytosis in cultured human monocytes. J Exp Med 1983; 158:1338–1343.
13. Betjes MGH, Tuk CW, Visser CE, Krediet RT, Arisz L, Beelen RHJ. Analysis of the peritoneal cellular immune system during CAPD and shortly before a clinical peritonitis. Nephrol Dial Transplant 1994; 9:684–692.
14. Cichocki T, Hanicki Z, Sulowicz W, Smolenski O, Kopec J, Zembala M. Output of peritoneal cells into peritoneal dialysate. Nephron 1983; 35:175–182.
15. Rubin J, Rogers WA, Taylor HM, et al. Peritonitis during continuous ambulatory peritoneal dialysis. Ann Intern Med 1980; 92:7–13.

16. Vassa N, Nolph KD, Prowant BF, Moore HL, Khanna R, Twardowski ZJ. Comparison of leukocyte counts in blood and dialysate in chronic peritoneal dialysis patients with peritonitis. Perit Dial Int 1995; 15(1):S34.
17. Boner G, Mhashilkar AM, Rodriguez-Ortega M, Sharon N. Lectin-mediated, nonopsonic phagocytosis of Type I *Escherichia coli* by human peritoneal macrophages of uremic patients treated by peritoneal dialysis. J Leuk Biol 1989; 14:239–245.
18. Ross GD, Vetvicka V. CR3 (CD11b/CD18): a phagocyte and NK cell membrane receptor with multiple ligand specificities and functions. Clin Exp Immunol 1993; 92:181–184.
19. Wright SD, Detmers PA. Adhesion-promoting receptors on phagocytes. J Cell Sci 1988; 9:99–120.
20. Arnaout MA, Gupta SK, Pierce MW, Tenen DG. Amino acid sequence of the alpha sub-unit of human leukocyte adhesion receptor Mo 1 (complement receptor type 3). J Cell Biol 1988; 106:2153.
21. Graham IL, Gresham HD, Brown EJ. An immobile subset of plasma membrane CD11b/CD18 is involved in phagocytosis of targets recognized by multiple receptors. J Immunol 1989; 142:2352–2358.
22. Graham IL, Brown EJ. Extracellular calcium results in a conformational change in Mac-1 (CD11b/CD18) on neutrophils. J Immunol 1992; 146:685–691.
23. Petersen MM, Steadman R, Williams JD. Human neutrophils are selectively activated by independent ligation of the subunits of the CD11b/CD18 integrin. J Leuk Biol 1994; 56:708–713.
24. Lewis SL, Norris PJ. Monocyte/macrophage function in continuous ambulatory peritoneal dialysis patients. In: Coles GA, Davies M, Williams JD, eds. CAPD: Host Defense, Nutrition and Ultrafiltration. Basel: Karger, 1989:1–9.
25. McGregor SM, Topley N, Jörres A, et al. Longitudinal evaluation of peritoneal macrophage function and activation during continuous ambulatory peritoneal dialysis: maturity, cytokine synthesis and arachidonic acid metabolism. Kidney Int. In press.
26. Goldstein CS, Bomalaski JS, Zurier RB, Neilson EG, Douglas SD. Analysis of peritoneal macrophages in continuous ambulatory peritoneal dialysis patients. Kidney Int 1984; 26:733–740.
27. McGregor SM, Topley N, Jörres A, et al. Long term changes in cytokines and arachidonic acid metabolism during CAPD. Nephrol Dial Transplant 1993; 8: 1025.
28. Springer T. Traffic signals for lymphocyte recirculation and leukocyte emigration: the multistep paradigm. Cell 1994; 76:301–314.
29. Liberek T, Topley N, Luttman W, Witowski J, Coles GA, Williams JD. ICAM-1 mediates the adherence of PMN to cytokine activated human peritoneal mesothelial cells. Nephrol Dial Transplant 1993; 8:1024.
30. Jonjic N, Peri G, Bernasconi S, et al. Expression of adhesion molecules and chemotactic cytokines in cultured human cells. J Exp Med 1992; 1165–1174.
31. Sironi M, Sciacca FL, Matteucci C, et al. Regulation of endothelial and mesothelial cell function by interleukin 13: selective induction of vascular cell adhesion molecule-1 and amplification of interleukin 6.

32. Cannistra SA, Ottensmeier C, Tidy J, DeFranzo B. Vascular cell adhesion molecule-1 expressed by peritoneal mesothelium partly mediates the binding of activated human T lymphocytes. Exp Haematol 1994; 22:996–1002.
33. Suassuna JHR, Das Neves FC, Hartley RB, Ogg CS, Cameron JS. Immunohistochemical studies of the peritoneal membrane and infiltrating cells in normal subjects and in patients on CAPD. Kidney Int 1994; 46:443–454.
34. Hemler ME, Elices MJ, Parker C, Takada Y. Structure of the integrin VLA-4 and its cell-cell and cell-matrix adhesion functions. Immunol Rev 1990; 114: 45–65.
35. Hession C, Tizard R, Vassallo C. Cloning of an alternative form of vascular cell adhesion molecule-1 (VCAM-1). J Biol Chem 1991; 266:6682–6685.
36. Holmes CJ. CAPD associated peritonitis: an immunological perspective on causes and interventions. In: Gurland H, Wetzels E, eds. Immunologic Perspectives in Chronic Renal Failure. Basel: Karger, 1990:73–90.
37. Topley N, Williams JD. Role of the peritoneal membrane in the control of inflammation in the peritoneal cavity. Kidney Int 1994; 46(suppl 48):S71–S78.
38. MacKenzie RK, Coles G, Williams JD. Eicosanoid synthesis in human peritoneal macrophages stimulated with *S. epidermidis*. Kidney Int 1990; 37:1316–1324.
39. Visser CE, Brouwer-Steenbergen JJE, Boorsma DM, Willemze R, Krediet RT, Beelen RHJ. Chemokines expressed by human mesothelial cells: interleukin 8, GRO, MCP, and IP-10. Perit Dial Int 1995; 15(suppl 1):S34.
40. Li F-k, Loetscher P, Williams JD, Topley N. Human peritoneal mesothelial cells synthesize MCP-1 and RANTES: induction by IL-1β and TNFα. Nephrol Dial Transplant. In press.
41. Forrest MJ, Jose PJ, Williams TJ. Kinetics of the generation and action of chemical mediators in zymosan-induced inflammation of the rabbit peritoneal cavity. Br J Pharmacol 1986; 86:719–730.
42. Muijsken MA, Heezius HJCM, Verhoef J, Verbrugh HA. Role of mesothelial cells in peritoneal antibacterial defense. J Clin Pathol 1991; 44:600–604.
43. Carlos TN, Harlan JM. Leukocyte-endothelial adhesion molecules. Blood 1994; 7(1):2068–2101.
44. Zemel D, Imholz ALT, de Waart DR, Dinkla C, Struijk DG, Krediet RT. Appearance of tumor necrosis factor-α and soluble TNF-receptors I and II in peritoneal effluent of CAPD. Kidney Int 1994; 46:1422–1430.
45. Zemel D, Krediet RT, Koomen GCM, Kortekaas WMR, Geertzen HGM, ten Berge RJM. Interleukin-8 during peritonitis in patients treated with CAPD; an in-vivo model of acute inflammation. Nephrol Dial Transplant 1994; 9:169–174.
46. Lin CY, Huang TP. Gene expression and release of interleukin-8 by peritoneal macrophages and polymorphonuclear leukocytes during peritonitis in uremic patients on continuous ambulatory peritoneal dialysis. Nephron 1994; 65:437–441.
47. Brauner A, Hylander B, Wretlind B. Interleukin-6 and interleukin-8 in dialysate and serum from patients on continuous ambulatory peritoneal dialysis. Am J Kidney Dis 1993; 22(3):430–435.

48. Topley N, Brown Z, Jorres A, et al. Human peritoneal mesothelial cells synthesize interleukin-8. Am J Pathol 1993; 142:1876–1886.
49. Betjes MGH, Tuk CW, Struik DG, et al. Interleukin-8 production by peritoneal mesothelial cells in response to tumor necrosis factor-α, interleukin-1, and medium conditioned by macrophages cocultured with *Staphylococcus epidermidis*. J Infect Dis 1993; 168:1202–1210.
50. Zeillemaker AM, Mul FPL, Hoynck Van Papendrecht AAGM, et al. Polarized secretion of interleukin-8 by human mesothelial cells: a role in neutrophil migration. Immunology 1995; 84:227–232.
51. Davenport A, Topley N, Williams JD. Control of polymorphonuclear leukocyte migration across human peritoneal msothelial cell monolayers. Proceedings of XXXIst European Dialysis Transplantation Association, 1955. In press.
52. Webb LMC, Ehrengruber MU, Clark-Lewis I, Baggiolini M, Rot A. Binding to heparan sulphate or heparin enhances neutrophil responses to interleukin 8. Proc Natl Acad Sci USA 1993; 90:7158–7162.
53. Davies M, Stylianou E, Yung S, Thomas GJ, Coles GA, Williams JD. Proteoglycans of CAPD-dialysate fluid and mesothelium. In: Coles GA, Davies JD, Williams JD. CAPD: Host Defense, Nutrition and Ultrafiltration. Basel: Karger 1990:134–141.
54. Tanaka Y, Adams DH, Hubscher S, Hirano H, Siebenlist U, Shaw S. T-cell adhesion induced by proteoglycan-immobilized cytokine MIP-1β. Nature 1993; 361(6407):79–82.
55. Harlan JM, Winn RK, Vedder NB, Doerschuk CM, Rice CL. In vivo models of leukocyte adherence to endothelium. In: Harlan JM, Liu DY, eds. Adhesion: Its Role in Inflammatory Disease. New York: Freeman, 1992:117–150.
56. Winn RK, Harlan JM. CD 18-independent neutrophil and mononuclear leukocyte emigration into the peritoneum of rabbits. J Clin Invest 1993; 93:168–1173.
57. Mileski W, Harlan J, Rice C, Winn R. *Streptococcus pneumoniae*-stimulated macrophages induce neutrophils to emigrate by a CD 18-independent mechanism of adherence. Circ Shock 1990; 31:259–267.
58. Collins PD, Jose PJ, Williams TJ. The sequential generation of neutrophil chemoattractant proteins in acute inflammation in the rabbit in vivo. J Immunol 1991; 2:677–684.
59. Beaubien BC, Collins PD, Jose PJ, et al. A novel neutrophil chemoattractant generated during an inflammatory reaction in the rabbit peritoneal cavity in vivo. Biochem J 1990; 271:797–801.
60. Conlan JW, Noth RJ. Listeria monocytogenes, but not *Salmonella typhimurium*, elicits a CD 18-independent mechanism of neutrophil extravasation into the murine peritoneal cavity. Infect Immun 1994; 62(7):2702–2706.
61. Mulligan MS, Varani J, Dame MK, et al. Role of endothelial-leukocyte adhesion molecule 1 (ELAM-1) in neutrophil-mediated lung injury in rats. J Clin Invest 1991; 88:1396–1406.
62. Sharar SR, Sasaki SS, Flaherty LC, Paulson JC, Harlan JM, Winn RK. P-selectin blockade does not impair leukocyte host defense against bacterial

peritonitis and soft tissue infection in rabbits. J Immunol 1993; 151:4982–4988.
63. Mayadas TN, Johnson RC, Rayburn H, Hynes RO, Wagner DD. Leukocyte rolling and extravasation are severely compromised in P selectin-deficient mice. Cell 1993; 74:541–554.
64. Bosse R, Vestweber D. Only simultaneous blocking of the L- and P-selectin completely inhibits neutrophil migration into mouse peritoneum. Eur J Immunol 1994; 24:3019–3024.
65. Arbones ML, Ord DC, Ley K, et al. Lymphocyte homing and leukocyte rolling and migration are impaired in L-selectin-deficient mice. Immunity 1994; 1: 247–260.
66. Watson SR, Fennie C, Lasky LA. Neutrophil influx into an inflammatory site inhibited by a soluble homing receptor-IgG chimera. Nature 1991; 349: 164–167.
67. Bullard DC, Qin L, Lorenzo I, et al. P-selectin/ICAM-1 double mutant mice: acute emigration of neutrophils into the peritoneum is completely absent but is normal into pulmonary alveoli. J Clin Invest 1995; 95:1782–1788.
68. Sligh JE, Ballantyne CM, Rich SS, et al. Inflammatory and immune responses are impaired in mice deficient in intracellular adhesion molecule-1. Proc Natl Acad Sci USA 1993; 90:8529–8533.
69. Leak LV. Interaction of mesothelium to intraperitoneal stimulation. I. Aggregation of peritoneal cells. Lab Invest 1983; 48:479–491.
70. Andreoli SP, Mallett C, Williams K, McAteer JA, Rothlein R, Doerschuk CM. Mechanisms of polymorphonuclear leukocyte mediated peritoneal cell injury. Kidney Int 1994; 46:1100–1109.
71. Liberek T, Topley N, Luttman W, Williams JD. Adherence of neutrophils to human peritoneal mesothelial cells: role of intracellular adhesion molecule-1. J Am Soc Nephrol. In press.
72. Parkos CA, Delp C, Arnaout MA, Madara JL. Neutrophil migration across a cultured intestinal epithelium: dependence on a CD11b/CD18-mediated event and enhanced efficiency in the physiologic direction. J Clin Invest 1991; 88: 1605–1612.
73. Colgan SP, Parkos CA, Delp C, Arnaout MA, Madara JL. Neutrophil migration across cultured intestinal epithelial monolayers is modulated by epithelial exposure to IFNγ in a highly polarized fashion. J Cell Biol 1993; 120:785–798.
74. Davenport A, Topley N, Williams JD. Neutrophil (PMN) transmigration across human peritoneal mesothelial cell monolayers is both ICAM-1 and IL-8 dependent. J Am Soc Nephrol 1994; 5:181–217.
75. Topley N, Jörres A, Luttman W, et al. Human pewritoneal cells synthesize interleukin-6: Induction by IL-1β and TNFα. Kidney Int 1993; 43:226–233.
76. Douvdevani A, Rapoport J, Konforty A, Argov S, Ovnat A, Chaimovitz C. Human peritoneal mesothelial cells synthesize IL-1α and β. Kidney Int 1994; 46:993–1001.

Index

About the Editors

LEENDERT C. PAUL is the Incumbent of the Keenan Chair in Medicine as well as a Professor in the Department of Medicine, University of Toronto at St. Michael's Hospital, Ontario, Canada. The editor or coeditor of five books, including *Organ Transplantation: Long-Term Results* (Marcel Dekker, Inc.), and the author or coauthor of over 110 book chapters and journal publications, he is a Fellow of the Royal College of Physicians and Surgeons of Canada and a member of the American Society of Nephrology and the American Society of Transplant Surgeons, among others. Dr. Paul received the M.D. degree (1969) and Ph.D. degree (1979) in transplantation immunology from the State University at Leiden, The Netherlands.

THOMAS B. ISSEKUTZ is a Professor in Medicine and Pediatrics at the University of Toronto, Toronto, and at The Toronto Hospital, Ontario, Canada. The author or coauthor of over 90 journal publications and book chapters, he is a member of the American Association of Immunology, the American Association of Pathology, and the American Society for Clinical Investigation, among others. Dr. Issekutz received the M.D. degree (1976) from Dalhousie University, Halifax, Nova Scotia, Canada.